Advances in Intelligent Systems and Computing

Volume 543

Series editor

Janusz Kacprzyk, Polish Academy of Sciences, Warsaw, Poland
e-mail: kacprzyk@ibspan.waw.pl

About this Series

The series "Advances in Intelligent Systems and Computing" contains publications on theory, applications, and design methods of Intelligent Systems and Intelligent Computing. Virtually all disciplines such as engineering, natural sciences, computer and information science, ICT, economics, business, e-commerce, environment, healthcare, life science are covered. The list of topics spans all the areas of modern intelligent systems and computing.

The publications within "Advances in Intelligent Systems and Computing" are primarily textbooks and proceedings of important conferences, symposia and congresses. They cover significant recent developments in the field, both of a foundational and applicable character. An important characteristic feature of the series is the short publication time and world-wide distribution. This permits a rapid and broad dissemination of research results.

More information about this series at http://www.springer.com/series/11156

Roman Szewczyk · Małgorzata Kaliczyńska
Editors

Recent Advances in Systems, Control and Information Technology

Proceedings of the International Conference SCIT 2016, May 20–21, 2016, Warsaw, Poland

Editors
Roman Szewczyk
Industrial Research Institute Automation and Measurements PIAP
Warsaw
Poland

Małgorzata Kaliczyńska
Industrial Research Institute for Automation and Measurements PIAP
Warsaw
Poland

ISSN 2194-5357 ISSN 2194-5365 (electronic)
Advances in Intelligent Systems and Computing
ISBN 978-3-319-48922-3 ISBN 978-3-319-48923-0 (eBook)
DOI 10.1007/978-3-319-48923-0

Library of Congress Control Number: 2016957850

Printed on acid-free paper

This Springer imprint is published by Springer Nature
The registered company is Springer International Publishing AG
The registered company address is: Gewerbestrasse 11, 6330 Cham, Switzerland

Preface

Today, we are the witnesses of the most important revolution in global industry during the twenty-first century. Conversion of the business model of global producers towards the idea of INDUSTRY 4.0 will change the global manufacture into flexible, robust and cost-effective systems. Production will be more effective due to shifting decision systems in field, building intelligent middleware structures, as well as implementing different level of intelligent sensors and control in robotic setups. As a result, the work of engineers and managers will become safer, easier and more prosperous. Moreover, the quality of life of all citizens of the globalized world will increase significantly.

This book presents results of discussion during the International Conference on Systems, Control and Information Technologies 2016. It presents research results from top experts in the fields connected with the idea of INDUSTRY 4.0 and determining the efficiency of its implementation, especially intelligent systems, advanced control, information technologies, industrial automation, robotics, intelligent sensors, metrology as well as new materials. Each chapter presents a deep analysis of a specific technical problem which is in general followed by a numerical analysis and simulation, and results of solution implementation for real-world problems.

We strongly believe that the presented theoretical results, practical solutions and guidelines will be useful for both researchers and engineers working in the area of transformation of global industry towards INDUSTRY 4.0 approach.

Warsaw
July 2016

Roman Szewczyk
Małgorzata Kaliczyńska

Contents

Systems and Control

Tender Participants Selection Based on Artificial Neural Network Model for Alternatives Classification 3
Tetiana Kolpakova and Valerii Lovkin

The Dimensionality Reduction Methods Based on Computational Intelligence in Problems of Object Classification and Diagnosis 11
Sergey A. Subbotin and Andrii A. Oliinyk

The Selection Methods for Multisensor System Elements of Drone Detection 20
Igor Korobiichuk, Michał Nowicki, Yuriy G. Danik, Sergey Dupelich, and Samchyshyn Oleksyj

Reduction of Uneven Pace of Internal Combustion Engine FP10C by Flywheel Design Improvement 27
Andrii Ilchenko and Volodymyr Lomakin

Synergetic Control of Social Networking Services Actors' Interactions 34
Ruslan Hryshchuk and Kateryna Molodetska

Artificial Neural Network as a Basic Element of the Automated Goniometric System 43
Irina Cherepanska, Elena Bezvesilna, and Artem Sazonov

The Program of Calculation of Rate of Movement of Cleaning Device for the Main Gas Pipeline 52
Volodymyr Kvasnikov and Oleksandr Stashynsky

Concept, Definition and Use of an Agent in the Multi-agent Information Management Systems at the Objects of Various Nature 59
Regina Boyko, Dmitriy Shumyhai, and Miroslava Gladka

Modelling of Double-Pendulum Based Energy Harvester for Railway Wagon 64
Vytautas Bučinskas, Andrius Dzedzickis, Nikolaj Šešok, Ernestas Šutinys, and Igor Iljin

Information Systems

Error Analysis of the Finite Element Method Calculations Depending on the Operating Range 75
Paweł Nowak and Roman Szewczyk

Models of Magnetic Hysteresis Loops Useful for Technical Simulations Using Finite Elements Method (FEM) and Method of the Moments (MoM) 82
Roman Szewczyk, Michał Nowicki, and Katarzyna Rzeplińska-Rykała

Parallel Computer System Resource Planning for Synthesis of Neuro-Fuzzy Networks 88
Andrii Oliinyk, Stepan Skrupsky, and Sergey A. Subbotin

The Sample and Instance Selection for Data Dimensionality Reduction 97
Sergey Subbotin and Andrii Oliinyk

Protection of University Information Image from Focused Aggressive Actions 104
Roman Korzh, Andriy Peleshchyshyn, Solomia Fedushko, and Yuriy Syerov

Development of the System for Detecting Manipulation in Online Discussions 111
Andrij Peleshchyshyn and Zoriana Holub

Capabilities of an Open-Source Software, Elmer FEM, in Finite Element Analysis of Fluid Flow 118
Marcin Safinowski, Maciej Szudarek, Roman Szewczyk, and Wojciech Winiarski

The Use of ICT in Formation of Professional Competence of Nurses in Ukraine 127
Natalia Shygonska

FEM Modelling and Thermography Validation of Thermal Flow 135
Alicja Praczukowska, Michał Nowicki, and Jakub Pełka

Numerical Analysis and Validation of the Human Impact on the Conditions in Model Chamber 144
Weronika Radzikowska-Juś, Maciej Szudarek, Andrzej Juś, and Stefan Owczarek

Design of an Interactive GUI for Multimedia Data Exchange Using SUR40 Multi-touch Panel 155
Rafał Kłoda, Jan Piwiński, and Aleksandra Nowak

Automatic Subtitling for Live 3D TV Transmissions by Real-Time Analysis of Spatio-Temporal Depth Map of the Scene 163
Konrad Bojar

Advances in FEM Based Modeling of Waveguide and Waveguide Systems for Microwave Applications, Using Newly Developed Open Source Software 172
Jakub Szałatkiewicz, Roman Szewczyk, Eugeniusz Budny, Mateusz Kalinowski, Juhani Kataja, Peter Råback, and Juha Ruokolainen

Industrial Automation

How to Increase Efficiency of Automatic Control of Complex Plants by Development and Implementation of Coordination Control System 189
Igor Korobiichuk, Anatoliy Ladanyuk, Dmytro Shumyhai, Regina Boyko, Volodymyr Reshetiuk, and Marcin Kamiński

Formalization of Energy Efficiency Control Procedures of Public Water-Supply Facilities 196
Liudmyla Davydenko, Viktor Rozen, Volodymyr Davydenko, and Nina Davydenko

The Measuring System for Bee Hives Environmental Monitoring 203
Jerzy Niewiatowski and Paweł Nowak

A Formal Model in Control Systems Design 211
Małgorzata Kaliczyńska, Stanisław Lis, Marcin Tomasik, and Tomasz Dróżdż

Automation of Evaporation Plants Using Energy-Saving Technologies 220
Anatoliy Ladanyuk, Olena Shkolna, and Vasil Kyshenko

Adaptive Control of Dynamic Load in Blooming Mill with Online Estimation of Process Parameters Based on the Modified Kaczmarz Algorithm 227
Vadim Kharlamenko, Sergii Ruban, Igor Korobiichuk, and Oleg Petruk

A Programmable Logic Controller (PLC); Programming Language Structural Analysis 234
Yulia Kovalenko

Energy-Efficient Electrotechnical Complex of Greenhouses with Regard to Quality of Vegetable Production 243
Igor Korobiichuk, Vitaliy Lysenko, Volodymyr Reshetiuk, Taras Lendiel, and Marcin Kamiński

Method of Control for Gasification Reactor in Ecological Technology of Biomass Waste Utilization 252
Jakub Szałatkiewicz, Grzegorz Zielono, Łukasz Obrzut, and Marzena Szałatkiewicz

Robots and Robotic Systems

Coordination and Cooperation Mechanisms of the Distributed Robotic Systems 263
Igor Korobiichuk, Yuriy Danik, Pavlo Pozdniakov, and Dorota Jackiewicz

Selected Aspects of Implementation of "EU Occupational Safety and Health (OSH) Strategic Framework 2014–2020" Connected with Automation & Robotics 273
Marcin Słowikowski, Jacek Zieliński, Zbigniew Pilat, Michał Smater, and Wojciech Klimasara

Diagnostic Systems of Mobile Robot Technical State 281
Svitlana Marchenkova

Positioning of Industrial Robot Using External Smart Camera Vision 288
Jacek Dunaj

Application of Direct Metal Laser Sintering for Manufacturing of Robotic Parts 312
Maciej Cader and Dominik Wyszyński

Inspection Robots in Hard Coal Mines 327
Maciej Cader and Leszek Kasprzyczak

Drop Test of Tactical Mobile Robot 345
Bartosz Blicharz and Maciej Cader

Human-Robot Interaction in the Rehabilitation Robot Renus-1 358
Jacek Dunaj, Wojciech J. Klimasara, and Zbigniew Pilat

Soft Flexible Gripper Design, Characterization and Application 368
Jan Fraś, Mateusz Maciaś, Filip Czubaczyński, Paweł Sałek, and Jakub Główka

Dynamics Model of a Three-Wheeled Mobile Robot Taking into Account Slip of Wheels 378
Maciej Trojnacki

Development of the Human-Robot Communication in Welding Technology Manufacturing Cells 387
Zbigniew Pilat, Jozef Varga, and Mikulas Hajduk

Modular Robotic Toolbox for Counter-CBRN Support 396
Grzegorz Kowalski, Adam Wołoszczuk, Agnieszka Sprońska, Damian Buliński, and Filip Czubaczyński

Tire Models for Studies of Wheeled Mobile Robot Dynamics on Rigid Grounds – A Quantitative Analysis for Longitudinal Motion 409
Przemysław Dąbek and Maciej Trojnacki

Sensors and Metrology

New Method for Calculation the Error of Temperature Difference Measurement with Platinum Resistance Thermometers (PRT) 427
Tadeusz Goszczyński

Measurement of the Number Servings of Milk and Control of Water Content in Milk on Stall Milking Machines 435
Volodymyr Kucheruk, Pavel Kulakov, and Natalia Storozhuk

Investigation of the Appropriate Method of Mounting Tested Elements in the Test Stand for Temperature Characteristics of Ultra-Precise Resistors 448
Andrzej Juś, Paweł Nowak, Roman Szewczyk, and Weronika Radzikowska-Juś

Acoustic Radiation Energy Focus in a Shell with Liquid 459
Volodimir Karachun and Viktorij Mel'nick

Torque and Capacity Measurement in Rotating Transmission 464
Ivan Grabar, Igor Korobiichuk, and Oleg Petruk

Analysis of the Phenomena Occurring During Initial Phase of Resistors Thermal Characteristics Measurement 473
Andrzej Juś, Paweł Nowak, Roman Szewczyk, and Weronika Radzikowska-Juś

Two-Channel MEMS Gravimeter of the Automated Aircraft Gravimetric System 481
Igor Korobiichuk, Olena Bezvesilna, Maciej Kachniarz, Andrii Tkachuk, and Tetyana Chilchenko

Measurement Setup for the Thermal and Line Regulation Characteristics of Reference Voltage Sources 488
Paweł Nowak, Andrzej Juś, and Roman Szewczyk

Development of Graphene Based Leak Detector 495
Marcin Safinowski, Krzysztof Trzcinka, Cezary Dziekoński, Andrzej Juś, Maciej Kachniarz, Roman Szewczyk, and Wojciech Winiarski

Earth Remote Sensing Satellite Navigation Based on Optical Trajectory Measurements 504
Ruslan Hryshchuk and Andriy Zavada

Time-Domain Reflectometry (TDR) Square and Pulse Test Signals Comparison 512
Sylwester Kostro, Michał Nowicki, Roman Szewczyk, and Katarzyna Rzeplińska-Rykała

Transforming the Conversion Characteristic of a Measuring System Used for Technical Control 524
Eugenij Volodarsky, Zygmunt Warsza, Larysa A. Kosheva, and Adam Idźkowski

Factors of AC Field Inhomogeneity in Impedance Measurement of Cylindrical Conductors 535
Aleksandr A. Mikhal, Aleksandr I. Glukhenkyi, and Zygmunt L. Warsza

Radio Electronic System Elements Diagnostics by Means of Lissajous Curves with the Extended Database 546
Serhii Yehorov

Investigation of Temperature Measurement Uncertainty Components for Infrared Radiation Thermometry 556
Nataliya Hots

Accuracy of Reconstruction of the Spatial Temperature Distribution Based on Surface Temperature Measurements by Resistance Sensors 567
Mykhaylo Dorozhovets, Mariana Burdega, and Zygmunt L. Warsza

Possibility of Sensors Application of 2714A Type Amorphous Alloys 577
Jacek Salach, Dorota Jackiewicz, and Magdalena Krześniak

New Type of the Test Stand for Surfaces and Lubricant Tribological Properties Test 584
Marcin Kamiński, Dawid Pogorzelski, Andrzej Juś, Tadeusz Missala, Roman Szewczyk, Wojciech Winiarski, Marcin Safinowski, and Marek Hamela

Analysis of Automated Ferromagnetic Measurement System 593
Tomasz Charubin, Michał Urbański, and Michał Nowicki

Experimental Research of Improved Sensor of Atomic Force Microscope 601
Vytautas Bučinskas, Andrius Dzedzickis, Ernestas Šutinys, Nikolaj Šešok, and Igor Iljin

Survey on River Water Level Measuring Technologies: Case Study for Flood Management Purposes of the C2-SENSE Project 610
Anna Bączyk, Jan Piwiński, Rafał Kłoda, and Mateusz Grygoruk

Error Ratio of a Measuring Instrument Under Calibration and the Reference Standard: Conditions and Possibilities of Decrease 624
Valerii A. Granovskii and Mikhail D. Kudryavtsev

Generalized Description of the Frequency Characteristics of Resistors 630
Stefan Kubisa and Zygmunt L. Warsza

Engineering Materials

A New Magnetic Method for Stress Monitoring in Steels 647
Evangelos V. Hristoforou, Aphrodite Ktena, Polyxeni Vourna, Eleni Mangiorou, and Stelios Mores

Study of Ultrasonic Characteristics of Ukraine Red Granites at Low Temperatures 653
Valentyn Korobiichuk

Study of the Durability of Reinforced Concrete Structures of Engineering Buildings 659
Lyudmyla Kuzmych and Volodymyr Kvasnikov

Comparison and Analysis of Steel and Tungsten Carbide Rockwell B Hardness Ball Indenters Utilizing a General Purpose Finite Element Approach 664
Vladimir Skliarov and Maxim Zalohin

Modeling of Freeform Reflecting Surface for LED Device 674
Natasha Kulik

Investigation on Functional Properties of Hall-Effect Sensor Made of Graphene 682
Oleg Petruk, Maciej Kachniarz, Roman Szewczyk, and Adam Bieńkowski

Possibilities of Application of the Magnetoelastic Effect for Stress Assessment in Construction Elements Made of Steel Considering Rayleigh Region 689
Dorota Jackiewicz, Maciej Kachniarz, Adam Bieńkowski, and Roman Szewczyk

Influence of Temperature and Magnetizing Field on the Magnetic Permeability of Soft Ferrite Materials 698
Maciej Kachniarz and Jacek Salach

Selected Trends in New Rapidly Quenched Soft Magnetic Materials 705
Peter Švec, Irena Janotová, Juraj Zigo, Igor Matko, Dušan Janičkovič, Jozef Marcin, Ivan Škorvánek, and Peter Švec, Sr.

The Comparison of Rapidly Quenched Co-Sn-B and Fe-Sn-B Alloys 713
Irena Janotová, Peter Švec, Sr., Peter Švec, Igor Mako, Dušan Janičkovič, Juraj Zigo, Jozef Marcin, and Ivan Škorvánek

General Science and Technology

Determining Prospects of European Countries' Positive Experience Implementation into the Water Consumption Process in Ukraine 723
Valerij Shygonskyy

Diesel Exhaust Gases Centrifugal-Jet Filter-Converter 734
Andrii Ilchenko, Vladislav Balyuk, and Neonila Kosnitskaya

Precautionary Statistical Criteria in the Monitoring Quality of Technological Process 740
Eugenij Volodarsky, Zygmunt Warsza, Larysa A. Kosheva, and Adam Idźkowski

Principles of Implementing an Electronic Progress Log at "Zhytomyr Nursing Institute" KVNZ 751
Svetlana Gordiichuk

Chi-Squared Goodness-of-Fit Tests: The Optimal Choice of Grouping Intervals 760
Ekaterina V. Chimitova and Boris Yu. Lemeshko

Homogeneity Tests for Interval Data 775
Stanislav S. Vozhov and Ekaterina V. Chimitova

On the Properties and Application of Tests for Homogeneity of Variances in the Problems of Metrology and Control 784
Boris Yu. Lemeshko and Tatyana S. Sataeva

E2LP Remote Laboratory: Evolution of the System and Lessons Learned 799
Rafał Kłoda and Jan Piwiński

Modern Business Internet Technology: Trends, Perspectives and Risks 810
Svetlana Ķumova and Olga Toropova

Quaternion Based Dynamics for Servicing Satellite with Mission Protocol Description 816
Elżbieta Jarzębowska and Michał Szwajewski

Author Index 827

About the Editors

Prof. Roman Szewczyk received both his PhD and DSc in the field of mechatronics. He specializes in the modelling of properties of magnetic materials as well as in sensors and sensor interfacing, in particular magnetic sensors for security applications. He leads the development of a sensing unit for a mobile robot developed for the Polish Police Central Forensic Laboratory and of methods of non-destructive testing based on the magnetoelastic effect. Professor Szewczyk has been involved in over 10 European Union funded research projects within the FP6 and FP7 as well as projects financed by the European Defence Organization. Moreover, he was leading two regional and national scale technological foresight projects and was active in the organization and implementation of technological transfer between companies and research institutes. Roman Szewczyk is Secretary for Scientific Affairs in the Industrial Research Institute for Automation and Measurements (PIAP). He is also Associate Professor at the Faculty of Mechatronics, Warsaw University of Technology and a Vice-chairman of the Academy of Young Researchers of the Polish Academy of Sciences.

Dr. Małgorzata Kaliczyńska received her M.Sc. Eng. degree in Cybernetics from the Faculty of Electronics, Wrocław University of Technology, and her Ph.D. degree in the field of fluid mechanics from the Faculty of Mechanical and Power Engineering in this same university. Now she is Assistant Professor in the Industrial Research Institute for Automation and Measurement (PIAP) and Editor of the scientific and technological magazine "Measurements, Automation, Robotics". Her areas of research interest include distributed control systems, Internet of Things, information retrieval and webometrics.

Systems and Control

Tender Participants Selection Based on Artificial Neural Network Model for Alternatives Classification

Tetiana Kolpakova(✉) and Valerii Lovkin

Zaporizhzhia National Technical University, Zaporizhzhya, Ukraine
t.o.kolpakova@gmail.com, vliovkin@gmail.com

Abstract. Business-to-business, business-to-government, business-to-consumer tender support is considered. The problem of tender participants selection is stated as a separate stage of tendering process. It is the problem of tender alternatives classification on the tender offer demands accordance. The method of tender participants selection was proposed for the problem solving. The method is based on artificial neural networks application. It is proposed to use feedforward neural networks with a hidden layer. The results of expert evaluation of tender alternatives (project value, project due date, technical parameters etc.) are used as neural networks inputs. Neural network models configuration is realized based on evolutionary modeling heuristic approach which allows to use this method for tendering process support in different subject fields. Application of the proposed method allows to select the set of tender alternatives which should be financed together or should be used interchangeably if only one tender alternative is financed. Information technology in the form of web-based system was developed on the basis of client-server architecture. The experimental investigation of the proposed method was conducted for the tender participants selection problem solving in the building projects realization support. The received experimental results allow to recommend the proposed method for use in practice.

Keywords: Tendering process · Artificial neural network · Genetic algorithm · Classification problem

1 Introduction

Tenders are based on competitive placing of orders for works by the terms predeclared in documentation, within the specified time limits on equity and effectiveness bases. Tenders are aimed at selection of such a product (service) supplier which is able to meet all the customer requirements (costs, delivery time, goods quality, etc.) [1]. Tenders are realized with finite resources in the following forms: business-to-business (B2B), business-to-government (B2G), business-to-consumer (B2C).

The sequence of tendering process stages could differ depending on subject field or conducting country [2, 3].

Before the interaction with tender participants significant organizational efforts to select experts and systematize requirements including transformation of verbal customer requests to formal representation should be done. Preliminary selection and registration

R. Szewczyk and M. Kaliczyńska (eds.), *Recent Advances in Systems, Control and Information Technology*, Advances in Intelligent Systems and Computing 543, DOI 10.1007/978-3-319-48923-0_1

is an optional but eligible stage for complex (for example building) projects [4]. These stages demand confirmative documents package from participants. The package is audited estimating reputation, financial conditions, opportunities and professional qualifications of every participant. Participants are divided into groups on the audit results basis and classified participants are rated.

The model of electronic tendering process was proposed in [5].

The purpose of this paper is to investigate tender participants selection problem.

2 The Problem Statement

New conditions which are unacceptable for one of the parties may appear during contract conditions specification stage. In this case it is necessary to apply to the participant which is placed next in the rating or to hold tender again. Besides tender organizers can make contracts with several suppliers which tender analogous products and services. It helps to avoid the risks of supply delays because of diversification. So it is necessary not only to rate alternatives but also to define alternatives which correspond to the main customer demands.

The problem of tender participants selection is a classification problem where alternatives $M = \{M_i \mid i = 1..N\}$ should be assigned to one of two classes $Y = \{y_0, y_1\}$:

- y_0 is corresponding to the tender task;
- y_1 isn't corresponding to the tender task.

Every alternative M_i is represented as a vector $X = \{x_k\}$. Components x_k of this vector represent different characteristics of i-th alternative which influence on the given sample classification decision making. The set of characteristics is defined by the tender task. Set of values for every criterion from tender offer could form the vector for the given alternative. The task of the classifier is to relate the sample to the class from set Y in accordance with the appointed partition of K-dimensional input space, so the dimension of the space corresponds to the number of vector components [6–9].

The model of tender participants selection problem can be represented by the following expression:

$$Y = f(X) = f\{x_k \mid k = 1..K\}, X \in M. \tag{1}$$

The level of M_i alternative's belonging to the class y_1 could be considered as risk of alternative's tender task mismatch.

At present tender participants classification isn't defined as a separate tender stage, so all the selected tender alternatives are considered in rating. If alternatives are classified it is made by decision maker manually based on personal experience without mathematical and technical support.

The proposed tender participant selection problem statement allows to exempt decision maker from manual analysis and allows to take founded decision based on the previous experience. So the presented problem is actual.

Every alternative M_i could be considered as a project which is a temporary endeavor undertaken to create a unique product or service [10]. Such endeavor is undertaken within the given time and budget. Tender task accordance estimation could be considered as estimation of project accordance to the expected result (product), that is a risk which belongs to quality risks. So if the given criterion isn't met completely, then realization of this project cannot result in the required product creation. When experts evaluate alternatives, this risk can be considered only as one criterion, so its influence can be decreased by other criteria. When alternatives classification is a separate stage, all alternatives, which are selected in this stage, can be financed.

3 The Method of Tender Participants Selection Based on Artificial Neural Network Model for Alternatives Classification

In the paper it is proposed to solve the tender participants selection problem (1) using artificial neural networks (ANN) [6, 7]. Feedforward ANNs are universal tools for function approximation and can be trained, so ANNs can be efficiently used for solving of classification problems [11–13]. This approach is especially effective for problems of expert judgment because ANNs allow to process huge number of factors which belong to different types (quantitative or qualitative).

The proposed architecture of ANN should reflect tender participants estimation problem decomposition. It is proposed to use three-layered feedforward ANN with one hidden layer (Fig. 1). In the proposed ANN vector $X = \{x_k \mid k = 1, \ldots, K\}$ is transformed by the hidden layer to the space with dimension of $h = 1, \ldots, H$ and then hyperplanes which correspond to the output layer neurons divide it into the classes.

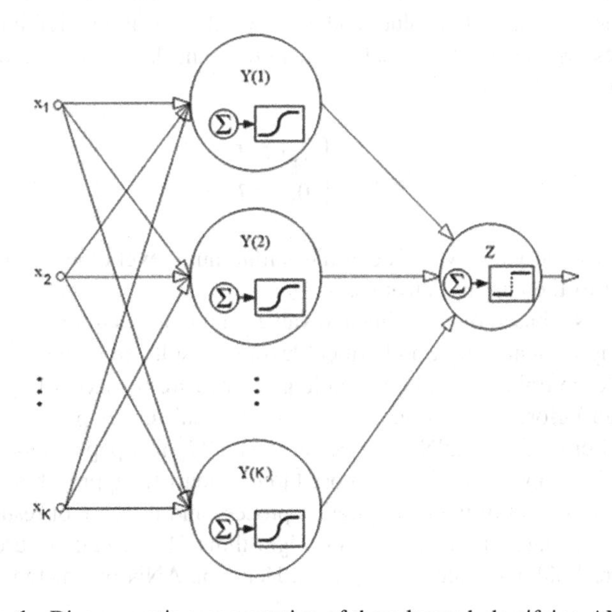

Fig. 1. Diagrammatic representation of three-layered classifying ANN

Every input neuron value x_k should be valued at $W_{alt}(i,k)$ which is absolute estimate of customer demands correspondence level for i-th alternative based on Cr_k criterion:

$$x_k = W_{alt}(i,k), x_k \in [0,1]. \tag{2}$$

For every neuron of the hidden layer the weighted sum of the input layer values should be calculated for every $k = 1, \ldots, K$, using neuron threshold value b_k:

$$Y_k = \sum_{q=1}^{K} x_q w_1(q,k) + b_k. \tag{3}$$

Sigma function should be used as transfer function of the hidden layer neurons, so the values of neuron outputs aren't binary and belong to interval [0, 1]:

$$f(x) = \frac{1}{1+e^{-x}}. \tag{4}$$

Adder function of the third layer neuron Z accumulates values from the hidden layer neurons weighted by w_2:

$$Z = \sum_{k=1}^{K} Y_k w_2(k),\ k = 1, \ldots, K. \tag{5}$$

Z is a risk of tender task demands mismatch of the given alternative.

Output layer gets calculated values and executes classification, dividing alternatives into two classes: y_0 and y_1. So transfer function of the last layer neuron should be threshold function:

$$f(x) = \begin{cases} 1, x \geq T, \\ 0, x < T, \end{cases} \tag{6}$$

where T is a threshold value which determines minimum level of estimated alternative correspondence to the current demands.

Value of T is set based on experimental values and belongs to the interval from 0.4 to 0.7 depending on tender task and critical level of results consequences.

ANN should be trained based on sample generated from successfully held tenders participants data before application. But every practical tender problem has its own alternatives and criteria, so ANN must be set manually. The papers [14, 15] consider alternative training approaches. It is proposed to use heuristic approach to evolutionary modeling for automation of the configuration process and increase of results reliability [15]. For the considered problem genetic algorithms [17] could be used for ANN weights and threshold automated setting [18]. Heuristic ANN parameters setting helps to get sufficiently precise results during reasonable time. Application of the combined neuro-evolutionary approach allows to unify tender participants classification problem solving.

For evolutional setting of ANN model neuron weights and thresholds are encoded into individual chromosome and then its fitness is estimated based on training results [16]. Crossing should be realized in such a way: new individuals should inherit connection weights which are presented in both parents and values which are different should be chosen randomly. Mutation is implemented by random modification of weights of one or several connections in individual chromosome.

ANN training based on heuristic approach is realized during the following stages.

At the initial stage input data are prepared. Test data are processed, generating samples $x(j, k), \ldots, x(j, 0), j = 1, \ldots, K, k = 1, \ldots, K$.

ANN architecture is chosen based on Fig. 1.

It is proposed to initialize ANN weights and thresholds for initial population by the following values: $w_1(q,k) = \frac{1}{K} + R$, $w_2(p) = \frac{1}{K} + R$, $b_k = R$, where $q = 1, \ldots, K$, $k = 1, \ldots, K$, $p = 1, \ldots, K$ and for every value of w_1, w_2 R is a random value from the interval [0, 1]. Variety of individuals in initial population is provided by random component.

Then optimal weights and thresholds should be calculated using genetic algorithm. It is proposed to implement mutation for considerable part of population to provide variety of genotypes on every step.

For every individual its fitness level should be calculated as classification error value β, where β is calculated by deviation of test sample (tender alternatives set) classification results from actual results, so β is determined by alternatives which correspond to the tender task but were classified as alternatives which don't correspond to the tender task and vice versa:

$$\beta = \frac{\sum_{i=1}^{n} |f(Z) - f(Z_0)|}{n}, \tag{7}$$

where n is a number of alternatives in test sample T; $f(Z)$ and $f(Z_0)$ are current alternative classification result value received by ANN and from the sample accordingly.

Population should be sorted by decrease of β and after that selection is implemented. Based on elitist selection principle 10% of the most adapted individuals of the current population should be selected without modification, 50% of individuals should be generated by crossing and 40% – by mutation. For genetic algorithm implementation individuals should selected not only from the best individuals. It should help to avoid early algorithm convergence to local optimum.

It is proposed to set training error level β_0 as algorithm termination condition. It is necessary to continue training while the fittest individual corresponds the condition $\beta \leq \beta_0$. Training time should be limited for situations where the level β_0 is unachievable.

The genotype with training error β_0 could be considered as acceptable and could be used for tender participants selection problem solving.

ANN classifies all tender alternatives from set M based on estimates of every expert, generating local lists: list of alternatives which correspond to the tender task L^1 and list of alternatives which don't correspond to the tender task L^0:

$$f : M \rightarrow \{L^1, L^0\}. \quad (8)$$

Based on individual results every alternative should be estimated according to the number of inclusions to the lists L^0. Every alternative should be as high in the rating as low its estimate is. If threshold value E is achieved, the given alternative should be excluded from rating because it doesn't correspond to the tender task.

The method of tender participants selection based on ANN model for alternatives classification can be presented by the following stage sequence:

- ANN architecture is chosen based on tender participants selection problem characteristics: feedforward ANN with a hidden layer, K neurons on the input and hidden layers and one neuron on the output layer;
- Generation of sample T based on successfully held tenders with analogue tender task;
- ANN training based on genetic algorithm for weight and threshold setting;
- Generation of alternatives set M based on current tender and expert estimation results on K criteria;
- Classification of alternatives M_i, $i = 1..N$ based on trained ANN;
- Aggregation of classification results received on the basis of individual expert estimates and generation of lists L^1 and L^0.

4 Experimental Investigation

The proposed method was implemented in tender support software which was developed based on client-server architecture using web-technologies. Server supports centralized management of all company tenders, storage of all tender data, system analysis of tender materials by experts and expert questioning when it is comfortable.

Historical data of tender tasks, participant offers and tender results accumulated by design companies LLC «Inzhenernye systemy» and «VST» and expert estimates were proposed for experimental investigation of the tender participants selection problem. These companies used price estimation or verbal discussions for holding tenders.

It's important to note that tender results can be observed in long time horizon, so evaluation of tender efficiency can be made after a period from one to six months after it finishes.

75% of historical data were used for generating samples T for training, residual data were used for generating set C which was used for classification based on trained ANN. This process was repeated iteratively and data were divided into sets T and C randomly at every iteration. Classification using set C and described approach by the method of tender participants selection had an error of 11.17%.

The proposed method was applied for tenders from set C based on comparison with held tenders. The comparison of the proposed method with current approach of design companies showed decrease of actual building projects costs for 4.7% and actual duration – for 9.2%.

In the following part the method of tender participants selection is applied for new building tender. At the beginning expert estimation was implemented, alternatives were rated and then the proposed method was applied. In this tender on supply of materials and equipment for installation of commercial centres building frames the following companies took part: "Master Profi" (21.31%), "Remko" (22.4%), "Atlas Word" (24.35%), "Askon" (11.81%), "IBT" (20.13%). For every company expert estimates are marked.

The following criteria which formed set X for every alternative were used for alternatives estimation during this tender: building frame cost, delivery date, additional constructions (rooftop unit frame, two-storeyed mezzanine), additional and installation works (costs and terms accounting), design works accounting, terms of delivery, terms of payment, productive capacity, availability and terms of materials supply.

After tender alternatives were estimated by experts of tender committee, the received estimates were used by ANN for classification by the method of tender participants selection. The following results were received: "Master Profi", "Atlas Word", "IBT" – correspond, "Remko", "Askon" – don't correspond to the tender task.

It is significant that tender offer by "Remko" company was highly rated by experts because of the low project costs and relatively early due date but system recommendations showed that interaction with this company isn't desirable, because there are significant tender task mismatch and some specific costs aren't taken into account in project costs. So in this case risk of tender task demands mismatch for the given alternative was higher than acceptable level and "Remko" company shouldn't be selected for the next tendering process stage. It confirms hypothesis which was moved in the result of problem statement.

5 Conclusion

The tender participants selection problem is stated. The problem is a classification problem. Tender alternatives proposed by participants are classified by the tender task accordance based on values of its characteristics.

The method of tender participants selection based on ANN model for alternatives classification was proposed for the problem solving. This method enables to exclude participants which offers don't correspond tender task from rating. This possibility differs this method from the existent alternative selection approach. It allows to avoid problems which are nontrivial at the problem estimation stage during interaction with the selected participants. It is proposed to use feedforward ANNs with a hidden layer in the method of tender participants selection. ANNs are configured and trained using the combined neuro-evolutionary approach.

Experimental investigation confirmed efficiency of the proposed method application for building project tenders.

References

1. Hashim, F., Alam, G.M., Siraj, S.: Information and communication technology for participatory based decision-making-E-management for administrative efficiency in higher education. Int. J. Phys. Sci. **5**(4), 383–392 (2010)
2. Du, R., Foo, E., Boyd, C., Fitzgerald, B.: Defining security services for electronic tendering. In: The Australasian Information Security Workshop: Conferences in Research and Practice in Information Technology, Dunedin, New Zealand, vol. 32, p. 10 (2004)
3. Yoong, N., Omran, A., Othman, O., Ramli, M., Bakar, H.A.: Contractor business strategy decision in competitive bidding: case studies. In: The International Conference on Economics and Administration, Bucharest, Romania, pp. 273–285 (2009)
4. Anagnostopoulos, K.P., Vavatsikos, A.P.: An AHP model for construction contractor prequalification. Oper. Res. **6**(3), 333–346 (2006)
5. Kolpakova, T.: Modelling of the tendering process. Cent. Eur. Res. J. **1**(1), 50–54 (2015)
6. Haykin, S.: Neural Networks and Learning Machines. Prentice Hall, Upper Saddle River (2008)
7. Korobiichuk, I., Bezvesilna, O., Ilchenko, A., Shadura, V., Nowicki, M., Szewczyk, R.: A mathematical model of the thermo-anemometric flowmeter. Sensors **15**, 22899–22913 (2015). doi:10.3390/s150922899
8. Oliinyk, A., Zaiko, T., Subbotin, S.: Training sample reduction based on association rules for neuro-fuzzy networks synthesis. Opt. Mem. Neural Netw. (Inf. Opt.) **23**(2), 89–95 (2014)
9. Oliinyk, A.O., Olijnyk, O.O., Subbotin, S.A.: Agent technologies for feature selection. Cybern. Syst. Anal. **48**(2), 257–267 (2012)
10. A Guide to the Project Management Body of Knowledge: PMBOK Guide, 5th edn. (2013)
11. Kar, A.K.: Using Fuzzy Neural Networks and Analytic Hierarchy Process for Supplier Classification in e-Procurement. Sprouts: Working Papers on Information Systems. No. 9 (2009)
12. Oliinyk, A.O., Zayko, T.A., Subbotin, S.A.: Synthesis of neuro-fuzzy networks on the basis of association rules. Cybern. Syst. Anal. **50**(3), 348–357 (2014)
13. Oliinyk, A., Subbotin, S.A.: The decision tree construction based on a stochastic search for the neuro-fuzzy network synthesis. Opt. Mem. Neural Netw. (Inf. Opt.) **24**(1), 18–27 (2015)
14. Oliinyk, A.O., Skrupsky, S. Yu., Subbotin, S.A.: Using parallel random search to train fuzzy neural networks. Autom. Control Comput. Sci. **48**(6), 313–323 (2014)
15. Oliinyk, A.O., Skrupsky, S. Yu., Subbotin, S.A.: Experimental investigation with analyzing the training method complexity of neuro-fuzzy networks based on parallel random search. Autom. Control Comput. Sci. **49**(1), 11–20 (2015)
16. Tsoy, Y.R., Spitsyn, V.G.: Using genetic algorithm with adaptive mutation mechanism for neural networks design and training. Opt. Mem. Neural Netw. **13**(4), 225–232 (2004)
17. Chambers, L.D. (ed.): The practical Handbook of Genetic Algorithms: Applications, 2nd edn. Chapman & Hall, CRC, Boca Raton (2000)
18. Korobiichuk, I.: Mathematical model of precision sensor for an automatic weapons stabilizer system. Measurement **89**, 151–158 (2016). doi:10.1016/j.measurement.2016.04.017

The Dimensionality Reduction Methods Based on Computational Intelligence in Problems of Object Classification and Diagnosis

Sergey A. Subbotin(✉) and Andrii A. Oliinyk

Zaporizhzhya National Technical University, Zaporizhzhya, Ukraine
subbotin@zntu.edu.ua, olejnikaa@gmail.com

Abstract. The set feature selection methods based on the paradigms of computational intelligence (evolutionary search and swarm intelligence) is proposed. Proposed methods speed up the search through the creation of special operators, taking into account a priori information about the data sample and concentrating search on the most perspective solution areas. This allows preserving the stochastic nature of the search to accelerate the obtainment of acceptable solutions through the introduction of deterministic component in the search strategy. The theoretical estimates of the computational (temporal) and spatial complexity of the developed methods are obtained. The proposed methods are experimentally studied on a set of problems of automatic object classification, technical and medical diagnosis. On the results of experiments the comparative characteristics and recommendations for the use of the proposed methods are given.

Keywords: Dimensionality reduction · Diagnosis · Classification · Feature selection

1 Introduction

The traditional tool for data sample dimensionality reduction is a feature selection methods [1–3], that select from a given set of input features a subset of the most significant features for computing the output feature value. According to the type of procedure of new solution formation on the basis of existing decisions they are divided on exhaustive search methods [1, 4] (implement deterministic transition from one decision to another, are highly complex and time-consuming) and stochastic search methods [2, 5] (implement random transition from the current to new solutions, but probabilistically taking into account information about previous solutions, avoid the falling into local minima, and find suboptimal solutions for a limited time).

The usage of known exhaustive methods [1, 4, 6] requires a large number of evaluations of feature combinations, produced from the initial set of features, which makes impossible to use this approach for a big number of features in the initial set, since it has a huge computational cost. Therefore, for feature selection it is reasonable to use the intelligent stochastic search methods [2, 3, 7], because they are more suited to finding new solutions by combining the best solutions obtained at different iterations

R. Szewczyk and M. Kaliczyńska (eds.), *Recent Advances in Systems, Control and Information Technology*, Advances in Intelligent Systems and Computing 543, DOI 10.1007/978-3-319-48923-0_2

and does not require a search of all solutions. However, known intelligent search methods are slow because they do not use the problem specific information in the process of a search.

Therefore, the aim of this work is to create a complex of Computational Intelligence methods for solving problems of feature selection making possible to accelerate the obtaining of solutions.

2 Formal Problem Statement

Let $<x, y>$ is a sample of precedents (cases, instances, exemplars), where $x = \{x^s\}$, $x = \{x_j\}$, $x^s = \{x_j^s\}$, $x_j = \{x_j^s\}$, $y = \{y^s\}$, $s = 1, 2, \ldots, S$, $j = 1, 2, \ldots, N$; x_j^s is a value of j-th input feature x_j of exemplar x^s, y^s is an output for precedent x^s (for the classification problems $y^s \in \{1, 2, \ldots, K\}$, $K > 1$, K is the number of classes), S is a number of precedents, N is a number of input features. Denote as $<x', y'>$ a fragment of $<x, y>$, as N' a number of features in $<x', y'>$, as $f()$ a user criterion describing the quality of argument according to the decided task, as *opt* an optimal desirable or acceptable value of the functional $f()$ for the problem. Then the problem of feature selection is for given $<x, y>$ find $<x', y>$, $x' \subset \{x_j\}$, $N' < N$, $S' = S$, $f(<x', y>, <x, y>) \rightarrow opt$.

3 Background of Computational Intelligence Based Search

Among the stochastic search methods the wide usage have evolutionary [3, 7, 8] and multi-agent [9–11] methods.

The evolutionary methods based on the idea of the evolution of solutions through simulation of natural selection. At the beginning the epoch counter setted as $t = 0$ and method form an initial population of solutions $P_t = \{H_j\}$, where $j = 1, 2, \ldots, N$, N is a number of population solutions, $H_j = \{h_{ij}\}$ is a j-th solution, h_{ij} is a value of i-th bit (gene) of j-th solution, $i = 1, 2, \ldots, L$, where L is a length of the solution. Then performed a cyclic replacement of one population solutions by the next more adapted. For this the solutions fitness is estimated for current t-th population $\{f(H_j)\}$. Then the checking of search stop conditions (reaching the epochs limit or a reasonable value of the objective function) is performed. If the conditions are satisfied then the search will stop, and otherwise for the current population of solutions method execute selection operator (on the basis of $f(H_j)$ values it probabilistically select solutions for crossover and mutation, with a preference for more adapted solutions), crossover operator (forming parent pairs from which create new solutions) and mutation operator (made spontaneous changes in decisions to increase the diversity of solutions).

On the basis of produced solutions a new generation of is formed, which includes new solutions, as well as the best solutions, the objective function values of which are the best in the population. Crossover of the fittest solutions results in that the most perspective search space parts are explored. Finally, the population of solutions will converge to an optimal task solution.

Multi-agent methods [3, 9–11] is a group of multi-dimensional stochastic search methods based on modeling the behavior of self-organizing agent colonies having the collective nature and provided by coordination (controls the spatial-temporal agent placement), specialization (distributes agent actions and form the spatial, temporal and social relations arising in the colony's work), cooperation (agents association mechanisms) and collective decision-making of agents (colony allows agents to react on changes in the external environment). These methods includes indirect agent communication methods (ant colony optimization (ACO) [3, 7], particle swarm optimization (PSO) [11], bacteria foraging optimization (BFO) [9] and direct communication methods (Bee Colony Optimization (BCO) [10]).

The methods of stochastic search are able to optimize the functions of any complexity without additional requirements on the objective function (unimodality, continuity, smoothness, monotonicity, differentiability). Thus, the use of stochastic search techniques is appropriate in the problems of feature and sample selection.

Despite these advantages the intelligent methods based on stochastic search has such common disadvantages as high iteratively and dependability of the methods speed from the choice of the search starting point (usually defined randomly) and a priori uncertainty of the time of convergence of methods due to the stochastic nature of the search, as well as neglect of search operators in available information from sample. Therefore, the improvement of the speed of intelligent stochastic search methods by reducing these disadvantages is urgent.

4 The Feature Selection Based on the Intelligent Stochastic Search

To speed up the feature selection based on stochastic search we propose to retain their stochastic nature add them deterministic components by modifying the search operators, under which they will take into account a priori information about the data sample. As such information we propose to use the individual evaluations of feature informativity [3, 11]. Since proposed methods are based on classical methods of intelligent stochastic search further we shall only present their key differences from the known basic methods.

4.1 Feature Selection Method Using the Entropy

This method as a search model use a canonical evolutionary search [8] with improvements listed below. Solution forming stage: set the probability of inclusion of i-th feature to the solution: $P_i = 1 - e_i$, where e_i – entropy [9, 10] of i-th feature. Solution crossover stage: use mask $h^* = \{h_i^*\}$, where $h_i^* = 1$, if $e_i < e_{\text{п.}}$ ($e_{\text{п.}}$ – threshold), $h_i^* = 0$, otherwise. Solution mutation stage: the mutation probabilities P_{M} of features with high e_i are increased, and with low e_i are decreased. Solution selection stage: the goal function is defined as

$$f(H_j) = \left(1 + \left(\sum_{i=1}^{N} h_{ij}\right)^{-1} \left(\sum_{i=1}^{N}\sum_{k=1}^{N} h_{ij}h_{ik}d_{ik}\right)E_j\right)^{-1} \left(\sum_{i=1}^{N} e_i \cdot h_{ij}\right)^{-1}, \; E_j \geq 0, \quad (1)$$

where H_j is a j-th solution, h_{ij} is an i-th digit of j-th solution, E_j is a model error for j-th solution, d_{ik} is an individual informativity value of i-th feature relative to k-th feature.

4.2 Feature Selection Method with Feature Grouping

This method as a search model use a canonical evolutionary search [3, 8] with improvements listed below. Solution forming stage: the set of solutions $\{H_j\}$, $H_j = \{h_{ij}\}$ is formed, where h_{ij} is a digit specifying usage of i-th feature taking into account its individual informativity value I_i. Solution crossover stage: using mask $h^* = \{h_i^*\}$, where $h_i^* = 1$, if $I_i > I_{\text{tr.}}$, ($I_{\text{tr.}}$ is a previously specified threshold value), $h_i^* = 0$, otherwise. Solution mutation stage: the mutation probabilities P_{M} of features with high I_i are decreased, and with low I_i are increased. Solution selection stage: the goal function is defined as:

$$f(H_j) = \left(\sum_{i=1}^{N} I_i h_{ij}\right)\left(1 + \left(\sum_{i=1}^{H} h_{ij}\right)\left(\sum_{i=1}^{N}\sum_{k=1}^{N} h_{ij} h_{ik} d_{jk}\right) E_j\right)^{-1}, E_j \geq 0. \qquad (2)$$

4.3 Evolutionary Feature Selection Method with Clustering

This method as a search model use a canonical evolutionary search [3, 8] with improvements listed below.

Initialization stage: the individual feature informativity values $\{I_i\}$ determined relatively to outputs, and the feature relations indicators are determined. The similar features are grouped. The most informative feature in each group is selected, that is a center of a cluster C_q.

Solution forming stage: the probability of i-th feature inclusion to the solution is determined on the Euclidean distance basis from it to the cluster center $d(x_i,C_q)$, also as the feature and cluster center individual informativities I_i and I_{Cq}, then $P_i = I_i + (I_i - I_{Cq})\ \{d(x_i, C_q)/d_{q\max} \mid x_i \in C_q, q = 1, 2, \ldots, Q\}$, where $d_{q\max}$ is a maximal distance in q-th cluster.

4.4 Feature Selection with Fixing of the Search Space Part

This method at initialization stage order features by I_i growing, then first $\alpha\eta N$ features are deleted, where α is a given coefficient, $0 < \alpha \leq 1/\eta$, η is a search space reduction coefficient:

$$\eta = N^{-1} \sum\nolimits_{i=1}^{N} \{1 | I_i < I_{\text{avg.}}\}, \qquad (3)$$

where $I_{\text{avg.}}$ is an average feature informativity value. Then method perform canonical evolutionary search for reduced feature set.

4.5 Multi-agent Feature Selection Method with Representation of the Destination Points by Features

This method as a search model use a multi-agent search with indirect communication on the basis of ACO model [3, 13] with such improvements. Search space forming: vertices of the search graph (destination points) represented by features. Search strategy: agent must pass the way to the specified number of destinations N', which determines the number of features that must be leave. Solutions encoding: path traveled by the j-th agent is a feature set (decision) H_j, on which the model is constructed: $H_j = \{h_{ij}\}$, $h_{ij} = 1$, if the point i is included in the j-th agent path, $h_{ij} = 0$ – otherwise.

4.6 Multi-agent Feature Selection Method Based on BFO Model

This method as a search model use multi-agent search with indirect agent communication based BFO model [9] with following improvements. Search space: each point in a search space represented by binary string, which consists of feature informativities as digits. Search strategy: the BFO model is hybridysed by using evolutionary operators of proportional selection and simple mutation. Search speed increasing: at the initialization stage for each i-th agent, $i = 1, 2, \ldots, H$, set initial position: $h_{ji} = 1$, if $rand_{ji} < I_j\left(\sum_{q=1}^{N} I_q\right)^{-1}$, $h_{ji} = 0$, otherwise, where $rand_{ji} \in [0; 1]$ is random numbers, I_j is an individual informativity value of j-th feature, $j = 1, 2, \ldots, N$.

4.7 Multi-agent Feature Selection Method with Representation of the Destination Points by Feature Informativities

This method as a search model use a multi-agent search with indirect communication on the basis of ACO model [3, 13] with listed below enhancements. Search space forming: search graph vertices (destination points) represented by randomly formed binary set $B = \{b_i\}$, $b_i = \{0;1\}$, $i = 1, 2, \ldots, N$, which contains N' element equal to one. Search strategy: each agent must pass the way to all destination points. Solutions encoding: the path traveled by the j-th agent by destination points is a binary set of informative features (decision) B used for model construction.

4.8 Multi-agent Feature Selection Method with Direct Agent Communication

This method as a search model use a multi-agent search with direct agent communication based BCO model [10, 14] with such improvements. Search space: $(N' \times N)$ – dimensional space, the feature informativity is a resource collected by the agent. Solution evaluation: based on agent stay on the iteration t in source h, in which $H^h(t)$ agents are located: $I^h(t) = a_h/H^h(t)$, $h = 1, 2, \ldots, N' \times N$, where $a_h = \varepsilon^*/E_h$, E_h is a model error for source h, ε^* is a reduction factor. Search speed increasing: when the agents-scouts starts the one part of scouts randomly placed in the search space (a condition for global search), and the other part is placed in the sources of resources proportionally the values of their informativity (accounting a priori information).

5 Complexity Analysis

For the developed feature selection methods at the sequential implementation of computations we obtain analytical assessments of temporal (computational) and spatial complexities (Table 1). Table 1 use the following notation: H is a number of agents (for multi-agent search) or solutions (for the evolutionary search), F is a complexity of the simulation of neuro-fuzzy network based model and the complexity of the error calculation on its base for the learning sample, T is a number of iterations taken by the method for search. The precise estimates of the complexity in the Table 1 are given in the soft form (without suppressing of members with less powers by members with greater powers). Taking $n = NS \approx N^2$ we present rough estimates in hard form (members with greater powers suppress members with a lesser powers).

Table 1. Estimates of the complexity of feature selection methods based on stochastic search

Method	Time complexity		Spatial complexity	
	Accurate	Rough	Accurate	Rough
4.1	$O(12NS + N^2 + N + THF)$	$O(n^2\sqrt{n})$	$O(13NS + 0.0625HFN)$	$O(n^2)$
4.2	$O(12SN(N + 1) + N + THF)$	$O(n^2\sqrt{n})$	$O(N(S + N + 2) + 0.0625HFN)$	$O(n^2)$
4.3	$O(12NS + N + THF)$	$O(n^2\sqrt{n})$	$O(NS + 2N + 0.0625HFN)$	$O(n^2)$
4.4	$O(1.5S(N - 1) + C + 5N + TFH)$	$O(n^2\sqrt{n})$	$O(NS + 3N + 0.0625HFN + C)$	$O(n^2)$
4.5	$O(TH(N^2 + 4N + H + F))$	$O(n^2\sqrt{n})$	$O(NS + H(3N + F + H))$	$O(n\sqrt{n})$
4.6	$O(TH(N^2 + F + 4N))$	$O(n^2\sqrt{n})$	$O(NS + 3HN + HF)$	$O(n\sqrt{n})$
4.7	$O(8H^2 + 6H^2N + HF + N + TH(2F + H + N + 4))$	$O(n^2\sqrt{n})$	$O(NS + HN + 3HF)$	$O(n\sqrt{n})$
4.8	$O(H + TH(F + 9 + HN - H))$	$O(n^2\sqrt{n})$	$O(NS + 5H + HN^2 - HN + FH)$	$O(n\sqrt{n})$

6 Experiments and Results

To study the proposed methods they have been implemented as software. The experimental investigation of the proposed methods and software was performed by solving practical problems, which sample characteristics shown in Table 2.

The results of experimental study are shown in Table 3. Here t is a time of feature selection, Nm is a computer memory volume used by the method, N'/N is a the proportion of selected features in original feature set.

Our experiments confirmed the efficiency of the developed methods. The proposed feature selection methods allow to significantly increase the speed of work, compared with the canonical evolutionary search [3, 8], because they use fewer references to the objective function and, consequently, require fewer model constructs.

The proposed methods based on evolutionary search can be recommended for use in cases where there are no significant limitations on the available computer memory, and methods based on multi-agent search – when computer memory is limited.

Table 2. The practical tasks collection for experiments

Task	Code	N	S
Diagnosis of air-engine blades [3]	SIGNAL	10240	32
Simulation the total index of quality of life of patients sick of a chronic obstructive bronchitis [11]	KOG	106	86
Diagnosis of chronic obstructive bronchitis [12]	HBR	28	205
Modeling of dependence of state of children's health from environmental pollution [11]	ECO	43	954
Automatic classification of vehicles [11]	AUTO	26	1062
Automatic classification of agricultural plants [11]	PLNT	55	248

Table 3. Averaged experimental performance evaluation of feature selection methods

Task	Canonical evolutionary method			Method with fixing of the search space part			Method with feature grouping		
	t, sec	$N_{m.}$, M	N'/N	t, sec	$N_{m.}$, M	N'/N	t, sec	$N_{m.}$, M	N'/N
SIGNAL	10317	7.22	0.113	3317.6	8.19	0.115	3317.6	8.19	0.115
KOG	21909	10.00	0.132	6984.1	10.10	0.132	6984.1	10.10	0.132
HBR	17150	14.12	0.786	5429.2	14.75	0.679	5429.2	14.75	0.679
ECO	43335	430.40	0.651	13832	487.84	0.628	13832	487.84	0.628
AUTO	35784	344.43	0.500	10518	373.90	0.500	10518	373.90	0.500
PLNT	4714	40.74	0.200	1538.2	45.09	0.218	1538.2	45.09	0.218
	Method using entropy			Method with feature clustering			BCO based method		
	t, sec	N_m, M	t, sec	t, sec	$N_{m.}$, M	N'/N	t, sec	$N_{m.}$, M	N'/N
SIGNAL	2947.1	7.73	0.113	2853	7.74	0.115	6008.5	3.06	0.115
KOG	5864	10.65	0.113	6686.6	9.55	0.132	13558	3.73	0.113
HBR	4700.1	14.46	0.714	4591.6	13.86	0.750	9822.1	5.30	0.750
ECO	12867	500.39	0.651	14764	472.47	0.628	28690	168.21	0.628
AUTO	10065	364.00	0.500	10400	355.00	0.577	22164	129.84	0.577
PLNT	1334.2	45.94	0.200	1436.6	41.83	0.164	3072.3	15.91	0.200
	ACO with features			ACO with feature informativities			BFO based		
	t, sec	$N_{m.}$, M	N'/N	t, sec	$N_{m.}$, M	N'/N	t, sec	$N_{m.}$, M	N'/N
SIGNAL	5748.5	0.89	0.115	5855.3	0.93	0.111	5946.2	1.21	0.113
KOG	12282	1.34	0.123	12792	1.25	0.132	14033	1.66	0.142
HBR	9327.3	1.96	0.750	9739.1	1.72	0.786	9394.1	2.28	0.714
ECO	28000	57.60	0.651	25507	57.30	0.651	27119	76.35	0.651
AUTO	19693	43.18	0.500	19378	43.89	0.538	21036	56.94	0.538
PLNT	2922.6	4.98	0.164	2814.9	4.68	0.200	2894.8	6.52	0.200

7 Conclusion

The problem of data dimensionality reduction have been studied. The dimensionality reduction methods based on intelligent stochastic search including feature selection methods and sample selection methods have been proposed. The experiments on study of proposed methods are conducted. They show that proposed set of methods allow to significantly reduce the data dimensionality.

Acknowledgements. This paper is prepared with partial support of "Centers of Excellence for young RESearchers" (CERES) project (Reference Number 544137-TEMPUS-1-2013-1-SK-TEMPUS-JPHES) of Tempus Programme of the European Union.

References

1. Dash, M., Liu, H.: Feature selection for classification. Intell. Data Anal. **1**, 131–156 (1997)
2. Jensen, R., Shen, Q.: Computational Intelligence and Feature Selection: Rough and Fuzzy Approaches. John Wiley & Sons, Hoboken (2008)
3. Boguslayev, A.V., Oleynik, A.A., Oleynik A.A. et al.: Progressivnyye tekhnologii modelirovaniya, optimizatsii i intellektual'noy avtomatizatsii etapov zhiznennogo tsikla aviatsionnykh dvigateley: monografiya. Motor Sich, Zaporozhye (2009). (in Russian)
4. Subbotin, S.A.: Methods of sampling based on exhaustive and evolutionary search. Autom. Control Comput. Sci. **3**(47), 113–121 (2013)
5. Korobiichuk, I., Podchashinskiy, Y., Shapovalova, O., Shadura, V., Nowicki, M., Szewczyk, R.: Precision increase in automated digital image measurement systems of geometric values. In: Jabłoński, R., Brezina, T. (eds.) Advanced Mechatronics Solutions. Advances in Intelligent Systems and Computing, vol. 393, pp. 335–340. Springer, Heidelberg (2016). doi:10.1007/978-3-319-23923-1_51
6. Korobiichuk, I., Bezvesilna, O., Ilchenko, A., Shadura, V., Nowicki, M., Szewczyk, R.: A mathematical model of the thermo-anemometric flowmeter. Sensors **15**, 22899–22913 (2015)
7. Engelbrecht, A.: Computational intelligence: an introduction. Wiley, Sidney (2007)
8. Ruan, D.: Intelligent Hybrid Systems: Fuzzy Logic, Neural Networks, and Genetic Algorithms. Springer, Berlin (2012)
9. Liu, Y., Passino, K.M.: Biomimicry of social foraging bacteria for distributed optimization: models, principles, and emergent behaviors. J. Optim. Theory Appl. **3**, 603–628 (2002)
10. Karaboga, D., Akay, B.: A survey: algorithms simulating bee swarm intelligence. Artif. Intell. Rev. **31**, 61–85 (2009)
11. Subbotin, S.A., Oleynik, A.A., Gofman, Y.A. et al.: Intellektual'nyye informatsionnyye tekhnologii proyektirovaniya avtomatizirovannykh sistem diagnostirovaniya i raspoznavaniya obrazov: monografiya. SMIT Co., Kharkov (2012)
12. Subbotin, S., Oleynik, A.: Entropy based evolutionary search for feature selection. In: Proceedings 9th International Conference (CADSM-2007), The Experience of Designing and Application of CAD Systems in Microelectronics, pp. 442–443. IEEE Press, Lviv (2007)

13. Subbotin, S., Oleynik, A.: Modifications of ant colony optimization method for feature selection. In: Proceedings 9th International Conference (CADSM-2007), The Experience of Designing and Application of CAD Systems in Microelectronics, pp. 493–494. IEEE Press, Lviv (2007)
14. Oliinyk, A.O., Oliinyk, O.O., Subbotin, S.A.: Agent technologies for feature selection. Cybern. Syst. Anal. **2**(48), 257–267 (2012)

The Selection Methods for Multisensor System Elements of Drone Detection

Igor Korobiichuk[1(✉)], Michał Nowicki[1], Yuriy G. Danik[2], Sergey Dupelich[2], and Samchyshyn Oleksyj[2]

[1] Industrial Research Institute for Automation and Measurements PIAP, Warsaw, Poland
ikorobiichuk@piap.pl, nowicki@mchtr.pw.edu.pl
[2] Zhytomyr Military Institute Named After S. Korolyov, Zhytomyr, Ukraine
{zhvinau, dypelych_sergey, samyj123}@ukr.net

Abstract. The application of the existing means for drones detection is rather troublesome due to the miniaturization of board equipment, usage of composite materials and special paint. The possible way to resolve this problem is to combine the technical solution means which function on the basis of physical phenomena and which are crucial for drones detection. The partial indexes and criteria of the effective functioning of multisensory systems of the surrounding area control are determined. The selection method of means for multisensory drone detection system is developed.

Keywords: Multisensory system · Monitoring · A drone · Effectiveness · Multicriterion optimization

1 Introduction

The main challenges we are facing nowadays have shifted to air sector and IT sphere. The rivalry in the sphere of technology development and the most modern means production has become one of the most determinative directions of competition among countries all over the world. At the same time, dual application drones (D) are developing extremely fast. The main approaches applied to drones detection by either active or passive means are investigated in [1–3]. But, the application of the existing means for drones detection is rather troublesome due to the miniaturization of board equipment [4–8], usage of composite materials and special paint.

The possible way to resolve this problem is to combine the technical solution means which function on the basis of physical phenomena and which are crucial for drones detection. The researches [9, 10] consider the complex approaches of drones detection and maintenance. Their common drawback is the absence of the selection method of surveillance means depending on the monitoring object class. Taking into account the mentioned above, there is the task of current interest to develop the selection method for creating the system optimum version to detect different drones-types at certain conditions and restrictions.

R. Szewczyk and M. Kaliczyńska (eds.), *Recent Advances in Systems, Control and Information Technology*, Advances in Intelligent Systems and Computing 543, DOI 10.1007/978-3-319-48923-0_3

2 Multisensory System Functioning Effectiveness at Drones Detection

In order to solve the task of grounding the technical system needed elements, there is the necessity to estimate the opportunities of system design of any version. Here, it is important to determine, as well as to analyze the effectiveness indexes of such system functioning. Such indexes, as a rule, possess different physical nature. In order to conduct the comparative analysis of complex technical systems, the determined indexes are expressed in n-dimensional vector form.

The basic effectiveness indexes of any multisensory system functioning used for drones detection are directly connected with the elements of the tactical performance specification features which compose its structure, are determined as [11]:

- range of operation d – is the space area, restricted by the controlled area;
- detection parameter x – is the value, which determines monitoring device resolution;
- setting distinction s – is the value, which determines the mathematical expectation of the feature applied to drones detection;
- coordinate determination precision σ_d – is rms (root mean square) error at drones location determination.

The determined indexes of multisensory system functioning effectiveness at drones detection for every single version of its design are comfortable to be presented in the form of matrixes $D = [d_{ijk}]$, $X = [x_{ijk}]$, $S = [s_{ijk}]$, $\Sigma = [\sigma_{ijk}]$. Vector-lines $d_i, x_i, s_i, \sigma_i, i = \overline{1 \ldots n}$ show the value of these indexes for every j component of the system design of the determined version M_j, $j = \overline{1 \ldots m}$ in order to provide its functioning when detecting different D types V_k, $k = \overline{1 \ldots v}$, where k – are the equipment elements.

The investigations focused on the grounding of the composition of complex technical systems have obtained the triad of criteria: effectiveness-cost-time [11]. System E_C effective functioning is used as the main criterion at its cost limitation c_C and time required to perform the task t_C:

$$\begin{array}{l} E_C : \{\{M\}, G\} \rightarrow M^* \\ t_C \leq t_{max} \\ c_C \leq c_{max} \end{array}, \tag{1}$$

where

$\{M\}$ is the determined set of options for the system design;
G is the formalized estimation criterion of the system design effectiveness version;
M^* is the optimum version of system design;
t_{max} is the maximum allowable time for the system to perform the tasks;
c_{max} is the most allowable system cost

Every single set of elements version of drones detection multisensory system is recommended to represent time t_C in the form of matrix $T = [t_{ijk}]$, where vector-lines $t_i, i = \overline{1 \ldots n}$ contain the quantitative value of time required for the tasks to be performed by every version of the system design M_j, and cost c_C – is the matrix $C = [c_{ijk}]$, where vector-lines $c_i, i = \overline{1...n}$ contain the quantitative value of the expenses to apply such version of the system design in order to provide its functioning for definite drones types V_k. Here t_{max} depends on D type and c_{max} is determined by the system of higher level.

Complex mathematical calculations are required to attain effect G based on the numerical characteristics which have been considered by vector indexes for the systems consisting of different types. Besides, it is not objective because of various approaches to the task solution for every single type of method. In order to fix the mentioned above drawbacks, it is recommended to establish fee Q as a unit of failure for the system to perform a single task. So, the quality of the solution is estimated due to the partial criteria which form s-dimensional vector $Q = (Q_1, Q_2, \ldots, Q_s)$, determined on the multitude M. The formalized principle of alternative selection from the determined multitude for drones detection multisensory system can be presented with s partial criteria defined by functionals:

$$Q_b = f_b[d_i(y), x_i(y), s_i(y)],\ b = \overline{1...s},\ i = \overline{1...n}, \tag{2}$$

where f_b are the functions which possess the continuous partial derivatives of d_i, x_i, s_i.

According to the purpose of the system, the first partial system criterion Q_1 characterizes a fee for D pass and is determined by the pass average probability for the maximum allowable time required for the tasks to be performed by the system:

$$Q_1 = 1 - \int_0^{t_{max}} P_в(t)dt, \tag{3}$$

where $P_в(t)$ is D detection conditioned probability.

D detection probability is estimated by the device resolution and is described as $P_в(t) = 1 - e^{-x(y)}$. So, the partial criterion Q_1 is determined:

$$Q_1 = 1 - \int_0^{t_{max}} 1 - e^{-x(y)}dt, \tag{4}$$

where $x(y)$ is the detection parameter.

The second partial parameter Q_2 characterizes a fee for the false drones detection and is determined by the average probability of the false detection for the maximum allowable time required for the tasks to be performed by the system:

$$Q_2 = 1 - \int_0^{t_{max}} P_p(t)dt, \tag{5}$$

where $P_p(t)$ is the correct probability of drones detection by a device with the defined indexes.

The partial criterion Q_2 used for the preset conditions is known as Newman-Pierson criterion [11] and requires finding the solution which is able to provide the minimum value of the conditioned probability of a target pass at the limitation of the conditioned probability of the false alarm F_0:

$$Q_2 = \int_0^{t_{max}} e^{-\frac{s^2(y) - 2z(y)s(y)}{2\sigma_s^2}}, \tag{6}$$

where $s(y)$ is the distribution density of the distinction parameter; $z(y)$ is the distribution density of noise (obstacle); σ_s – the rms deviation of distinction parameter.

The third partial criterion Q_3 characterizes a fee for drones maintenance breakdown and is determined by the average probability of the maintenance breakdown for the maximum allowable time for the tasks to be performed by the system. The maintenance breakdown can be estimated by the probability of drones going beyond the radius limits d:

$$Q_3 = \int_0^{t_{max}} 1 - e^{-\left(\frac{d(y)}{2\sigma_d}\right)^2} dt, \tag{7}$$

where $d(y)$ is the trajectory of drones maintenance.

The task of optimum selection of the design for drones detection multisensory system M consists of vectors $M = (M_1, M_2, \ldots, M_n)$, n – dimensional Euclidean space and the partial criteria are the components of s – dimensional vector $Q = (Q_1, Q_2, \ldots, Q_s)$ determined on the multitude M. So, there is the necessity to find such solution $M^* \in M$ which is able to optimize the effectiveness vector Q:

$$E : \{\{M\}, \{Q\}\} \to M^*, 0 \le Q_s \le 1. \tag{8}$$

at the preset conditions and limitations.

The optimum version of drones detection multisensory system can be found if the partial criteria $Q_1 - Q_3$:

$$\begin{array}{l} Q_1 \to min \\ Q_2 \to \underset{F_0}{max} \\ Q_3 \to \underset{d}{min} \end{array}. \tag{9}$$

are distributed.

Taking into account all the determined criteria of the estimation of effectiveness of drones detection multisensory system there is the task of means selection. This task has to be considered as the multicriterion task for an alternative selection. The solution to this task either directly or indirectly leads to the case when the partial criteria are combined into the main optimization criterion. The optimization criterion definition is advisable to be based on A.M. Voronin convolution [12] that has forecast the introduction of the significance coefficient of every single partial criterion. This has to be done, taking into consideration the necessity to determine the version of detection system design depending on drones type and preset limitations in either cost or time.

3 The Method for Drones Detection Multisensory System

Taking into account the determined partial criteria, the selection method for drones detection multisensory system is:

1. The establishment of area of conditions *C*, *T*.
2. The normalization of the determined partial criteria Q_b. In order to obtain this, it is recommended to find supremums of the preset partial criteria which are determined within the solution area $Q \in M$:

$$Q_b^{max} = \left\{ \sup_{Q \in M} Q_b \right\}_{b=1}^{s}. \tag{10}$$

3. The application of scalar convolution method.

A.M. Voronin's scalar convolution for the preset indexes of the effectiveness of drones detection multisensory system functioning obtains the form of:

$$M^* = \arg\min_{Q \in M} \sum_{b=1}^{s} \gamma_b \left[1 - Q_b^{norm}\right]^{-1}, \tag{11}$$

where:

Q is the optimization parameter;
M is feasible solution area;
γ_b is significance coefficient of b-partial criterion;
$Q_b^{norm} = \frac{Q_b}{Q_b^{\max}}$ is normalized function of b-partial criterion;
$b = \overline{1 \ldots s}$ is the coefficient of summing based on the number of partial criteria;
M^* is the selected version of system design;

4. The determination of convolution minimum by means of equating its partial derivatives on varied parameters to zero:

$$\sum_{b=1}^{s} \frac{\frac{Q_b}{\partial m}}{\left(1 - Q_b^{norm}\right)} = 0. \tag{12}$$

5. Testing Pareto-optimality of the solution $m*$ of convolution minimization:

$$M^* \in M^K = \left\{ m^* \middle| M^* \in M; \forall Q \in M : Q_b' \leq Q_b, b = 1 \ldots s \right\}, \tag{13}$$

where M^K is the area of compromises;
Q_b' are minimized criteria.

6. The selection of optimum version of multisensory system for drones detection in accordance with the preset conditions and limitations.

4 Conclusions

The selection of indexes and partial criteria of the effectiveness of multisensory system for drones detection is done as a result of the conducted investigations. The selection method of optimum version of the design of multisensory system to provide drones detection of different types at the preset conditions and limitations is proposed. It is done on the basis of multicriterion task of alternative selection.

The conduction of experimental calculations is the promising direction of the further research. These calculations are targeted on various versions of the design of multisensory system for drones detection depending on the class of monitoring objects in order to estimate the effectiveness of these version functioning and, finally, to select the optimal one.

References

1. Tkachenko, V.I., Dannyk, Y.G., Drobaha, G.A., Karpenko, V.I., Pashenko, R.E., Smirnov, E.B.: Theory and technique of resistance to drones of air attack, Application and Perspectives of Development, Detection of Undistinguished Means of Air Attack, 220 p. HMU, Harkiv (2002)
2. Moses, A., Rutherford, M.J., Valavanis, K.P.: Radar-based detection and identification for miniature air vehicles. In: IEEE International Conference on Control Applications, Denver, CO, USA, pp. 933–940, 28–30 September 2011
3. Saravanakumar, A., Senthilkumar, K.: Exploitation of acoustic signature of low flying aircraft using acoustic vector sensor. Defence Sci. J. **64**(2), 95–98 (2014)
4. Korobiichuk, I., Podchashinskiy, Y., Shapovalova, O., Shadura, V., Nowicki, M., Szewczyk, R.: Precision increase in automated digital image measurement systems of geometric values. In: Jabłoński, R., Brezina, T. (eds.) Advanced Mechatronics Solutions. AISC, vol. 393, pp. 335–340. Springer, Heidelberg (2016). doi:10.1007/978-3-319-23923-1_51
5. Korobiichuk, I.: Analysis of ways to increase accuracy of aviation gravimeters. Int. J. Sci. Eng. Res. **6**(10), 1162–1164 (2015)
6. Korobiichuk, I., Bezvesilna, O., Tkachuk, A., Nowicki, M., Szewczyk, R.: Piezoelectric gravimeter of the aviation gravimetric system. In: Szewczyk, R., Zieliński, C., Kaliczyńska, M. (eds.) Challenges in Automation, Robotics and Measurement Techniques. AISC, vol. 440, pp. 753–761. Springer, Heidelberg (2016). doi:10.1007/978-3-319-29357-8_65
7. Korobiichuk, I., Bezvesilna, O., Tkachuk, A., Chilchenko, T., Nowicki, M., Szewczyk, R.: Design of piezoelectric gravimeter for automated aviation gravimetric system. J. Autom. Mob. Rob. Intell. Syst. (JAMRIS) **10**(1), 43–47 (2016)
8. Korobiichuk, I.: Mathematical model of precision sensor for an automatic weapons stabilizer system. Measurement **89**, 151–158 (2016). doi:10.1016/j.measurement.2016.04.017147
9. Shi, W., Arabadjis, G., Bishop, B., Hill, P.: Detecting, tracking and indentifying airborne threats with netted sensor fence. In: Sensor Fusion – Foundation and Applications, pp. 139–158. InTech Europe, Rijeka (2001)
10. Godunov, A.I., Shishkov, S.V., Yurkov, N.K.: The complex for detection and resistance to compact drones. Reliab. Qual. Complex Syst. **2**(6), 62–70 (2014)

11. Toropchyn, A.Y., Romanenko, I.O., Dannyk, Y.G., Pashenko, R.E. et al.: Air Defense Reference, 368 p. H.: HMU, K.: MO of Ukraine (2003)
12. Voronin, A.V., Ziadtinov, Y.K., Klimova, A.S.: Information Systems of Decision-Making: Tutorial, 136 p. Publishing House of Nationa Aviation University «HAU-print», K. (2009)

Reduction of Uneven Pace of Internal Combustion Engine FP10C by Flywheel Design Improvement

Andrii Ilchenko(✉) and Volodymyr Lomakin

Zhytomyr State Technological University, Zhytomyr, Ukraine
avi_7@ramble.ru, rootsymbol@gmail.com

Abstract. The kinematics of operation of the crank mechanism elements gives rise to change of the presented moment of the internal combustion engine (ICE) inertia during the turn, which is often neglected. However, there is a number of tasks where it is required to take into account the change of the presented moment of the crank mechanism inertia. The authors proposed a design of the flywheel with a variable inertia moment for FP10C engine, which allows reducing the uneven pace significantly. As a result of experimental studies of a single-cylinder engine FP10C with a variable inertia moment flywheel and with a constant inertia moment flywheel, it was found that the variable inertia moment flywheel can significantly reduce the unevenness of the engine pace (up to 37.5%) and improve its fuel economy (up to 9.4%). This significantly reduced the energy of the crankshaft vibration processes, which results in the overall reduction of energy consumption.

Keywords: Uneven pace of engine · Flywheel · Fuel consumption

1 Introduction

Lots of theoretical and experimental studies, especially in the period 1940–1980, were devoted to the theory of oscillations associated with the internal combustion engine (ICE) [1]. But the problem of the study of oscillatory processes occurring in internal combustion engines is quite complex. This is due to the ambiguity of the choice of calculation models and calculation systems with many degrees of freedom, as well as the interdependence of certain forms of oscillations. Besides, driving forces of oscillation disturbance have a wide frequency range and have not only periodic but also casual character [1–4]. In most works on a dynamic synthesis the parameters of the oscillating systems are considered as linear models, to a great extent conditioned by bulkiness and labour-intensiveness of solving the tasks of analysis for the nonlinear systems.

In the modern terms of engine-building development, more and more attention is paid to seemingly insignificant processes. On the one hand, the mechatronics achievements made it possible, and on the other hand, rationalization of the use of resources does not give us any choice [5]. It results in the necessity of efficiency coefficient increase, improvements of ecological compatibility and safety of ICE operation [6, 7].

R. Szewczyk and M. Kaliczyńska (eds.), *Recent Advances in Systems, Control and Information Technology*, Advances in Intelligent Systems and Computing 543, DOI 10.1007/978-3-319-48923-0_4

Therefore, it is high time to pay attention to such phenomenon, as influence of the equivalent moment of inertia (MI) of the crank-and-rod mechanism on the engine operation. This influence is insignificant in the engines with more than 4 cylinders, and in such types of ICE it can be not taken into account. It can also be substantially decreased by the rational choice of mass-geometrical parameters of the crank-and-rod mechanism of ICE, but that is, unfortunately, not always possible.

It is necessary to note, that the distinctive feature of engine-building development of the recent years is diminishing of the engines swept volumes, their speedup, diminishing of the number of cylinders. As a result, some, very popular at one time engines, for example V-type 8-cylinder engines, are used rarer and rarer. It results in considerable increase in manufacturing of 4- and fewer cylinder engines. 2- and 3-cylinder engines, which were not used for a long time, are widely-spread now. And it means that the processes, related to the change of the equivalent MI of the crank-and-rod mechanism, must be taken into account. Unevenness of the motion even of a 4-cylinder engine at the idling mode can reach 23% and more. Certainly, everything depends on the construction of an engine, but it is possible to define and eliminate some tendencies at the stage of designing.

2 Analysis of the Recent Research and Publications

Technological food industry facilities operate under uncertainties of diverse nature. In the diagnosis tasks, as well as for simplifying, the equivalent MI of the crank-and-rod mechanism is considered constant [8, 9]. It means that the dependence of the effect of moving masses of modern internal combustion engines on mass-geometry parameters and layout diagrams has not been properly investigated yet.

To compensate for changes in the equivalent MI of the crank-and-rod mechanism for its turn, it is necessary to use special devices, such as flywheels of alternating MI. Despite the fact that the flywheel has long been one of the simplest elements of the crank-and-rod mechanism, this device is now subjected to considerable changes.

Nowadays, there are quite successful designs of flywheels [10] that reduce weight, improve the internal combustion engine injectivity and facilitate its start. However, these known designs do not compensate for the impact of changes in the equivalent MI of the crank-and-rod mechanism for its turnover.

Of special note is the development of ICE crankshaft oscillation dampers, which functions are performed by dual mass flywheels [11]. These devices can help significantly reduce the crankshaft oscillations such as of a gear box of the car, but they have little effect on the uneven pace of the internal combustion engine. Changes effect compensation in the equivalent MI of the crank-and-rod mechanism for its turn reduces the unevenness of the internal combustion engine pace.

The aim is to decrease the pace unevenness of FP10C engine using alternating MI flywheel.

3 The Results of the Research

FP10C engine is single-cylinder, four-stroke, the main characteristics of which are given in [10]. For this engine, the authors developed an alternating MI flywheel (Fig. 1) and the hardware and software system for the analysis of its work [12].

The study of uneven pace, as well as fuel consumption measurements were performed at idling, full-load and mixed (5 s of idling, then 5 s of full load) modes.

Each of the modes described lasted for at least 5 min. Results repeatability was ensured by the necessary number of measurements (standard deviation was not higher than 0.05), the relative error of measured values in all measurements did not exceed 8%. First, the studies were conducted with constant MI flywheel, and then – with alternating MI flywheel (Table 1).

Analyzing the data in Table 1, it can be concluded that the alternating MI flywheel can significantly reduce the unevenness of the engine pace (to 37.5%) and improve fuel economy (up to 9.4% at idling mode).

For further analysis it is necessary to assess the energy component of oscillatory processes that occur when using alternating and constant MI flywheels. It can be done by applying the energy conservation law [13].

Based on this law, at the rotational motion of the body with alternating MI, the total moment of the external forces can be expressed:

$$M = I(\varphi)\omega\frac{d\omega}{d\varphi} + \frac{\omega^2}{2}\frac{dI(\varphi)}{d\varphi}, \tag{1}$$

where $I(\varphi)$ is alternating MI of the crank-and-rod mechanism, kg·m^2; φ is crankshaft rotation angle, degrees; ω is rotational rate, rad/s.

Fig. 1. Alternating MI flywheel for FP10C engine

Table 1. The results of an experimental study of uneven pace and fuel consumption by FP10C engine

Index		Engine behavior		
		Idling	Full load	Mixed (5 of idling, then 5 s of full load)
Average engine speed	With the constant MI flywheel, rev/min	2272	5288	–
	With the alternating MI flywheel, rev/min	2891	5381	–
Uneven engine pace	With the constant MI flywheel, δ_{const}, %	1.14	0.56	–
	With the alternating MI flywheel, δ_{var}, %	0.74	0.35	–
Relative decrease in the uneven pace, %	$100\% - \frac{\delta_{var}}{\delta_{const}} \times 100\%$	35.1	37.5	–
Fuel consumption for 5 min, g	With the constant MI flywheel, G_{const}, g/h	263.6	695.7	472.5
	With the alternating equivalent MI flywheel, G_{var}, g/h	238.4	651.3	437.2
Relative reduction in fuel consumption, %		9.6	6.4	7.5

Table 2. The results of the coefficients calculation of the regression model of FP10C engine rotational frequency

	Average frequency of the crankshaft/δ (uneven pace)	β_0	β_1	β_2	β_3	β_4
Full-load mode						
With the alternating MI flywheel	5077 rev/min/0.35%	5077.1	–0.08	3.68	0.28	1.13
With the constant MI flywheel	5288 rev/min/0.56%	5287.4	0.41	–4.14	0.54	–2.39
Idling						
With the alternating MI flywheel	2891 rev/min/0.74%	2890.0	4.70	1.27	–0.46	0.63
With the constant MI flywheel	2272.38 rev/min/1.14%	2272.4	3.68	–2.83	–0.15	–1.20

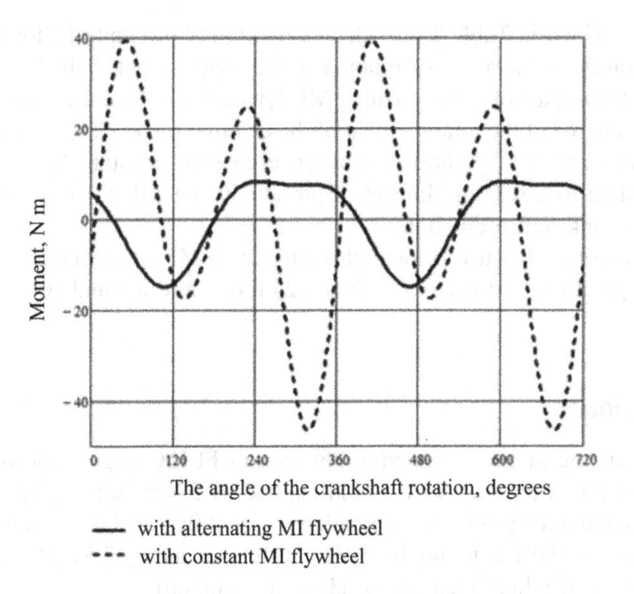

Fig. 2. Changes in the moment of the external forces with alternating and constant MI flywheels for FP10C engine at idling

To determine crankshaft rotational rate, the method of least squares for 240 consecutive engine operating cycles was used. As a function, the first 5 elements of the Fourier series were taken $\beta_0 + \beta_1 \cos(x) + \beta_2 \sin(x) + \beta_3 \cos(2x) + \beta_4 \sin(2x)$ (Table 2).

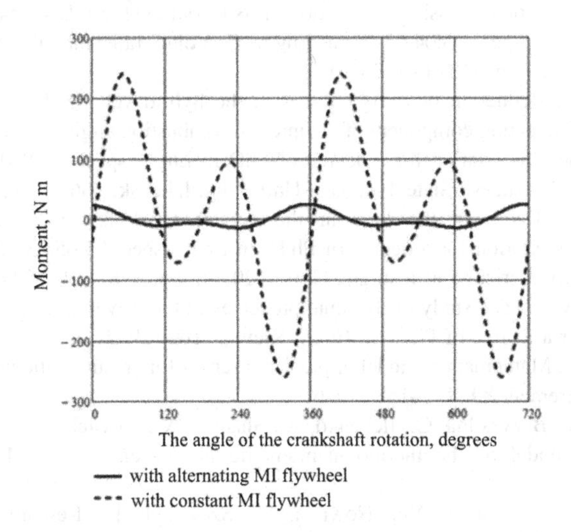

Fig. 3. Changes in the moment of the external forces with alternating and constant MI flywheels for FP10C engine at full load

The data obtained in Table 2 provide an opportunity to graph the total moment of the external forces on the angle of rotation of the engine crankshaft (Figs. 2 and 3).

To assess the impact of the alternating MI flywheel, compare the values of the area limited by the curves of the total moment of the external forces *M,* defined for the case of use of alternating MI flywheel, and when used with constant MI flywheel. This makes it possible to assess the impact of the alternating MI flywheel on the kinetic energy of the crankshaft oscillations.

As the result of studies it was found that alternating MI flywheel reduces the energy of oscillatory processes of the crankshaft at 2.24 times at idling and 10.83 times at full load.

4 Conclusions

1. Using alternating moment of inertia flywheel for FP10C engine allowed to reduce the unevenness of pace by 35.1% at idling and 37.5% at full load modes.
2. It was found that the use of the alternating moment flywheel allowed to reduce fuel consumption by 9.6% at idling, by 6.4% at full load, and by 7.5% at mixed mode of operation (5 s of idling, then 5 s of full load operation).
3. Application of an alternating moment of inertia flywheel reduced the oscillation kinetic energy at 2.24 times at idling and 10.8 times at full load.

References

1. Dolgov, K.O.: Improving of balance, mass overall indices and characteristics of the piston engine oscillations on the basis of perfection of its layout diagram. Dissertation for a degree of Ph.D. in Engr.: spec. 05.04.02 "Heat Engines", Volgograd State Technical University, Volgograd, 185 p., pp. 167–180 (2006)
2. Vasilyev, V.A.: Reduction of energy losses in the hybrid vehicle drive by reducing the impact of the fluctuating component of the internal combustion engine torque and applying a rational scheme. Dissertation for a degree of PhD in Engr.: spec. 05.05.03 "Wheeled and tracked vehicles", Izhevsk State Technical University, Izhevsk, 166 p., pp. 131–144 (2007)
3. Gudkov, A.V.: Reducing the load in the vehicle transmission caused by torsional oscillations. Dissertation for a degree of Ph.D. in Engr.: spec. 05.05.03 "Automobiles and Tractors", Gorky Agricultural Institute, Gorky, 201 p., pp. 158–174 (1984)
4. Savustyanov, V.V.: The study of dynamic processes in the driveline of heavy motorcycles. Dissertation for a degree of Ph.D. in Engr., Kiev, p. 169 (1982)
5. Korobiichuk, I.: Mathematical model of precision sensor for an automatic weapons stabilizer system. Measurement **89**, 151–158 (2016)
6. Korobiichuk, I., Bezvesilna, O., Ilchenko, A., Shadura, V., Nowicki, M., Szewczyk, R.: A mathematical model of the thermo-anemometric flowmeter. Sensors **15**, 22899–22913 (2015)
7. Korobiichuk, I., Shavursky, Yu., Nowicki, M., Szewczyk, R.: Research of the thermal parameters and the accuracy of flow measurement of the biological fuel. J. Mech. Eng. Autom. **5**, 415–419 (2015)

8. Grebennikov, A.S.: Diagnosis of automotive engines by intracyclic changes of the angular velocity of the crankshaft: methods, tools, technologies. Dissertation for a degree of Ph.D. (Doctor of Science) in Engr.: spec. 05.20.03 "Technologies and means of maintenance in agriculture", Saratov State Technical University, Saratov, 292 p., pp. 238–275 (2002)
9. Borshchenko, Y.A.: Development of the method of diagnosing of automobile diesels on the crankshaft rotation unevenness. Abstract of Dissertation for a degree of Ph.D. in Engr.: spec. 05.22.10 "Operation of motor transport", Kurgan State University, K., 21 p. (2003)
10. Kimura, Y., Tsuboi, T., Endoh, T.: Variable Flywheel Mechanism and Flywheel Apparatus: Patent Application Publication United States US 2007/0179012 A1, МПК F16H57/08. No. P2006-022445, Filed 31.01.2006, Publish 02.08.2007, Appl. No. 11/699, 368
11. Lee, C.G.: Dual mass flywheel. Patent Application Publication United States US 2014/0144284 A1, МПК F16F15/31. Pyeong Hwa Clutch Industry Co. Ltd., Filed 16.07.2013, Publish 29.05.2014, Appl. US13/943, 567
12. Lomakin, V.A., Ilchenko, A.V.: Alternating moment of inertia flywheels for the piston engines (theoretical basis of the development and practical application). 112 p., pp. 94–103. LAP Lambert Academic Publishing, Germany (2015). Monograph, ISBN-13 978-3-659-71200-5
13. Weitz, V.L., Kochura, A.E.: Dynamic submachine units with internal combustion engines. L.: Engineering, 352 p. (1978)

Synergetic Control of Social Networking Services Actors' Interactions

Ruslan Hryshchuk[1(✉)] and Kateryna Molodetska[2]

[1] Cybersecurity Department of the Research Center,
Sergey Korolyov Zhytomyr Military Institute, Zhytomyr, Ukraine
dr.hry@i.ua

[2] IT and Simulation Department,
Zhytomyr National Agro-Ecological University, Zhytomyr, Ukraine
kmolodetska@gmail.com

Abstract. The evolutionary process of social networking is a transition from one state to another through chaos that features high system sensitivity to external disturbances. Thus, a system may be in a certain stable state called an attractor adversely affected by a potential threat to social networking actors. So let us have synergetic control conceptualized here in terms of social network actors' interactions control to ensure national information security. The design of a synergetic system to control self-organizing virtual communities enables a chaos-control transition thus achieving the predicted result of their actors' interactions. There are some models introduced.

Keywords: Social networking service · Actors' interactions · Chaotic dynamics · Synergetic control · Information security

1 Introduction and Analysis of the Latest Studies and Printed Works

The communications being as sophisticated as they are for now make of social networking services (SNS) a greatest contributor to the socialization process [1–4]. People called the actors use SNS to satisfy their individual and collective wants within the cyberspace. Those services are known to belong to nonlinear dynamical systems [5, 6].

One of SNS properties is randomness of a social network actors' behavior in a system. This is due to the system's sensitivity to interferences and to its transition into an uncontrolled chaotic state. The actors' chaotic behavior will eventually get under control, and in case of their interaction and certain compliance, it will unlock synergies. Those synergies originate new SNS features also known as emergent ones within the high order systems of inter-hierarchical interaction. The on time determination of the nature and intention of the synergies in SNS alongside with timely identification and prediction is a great matter of individual, collective and national information security.

The comprehensive case study reveals a number of key factors of social networking [4, 7–9]. Those are sociability and dissipativity of SNS actors' interactions alongside with their associational capability. A chaotic system is highly sensitive both to insignificant perturbation of the system parameters and to initial conditions rendering

R. Szewczyk and M. Kaliczyńska (eds.), *Recent Advances in Systems, Control and Information Technology*, Advances in Intelligent Systems and Computing 543, DOI 10.1007/978-3-319-48923-0_5

its behavior impossible to predict and control. Further study of SNS actor's interactions requires application of the chaotic dynamics theory to control the system behavior by controlling its chaotic dynamics [10].

The optimal chaotic dynamics control solution is to choose a controlling action to stabilize the desired trajectory in a system of chaotic behavior [5]. Such an action has certain disadvantages sufficiently affecting the system dynamics. The desired behavior of a system is known to result from its self-organization capability that needs some further study [6, 10]. Synergetic suggests decentralization and state of equilibrium as a prerequisite to the emergence of self-organization. The self-organization process narrows down variables or parameters to a few which determine system dynamics and are termed as "order parameters". Thus under small disturbances a system gains stability due to its dynamics modifications. That is why solution of the synergetic control design problem aimed at predicted SNS self-organization to trigger a desired system transition to a controlled state remains a live issue today.

The paper challenges a generalized conceptual design of SNS actors' interactions control.

2 Concept Presentation

The modern synergetic control theory [11–13] referring to technical, social and economic nonlinear systems is based on the analytical approach to the nonlinear aggregated controllers design [14]. The SNS actors' expected self-organization control design based on the analytical approach to the nonlinear aggregated controllers design is the core of the concept just developed. Its posit shall be a composition method [15] that will not conflict with other proprietary ones [10, 12]. The problem of SNS formalization should have a solution at the first stage.

Phase 1. SNS formalization. The first step is to set the simultaneous nonlinear differential equations that represent the SNS actors' interactions control flow. For instance to control the SNS actors' market for information is to control the number of the enlisted in a cause. Thus, the resulting set of nonlinear differential equations takes the general form

$$\begin{cases} \frac{dx_i(t)}{dt} = f_i\big(x_1(t),\ldots,x_\lambda(t),y_{\lambda+1}(t),\ldots,y_\mu(t)\big); \\ \frac{dy_j(t)}{dt} = f_j\big(x_1(t),\ldots,x_\lambda(t),y_{\lambda+1}(t),\ldots,y_\mu(t),u_1(t),\ldots,u_\gamma(t)\big), \end{cases} \quad (1)$$
$$x_i(t_0) = x_i^0, \quad y_i(t_0) = y_i^0,$$

where $x_i(t)$, $y_j(t)$ stand for SNS actors' interaction indicators $i = 1, 2, \ldots, \lambda$, $j = \lambda + 1, \lambda + 2, \ldots, \mu$; $u_\gamma(t)$ stands for SNS synergetic control feedback; $x_i(t_0) = x_i^0$, $y_i(t_0) = y_i^0$ - initial conditions.

Phase 2. Validation and selection of the order parameters that determine the SNS actors' interaction dynamics. As the order parameter that generally determines the SNS actors' interaction dynamics serves a macrovariable in the form of $\psi_\upsilon(x_i, y_j) = 0$, $\upsilon = 1, 2, \ldots$. It ensures self-organization of the system designed. The macrovariable selected presents the collective properties of the system designed that carry synergetic information on SNS actors' interactions.

With this in view, let us introduce some certain dynamic invariants aka attractors to the system structure that are based on natural features of SNS. The variability invariants chosen shall represent conservative (conservation laws) and dissipative (regulation and self-organization laws) properties of the system. The resulting macrovariable takes the form of

$$\psi_{\upsilon} = \psi_k(x_1, y_1, \ldots, x_\lambda, y_\mu) + \psi_d(x_1, y_1, \ldots, x_\lambda, y_\mu), \tag{2}$$

where $\psi_k(x_1, y_1, \ldots, x_\lambda, y_\mu)$ - conservative component or control aspect of SNS actor's interactions, $k = 1, 2, \ldots$; $\psi_d(x_1, y_1, \ldots, x_\lambda, y_\mu)$ - dissipative component to determine the form of a desired attractor, $d = 1, 2, \ldots$.

The dissipative component of the macrovariable in (2) determines the SNS actors' interaction response to the invariant given.

Note 1. If a macrovariable (2) chosen, its dissipative component allows Lyapunov function in a system designed.

Description. The designed system state in the phase space shall be a point called *a synergy splash point.* However, its trajectory in the phase space shall match the equation

$$T_{\upsilon} \frac{d\psi_{\upsilon}(t)}{dt} + \psi_{\upsilon}(t) = 0, \tag{3}$$

where T_{υ} stands for the period of all the transitions triggered with the SNS actors' interaction synergetic control.

Phase 3. SNS actors' interactions synergetic control design. This step represents the control design $u_\gamma(x_1, y_1, \ldots, x_\lambda, y_\mu)$ substituting $\frac{d\psi_{\upsilon}(t)}{dt}$ in (3). Thus, the synergetic control encourages a system to have a stabilizing invariant $\psi_{\upsilon}(t) = 0$ of the SNS actors' interaction process. Then, implementing a well-known technique [12–14], we design an equation of the system motion on the invariant manifold $\frac{d\psi_{\upsilon}(t)}{dt}$. Subsequently we determine the stationary values of the SNS actors' interaction indicators $x_{i\upsilon}$, $y_{j\upsilon}$ the dynamic system moves towards.

Note 2. SNS actors' interaction indicators $x_{i\upsilon}$, $y_{j\upsilon}$ under synergetic control $u_\gamma(x_1, y_1, \ldots, x_\lambda, y_\mu)$ are time argument-independent and based on internal system parameters (1) and a macrovariable chosen (2).

Note 3. The value change of SNS interaction indicators $x_{i\upsilon}$, $y_{j\upsilon}$ under synergetic control $u_\gamma(x_1, y_1, \ldots, x_\lambda, y_\mu)$ results from the expected synergy splash points accessed by the system (1) in the phase space.

Phase 4. SNS actors' interaction synergetic control modelling. SNS modelling involves the system introduced with nonlinear differential Eq. (1) of the synergetic control $u_\gamma(x_1, y_1, \ldots, x_\lambda, y_\mu)$ designed at the previous stage. Thus the SNS actors' interaction control design $u_\gamma(x_1, y_1, \ldots, x_\lambda, y_\mu)$ maintains the desired interaction indicators $x_{i\upsilon}$, $y_{j\upsilon}$ and assures controlled self-organization upon attaining the predicted synergy splashes.

3 Conceptual Flowchart

Figure 1 represents the concept as a flowchart.

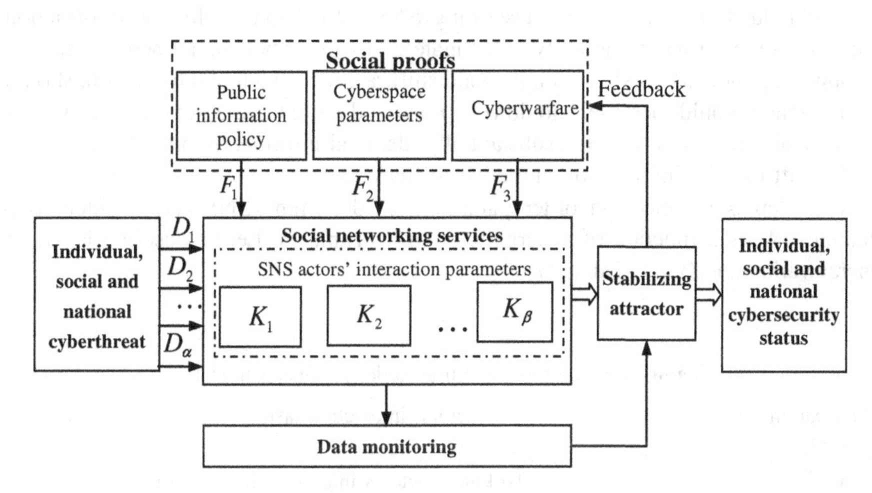

Fig. 1. SNS actors' interaction synergetic control flowchart

Cyberthreats build the vector of potential threats $D = \{D_1, D_2, \ldots, D_\alpha\}$ to SNS actors [15]. This results in altering SNS actors' interaction vector values $K = \{K_1, K_2, \ldots, K_\beta\}$ and disseminating certain information on SNS. The constant data monitoring establishes a control base for detecting both destructive social proof and control-demanding SNS actors' interaction parameters. The vector $K = \{K_1, K_2, \ldots, K_\beta\}$ values alteration aims at SNS actors' cue ruggedization. This aim is achieved by selecting a stabilizing attractor to implement the synergistically controlled self-organization in SNS.

The synergetic control vector $F = \{F_1, F_2, F_3\}$ triggers transition of a system to the state of equilibrium. This vector is built by means of the public communications policy, perturbations of cyberspace parameters and the consequences of a cyberwar between its actors. Thus with SNS actors' interactions being under synergetic control $u_\gamma(x_1, y_1, \ldots, x_\lambda, y_\mu)$ and the indicator values being $x_i(t)$, $y_j(t)$, the cybersecurity status of the virtual community can be obtained as a desired state in the stabilizing attractor ψ_υ.

4 Modelling

4.1 Modelling Practice

Let us put the concept into modelling practice. To do this let us consider the actors' interaction synergetic control problem in a certain SNS. Suppose the formalized SNS actors' interaction at the conceptual phase is governed by a set of differential equations of the form [13]

$$\begin{cases} \frac{dx(t)}{dt} = ax - xy - bx^2; \\ \frac{dy(t)}{dt} = -cy + xy, \end{cases} \tag{4}$$

where $x(t)$ stands for the process describing SNS actors' market for the information attractive to the virtual community investigated; $y(t)$ describes the attractive information supply; a indicates SNS actors' demand shift rate for attractive information, should $a > 0$ demand would increase, should $a < 0$ demand would decrease; b indicates SNS actors' rivalry in response to the substantially identical information posted; c indicates supply shift rate for information attractive to SNS actors.

Next step is to select an order parameter to determine SNS actors' interaction dynamics (4) as a function of synergetic control (Table 1). Let us describe it with a general variable $\psi_\upsilon(x, y)$ (Table 1).

Table 1. Output to discern an order parameter model

Order parameter $\psi_{\upsilon z}(x,y),\ z = 1,2$	SNS actors' interaction task
$\psi_{\upsilon 1}(x,y) = \varepsilon_1 x - \varepsilon_2 y$, where ε_1, ε_2 – attractive information market coefficient	To kindle actor's interest in information
$\psi_{\upsilon 2}(x,y) = \varphi_1 x + \varphi_2 \left(1 - \frac{y}{N}\right)$, where φ_1, φ_2 – attractive information market coefficient; N – attractive information supply rate subject to its value	To control SNS actors' interaction with the attractive information value shift factored in

At phase 3 of designing SNS interactions synergetic control $u_{\gamma z}(x, y)$ let us define each order parameter $\psi_{\upsilon z}(x, y)$ at its splash point $(x_{\upsilon z}, y_{\upsilon z})$ (Table 2).

Table 2. SNS actors' interaction control effects

Order parameter, $\psi_{\upsilon z}(x,y)$	Synergetic control, $u_{\gamma z}(x,y)$	Synergy splash point, $(x_{\upsilon z}, y_{\upsilon z})$
$\psi_{\upsilon 1}(x,y)$	$u_{\gamma 1}(x,y) = \frac{\varepsilon_2}{\varepsilon_1} x \left(a - y - bx + \frac{1}{T}\right) - y\left(-c + x + \frac{1}{T}\right)$	$x_{\upsilon 1} = \frac{\varepsilon_1 a}{\varepsilon_1 b + \varepsilon_2},\ y_{\upsilon 1} = \frac{\varepsilon_1 \varepsilon_2}{\varepsilon_1 b + \varepsilon_2}$
$\psi_{\upsilon 2}(x,y)$	$u_{\gamma 2}(x,y) = \frac{\varphi_1}{\varphi_2} N\left(ax - xy - bx^2\right) + \frac{1}{\varphi_2 T} N \psi_{\upsilon 2}(x,y) + cy - xy$	$x_{\upsilon 2} = \frac{a-N}{\frac{\varphi_1}{\varphi_2}N + b},\ y_{\upsilon 2} = \frac{\varphi_1}{\varphi_2} N \frac{a-N}{\frac{\varphi_1}{\varphi_2}N + b} + N$

To visualize the effects the output model (4) in Fig. 2 is given a phase portrait both before (Fig. 2a) and after (Fig. 2b and c) the SNS actors' interaction synergetic control effect.

4.2 Evaluating Modelling Performance

First, the SNS interaction system (Fig. 2a) will stop behaving chaotically (Fig. 2b) being under synergetic control. Second, the phase trajectories convergence on the manifold $\psi_{\upsilon z} = 0$ to the synergy splash point, thus being an attractor, guarantees the desired actors' interactions by means of synergetic control.

Thereby, the simulation practical effect is to acknowledge the truth of SNS actors' interactions synergetic control concept.

4.3 Experiment

The experiment is aimed at a qualitative analysis of the SNS actors' interaction synergetic control designed for outreaches destructive in their content. With this in view, *Yandex Wordstat* was used to define the dynamics of the SNS actor's average number of queries including "crucified boy". The test observation period is 01.01.2015 – 12.31.2015.

Russia's mass media are known to start broadcasting news about a boy nailed by a Ukraine's army official to a wooden board "right before his mother's eyes" in July 2014. That destructive information hotly debated and actively reposted was a smear campaign against the Army Forces of Ukraine.

The operational characteristics of the non-linear differential equation system (4) are found to take values: content market change rate $a = 0.5$; rivalry change rate $b = 0.45$ and content supply change rate $c = 0.3$. The attractor $\psi_{\upsilon 2}(x, y)$ is applied to regulate the SNS content market for the "crucified boy" query and the synthesized control $u_{\gamma 2}(x, y)$ is designed with the following parameters set: content supply level $N = 0.5$; regulation parameters $\varphi_1 = 1.1$ and $\varphi_2 = 0.9$. The functions of synergetic control and market regulation accordingly will change under the controlled self-organization forced for the actors in a virtual community as shown in Fig. 3.

Figure 3 shows that the chaotic dynamics is under control in a complex system that is proved to transfer into a controlled state. That could substantially suppress the illegal outreach activities which create a hazard for the noosphere and pose a great cyber security threat. The study proves that unlike any other method, the introduced one factors in the SNS actors' natural behavior in virtual communities.

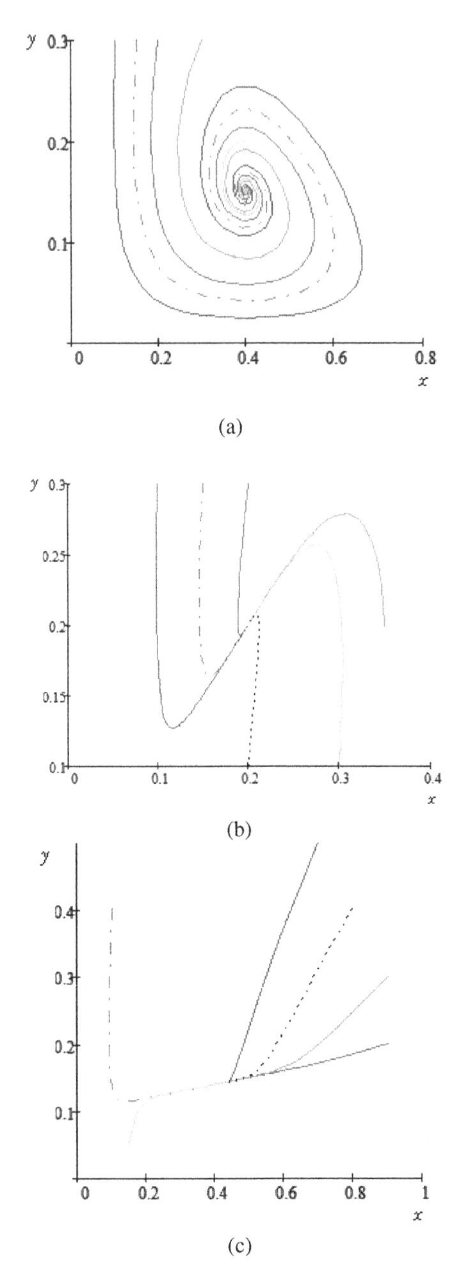

Fig. 2. SNS actors' interaction phase portraits: dynamic system in a chaotic state $a = 0.25$, $b = 0.25$, $c = 0.4$, $T = 1$ (a); synergetic control $u_{\gamma 1}(x, y)$ for $\psi_{\upsilon 1}(x, y)$, $\varepsilon_1 = \varepsilon_2 = 0.1$ (b); synergetic control $u_{\gamma 2} = (x, y)$ for $\psi_{\upsilon 2} = (x, y)$, $\varphi_1 = \varphi_2 = 0.3$, $N = 0.1$ (c)

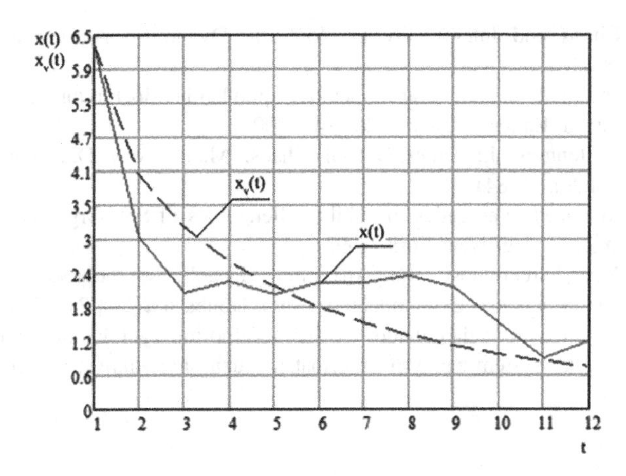

Fig. 3. Content market behavior dynamics for the "crucified boy" query: content market, thousand queries $x(t)$; content market under synergetic control, thousand queries $x_v(t)$

5 Conclusions and Further Studies

The concept of SNS actors' interaction synergetic control is designed to generalize the known methods of SNS actors' interaction control and to classify them as nonlinear systems described in the theory of chaotic dynamics. The concept is meant for application in the short term to put SNS actors' interactions under synergetic control underpinning their market for attractive information.

References

1. Horbulin, V.P., Dodonov, O.H., Lande, D.V.: Informatsiini operatsii ta bezpeka suspilstva: zahrozy, protydiia, modeliuvannia. Intertekhnolohiia, Kyiv (2009)
2. Tatnall, A.: Actor-Network Theory and Technology Innovation: Advancements and New Concepts. Information Science Reference, New York (2010)
3. Rogers, E.M.: Diffusion of Innovations. Free Press, New York (2003)
4. Castells, M., Cardoso, G.: The Network Society: From Knowledge to Policy. Johns Hopkins Center for Transatlantic Relations, Washington (2005)
5. Epstein, J.M.: Generative Social Science: Studies in Agent-Based Computational Modeling. Princeton University Press, Princeton (2012)
6. Epstein, J.M.: Nonlinear Dynamics, Mathematical Biology, and Social Science: Lecture Notes. Addison-Wesley Publishing Company, Massachusetts (1997)
7. Barrett, C., Eubank, S., Marathe, M.: Modeling and simulation of large biological, information and socio-technical systems: an interaction based approach. In: Goldin, D., Smolka, S.A. (eds.) Interactive Computation, pp. 353–392. Springer, Heidelberg (2006)
8. Wasserman, S., Faust, K.: Social Network Analysis: Methods and Applications. Cambridge University Press, Cambridge (1994)
9. Missaoui, R., Sarr, I.: Social Network Analysis - Community Detection and Evolution. Springer International Publishing, Switzerland (2014)

10. Tabor, M.: Chaos and Integrability in Nonlinear Dynamics: An Introduction. Wiley, Michigan (1989)
11. Kolesnikov, A.A.: Sinergeticheskoe metody upravlenija slozhnymi sistemami: teorija sistemnogo sinteza. Editorial URSS, Moskow (2005)
12. Prigogine, I., Stengers, I.: Order Out of Chaos. Man's New Dialogue with Nature. Heinemann, London (1984)
13. Haken, H.: Advanced Synergetics: Instability Hierarchies of Self-Organizing Systems and Devices. Springer-Verlag, New York (1993)
14. Serikov, A.V.: Jeffektivnost' hozjajstvennoj dejatel'nosti: opredelenie, izmerenie, sinergeticheskoe upravlenie. Ekonomichnyi visnyk Donbasu **2**(24), 212–219 (2011)
15. Hryshchuk, R.V.: Kontseptsiia pobudovy dyferentsialno-ihrovykh harantovano zakhyshchenykh rozpodilenykh system zakhystu informatsii. Suchasnyi zakhyst informatsii **1**(6), 4–9 (2011)

Artificial Neural Network as a Basic Element of the Automated Goniometric System

Irina Cherepanska(✉), Elena Bezvesilna, and Artem Sazonov

Zhytomyr State Technological University, Zhytomyr, Ukraine
cheri_ko@mail.ru, artyomsazonov@mail.com

Abstract. The approach to automatic definition of components of the systematic error and the sources of its appearing in the automated goniometric system that is based on artificial neural networks are proposed in the article. In particular, the input and output vectors and the structure of artificial neural network are defined. For this propose systematic error of automated goniometric system is presented as a totality of instrumental, methodic and subjective components, and each of them has defined primary components. These components form the structure and content of the artificial neural network input vector. The structure and the content of the output vector allow to detect the causes of errors and to correct measuring result in future. Generalized methodics of the proposed "back-propagation" neural network is given. The last one will be trained by the supervised learning.

Keywords: Goniometers · Artificial neural networks · Automated measuring · Goniometric systems · Precise measurements · Measurement errors

1 Introduction

Development of the precise automated goniometric systems with data processing in real-time mode is the most important task because these systems are used for contactless angles measuring in different fields: industry, navigation, orientation, etc.

The analysis of known investigations and publications [1–6] has shown that the task of goniometric systems synthesis singled out in separate and topical research field.

High accuracy, possibility of the measurement automation and data processing automation, and also high reliability are the main requirements that are imposed on the modern goniometric systems. It can be partially provided by the artificial intelligence usage e.g. artificial neural networks (ANN). The last ones are successfully used to solve different problems of data analysis and processing in condition of incompleteness, inconsistency and dynamism of the input information [5]. Thus the above mentioned specifies the topicality of this theme.

Based on the above mentioned, the ANN as a component of the precise automated goniometric system structure synthesis is the aim of the article. It will allow to analyze errors and to define its value, and also compensate them in real-time mode in the sequel. Particularly in [7–11] the problem of accuracy improvement is solved with the help of artificial neural networks.

R. Szewczyk and M. Kaliczyńska (eds.), *Recent Advances in Systems, Control and Information Technology*, Advances in Intelligent Systems and Computing 543, DOI 10.1007/978-3-319-48923-0_6

For example, in [7] the results of development and investigation of test bench equipment to account the errors of electronic goniometric tools which are used in geodesy at the basis of neural network algorithms are presented. It is shown that neural network algorithms usage allows to decrease the time of systematic error detection in calibration of geodesic goniometric tools and also makes this procedure simpler.

The problem of accuracy and stability improvement of coordinate and time ensuring of users in differential navigating system on the basis of software and algorithmic redundancy in neural network algorithm usage to process the control station measurement results are solved in [8].

The materials given in [9] concern the learning of newest principles of error's measurement and account on the basis of application of artificial neural networks.

The model of neural network measuring transducer for angular measurements is presented in [10]. Neural network is used to pick out and estimate the errors. The results of the error estimation of neural network model of nonlinear measurement transducer on the basis of three-layer perceptron are presented in [11].

2 The Main Part

The Automated Goniometric System is shown as a multilevel structure. Its block diagram is presented in Fig. 1.

The lower level of the system (*0-level*) is the level of the input measurement signal forming α_{in} and it is presented as a *subsystem of the angle high-precision measuring* (*SAM*). It is based on the high-precision laser goniometer.

1-level is presented as a *subsystem of signal preparing* (*SSP*). It is used for preprocessing of input analog signal α_{in}^{A} from SAM (matching, amplification, filtering) and its conversion into digital form α_{in}^{D} with the following processing.

2-level of the Automated Goniometric System is presented as a *subsystem of signal processing and displaying* (*SSPD*). SSPD is used for processing of digital signal from SSP α_{in}^{D} and its presentation in convenient form for the following visualization and its perception by the viewer and/or the following automated processing in real-time mode by means of artificial intelligence usage, e.g. neural networks.

It is obvious, that the systematic and random errors appear at the angular measuring. Imperfection of the measurement tools or so-called *instrumental error* Δy_I; imperfection of the measuring method or so-called *truncation error* Δy_M; insufficient skills of the operator, so-called *subjective error* Δy_S are the main sources of the systematic errors. Appearing of the random errors is conditioned by the common influence of the many random factors on the measurement tool and object.

Angular measurement accuracy improving of the proposed automated goniometric system can be reached by means of automated analysis and detection of systematic errors sources. Artificial neural networks (ANN) as a set of mathematic and algorithmic methods to solve broad spectrum of tasks of data analysis and processing can play an important role in it. In the context of solving tasks ANN allow to automate the processing and assessment of the informative parameters of all components of systematic error with parameters set usage in real-time mode. Furthermore, operational efficiency, probability of correct data processing in conditions of its incompleteness and

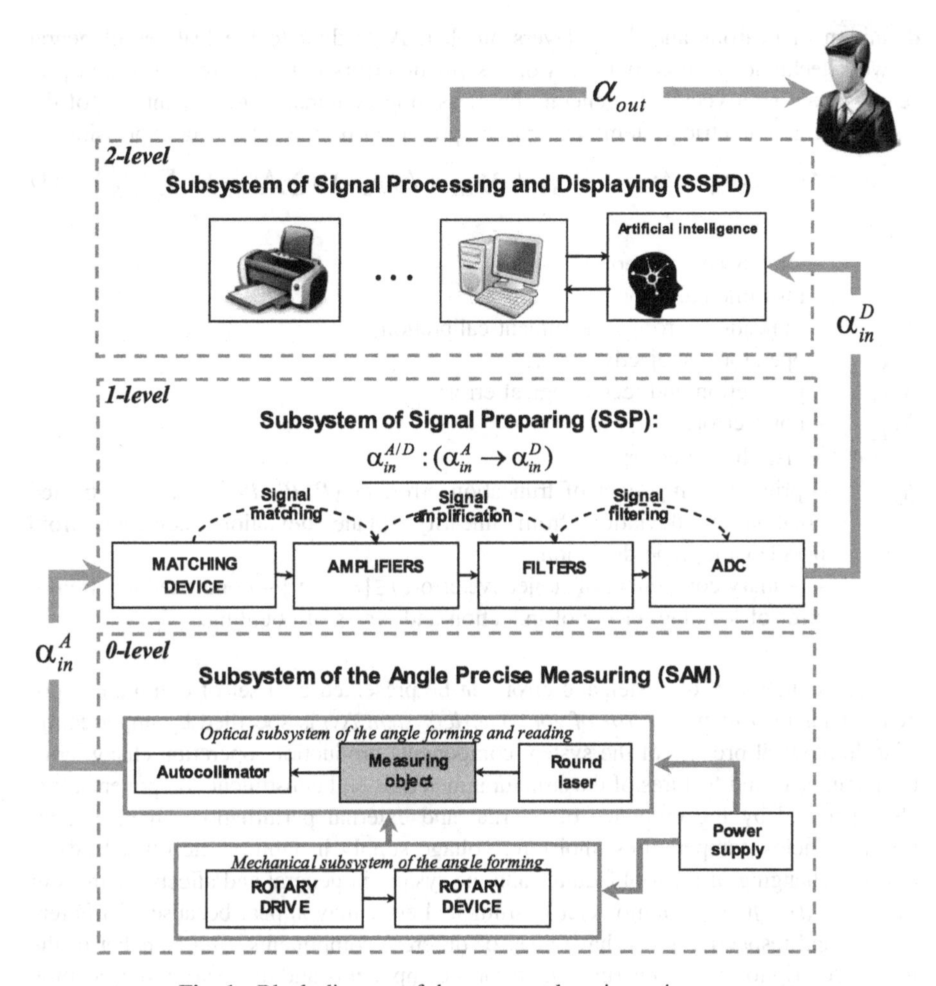

Fig. 1. Block diagram of the automated goniometric system

contradictions, as well as training and retraining simplicity allow to change the new kinds of tasks timely in case of external factors change.

For the above-mentioned it is proposed to use Kohonen ANN that allows to change internal and external data structure dynamically. It is determined by constant development and complication of goniometric complexes and systems, nomenclature of measuring objects, also means and methods of angles measuring. This distinguishes it from the other ANNs, e.g. ANN Hopfield, ANN Hamming, Multilayer Perceptron, RBF, etc.

Concrete type of data conversion performed with ANN is defined not only by its artificial neurons but also specifics of its architecture, i.e. by certain subsets of internal, external and hidden neurons. Thereby the synthesis of ANN for systematic error automated analysis of the proposed goniometric system intends to solve previously the tasks of structure and both internal and external vectors definition with the following definition of neurons and ANN layers number. According to the features of neural network technology all components of systematic errors must be predefined and presented as an input vector. In general, the measuring systematic component Δy_Σ of the automated goniometric system error can be presented by the following expression:

$$\Delta y_\Sigma = \Delta y_I + \Delta y_M + \Delta y_S = (\Delta y_{BT} + \Delta y_{Ш} + \Delta y_{F,F'}) + (\sum \Delta y_M^k) + (\sum \Delta y_S^j) \quad (1)$$

where

Δy_Σ – total systematic error;
Δy_I – instrumental error;
Δy_M – truncation error of instrument calibration;
Δy_S – operator's subjective error;
Δy_{BT} – production and technological error;
$\Delta y_{Ш}$ – noise error;
$\Delta y_{F,F'}$ – disturbance error;
Δy_M^k – primary component of truncation error; $k \in \{P, Pl, Pr, Rot\}$ – contracted notation of deviation from linearity, plane deviation, deviation from parallelism, angle deviation;
Δy_S^j – primary component of subjective error; $j \in \{\delta, n, \nu\}$ – contracted notation of error of locating, errors of induction and errors of counting.

Each component of systematic error can be presented as a set of definite components. Thus *instrumental error of goniometric system* Δy_I is specified by imperfection of technological process of the system components production, operation changing of the parameters and features of the structural materials and constituent components. It is also specified by the influence of internal and external perturbations, (e.g. electromagnetic fields, temperatures, moisture, voltage supply instability, etc.) that leads to periodic changing of the tool features and the system in general and affects the transfer function $Y(t) = f(X(t))$. Goniometer instrumental error may appear because of different reasons and respectively can have a set of different components. The alphabet of the formal description of the instrumental error's components and its vector representation are shown in Table 1.

The structure of the input vector should correspond to the structure of systematic error. According to the expression (1) and Table 1 the input ANN vector is presented by the following expression:

$$\mathrm{X} = \{x_1^1, x_2^1, x_2^2, x_2^3, x_2^4, x_2^5, x_2^6, x_3^1, x_3^2, x_4^1, x_4^2, x_4^3, x_4^4, x_5^1, x_6^1, x_7^1\}, \quad (2)$$

Where x_1^1 is an ANN input signal, corresponding to the production and technological errors of metrological parameters from nominal values; x_2^1 – ANN input signal corresponding to the quantization noise existence; x_2^2 – ANN input signal corresponding to the noise from electronic components; x_2^3 – ANN input signal corresponding to the existence of parasitic capacity and inductive connections; x_2^4 – ANN input signal corresponding to the leakage resistance existence; x_2^5 – ANN input signal corresponding to the existence of parasitic thermal e.m.f.; x_2^6 – ANN input signal

Table 1. The alphabet and vector representation of goniometric system systematic error according to the features of neural network technology

Components of systematic error	Elements (reasons of appearing) of systematic error components	Vector alphabet	
		Elements of input vector	Input signal maximum value
Production and technological error Δy_{BT}	Production and technological deviation of metrological parameters A_i of the tool from nominal value	x_1^1	1
Noise error $\Delta y_{ш}$,	Quantization noise of output signal $\Delta y_{ш}^{K}$,	x_2^1	1
	Noise of electronic components $\Delta y_{ш}^{e}$	x_2^2	1
	Parasitic capacity and inductivity connections $\Delta y_{ш}^{CL}$	x_2^3	1
	Leakage resistance $\Delta y_{ш}^{r}$	x_2^4	1
	Parasitic thermal e.m.f $\Delta y_{ш}^{tE}$	x_2^5	1
	Photon noise $\Delta y_{ш}^{\Phi}$	x_2^6	1
Disturbance error $\Delta y_{F,F'}$	External disturbance $F(t)$	x_3^1	1
	Internal disturbance $F'(t)$	x_3^2	1
Error of measuring object's parameters and geometric sizes	Deviation from linearity Δy_M^P	x_4^1	1
	Deviation from flatness Δy_M^{Pl}	x_4^2	1
	Deviation from surfaces parallelism Δy_M^{Pr}	x_4^3	1
	Angle deviation Δy_M^{Rot}	x_4^4	1
Operator's error	Error of location Δy_S^{δ}	x_5^1	1
	Error of induction Δy_S^{n};	x_6^1	1
	Error of counting Δy_S^{v}	x_7^1	1
General structure of the input vector X: $\{x_1^1, x_2^1, x_2^2, x_2^3, x_2^4, x_2^5, x_2^6, x_3^1, x_3^2, x_4^1, x_4^2, x_4^3, x_4^4, x_5^1, x_6^1, x_7^1\}$			

corresponding to photonic noise existence; x_3^1 – ANN input signal corresponding to the existence of external disturbance; x_3^2 – ANN input signal corresponding to the existence of the internal disturbance; x_4^1 – ANN input signal corresponding to existence of deviation from linearity of; x_4^2 – ANN input signal corresponding to existence of deviation from flatness of the measured object; x_4^3 – ANN input signal corresponding to existence of deviation from parallelism of the measured object surfaces; x_4^4 – ANN input signal corresponding to existence of angle deviation; x_5^1 – ANN input signal that registers inaccuracy or error of the measured object location. x_6^1 – ANN input signal that registers inaccuracy of induction; x_7^1 – ANN input signal corresponding to the counting inaccuracy.

Vector Y is formed at the ANN output according to the requirements of the ANN technology (Table 1) and based on the vector description of the systematic error components. This vector contains 6 components that define the content of the systematic error and its value (Table 2).

Table 2. Output vector characteristic $Y = \{y_k | k = \overline{1,6}\}$

Content of systematic error	Meaning and value of systematic error	Vector alphabet	
		Output vector elements	Maximum value of the output signal
Instrumental error Δy_I	visible	y_1	1
	invisible		0
	max	y_2	1
	min		0
Methodology error Δy_M	visible	y_3	1
	invisible		0
	max	y_4	1
	min		0
Subjective error Δy_S	visible	y_5	1
	invisible		0
	max	y_6	1
	min		0
Common structure of the output vector Y: $\{y_1, y_2, y_3, y_4, y_5, y_6\}$			

Input and output vectors allow synthesis of the ANN schematic model to analyze systematic error of the goniometric system (Fig. 2), and this ANN is a three-layer perceptron. The first layer, so-called “input”, is formed by the input neurons and is intended to receive the input information in the form of an input vector. In this layer the computational procedures are not performed and information is transmitted from the neuron’s input to output by changing its activation. Intermediate information preprocessing is carried out in so-called “hidden” layer and as a result of this action the linearly disjoint sets are fed into the neuron’s output layer. The “output” layer is formed by the output neurons. Their output values are the out as a network output vector.

The experimental investigation of ANN has been carried out on pre-generated training set. The research has been conducted to confirm possibility of its using in solving assigned tasks as well as structure correction of ANN (i.e. changing the number of neurons in hidden layer) in case of receiving unsatisfactory results. The ANN has been trained with back propagation errors method. The step is 0.1; the moment is 0.9. Optimization criterion of training algorithm is the mean-square error E = 0.05; activation function slope –1.

The results of experimental investigation of proposed ANN are presented on plots as changing of mean-square error depending on hidden layer dimension (Fig. 3). It shows the deviation of desired input value of relevant ANN from the actual one. The results given can be treated as follows:

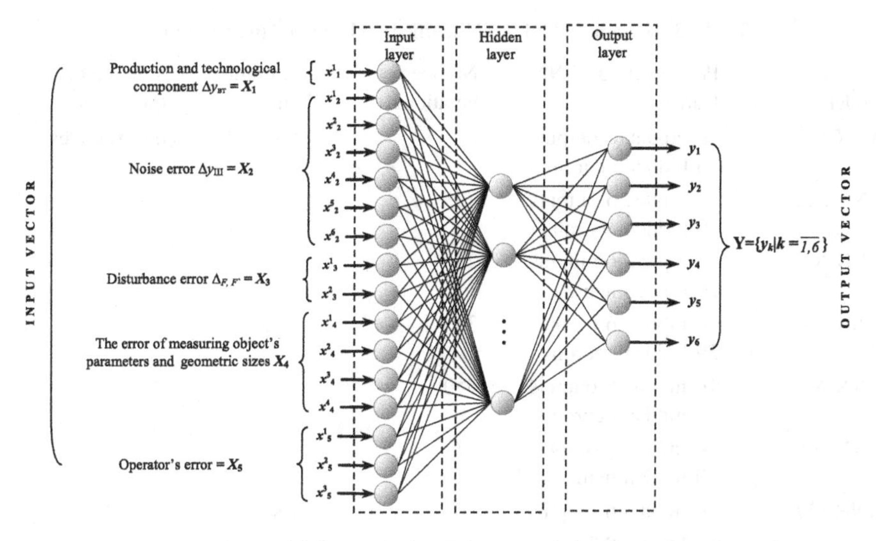

Fig. 2. ANN schematic model for analysis of the systematic error of goniometric system

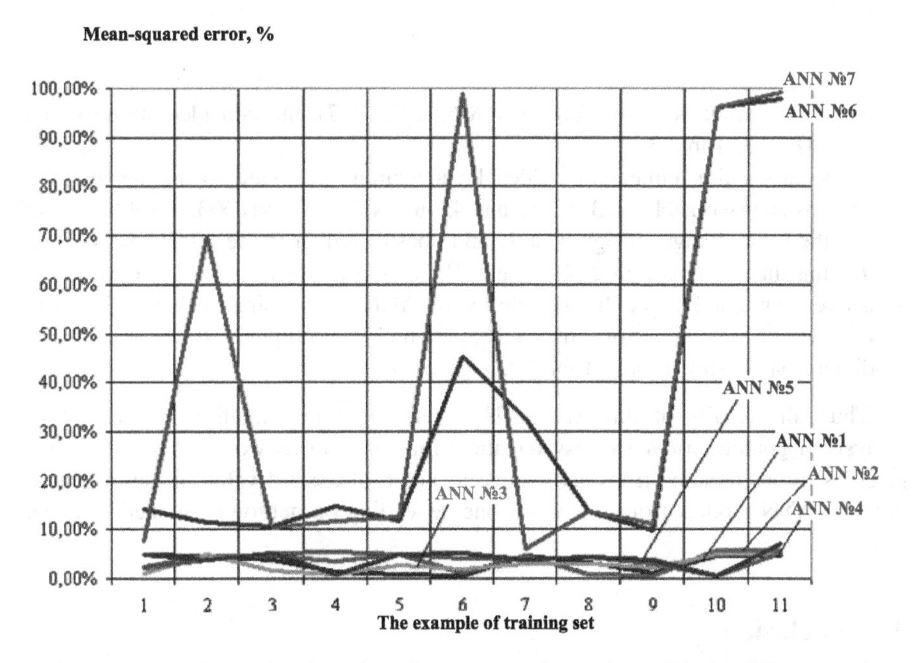

Fig. 3. The graph of changing the value of mean-squared error for ANN with different numbers of neurons in hidden layer (Table 3)

– to solve the assigned task the work of 7 ANNs with different dimensions of hidden layer have been learned (Table 3). The results of ANN's training in regard to Fig. 3, Table 1 showed that when the number of neurons in hidden layer is less than the

Table 3. Parameters of ANN learned and time of their training

№ of ANN model	Parameters of ANN learned	Number of iterations	Training time	Volume of training set
ANN №1	16 inputs, 6 outputs, 16 hidden neurons	52	1 s	66 examples
ANN №2	16 inputs, 6 outputs, 19 hidden neurons	72	2 s	
ANN №3	16 inputs, 6 outputs, 29 hidden neurons	284	4 s	
ANN №4	16 inputs, 6 outputs, 39 hidden neurons	310	17 s	
ANN №5	16 inputs, 6 outputs, 58 hidden neurons	350	21 s	
ANN №6	16 inputs, 6 outputs, 10 hidden neurons	43	1 s	
ANN №7	16 inputs, 6 outputs, 5 hidden neurons	40	1 s	
Measurement range of hidden layer value [12]		From 2 to 58 neurons		

number of input neurons (for ANN №6, ANN №7), the considerable errors are made (Fig. 3, Table 3).

- in increasing the number of hidden layer neurons in regard to number of input neurons approximately on 3, 13, 23 and 42 in ANN №2, ANN №3, ANN №4, ANN №5 the error changes insignificantly and doesn't influence the results. In so doing training time increases in 2, 4, 17 and 21 times correspondingly (Table 3);
- above mentioned argue for possibility of ANN № 1 using with the following parameters: the number of inputs – 16; the number of outputs – 6; the hidden layer dimensions – 16 neurons (Table 3).

Thus, the results of different ANN experimental investigations conducted for analysis of goniometric system systematic error points to its workability. The results given as mean-squared error curve show deviation of desired value of the outputs of different ANN models from the actual one depending on neuron's number in hidden layer.

3 Conclusions

Thus, the results of experimental investigations of different ANN models for the analysis of systematic error of goniometric system points to its workability. It is supposed that ANN shown by the authors allows to analyze systematic error and to define its values with the following compensation in real-time mode.

It should be noted that a number of additional tasks that appear in development and synthesis of the proposed ANN have been solved by the authors. In particular, the tasks of changing the input, output, weight factors, ultimate levels and other parameters of the proposed ANN, as well as the values defining and setting to the last one. The tasks of testing and training bases forming and ANN training should be also solved. The last one requires an additional research in this direction.

Besides, the presented research points to perspectives of ANN usage as a part of automated goniometric system to improve its accuracy.

References

1. Yu, A.M.: Development and research of goniometric systems for angle transducers check. Ph.D. dissertation abstract on the specialty “Information and measuring checked systems (instrument making)”, Saint-Petersburg, p. 17 (2009)
2. Aksenenko, V.D.: Automated correction of the angular sensors error. Aerospace instrument making, vol. 6, pp. 2–7 (2003)
3. Barinova, E.A.: Development and research of the goniometric systems accuracy improving methods. Ph.D. dissertation on the specialty “Information and measuring checked systems (instrument making)”, Saint-Petersburg, p. 167 (2009)
4. Bereza, B.V.: Development and research of the laser goniometric system for checking measuring tools of angular movements : Ph.D. dissertation abstract on the specialty “Systems of data processing and control”, Saint-Petersburg, p. 20 (1993)
5. Pavlov, P.A.: Development and research of the precise laser goniometric systems. Doctoral dissertation on the specialty “Information and measuring checked systems (instrument making)”, Saint-Petersburg, p. 280 (2008)
6. Korotaev, V.V.: Accuracy of the measuring optical-electronic tools and systems. Manual, Saint-Petersburg, p. 42 (2011)
7. Hinoeva, O.B.: Developing and application of neural network algorithms of etalon tools errors accounting in calibration of goniometric geodesic equipment: abstract of a Ph.D. thesis on the specialty 25.00.32 “Geodesy”/Olga Borisovna Hinoeva – Moscow, p. 24 (2007)
8. Kashaev, I.A.: Neural network technologies usage to process the information in control station of in differential navigation system. In: Kashaev, I.A., Podorozhniak, A.A., Dejneko, V.N. (eds.) System of Data Processing, vol. 16(6), pp. 229–233 (2001)
9. Shevchuk, V.P.: Metrology of intelligent measurement system: monograph. In: Shevchuk, V.P., Drop, V.I., Zheltonogov, A.P., Lyasin, D.N., (eds.) (generally edited by academician of Metrological Academy of Russia, Prof. Shevchuka, V.P.) VolgGTU – Volgograd, p. 210 (2005)
10. Skvortsov, M.G.: Neural network measuring transducer. In: Skvortsov, M.G. (ed.) Proceedings of VolgGTU, vol. 6(5), pp. 88–90 (2011)
11. Zaporozhets, O.V.: The research of the error of neural network model of nonlinear measuring transducer. In: Zaporozhets, O.V., Ovcharova, T.A. (eds.) Proceedings of the Conference “Informatics, Simulation and Economics”, vol. 1, pp. 60–66, Smolensk, April 23–25 (2014)
12. Diakonov, V.P.: Matlab 6.5 SP1/7/7 SP1/7 SP2 + Simulink 5/6. Tools of artificial intelligent and bioinformatics. In: Diakonov, V.P., Kruglov, V.V. (eds.), p. 456. Solon-Press, Moscow (2006)

The Program of Calculation of Rate of Movement of Cleaning Device for the Main Gas Pipeline

Volodymyr Kvasnikov[1(✉)] and Oleksandr Stashynsky[2]

[1] National Aviation University, Kiev, Ukraine
kvp@nau.edu.ua
[2] Gas Mains Administration (GMA) «CHERKASYTRANSGAZ», Cherkassy, Ukraine
stashinskiy-ap@utg.ua

Abstract. This work is devoted to the decision of problem of cleaning of internal cavity of main gas pipeline by means of cleansing devices. Realization of the high-quality cleaning is executed with the purpose of increase of hydraulic efficiency and carrying capacity of gas pipeline to the project values. We know that the rate of movement of cleansing device on length of gas pipeline grows from the beginning of area and to the end with every kilometer. Sometimes speed of cleansing device at the end of area of gas pipeline can attain a size in 1.5 times more than the size at the beginning of area of gas pipeline. In result, the cleansing device moving with speed more than his optimal set speed, diminishes the efficiency of work and does not provide the complete cleaning of gas pipeline. The methods offered by an author consist in providing of optimal rate of movement of cleansing device on all length of area of gas pipeline. It will provide more effective cleaning, without additional expenses.

Keywords: Cleansing device · Compressor stations · Main gas pipeline · Gas flow · Mathematical model · Information technologies

1 Introduction

During operation the main gas pipeline, in its internal cavity always accumulate contaminants: soil, rocks, pieces of welding electrodes, etc. These contaminants are reducing the cross section of the pipeline, and reducing its capacity. To determine the capacity perform. To determine the capacity of pipeline must perform assessment of the internal cavity. This assessment is characterized by a coefficient of hydraulic efficiency of main gas pipeline

$$E = \sqrt{\frac{\lambda_m}{\lambda_f}} = \frac{q_{\mathrm{f}}}{q}, \tag{1}$$

where, λ_m, λ_f – theoretical and factual coefficients of hydraulic resistance; q_f, q – theoretical and factual carrying capacity of gas pipeline.

R. Szewczyk and M. Kaliczyńska (eds.), *Recent Advances in Systems, Control and Information Technology*, Advances in Intelligent Systems and Computing 543, DOI 10.1007/978-3-319-48923-0_7

Cleaning of internal cavity of gas pipeline is executed for the increase of its hydraulic efficiency and carrying capacity to the project values.

Now a few methods of cleaning of pipeline are known. To them belong mechanical, vibromechanical, ultrasonic, chemical and others. On main gas pipelines most widespread is a method of the mechanical cleaning by means of scrapers, delimiters, pistons etc.

The rate of movement of cleansing device on length of gas pipeline grows from the beginning of area and to the end with every kilometer and depends on the change of basic parameters of the mode of operations of gas pipeline: sizes of pressure of gas, temperature, and expense. Sometimes speed of cleansing device at the end of area of gas pipeline can attain a size in 1.5 times more than the size at the beginning of area of gas pipeline (Fig. 1).

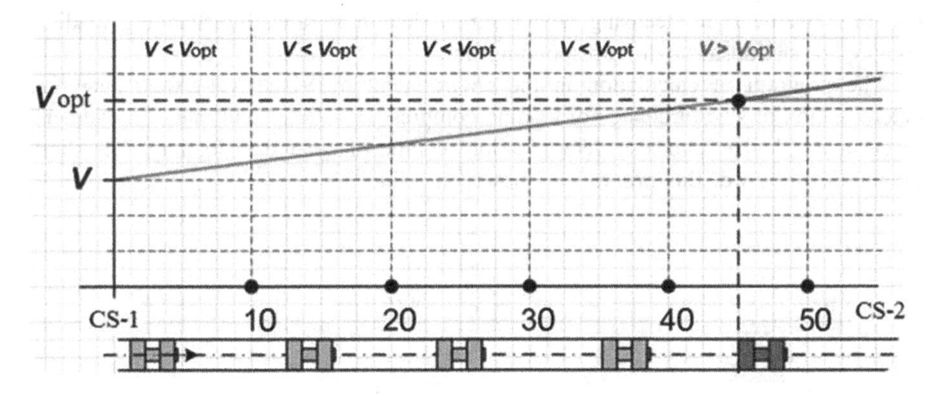

Fig. 1. The increase of speed of cleansing device along pipeline, where V – factual speed of cleansing device; V_{opt} – optimal speed of cleansing device; CS-1 – compressor station that launch the cleansing device; CS-2 – compressor station that receive the cleansing device; 10, 20, 30, …, – kilometer of gas pipeline

In result, the cleansing device moving with speed more than his optimal set speed, diminishes the efficiency of work and does not provide the complete cleaning of gas pipeline [1]. Also it is possible destruction of structural elements of cleansing device (Fig. 2).

Fig. 2. Destruction of structural elements of cleansing device

Aim of the Work. Develop methods, which consist in providing of optimal rate of movement of cleansing device on all length of area of gas pipeline.

2 Materials and Results of the Research

Thus, it is necessary to support the set rate of movement permanent on all length of area. For this purpose with every change of basic parameters of work of gas pipeline, it is necessary to compensate their value according to the set rate of movement of cleansing device [3, 4]. It will make the change of loading compressor stations (CS) during motion between them of cleansing device [5–7].

So we change the mode of operations of area of gas pipeline, and regulate the set rate of movement of cleansing device on all length of area by changing loading compressor station through corresponding time.

The program of calculation in the environment of Visual C++ was made for determination of size of change of loading compressor stations, and rate of movement of cleansing device, and also determination of time of change of loading. The short graphic algorithm of this program is shown on a Fig. 3.

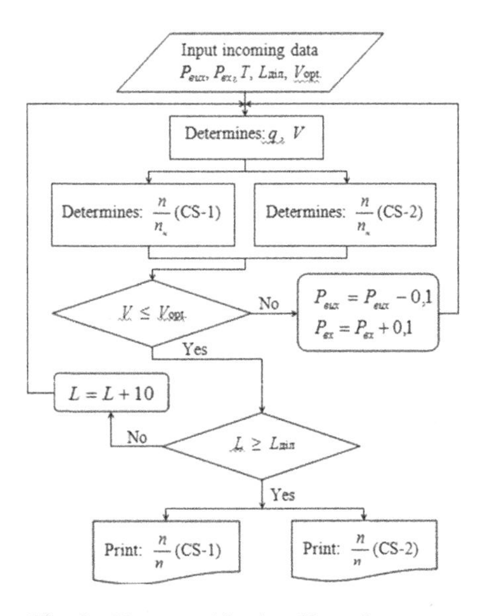

Fig. 3. Short graphic algorithm of program

Essence of calculation consists in the following.

1. Input incoming data:
   - entrance and initial pressures and temperatures of gas are in a CS-1. Compressor station that launch the cleansing device;

- entrance and initial pressures and temperatures of gas are in a CS-2. Compressor station that receive the cleansing device;
- length of the pipeline;
- optimal rate of movement of cleansing device.

2. Determine the carrying capacity of gas pipeline – q, between two compressor stations [4]:

$$q = 3,26 \cdot 10^{-7} \cdot d^{2.5} \sqrt{\frac{P_n^2 - P_\kappa^2}{\lambda \cdot Z_{cp} \cdot \Delta \cdot T_{cp} \cdot L}}, \tag{2}$$

where P_κ, P_n – finite and initial pressures of gas are in a gas pipeline, λ – coefficient of hydraulic resistance; T_{cp} – mean value of temperature of gas, Δ – relative gas density, L – length of the pipeline, Z_{cp} – compressibility factor.

3. Divide the area of pipeline into parts through 10 km (Fig. 4), and determine middle rate of movement of cleansing device on each parts.

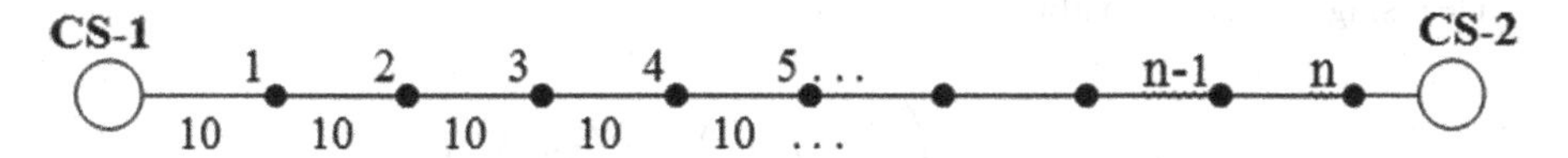

Fig. 4. Scheme of the pipeline between CS-1 and CS-2

For part of CS-1-1:

$$V^{CS1-1} = 0,0052 \frac{T_{cp}^{CS1-1} \cdot z_{cp}^{CS1-1} q}{d^2 \cdot P_{cp}^{CS1-1}}, \tag{3}$$

where, $T_{cp}^{CS1-1}, z_{cp}^{CS1-1}, P_{cp}^{CS1-1}$ – mean value of gas pressure, mean value of gas temperature, coefficient of gas compressibility at part of CS-1-1, d – internal diameter of gas pipeline.

To find these values is necessary to determine the pressure and temperature of gas at point 1.

Temperature of gas at point 1:

$$T_1 = T_{gp} + \left(T_{CS-1} - T_{gp}\right) \cdot e^{-ax}, \tag{4}$$

where T_{gp} – soil temperature, T_{CS-1} – temperature of gas at the beginning of part CS-1-1, x – the length of part of CS-1-1 = 10 km, a – conditional parameter.

Mean value temperature of gas:

$$T_{cp}^{CS-1} = T_{gp} + \frac{T_{CS-1} - T_1}{ax}. \tag{5}$$

Pressure of gas at point 1:

$$P_1 = \sqrt{P^2_{CS-1} - \frac{\lambda \cdot \Delta \cdot z^{CS1-1}_{cp} \cdot T^{CS1-1}_{cp} \cdot x \cdot q^2}{(3,26 \cdot 10^{-7})^2 \cdot d^5}}, \quad (6)$$

where, P_{CS-1} – pressure of gas at the beginning of part CS-1-1.

Mean value pressure of gas:

$$P^{CS-1}_{cp} = \frac{2}{3}\left(P_{CS-1} + \frac{P_1^2}{P_{CS-1} + P_1}\right) \quad (7)$$

Substitute T^{CS1-1}_{cp} and P^{CS1-1}_{cp} into (3) and determines middle rate of movement of cleansing device V^{CS1-1}. For other parts of pipeline the calculation is similar, only initial values of gas parameters at the next part are equal the finite values of gas parameters at the previous part.

4. Determines loading of the compressor stations, on which a start and reception of cleansing device is conducted:

$$\frac{\varepsilon^{\frac{k-1}{k}} - 1}{\frac{n}{n_n}\sqrt{\frac{z_{zb} \cdot R_{zb} \cdot T_{zb}}{z_{bc} \cdot R \cdot T_{bc}}}} = \left(a + b\left(\frac{Q_{bc}}{\frac{n}{n_n}}\right) + c\left(\frac{Q_{bc}}{\frac{n}{n_n}}\right)^2\right)^{\frac{k-1}{k}} - 1 \quad (8)$$

where, $\frac{n}{n_n}$ – loading of the compressor stations, %, ε – degree of increase of pressure on compressor stations

$$\varepsilon = \frac{P_{out}}{P_{in}}; \quad (9)$$

P_{out}, P_{in} – entrance and initial pressures of gas are in a compressor stations, $Q_{вс}$– carrying capacity of compressor stations, million m^3/24 h, z_{zb}, R_{zb}, T_{zb} – erected parameters of gas, $z_{вс}$, R , $T_{вс}$ – parameters of gas on the entrance of compressor stations, a, b, c – coefficients of mathematical model of descriptions of compressors of compressor stations.

5. Farther the sizes of the found middle rate of movement of cleansing device made on every area with the set optimal speed V_{opt} (Fig. 1) are compared by means of logical operators of the program. If middle speed will exceed the value of the set optimal speed on a 1 m/s, the program diminishes the value of actual speed by means of change of sizes of pressures of gas at the beginning and at the end of area of gas pipeline (at the beginning of area – pressure falls down, in the end – rises on 0.1 kgf/cm^2).

The values of incoming data of calculation change accordingly, the calculation of values of carrying capacity, loading of CS is conducted on new, and sizes of middle rates of movement of cleansing device on length of area through each 10 km, so an analogical calculation is conducted, only with the incoming data which are set by the program. If the rate of movement of cleansing device on an area did not diminish on 1 m/s from the set speed, then the program changes the value of pressures of gas on 0.1 kgf/cm^2, and a calculation recurs again. It will be carried out until the above-mentioned condition is executed. And farther a calculation is conducted for a next area like previous.

The time of change of loading CS is determined in dependence on the rate of movement of cleansing device, and area of gas pipeline passed by it. Wave of pressure, which appears at the change of entrance and initial pressure of gas in a gas pipeline is moving at the speed of sound:

$$c = \sqrt{k \cdot z \cdot R \cdot T} \tag{9}$$

From data of experimental calculation (Fig. 5), between the fiftieth and the sixtieth kilometer of gas pipeline the rate of movement of cleansing device becomes more on a 1 m/s from the set speed. Farther the values of initial and finite pressure of gas begin to deviate on an area, and all modes of operations of gas pipeline change, that causes to the change of loading on CS. Then after the sixtieth kilometer there is a permanent change of all parameters of the mode of gas pipeline which supports the stable rate of movement of cleansing device, so its traffic regulation is executed by the change of loading compressor stations.

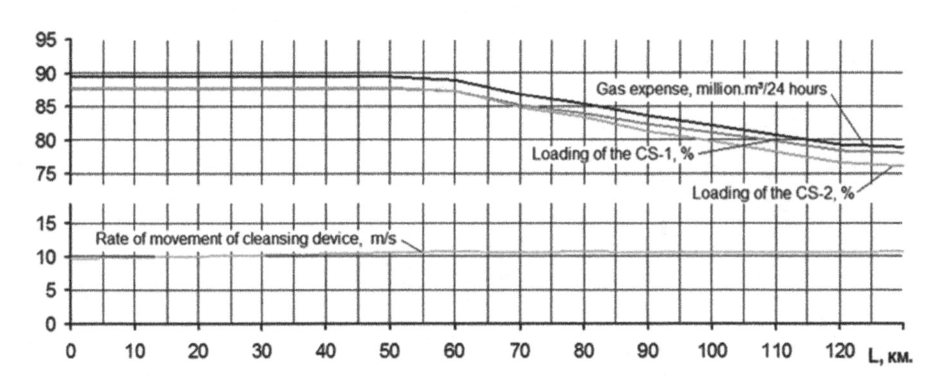

Fig. 5. Dependences of parameters of the mode of operations of gas pipeline

3 Conclusions

Analyzing a programmatic calculation it is possible to say, that it is really possible to regulate the rate of movement of cleansing device by changing the loadings of the compressor stations on all length of area of gas pipeline. It is not advantageous on one side, as a mode of operations of gas pipeline breaks, but on the other side, this method is advantageous enough, as it will provide optimal motion of cleansing device, and thus high-quality cleaning of gas pipeline.

References

1. Pankratov, V.S., Dubins'kiy, A.V., Sipershteyn, B.I., Nedra, L.: Informatsionno-vychislitel'nyye sistemy v dispetcherskomu pravlenii gazoprovodami, p. 246 (1988)
2. McAllister, W.E.: Pipeline Rules of Thumb Handbook: Quick and Accurate Solutions to Your Everyday Pipeline Problems, 632 p. Heinemann (2002)
3. Korobiichuk, I., Bezvesilna, O., Ilchenko, A., Shadura, V., Nowicki, M., Szewczyk, R.: A mathematical model of the thermo-anemometric flowmeter. Sensors **15**, 22899–22913 (2015)
4. Korobiichuk, I., Shavursky, Yu., Nowicki, M., Szewczyk, R.: Research of the thermal parameters and the accuracy of flow measurement of the biological fuel. J. Mech. Eng. Autom. **5**, 415–419 (2015)
5. Larock, B.E., Jeppson, R.W., Watters, G.Z.: Hydraulics of Pipeline Systems, 533 p. CRC Press (2000)
6. Lur'ye, M.V.: Matematicheskoye modelirovaniye protsessov truboprovodnogo transporta uglevodorodov. M.: GUP "Neft' i gaz" RGU neftiigazaim. I.M. Gubkina (2002)
7. Karachun, V., Mel'nick, V., Korobiichuk, I., Nowicki, M., Szewczyk, R., Kobzar, S.: The additional error of inertial sensor induced by hypersonic flight condition. Sensors **16**(3) (2016). doi:10.3390/s1603029

Concept, Definition and Use of an Agent in the Multi-agent Information Management Systems at the Objects of Various Nature

Regina Boyko, Dmitriy Shumyhai(✉), and Miroslava Gladka

National University of Food Technologies, Kiev, Ukraine
{rela,mira}@ukr.net, shumyhai@gmail.com

Abstract. One of the most effective methods in information systems management is a multi-agent approach and its implementation in distributed structures, especially in the control under uncertainty. In the article is shown the basic concepts that make it possible to create multi-agents control systems at various levels and purposes, first of all this concerns an organizational and technological (technical) systems, basic terms and concepts which need clarification allowing to the system's feature, such as common approach to technological complex and enterprises in general, which have different structures and the nature of the solved problems.

Keywords: Information · Agent · System · Object · Multi-agent

1 Introduction

Almost all the works, which provides definitions of what agent is and what its basic properties, the general, is the lack of consensus on the matter. In fact, using the term «agent», every author defines «agent» as its own specific set of properties. The concept of «agent» is used in various fields, for example, at manufacturing «agent» may be called a robot, in the telecommunications – program, etc. As a result, depending on the agent's habitat it has different properties [1]. Therefore, in the development and implementation of systems at this area a lot of agent's types were appeared, such as: autonomous agents, mobile agents, intelligent agents, social agents, etc. So, instead of a single basic definition of «agent» there are a lot of definitions of derivative types.

In one of the most influential works of modern artificial intelligence, issued by S. Russell and P. Norvig [2] «agent» means «any entity that is in some environment, sees it through sensors, getting data which reflect events that occur in the environment, interprets this data and act on the environment using effectors». As follows, it distinguishes four input factors of agent's forming – environment, perception, interpretation, action. According to Mr. Maes [3], «autonomous agents – computer systems that operate in a complex, dynamic environment, able to feel and act autonomously in this environment and thus perform many tasks for which they are intended» . Two restrictions on agent's environment are purposed: complex and dynamic.

R. Szewczyk and M. Kaliczyńska (eds.), *Recent Advances in Systems, Control and Information Technology*,
Advances in Intelligent Systems and Computing 543, DOI 10.1007/978-3-319-48923-0_8

2 The Formulation of the Task of the Multi-agents

Often agents are understood as computational units supporting state and local parallel computing, as well as communication processes capable to achieve the status of other agents and automatically perform actions in certain environmental conditions [4]. Agent can be considered as a software object representing perception of environment and environmental action in order to achieve a certain goal. Perception of an environment by «agent» can be made by perception of the physical environment which consists of system components (typically represented as a database) and the perception of a multi-agent environment which consists of other agents [5].

3 Definable Description of MAS

The complete sequence of the perception of the physical environment at time $\boldsymbol{t}$ denoted as ENV_t, multi-agent environment – MAE. The complete sequence of perception to date consists of individuals and multi. Designate $\boldsymbol{h}$ as the present point in time, the complete sequence of perception can be represented as:

$$P' = ENV_h uMAS \tag{1}$$

Agent usually has some knowledge about the world, encoded as rules, these rules are reflected to as encoded $\boldsymbol{k}$. This skills include general knowledge, that is, those that are available to all agents, and knowledge which is unique to each agent separately. In our environment general knowledge may include algorithms for problems to be solved in the management systems of technological complex and other publicly and procedures. As long as agent is not activated, it can update its internal state based on new perceptions and own knowledge of the world; adjusting function which generates a new states $\boldsymbol{S}$ is indicated as follows:

$$SG : (\frac{P}{k}) - S^* \tag{2}$$

Therefore, «agent» is able to distinguish some plight of the environment, in other words, «agent» has the same priority for some plights of the environment. It is essential in complex environments, such as power systems, which can have plurality of plights, but allow to individual agents deal with a limited number of plights. As a rule, «agent» has set the following properties:

- adaptability: the agent has the ability to learn;
- autonomy: agent works as an independent program, setting itself goals and carry out actions to achieve these goals;
- communication: agent can communicate with other agents;
- permeability: agent can communicate with other agents in several ways, for example, playing the role of the supplier/customer information, or both at the same time the role and the role of the intermediary;

- the ability to reasoning: agent may have a partial knowledge or output mechanisms, such as knowledge how to bring data from different sources to the same species, agent may specialize in specific subject area;
- mobility: agent's ability to transfer the code from one data source to another.

From the perspective of developers of the information systems «agent» is the module (component) software, running on a particular platform, commits certain acts, such as the display and input of information, exchanging messages with other agents or person [6]. Author in this sequel will adhere to this definition.

Agent's programming languages. Currently, there is no programming languages that fully would answer the needs of multi-agent system technology. Currently developing agent's systems use a wide of different base languages, but, unfortunately, none of them can be regarded as a true «agent-oriented». There are attempts to extend existing languages, and attempts to use traditional programming languages. There are a number of projects to develop new specialized agent's languages.

Below is shown the classification of the languages, which most commonly used in the of intelligent agent technology.

1. Universal programming language (Java);
2. Languages which are focused on knowledge;
   (a) knowledge representation languages (KIF);
   (b) the language of negotiations and of sharing knowledge (KQML, AgentSpeak, April);
   (c) language of agent's specifications.
3. Specialized programming language agents (TeleScript).
4. Scripting languages (Tcl/Tk).
5. Symbolic language and logic programming languages (Oz, Prolog).

If we talk about the requirements that apply to programming languages of agent-based systems, they can be formulated as follows:

1. ***Ensuring portability on various platforms.*** This requirement arises whenever necessary to ensure an agent's mobility. To ensure the mobility agent, the language must support sending mechanism, transmitting, receiving and executing code containing agents. There are two different approaches to solve the mobility problem. The first is an agent's transfer in text form, as a special script followed the interpretation of this scenario on the receiving computer, the second – agent's transfer into the form of machine-independent byte code. Both of these methods have their advantages and disadvantages. This byte-code compiler generated during the creation of agency system, is sent over the network and runs the bytecode interpreter on the receiving computer.
2. ***Availability on multiple platforms.*** This requirement follows directly from the previous one. Intelligent agents have to work in heterogeneous computing environment. Any computer that receives the agent should be able to accept and execute it.
3. ***Support networking***. Support for network services may include family appropriate programming interfaces (APIs) such as: sockets, interfaces to databases, interfaces

interaction of objects (CORBA, OLE, ActiveX, etc.), special mechanisms are built into the language (such as Remote Method Invocation in the language Java), special primitives to implement language negotiation agent, etc.

4. ***Multithread processing («Multithreading»).*** Agent can perform certain actions at one time. Thus, the programming language agents include support for parallel execution of various functions of an agent (such as «threads») and various synchronization primitives (semaphores, monitors, critical sections, etc.). Moreover, the process that performs all agents (and may in fact be viewed as a meta-agent) have to support parallel execution agents. The latter can be achieved through a separate virtual machine implemented preemptive multitasking mode and own strategy of time-sharing.
5. ***Support for symbolic computations***. Since under the current views agent have to use actively the achievements and methods of artificial intelligence, it would be helpful to have support for symbolic computation and, possibly, logic programming, built-in language (like PROLOG and LISP), as well as having a built-in terminals mechanism involving various strategies search solutions. Automatic memory management and garbage collection - standard tools for these languages.
6. ***Safety in particular to protect the system from unauthorized access or «bad codes»***. This is an important for many reasons. Mobile agents who comes from the network can carry many dangers to the host machine, as they're performed in its address space. In order to ensure security in the transmission control each agent have to perform the authorization process agent, i.e. to check whether he is registered if the appropriate authority (privileges) to perform an action or access certain resources. The security system should prevent any unauthorized actions of agent.
7. ***Object orientation***. Language should have mechanisms of inheritance, consider the procedure calls as "message sent" from one object to another, include the possibility of synchronous and asynchronous communication facilities as well as to allow parallel inside the object.
8. ***Language support agent's properties***. It would probably be convenient to have the support of specific agent's properties, built into the language at the syntax level, so that, for example, properties such «beliefs-desires-intentions» (BDI), guidelines for negotiations and mobility, meeting place, etc. could be expressed by using the appropriate language primitives [7].

4 Conclusions

It is shown that the practical use of multi-agent approach, above all, necessary to define the basic concepts and their application features that are different features of operation, structure, power, duration of work, and etc. specific object.

Multi-agent approach to information management systems based on intelligent methods are effective in supporting modes of operation, such as complex technological components and systems that operate under uncertainty.

References

1. Korobiichuk, I., Podchashinskiy, Y., Shapovalova, O., Shadura, V., Nowicki, M., Szewczyk, R.: Precision increase in automated digital image measurement systems of geometric values. In: Jabłoński, R., Brezina, T. (eds.) Advanced Mechatronics Solutions. AISC, vol. 393, pp. 335–340. Springer, Heidelberg (2016). doi:10.1007/978-3-319-23923-1_51
2. Russell, S., Norvig, P.: Artificial Intelligence: A Modern Approach, 2nd edn., 1408 p. Publishing House "Williams" (2006). Trans. from English
3. Maes, P.: Artificial life meets entertainment: life like autonomous agents. Commun. ACM **38** (11), 108–114 (1995)
4. Korobiichuk, I., Bezvesilna, O., Ilchenko, A., Shadura, V., Nowicki, M., Szewczyk, R.: A mathematical model of the thermo-anemometric flowmeter. Sensors **15**, 22899–22913 (2015)
5. Gorodetsky, V.: Multiagent system: current state and prospects of research. News Artif. Intell. (1), 44–59 (1996)
6. Tarasov, V.: The agents, multi-agent system, virtual communities: strategic direction in computer science and artificial intelligence. News Artif. Intell. (3), 5–54 (1998)
7. Gorodetsky, V., Grushinskiy, M., Khabalov, A.: Multi-agent systems (review). News Artif. Intell. (1998)

Modelling of Double-Pendulum Based Energy Harvester for Railway Wagon

Vytautas Bučinskas(✉), Andrius Dzedzickis, Nikolaj Šešok, Ernestas Šutinys, and Igor Iljin

Vilnius Gediminas Technical University, Vilnius, Lithuania
{vytautas.bucinskas,andrius.dzedzickis,nikolaj.sesok, ernestas.sutinys,igor.iljin}@vgtu.lt

Abstract. Powering various electronic devices using mechanic energy harvester became ordinary solution for remotely installed ones. Recent installation of harvesters on surfaces with steady-state harmonic vibrations is well known. Harvesters, utilizing chaotic vibration for energy gaining, require define design and its dynamic parameters. Implementation of energy harvester on railroad cargo wagon for signaling and diagnostic information transfer requires special characteristics of such harvester. In such case, mostly vibrating surface in the cargo wagon is bogy, therefore vibration data for the harvester taken from real measurements of the bogy vibration on real railroad. In order to solve problem of harvester parameters, modelling using Simulink software performed. Dynamic model of harvester contains horizontal pendulum and electrical elements, serving of mechanical damper – electric energy transfer. Excitation of this system applied as kinematic excitement of pendulum pins, connected with harvester body. This model build using II type of Lagrange equation, pendulums built using special supporting springs, limiting pendulum active angles. Simulation of this model performed for 1 km of drive using measurements of newly build railway. Results of the modelling presented graphically and amount of gained energy evaluated by integrating resulting vibration. On basis of the results, conclusions are drawn and recommendations given.

Keywords: Double-pendulum · Energy harvester · Railway wagon

1 Introduction

Harvesting of mechanical energy there is well known and widely used process in small energy applications. It is necessary to state that mostly these devices are efficient in harmonic vibration cases and they are tuned to single frequency. Non-tuned harvesters still are under research and harvesting of energy from chaotic vibrations still is a challenge [1, 2].

In XXI century technologies such as microelectromechanical system (MEMS), mechanical system (MS) and wireless systems, monitoring and supporting systems are increasingly penetrating to human is life. All these technologies requires source of energy, which typically are batteries of different kind. Significant amount of different batteries in these devices requires monitoring and change of these batteries on time.

R. Szewczyk and M. Kaliczyńska (eds.), *Recent Advances in Systems, Control and Information Technology*, Advances in Intelligent Systems and Computing 543, DOI 10.1007/978-3-319-48923-0_9

In order to make systems more reliable and free from taking care of these sources of energy, there are possible to use mechanical energy for fulfilling of such small energy needs [1–6].

In recent years, many research activities have been carried out in the area of micro-energy harvesting with the aim to find alternative solutions to batteries, which require regular recharging or replacement, as power supplies of low-power devices. The basic idea behind micro-energy harvesting is to capture free environmental energy and convert it into useful electrical energy, which can be used to power micro devices such as temperature sensors, pressure sensors, gas sensors, humidity sensors or micro-actuators [5].

Usually energy-harvesting operations based on the collection and conversion of kinetic energy to the electricity or other types of energy. Focusing on external kinetic energy, it represents undoubtedly a very convenient solution widely investigated for applications in both macro scale and integrated systems [7].

Kinetic energy is very often present in nature and abundant. Typically, human motions, vehicles vibrations, noisily environments, appear in the form of vibrations, random displacements or forces. In order to generate electrical energy from motion, a mechanical coupling between the moving body and a physical device capable to generate electricity is required. In this context, several transduction mechanisms have been presented in recent years on harvesting energy using piezoelectric, electrostatic, electro-magnetic or polymeric approaches, with sufficient energy harvested to enable communication of Wireless Sensor Networks (WSNs), to supply smart node or wearable devices [7].

The piezoelectric conversion mechanism exploits the deformation within a piezoelectric material whilst the electrostatic mechanism utilizes the relative motion of a capacitor plate to generate electricity. Piezoelectric and electrostatic conversion mechanisms are widely used since they are more compatible with prevailing MEMS-fabrication processes [7].

This paper is dedicated to some aspects of building wagon frame generated mechanical energy harvester, based on electromagnetic induction effect. Wagon frame motion has low frequency (< 20 Hz) and chaotic directions. Classic vibration harvesters with low resonance frequency will have big mass and size, so there exist mechanical problem to build a sensitive dynamic system, consisting from high-frequency energy transforming systems and sensitive to low frequency mechanical system. In order to create a harvester system is necessary to research dynamical properties of such system. Research here is performed in few steps using dynamic and mathematical model of the system. Use of such model allows finding values of the system stiffness, damping ratio, which will allow maximize amount of generated power from railway wagon vibrations.

This paper is an attempt to power railway wagon sensor feeding device, which can give autonomy for sensors and other information transmitting devices.

2 Theoretical Research

The pendulum revolves freely around suspension point, which is excitation aroused by coordinates x_0 and y_0 (see Fig. 1). At a time, when the pendulum touches the lower spring, the system get kinematic excitation y_0 from the spring in vertical direction. The movements of the pendulum are described by coordinate θ_1 [1].

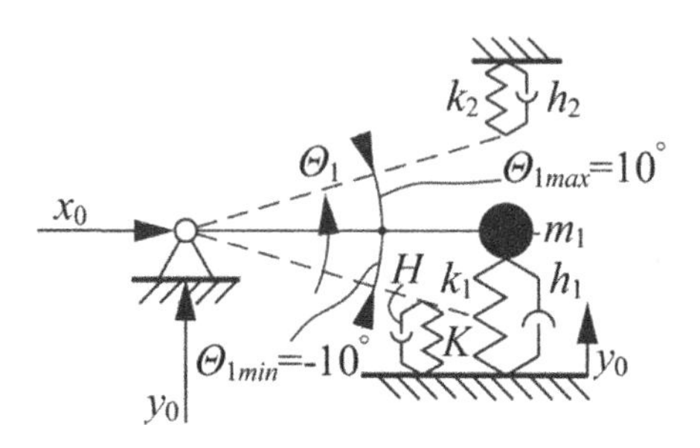

Fig. 1. The horizontal pendulum revolves freely around suspension point

The coordinates of the pendulum concentrated mass:

$$x_1 = x_0 + l_1 \cdot \cos\theta_1;\ y_1 = y_0 + l_1 \cdot \sin\theta_1. \tag{1}$$

The systems kinematic energy is [1]:

$$T = \frac{1}{2}\left(m_1\dot{x}_1^2 + m_1\dot{y}_1^2\right) \tag{2}$$

Considering that x_0, y_0 and θ_1 is function of time [1]:

$$\dot{x}_1 = \dot{x}_0 - l_1\dot{\theta}_1 \cdot \sin\theta_1;\ \dot{y}_1 = \dot{y}_0 + l_1\dot{\theta}_1 \cdot \cos\theta_1. \tag{3}$$

In this way, the kinematic energy expression, which is applied to any state of the system [1]:

$$T = \frac{1}{2}m_1\left(\dot{x}_0 - l_1\dot{\theta}_1 \cdot \sin\theta_1\right)^2 + \frac{1}{2}m_1\left(\dot{y}_0 + l_1\dot{\theta}_1 \cdot \cos\theta_1\right)^2. \tag{4}$$

The Lagrange equation of second type created for each state of the system is [1]:

$$\frac{d}{dt}\left(\frac{\partial T}{\partial\dot{\theta}_1}\right) - \frac{\partial T}{\partial\theta_1} + \frac{\partial\phi}{\partial\dot{\theta}_1} + \frac{\partial\Pi}{\partial\theta_1} = 0. \tag{5}$$

There are three state of the object in question:

(1) The horizontal pendulum lying on the bottom affixed springs and when $\Theta_1 < 10^{\circ}$ (see Fig. 2):

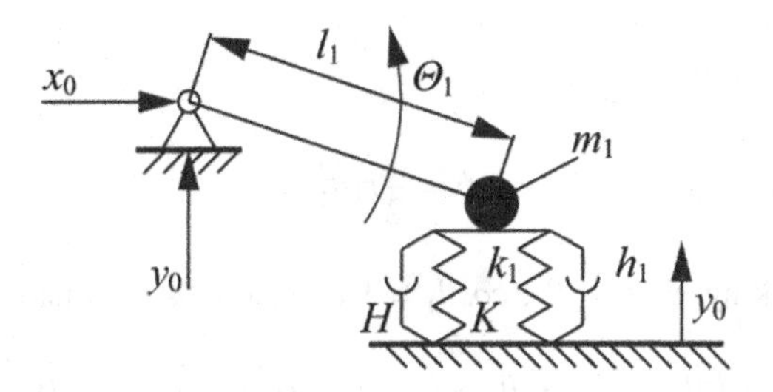

Fig. 2. The horizontal pendulum in the lower position

In this case, the potential energy is:

$$\Pi = m_1 g y_1 + \frac{1}{2}(k_1 + K)\Delta_{11}^2; \Delta_{11} = l_1 \cdot \theta_1 - y_0 \tag{6}$$

Dissipation function is:

$$\phi = \frac{1}{2} r_1 \dot{\theta}_1^2 + \frac{1}{2}(h_1 + H)\dot{\Delta}_{11}^2; \dot{\Delta}_{11} = l_1 \cdot \dot{\theta}_1 - \dot{y}_0 \tag{7}$$

In this case, the equation corresponds to this condition:

$$\begin{aligned} m_1 l_1^2 \ddot{\theta}_1 - m_1 l_1 \ddot{\mathbf{x}}_0 \cdot \sin\theta_1 + m_1 l_1 \ddot{\mathbf{y}}_0 \cdot \cos\theta_1 + \left(r_1 + (h_1 + H) l_1^2\right)\dot{\theta}_1 + m_1 g l_1 \cdot \cos\theta_1 \\ + (k_1 + K) l_1^2 \theta_1 = (k_1 + K) l_1 y_0 + (h_1 + H) l_1 \dot{y}_0. \end{aligned} \tag{8}$$

Or taking into consideration that $sin\theta_1 = \theta_1$; $cos\theta_1 = 1$.

$$\begin{aligned} m_1 l_1^2 \ddot{\theta}_1 - m_1 l_1 \ddot{\mathbf{x}}_0 \theta_1 + m_1 l_1 \ddot{\mathbf{y}}_0 + \left(r_1 + (h_1 + H) l_1^2\right) \cdot \dot{\theta}_1 + m_1 g l_1 + (k_1 + K) l_1^2 \theta_1 \\ = (k_1 + K) l_1 y_0 + (h_1 + H) l_1 \dot{y}_0. \end{aligned} \tag{9}$$

(2) Horizontal pendulum raised up from the rest position, when $\Theta_1 > 0^{\circ}$, $\ddot{\Theta}_1 l_1 > g$, that $K = 0$ and $k_2 = 0$, as shown in Fig. 3:

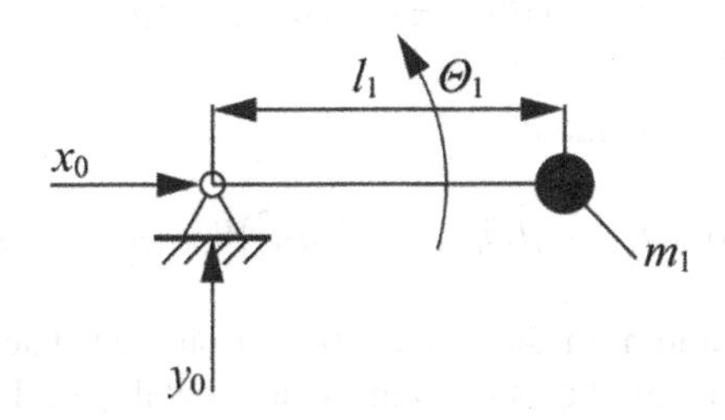

Fig. 3. The horizontal pendulum intermediate position

In this case, the potential energy is:

$$\Pi = m_1 g y_1 \tag{10}$$

Dissipation function is:

$$\phi = \frac{1}{2} r_1 \dot{\theta}_1^2 \tag{11}$$

This state of taking $sin\theta_1 = \theta_1$; $cos\theta_1 = 1$ corresponds to equation:

$$m_1 l_1^2 \ddot{\theta}_1 - m_1 l_1 \ddot{x}_0 \theta_1 + m_1 l_1 \ddot{y}_0 + r_1 \dot{\theta}_1 + m_1 g l_1 = 0. \tag{12}$$

(3) Horizontal pendulum raised to the top for some reason and it compresses on the top the protective spring when $\Theta_1 > 10^{\circ}$ (see Fig. 4).

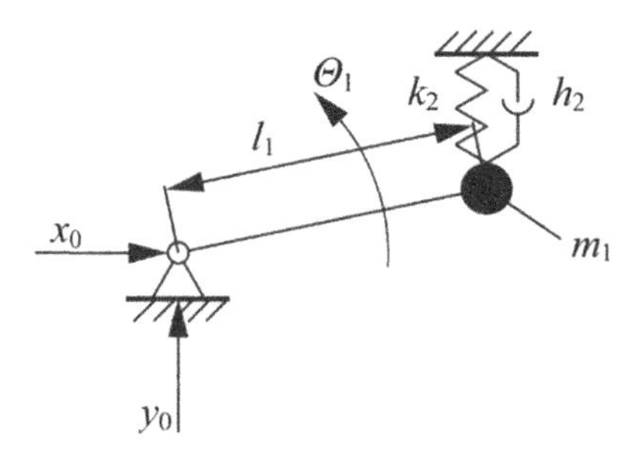

Fig. 4. The horizontal pendulum in the top position

The potential energy is:

$$\Pi = m_1 g y_1 + \frac{1}{2} k_2 \Delta_{12}^2; \Delta_{12} = l_1 \theta_1. \tag{13}$$

Dissipation function is:

$$\phi = \frac{1}{2} r_1 \dot{\theta}_1^2 + \frac{1}{2} h_2 \dot{\Delta}_{12}^2; \dot{\Delta}_{12} = l_1 \dot{\theta}_1 \tag{14}$$

Corresponding to the equation:

$$m_1 l_1^2 \ddot{\theta}_1 - m_1 l_1 \ddot{x}_0 \theta_1 + m_1 l_1 \ddot{y}_0 + \left(r_1 + h_2 l_1^2\right) \dot{\theta}_1 + m_1 g l_1 + k_2 l_1^2 \theta_1 = 0 \tag{15}$$

The horizontal pendulum transition conditions from one state to another.

The starting position of the pendulum is horizontal (see Fig. 2). The mass of pendulum m_1 lying on the two springs (k_1, h_1, K, H). Whereas the lower springs

(k_1, h_1, K, H) bottom induction operates in a known oscillation y_0, is switching to another condition (the horizontal pendulum intermediate position see Fig. 3) consider the condition (the lower spring not compressed and in which case it is disable with):

$$L_1 \cdot \theta_1 - y_0 > 0.$$

The transition to another condition (the pendulum rise to achieve and presses the spring (k_2, h_2) (see Fig. 4) consider the condition:

$$\theta_1 > \theta_{1\max}$$

It is assumed that $\theta_{1max} = 10°$.

Equations are solved using methodology described in [8]. Equations are rewritten in operational form after that in virtual MATLAB environment is created SIMULINK model. In both system states excitation coordinates is x_0, y_0.

The system of horizontal pendulum is described by coordinate θ_1. Pendulum oscillation speed is $\dot{\theta}_1$ kinetic energy of horizontal pendulum is T_1:

$$T_1 = \frac{m_1 \cdot L_1^2 \cdot \dot{\theta}_1^2}{2}. \tag{16}$$

Simplified schematic of energy harvester with horizontal pendulum is presented in Fig. 5.

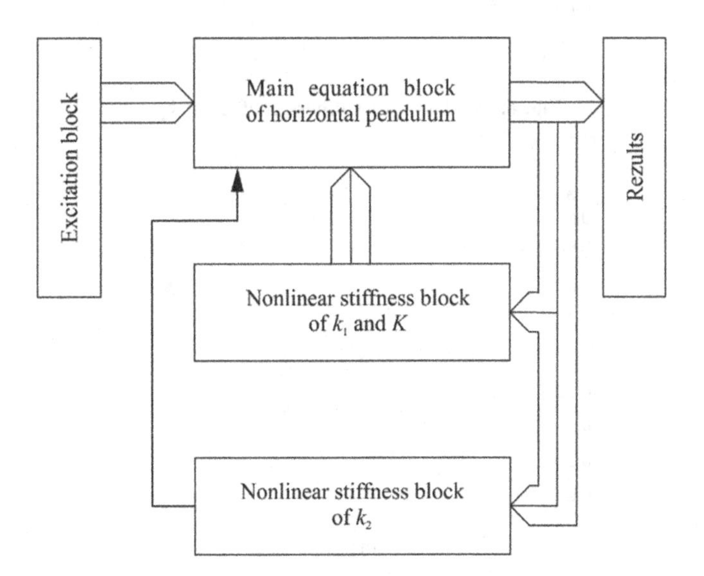

Fig. 5. The simplified block diagram of horizontal vibration harvester

Electromagnet, which is used in our system, creates nonlinear coefficients of resistance. $r_1 = f\left(\theta_1; \dot{\theta}_1\right)$. In simplified diagrams this coefficients are represented by nonlinear damping blocks.

3 Results of Research

In order to determine optimal parameters of pendulum it was decided to change oscillating mass m and coefficient of stiffness k. For the theoretical research was used these coefficient: m_1 = 0.050 kg; k_1 = 10000 kg/m; k_2 = 100000 kg/m; K = 100000 kg/m; h_1 = 0.1 Ns/m; h_2 = 0.1 Ns/m; H = 0.15 Ns/m; l_1 = 0.1 m. The bests result obtained solving combination m_1k_3, these results represented in Figs. 6, 7 and 8.

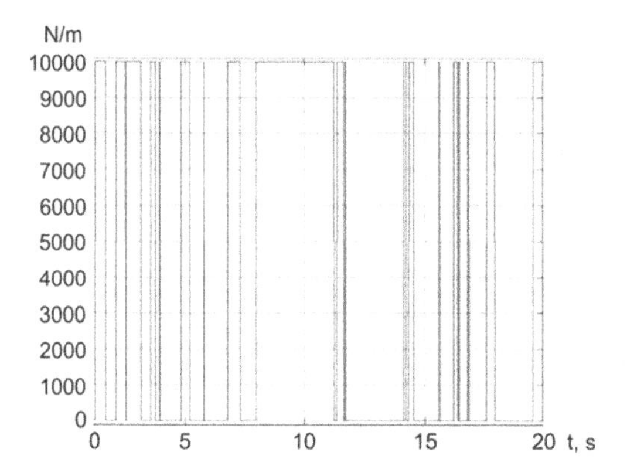

Fig. 6. Variation of stiffness coefficient k_1

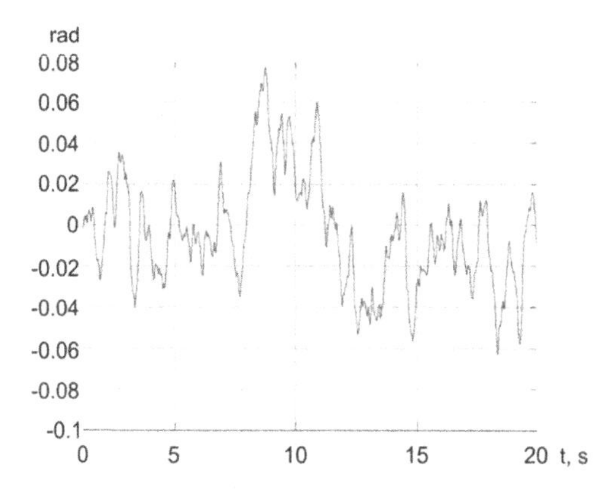

Fig. 7. Variation of coordinate Q_1

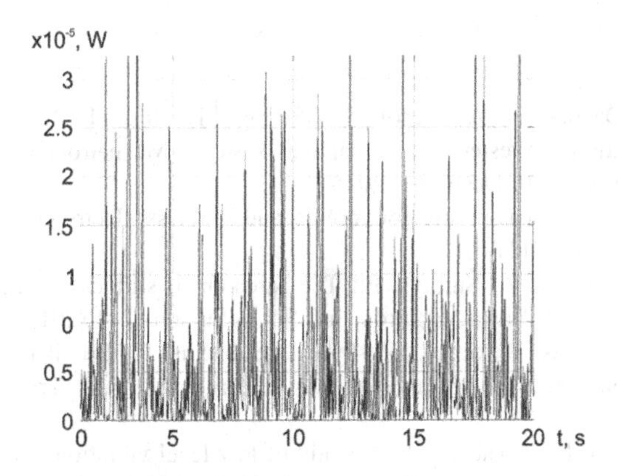

Fig. 8. Dependency of kinetic energy in respect to time

Figure 6 show how variation of stiffness coefficient k_1 in respect to time. When stiffness coefficient $k_1 = 0$, horizontal pendulum is flaying between two springs. When horizontal pendulum is flaying between two springs is used a pendulum restrictions $\ddot{\Theta}_1 l_1 > g$ and $\ddot{\Theta}_1 l_1 > 0$.

From Figs. 7 and 8 it is seen, then value of horizontal pendulum rotation angle θ_1 is negative (horizontal pendulum is in the lower position, but not in contact with two lowers springs), the kinetic energy of horizontal pendulum is the biggest.

4 Conclusions

Performed theoretical research and modelling of horizontal pendulum of one axis harvester behavior from kinematic excitation brings interesting and useful results. This pendulum can be excited from movement of real transport mean suspension parts, railroad wagon or bogie frame, for example. Such pendulum system can realize movement of wagon for energy harvesting in efficient way. Given results allows drawing these conclusions:

- From results seen, that system efficiently excited from existing movement of railway wagon frame and bogie and horizontal pendulum began oscillate and its kinetic energy can be converted to electrical one.
- Obtained results seems promising and research on this field will be continued.

References

1. Bučinskas, V., Dzedzickis, A., Šešok, N., Šutinys, E., Iljin, I., Kazickij, A.: Two-axis mechanical vibration harvester. In: Dynamical systems Mechatronics and Life Sciences, Lodz, Poland, pp. 99–110, December 2015
2. Paradiso, J.A.: Energy scavenging for mobile and wireless electronics. J. IEEE Pervasive Comput. **4**, 312–318 (2011)
3. Kazickij, A., Bučinskas, V., Šutinys, E.: The research of stiffness characteristics of active elements of harvester. J. Electr. Electron. Inf. Sci. (eStream) 1–4 (2015)
4. Kazickij, A., Bučinskas, V., Šešok, N., Iljin, I., Subačius, R., Bureika, G., Dzedzickis, A.: Research on implementation of harvester on the cargo wagon. J. Transp. Probl. 247–256 (2015)
5. Roundy, S., Wright, P.K., Rabaey, J.: A study of low level vibrations as a power source for wireless sensor nodes. J. Comput. Commun. **26**, 1131–1144 (2003)
6. Bučinskas, V., Šutinys, E., Augustaitis, V.K.: Experimental research of steel rope integrity problem. J. VibroEng. **13**(2), 312–318 (2011)
7. Bouendeu, E., Greiner, A., Smith, P.J., Korvink, J.G.: An efficient low cost electromagnetic vibration harvester. In: Power MEMS, Washington DC, USA, pp. 320–324, December 2009
8. Augustaitis, V.K., Gichan, V., Sheshok, N., Iljin, I.: Computer-aided generation of equations and structural diagrams for simulation of linear stationary mechanical dynamic systems. Mechanika **17**, 255–263 (2011)

Information Systems

Error Analysis of the Finite Element Method Calculations Depending on the Operating Range

Paweł Nowak[1(✉)] and Roman Szewczyk[2]

[1] Industrial Research Institute for Automation and Measurements, Al. Jerozolimskie 202, 02-486 Warsaw, Poland
pnowak@piap.pl
[2] Faculty of Mechatronics, Warsaw University of Technology, sw. A. Boboli 8, 02-525 Warsaw, Poland

Abstract. Paper presents an analysis of the FEM modeling error in the function of the simulation range. Test was conducted on the open-source Finite Element Method software, which utilizes Whitney edge elements for solving Maxwell's equations. Simulations focused on the magnetic flux distribution around the Helmholtz coils setup. Due to the utilization of analytically solvable example, comparison of simulation results with reliable reference result was possible. Simulations were conducted on a typical Helmholtz coils setup powered by constant current. In order to simulate electromagnetic phenomenon in the air surrounding the setup, both coils were placed in sphere, which provided proper simulation space. Also, the external surface of the sphere is utilized for application of boundary conditions for the differential equations solved during the modeling. Thus, proper selection of the simulation range is crucial for obtaining reliable results. On the other hand, utilization of excessively large simulation space significantly increases computation cost. This paper focuses on the optimization of a modeling range in the function of obtaining reliable and accurate results by minimization of modeling error, as well as on minimization of computation time.

Keywords: Elmer FEM · Error analysis · Finite Element Method

1 Introduction

Finite Element Method is commonly used during scientific and engineering research as a useful tool for solving complicated problems described by the partial differential equations. It can be utilized for solving Navier-Stokes equations in the fluid mechanics, as well as in many different applications.

Each Finite Element Method solver requires proper differential equations, describing physical phenomenon [1] which can be linearized in the mesh of finite elements and after solving them in small subdomains approximate solution of global problem can be obtained.

R. Szewczyk and M. Kaliczyńska (eds.), *Recent Advances in Systems, Control and Information Technology*, Advances in Intelligent Systems and Computing 543, DOI 10.1007/978-3-319-48923-0_10

During modeling of one-element problems, such as stresses calculation [2] modeled elements contain only analyzed object. On the other hand, utilization of FEM for soling electromagnetic problems [3–5] often requires placing analyzed model in some kind of space which, properly meshed, allows to calculate the electromagnetic field distribution around the object.

This paper focuses on the proper selection of a modeling range in order to minimize both modeling error as well as its computation cost.

2 Helmholtz Coil Model

During modeling two universal models of Helmholtz coil was developed. Both consisted two coaxial rings which represented one turn of solenoid. Those rings had average distance between them equal to the ring radius, which is crucial requirement for Helmholtz coil setup [6]. Created models differed with the radius of the coil, as well as the distance between the rings, but the requirement was fulfilled. Coil setup was placed in a sphere with the midpoint between the coils. As mentioned before, sphere was utilized to limit the space for FEM calculations, as well as for applying proper boundary conditions for the solver. During the calculations value of potential vector [7] on the external surface of the sphere was set to zero.

Two developed models of Helmholtz coil utilized for simulations varied with the radius of the coils. Also the radiuses of spheres utilized for the simulations varied depending on the coils radius. First created model consisted coils with average 0.9 m radius, placed in spheres of radius from 1.2 m to 20 m. Second model consisted coils with average 0.5 m radius placed in spheres with a radius of from 0.67 m to 11.11 m. The ratio of sphere and coil radius has been maintained during the modeling in order to obtain reliable data about proper designation of simulation space simulation depending on the model size.

Helmholtz coil model was selected for the simulations due to the fact, that it is analytically solvable. The value of magnetic flux in the midpoint of the coils setup can be obtained from Maxwell equations [6] and equals:

$$B = \frac{4^{\frac{3}{2}}}{5} \frac{\mu_0 n I}{R} \tag{1}$$

where B is for the magnetic flux in the center point, μ_0 is free space permeability, n is the number of the turns in wire, I is the current powering the coils and R is their radius and distance.

During the simulations coils were formed by one wire turn and were powered with constant current. For the 0.9 m radius coil the value of the current equals 0.04 A, and for the 0.5 m radius coil the value of the powering current equals 0.01 A. Thus the reference values of the magnetic flux in the analyzed point equal respectively $3.99 \cdot 10^{-8}$T and $1.79 \cdot 10^{-8}$T.

3 Modeling Description

As mentioned before, the object of the study was to analyze the influence of the sphere radius on the modeling error, caused mostly by the unrealistic application of the boundary conditions. In order to do that, same simulation had to be conducted for a different sphere radius. For each radius different modeling mesh was prepared in Netgen software. Due to the utilization of Delaunay algorithm [8], size of the elements varied depending on the distance from the coil-air boundary. Simulations were conducted for extremely small sphere in which the coils barely fitted as well as for excessively large space. Obtained magnetic flux distribution for average sphere is consisted with the literature data [6] and is presented in Fig. 1.

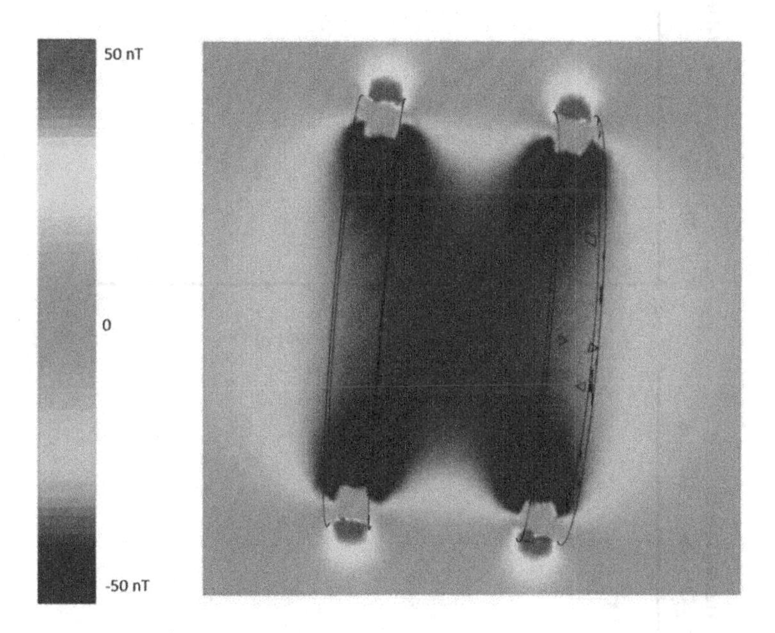

Fig. 1. Exemplary visualization of magnetic flux distribution B around the coils

4 Result Analyze

Results of the simulations, compared with the radius of the operating sphere are presented in Fig. 2. In order to compare results between two models, relative modeling error was compared with the ratio between the coil radius and sphere radius. Results of comparison are presented in Fig. 3. As presented, relative modeling error significantly decreases with the increment of the simulation range. For small sphere radius, error is caused by unrealistic setting of boundary condition for the Finite Element Method solver. For bigger radius, after the sphere radius is 5 times bigger than coil radius, modeling error is more constant. Figure 4 presents closer look to the modeling error for bigger spheres. It is clearly visible, that the error is smaller for bigger Helmholtz coil setup model. Typical error does not exceed 0.5 %, when average error for smaller model equals 1.8 %. This may be caused by the wrong scaling of minimal size of mesh elements, which influence on the modeling accuracy has been confirmed previously [9]. Generally modeling error may be caused by the finite dimensions of the coil, which were not took into account in analytical solution [6].

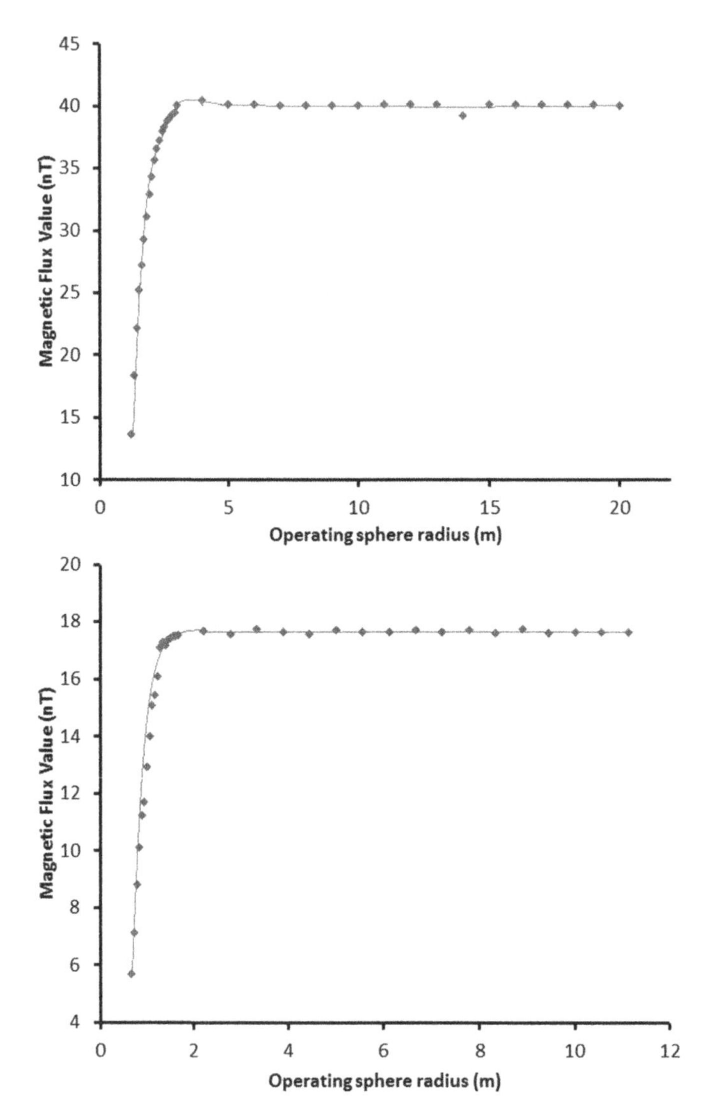

Fig. 2. Absolute values of magnetic flux in the midpoint of Helmholtz coils setup in the function of operating sphere radius

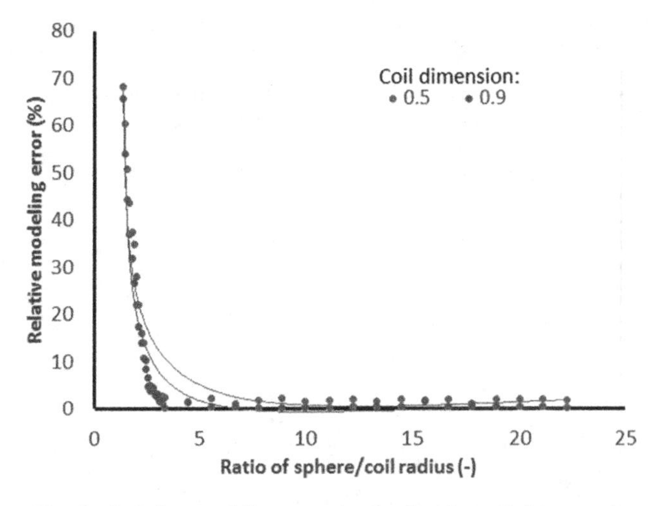

Fig. 3. Relative modeling error in the function of objects ratio

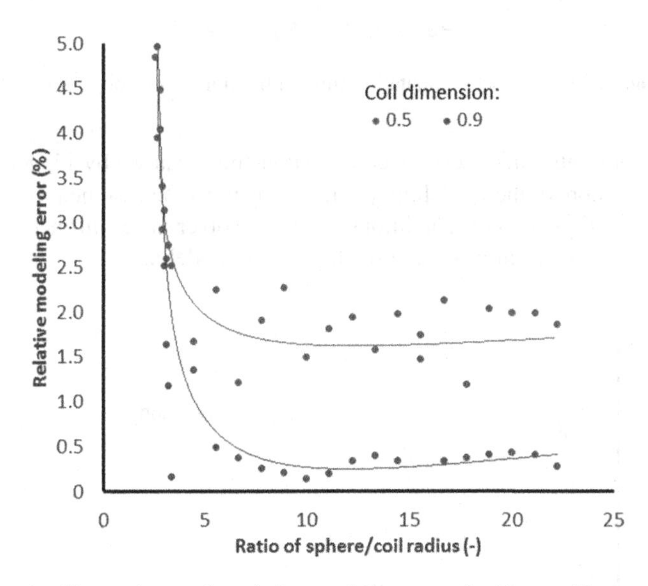

Fig. 4. Closer view to the relative modeling error for bigger objects ratio

Despite that, the influence of simulation space on modeling accuracy is confirmed. On the other hand, increment of the modeling sphere increases the number of elements in the mesh which increases the computation cost. Analysis of computation time in the function of the ratio between the coil radius and sphere radius is presented in Fig. 5. This data confirms, that increment of sphere radius significantly influences the modeling time. Also, the disproportion of mesh elements size between two models is confirmed. In order to select optimized ratio of the sphere radius, modeling error was compared with the required computation time.

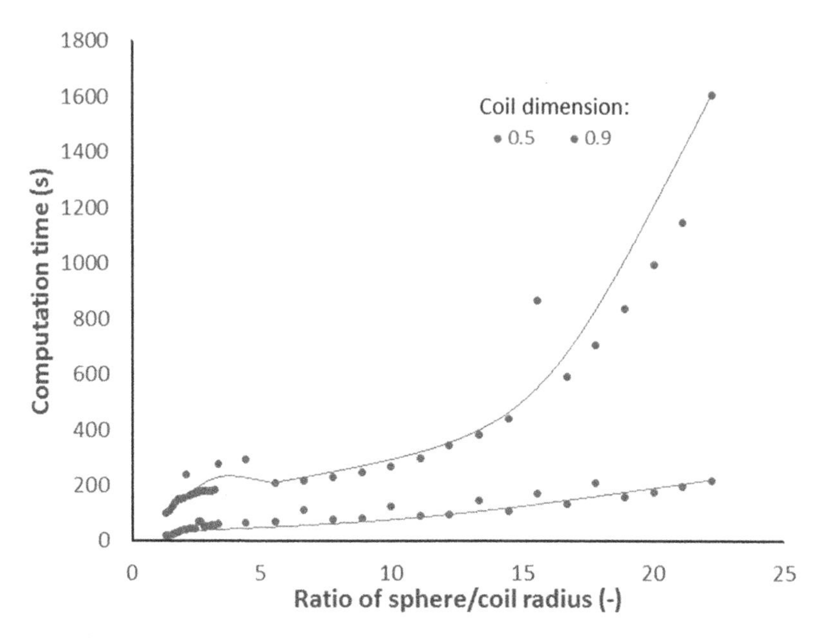

Fig. 5. Analysis of computation time in the function of objects ratio

As Fig. 6 presents, increment of computation time, caused by bigger simulation range, does not improve the modeling quality. Thus it can be assumed, that after some point, when realistic boundary conditions for FEM solver is set, further increment of modeling space does not improve the quality of the modeling.

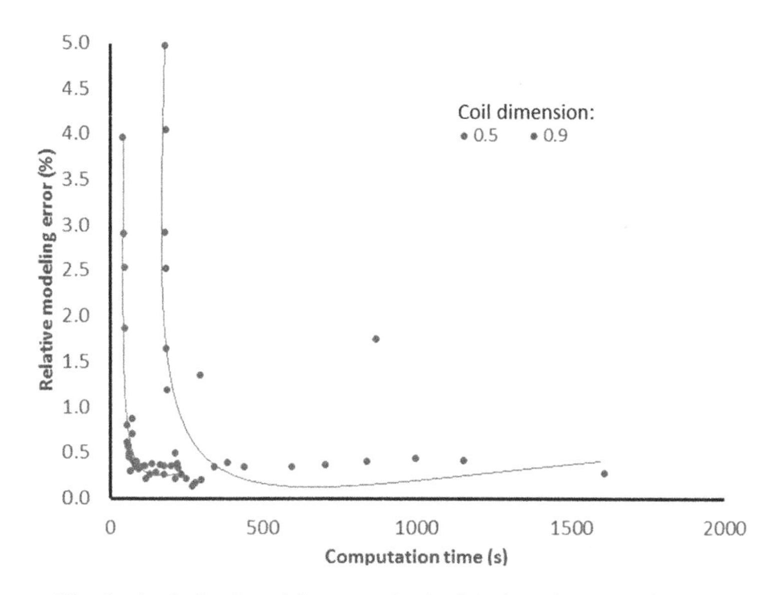

Fig. 6. Analysis of modeling error in the function of computation time

5 Conclusion

Presented tests provided useful and practical data about proper selection of modeling space for electromagnetic calculations utilizing open-source Finite Element Method solver. Analysis showed that too small simulation space causes unacceptable modeling error reaching up to 70 % in worst cases. This is caused by the wrong presumption of boundary condition for partial differential equations describing the phenomenon. Creation of bigger modeling space results with more realistic setting of boundary conditions and then modeling error decreases logarithmically to 2 %.

On the other hand utilization of too big modeling space significantly increases computation cost, without the improvement of the modeling quality. It has been shown, that optimal ratio of the simulation space to the modeled object, during electromagnetic calculations, is 5. This ratio results with the smallest achieved modeling error (up to 2 % for smaller coil and 0.5 % for bigger) and has low computation cost. It is important to notice, that other conditions, such as finite dimensions of coils (which were not took into account for analytical solution represented by (1)) may be the cause of the resulting modeling error. Further tests will consider this issue.

References

1. Szałatkiewicz, J., Szewczyk, R., Budny, E., Missala, T., Winiarski, W.: Measurement and control system of the plasmatron plasma reactor for recovery of metals from printed circuit board waste. In: Szewczyk, R., Zieliński, C., Kaliczyńska, M. (eds.) Recent Advances in Automation, Robotics and Measuring Techniques. AISC, vol. 267, pp. 687–696. Springer, Heidelberg (2014)
2. Jackiewicz, D., Szewczyk, R., Salach, J., Bieńkowski, A., Wolski, K.: Investigation method for the magnetoelastic characteristics of construction steels in nondestructive testing. In: Březina, T., Jabłoński, R. (eds.) Mechatronics 2013, pp. 479–486. Springer, Heidelberg (2014). doi:10.1007/978-3-319-02294-9_61
3. Nowicki, M., Szewczyk, R.: Modelling of the magnetovision image with the finite element method. In: Vajda J., Jamnicky I. (eds.) PROCEEDINGS of the 20th International Conference on Applied Physics of Condensed Matter (APCOM2014) (2014)
4. Szewczyk, R., et al.: Forward transformation for high resolution eddy current tomography using Whitney elements. In: Vajda J., Jamnicky I. (eds.) PROCEEDINGS of the 20th International Conference on Applied Physics of Condensed Matter (APCOM2014) (2014)
5. Bossavit, A.: Whitney forms: a class of finite elements for three-dimensional computations in electromagnetism. In: IEE Proceedings A (1988)
6. Cacak, R.K., Craig, J.R.: Magnetic field uniformity around near-Helmholtz coil configurations. Rev. Sci. Instrum. **40**(11), 1468–1470 (1969)
7. Raback, P. et al.: Elmer Models Manual, CSC – IT Centre for Science, Finland (2014)
8. Schöberl, J.: NETGEN – 4.3, Department of Computational Mathematics and Optimization, University Linz, Austria (2003)
9. Nowak, P., Szewczyk, R.: Error analysis of the FEM calculations depending on the mesh density. Int. J. Sci. Eng. Res. **6**(10) (2015)

Models of Magnetic Hysteresis Loops Useful for Technical Simulations Using Finite Elements Method (FEM) and Method of the Moments (MoM)

Roman Szewczyk(✉), Michał Nowicki, and Katarzyna Rzeplińska-Rykała

Industrial Research Institute for Automation and Measurements,
Al. Jerozolimskie 202, 02-486 Warsaw, Poland
szewczyk@mchtr.pw.edu.pl

Abstract. The paper provides the analyse of three simplified models of hysteresis loop suitable for technical purposes. Models consider the coercive field and utilize linear approximation with saturation, Langevin equation as well as arcus tangent functions. Validation of the models was done on the experimental data from measurements of magnetic hysteresis loops of four different materials. Accuracy of the models is assessed quantitatively. Finally, the parameters for practical application of the models are presented from the point of view different magnetic materials used in modelling by the finite elements method or the method of the moments.

1 Introduction

Quantitative description of magnetic hysteresis loop is the one of the most sophisticated problem connected with contemporary physics of magnetic materials. Among recently developed models of the magnetic hysteresis, the most effective are Jiles-Atherton [1] model and Preisach model [2]. Both these models require solving of sophisticated equations. Moreover, the Jiles-Atherton model requires numerical integration in the case of anisotropic materials [3] as well as solving of ordinary differential equations (ODE), which may lead to different numerical problems, especially for high-permeability materials [4].

On the other hand, technical simulations oriented on finite element method or method of the moments don't require sophisticated analyses of the shape of the hysteresis loops. To be useful for technological simulations, the model of the magnetic hysteresis loop should provide fast and reliable reproduction of the shape of saturated magnetic hysteresis loops.

Approximation of magnetic hysteresis loops by the different mathematical functions were presented previously for specific cases [5, 6]. However, quantitative comparative analyses of efficiency of such approximation for modern magnetic materials was not presented. This paper is filling this gap, to enable effective application of simplified models of magnetic hysteresis loop in magnetostatic and magnetodynamic systems modelling for technical purposes.

R. Szewczyk and M. Kaliczyńska (eds.), *Recent Advances in Systems, Control and Information Technology*, Advances in Intelligent Systems and Computing 543, DOI 10.1007/978-3-319-48923-0_11

2 Proposed Simplified Models of Magnetic Hysteresis

For technical purposes, three models are proposed: linear model with saturation, model based on the Langevin equation as well as model utilizing the arcus tangent functions. In all three models, the coercive field H_c is considered in saturation hysteresis loops.

Linear model with saturation utilizes three parameters: coercive field H_c, relative permeability μ and saturation flux density B_s. This model is given by the set of following equations:

$$B(H) = \begin{cases} B_s \, when \, \mu\mu_0(H \pm H_c) > B_s \\ \mu\mu_0(H - H_c) \, when \, \frac{dH}{dt} > 0 \\ \mu\mu_0(H + H_c) \, when \, \frac{dH}{dt} < 0 \\ -B_s \, when \, \mu\mu_0(H \pm H_c) < B_s \end{cases} \tag{1}$$

where μ_0 is magnetic constant.

Langevin function based model also uses three parameters: H_c, B_s and a. This sigmodal-shaped function is given by the following set of equations [7]:

$$\mathrm{B(H)} = \begin{cases} \mathrm{B_s}\left(\coth\left(\frac{\mathrm{H-H_c}}{\mathrm{a}}\right) - \left(\frac{\mathrm{a}}{\mathrm{H-H_c}}\right)\right) \text{ when } \frac{\mathrm{dH}}{\mathrm{dt}} > 0 \\ \mathrm{B_s}\left(\coth\left(\frac{\mathrm{H+H_c}}{\mathrm{a}}\right) - \left(\frac{\mathrm{a}}{\mathrm{H+H_c}}\right)\right) \text{ when } \frac{\mathrm{dH}}{\mathrm{dt}} < 0 \end{cases} \tag{2}$$

In opposite to linear function and Langevin function based models, the arcus tangent function based model doesn't explicit specify saturation flux density. However, it is also based on the three parameters: H_c, μ and k. Model with hysteresis is specified by the following set of equations [8]:

$$B(H) = \begin{cases} \frac{\mu_0\mu}{k} \operatorname{atan}(H - H_c) \, when \, \frac{dH}{dt} > 0 \\ \frac{\mu_0\mu}{k} \operatorname{atan}(H + H_c) \, when \, \frac{dH}{dt} < 0 \end{cases} \tag{3}$$

As it can be seen, each model utilizes three parameters. However, the physical background of each parameter is different in the case of each models. Moreover, some of parameters, such as a in the arcus tangent model, doesn't have physical explanation and should be determined experimentally for each shape of hysteresis loop.

3 Tested Materials and Measuring Method

Experiments were performed on four different magnetic materials:

- anisotropic electrical steel M130-27 s magnetized in the easy axis direction. Sample was in the form of the Epstein frame [9],
- martensitic, stainless steel 3H13 for energetic purposes in the form of frame-shaped samples [10],
- manganese-zinc high permeability ferrite F801 in the form of ring-shaped samples,
- Fiemet-type nanocrystalline alloy in the form of ring-shaped samples.

Magnetic hysteresis loops were measured in quasi-static conditions by computer controlled hysteresis graph. Measurements were carried out in the room temperature.

4 Results of Modelling of Hysteresis Loops

Presented models were implemented with use of open-source OCTAVE 4.0 software. Parameters of them models were initially approximately determined on the base of its physical meaning (such as value of saturation flux density B_s or relative permeability μ). Next, the parameters of the models of hysteresis loops were identified during the optimisation process using a derivative-free Nelder and Mead simplex algorithm [11]. The target function F for optimization process was given by the following equation:

$$F = \sum_{i=1}^{n} \left(B_{model}(H_i) - B_{meas}(H_i)\right)^2 \tag{4}$$

where $B_{model}(H_i)$ were the results of the modelling and $B_{meas}(H_i)$ were the results of the experimental measurements of hysteresis loops, both for the value H_i of magnetizing field.

Results of the modelling of magnetic saturation hysteresis loops of four materials described in the Sect. 3 with use of linear function based model, Langevin function based model and arcus tangent function based model are presented in the Figs. 1, 2 and 3 respectively. Parameters of the models and assessment of accuracy of modelling is presented in the Table 1.

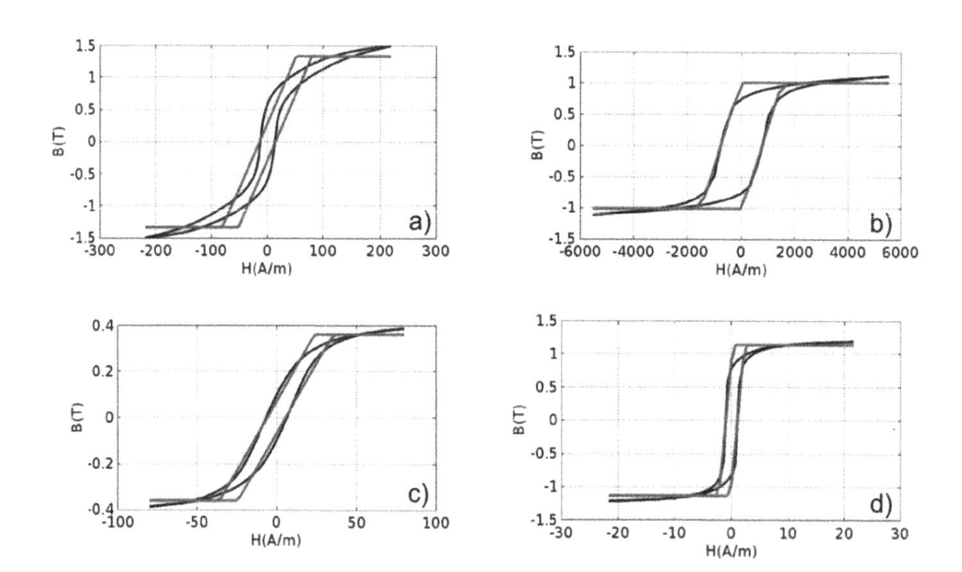

Fig. 1. Results of the modelling of saturation magnetic hysteresis loops using linear function based model for: (a) electrical steel M130-27 s, (b) martensitic steel 3H13, (c) Mn-Zn ferrite, (d) Finement-type nanocrystalline alloy

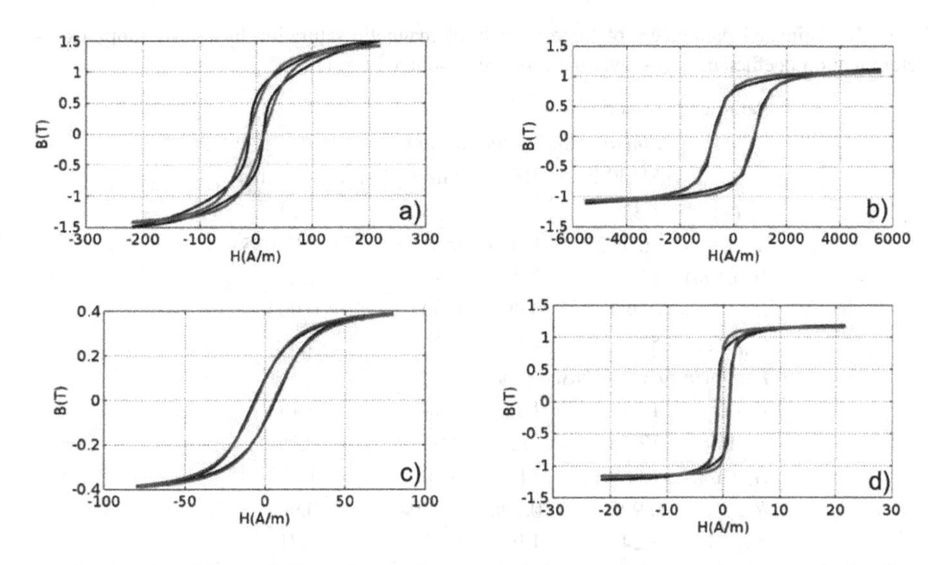

Fig. 2. Results of the modelling of saturation magnetic hysteresis loops using Langevin function based model for: (a) electrical steel M130-27 s, (b) martensitic steel 3H13, (c) Mn-Zn ferrite, (d) Finement-type nanocrystalline alloy

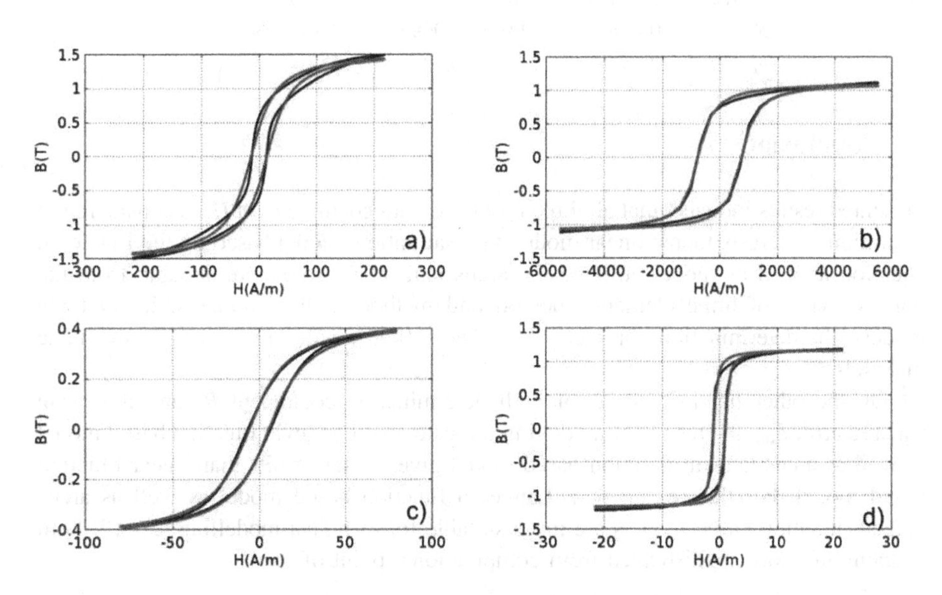

Fig. 3. Results of the modelling of saturation magnetic hysteresis loops using arcus tangent function based model for: (a) electrical steel M130-27 s, (b) martensitic steel 3H13, (c) Mn-Zn ferrite, (d) Finement-type nanocrystalline alloy

Table 1. Estimated parameters of three models of magnetic saturation hysteresis loops (R^2 – determination coefficient, e_{std} – average root mean squared error)

Parameter	Materials			
	Linear function based model			
	M130-27 s	3H13	Mn-Zn ferrite	Fimemet
B_s (T)	1.33	1.01	0.36	1.13
μ	16 112	1 099	9 549	792 085
H_c (A/m)	14.19	768	5.77	1.01
R^2	0.983	0.991	0.994	0.993
e_{std} (%)	5.6	4.6	3.2	5.3
Langevin function based model				
B_s (T)	1.54	1.11	0.44	1.18
a (A/m)	17.16	212	8.95	0.26
H_c (A/m)	13.89	747	5.80	1.02
R^2	0.995	0.998	0.9994	0.998
e_{std} (%)	3.24	1.91	0.77	2.71
Arcus tangent function based model				
μ	25 589	1 506	13 567	1 282 538
k	0.032	0.0026	0.0599	2.14
H_c (A/m)	13.84	748	5.81	1.02
R^2	0.996	0.9988	0.9993	0.9986
e_{std} (%)	2.97	1.70	0.93	2.48

5 Conclusions

Presented results indicate, that all three proposed with coercive field H_c is considered in saturation hysteresis loops: linear model with saturation, model based on the Langevin equation as well as model utilizing the arcus tangent functions are suitable from the point of view of finite elements method and method of the moments. In all three models, the determination parameter R^2 exceeds 0.98 for different types of magnetic materials.

On the other hand, in terms of both determination coefficient R^2 and the mean squared error e_{std}, the results of modelling indicate, that Langevin function based model as well as arcus tangent function based model gives better results than linear function based model. For this reason both Langevin function based model as well as arcus tangent function based models are more suitable for technical modelling, even if both of them are more sophisticated from computational point of view.

References

1. Jiles, D.C., Atherton, D.L.: Theory of ferromagnetic hysteresis. J. Magn. Magn. Mater. **61**, 48–60 (1986)
2. Liorzou, F., Phelps, B., Atherton, D.L.: Macroscopic models of magnetization. IEEE Trans. Magn. **36**, 418 (2000)
3. Ramesh, A., Jiles, D.C., Roderik, J.: A model of anisotropic anhysteretic magnetization. IEEE Trans. Magn. **32**, 4234–4236 (1999)
4. Szewczyk, R.: Validation of the anhysteretic magnetization model for soft magnetic materials with perpendicular anisotropy. Materials **7**, 5109–5116 (2014)
5. Barton, J.P.: Empirical equations for the magnetization curve. Trans. Am. Inst. Electr. Eng. **52**, 659 (1933)
6. Maxim, A., Andreu, D., Boucher, J.: A new analog behavioral SPICE macromodel of magnetic components. In: Proceedings of the IEEE International Symposium on Industrial Electronics ISIE 1997, p. 183 (1997)
7. Jiles, D.C., Atherton, D.L.: Ferromagnetic hysteresis. IEEE Trans. Magn. **19**, 2183 (1983)
8. Ponjavic, M.M., Duric, M.R.: Nonlinear modeling of the self-oscillating fluxgate current sensor. IEEE Sens. J. **7**, 1546 (2007)
9. Szewczyk, R.: Application of Jiles-Atherton model for modelling magnetization characteristics of textured electrical steel magnetized in easy or hard axis. In: Szewczyk, R., Zieliński, C., Kaliczyńska, M. (eds.) Progress in Automation, Robotics and Measuring Techniques. Advances in Intelligent Systems and Computing, vol. 350, pp. 293–302. Springer, Heidelberg (2015)
10. Jackiewicz, D., Kachniarz, M., Rożniatowski, K., Dworecka, J., Szewczyk, R., Salach, J., Bieńkowski, A.: Temperature resistance of magnetoelastic characteristics of 13CrMo4-5 constructional steel. Acta Phys. Pol., A **127**, 614–616 (2015)
11. Lagarias, J.C., Reeds, J.A., Wright, M.H., Wright, P.E.: Convergence properties of the Nelder-Mead simplex method in low dimensions. SIAM J. Optim. **9**, 112–147 (1998)

Parallel Computer System Resource Planning for Synthesis of Neuro-Fuzzy Networks

Andrii Oliinyk(✉), Stepan Skrupsky, and Sergey A. Subbotin

Zaporizhzhya National Technical University, Zaporizhzhya, Ukraine
olejnikaa@gmail.com, subbotin@zntu.edu.ua

Abstract. The problem of parallel computer systems resource planning in the synthesis of neuro-fuzzy networks (NFN) is considered. The model of parallel computer system resource planning for the synthesis of the NFN is developed. The developed model takes into account the following parameters: type of computer system, the number of processes that are running problem, the bandwidth of data network, the number of possible solutions to be processed in the method, the parameters of solved applied problem (the number of instances and the number of features in a given training sample). The synthesized model enables to estimate the time required parallel system to perform the method of synthesis of neuro-fuzzy networks.

Keywords: Cluster · Neural network · Parallel computing · Resource planning · Training sample

1 Introduction

Modern time-consuming tasks are usually realized using parallel calculations on clusters and graphical processors (GPU) [1]. The use of such resources is expensive and unavailable for most users. For effective use of parallel computer systems first have to perform resource pre-planning, which can be accomplished by methods of mathematical modeling of system behavior in solving relevant problems [2–7]. In this case it is advisable to apply methods of neural network modeling [8–10], because they allow the identification of non-linear relationships among characteristics of a parallel computer system, parameters of used mathematical software, parameters of solved applied problem and time spent by the system solving applied problem.

A method of neuro-fuzzy networks synthesis based on parallel calculations, solving applied problems of pattern recognition, technical and biomedical diagnostics, suggesting the need to build models based on different data samples [11–14], has been proposed in the work [11]. Implementation of this method consumes time-and compute resources, which necessitates the use of parallel computer systems. Therefore, for rational use of these systems in the implementation of the method [11], it is necessary to perform resource pre-planning. The purpose of the work is a construction of the model of parallel computing systems' resources planning performing solution of applied problems based on the parallel method of NFN synthesis [11].

R. Szewczyk and M. Kaliczyńska (eds.), *Recent Advances in Systems, Control and Information Technology*,
Advances in Intelligent Systems and Computing 543, DOI 10.1007/978-3-319-48923-0_12

2 Problem Statement

As noted above, in the planning of the resources of a parallel system, important characteristics that determine its effectiveness and speed of obtaining results are the following parameter groups: technical characteristics of a parallel system, the parameters of used mathematical software and the features of solved applied problem [2].

The main characteristics of the system that affect the time of solving the practical problem are: x_1 is a type of system type (cluster of CPU or GPU), x_2 is a number of processes on which problem is executed (N_{pr}), x_3 is a network bandwidth (V), Gb/s. The main parameter of used mathematical software (in this case, method of NFN synthesis [11, 14]): x_4 is a number of possible solutions N_χ [11], which system operates on each iteration of the method. As parameters of an applied problem significantly affects the speed of work of a parallel system for the NFN syn`thesis, we can use: x_5 is a number of cases (Q) in a given training sample observations (S), x_6 is a number of features (M) in a variety of observations S.

Thus, to estimate the operating time (t) of a parallel computer system for the NFN synthesis need to build a model of the form

$$t = t(type, N_{pr}, V, N_\chi, Q, M), \quad (1)$$

allowing to execute the prediction of time spent to perform the parallel method of NFN synthesis depending on the characteristics of the system, the parameters of software and features of solved applied problem.

3 Resource Planning Model for Parallel Computer Systems for NFN Synthesis

The proposed method [11] has been applied on a cluster CPU and on the GPU at the data processing from a public repository [15]. Characteristics of processed data sets are shown in Table 1.

As a result of data processing [12], the training set (2) included 1034 results of the method [8] execution was formed: $D = \langle X, T \rangle$, where $X = \{x_1, x_2, x_3, x_4, x_5, x_6\}$, $x_i = \{x_{i1}, x_{i2}, \ldots, x_{iN}\}$, $N = 1034$, $T = \{t_1, t_2, \ldots, t_N\}$. Thus the training set is a table consisting of 1034 rows and seven columns containing values of six input features and one output (running time of the method) for each application of the method [11] in the parallel system. Fragment of the training sample is given in Table 2.

As a result, the range of feature values has been reduced to a single interval [0, 1]. As a basis for model (1) construction the feed forward neural network [16] have been used, allowing approximate complex nonlinear dependence with high accuracy. Model (1) was synthesized in the form of a three-layer perceptron [16], the first layer contains three neurons, the second layer – two neurons, the third layer – one neuron. All neurons have sigmoid activation function. As a discriminant function of neurons we used a weighted sum.

For neural model synthesis and determining of its parameters (weights and biases of each neuron) on its inputs fed of normalized features, the output is the value of the

Table 1. Characteristics of processed data sets

№	Name	Attribute types	Number of instances Q	Number of attributes M
1	Auto MPG	Categorical, Real	398	8
2	Automobile	Categorical, Integer, Real	205	26
3	Computer hardware	Integer	209	9
4	Housing	Categorical, Integer, Real	506	14
5	Servo	Categorical, Integer	167	4
6	Solar flare	Categorical	1389	10
7	Forest fires	Real	517	13
8	Concrete compressive Strength	Real	1030	9
9	Communities and Crime	Real	1994	128
10	Parkinsons Telemonitoring	Integer, Real	5875	26
11	Energy efficiency	Integer, Real	768	8

Table 2. Fragment of the training sample

Feature values						T
Type	N_{pr}	V	N_{χ}	Q	M	
0	1	20	50	209	9	86
0	5	20	50	209	9	22.30
0	10	1	100	209	9	50.60
1	120	32	50	209	9	24.54
1	260	64	50	209	9	25.30
0	25	20	100	506	14	17.94
1	60	32	100	506	14	77.26
1	200	64	50	1389	10	27.73
0	32	20	100	1994	128	21.11
...	...	...	...	...	...	...
1	260	64	100	5875	26	73.49

run-time of method [11] in a parallel system. As the objective function of neural model training the minimum of mean square error has been used. Neural network trained on the basis of back-propagation error algorithm [16]. Considered acceptable level of mean square error was 10^{-4}.

After substituting of the obtained weighs and biases of neural network and using the activation function and discriminant function we obtain the mathematical description of the synthetic neural network model (3), which describes the relationship between the characteristics of the parallel computer system, parameters of the investigated method and the time spent to perform the process of NFN synthesis based on method [11]. Graphic interpretation of synthetic model is shown in Fig. 1.

$$\begin{cases} Y_{NN} = \psi_{(3,1)} = \left(1+e^{-(-9{,}36+27{,}73\psi_{(2,1)}-4{,}83\psi_{(2,2)})}\right)^{-1}; \\ \psi_{(2,1)} = \left(1+e^{-(25{,}59+0{,}84\psi_{(1,1)}+0{,}42\psi_{(1,2)}-27{,}02\psi_{(1,3)})}\right)^{-1}; \\ \psi_{(2,2)} = \left(1+e^{-(0{,}36-18{,}17\psi_{(1,1)}+4{,}85\psi_{(1,2)}-6{,}24\psi_{(1,3)})}\right)^{-1}; \\ \psi_{(1,1)} = \left(1+e^{-(-12{,}42-0{,}42type+0{,}29N_{pr}-1{,}8V+11{,}68N_{\chi}+0.26Q+0.08M)}\right)^{-1}; \\ \psi_{(1,2)} = \left(1+e^{-(-23{,}41+20{,}9type-0{,}13N_{pr}+6{,}06V+0{,}41N_{\chi}+0.95Q+0.06M)}\right)^{-1}; \\ \psi_{(1,3)} = \left(1+e^{-(3{,}81-9{,}94type+50{,}58N_{pr}+0{,}64V+0{,}08N_{\chi}-0.11Q-0.03M)}\right)^{-1}. \end{cases} \quad (3)$$

Thus, synthesized neural network model is a hierarchical structure containing neurons and allows to calculate the runtime of a parallel computer system for predicting the time spent to perform the process of NFN synthesis based on method [11]. The value of mean square error of the model is $9{\cdot}10^{-4}$, which is acceptable for this kind of problems solved using the synthesized model.

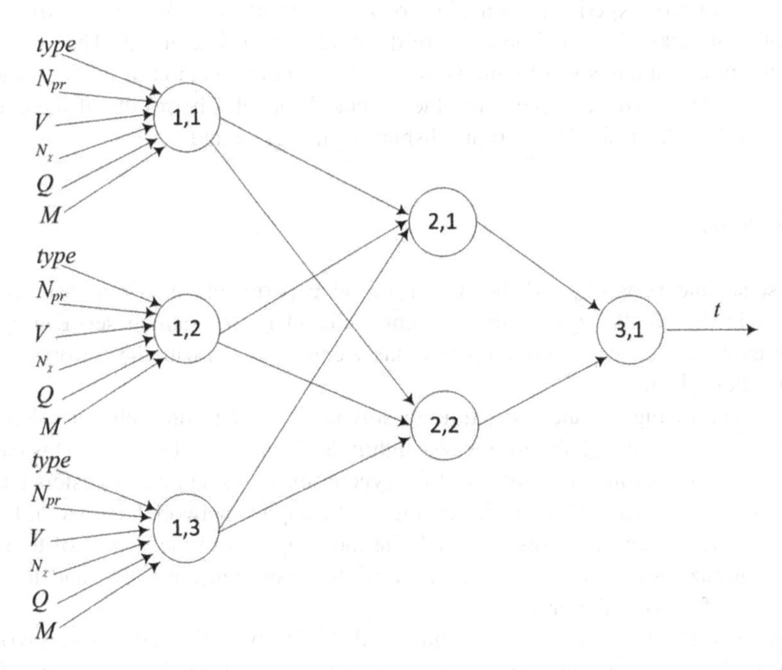

Fig. 1. Synthesized neural network model

4 Experiments

To provide the experimental researches of the considered method and the proposed neural network model the following computer systems have been used:

- the cluster of Pukhov Institute for Modelling in Energy Engineering NAS of Ukraine (IPME, Kyiv, Ukraine): processors Intel Xeon 5405, RAM – 4 × 2 GB DDR-2 for each node, communication environment InfiniBand 20 Gb/s, middleware Torque and OMPI [14];
- the cluster of Zaporizhzhya National Technical University (ZNTU, Zaporizhzhya, Ukraine): processors Intel E3200, RAM – 1 GB DDR-2 for each node, communication environment Gigabit Ethernet 1 Gb/s, middleware Torque and MPICH;
- GPU NVIDIA GTX 285 + 240 CUDA nodes;
- GPU NVIDIA GTX 960 1024 CUDA nodes.

During experiments the number of processes involved x_2 varied from 1 to 32 for clusters and from 60 to 260 for GPUs. Network bandwidth x_3 – from 1 to 20 Gb/s, the number of possible solutions to be processed in the method x_4 – from 50 to 100.

To perform the experiments the software, based on C language with the MPI library [18], was developed.

5 Results

The results of the experiments on clusters ZNTU (with $x_4 = 50$) and IPME (with $x_4 = 100$) for tasks № 3, 4 (Table 1) are displayed in Figs. 2 and 3. The solid line shows the time actually spent by the system to perform the method of NFN synthesis, dotted line is the predicted time using the proposed model. The results of the experiments on GPUs for tasks № 9, 10 are displayed in Figs. 4 and 5.

6 Discussion

The test sample consisting of the results of 69 experiments included instances of solutions in the parallel system that are not included in the training set. During the experiments the adequacy of complexity of tasks corresponded with the performance of used parallel systems.

As shown in Figs. 2 and 3 the time of solving a problem on a cluster calculated using the proposed model almost always slightly less than the actual time. This can be explained by the fact that the time spent for synchronizations and data transfers among processes of cluster varied greatly depending on the applied network bandwidth. In this case, the more cluster processes involved, the more significant the impact of transfers and synchronizations, and the greater the deviation between the actual and the predicted time of solving the problem.

Figures 4 and 5 provide insights that on the GPU with the number of involved threads to 140 inclusive there is a decrease in time spent on the execution of the method. Time predicted using the proposed model behaves in a similar way. With further increase of the number of involved GPU threads the impact of overhead grows significantly, the proportion of transfers and synchronizations starts to exceed the proportion of the target computations, so the time actually spent executing the method starts to increase. The time predicted using the proposed model also increases, but

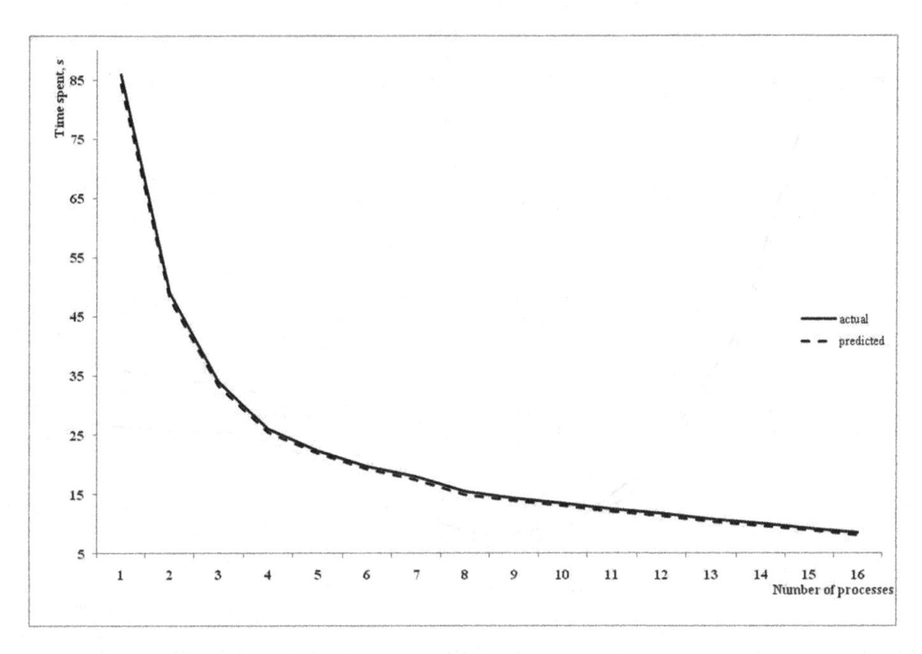

Fig. 2. The results of the experiments on cluster IPME solving Computer Hardware task ($N_\chi = 100$, $Q = 209$, $M = 9$)

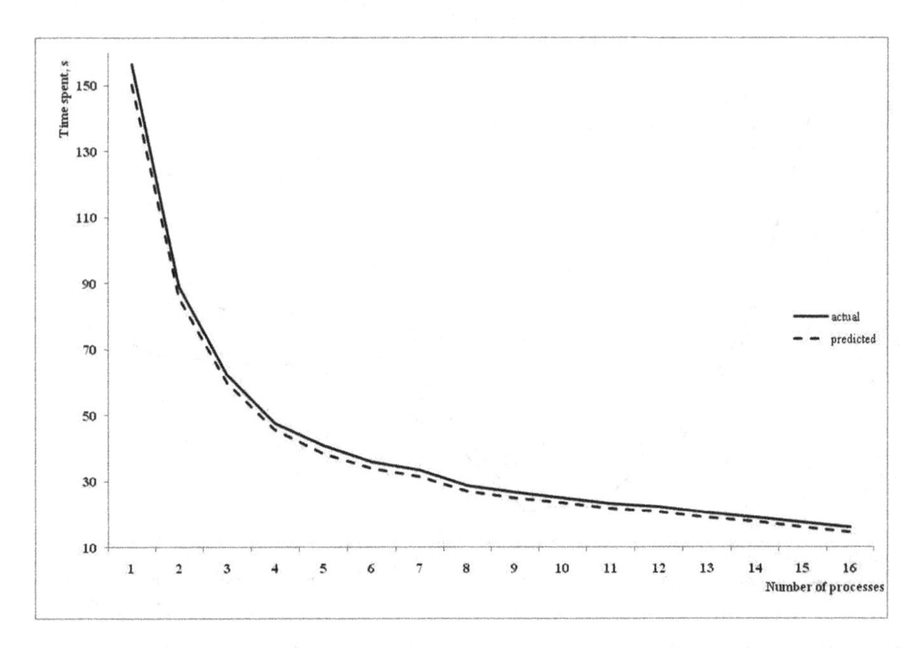

Fig. 3. The results of the experiments on cluster ZNTU solving Housing task ($N_\chi = 50$, $Q = 506$, $M = 14$)

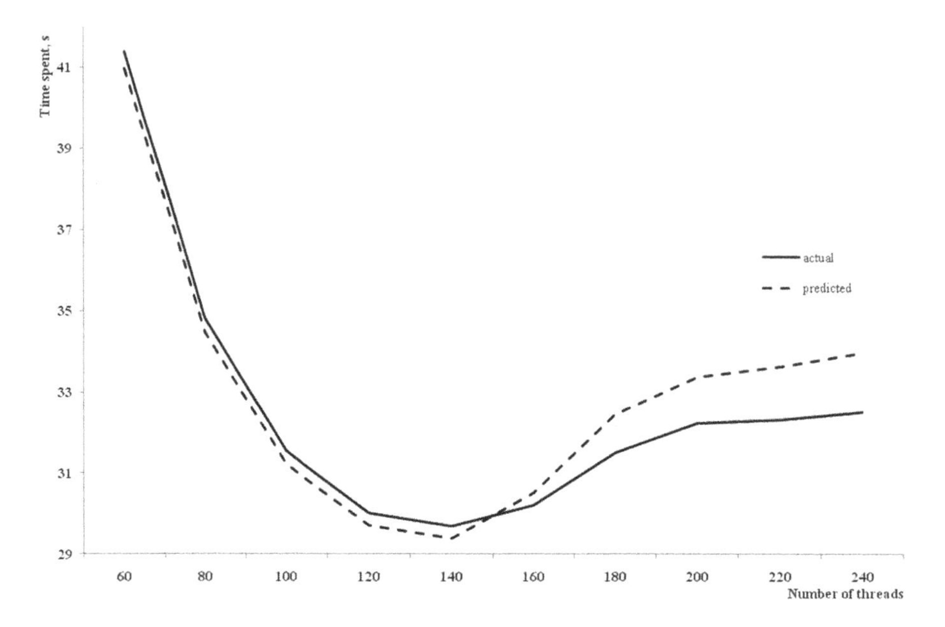

Fig. 4. The results of the experiments on GPU NVIDIA GTX 285 + solving Communities and Crime task ($N_\chi = 50$, $Q = 1994$, $M = 128$)

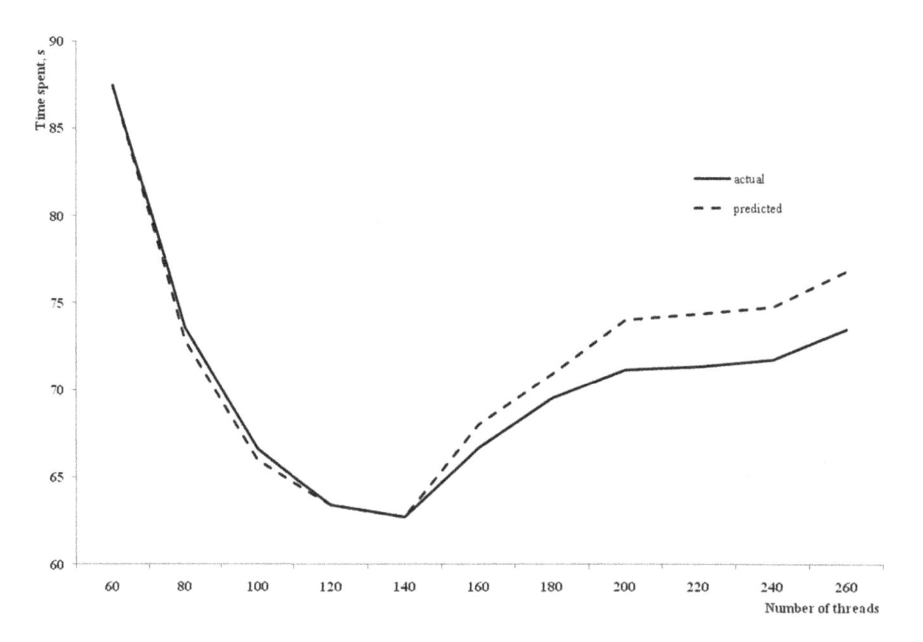

Fig. 5. The results of the experiments on GPU NVIDIA GTX 960 solving Parkinsons Telemonitoring task ($N_\chi = 100$, $Q = 5875$, $M = 26$)

slightly faster. Thus, the proposed model takes into account an impact of overhead of the computational process in a parallel system.

The experimental results give evidence of the fact that regardless of choice of the particular application (Table 1) within the method of NFN synthesis the proposed model behaves adequately. The value of the mean square error on the test sample amounted to $2.33 \cdot 10^{-3}$, which is much worse than in the training set, but remains within acceptable values and allows to recommend the proposed model in practice for the rational choice of computer system resources in the implementation of the parallel method of NFN synthesis.

7 Conclusion

In this paper we solve the problem of parallel computing systems resource planning for synthesis of neuro-fuzzy networks.

Scientific novelty lies in the fact that parallel computing systems resource planning model for synthesis of neuro-fuzzy networks has been developed. It takes into account the type of computer system, the number of processes that are running problem, the bandwidth of data network, the number of possible solutions to be processed in the method, the parameters of solved applied problem (the number of instances and the number of features in a given training sample). The synthesized model enables to estimate the time required parallel system to perform the method of synthesis of neuro-fuzzy networks.

The practical value of the results is the developed software which implements the proposed model allowing rational planning of selection of computer system resources for synthesis of neuro-fuzzy networks to solve practical problems of pattern recognition, diagnosing and prediction.

Acknowledgements. This paper is prepared with partial support of "Centers of Excellence for young RESearchers" (CERES) project (Reference Number 544137-TEMPUS-1-2013-1-SK-TEMPUS-JPHES) of Tempus Programme of the European Union.

References

1. [https://www.cs.utexas.edu/~pingali/CS378/2015sp/lectures/IntroGPUs.pdf] – Introduction to GPUs, (in Russian)
2. Sulistio, A., Yeo, C.S., Buyya, R.: Simulation of parallel and distributed systems: a Taxonomy and survey of tools. Int. J. Softw. Pract. Exp. 1–19 (2002)
3. Yu, S.S., Kudermetov, R.K.: Imitacionnye modeli raspredelennyx sistem kompressii videoinformacii. Naukovi pratsi Donetskoho natsionalnoho tekhnichnoho universytetu. Seriia: Informatyka, kibernetyka ta obchysliuvalna tekhnika **15**(203), 190–202 (2012). (in Russian)
4. Praczukowska, A., Nowicki, M., Korobiichuk, I., Szewczyk, R., Salach, J.: Modeling and validation of magnetic field distribution of permanent magnets system. Int. J. Sci. Eng. Res. **6/5**(78), 4–11 (2015)

5. Korobiichuk, I.: Mathematical model of precision sensor for an automatic weapons stabilizer system. Measurement **89**, 151–158 (2016). doi:10.1016/j.measurement.2016.04.017
6. Buyya, R., Murshed, M.: Gridsim: a toolkit for the modeling and simulation of distributed resource management and scheduling for grid computing. Concurrency Comput. Pract. Exp. **14**, 1175–1220 (2002)
7. Ponomarenko, V.S., Listrovoj, S.V., Minuhin, S.V., Znahur, S.V.: Metody i modeli planirovaniya resursov v GRID-sistemax: monografiya. INZhE'K, Kharkov (2008). (in Russian)
8. Oliinyk, A., Zaiko, T., Subbotin, S.: Training sample reduction based on association rules for neuro-fuzzy networks synthesis. Opt. Memory Neural Netw. (Inf. Opt.) **2**(23), 89–95 (2014)
9. Oliinyk, A., Skrupsky, S., Subbotin, S.: Experimental investigation with analyzing the training method complexity of neuro-fuzzy networks based on parallel random search. Autom. Control Comput. Sci. **1**(49), 11–20 (2015)
10. Korobiichuk, I., Bezvesilna, O., Ilchenko, A., Shadura, V., Nowicki, M., Szewczyk, R.: A mathematical model of the thermo-anemometric flowmeter. Sensors **15**, 22899–22913 (2015). doi:10.3390/s150922899
11. Oliinyk, A., Skrupsky, S., Subbotin, S.: Using parallel random search to train fuzzy neural networks. Autom. Control Comput. Sci. **6**(48), 313–323 (2014)
12. Oliinyk, A., Zaiko, T., Subbotin, S.: Factor analysis of transaction data bases. Autom. Control Comput. Sci. **2**(48), 87–96 (2014)
13. Oliinyk, A., Subbotin, S.A.: The decision tree construction based on a stochastic search for the neuro-fuzzy network synthesis. Opt. Memory Neural Netw. (Inf. Opt.) **1**(24), 18–27 (2015)
14. Subbotin, S., Oliinyk, A., Skrupsky, S.: Individual prediction of the hypertensive patient condition based on computational intelligence. In: International Conference on Information and Digital Technologies (IDT 2015), pp. 336–344. IEEE Press, Zilina (2015)
15. UCI Machine Learning Repository. http://archive.ics.uci.edu/ml/
16. Rassel, S., Norvig, P.: Iskusstvennyj intellekt: sovremennyj podxod. Vil'yams, Moscow (2006). (in Russian)
17. [http://www.ipme.kiev.ua/ukr/grid_vuzol/charakter-g.html] – Kharakterystyky GRID-vuzla NANU, (in Russian)
18. Quinn, M.J.: Parallel Programming in C with MPI and OpenMP. McGraw-Hill, New York (2004)

The Sample and Instance Selection for Data Dimensionality Reduction

Sergey Subbotin(✉) and Andrii Oliinyk

Zaporizhzhya National Technical University, Zaporizhzhya, Ukraine
subbotin@zntu.edu.ua, olejnikaa@gmail.com

Abstract. The paper proposes tools for data dimensionality reduction containing sample selection method and instance informativity indicators based on the evolutionary search, which is modified to speed up the search through the creation of special operators, taking into account a priori information about the data sample and concentrating search on the most perspective solution areas. This allows preserving the stochastic nature of the search to accelerate the obtainment of acceptable solutions through the introduction of deterministic component in the search strategy. The proposed methods are experimentally studied. On the results of experiments the comparative characteristics and recommendations for the use of the proposed methods are given.

Keywords: Dimensionality reduction · Diagnosis · Classification · Sample selection · Instance selection

1 Introduction

For the decision making in technology and medicine, where the expert knowledge is missing or insufficient and for object classification usually we need to build a recognition model on sample of precedents (cases, observations, instances, exemplars).

If the sample characterized by a big number of instances then it have a big dimensionality. The usage of big samples for model building, as a rule, increases the time on the model construction and the volume of required computer memory.

To speed-up model training process we need to reduce the data dimensionality, for example, using sample selection [1, 2].

The sample selection methods [1–4] are extracted from the original sample of a large volume the sub-sample of a smaller volume, trying to ensure the preservation of the most important properties of the original sample in the subsample. These methods are divided into deterministic and probabilistic.

Deterministic methods of sample selection [5–12] provide selection of instances on the basis of assumptions about their usefulness (information content), wherein some examples cannot be selected, or the probability of their the selection cannot be accurately determined. They are usually based on cluster analysis and seek to ensure topological similarity of the original sample. The disadvantage of these methods is the impossibility of error estimation for generated samples. The advantage of deterministic methods is that they can identify the most important precedents for diagnostic model

R. Szewczyk and M. Kaliczyńska (eds.), *Recent Advances in Systems, Control and Information Technology*, Advances in Intelligent Systems and Computing 543, DOI 10.1007/978-3-319-48923-0_13

synthesis problem solving, which could also be used to initialize the recognition model and accelerate the model learning process.

Probabilistic methods [1, 2, 4, 13, 14], suppose a random selection of a set of instances from the original sample, wherein each instance of the original sample has a non-zero probability, which can be precisely defined, to be included into the formed sample. The advantages of probabilistic methods of sample selection are their relative simplicity and the possibility of sample error estimating. The disadvantage of probabilistic methods is that they do not guarantee that formed the sample of small volume will be good reflect the properties of the original sample, and will not be excessive and will not be artificially simplify the task.

Thus, we need to combine probabilistic and deterministic approaches for sample selection. The evolutionary search as probabilistic method provide a global search, does not provide specific requirements for the objective function, take into account the experience gained at the previous search iterations. Therefore, it is a good basis for sample formation. However, it is slow because they do not use the problem specific information in the process of a search.

Therefore, the aim of this work is to create an evolutionary method for sample selection making possible to accelerate the obtaining of solutions.

2 Formal Problem Statement

Let $<x, y>$ is a sample of precedents (cases, instances, exemplars), where $x = \{x^s\}$, $x = \{x_j\}$, $x^s = \{x_j^s\}$, $x_j = \{x_j^s\}$, $y = \{y^s\}$, $s = 1, 2,\ldots, S$, $j = 1, 2,\ldots, N$; x_j^s is a value of j-th input feature x_j of exemplar x^s, y^s is an output for precedent x^s (for the classification problems $y^s \in \{1, 2,\ldots, K\}$, $K > 1$, K is the number of classes), S is a number of precedents, N is a number of input features. Denote as $<x',y'>$ a subsample of $<x, y>$, as S' a number of precedents (examples) in $<x',y'>$, as $f()$ a user criterion describing the quality of argument according to the decided task, as *opt* an optimal desirable or acceptable value of the functional $f()$ for the problem.

The problem of training sample selection is for given: $X = <x, y>$ find: $X' = <x',y'>$, $X' \subset X$, $y^* = \{y^s|x^s\hat{I}x'\}, S' < S, N'' = N, f(<x',y'>,<x,y>) \rightarrow opt$.

3 Sampling Method Based on Evolutionary Search

To reduce the number of matching combinations in the search process it is rationally to ensure efficient use of the information about before analyzed solutions for the transition to the consideration of new solutions, similar to the previously discussed. It is also necessary to ensure the chances for each of the possible solutions to be considered.

The proposed sampling method based on evolutionary stochastic search with pseudo clustering includes the following steps.

Initialization stage: set the original sample $<X, Y>$ of S instances, the maximum number S_f of formed sample $<x, y>$, the parameters of evolutionary search, the mutation probability $P_{M.}$, as well as an acceptable quality criterion value I^*. Calculate the quality criterion value of the original sample I.

Stage of pseudo-clustering of solutions. For each i-th feature, $i = 1, 2, \ldots, N$, sort instances of the original sample in a decreasing order of i-th feature values, then looking through an ordered set of instances along the axis of the i-th feature from smaller to larger values pairwise for every two adjacent instances, both instances are included in the sample $<X', Y'>$, if they belong to the different classes. The left and right instances along the axis values of i-th feature also included to the sample $<X', Y'>$.

Stage of formation of the initial population of H binary decisions: for $k = 1, 2, \ldots, H$, $s = 1, 2, \ldots, S$, set the probabilities of instance inclusion to the k-th solution as:

$$P(h_s^k) = \begin{cases} \frac{\lambda + \text{rand}}{2}, X^s \in <X', Y'>; \\ 0,5\text{rand}, X^s \notin <X', Y'>, \end{cases} \quad (1)$$

$$\lambda = \begin{cases} 1, S' \leq S_f; \\ \min\left\{0,5; \frac{S_f}{S'}\right\}, S' > S_f. \end{cases} \quad (2)$$

where h_s^k is an s-th digit (bit) of solutions h^k, rand is a function returning random numbers.

Stage of test on end of the search. For each k-th solution the sample is formed, for which the $I(k)$ estimated. If it is performed more than the T iterations, or there is exists a solution with the number k: $I(k) \geq I^*$, then stop the search and return as a result the sample with the highest value of $I(k)$.

Stage of the decision selection: according to the goal $I(k) \rightarrow \max$ based on the "roulette" rule the parental pairs of solutions are formed [15, 16].

Stage of solution crossing: new solutions are generated on the basis of a one-point crossover [15, 16].

Stage of solution mutation: for each solution randomly inverted no more than round $(P_M S)$ bits, and for decisions in which the number of bits equal to one, more than S_ϕ, randomly invert the remaining bits equal to one. The previously encountered solutions are excluded from the population. Go to the stage of test on end of the search.

This method combines the idea of random sampling and the determined search for the best solutions. It starts with the selection of the most promising instances to be included in solutions, but retains the chances of remaining instances to be included in the generated sample, and in the course of its work purposefully improves considered solutions. This method ensures that each of the examined samples will have a volume of no more than S_f.

4 Instance Selection

For the selection of instances in the methods of the sampling is proposed to use the following individual instance informativeness indicators.

The indicator of the individual informativeness of the s-th instance of the sample relative to the internal boundary of the classes:

$$I_G^s = \frac{1}{S-1}\sum_{p=1}^{S}\left\{ e^{-\alpha_{y^s,y^p}\sum_{j=1}^{N}(x_j^s - x_j^p)^2} \middle| s \neq p,\ y^s \neq y^p \right\}, \tag{3}$$

where $\alpha_{k,q} = \frac{1}{\frac{1}{S^k S^q}\sum_{s=1}^{S}\sum_{p=1}^{S}\left\{\sum_{i=1}^{N}(x_i^s - x_i^p)^2 \middle| (y^s = k, y^p = q) \vee (y^s = q, y^p = k)\right\}}$.

This indicator will take values between 0 and 1; i.e., its value will be the greater the closer the corresponding instance is to the boundary between the different classes.

The indicator of the individual informativeness of the s-th instance of the sample relative to its distance from the class boundary:

$$I_U^s = 1 - \min_{p=1,2,\ldots,S}\left\{ e^{-\alpha_{y^s}\sum_{j=1}^{N}(x_j^s - x_j^p)^2} \middle| s \neq p, y^s = y^p \right\}, \tag{4}$$

where $\alpha_k = \frac{1}{\max\limits_{\substack{s=1,2,\ldots,S;\\ p=s+1,\ldots,S}}\left\{\sum_{j=1}^{N}(x_j^s - x_j^p)^2 \middle| y^s = k, y^p = k\right\}}$;

This indicator will take a value between zero and one; i.e., its value will be the greater the closer the instance is to the "core" of the class.

Indicator of the individual informativeness of the s-th instance of the sample relative to its proximity to the center of the class (cluster):

$$I_O^s = \frac{1}{S^{y^s}-1}\sum_{p=1}^{S}\left\{ e^{-\sum_{j=1}^{N}(x_j^s - x_j^p)^2} \middle| s \neq p,\ y^s = y^p \right\}. \tag{5}$$

This indicator will take values between zero and one; i.e., its value will be the greater the farther the instance is in relation to other instances of the class or sample; i.e., it will identify rare atypical cases and instances located on the external borders of the borders of classes.

The integral indicator of the individual informativeness of the instance x^s: $I(x^s) = \max\{I_G^s, I_O^s, I_U^s\}$. This indicator will take a value between zero and one; i.e., the greater its value, the more important the s-th instance is for the construction of the model, since it is either located on the border between the classes, it is a unique observation, it is on the outer border of the class, or it corresponds to the "core" of the class (close to the center of the cluster).

The indicator of the group informativeness of the instances of the sample:

$$\bar{I} = \max_{x^s \in X} \{I(x^s)\}.$$

5 Experiments and Results

To study the proposed methods they have been implemented as software.

An important characteristic of the sampling methods is the speed of processing of the search space, which can be estimated by the λ – a proportion of considered or excluded decisions regarding the entire set of possible solutions in one iteration (epoch). For the developed sampling method by averaging the observations during the experiment, the dependence of λ from the epoch number have been studied.

Figure 1 demonstrates obtained dependences of λ from number of epochs in semi-logarithmic coordinates with different values of number of iterations T and H comparing to reduced exhaustive search [14].

The proposed indicators for instance informativity have been studied for the Fisher IRIS problem [17] to select most informative instances. The fragment of obtained results is shown on Fig. 1. Here three different plant classes shown by ".", "x" and " + " markers. Circle markers "o" show the exemplars with values greater than average value for each indicator.

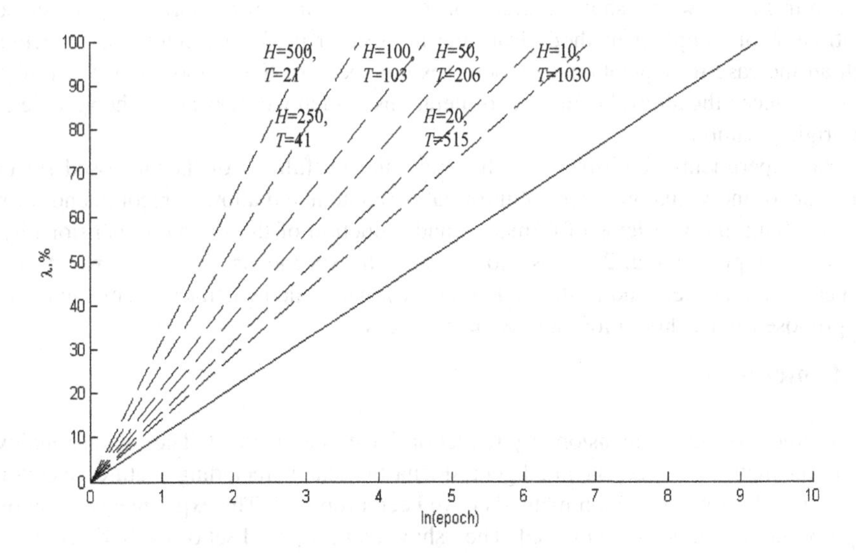

Fig. 1. Graphs of dependences of λ from the epoch for the sampling methods based on reduced search (-) and evolutionary search (–)

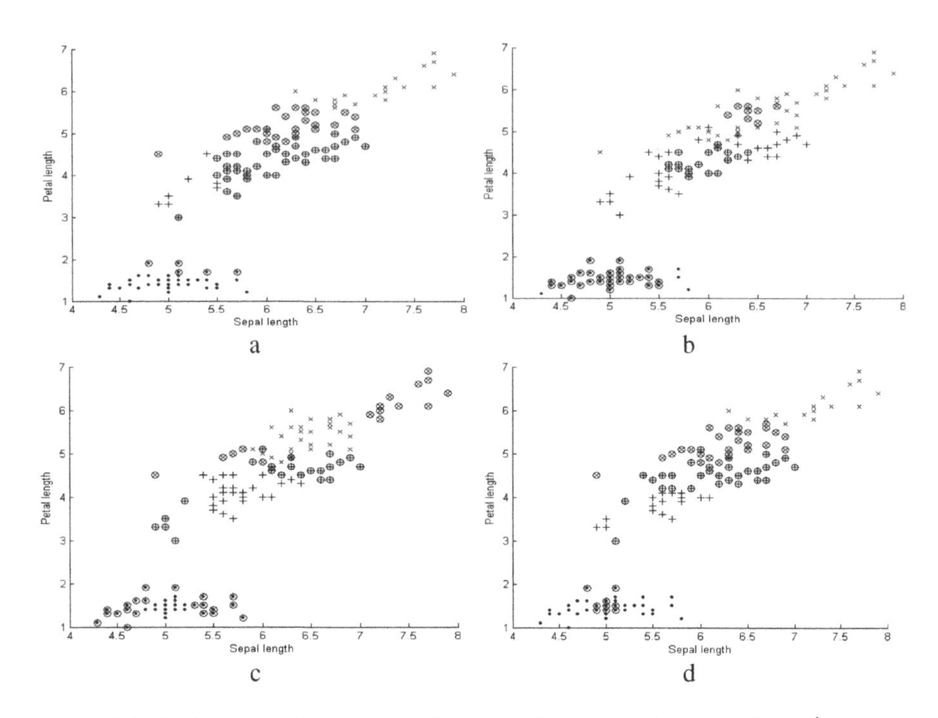

Fig. 2. Instance selection for IRIS problem for: (a) $I^s_{\bar{G}}$, (b) I^s_O, (c) I^s_U, (d) $\hat{I}^s$

6 Discussion

The study of methods of formation samples have revealed their performance. As can be seen from Fig. 2, which shows a graph of the dependence of proportion of processed solutions λ of sampling methods from the number of performed iterations (epochs), with an increase in population size increases the speed of the method, but it will also greatly reduced the allowable number of iterations, which will depend on the volume of the original sample.

The experiments confirmed the efficiency and usefulness of the proposed set of indicators of individual instances of information content and allow to recommend it for use in solving the problems of formation and reduction of the sample dimensionality.

For example, in Fig. 2 is easy to see that the specimens showed signs of "o", located on the borders and in the middle classes, which demonstrates the efficiency of the proposed individual informativity of instances.

7 Conclusion

The problem of data dimensionality reduction have been studied. The dimensionality reduction methods based on intelligent stochastic search including feature selection methods and sample selection methods have been proposed. The experiments on study of proposed methods are conducted. They show that proposed set of methods allow to significantly reduce the data dimensionality.

Acknowledgements. This paper is prepared with partial support of "Centers of Excellence for young RESearchers" (CERES) project (Reference Number 544137-TEMPUS-1-2013-1-SK-TEMPUS-JPHES) of Tempus Programme of the European Union.

References

1. Lavrakas, P.J.: Encyclopedia of Survey Research Methods. Sage Publications, Thousand Oaks (2008)
2. Korobiichuk, I., Podchashinskiy, Y., Shapovalova, O., Shadura, V., Nowicki, M., Szewczyk, R.: Precision increase in automated digital image measurement systems of geometric values. In: Jabłoński, R., Brezina, T. (eds.) Advanced Mechatronics Solutions. AISC, vol. 393, pp. 335–340. Springer, Heidelberg (2016). doi:10.1007/978-3-319-23923-1_51
3. Hansen, M.H., Hurtz, W.N., Madow, W.G.: Sample Survey Methods and Theory. Wiley, New York (1953)
4. Bernard, H.R.: Social Research Methods: Qualitative and Quantitative Approaches. Sage Publications, Thousand Oaks (2006)
5. Korobiichuk, I., Bezvesilna, O., Ilchenko, A., Shadura, V., Nowicki, M., Szewczyk, R.: A mathematical model of the thermo-anemometric flowmeter. Sensors **15**, 22899–22913 (2015). doi:10.3390/s150922899
6. Subbotin, S.: The instance and feature selection for neural network based diagnosis of chronic obstructive bronchitis. In: Bris, R., Majernik, J., Pancerz, K., Zaitseva, E. (eds.) Applications of Computational Intelligence in Biomedical Technology. SCI, vol. 606, pp. 215–228. Springer, Heidelberg (2016). doi:10.1007/978-3-319-19147-8_13
7. Ghosh, S.: Multivariate Analysis, Design of Experiments, and Survey Sampling. Marcel Dekker Inc., New York (1999)
8. Subbotin, S.A.: The training set quality measures for neural network learning. Opt. Memory Neural Netw. (Inf. Opt.) **2**(19), 126–139 (2010)
9. Smith, G.: A deterministic approach to partitioning neural network training data for the classification problem: dissertation. Virginia Polytechnic Institute & State University, Blacksburg (2006)
10. Subbotin, S.A.: The sample properties evaluation for pattern recognition and intelligent diagnosis. In: 10th International Conference on Digital Technologies, pp. 321–332. IEEE, Zilina (2014)
11. Plutowski, M.: Selecting training exemplars for neural network learning: dissertation. University of California, San Diego (1994)
12. Oliinyk, A., Zaiko, T., Subbotin, S.: Training sample reduction based on association rules for neuro-fuzzy networks synthesis. Opt. Memory Neural Netw. (Inf. Opt.) **2**(23), 89–95 (2014)
13. Chaudhuri, A., Stenger, H.: Survey Sampling Theory and Methods. Chapman & Hall, New York (2005)
14. Subbotin, S.A.: Methods of sampling based on exhaustive and evolutionary search. Autom. Control Comput. Sci. **3**(47), 113–121 (2013)
15. Tenne, Y., Goh, C.-K.: Computational Intelligence in Expensive Optimization Problems. Springer, Berlin (2010)
16. Talbi, E.: Metaheuristics: from Design to Implementation. Wiley, Hoboken (2009)
17. Iris Data Set. https://archive.ics.uci.edu/ml/datasets/Iris

Protection of University Information Image from Focused Aggressive Actions

Roman Korzh(✉), Andriy Peleshchyshyn, Solomia Fedushko, and Yuriy Syerov

Lviv Polytechnic National University, Lviv, Ukraine
korzh@lp.edu.ua, {apele,syerov}@ridne.net, felomia@gmail.com

Abstract. The article considers an actual problem of protecting university information image from focused aggressive actions. It also studies the protection of university information image in the context of its formation. There are two duration-based types of aggression (short-term aggression: less than one cycle of university information activity; permanent or long-term aggression: more than one cycle of university information activity). The paper considers processes that provide formation of information image of Internet social media, which greatly differ from traditional World Wide Web development processes. This raises the problem of forming university information image in social media, which is autonomous and requires separate study. There are phases of complex multi-stage process of protecting university information image from aggression (identification of aggression, analysis and prioritization of areas of aggression, resistance to aggressive actions, monitoring consequences of aggressor's confrontation and attacks). The paper presents systemic features of aggression in respect to information image in the field of education. Motivation and actions of aggression coordinator and direct perpetrators have also been analyzed in the paper. Such separation is important from the standpoint of system, as it helps to better understand the nature of threats, their consequences and, therefore, consider them in further development of threat avoidance methods.

Keywords: Internet social media · Monitoring · Aggression in internet social media · Education

1 Introduction

Internet social media (ISM) provide significant opportunities for universities related to their public promotion and public communication by creating a positive attitude of online communities to an educational institution [1]. On the other hand, some ISM users and their communities can discredit universities and inflict other types of damage. Analysis of the entire complex of reasons leading to aggression towards a university goes beyond this study, but this does not deny the necessity to examine practical threats to universities in the ISM and ways of combating them. Anyway, great significance of universities as socially important institution inevitably gives rise to threats and need for active protection from them. Destructive-minded social groups, unfair competitors,

R. Szewczyk and M. Kaliczyńska (eds.), *Recent Advances in Systems, Control and Information Technology*, Advances in Intelligent Systems and Computing 543, DOI 10.1007/978-3-319-48923-0_14

influencers of hostile intelligence agencies and others can be possible organizers of aggression towards a university as a public institution. Formation of communities with unverified data [2–5] and biased negative attitude to universities is often a logical consequence of passivity of educational institutions. A positively-motivated interested user who has no direct contact with competent employees of universities in the community cannot meet his/her information needs, and therefore is not actively involved in the community. However, absence of university representation allows free expression of negative thoughts that attracts negatively-motivated users and thereby ensures necessary human resources for aggression organizers. Improvement of this situation should be a priority in the information field of a modern university. Universities' information activities in social media, which are performed regular, have clear goals, objectives, performance assessment criteria and appropriate financial and organizational support, significantly increases level of universities' information image protection from possible aggression.

2 System Features of Aggression Towards an Information Image in the Field of Education

In our time scientific and practical research in the field of protection from this type of actions is actively carried out in the areas of political PR, information protection of web-community user [5–7], library computerization [8], cyber security [9], internet marketing [10–12] and media strategic management [13, 14].

Let's consider briefly the features of each of these areas compared to the aggression towards universities and protection therefrom. Table 1 shows a comparison of aggression features in each of the areas: commercial, political and educational.

Let's analyze the details of Table 1. It is important to note that characteristics of attacks are the basis for determining main activities on protection of universities' information image from aggression in social media. The table separately analyzes motivation and actions of aggression coordinators and direct perpetrators. This separation is important in terms of a system, as it helps to better understand the essence and nature of threats and their consequences and therefore to consider them in further development of methods to combat threats.

1. An important difference of commercial and political aggression from aggression in the field of education is a higher role of coordination centre and its motivation. Generally, all collective aggressive actions of commercial and political nature have coordinators, who have material resources, strategy and possibility to motivate perpetrators financially and control their activities. Similarly, perpetrators are motivated and accountable. In case of attacks on information image of a university, the role and capabilities of coordinator is much smaller. Typically, they are reduced to the initiation of certain aggressive actions. Perpetrators usually have not direct financial motivations. These differences are caused by the nature of aggression in various fields. Commercial aggression is cause by competitors, political aggression is caused by political opponents, and educational aggression is mainly induced by consumers of educational services. The likelihood of commercialized non-contextual aggressive actions of

Table 1. Comparison of aggressions by area

		Internet marketing	Political activity in internet	Information activity of universities in internet
Motivation of coordinator	Commercial	High	High	Low
	Conviction	Low	High	High
Motivation of perpetrators	Commercial	High	High	Low
	Conviction	Low	High	High
Capacity of coordination influence		High	Low	Low
Nature of perpetrator's conviction		Impulse	Firm	Impulse
Perpetrator's conviction perseverance		Low	High	Low
Community coverage		Selective	Selective, wide	Wide
Perpetrator's ethic level		High	Low	Low
Perpetrator's ability to support discussions		Low	High	Low
Perpetrator's communication competence level		Average	Low	High
Influence resistance of audience		Low	High	Low

competitors is quite low for certain obvious reasons. On the one hand, the difference in threats simplifies the counteraction, because weakness of coordination centre significantly reduces effectiveness of targeted actions. *On the other hand, absence of purposeful coordination and financial motivation increases social prestige of a campaign, reduces society's criticism towards aggression and creates illusion of objectivity.* Besides, principles of counteraction also changes, as it becomes inefficient in respect to a coordinator. High degree of non-commercial motivation in aggression also provides additional resistance of aggressive action to a lack of financial resources and pressure on perpetrators.

2. Perpetrators' convictions about information image of a university are characterized by occurrence impulsion and relatively low perseverance. Impulse nature of conviction occurrence is usually caused by specific training and examinational situations, which give a student a negative impression. Other cases, such as failure to enter the university and organizational problems, also lead to impulse short-term motivation. Perseverance of such convictions is rather weak, i.e. correct explanation of situation, its improvement or opinion of other fellow students can often change opinion of a perpetrator. Otherwise, motivation can be reduced to a level insufficient to maintain aggressive information activities. On the other hand, impulsiveness of aggressor leads to defamatory, obscene, exaggerated and generalized posts which greatly harm information image of a university.

3. Communication characteristics of perpetrator and environment. As has been said, due to impulse motivation aggressor can actively use profanity, insults and slander in his/her efforts. However, a typical performer of aggressive actions toward university's information image is characterized by high computer and network competence, ability to communicate in online communities, creativity, attractiveness, ability to use multimedia element, etc. All this makes an aggressor a tricky opponent to conduct discussions in online communities. Aggressive actions are carried out in a wide range of communities, i.e. not only in a predefined set of authoritative websites, but on websites of multifarious subjects, on unpopular websites and in closed communities. This complicates to a large extent a search for sensitive information and increases importance of problem of determining priorities in combating aggression.

3 Protection of Information Image of Universities in the Context of Its Formation

Protection of information image of universities belongs to a class of problems including protection of a facility from discreditation. Due to a number of factors, necessary part of information image protection is the early formation of image resistant to misconduct, constant monitoring of environment to identify threats at an early stage and effective rapid response.

Protection from aggression is a complex multistage process, in which we distinguish the following phases: identification of aggression; analysis and prioritization of aggression areas; resistance to aggressive actions; monitoring of adverse effects of conflict and attacks of aggressor. It should be studied how the specified process corresponds to the above-mentioned (Fig. 1) multistage process of formation of university's information image. This allows comparison of certain elements of the processes taking formation of university's information image as a basic process, and, accordingly,

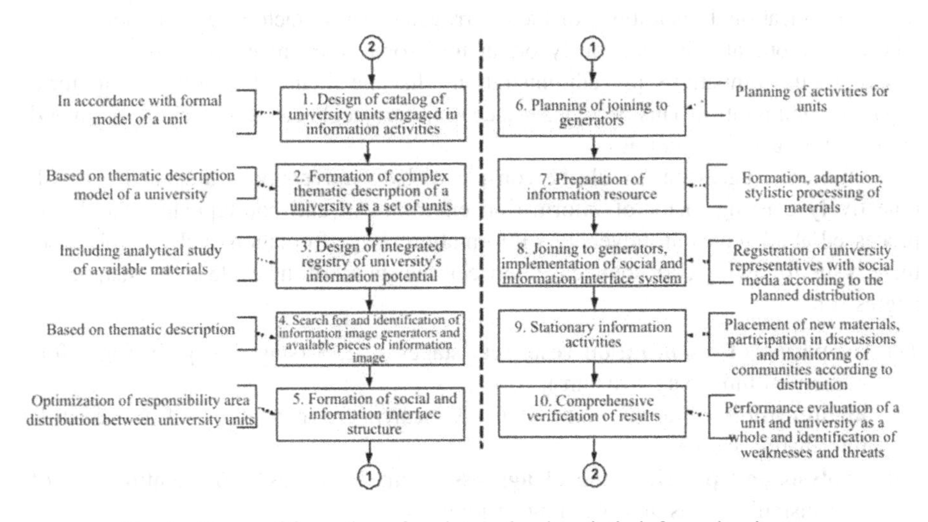

Fig. 1. Stages of formation of an integral university's information image

is a theoretical basis for planning measures for protection from aggression. There are two classes of aggression in terms of duration: short-term (less than one cycle of university's information activity); permanent or long-term (more than one cycle of university's information activity).

Short-term aggression lasts less than one cycle of information activities, so it is impossible to compare stages of protection with stages of formation of university's information image. Typically, short-term aggression occurs as a reaction of a group of people to actions of university employees or managers in a particular situation. In practice, this means that a full range of protection measures is carried out within the available ISM system on the stage of stationary information activities. Analysis of success in attack defeat is carried out on the stage of comprehensive verification of results. Considering time limit, it is difficult to restructure available resources in units in case of short-term aggression, so basic organizational measures are as follows: additional unit staff mobilization (within the principle of a social network [15, 16]); involvement of experts from information and administration units.

Long-term aggression can last several cycles of information activities or be a permanent factor of external pressure on university's information image. In this case, resistance to aggression becomes a part of the overall 'regular' information activity of a university. Long-term aggression is characterized by generally societal factors such as social tension, illegal methods of communities' monetization and unfair competition. Social tension is manifested in the presence of a significant proportion of society which is not satisfied with the situation and directs this dissatisfaction on the higher education system as a whole or specific universities in particular. Illegal methods of communities' monetization lies in availability of ways to profit from communities formed around negative social ideas. Profits can be payments for abstract ad view and for placing discrediting customized materials. Separately, owners of this type of communities can profit from brokerage in activities which are suspicious in terms of law [17, 18] or ethics. Unfair competition is implemented in constant systematic deterioration of information image of a specific university or a group of universities. Competitors may be other educational institutions or their surrogates. These factors are complementary and usually operate simultaneously or, in the worst case, in reverse order. That is, unscrupulous competitors use administrations that are financially motivated to form negative communities. This becomes a successful project because of social tension and demand for such communities.

Long-term aggression should be considered by a university as a special and still objectively existing factor of information environment and consequently should be processed at all stages of university's information image formation cycle. In this case, there is the following correspondence between the stages of the cycle and resistance to aggression:

(1) **formation of mobilization resource:** stages 1, 2, 3 (stages of gathering information on university structure);
(2) **identification of aggression:** stage 4, search for generators of university's information image;
(3) **analysis and prioritization of aggression areas:** stages 5, 6 (identification of responsibility areas and planning of joining);

(4) **resistance to aggressive actions:** stages 7, 8, 9 (immediate preparation for information activities in communities and stationary information activities);
(5) **monitoring of adverse effects of conflict and attacks of aggressor:** stage 10 (integrated verification of results).

Admittedly, long-term aggression occurs on a certain cycle of university's information activities. Within this cycle aggression is considered as a short-term, and respectively all necessary operational measures are taken. If comprehensive analysis of results at the end of the cycle shows that aggression is not eliminated, it is assigned the status of long-term. In this case, the stage of comprehensive verification implies check of completion of all available aggressions. The first stage of resistance to long-term aggression implies formation of a certain mobilization resource among workers and students of a university for their rapid involvement in case of intensification of aggressive actions.

4 Conclusions

Formalization of university's information activities, as well as methods and means of university's information activities improvement proposed in the paper involve improvement of resistance of university's information activities to external influences by establishing a system of active representation of universities in main communities. Aspects of protection of university's information image from focused deliberate aggressive actions which are possible in internet social media require a thorough study. This paper examines system features of aggression towards the information image in the field of education and presents a formal description of aggression in internet social media. It also describes a method of protection of university's information image in the context of its formation.

References

1. Korzh, R., Peleschyshyn, A., Syerov, Y., Fedushko, S.: The cataloging of virtual communities of educational thematic. Webology **11**(1), 1–16 (2014). http://webology.org/2014/v11n1/a117.pdf
2. Peleschyshyn, A., Fedushko, S., Peleschyshyn, O., Syerov, Y.: The verification of virtual community member's socio-demographic characteristics profile. Adv. Comput. Int. J. **4**(3), 29–38 (2013). doi:10.5121/acij.2013.4303
3. Syerov, Y., Peleschyshyn, A., Fedushko, S.: The computer-linguistic analysis of socio-demographic profile of virtual community member. Int. J. Comput. Sci. Bus. Inf. **4** (1), 1–13 (2013)
4. Fedushko, S.: Development of a software for computer-linguistic verification of socio-demographic profile of web-community member. Webology **11**(2), 1–14 (2014). http://www.webology.org/2014/v11n2/a126.pdf
5. Fedushko, S.: Peculiarities of definition and description of the socio-demographic characteristics in social communication. Comput. Sci. Inf. Technol. **694**, 75–85 (2011)

6. Biluschak, H.I., Fedushko, S.S.: Formation of linguistic and communicative indicators of socio-demographic characteristics of the web-community member. Manag. Complex Syst. **18**, 112–122 (2014)
7. Fedushko, S., Syerov, Y.: Design of registration and validation algorithm of member's personal data. Int. J. Inf. Commun. Technol. **2**(2), 93–98 (2013)
8. Andrukhiv, A., Sokil, M., Fedushko, S.: Integrating new library services into the University Information System. Library Management. Nr1 (6), Gdańsk, pp. 79–87 (2014)
9. Fedushko, S., Bardyn, N.: Algorithm of the cyber criminals identification. Global J. Eng. Des. Technol. **2**(4), 56–62 (2013)
10. Fedushko, S., Melnyk, D., Syerov, Y.: Analysis of the function and organization of virtual communication environments of Ukrainian scientists. Comput. Sci. Inf. Technol. **732**, 293–305 (2011)
11. Bekesh, Y., Fedushko, S.: The approaches to attract and inform customers of travel agencies in social networks. In: International Academic Conference on Information Communication Sociology, pp. 216–217 (2014)
12. Fedushko, S.: Disclosure of web-members personal information in Internet. In: Conference of Modern Information Technologies in Economics, Management and Education, Lviv, pp. 163–165 (2010)
13. Peleschyshyn, A., Fedushko, S.: Gender similarities and differences in online identity and Internet communication. In: International Conference of Computer Science and Information Technologies, Lviv, pp. 195–198 (2010)
14. Korzh, R., Fedushko, S., Peleshchyshyn, A., Syerov, Y.: Verification of socio-demographic characteristics of virtual community members. In: Proceedings of the XII International Conference on Modern Problems of Radio Engineering, Telecommunications and Computer Science, p. 632 (2014)
15. Fedushko, S.: Analysis architecture and modern trends of virtual communities' development. Inf. Syst. Netw. **699**, 362–375 (2011)
16. Syerov, Y., Fedushko, S., Peleschyshyn, A.: Definition methods of socio-demographic characteristics of the users' social communications. In: 7th International Academic Conference on Computer Science and Engineering, pp. 358–361 (2011)
17. Fedushko, S., Biluschak, H., Syerov, Y.: Statistical methods of virtual community users age verification. Int. J. Math. Comput. Sci. **1**(3), 174–182 (2015)
18. Biluschak, H., Fedushko, S., Syerov, Y.: Investigations of web-community members age verification by statistical methods. Int. J. SocNet.&Vircom. **3**(1) (2014) http://www.iaesjournal.com/online/index.php/VirCom/article/view/5909/0

Development of the System for Detecting Manipulation in Online Discussions

Andrij Peleshchyshyn and Zoriana Holub(✉)

Social Communication and Information Science Department,
Lviv Polytechnic National University, S. Bandery str. 12, Lviv, Ukraine
apele@ridne.net, zorianaholub@gmail.com

Abstract. An online discussion site is a constituent of numerous internet platforms. A large number of visitors, simplicity of publishing information as well as high level of trust in the posted information create an environment vulnerable to mind manipulation processes. This research is devoted to the development of the system for detecting manipulation in online discussions. Considering peculiarities and tendencies of communication in online discussions, the architecture of the system for detecting manipulation was designed. Peculiarities and tendencies were classified according to temporal type and level. The system for detecting manipulation consists of two components: the system of filters, which outputs the set of profiles, whose activities should be monitored; and a tool for tracking users` actions in real time. The latter is aiming at detecting initiated manipulations. The system of filters is constructed on the basis of profile features, which belong to the static temporal type. Static features provide a general impression of user`s activities in online discussions. The tool for monitoring activities operates on features of the dynamic temporal type. Piecewise-linear aggregates are used for modelling the tool. Schemes of manipulations are created by means of transactional analyses and formal tools of neurolinguistics programming.

Keywords: Manipulation · Online discussion · Piecewise-linear aggregate · Psychological state · Manipulation method

1 Introduction

Informational and psychological manipulations pose a threat to a community member as well as to social institutions on the whole. Therefore finding a counteraction to mind manipulation is a pressing problem. Instant detection of manipulation is a prerequisite to successful fighting manipulation.

Taking into consideration the very dynamic nature of socializing in online discussions, it is of utmost importance to detect facts of manipulation as soon as possible. To achieve this, information in discussions should be processed in real time. General features of information flow as well as high updating frequency, requires formalized approaches to detecting manipulation in large communities in the national network segment [1].

R. Szewczyk and M. Kaliczyńska (eds.), *Recent Advances in Systems, Control and Information Technology*, Advances in Intelligent Systems and Computing 543, DOI 10.1007/978-3-319-48923-0_15

Existing means for fighting manipulation are not comprehensive and perform highly-specified tasks. Traditional protection from manipulation was focused on speeches in mass media [2, 3]. Thus, designing program and algorithm means for detecting manipulation is a topical task.

Manipulative operations in the internet combine features of interpersonal (dialogic, interactive) and mass communication (large unspecialized audience) [4] and discussions. In light of danger, dynamics and peculiar nature of internet, it is called for an algorithmic and program complex for manipulation detecting. The complex should embody method for detecting and fighting manipulation in the traditional environment as well as specific for the internet signal features of a manipulative operation.

2 Determining Criteria for Detecting Manipulation

The algorithmic and program complex should have a two-level structure: the system of filters and tools for tracking users actions in real time. In the beginning the system of filters is to make a list of suspicious profiles, after that these profiles are going be monitored according to definite methodology by means of the tool. In this way the complex efficiency is significantly increased.

The system of filters and tools is based on the features of different nature. The system of filters is aimed to process general characteristics of profiles activities during the definite time period. In contrast, the tool for tracking manipulations in the real time is analyzing verbal and nonverbal data in a message, as well as their sequence. Thus, the classification of signal features on dynamic and static is suggested.

Additionally, while designing the algorithmic and program complex, multi-level structure of the internet platforms that contain discussions is taken into consideration [5].

Table 1. Examples of criteria for detecting manipulation

Message	Discussion	Community	Internet
Likes count of a message	Frequency of posting links	Community attendance frequency	Quality of user's account in social networks
Number of characters in a message	Ration of the user`s messages to all messages in the discussion	Time period characterized by user`s peak activity	Participants in other communities
Number of punctuation marks	Percentage of answers on other user's messages	Number of initiated discussions	Existence of private blog
Number of hashtags	Several sensory modalities (digital, visual, audial, kinesthetic) [8]	Existence of users with a similar profile in the community	Number of platforms, where a user participates in discussions

Identifiers of manipulation differentiate on each level. Therefore, organizational and structural levels as well as specific for each level criteria for detecting manipulation are determined.

Having applied manipulation method, manipulator leaves trail [6]. This trail could be identified on each level by means of different criteria (Table 1). Therefore, manipulation should be scrutinized from the perspective of organizational and structural level of the internet, community, discussion, and message. Communities differentiate in rules, customs and styles [7]. Therefore manipulation identifiers of each criterion should be defined in the light of community's peculiarities.

3 System of Filters for Identifying Suspicious Profiles

Criteria for detecting manipulation on discussion, community and internet levels are defined grounding on the communicative activities during the certain time period. These criteria are called static. The system of filters for detecting manipulation is functioning on the basis of these criteria.

As an example operation of the filter that filters out quality and developed profiles is illustrated in the Fig. 1.

The filter is based on the observed tendency, that in order to accomplish their task manipulators often use fake profiles. One of characteristic features of a fake profile is a small number of friends. Popular among manipulators way to increase a number of friends is to create and add to friends fake profiles. These fake friends usually have no avatars. In addition, aiming to recognize each other manipulators engaged in the same operation often have similar nicknames. Described above identifiers of manipulation are utilized in the filter for sorting out genuine and fake profiles.

For each quantitative identifiers of manipulation is set a threshold value. In the filtering process, the feature of user`s activities is compared with a threshold value. If the feature surpasses its threshold value, the profile added to the list of suspicious. For various discussions different threshold value is determined. Precision of the filter heavily depends on the correctly determined threshold value.

The chance of filtering errors cannot be denied. However, the error would underlie in adding ordinary users to the list of suspicious profiles, overlooking a manipulator is less likely.

In the considered algorithm ratio of friends having no avatar ($n^{friendsNoAvatar}$) to friends total ($n^{friends}$) is compared to the predetermined for a particular discussion threshold value ($n^{threshold}$). String variables $Nickname^{user}$ and $Nickname^{friend}$, containing users' nickname and friend's nickname, respectively, are compared in order to identify profiles with similar nicknames. If a profile fails to pass either of the checks, it is claimed to be suspicious and its activities are to be monitored in real time.

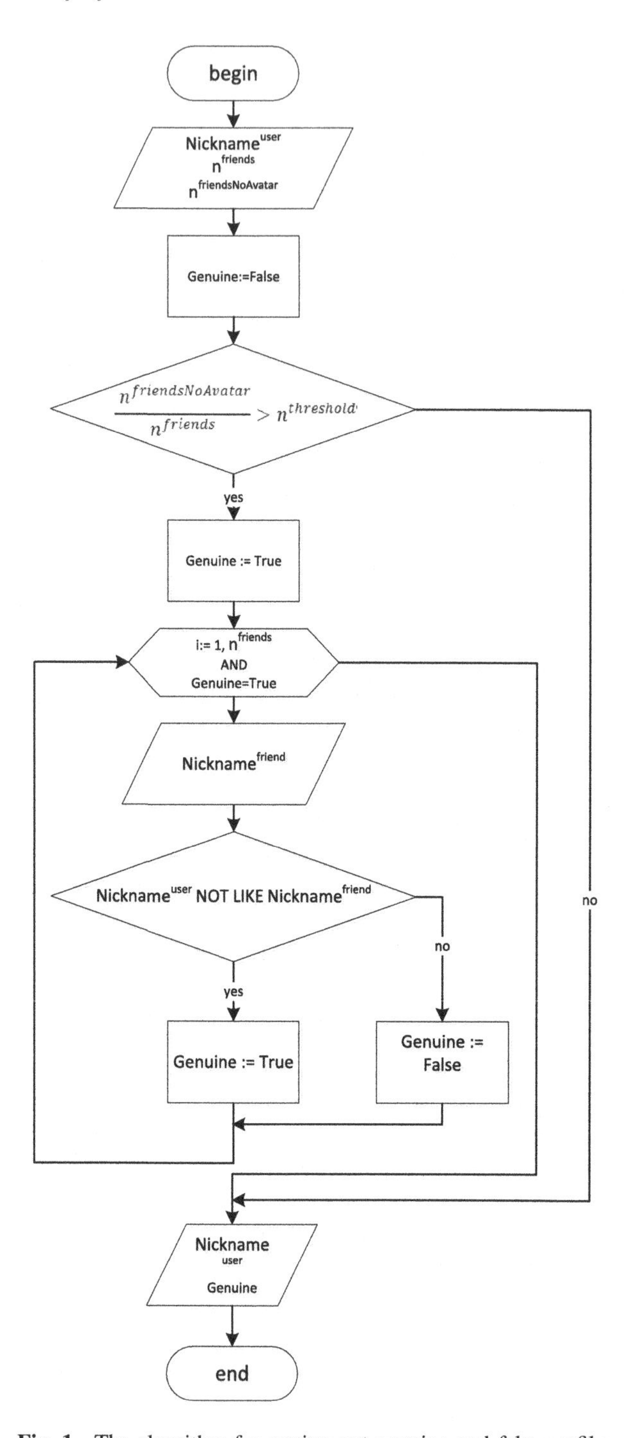

Fig. 1. The algorithm for sorting out genuine and fake profiles

4 Monitoring Activities of Suspicious Profiles

Dynamic criteria of manipulation are semantic, metagraphemic and nonverbal criteria of another temporal type associated with a certain message. Exploiting the latter in detecting manipulation, it is imperative to consider their sequence. As long as identified using criteria manipulation methods constitute manipulation tactics, it is reasonable to model the latter by means of finite-state machine and piecewise-linear aggregate [9]. Having detected several identifiers of a manipulation method in a discussion, it is necessary to compare their appearance sequence with sequence of method in manipulation tactic. If they match, the manipulation is deployed in the discussion.

Finite-state machine and piecewise-linear aggregate are apt for modeling changes of psychological states effected by manipulative acts. For the reason that finite-state machine is used for modeling dependences of the system states on exterior environments influences.

The tactic of attitudes manipulation is depicted below by means of a Moore machine. It consists of a set of psychological states (1) and a set of manipulative methods (2) that are applied in aforementioned tactic [10].

$$State = \{S_1, S_2, S_3, S_4\} \tag{1}$$

$$Method = \left\{ \begin{array}{l} Rhetorical\_question, Imitating\_recipients\_style, \\ Provocativ\_futuristic\_statement, Pseudologic, \\ Pseudolink, Card\_stacking, Evoking\_strong\_emotions \end{array} \right\} \tag{2}$$

To reach a target manipulator, as a rule, repeats a method several times, however each time it has different surface representation (Fig. 2).

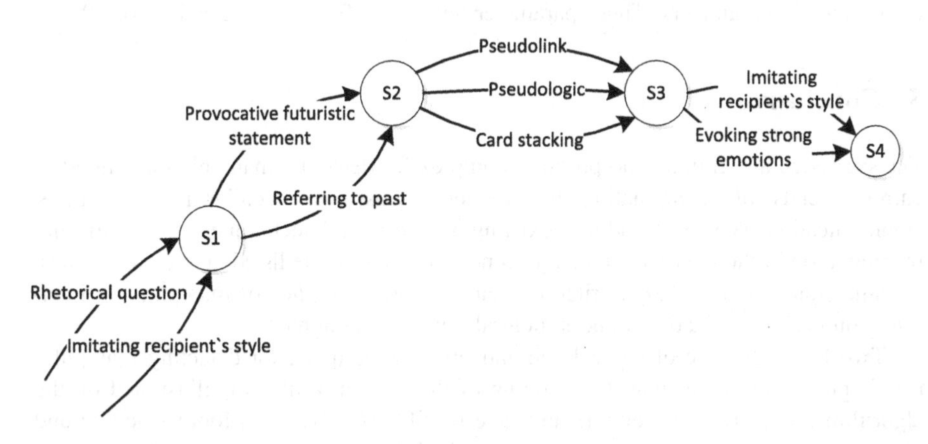

Fig. 2. The tactic of attitudes manipulation depicted utilizing a Moore machine

If it is required to detail a model, piecewise-linear aggregate is to be applied. The latter gives information about exact number of reiterations of a method that are necessary to achieve a state.

Piecewise-linear aggregate, like a Moore machine, consists of a set of psychological states (3) and manipulation methods (4):

$$State = \{State_i\}_{i=1}^{n^{State}}, \text{ where } n^{State} \text{ is a number of possible states} \tag{3}$$

$$Method = \{Method_j\}_{j=1}^{n^{Method}}, \text{ where } n^{Method} \text{ is a number of possible methods} \tag{4}$$

However the difference underlies in the fact, that states and methods are concretized by a parameters vector. Vector size is predetermined for each manipulation tactic. Methods (5) and states (6) are detailed in the way illustrated below.

$$Method_j = \left\langle idMethod^{(j)}, pMethod_1^{(j)}, \ldots, pMethod_k^{(j)} \right\rangle \tag{5}$$

Each state parameter $pState_k^{(i)}$ acquires discrete value from a definite interval $pState_k^{(i)} \in \left[\min^{(i)}, \max^{i}\right]$.

$$State_i = \left\langle idState^{(i)}, pState_1^{(i)}, \ldots, pState_k^{(i)} \right\rangle \tag{6}$$

If a state parameter's value goes beyond the set interval, a psychological state of a recipient changes. This change is formalized in the following way (7).

$$State_{i+1} = \left\langle idState_{i+1}, pState_1^{(i)} + pMethod_1^{(j)}, \ldots, pState_k^{(i)} + pMethod_k^{(j)} \right\rangle \tag{7}$$

The tactic of attitudes manipulation could be detailed by means of $Doubt^{(i)}$ and $Involvement^{(i)}$ parameters. These parameter values are from the interval [–10; 10].

5 Conclusion

The suggested algorithmic and program complex for detecting manipulation embodies current trends of manipulation in internet discussions, peculiarities of online-communication as well as adopts existing methods and techniques for identifying manipulation in the interpersonal environment. Basing on the listed above approaches, classification of manipulation criteria is carried out. Difference of manipulation identifiers on each organizational and structural level is highlighted.

Two-level structure of algorithmic and program complex for detecting manipulation is proved to be efficient. Functioning of the system of filters is illustrated on the algorithm for sorting out genuine and fake profiles. Applying a Moore machine and piecewise-linear aggregate for modeling manipulation tactics is scrutinized.

Suggested formal methods and tools for detecting manipulation can be utilized not only when analyzing communication in internet discussions by means of transactional analyzes and tools of neurolinguistics programming, yet other psychological methods as well.

References

1. Fedushko, S., Biluschak, H., Syerov, Y.: Statistical methods of virtual community users age verification. Int. J. Math. Comput. Sci. **1**(3), 174–182 (2015)
2. Syerov, Y., Peleschyshyn, A., Fedushko, S.: The computer-linguistic analysis of socio-demographic profile of virtual community member. Int. J. Comput. Sci. Bus. Inform. **4**(1), 1–13 (2013)
3. Peleshchyshyn, A., Trach, O.: The relevance of the life cycle of virtual community. In: Materials 4th International Science Conference on Information, Communication, Society, ICS - 2015, National University of "Lviv. Polytechnic", Lviv, Ukraine, Slavske, 20–23 May 2015
4. Korzh, R., Peleschyshyn, A., Holub, Z.: Analysis of integrity and coverage completeness of the informational image of a higher education institution. In: Proceedings of the XIII International Conference on Modern Problems of Radio Engineering, Telecommunications and Computer Science, TCSET 2016, pp. 825–827. Publishing House of Lviv Polytechnic, Lviv–Slavske, Ukraine, 23–26 February 2016
5. Fedushko, S.: Development of a software for computer-linguistic verification of socio-demographic profile of web-community member. Webology **11**(2), 1 (2014). http://www.webology.org/2014/v11n2/a126.pdf
6. Biluschak, H.I., Fedushko, S.S.: Formation of linguistic and communicative indicators of socio-demographic characteristics of the web-community member. Manag. Complex Syst. **18**, 112–122 (2014)
7. Peleschyshyn, A., Fedushko, S., Peleschyshyn, O., Syerov, Y.: The verification of virtual community member's socio-demographic characteristics profile. Adv. Comput. Int. J. **4**(3), 29–38 (2013)
8. Berne, E.: Games People Play – The Basic Hand Book of Transactional Analysis, p. 81. Ballantine Books, New York (1964)
9. Rodger, S., Finley, T.: JFLAP: An Interactive Formal Languages and Automata Package, 1st edn. Jones and Bartlett, Sudbury, MA (2006)
10. Wake, L.: Neurolinguistic Psychotherapy: A Postmodern Perspective. Routledge, London (2008)

Capabilities of an Open-Source Software, Elmer FEM, in Finite Element Analysis of Fluid Flow

Marcin Safinowski[1(✉)], Maciej Szudarek[2], Roman Szewczyk[2], and Wojciech Winiarski[1]

[1] Industrial Research Institute for Automation and Measurements, Al. Jerozolimskie 202, 02-486 Warsaw, Poland
msafinowski@piap.pl

[2] Faculty of Mechatronics, Warsaw University of Technology, sw. A. Boboli 8, 02-525 Warsaw, Poland
mszudarek@gmail.com

Abstract. Computational Fluid Dynamics (CFD) is widely used to model fluid flow and optimize industrial processes. The main obstacle to implement these methods to small businesses might be the price of commercial software. However, many cases do not necessarily require state-of-the-art algorithms and use of open-source software might be the best solution. The aim of the article was to study and present the capabilities of a solver Elmer FEM. Meshes were generated with Gmsh software and data visualization was done with the use of application ParaView, all of which are open-source software. The chosen software satisfied criteria of being simple to learn, being compatible with widely used file formats and having an active community. What is more, it allows the user to solve multiphysics problems. The object of simulations was a prototype of a graphene flow meter, developed by Industrial Research Institute for Automation and Measurements. The article focused on a key issue in most of the CFD simulations, which is turbulence modelling. Both RANS-based and Large Eddy Simulation models were tested. The process of setting up simulations and its results are presented.

Keywords: Computational fluid dynamics · CFD · Flow measurement · Graphene open-source · Turbulence

1 Introduction

Computational Fluid Dynamics is used widely in numerous branches of science. It opens up opportunities to optimize geometry, visualize fluid flow and evaluate its parameters. Unfortunately, high prices of most of the commercial software have a deterrent effect. On the other hand, simple cases do not necessarily require state-of-the-art algorithms and use of open-source software might be the best solution. The aim of the article was to study and present the capabilities of a solver Elmer FEM in finite element analysis of fluid flow. The article focused on a key issue in most of the CFD simulations, which is turbulence modelling. The object of simulations was a

R. Szewczyk and M. Kaliczyńska (eds.), *Recent Advances in Systems, Control and Information Technology*, Advances in Intelligent Systems and Computing 543, DOI 10.1007/978-3-319-48923-0_16

prototype of a graphene flow meter, developed by Industrial Research Institute for Automation and Measurements.

2 Used Software

There is a wide variety of free software for flow simulation available to choose from, such as OpenFOAM, UFO-CFD, Helix-OS or Elmer FEM. Upon making a decision, one has not only to keep in mind its capabilities, but also costs related to time spent for learning how to use it. Although Elmer FEM is based on finite element method, which requires more memory and has slower solution times that the finite volume method, it is easy to use and allows the user to solve multiphysical problems, involving fluid mechanics, mechanics, electromagnetics, heat exchange, acoustics, etc.

Meshes were prepared with Gmsh and data visualization was done with the use of ParaView. Both of these are also an open-source software.

3 Test Stand

Modelled stand comprised of a tank, an electrode and a sensor. The tank had a cylindrical shape with the inner radius r_i = 5.75 cm and the outer radius r_o = 11.325 cm. The water level was 12 cm. The electrode, which was a 0.8 mm diameter rod, and the sensor were dipped in the tank. The sensor was a layer of graphene on a silicon plate. Such a plate was mounted on the both sides of a plastic support. The angular velocity was controlled through a DC motor.

4 Simulation

4.1 Theoretical Model

The flow of a viscous fluid confined in the gap between two rotating cylinders is known in fluid dynamics as the Taylor-Coutte flow [1]. To predict flow instabilities Taylor number is used.

$$T = Re^2 \frac{h}{R_0} \tag{1}$$

where *Re* is Reynolds number based on the difference between cylinders' diameters *h* and the angular velocity of the inner cylinder, R_o is a mean diameter of two cylinders.

Given that the inner cylinder is rotating in the same direction and with the same angular velocity as the outer cylinder, it is proved that the flow is laminar up to a critical value of Taylor number, equal to 1708. For higher values of *T* the flow is turbulent and its mean velocity is constant. The are no complex instabilities that are present when the cylinders' angular velocities are different [2].

To test turbulence models that are available in Elmer FEM, angular velocities were chosen accordingly to the given geometry, so that Taylor number exceeds the critical value and the flow is turbulent.

4.2 Available Turbulence Models

There are three main approaches to numerical simulation of turbulent flows. The most accurate, as it marginalizes the effect of models, is DNS (Direct numerical simulation). For most cases it requires a computational power that exceeds the capacity of the most powerful computers currently available, therefore it is useless for industrial applications.

Less accurate, but widely used are RANS-based turbulence models (Reynolds-averaged Navier-Stokes). They model a time-averaged solution. In Elmer FEM there is a couple of RANS-based turbulence models, one of which is *k*-epsilon, studied in the article as the most stable.

The last of common models is LES (Large Eddy Simulation), a technique that is a compromise between the previously mentioned. The smallest scales of the flow are removed through a filtering operation and only the largest scales of turbulence are resolved. Although the computational cost is greatly reduced in comparison to DNS, the mesh has to be dense enough to not omit the important data. Elmer FEM allows to implement LES model with VMS method (Variational multiscale).

4.3 Geometry and Mesh

Geometry was created with an open-source software, Gmsh. All of the dimensions were chosen accordingly to Fig. 1. In the region distant from the sensor and the electrode structured mesh was used. It allows to orientate mesh elements to predicted streamlines and leads to the converging solution faster and more accurately. However, this cannot be applied in proximity of the sensor and the electrode, as it is impossible to cover such a complicated geometry with structured mesh with decent mesh quality. Mesh quality is determined by skewness and proportions of the elements, as well by the rate of size change in the domain [3].

Caution must be taken while creating a hybrid mesh in Gmsh, as it is highly probable that nodes on the edge separating structured and unstructured mesh would overlap. Such an error is hard to spot, will not be automatically repaired while importing mesh to Elmer FEM and results in a failed simulation.

The VMS model is used in time dependent simulations. Time steps have to be small enough so that a convergent solution with a reasonable error can be achieved. Time step size is limited by the mesh size, as it cannot be larger than it takes for the fluid to move from one node to another. Because of small dimensions of both the electrode and the sensor, the mesh that would allow to capture the separation point and vorticity had to be very dense. It resulted in a small time step and very long computation time, exceeding 24 h. Due to lack of computational power, ILU (Incomplete LU factorization) algorithms, that would definitively speed up the process of achieving convergent

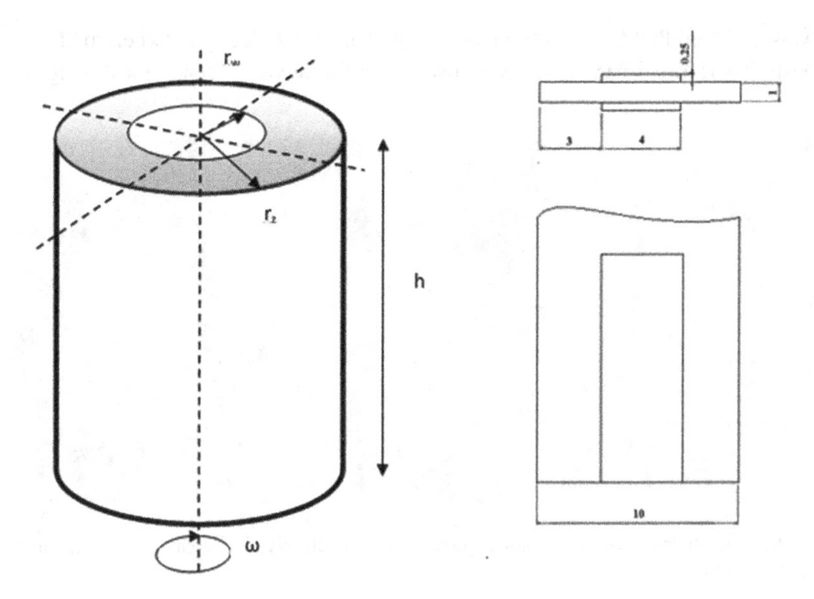

Fig. 1. Tank and a sensor, dimensions are in milimeters

results, could not have been used. Because of the issues presented above, it was decided to test turbulence models on a less computationally demanding 2-D model. 3-D model was used only to learn the principles of operation of the free surface module. Authors do realize that turbulence breaks any symmetries that would justify such a simplification, therefore 2-D model is merely a test case for turbulence models available in Elmer FEM.

4.3.1 Two-Dimensional Model

Simulations with the VMS model for multiple angular velocities were preceded by a mesh independence test. Multiple versions of mesh with varying number of elements were created to determine what is the minimal density of mesh that can be applied. That was especially important due to the number of time steps that reached up to several thousands.

To determine the effect of mesh density on result, extreme pressure and velocity values, as well as pressure and velocity distributions were compared. The test was performed for an angular velocity that would cause generation of Karman vortices behind the electrode, calculating from the time moment when the first instabilities would appear (Table 1).

Table 1. Mesh independence test

Mesh quality	Number of elements	Maximum velocity in m/s	Maximum pressure difference P_{max}–P_{min} in kPa
Fine	4 536	0.5789	0.2984
Medium	13 836	0.5808	0.3204
Coarse	36 788	0.6101	0.3202

It was assumed that flow parameters do not change drastically between medium and fine mesh. Therefore VMS model was used with the medium dense mesh (Fig. 2).

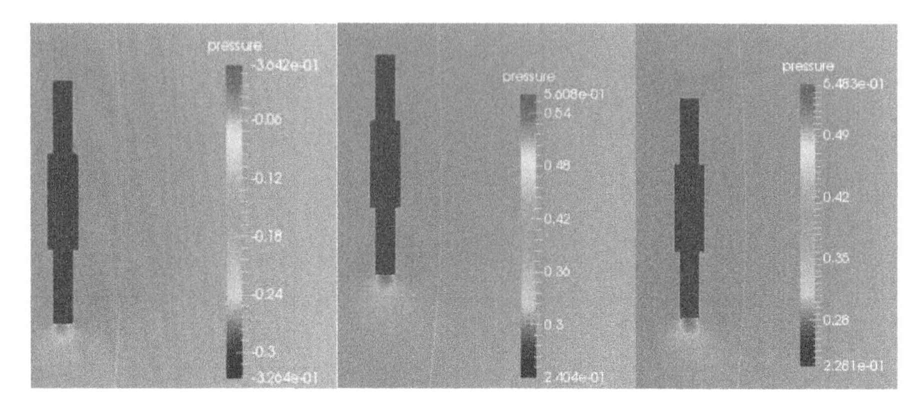

Fig. 2. Pressure distribution in sensor's proximity respectively for: coarse, medium and fine mesh, units are kPa

Different algorithm to determine mesh size was used with RANS-based turbulence model. As it is advantageous to use a wall function approximation in the closeness of a wall boundary rather than to compute it, Reichards Wall Law was used. In such a case a dimensionless wall distance y^+ [4] is used to estimate recommended mesh size near wall boundary.

$$\tau = 0,5 \cdot \rho \cdot 0,079 \cdot Re^{-0,25} \cdot w^2 \tag{2}$$

$$U_\tau = \left(\frac{\tau}{\rho}\right)^{\frac{1}{2}} \tag{3}$$

$$y = y^+ \cdot \frac{\mu}{\rho \cdot U_\tau} \tag{4}$$

where w is a mean velocity in m/s, ρ is density in kg/m^3, μ is dynamic viscosity in Pa·s, Re is Reynolds number, y is the distance of the first element from the wall, measured from its middle point.

Because the calculated y distances are roughly the size of the electrode and the sensor, the mesh in their proximity was generated as more dense (Table 2).

Table 2. Calculation of the recommended distance of the first element from the wall

Angular velocity in rad/s	y for y^+ = 30, in mm	y for y^+ = 300, in mm
5	0.9	9.2
15	0.3	3.7

4.3.2 Three-Dimensional Model

Large angular velocities result in drastic change of free surface shape and alters velocity gradient in the tank. However, free surface module available in Elmer FEM gives a possibility to take that into account. The principle of operation is as follows. Firstly, one has to choose the boundary that will be considered as a free surface and specify its initial shape, just like any other initial condition. Then, when the simulation begins, mesh would be updated to the new shape before the first Navier-Stokes equations iteration. The equations that describe the altering shape of the free surface are dependent upon flow equations and are updated with user-defined frequency. The mesh is then adapted to the surface equations (Fig. 3).

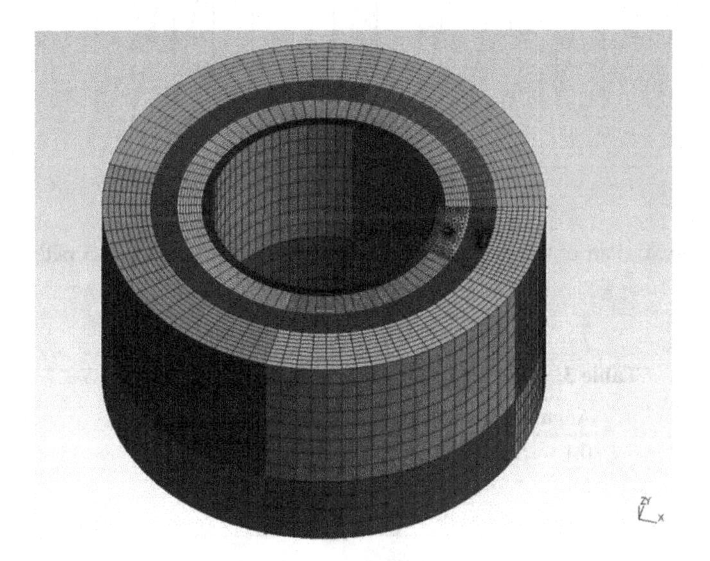

Fig. 3. 3-D model mesh

For the 3-D model a hybrid mesh containing 84 729 elements was created.

4.4 Simulation Settings and Results

The centrifugal force was included in the model. The angular velocity of the fluid was set as an initial condition and the boundary condition were rotating walls and no-slip condition on the electrode and the sensor. Different angular velocities required different duration of time steps. All of the simulations resulted in monotonically convergent results. Figure 4 presents a part of the convergence history for the angular velocity 5 rad/s. Each peak corresponds to the beginning of a new time step. For every other angular velocity the convergence history was almost identical (Table 3).

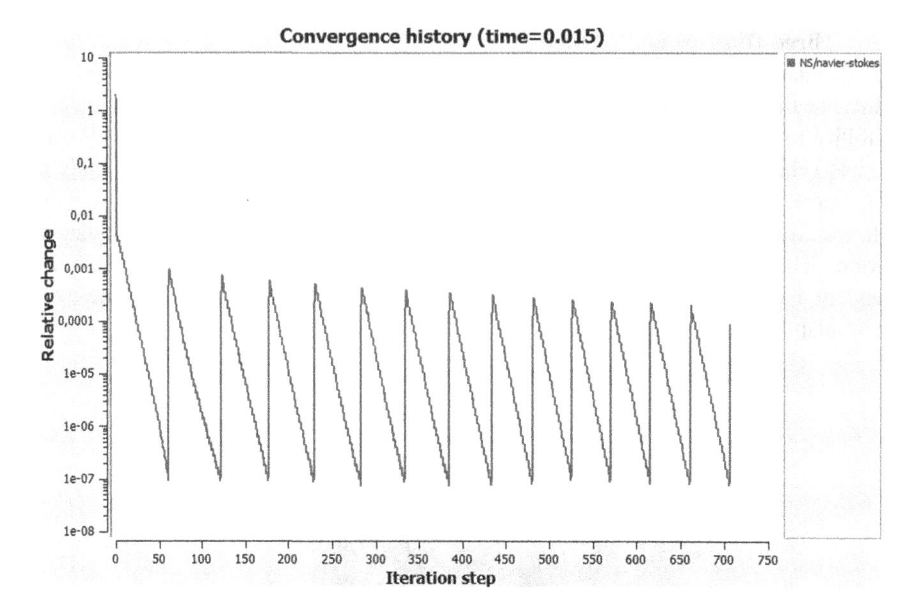

Fig. 4. Part of the convergence history for the angular velocity 5 rad/s

Table 3. Time step values for different angular velocities

Angular velocity in rad/s	Time step size in ms
0.1	100
5	1
15	1

In the Figs. 5 and 6 below sample results are presented in form of velocity and pressure distributions in the proximity of the sensor and the electrode.

With RANS-based turbulence k-epsilon model it was necessary to add initial and boundary conditions for the turbulent kinetic energy k and the turbulent dissipation ε. These were chosen iteratively, using the first approximation calculated with the equations given in [5].

The results were convergent until regular oscillations could be observed. This was an effect of Karman vortices that would generate behind the electrode. Both LES and RANS-based models generated realistic and comparable results.

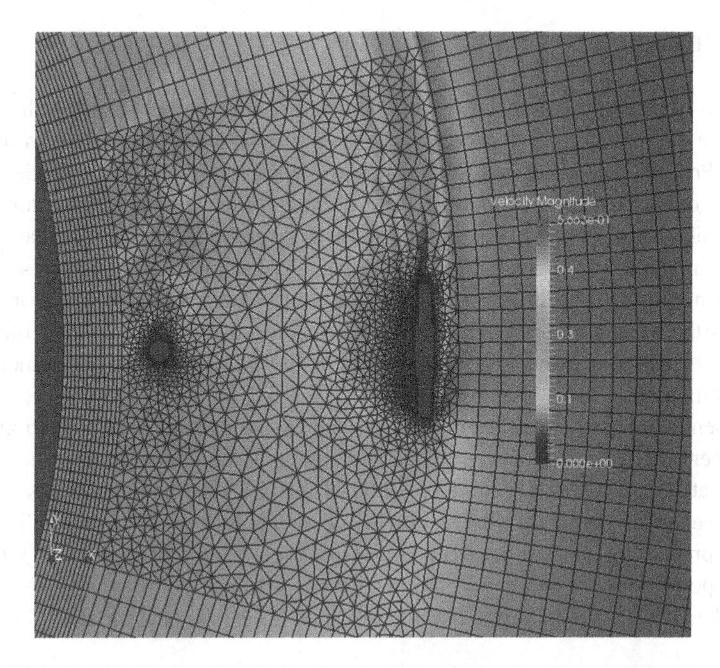

Fig. 5. Velocity distribution in m/s for the angular velocity 5 rad/s, mesh lines are visible

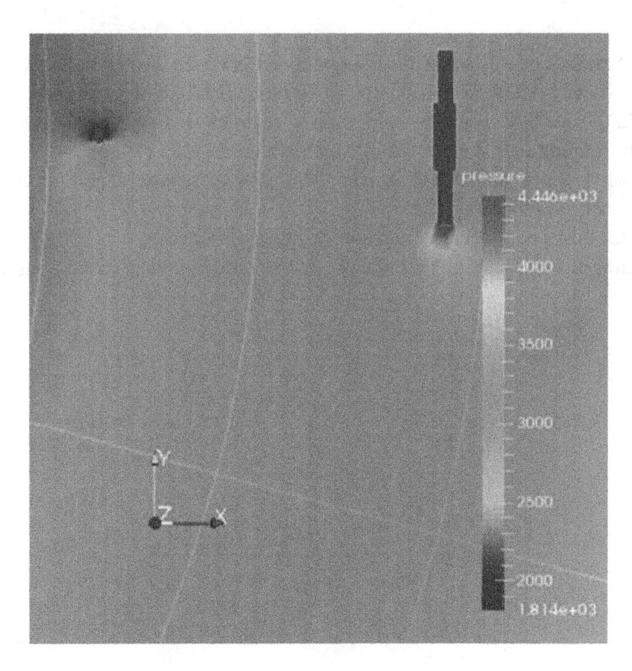

Fig. 6. Pressure distribution in Pa for the angular velocity 15 rad/s

5 Conclusion

The process of setting up simulations and achieved results are presented. Although the results were not validated experimentally, they are in line with expectations and calculations. Pressure and velocity distributions are alike regardless of the model used.

Despite of the inevitable flaws that come with the finite element method, turbulence models available in Elmer FEM allow to solve simple cases in a reasonable time. Software's modularity is a great advantage, especially because the process of setting up a simulation is quick and intuitive. The active users' community is also worth mentioning, as there exist instructions on how to develop user's own defined functions. That could be used for extending functionality of Elmer FEM with such elements as a roughness module.

The mentioned restrictions are mostly caused by the principles of operation of the finite element method. The number of used elements is limited as their size is significantly greater than in finite volume method. That is a source of problems with the convergence and stability of the results while using turbulence models. Therefore critical approach, or at best experimental validation, is required. Other issues are limited opportunities for mesh generation in comparison to commercial software and lack of other RANS-based turbulence models than k-epsilon or k-omega SST.

References

1. Taylor, G.I.: Stability of a Viscous Liquid contained between Two Rotating Cylinders. Phil. Trans. Royal Society **A223**, 605–615, doi:10.1098/rsta.1923.0008
2. Andereck, C.D., Liu, S.S., Swinney, H.L.: Flow regimes in a circular Couette system with independently rotating cylinders. J. Fluid Mech. **164**, 155–183 (1986)
3. Accuracy, Convergence and Mesh Quality. http://www.pointwise.com/theconnector/May-2012/Mesh-Quality.shtml
4. http://www.cfd-online.com/Wiki/Dimensionless_wall_distance
5. http://www.cfd-online.com/Wiki/Turbulence_free-stream_boundary_conditions

The Use of ICT in Formation of Professional Competence of Nurses in Ukraine

Natalia Shygonska(✉)

Zhytomyr Nursing Institute, Zhytomyr, Ukraine
shygonska@mail.ru

Abstract. Nurse training system in Ukraine is pending continuous alterations. The basic principles and areas of its operation are determined by the laws of Ukraine on Higher Education and on Scientific and Scientific-Technical Activities, emphasizing the importance of forming competitive experts. The objective of this article is to specify the information and computer technologies, the use of which is feasible in the course of the nurse training in Ukraine in medical colleges and institutes, and determination of their scopes of application. To achieve the scientific research objective, we used the methods of system analysis, synthesis and logic generalization to systematize the obtained information and subsequently use the same when developing the educational standards of nurse training in Ukraine. According to the Law of Ukraine on Higher Education, the nurse training in Ukraine will be implemented at three educational levels, Junior Bachelor, Bachelor and Master. The content of occupational training is focused on obtaining of programmed educational outcome, determined by the structure and content of the professional competence of nurses, in which both a comprehensive information competence and specialization component of the educational standard play an important role. In this case, the information and computer technologies become the means of forming a professional competence and curriculum outcome. Thus, the use of ICT in the system of nurse training in Ukraine will allow upgrading the training content and form of the training process organization substantially.

Keywords: Nurses · Professional competence · Information and computer technologies

1 Introduction

Ukrainian society is permanently changing and consequently these alterations touch upon all spheres of its functioning. Deep transformations have initiated elaborating new strategies within all educational areas [7, 8] aimed at forming qualitatively new educational standards and correspondingly competitive specialists [10, 11].

Completely new phenomenon for Ukrainian system of education has outlined such new brunch of study as "Nursing" that in its turn completely meets the European standards. We must acknowledge the promptness of Ministry of Education and Science of Ukraine in responding to innovative processes that occur not only in educational area but social life as well [8].

R. Szewczyk and M. Kaliczyńska (eds.), *Recent Advances in Systems, Control and Information Technology*, Advances in Intelligent Systems and Computing 543, DOI 10.1007/978-3-319-48923-0_17

Since November 11, 2015 the expert groups created by Ministry of Health Care of Ukraine have been persistently working on elaborating new Europe-oriented educational strategies and standards backed on B. Bloom's Taxonomy – Learning in Action [4, 13]. New educational standards propose to form a number of professional competences at future specialists' nurses. The students' curriculum outcome consists of two compound constituents: the basic and special competences. They make together the main planned outcomes [4, 13]. The leading position in this process is given to using the information and computer technologies as the means and result of forming nurses' professional competence and learning outcome [4].

The formulation of the task. The objective of this article is to specify the information and computer technologies, the use of which is feasible in the course of the nurse training in Ukraine in medical colleges and institutes, and determination of their scopes of application.

2 Materials and Methods

To achieve the scientific research objective, we used the methods of system analysis, synthesis and logic generalization to systematize the obtained information and subsequently use the same when developing the educational standards of nurse training in Ukraine.

3 Results and Discussion

According to the Law of Ukraine on Higher Education, the nurse training in Ukraine will be implemented at three educational levels, Junior Bachelor, Bachelor and Master [7]. The content of occupational training is focused on obtaining of programmed educational outcome, determined by the structure and content of the professional competence of nurses, in which both a comprehensive information competence and specialization component of the educational standard play an important role. In this case, the information and computer technologies become the means of forming a professional competence and curriculum outcome [2, 12].

The proposed perspective strategy of training future professionals on Nursing within healthcare system has supposed the direct use of the information and computer technologies [1, 7, 12]. At the first thought, it seems that there is nothing new in here We want to emphasize that the effective and efficient technologies have been sufficiently used in medical universities of Ukraine and consequently their implementation bears a systemic character either for students or for teaching staff proposing at the same time a wide use of internal monitoring possibilities: Horbachevsky Ternopil State Medical University [5], Bukovinian State Medical University [3], Vinnitza Pirogov National Medical University [9] while the educational institutions that have targeted their activity at training the specialists of middle rank – nurses, medical assistants etc. just now open the great variety of possibilities to use such technologies [14].

Using ICT in education is aimed at solving many complicated tasks among which:

- Creating educational environment;
- Making didactic software;
- Using ICT during academic projects;
- Implementing Multimedia;
- Elaborating distant learning courses;
- Managing educational institution activity;
- Making Web-sites;
- Making open-access journals, repositories;
- Creating occupational guidance for prospective students [2, 6, 16].

In general, there can be distinguished several main directions of implementing ICT by the educational providers (Table 1):

- Software product as the element of the inner system monitoring the general activity in the educational institution and, thus, providing the qualitative support of the specialists training;
- ICT for monitoring students' learning outcomes;
- Software for creating didactic products;
- Web-technologies support for academic process;
- Tele-Communication and Tele-Education.

Software products that serve as the structural elements for providing the inner system of monitoring the general activity of the educational institution, generalize the whole informational flow either from students and teachers or from academic department interpreting and making the secondary information product ready to its further use and final representation (terminal, annual reports, final academic results) – "Electronic journal", "Dekanat", etc.

Using ICT for monitoring students' learning outcomes implies monitoring the stream data of the students' curriculum and learning outcomes as well as the final assess. Here the academic tasks and tests are to be classified according to the student's

Table 1. Application types of information and computer technologies in academic process

Type	Characteristics
Software product as the element of the inner system monitoring	Contributes to support academic process within the educational institution
ICT for monitoring students' learning outcomes	Monitors the stream data of the students' curriculum and learning outcomes as well as the final assess
Software for creating didactic products	Supports creating academic tasks, tests, manuals, electronic textbooks, computer test systems, etc.
Web-technologies support for academic process	Information sites, search systems, catalogues, blogs
Tele-Communication and Tele-Education	Web and Skype conferences, Web-forums

knowledge level, course of the study, expected learning results and professional competences that would have been formed.

Software for creating didactic products is oriented at making a number of academic tasks, tests, manuals, electronic textbooks, computer test systems, electronic reference books, distant and laboratory courses, audio and video materials.

Wide-world internet penetration into all human activities initiated using so-called web-technologies aimed at supporting academic process within the institutions and making strong inter-institutional, interpersonal and international ties, thus, providing outer and inner mobility of the educational establishment (official information sites, personal accounts at various systems, search systems, catalogues, blogs, etc.).

The special attention should be paid to Tele-Communication and Tele-Education as the innovative methods in vocational training in Ukraine. Primarily their implementation proposes opportunities for enhancing scientific competence as well as information one [15]. Web conferences and Skype help to make the communication interactive and allow all the applicants to participate during process. Besides, a self analysis becomes the inherent element of interaction.

Bright illustration of providing the information and computer technology into the academic process of institutions training nurses is the system “Intranet”.

The program “Intranet” has been functioning for several years in Zhytomyr Nursing Institute and definitely proved its positive influence onto the learners’ outcome [14]. This educational provider is among the first in Ukraine among institutions training nurses that has offered a stable approach in delivering learning information for students and teaching staff based on introducing ICT as the component of inner quality system (Fig. 1). The system is oriented at delivering the full pack of academic materials to learners and there through give access to learning resources from anytime and anywhere just when students need it so to make and deliver tasks. The learners have

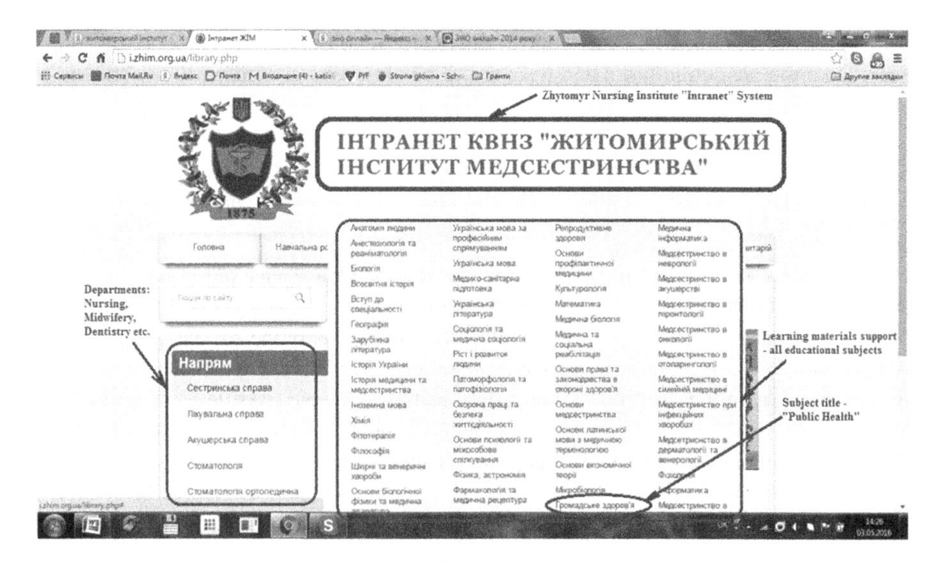

Fig. 1. The system “Intranet” – home page

possibility to complete everyday educational transactions at any convenient time. In such a way the given product helps to form both basic and special nurses' competences that result professional competence at future specialist.

The system is rather unsophisticated and easy for learners to use. The title of the system – Zhytomyr Nursing Institute "Intranet" System is at the center of the home page so to grasp and focus users' attention. Information is classified into thematic blocks: home page, academic programs, practice, electronic repository, regulatory work etc. to simplify information search. The freestanding filters are institution departments: Nursing, Midwifery, Dentistry, etc. They filter educational information according to the training area, thus, only the materials adequate to that study area. In the center, as the search result, the number of learning materials to every subject is given e.g. "Public Health" and many others.

Electronic repository (Fig. 2) comprises all electronic materials, handouts, text-books, course-books, laboratory manuals on certain subjects, thematic areas within peculiar specialty (Nursing, Midwifery, Dentistry, etc.), e.g. Anatomy, Physiology, Gynecology, Humanities, Medical Biology, Microbiology.

The next Fig. 3 presents the pack of Performance-Based Assessment Tests for students majoring at Master's degree programme as well as Nursing, Midwifery, Medical Care departments. They can be loaded or practiced on-line at home or at the institute. It helps to monitor students learning activity and outcomes. The resulted data are systematized and analyzed further by the Academic Affairs Department to get objective information; data based terminal and annual reports.

The Fig. 4 delivers general mechanism of loading necessary learning materials illustrated by working programme and learning pack. Rightward window presents the loaded pack. It consists of: situational tasks, practical algorithms, Nursing protocols and subject textbook.

Fig. 2. Electronic repository

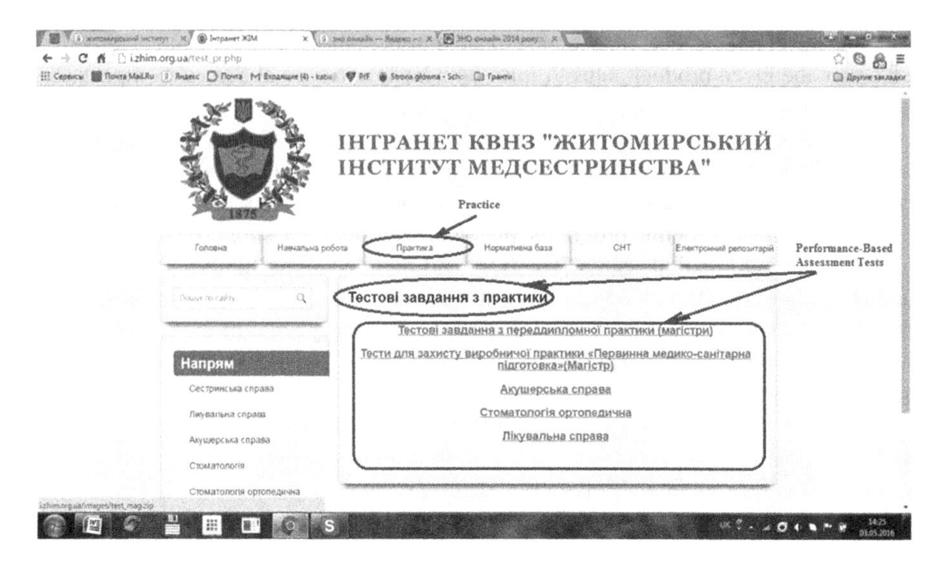

Fig. 3. Performance-based assessment tests

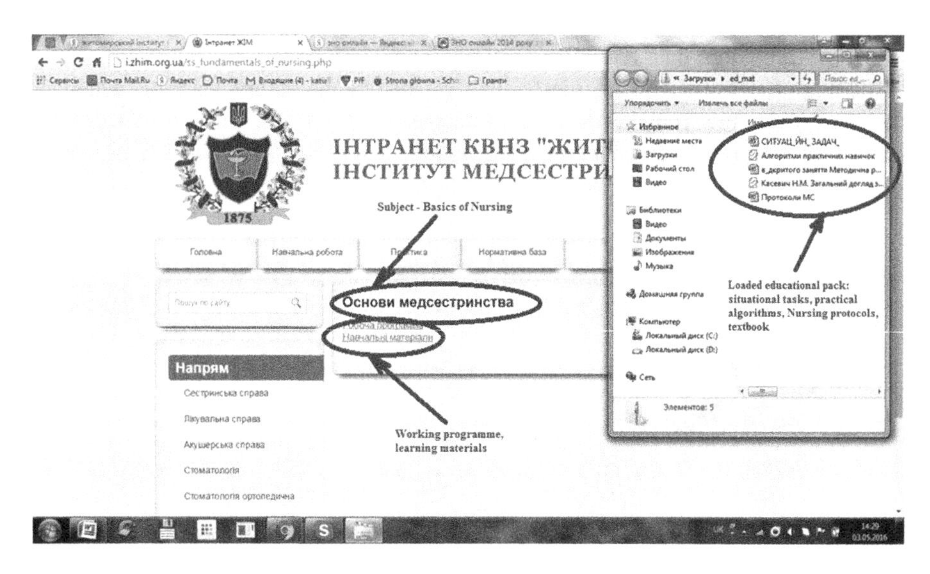

Fig. 4. The full-pack of learners' materials – subject "Basics of Nursing"

Submitted system is rather simple in users-friendly. It serves as the educational means and synchronously the learning outcome – skills of free work in the information systems and with the information and computer technologies.

Still, wide applying of ICT within the educational process, while forming nurses' professional competence, has a number of positive and negative characteristics.

The positive ones include:

- Individualization in learning;
- Information packing;
- Firm motivation;
- Close interconnection between practice and theory;
- Monitoring students' cognitive activity;
- Forming students' creative features;
- Training and learning activity differentiation;
- Modeling real situations;
- Creating necessary emotional state;
- Contribution to forming skills of individual work;
- Developing students' self-reflection and self analyses.

The negative features represent:

- Students' isolation;
- Sometimes absence of student-teacher interaction;
- Health and nervous breakdowns;
- Internet and computer addiction.

Unquestionably, the ICT in education have a good deal more advances but the negative ones set alarm bells ringing. Health and nervous breakdowns become the disturbing factors in increasing frequency. Students spend more and more time with the computers or M-devices surfing the internet and become internet and computer addicted.

4 Conclusions

Strategies for the use of information and communication technologies in academic process are aimed at supporting quality learning and quick delivering learning materials and, thus, facilitating student-teacher interaction.

The popularity and interactivity of information and computer technologies are among the factors that empower learners and transform the traditional learning into the untraditional one. The educational providers view ICT as an important part of their mission and commitment to addressing and supporting the students' educational needs.

Thus, the use of ICT in the system of nurse training in Ukraine will allow:

- upgrading the training content and form the training process organization substantially and prospectively;
- setting innovative educational goals;
- reaching them persistently and rapidly;
- making the interactive academic process;
- creating firm learners' motivation;
- developing students' self-analyses;
- contributing to quality learning.

References

1. Bates, D.W., Gawande, A.A.: Improving safety with information technology. New England J. Med. **348**, 2526–2534 (2003)
2. Bykov, V.Ju.: Theoretical and methodological basis of modeling learning environment of modern educational systems. In: Proceedings of Scientific Papers Information Technologies and Educational Methods APS of Ukraine, Kyiv, Atika (2005). (in Ukrainian)
3. Bukovinian State Medical University. http://www.bsmu.edu.ua/
4. Elaboration of the Educational Programmes: Methodological Recommendations. http://ihed.org.ua/images/biblioteka/rozroblennya_osv_program_2014_tempus-office.pdf. (in Ukrainian)
5. Horbachevsky Ternopil State Medical University. http://www.tdmu.edu.ua/en/
6. Kademiia, M.Ju., Shahina, I.Ju.: Information and Communication Technologies in the Academic Process. Planer, Vinnitsa (2011). (in Ukrainian)
7. Law of Ukraine, About Higher Education, 01.07.2014 №1556-VII. http://www-library.univer.kharkov.ua/pages/zakon/1556-VII.pdf. (in Ukrainian)
8. Order of the Cabinet of Ministers of Ukraine from 30.12.2015 №1187, On Approving License Requirements as for Terms of Introducing Educational, Activities by Educational Institutions. http://zakon3.rada.gov.ua/laws/show/1187-2015-%D0%BF. (in Ukrainian)
9. Vinnitza Pirogov National Medical University. http://www.vnmu.edu.ua/
10. Shygonska, N.: Educational administrative situations in the system of nursing management. In: Proceedings of the International Expert-Scientific Conference, pp. 296–310. Heraldika Kiado, Budapest (2015)
11. Shygonska, N.: Zhytomyr nursing institute within Ukrainian education system. In: Mroczkowska, R., Molka, W. (eds.) Współczesne Pielęgniarstwo Specjalistyczne: Wiedza, Kompetencje, Praktyka, Bytom, Poland, pp. 219–229 (2015)
12. Staggers, N., Thompson, C.B.: The evolution of definitions for nursing informatics: a critical analysis and revised definition. J. Am. Med. Inf. Assoc. **9**(3), 255–261 (2002)
13. TUNING – http://www.unideusto.org/tuningeu/. (in Ukrainian)
14. Zhytomyr Nursing Institute. http://zhim.org.ua/
15. Zollo, A.S., et al.: Tele-education in a telemedicine environment: implications for rural health care and academic medical centers. J. Med. Syst. **23**(2), 107–122 (1999)
16. Wutoh, R., Boren, S.A., Balas, E.A.: e-Learning: a review of internet-based continuing medical education. J. Continuing Educ. Health Prof. **24**(1), 20–30 (2004). doi:10.1002/chp.1340240105

FEM Modelling and Thermography Validation of Thermal Flow

Alicja Praczukowska[1](✉), Michał Nowicki[2], and Jakub Pełka[3]

[1] Industrial Research Institute for Automation and Measurements PIAP, Al. Jerozolimskie 202, 02-486 Warsaw, Poland
alicjapraczukowska@gmail.com
[2] Institute of Metrology and Biomedical Engineering, Warsaw University of Technology, Św. Andrzeja Boboli 8, 02-525 Warsaw, Poland
nowicki@mchtr.pw.edu.pl
[3] Institute of Micromechanics and Photonics, Warsaw University of Technology, Św. Andrzeja Boboli 8, 02-525 Warsaw, Poland

Abstract. Heat flow is one of the major technological issues encountered during the design, fabrication and exploitation of almost every element and device. The design process of such an item requires a proper modelling method in order to predict spatial and temporal heat distribution. The most suitable solution to resolving this issue is an open-source software: Elmer FEM. It utilizes Finite Element Method for multiphysics calculations and analysis, including heat distribution. The possibility of straightforward custom solver creation is remarkable for this free software. The natural consequence of applying self-designed solver algorithms is the necessity of result verification. The following paper presents a comparison of simulated and measured heat distribution. The primary test sample was a metal solid of a complicated geometry in which a significant heat flow was induced. The simulation data acquired via Elmer FEM is presented against the results given by a thermographic measurement utilizing high-resolution FLIR camera. After taking environment influence into account, the correlation of the results given by both methods justifies using Elmer FEM with proposed solver as a viable and credible solution for determining the distribution of heat flow.

Keywords: Finite element method · Heat flux · Temperature distribution · Infrared measurement · Elmer simulation

1 Introduction

Temperature plays a key role in many engineering applications. It changes most of the basic material properties, including stiffness, electrical resistance and general durability (see Fig. 1). For this reason, calculating accurate temperature distribution is vital during the design process and further exploitation. Unfortunately, in most cases manual calculations are either impossible or too complex to conduct due to differential character with many boundary conditions. An obvious solution would be a computer simulation. In order to use this tool however, we need to prepare a numerical model, recreate

R. Szewczyk and M. Kaliczyńska (eds.), *Recent Advances in Systems, Control and Information Technology*, Advances in Intelligent Systems and Computing 543, DOI 10.1007/978-3-319-48923-0_18

object's geometry and validate the results. This article shows how to conduct a validation process and how adequate is the model to the simulation.

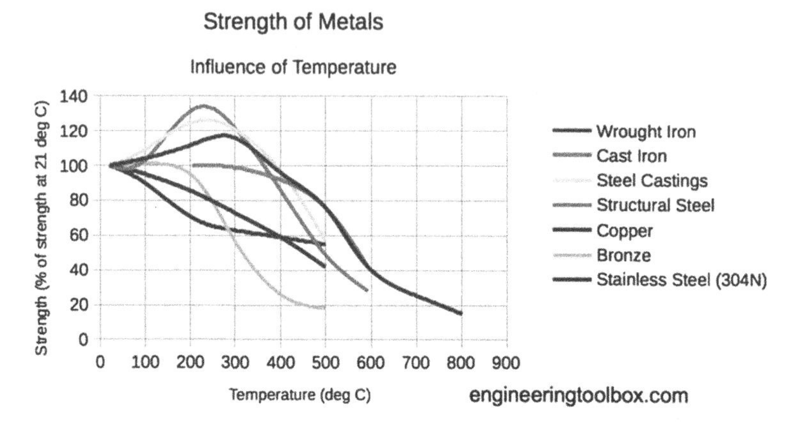

Fig. 1. Material strength change caused by temperature [1]

2 Experimental Setup

The first step of validation process was preparing the real object for measurement to which a simulation model would be compared. The examined sample was an aluminum plate of 20 mm × 200 mm × 300 mm dimensions (Fig. 2). Various circular and linear holes, as well as a 100-milimeter-long slit were drilled through it. The surface was scratched using sandpaper and then covered with uniform layer of black, dull, temperature-resistant paint.

Fig. 2. Aluminum plate with black matte coat

In order to achieve thickness independence, the heating element (in this case a resistance heating wire) was placed inside the internally insulated copper tube going through the plate. The heating wire was powered and controlled by LPS-305 power supply. For additional thermal insulation, the plate was placed on a Styrofoam block (see Fig. 3).

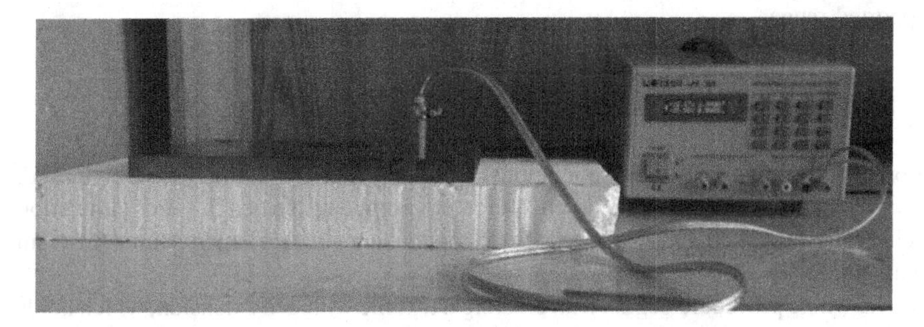

Fig. 3. Test stand

3 Thermography

Among various methods for measuring temperature distribution infrared cameras become most popular. They offer quick and relatively easy way of mapping temperature distribution in the element. However they require some additional preparation.

Infrared cameras do not measure temperature directly but rather a combination of multiple parameters. The detector receives IR radiation containing both object emission spectrum and some reflected or ambient radiation both of which are additionally filtered by air spectral absorption (see Fig. 4). The temperature is calculated using these values and known emissivity coefficient. For example, a smooth metal surface would act as a mirror and wouldn't allow for measuring its temperature that way. This is why we used a dull black paint to cover the examined surface.

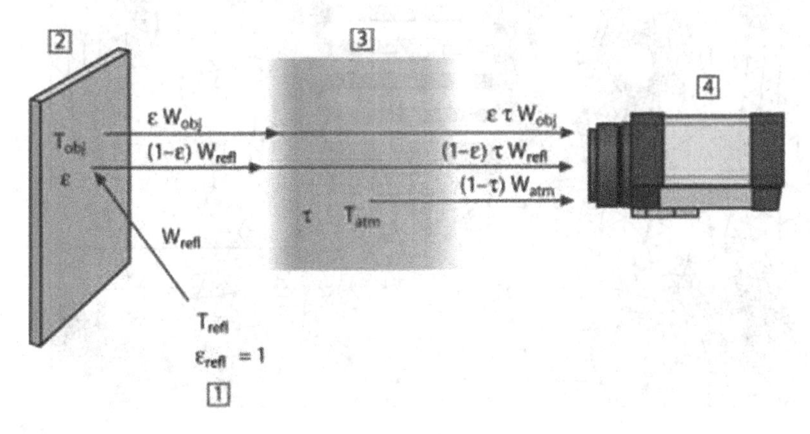

Fig. 4. General schematic of IR measurement: 1 – surrounding, 2 – object, 3 – atmosphere, 4 – IR camera [3]

Taking above into regard, IR Camera measurements tend to have relatively large errors due to detector noise, imprecise emissivity coefficient, element and background reflections and atmosphere spectral absorption [2].

In this case we used FLIR E60 IR Camera along with official firmware. It has an uncertainty of temperature measurement declared as up to 2 % of absolute temperature value, detector matrix of 320 × 240 pixels, sensitivity of 0.05 K and –20 °C–650 °C supported range.

4 Simulation

In order to conduct necessary simulations, an appropriate digital model was required. The meshes were created in the open-source environment: Netgen 5.3. For simulation itself Elmer FEM software was used. Both applications are free, open-source program compatible with all popular operating systems.

The whole procedure begins with creating a text file with.geo extension, which holds the definition of element/surrounding geometry. It is further used to generate meshes in Netgen. The surrounding was defined as a cylinder of 2 m radius and height of 2 meters.

After uploading the definition of geometrical boundary conditions the discretization of the model needs to be performed. In order to generate the mesh, Netgen uses the Delaunay algorithm [4]. The visible change in size of fragments shows that the density of the mesh increases proportionally to the radius of the corner (see Fig. 5). The number of elements was equal to 1563286. It should be noted that the simulation has been conducted with the precision set to 'very fine', what appeared to be sufficient in current examination.

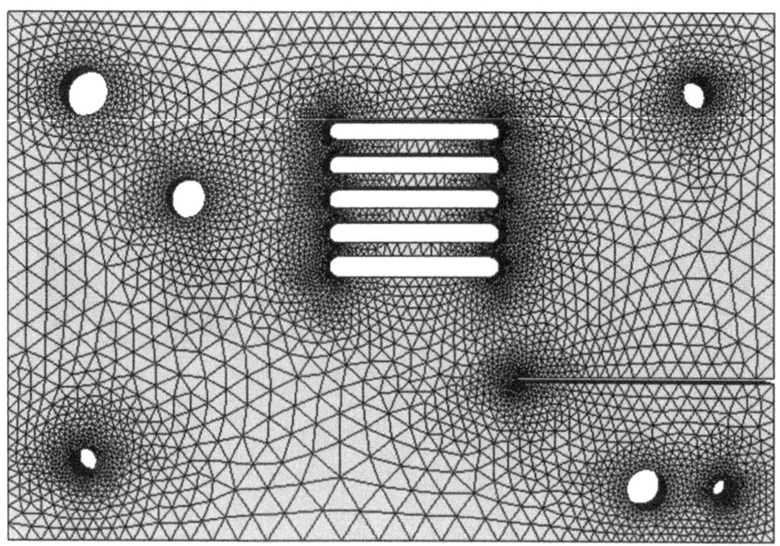

Fig. 5. Simulation model of examined plate generated in Netgen

Temperature distribution was calculated with the use of provided Heat Solver [5] applied to prepared mesh. The surrounding was defined as air of constant room temperature (20 °C) along its outer faces. The plate was made of aluminum which properties are shown on Fig. 6.

```
Material 1
  Name = "Air (room temperature)"
  Heat Conductivity = 0.0257
  Heat Capacity = 1005.0
  Density = 1.205
  Relative Permittivity = 1.00059
  Viscosity = 1.983e-5
  Sound speed = 343.0
  Heat expansion Coefficient = 3.43e-3
End

Material 2
  Name = "Aluminium (generic)"
  Heat Conductivity = 237.0
  Youngs modulus = 70.0e9
  Mesh Poisson ratio = 0.35
  Heat Capacity = 897.0
  Density = 2700.0
  Poisson ratio = 0.35
  Sound speed = 5000.0
  Heat expansion Coefficient = 23.1e-6
End
```

Fig. 6. Properties of air and aluminum

In order to perform the calculation, heat flux needed to be defined for aluminum-air contact faces. It was determined with partial help of FLIR measurement data along X axis and then verified with distribution along Y axis. Both axes were located in the center of plate and were parallel to the longest edges of test plate.

The simulation was conducted for two different heater locations. Heat flux needed to be calculated separately for each of them and varied.

5 Measurement Results

The images below present temperature distributions along X and Y axes on plate surface. The yellow line contains simulation data, while the blue one shows measured values (see Figs. 7, 8, 9 and 10). During the heating of upper hole its inner face temperature was 51 °C and during the heating of bottom hole –42 °C.

Heat flux was determined using the measured data along X axis (horizontal) and then verified with Y axis (vertical) results. In the first case, general heat flux was defined as –0.35 W/(mK) and for the other as –0.2 W/(mK).

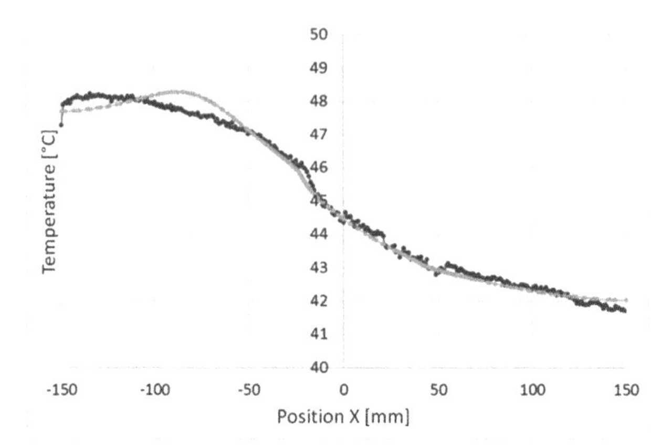

Fig. 7. Comparison of simulated and measured temperatures for top hole along X axis

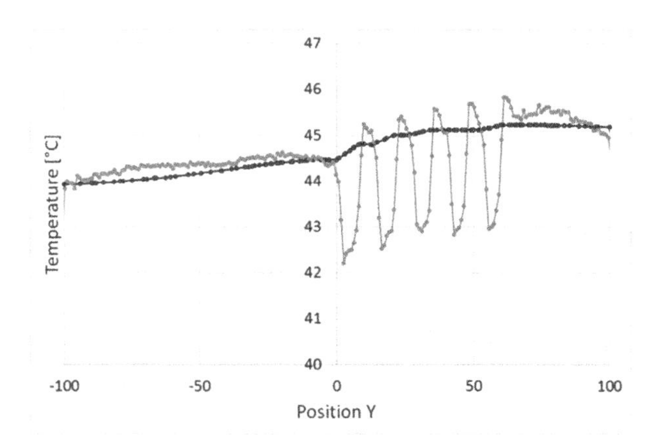

Fig. 8. Comparison of simulated and measured temperatures for top hole along Y axis

Both holes were heated separately. After each measurement the element was left to cool down until uniform room temperature (20 °C) was observed on the whole surface. Measured temperature values data were acquired with FLIR Tools application [6].

Temperature distribution maps received from Elmer and FLIR used slightly different color palettes, not corresponding to each other. In order to prevent misinterpretation, the simulation color map type was left unchanged and FLIR type was switched to iron scale. Temperature distribution maps are shown on Figs. 11, 12, 13 and 14. Due to different coloring each figure has a color scale included.

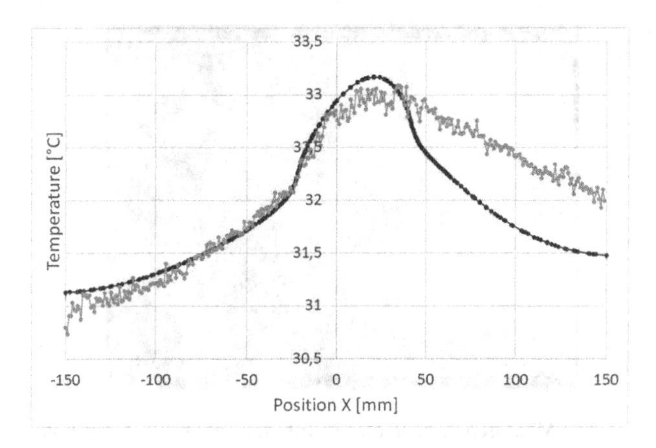

Fig. 9. Comparison of simulated and measured temperatures for bottom hole along X axis

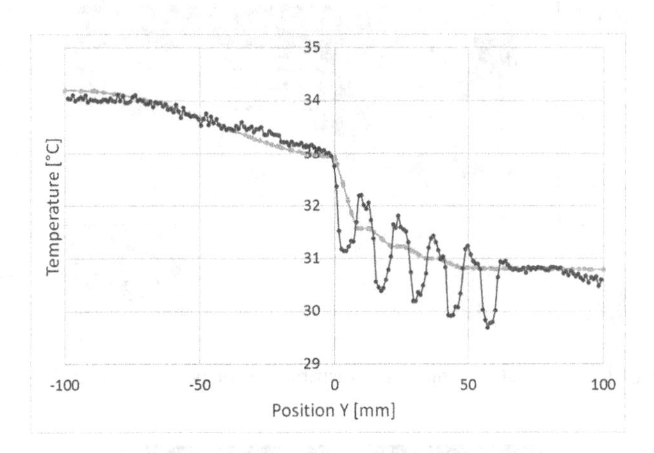

Fig. 10. Comparison of simulated and measured temperatures for bottom hole along Y axis

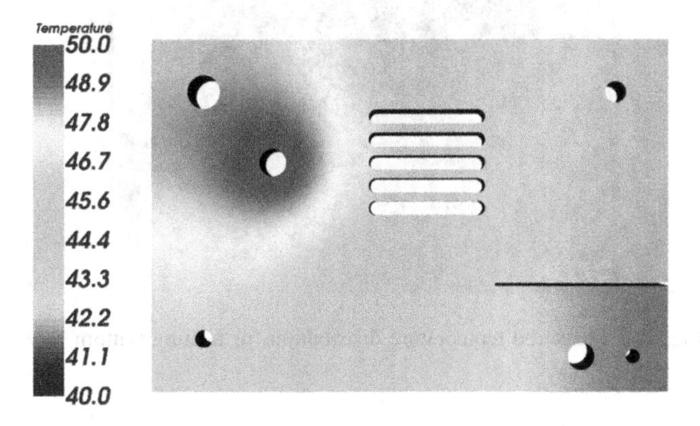

Fig. 11. Simulated temperature distribution for heating top hole

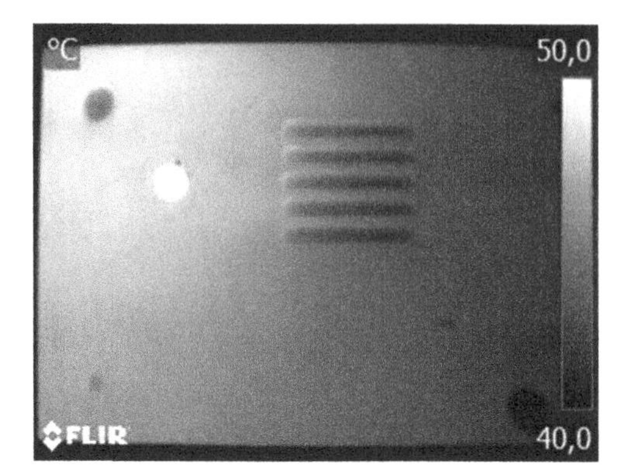

Fig. 12. Measured temperature distribution for heating top hole

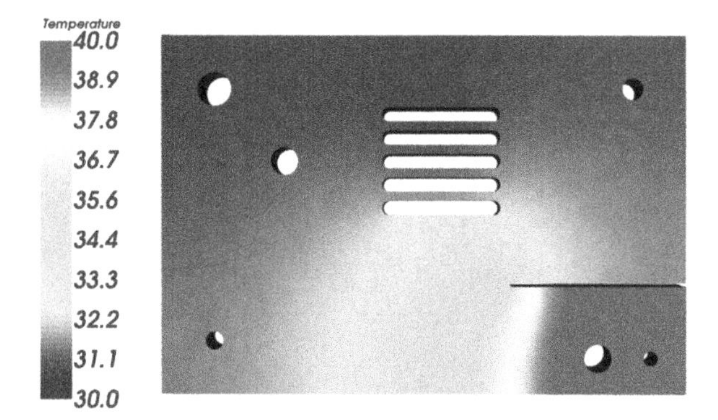

Fig. 13. Simulated temperature distribution for heating bottom hole

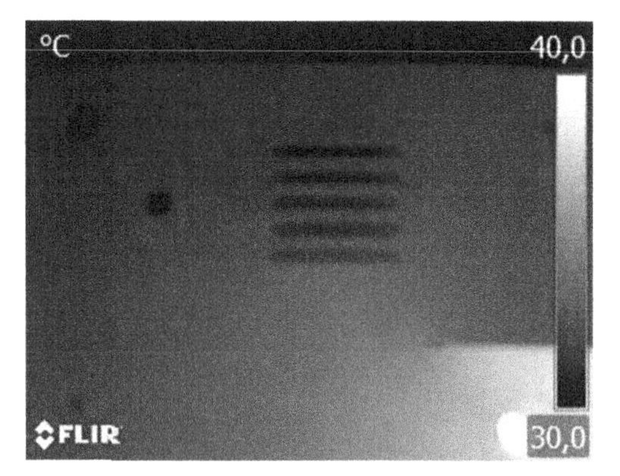

Fig. 14. Measured temperature distribution for heating bottom hole

6 Sources of Error and Signal Error Analysis

There were a few possible error sources:

- Imprecise simulation coefficient values like heat flux,
- Imprecise selection of X and Y axes, which need to be set manually in FLIR Tools,
- Measurement errors – assumed emissivity coefficient, slightly varying room temperature, photograph perspective, etc.
- Element geometry errors,
- Not ideal paint coating.

7 Conclusion

Measured temperature distribution on the surface is similar to simulated one. It applies in both simple and complicated geometry. Heat distribution near the narrow slit was marked as the potential source of the largest errors. In fact, its impact was smaller in realty than in simulation, what can be seen on Fig. 9.

It is important to notice irregular 'sawtooth' characteristic along Y axis. It is caused by IR camera which is unable to measure air temperature properly. Nevertheless, remaining part of the graph corresponding to material between holes, has regular shape adequate to simulation.

The general conclusion is that after initial difficulties with configuring the simulation, it can be easily applied to less complicated geometrical cases. Utilizing such a method for more complex issues requires detailed configuration, including precise calculations of heat flux density for each face, increasing simulation parameter values and mesh level of detail.

References

1. http://www.engineeringtoolbox.com/
2. Jozwicki, R., Wawrzyniuk, L.: Technika Podczerwieni. OWPW, Warszawa (2014)
3. User manual FLIR Exx Series
4. Jurczyk, T.: Generowanie niestrukturalnych siatek trójkątnych z wykorzystaniem triangulacji Delaunay'a. Praca magisterska. WEAIE AGH, Kraków (2000)
5. Raback P.: Elmer Models Manual. CSC 2014 (2014)
6. FLIR. http://www.flir.eu

Numerical Analysis and Validation of the Human Impact on the Conditions in Model Chamber

Weronika Radzikowska-Juś[1](✉), Maciej Szudarek[2], Andrzej Juś[3], and Stefan Owczarek[1]

[1] Faculty of Civil Engineering and Geodesy, Military University of Technology, gen. Sylwestra Kaliskiego 2, 00-908 Warsaw, Poland
weronika.radzikowska-jus@wat.edu.pl
[2] Faculty of Mechatronics, Warsaw University of Technology, sw. A. Boboli 8, 02-525 Warsaw, Poland
[3] Industrial Research Institute for Automation and Measurements, Al. Jerozolimskie 202, 02-486 Warsaw, Poland

Abstract. The paper presents issue of the human impact on the conditions in the room. It focused on temperature changes in time. The importance of the issue stems from the necessity of ensuring thermal comfort for human inside the room. So the theory of thermal comfort and the phenomena determining the pace and the way of heat exchange between man and the environment were the basis for discussion. After the analysis of the phenomena related to heat transfer the model of the heat exchange, created in ANSYS 13.0 academic research software based on FVM, was presented. In particular, the model is taking into account the way of heat exchange between man and the environment, and between the layers of the room (environment). Different properties in terms of thermal conductivity of various materials and convection phenomena were taken into account. To validate and optimize the developed model the phenomenological tests were conducted. The research relied on observation of the temperature changes in selected zones of the room after entry of a man into it. Presented model is the basis for further analysis of the human's impact on the conditions in the shelter according to the conditions beyond it.

Keywords: Heat transfer · Convection · FVM modelling · Thermal comfort

1 Introduction

In recent decades, the quality of life and the expectations of people, especially in the countries of high and very high Human Development Index [1, 2], quickly grew. Such growth of expectations also covered the requirements for conditions in indoor spaces, where nowadays people spend major part of their lives. However, up until now this process has not been applied a special type of indoor spaces, which are shelters. Normative documents concerning the requirements for the conditions in shelters are outdated, as often the objects themselves. That is why taking up the issue and redefining requirements, appropriate to the present time, is so important.

R. Szewczyk and M. Kaliczyńska (eds.), *Recent Advances in Systems, Control and Information Technology*, Advances in Intelligent Systems and Computing 543, DOI 10.1007/978-3-319-48923-0_19

The issue undertaken in the paper is influence of human on the conditions inside model chamber of small volume. Results will be useful for further work on the formulation of requirements for civilian shelters. For this purpose, a numerical model using software ANSYS 13.0 was developed and its results validation was performed.

2 Principles of the Phenomenon

2.1 Heat Transfer Principles

The main ways of heat transfer are convection, conduction and radiation. Heat conduction based on transfer of the heat by chaotic collisions of particles and by diffusion of free electrons. Heat conduction occurs mainly in solid bodies. Convection occurs only in fluids (mainly liquids and gases). It relies on moving heat energy by heat flow or mix of fluid streams. Convection may be forced by external influence or be natural (caused by gravitational force). Thermal radiation is the transfer of energy by photons in electromagnetic waves in the band of thermal energy transfer [3, 4].

Heat transfer may be steady or unsteady. In the first case, the temperature of the various points is steady over time opposed to unsteady heat flow [3, 5]. Heat transfer model based on the law of conservation of energy and Fourier's law to consider convection. Such equations make possible to derive the final form of Fourier-Kirchhoff's equation for systems without phase transition [5]:

$$\rho c_p \frac{DT}{D\tau} = \nabla \bullet (\lambda \ \nabla T) + \frac{Dp}{D\tau} + \Phi + \dot{q}_v \tag{1}$$

where:

ρ - density,
c_p - specific heat capacity,
T - temperature,
τ - time,
∇ - del operator (nabla) – vector differential operator,
λ - heat conductivity,
$\bullet$ - dot product,
P - pressure,
Φ - dissipation density over the volume,
$\dot{q}_v$ - volumetric efficiency of internal heat sources

When such equation is used for turbulent thermal flow the Reynolds assumptions on averaging and decomposition of instantaneous values have to be taken into account [5]. Obtained the following form of the equation [5]:

$$\rho c_p \frac{D\overline{T}}{D\tau} = \nabla \bullet \left(\lambda\ \nabla\overline{T}\right) + \frac{D\overline{p}}{D\tau} + \Phi + \dot{q}_v - \rho c_p \left(\frac{\partial \overline{w'T'}_x}{\partial x} + \frac{\partial \overline{w'T'}_y}{\partial y} + \frac{\partial \overline{w'T'}_z}{\partial z} \right) \quad (2)$$

where:
T' - fluctuational deviation from average value of temperature,
w' - fluctuational deviation from average value of velocity

Because numerical solver based on FVM method and geometry is axisymmetric these Eqs. (1, 2) are transformed to cylindrical coordinates and formulas are simplified according to the algorithm presented in [5, 6].

To model the heat transfer equation of mass balance and Navier-Stokes momentum equation are also used. However, because of the largest role of Fourier-Kirchhoff's Eq. (2) was presented above while description of the other two equations can be found in the literature [3–7].

2.2 Heat Exchange Between Human and Environment and the Issue of Thermal Comfort

Understanding of processes of heat exchange between a human body and the environment is the basis for design of the microclimate in rooms. Heat exchange in a human-environment relation stems from correlations between many factors involved in shaping of human bodies' heat balance. Its main task is to maintain a constant internal body temperature. Some of such factors are: energy expenditure, resistance of heat conduction through clothes, air temperature, average temperature of radiation, vapor pressure and air velocity, which is included in the thermal balance equation. But, especially when a man is dressed, great part of warm air is carried away from its vicinity by convection [8–10].

Thermal comfort is a state when a man feel satisfaction of surrounding environments thermal conditions, so the man feels neither too cold nor too warm [11]. Determination of thermal comfort generates a lot of problems, because the feeling of comfort is highly personalized (e.g. different for men and women) and subjective (e.g. depending on the psychical condition). Therefore it comes down to seek conditions, in which the greatest number of people feel comfortable [11, 12].

The publications [11–14] on issue of the thermal comfort of people, identify three different approaches to it. First one treats man as a physical object analogous to a machine generating heat and carrying out some actions. Second approach is called a psychological and it consists on different kinds of tests and questionnaires. The third approach is based on thermophysiology. In this case research methods are based on direct or indirect human calorimetry, which allow to precisely determine the states of thermal stresses.

2.3 Natural Convection and Rayleigh Number

As it was described in Sect. 2.2 convection is the main way of removing warm air from the vicinity of a man. If there is no ventilation or other external movement of air it is approximately natural convection. The term which determines if natural convection is turbulent or laminar is Rayleigh number, *Ra*. It is a dimensionless number defined as the product of the Grashof and Prandtl numbers. Grashof number describes the relationship between buoyancy and viscosity of a fluid. While Prandtl number describes the relationship between thermal and momentum diffusivities. Equation of Rayleigh number is [15]:

$$Ra = \frac{g\beta \Delta TL^3 \rho}{\mu \alpha} \tag{3}$$

where:

g - gravitational acceleration,
β - thermal expansion,
ΔT - temperature difference,
L - characteristic dimension,
ρ - density,
μ - kinematic viscosity,
α - thermal conductivity

If Rayleigh number is lower than 10^8 then flow is laminar. While for values over 10^8–10^{10} occurs turbulences [15].

During numerical solving of natural convection often it is appropriate to use Boussinesq Model. This is a simplification that excludes variability of fluid density except of the buoyancy term in the momentum equation. It can be used if differences of temperatures in object aren't large [16].

3 Developed Model and Results

3.1 Numerical Model

The aim of the simulation is testing the human impact on the conditions in the model room. Simulations are performed using ANSYS 13.0 software. Model chamber is cylinder of 140 cm diameter and 250 cm height. Inside it smaller cylinder of 50 cm diameter and 130 cm height is axially located – it represent a man. Because of axisymmetric of the geometry simulations were performed in a model 2D axisymmetric. Geometry was meshed of 11759 rectangular elements. Mesh was concentrated near borders of the objects. Schematic diagram of the geometry with marked boundary conditions is presented in Fig. 1.

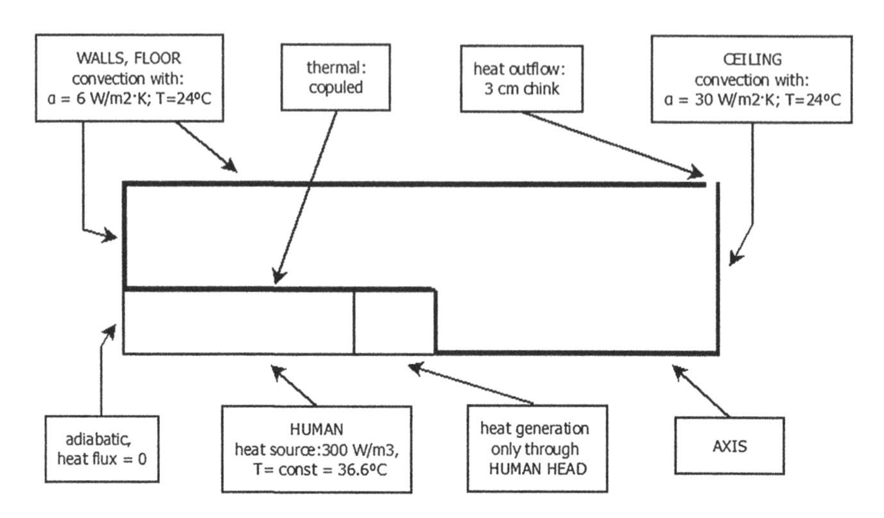

Fig. 1. Geometry with marked boundary conditions

In the simulations, it was assumed that the heat is generated through human head (this is justified because the man gives a lot of the heat by breathing). Furthermore, in the upper part of the chamber, near the ceiling, assumes the existence of a 3 cm chink. Such chink occurs during empirical tests.

To evaluate whether the heat flow is laminar or turbulent Rayleigh number *Ra* was determine according to the formula (3). To calculations assumes following values (correct for air) [17, 18]: g = 9.81 m/s^2, β = 0.003125 K^{-1}, ΔT = 36.6 °C – 24 °C = 12.6 °C (human body temperature minus ambient temperature), L = 1.56 m, ρ = 1.225 kg/m^3, μ = 1.7894·10^{-5} kg/m·s, α = 0.0242 W/m·K. Determine value of *Ra* is:

$$Ra = \frac{g\beta \Delta T L^3 \rho}{\mu \alpha} = \frac{9,81 \cdot 0,003125 \cdot 12,6 \cdot 3,848}{1,7894 \cdot 10^{-5} \cdot 0,0242} \approx 3,4 \cdot 10^6 < 10^8,$$

So heat flow is laminar.

In Table 1 summarized simulation settings. And in Fig. 2 shows the results of simulation after varying time of its launch.

Table 1. Summarization of simulation settings

Simulation settings	
Software	Fluent 13 academic research, double precision, unsteady solution, axisymmetric
Models	Viscous – laminar, Energy equation
Time step	0.4 s
Material 1, air	Convection modeled by Boussinesq approximation; Boussinesq temperature set as an average between human (36.6 °C) and air (24 °C), that is 30.3 °C

(*continued*)

Table 1. *(continued)*

Simulation settings	
Material 2, man	User-defined material [19]: - specific heat: 3470 J/kg °C - density: 1010 kg/m^3 - thermal conductivity: 0.40 W/m/°C
Gravitation	X = –9.81 m/s^2
Boundary conditions	As stated on Fig. 4
Relaxation factors other than default	Pressure: 0.7 Pa Momentum: 0.3 kg·m/s
Initial conditions	T = 24 °C, axial velocity = 0.01 m/s (for the convection to kick in), human body temperature 36.6 °C

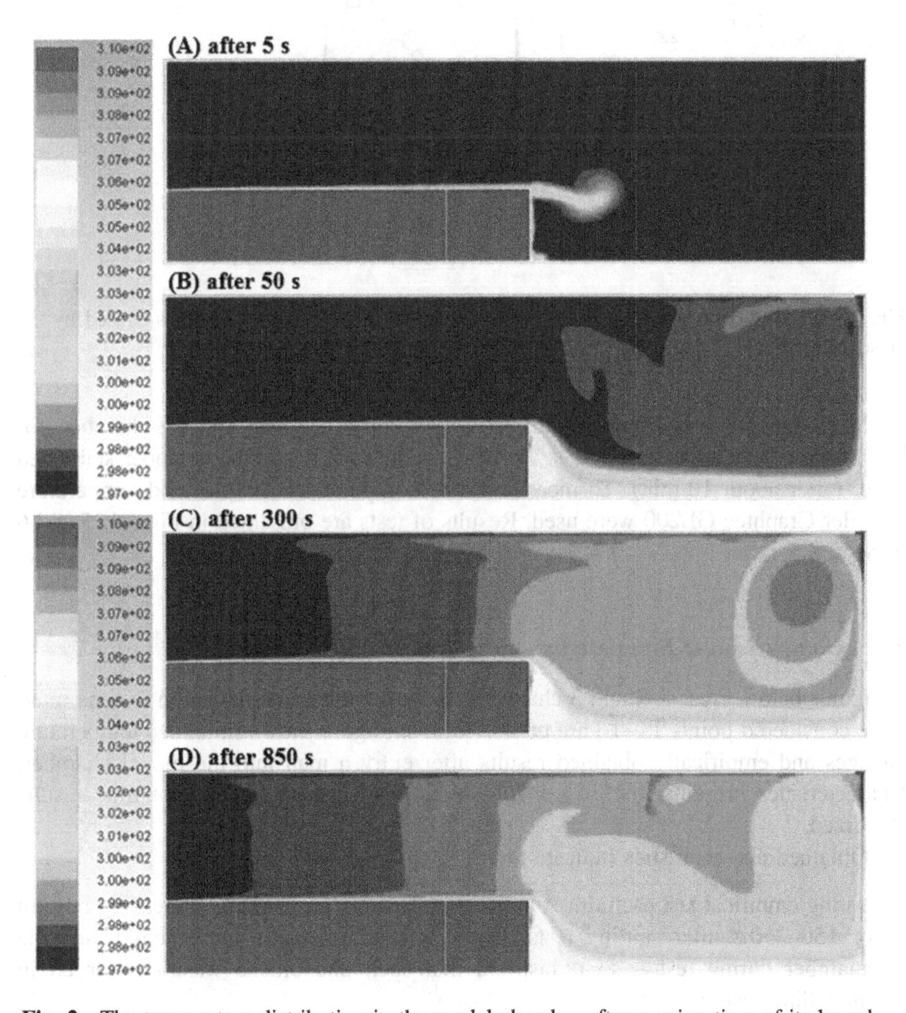

Fig. 2. The temperature distribution in the model chamber after varying time of its launch

3.2 Empirical Model

To validate the results of simulations presented in Sect. 3.1 performed empirical tests. Inside model chamber (its dimensions are same as in simulated geometry and are presented in Fig. 3) a man was located. It is sitting the central part of the chamber. Temperature measurements conducted in three points (T1–T3) on different heights: T1 – 220 cm, T2 – 160 cm, T3–70 cm, at a distance of 10 cm from the wall of the room.

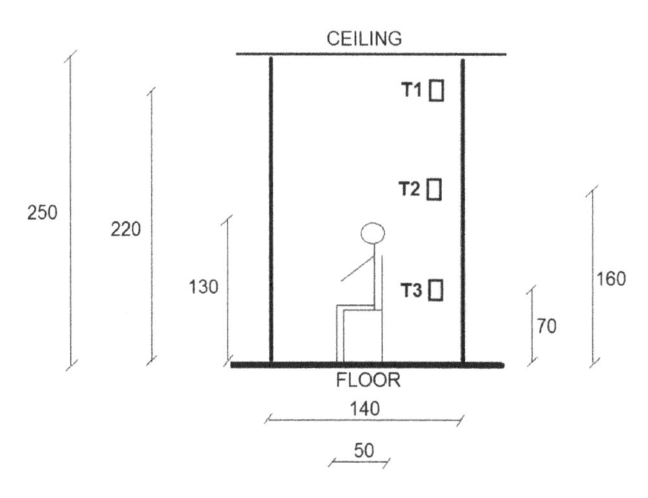

Fig. 3. Diagram of test bench during the measurement of temperature changes caused by the presence of man

Measurements was conducted, directly after entry of a man into the chamber, for half an hour with time step of 1 s. However conditions inside the chamber stabilized faster (after about 10 min). To measurements K-type thermocouples and temperature recorder Graphtec GL200 were used. Results of tests are presented in Figs. 4, 5 and 6 in Sect. 3.3.

3.3 Comparison of Simulations and Empirical Tests Results

In figures below Figs. 1, 2 and 3 characteristics of temperature deviation in time in all three considered points T1–T3 are presented. Each figure shows simulated temperature changes and empirically obtained results after entry a man into the model chamber. Characteristics presents first 15 min of the results – because after that time results stabilized.

Obtained characteristics indicate that:

- during empirical test each time observed collapse of temperature increasing in term of 150–250 s after starting tests. It was related to man's movement inside the chamber during test – so it justified that such phenomena wasn't observed in modelling results,

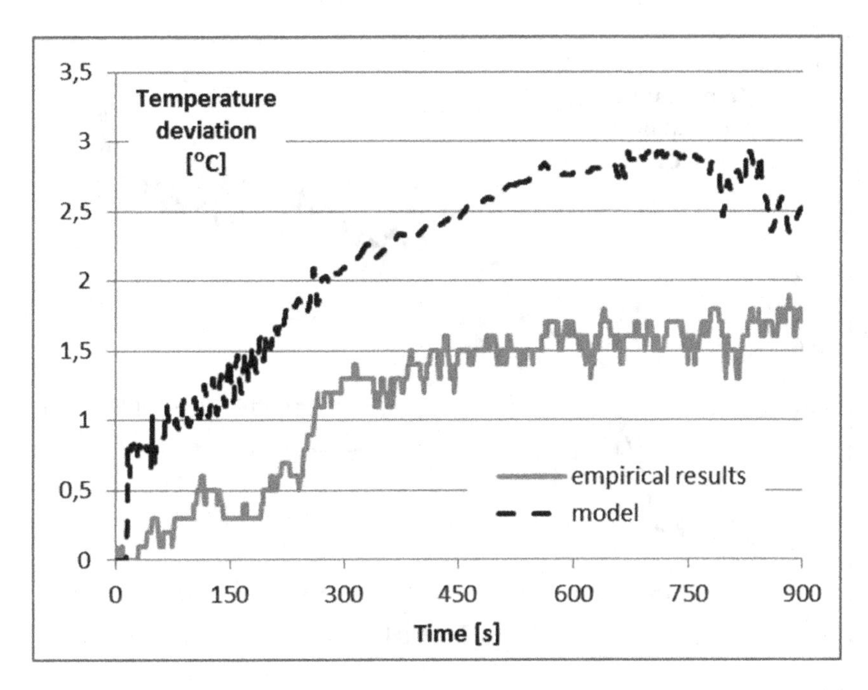

Fig. 4. Changes of temperature in time obtained directly after entry of a man into model chamber: empirical and model results in T1

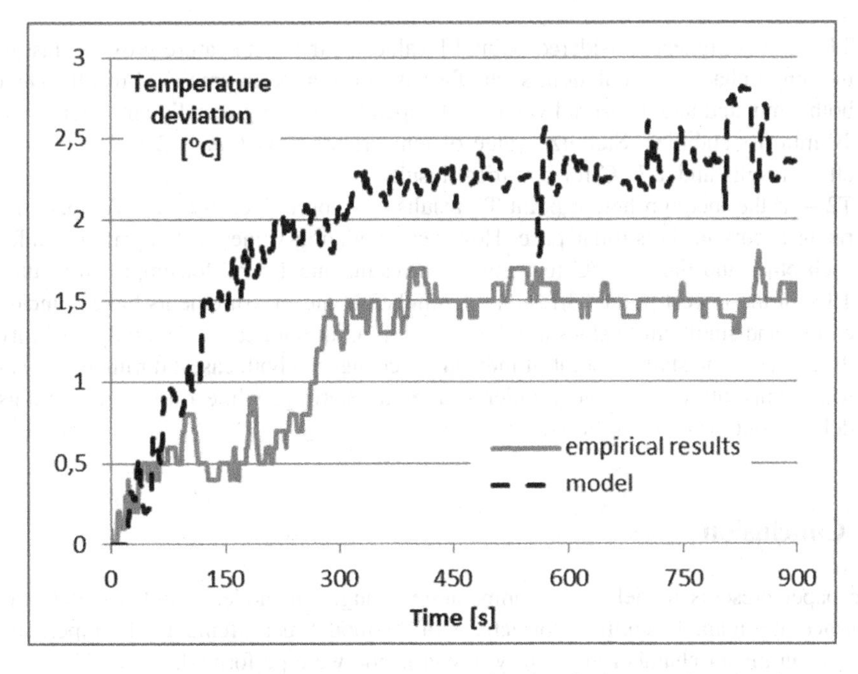

Fig. 5. Changes of temperature in time obtained directly after entry of a man into model chamber: empirical and model results in T2

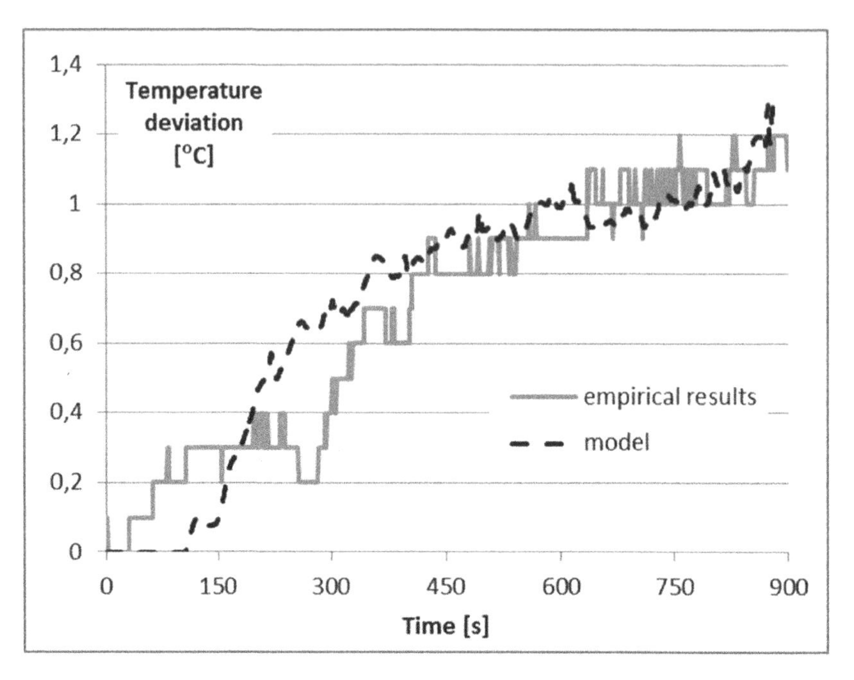

Fig. 6. Changes of temperature in time obtained directly after entry of a man into model chamber: empirical and model results in T3

- T1 – in the highest considered point T1 values of the temperature growing fast in the initial phase of simulation, such effect isn't observed in empirical results. Next both simulated and measured values of temperature rising gradually and after about 10 min it's stabilize. Stabilize value of temperature deviation is 2.8 °C for modelling results and 1.7 °C for empirical results,
- T2 – in the medium height point T2 results are closer to each other. Temperature rising occurs in the similar pace. However stabilizing values of temperature differ each other and they 2.3 °C for simulation results and 1.5 °C for empirical results,
- T3 – in the lowest point T3 results of simulation and measurements have coincide course and stabilizing values are the same 1.1 °C in both cases. The only visibility difference is the starting point of indications changes in both cases: during empirical tests temperature rising very quickly after its starting, while during simulations delay about 100 s is observed.

4 Conclusion

The paper presents model of the temperature changes in model chamber due to the presence of a man. To confirm correctness of the model measurements of temperature changes in model chamber after entry a man inside were performed.

Simulated and measured results have analogical course. So as stabilizing values of temperature, however in this case some discrepancies, especially in the upper part of chamber, are observed. This is due to the simplified nature of the model. It takes into account convection and conduction, but no a human body radiation, which is one of important routes of heat dissipation. A direction of air exhalation (practically in the direction of the chamber wall) isn't taken into account, too. As a result of these simplifications in model air movement takes place mainly up, and only after then it moves into the lower parts of the room displaced by warmer air. In real conditions air held up and move towards the edges of the chamber same time. Therefore heat dissipation through the walls of the room is faster and takes place in larger wall surface, so stabilize values of temperature differ each other (are little higher in model results).

Summing up developed model satisfactorily simulates the course of temperature changes in the model room after entry the man inside. Therefore it constitutes the starting point for development of model to rooms or objects of larger dimensions and with more people inside. In particular with a focus on massive structures - like civil shelters.

References

1. Anand, S.: Human Development Index: methodology and measurement. Human Development Report Office (HDRO), United Nations Development Programme (UNDP) (1994)
2. Human Development Report 2015: Work for Human Development. United Nations Development Programme (UNDP) (2015)
3. Kreith, F., Manglik, R.M., Bohn, M.S.: Principles of heat transfer. Cengage learning (2012)
4. Kostowski, E.: Przepływ ciepła. Wydawnictwo Politechniki Śląskiej (2006)
5. Szargut, J. (ed.): Modelowanie numeryczne pól temperatury: praca zbiorowa. Wydawnictwa Naukowo-Techniczne (1992)
6. Błażejczyk, K.: Wymiana ciepła pomiędzy człowiekiem a otoczeniem w różnych warunkach środowiska geograficznego. IGiPZ PAN (1993)
7. Acheson, D.J.: Elementary Fluid Dynamics. Oxford University Press, Oxford (1990)
8. Clark, R.P., Edholm, O.G.: Man and His Thermal Environment. Arnold, London (1985)
9. Sorensen, D.N., Voigt, L.K.: Modelling flow and heat transfer around a seated human body by computational fluid dynamics. Build. Environ. **38**(6), 753–762 (2003)
10. De Dear, R.J., et al.: Convective and radiative heat transfer coefficients for individual human body segments. Int. J. Biometeorol. **40**(3), 141–156 (1997)
11. Fanger, P.O.: Komfort cieplny. Arkady (1974)
12. Butera, F.M.: Principles of thermal comfort. Renew. Sust. Energ. Rev. **2**(1), 39–66 (1998)
13. Nicol, J.F., Humphreys, M.A.: A stochastic approach to thermal comfort – occupant behavior and energy use in buildings. ASHRAE J. **110**(2), 554–568 (2004)
14. Djongyang, N., Tchinda, R., Njomo, D.: Thermal comfort: a review paper. Renew. Sust. Energ. Rev. **14**(9), 2626–2640 (2010)
15. Oosthuizen, P.H., Naylor, D.: An Introduction to Convective Heat Transfer Analysis. McGraw-Hill Science, Engineering & Mathematics, New York (1999)
16. Tritton, D.J.: Physical Fluid Dynamics. Springer Science & Business Media (2012)
17. Szymczyk, T., Rabiej, S., Pielesz, A.: Tablice matematyczne, fizyczne, chemiczne, astronomiczne. PPU "PARK" (1999)

18. Webb, R.C. et al.: Thermal transport characteristics of human skin measured in vivo using ultrathin conformal arrays of thermal sensors and actuators. PloS One **10**(2) (2015). doi:10.1371/journal.pone.0118131
19. Vargaftik, N.B.: Handbook of Physical Properties of Liquids and Gases-Pure Substances and Mixtures. Springer, Heidelberg (1975)

Design of an Interactive GUI for Multimedia Data Exchange Using SUR40 Multi-touch Panel

Rafał Kłoda[1(✉)], Jan Piwiński[1], and Aleksandra Nowak[2]

[1] Industrial Research Institute for Automation and Measurements PIAP, Al. Jerozolimskie 202, 02-486 Warsaw, Poland
{rkloda, jpiwinski}@piap.pl

[2] Institute of Metrology and Biomedical Engineering, Warsaw University of Technology, Św. Andrzeja Boboli 8, 02-525 Warsaw, Poland
aleksandra.anna.n@gmail.com

Abstract. Designed interface is a proposed solution for the FP7 CARRE project. The project focuses on the development of medical experts supporting technologies. The most important part of this paper is the evolution of the developed interface, which allows to present and exchange multimedia data to the external devices. The design has been developed to ensure the convenience of usage for both, medical expert and the patient. The paper also presents basic design guidelines, as well as tools that are available for developers. Overview of components for application development gives an idea of the possibilities and is a good starting point for further exploration of issues associated with the selected multi-touch panel.

Keywords: Graphical User Interface · GUI · Design · SUR40 · Multi-touch panel

1 Introduction

Graphical User Interface (GUI) plays a key role in interaction between human and device. For a common user the quality of the interface is equivalent to quality of the whole product. He uses the interface without thinking about the complicated application architecture. Ergonomy of the interface is the main factor that determines whether the application is seen as useful. Designing the interface should be the first thing that is done while creating a program, since it is the most important element for the user. The user, however, as the recipient of the application, is the most important for the creator [1–4]. User experience is a new term that has been established in last few years and it is described as a person's total experience using an interactive product. Providing a positive experience while interacting with application became an additional challenge for the designers [2].

The way of designing interfaces changes with the constant development of technology. The invention of touch screens was revolutionary for interface design. Many elements of the ordinary interface can be replaced by gestures. Therefore, the aim is to

R. Szewczyk and M. Kaliczyńska (eds.), *Recent Advances in Systems, Control and Information Technology*, Advances in Intelligent Systems and Computing 543, DOI 10.1007/978-3-319-48923-0_20

minimize the interface and to concentrate on the content. Nowadays programs are more intuitive and ergonomic, because the intermediary items like keyboard or mouse are no longer needed.

Designing interfaces for multi-touch screens is a specific challenge. Multi-touch interface has to provide simultaneous access to content for many users. The way of accessing the content has to be intuitive and simple. When it comes to working with electronic devices, this type of interaction has a huge potential in terms of performance, usability and intuitiveness. There are many possible uses of multi-touch screens. Currently the most popular devices with multi-touch screens are tablets and smart phones. Touch screens for laptops and PCs are also being produced. Multi-touch screens can also be used as large-format interactive walls in stores or shopping centers. Touch screen used as an interactive table can significantly improve the cooperation of people working on a common project. Some of the technologies that are being developed, enable objects recognition using specially prepared tags. It gives endless possibilities of interacting with computers using physical objects. This kind of user experience is much more immersive and satisfying than the usual work with a keyboard and mouse.

2 SUR40 Multi-touch Panel

Samsung SUR40 is an example of a large-format multi-touch screen. It has been developed in cooperation by Samsung and Microsoft. It is provided with PixelSense technology. The device is the size of an average 40″ TV screen and thanks to the 360-degree interface lets a group of people use the SUR40 simultaneously. It is possible to create an application that allows one person to present the information to others, or allows a group of people to make a collaborative decision.

2.1 Unit Description

Samsung SUR40 with Microsoft PixelSense ships with the AMD Athlon II X2 2.9 GHz Dual Core processor and The AMD Radeon HD 6570M desktop graphics card, both of which deliver clear and vibrant visuals. The device has four built-in speakers. The Samsung SUR40 with MS PixelSense utilizes a Full HD LED display. Featuring a large 16:9 40″ design with 1920 × 1080 resolutions. Pixel size is 0.46125 mm × 0.46125 mm. The SUR40 includes the world's largest sheet of Gorilla Glass bonded to any display. The material of the protective layer is touted by the manufacturer as an extremely strong, lightweight and resistant to scratches. One-hour water ingress protection is also featured. The device is 4″ thin. It features four USB ports, HDMI port, a Wi-Fi 802.11n router and Bluetooth and Ethernet connections. Samsung SUR40 with MS PixelSense ships with Windows 7 operating system and additional component named Surface Shell. This component is responsible for working in Surface mode [5, 9, 10].

2.2 PixelSense Technology

PixelSense is a technology that allows each pixel on the screen to act as a sensor. This enables detection of objects that interact with the device. Integrated PixelSense sensors are built directly into the layers of the LCD screen. These sensors enable detection, identification and reaction to objects with predefined tags and to untagged objects. This technology allows simultaneous identification of 52 touch points [6, 8].

2.3 Microsoft Surface 2.0 SDK

Microsoft Surface 2.0 SDK is intended for developing applications based on Microsoft PixelSense platform. It comprises two development environments .NET 4.0 and XNA. The SDK provides two APIs. The first one is *Presentation Layer* integrated with Windows Presentation Foundation (WPF), which is based on .NET 4.0. Presentation Layer interface extends WPF, adding controls designed for Microsoft Surface multi-touch platform. GUI is designed in XAML markup language and the behavior of the application is programmed in the programming language C#. The second one API is *Core Layer* based on XNA Game Studio 4.0. This programming interface can be used to develop 3D graphic applications [8, 11, 12]. In order to program applications using the Surface SDK, the following software is required:

- Microsoft Visual C# 2010 Express Edition or Microsoft Visual Studio 2010,
- Microsoft .NET Framework 4.0,
- Microsoft XNA Framework Redistributable 4.0,
- Windows 7 operating system.

It should be mentioned that SDK allows writing applications not only for PixelSense devices, but also for computers with a touch-screen and a Windows7 operating system.

2.4 Tagged Objects and Blobs

Microsoft PixelSense technology enables Samsung SUR40 to recognize tagged and untagged objects. Tag is a marker located on the object has a special pattern (Fig. 1), which thanks to PixelSense technology, can be read using the infrared rays. Markers can be used to identify objects or people.

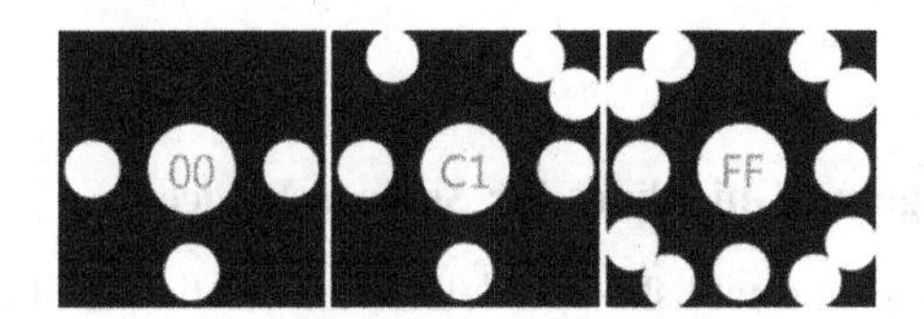

Fig. 1. Examples of tags [13]

Tags store 8 bits of information, so they can contain 256 different values. Each tag to work properly must have dimensions of exactly 0.75 × 0.75 inches. For the application to react properly to the markers, special control, *TagVisualizer* must be used. The application can be programmed to turn on itself when appropriate marker is placed on the screen. The object on which marker was placed should not reflect infrared rays, to prevent generating additional touch points.

Blobs are objects that do not have tags and are not fingers. These objects are unlabeled. The application can be programmed so that it behaves in a certain way when an unmarked object is detected, for example by displaying a visualization object on the screen [11–13].

3 General Guidelines for Interface Designing

The main goal in designing an interface is to find a way to explain to user how the application is working. This involves the creation of graphic elements, which are demonstratively illustrating what the application does and how it should be used. The interface should ensure proper communication between the user and the computer [2].

When designing the interface, it is crucial to determine for whom the product is intended. This determines the selection of graphics, colors, and arrangement of interface elements. The basic features of a well-designed interface are following:

- Intelligibility – the interface should be easy to use and understand. Users of intelligible interfaces are less likely to be confused and they work more efficiently.
- Brevity – it is important that the interface does not contain unimportant or repetitive content.
- Familiarity – the interface should use elements and symbols that are obvious to users.
- Accessibility – the interface should work quickly and provide feedback to ensure the user that he/she performed a proper operation.
- Consistency – maintaining consistency allows the user to quickly identify behavior patterns in each following application window.
- Aesthetics – although it is not necessary for the proper operation of the program, it is important because of its impact on the level of user satisfaction. Aesthetic plays an important role in creating a positive user experience.
- Efficiency – interface is designed to increase user productivity.
- Forbearance – the possibility of withdrawal from accidental or unintended actions [2, 4, 14, 15].

3.1 Interface Designing Guidelines for Surface Devices

Microsoft developers team established guidelines for creating GUI on Surface devices. They are designed to make application intuitive, engaging and visually appealing. Many of those guidelines refers to general rules of designing interface:

- Simple – the application must be clear. The way to use it should be obvious to the user. Excessive decorative elements should be avoided.
- Organized – elements should be arranged hierarchically and they should form a consistent structure.
- Content Oriented – information and data are always the core of the application, controls are secondary.
- Dynamic – it is important to take care of move and animation smoothness.

Samsung SUR40 device with PixelSense technology can be attached horizontally or vertically, depending on the intended use of the device. If SUR40 unit is placed horizontally, the interface has to provide possibility of using the application from all sides, it has to be capable of 360-degrees interaction. Therefore, the following guidelines should be taken into account:

- avoiding elements oriented along one edge of the display,
- enabling the change of elements' orientation by the user,
- orienting the newly opened content in the direction of the person who opened it,
- providing access to any element for every user, regardless of its position in relation to the table,
- ensuring readability of the contents from every side of the device.

The elementary rule of multi-touch interface designing is to use developed standards for operating applications using gestures (e.g. changing the size of the item with two fingers, dragging items with one finger).

4 SUR40 as the Interactive User Interface in CARRE Project

4.1 About CARRE Project

CARRE is an EU FP7-ICT funded project with the goal to provide innovative means for the management of comorbidities (multiple co-occurring medical conditions), especially in the case of chronic cardiac and renal disease patients or persons with increased risk of such conditions.

Sources of medical and other knowledge will be semantically linked with sensor outputs to provide clinical information personalized to the individual patient, to be able to track the progression and interactions of comorbid conditions. Visual analytics will be employed so that patients and clinicians will be able to visualize, understand and interact with this linked knowledge and take advantage of personalized empowerment services supported by a dedicated decision support system.

The ultimate goal is to provide the means for patients with comorbidities to take an active role in care processes, including self-care and shared decision-making, and to support medical professionals in understanding and treating comorbidities via an integrative approach [7].

4.2 Decision Support System in CARRE Project

Decision making in healthcare is a complex process in terms of number of parameters and variables, outcome possibilities and amount of information must be processed.

Decision support systems (DSS) can assist patients and provide to him advices, recommendations and diagnosis of problems in cardiorenal domain, where the optimal solutions for a given sort of data about the possible consequences are determined similar as human experts in the field.

A modern intelligent decision support system not only provides access to data and models. It is also a significant development in the field of analytical data processing, data warehousing and artificial intelligence-aided methods of knowledge discovery in databases i.e. data mining [7].

Our aim was to explore the possibilities of using the SUR40 device and its software as visualization platform in CARRE project.

4.3 Designed Interface

The interface was designed for CARRE project purposes and its goal is to improve the communication between patient and internist (or medical expert). The interface was equipped with set of visual elements, which enable the future end user to acquire the medical domain knowledge in more useful as well as user-friendly way which is shown in the Fig. 2 below. The designed interface can assist patients and doctors by providing advices, recommendations and diagnosis of problems in aging and growing population with chronic diseases, by means of interactive visualization interface, variables and recommendation to intuitive and user-friendly visualization in patient application.

Fig. 2. Designed interface presented on SUR40 device

The interface has scalable architecture, which enable further development. The interface has been created using rules and guidelines described in Subsect. 3.1. It is also possible to present multimedia data and to send it to an external device using Bluetooth protocol. Information about contents are accessible through QR codes generated by application. Designed interface provides to medical expert possibility review all types of data from the hospitals' records and the patient's records together with patient. They can sit side-by-side or across from each other viewing the same information using hand gestures to scroll through, open, zoom, rotate in 3D, push across the table, and drag and drop records from one storage repository to the other.

The proposed interface improve the doctor-patient communication by the use of following features:

- possibility of multiple users interaction with the one application,
- easy manipulation of the elements that display the contents,
- quick and easy access to information,
- transmission of multimedia data,
- intuitive navigation system.

5 Summary

Designed graphical interface supports the exchange of information using multi-touch Surface device. The designed GUI has been implemented for the SUR40 unit. The application has been tested and it is working correctly. It can be used in patient – doctor cooperation. In further development of our interface we consider to add following possibilities:

- animating transitions between windows to make interface more dynamic and consistent,
- enriching applications with sound effects that would indicate an action made by the user. Sounds could also provide feedback on whether an operation is executable,
- currently, the removal of elements is only possible directly from code, so that user cannot accidentally remove any important component. There is a possibility of providing extra fields for confirming the intentions of the user, therefore reducing the risk of removing the essential elements. The user will be able to use the interface more freely,
- allowing the user to undo last action,
- placing search engine in the application,
- extending application database.

Acknowledgment. This work was supported by the FP7-ICT project CARRE (No. 611140), funded in part by the European Commission.

References

1. Kuliński, M.: Obiektywne i subiektywne czynniki jakości użytkowej graficznych interfejsów użytkownika. Praca doktorska (2007)
2. Sikorski, M.: Interfejs użytkownika: od pracy, przez emocje, do relacje. Unpublished paper presented at Interfejs użytkownika - Kansei w praktyce Conference, Warszawa (2006)
3. Najmiec, A.: Ergonomia oprogramowania – od przepisów do praktyki. Bezpieczeństwo Pracy – Nauka i Praktyka, 5/2002
4. Fadeyev, D.: Interfejsy użytkownika w nowoczesnych aplikacjach webowych. The Smashing Book #1, Helion (2009)
5. http://www.samsung.com/
6. http://www.microsoft.com/en-us/pixelsense/
7. http://www.carre-project.eu/
8. Designing and Developing Microsoft Surface Application. http://www.microsoft.com/en-us/pixelsense/training20/index.html
9. Samsung SUR40 – Specification Sheet
10. Samsung SUR40 – User Manual
11. Developing Applications for Microsoft Surface 2.0 (2011)
12. Microsoft Surface 2.0 Design and Interaction Guide (2011)
13. Tagged Object Integration for Surface 2.0 (2012)
14. Inchauste, F.: Widoczne kontra niewidoczne. The Smashing Book #2. Helion (2011)
15. Chapman, C.: The Smashing Idea Book: From Inspiration to Application. Helion (2011)

Automatic Subtitling for Live 3D TV Transmissions by Real-Time Analysis of Spatio-Temporal Depth Map of the Scene

Konrad Bojar(✉)

Industrial Research Institute for Automation and Measurements PIAP,
al. Jerozolimskie 202, 02-486 Warsaw, Poland
kbojar@piap.pl

Abstract. In order to maximize experience and perception of the 3D TV transmission, there is a rule of thumb for the 3D video content subtitling which states that the subtitle should appear in front of the in-focus content at all times of the subtitle exposure. The main problem with live 3D transmissions containing subtitles, such as TV news or football matches, is that besides a pure text and a pair of video streams acquired by a stereo rig, there must be some additional information calculated which would allow to settle the correct subtitle depth. Therefore, either all set-top-boxes must determine this depth by themselves or the broadcaster must calculate and provide this information in the Disparity Signalling Segment (DSS). In this paper we present an algorithm for automatic subtitle depth estimation based on unsupervised spatio-temporal analysis of stereoscopic pair of compressed video streams. The proposed algorithm first analyzes the texture in the streams for left and right eye in the area where the subtitle should appear. The result of this analysis is a set of correspondences, that is pairs of points corresponding to the same single point in the scene. Every correspondence yields a stereoscopic parallax vector, and the magnitude of this vector is inversely proportional to the depth of point in the scene. It is shown how to effectively calculate the depth of the subtitle from depth maps for every stereoscopic pair of frames in which this subtitle should to appear. Also, latency problems and hardware aspects of low-cost FPGA implementation of the algorithm are discussed.

1 Introduction

The problem of efficient 3D content subtitling is not new and it has been addressed many times from the very beginning of 3D technology for theatrical and TV broadcasting. There are numerous solutions and systems, for example [4,18], for preparation of 3D subtitles, however they only aim at off-line preparation of content, while none of them is targeted at solving the task of real-time 3D subtitling. On the other side, the corresponding ETSI norm for 3D TV broadcasting [5] (dated 2012) and its newest part [6] (dated 2015) refer to the document [7] (dated 2011) in which one finds a special data structure called Disparity Signalling Segment (DSS) meant for setting depth during 3D subtitles rendering. Hence, the

R. Szewczyk and M. Kaliczyńska (eds.), *Recent Advances in Systems, Control and Information Technology*, Advances in Intelligent Systems and Computing 543, DOI 10.1007/978-3-319-48923-0_21

DVB standards are ready for broadcasting 3D subtitles, but the component for live positioning of these is missing. As a result, to the best of author's knowledge, currently no TV station broadcasts the DSS for 3D subtitles and no home cinema system offers variable-depth 3D subtitling when no disparity is provided alongside the 3D video stream. In this paper we are going to fill this gap by presenting an algorithm for automatic subtitle depth estimation. It can be applied either on the broadcaster side or on the receiver side.

Naturally, for the 3D content to be perceived satisfactorily, the subtitles must be correctly composed into the scene. This means that the depth of the subtitles must be adjusted in accordance to the depth of the region in which they should appear. In practice, the term of accordance is not defined in any exact manner. For example, in technical guidelines of ARTE television one can read that "all other burned-in content must be positioned in the comfort zone" [2, Par. 2.9.5] and this example is representative. In order to uncover meaning of this term, one has to pursue quality assessment rules of the 3D content [3,8,11]. Therefore we start this paper with a summary of 3D video subtitling rules (Sect. 2). Next, we present an outline of the algorithm (Sect. 3) and consider its computational complexity, parallelization potential and briefly discuss hardware implementation aspects (Sect. 4). Then we conclude the paper (Sect. 5).

2 3D Video Subtitling Rules

Since the DVB norm [7] imposes no content subtitling rules, these rules vary a little bit between broadcasters. However, general rules do not change. A representative example of such subtitling rules can be taken from technical guidelines of ARTE television [2]. Naturally, there is quite a number of rules, but only several of them are relevant from the point of view of the automatic subtitling algorithm structure, and only such rules are presented below.

- General rules for subtitling video content (2D and 3D)
  - G1 There should be no more than 40 characters per line.
  - G2 Double-height single-width font should be used.
  - G3 Consecutive subtitles should be separated by at least 5 frames.
  - G4 Each subtitle should be presented no shorter than 1 s and no longer than 10 s.
  - G5 When calculating shortest subtitle presentation time it is assumed that reading 15 characters takes 1 s.
  - G6 Subtitles should not overlap a shot change, but if it is necessary, subtitles should appear at least 1 s before and disappear at least 1 s after the shot change.
- Rules specific for subtitling 3D content
  - S1 Subtitles should appear in front of the content.
  - S2 Subtitles should not change its initial depth after a shot change.
  - S3 The depth of subtitles should not require sharp eye accommodation changes.

The above set of rules will be useful when pursuing desired algorithm structure and estimating resources needed for the algorithm to run.

3 The Algorithm

Before we start considerations related to the algorithm itself, we have to analyze the video processing ecosystem into which this algorithm has to fit. As we have mentioned earlier, our algorithm can be applied either on the broadcaster side or on the receiver side. However, the receiver side is much more demanding than the broadcaster side because it can be assumed that the broadcaster has practically unlimited processing resources which can be attributed to solve this problem, which is exactly opposite to situation on the receiver side. Hence, in further text we assume that our algorithm will operate on the receiver side, inside a set top box. After analyzing a typical 3D TV architecture [1], we come a to conclusion that the subtitling algorithm should fit the video processing ecosystem as shown in Fig. 1.

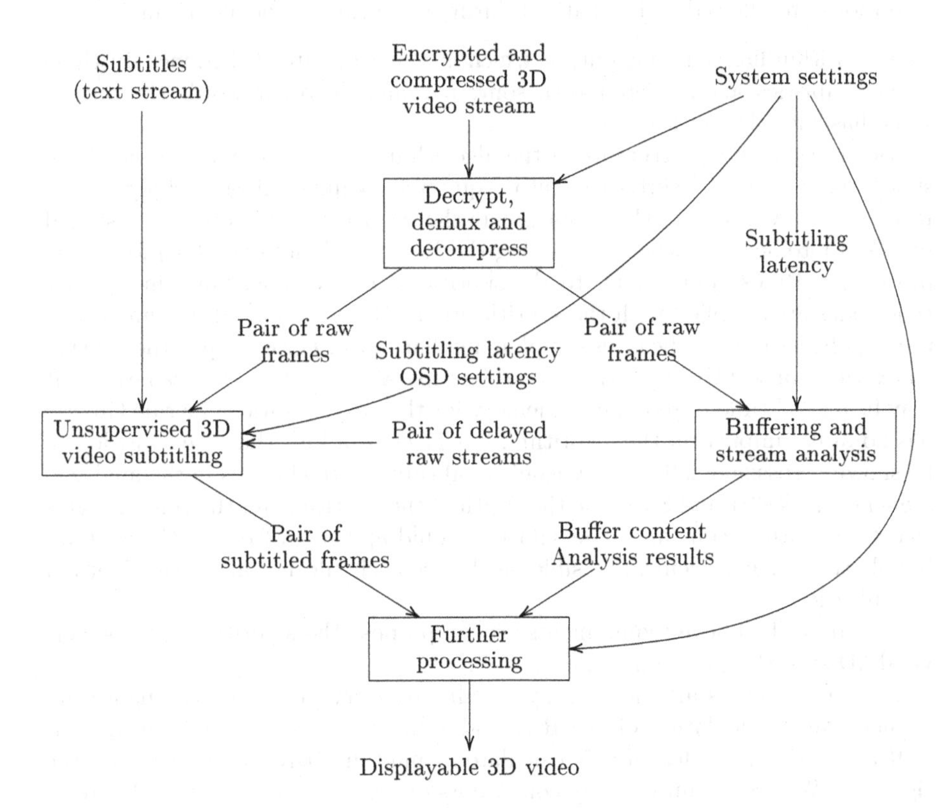

Fig. 1. 3D video subtitling algorithm as a part of the video processing ecosystem. Processing goes from the top to the bottom.

In this figure we see that inputs of the algorithm are

1. Subtitles in a form of a text stream. We implicitly assume that the stream is supplemented with time codes for each subtitle to be shown. For each subtitle there are two time codes, defining the first and the last frame in which the subtitle is presented.
2. Subtitling latency. This single numerical parameter defines time delay introduced by the subtitling algorithm.
3. Pair of current raw video frames. Here we assume that these frames are raw images stored in memory of the device running the algorithm.
4. Pair of delayed raw streams. These streams of raw video frames are assumed to be buffered sequences frames. Size of the buffer is defined by the subtitling latency.
5. OSD settings. This set of parameters carries current rendering options. These options are allowed to be changed during execution of the algorithm.

The algorithm has only one output which is a pair of subtitled frames. In these frames subtitles are rendered with some parallax; in other words the output video has subtitles burned in.

Before we give the structure of the algorithm, we have to analyze the above subtitling rules to convert them into requirements imposed on our algorithm. Rules G1 and G2 refer to the region where the subtitle should appear and should be taken into account when calculating the Region Of Interest (ROI) for image processing. Rules G3-G5 refer to the subtitle exposure time. Since in and out time codes are supplied with the subtitle stream, they have no direct impact on the algorithm, but can be of use to plan auxiliary jobs. For example, the rule G3 states that consecutive subtitles should be separated by at least 5 frames and it may be a good time to reorganize memory for the next subtitle. The rule G6 also has no direct impact on the algorithm, but gives us a hint that the subtitle can be shown across shots. This may result in abrupt depth changes in the analyzed sequence. Rules S1 and S3 define the depth of the subtitle, and the rule S2 states that this depth is constant. The subtitle should appear in front of the content, but short, transient elements visible on the scene should not affect the depth of the subtitle.

Based on the above requirements we can propose the algorithm for unsupervised 3D subtitling as seen in Fig. 2.

The algorithm starts identically as almost every stereo vision algorithm. Namely, after calculation of the ROI and subtitle exposition time, inside this ROI a set of texture features is calculated simultaneously for the left and the right eye. We do not impose any constraints on feature type to be used here as the correct choice is heavily dependent on hardware capabilities. Typically, in machine vision applications SIFT [17], SURF [15], ORB [12], or like features are used. Here, such features may turn out to be too complex. Probably, features like Harris corners [9] will be more suitable. Next, after two feature clouds are available, these clouds are matches against each other to find corresponding feature pairs. Again, we do not point here any specific algorithm. Depending on hardware architecture, a RANSAC-type stochastic algorithm, like ORSA [14],

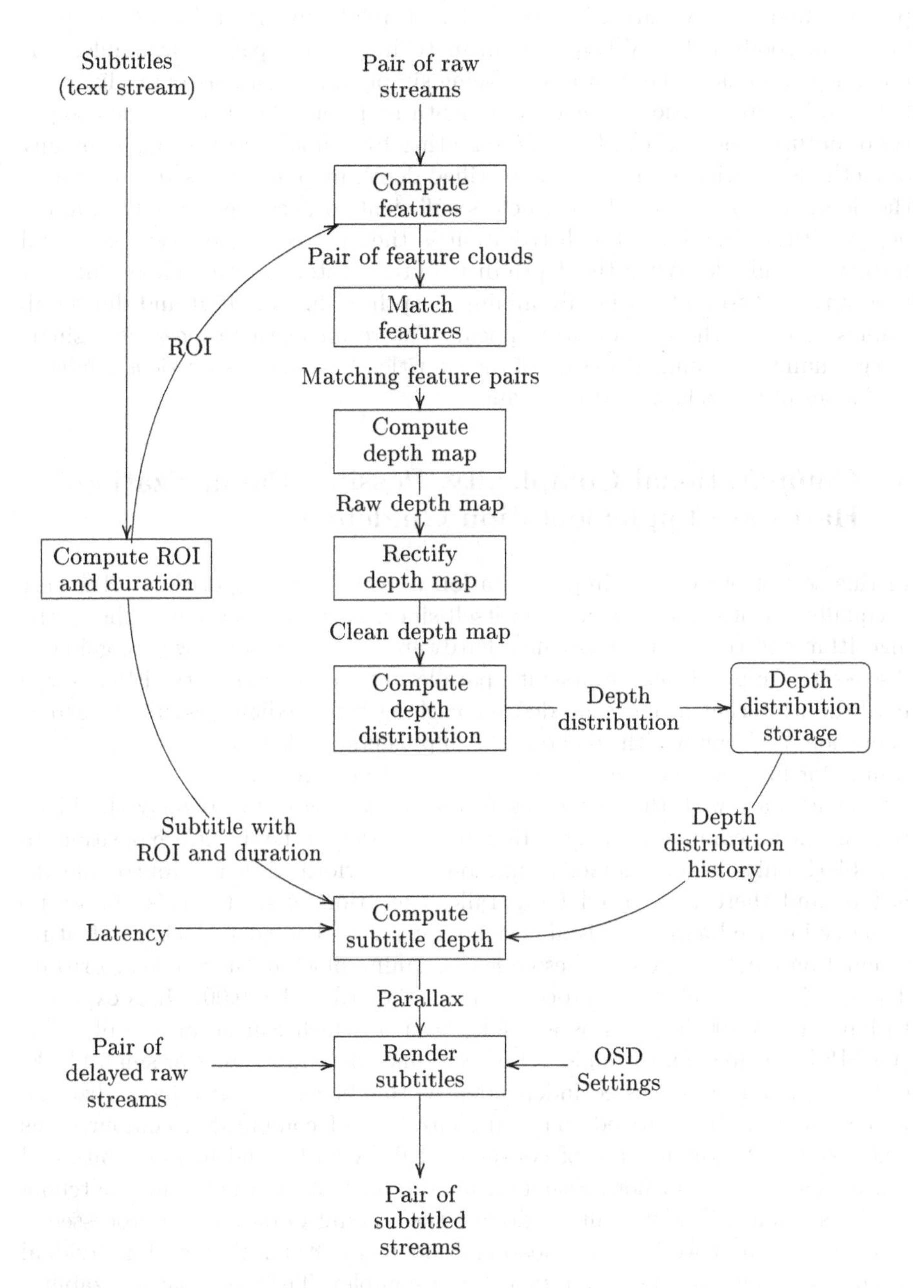

Fig. 2. Decomposition of the the unsupervised 3D video subtitling algorithm. Processing goes from the top to the bottom.

may be a good choice, but Kanade-Lucas-Tomasi feature tracker [16] may also perform well since we are going to deal with relatively small displacements in images of good quality. When a list of matching feature pairs is available, it is possible to calculate the depth map, being simply a map of horizontal distances between feature locations. Such a raw depth map must be then rectified to get rid of outliers and this can be achieved either by simple thresholding or by any of methods mentioned in [10]. The rectified depth map is then used to calculate the depth vs area distribution which is sufficient to decide essentially minimal depth within the ROI. This distribution is then stored for analysis when full history is available. When the depth distribution history is available in full, is it then analyzed to find essentially minimal depth within the ROI and during all frames in which the subtitle will appear. The result of this process is a single integer number defining disparity of the subtitle. At last, the subtitle is rendered on frames of the delayed video stream.

4 Computational Complexity, Possible Parallelization, Hardware Implementation Guidelines

In this section we discuss implementation aspects of the algorithm. This part is equally essential as the algorithm itself since, as it was mentioned above, the algorithm will run on a low resource hardware. For every block we are going to discuss its computational complexity, parallelization potential in the FPGA-type hardware, internal memory needed for caching intermediate results of current operation, and bandwidth to external components needed to store intermediate results for further processing or to pass the output stream.

Let us start with the first block in the processing chain, namely the block responsible for computation of the ROI and duration of the subtitle exposition. In this block only several simple calculations are performed, hence its complexity is low, and there is no need for parallel execution. Also, there is almost no memory involved and there is almost no data flow. The next block, the feature calculation block, is the most resource consuming block in the whole algorithm. The number of pixels to be processed is on the order of 100,000. It is expected to have up to 100 features as a result, each of which can be of size of 1 kbit (for SIFT features, for example). However, since individual lines or small blocks of the ROI can be processed independently, this block has large parallelization potential. It must be noted that parallelization of complicated computations will result in linear increase of consumed FPGA CLBs and memory bits and linear increase of data flow. Situation for the next block, the feature matching block, is similar. However, although here the amount of data to be processed is smaller, yielding lower data bus load and memory consumption, each individual feature pair matching decision is much more complex. Therefore, parallelizability potential is here smaller due to large size of a single processing block in the FPGA fabric. Passing to the depth map computation block we find that this operation involves analysis of feature locations only. Hence, it is computationally light and does not need any intermediate memory. Moreover, external memory usage can be avoided by pipelining results from the previous block. The next block, performing depth map rectification, is very similar from our current point of view, but the computation may be slightly more complicated, and the block for depth distribution computation is very similar even without the latter remark. The next element on the processing path is actually a data store for the depth distribution history, hence it will consume some of the internal memory bits. The size of this store is completely determined by maximum assumed system latency. The next processing element performs the subtitle depth calculation, which is yet another lightweight element of the algorithm, while the last block, the subtitles rendering block, will consume consume some of the cache memory for the font, some of the data bus bandwidth for ROI transfer and will require some of the FPGA fabric size for coding the bit-wise masking operation.

The table below summarizes our considerations on computational complexity, possible parallelization in FPGA-type hardware and hardware resources needed for implementation of respective algorithm blocks.

Block name	Complexity	Parallelizability	Cache memory	Bandwidth
Compute ROI and duration	Low	Low	Low	Low
Compute features	High	High	High	High
Match features	High	Medium	Medium	Medium
Compute depth map	Low	High	Low	Low
Rectify depth map	Medium	High	Low	Low
Compute depth distrib.	Low	High	Low	Low
Depth distribution storage	N/A	N/A	Medium	N/A
Compute subtitle depth	Low	High	Low	Low
Render subtitles	Medium	High	Medium	Medium

At last, we note that for architectures with low on-chip memory, the last two columns of the above table should be combined into one and the value should go one gradation up.

5 Summary

We have presented an algorithm for automatic subtitling of live 3D TV transmissions. The core of this algorithm is the analysis of spatio-temporal depth map of the scene which allows to determine the desired subtitle depth in real time. The algorithm is designed to fit low resource FPGA-type platforms which are now becoming common in home cinema systems. We have also vaguely discussed envisaged computational complexity and memory and bus bandwidth consumption. In future works we will synthesize this algorithm on a chosen FPGA evaluation board to test its performance in the target environment.

Acknowledgments. This work was supported by "Satellite mutual navigation system for on-orbit servicing and formation flying" project, realized for The National Centre for Research and Development, grant number PBS2/B3/17/2013.

References

1. Angueira, P., de la Vega, D., Morgade, J., Vélez, M.M.: Transmission of 3D video over broadcasting. In: Zhu, C., Zhao, Y., Yu, L., Tanimoto, M. (eds.) 3D-TV System with Depth-Image-Based Rendering: Architectures, Techniques and Challenges, pp. 299–344 (2013)
2. Complete technical guidelines ARTE G.E.I.E. (2015). http://wp.arte.tv/corporate/files/Complete-Technical-Guidelines-ARTE-GEIE-V1-04.pdf
3. Barkowsky, M., Brunnström, K., Ebrahimi, T., Karam, L., Lebreton, P., Le Callet, P., Perkis, A., Raake, A., Subedar, M., Wang, K., Xing, L., Iess, J.: Subjective and objective visual assessment in the context of stereoscopic 3D-TV. In: Zhu, C., Zhao, Y., Yu, L., Tanimoto, M. (eds.), 3D-TV System with Depth-Image-Based Rendering: Architectures, Techniques and Challenges, pp. 413–437 (2013)
4. Claydon, L.J., Gardner, J.F., Corne, R., Wang, R., McDermott, J.: System, apparatus and methods for subtitling stereoscopic content. US Patent Application No. 12/713,685 (2009)
5. Etsi, T.S.: 101 547–2: Digital video broadcasting (DVB); Plano-stereoscopic 3DTV; Part 2: Frame compatible Plano-stereoscopic 3DTV. ETSI, France (2012)
6. Etsi, T.S.: 101 547–4: Digital video broadcasting (DVB); Plano-stereoscopic 3DTV; Part 4: Service frame compatible Plano-stereoscopic 3DTV for HEVC coded services. ETSI, France (2015)
7. Etsi, E.N.: 300 743: Digital video broadcasting (DVB); Subtitling systems. ETSI, France (2015)
8. Gonzalez-Zuniga, D., Carrabina, J., Orero, P.: Evaluation of depth cues in 3D subtitling. Online J. Art Des. **1**(3), 16–29 (2013)
9. Harris, C., Stephens, M.: A combined corner and edge detector. In: Proceedings of Alvey Vision Conference, pp. 147–151 (1988)
10. Hodge, V., Austin, J.: A survey of outlier detection methodologies. Artif. Intell. Rev. **22**(2), 85–126 (2004)
11. Huynh-Thu, Q., Barkowsky, M., Le Callet, P.: The importance of visual attention in improving the 3D-TV viewing experience: overview and new perspectives. IEEE Trans. Broadcast. **57**(2), 421–431 (2011)
12. Kulkarni, A., Jagtap, J., Harpale, V.: Object recognition with ORB and its implementation on FPGA. Intl. J. Adv. Comput. Res. **3**(11), 164–169 (2013)
13. Liao, C.-K., Yeh, H.-C., Zhang, K., Vanmeerbeeck, G., Chang, T.-S., Lafruit, G.: Stereo matching and viewpoint synthesis FPGA implementation. In: Zhu, C., Zhao, Y., Yu, L., Tanimoto, M. (eds.) 3D-TV System with Depth-Image-Based Rendering: Architectures, Techniques and Challenges, pp. 69–106 (2013)
14. Moisan, L., Stival, B.: A probabilistic criterion to detect rigid point matches between two images and estimate the fundamental matrix. Int. J. Comput. Vis. **57**(3), 201–218 (2004)
15. Schaefferling, M., Kiefer, G.: Flex-SURF: A flexible architecture for FPGA-based robust feature extraction for optical tracking systems. In: Proceedings of International Conference of Reconfigurable Computing and FPGAs, pp. 458–463 (2010)
16. Tomasi, C., Kanade, T.: Detection and tracking of point features, Carnegie Mellon University Technical Report CMU-CS-91-132 (1991)

17. Yao, L., Feng, H., Zhu, Y., Jiang, Z., Zhao, D., Feng, W.: An architecture of optimised SIFT feature detection for an FPGA implementation of an image matcher. In: Proceedings of International Conference on Field-Programmable Technology, pp. 30–37 (2009)
18. Zhang, T.: System and method for combining 3D text with 3D content. US Patent Application No. 13/521,290 (2010)

Advances in FEM Based Modeling of Waveguide and Waveguide Systems for Microwave Applications, Using Newly Developed Open Source Software

Jakub Szałatkiewicz[1], Roman Szewczyk[2], Eugeniusz Budny[1], Mateusz Kalinowski[1(✉)], Juhani Kataja[3], Peter Råback[3], and Juha Ruokolainen[3]

[1] Industrial Research Institute for Automation and Measurements PIAP, Al. Jerozolimskie 202, 02-486 Warsaw, Poland
mkalinowski@piap.pl
[2] Warsaw University of Technology, Pl. Politechniki 1, 00-661 Warsaw, Poland
[3] CSC – IT Center for Science Ltd., P.O. Box 405, 02101 Espoo, Finland

Abstract. This article presents application of new open-source software for solving time-harmonic Maxwell's equations, allowing modeling of microwave waveguide lines. Paper presents investigated cases of 2.45 GHz and 5.8 GHz frequency waveguides lines. First is straight rectangular waveguide model, next are two coaxial to waveguide couplers models. On presented models microwave modes are shown with electric field magnitude resulted from physical geometry. Modeling results obtained from new Elmer Vectorial Helmholtz module show that application of open-source software can be effective in R&D works for high-tech small and medium enterprises involved in microwave technology.

Keywords: Microwaves · Waveguide · Resonant cavity · Open-source · Elmer · FEM · Time-harmonic maxwell equations · Electromagnetic waves

1 Introduction

It is possible to describe many physical phenomenon using Partial Differential Equations (PDE). However in most cases it is hard to compute PDEs and find theirs solutions. In those cases numerical methods are used to calculate simplified solutions. Among other, the most popular is Finite Elements Method. Broad application of numerical methods was possible due to rapid growth of computers computing power. To calculate one of the presented simple cases (mesh of 107k elements), it took about 2 min on test machine.

Nowadays there are several Computer Aided Engineering (CAE) programs for many applications, in example: structural analysis, fluids flow, electromagnetism, heat transfer and others. Companies like Siemens, ANSYS or Comsol have advanced and refined packages for numerical analysis, 3D modeling and pre/post processing of simulation. Next to above commercial software there are also open-source alternatives like Calculix, Elmer or OpenFOAM et al. Those are very powerful tools, however their

R. Szewczyk and M. Kaliczyńska (eds.), *Recent Advances in Systems, Control and Information Technology*, Advances in Intelligent Systems and Computing 543, DOI 10.1007/978-3-319-48923-0_22

usage is not always as intuitive as commercial software. Despite this issue, open-source free software have many advantages, among others: source code can be accessed and reviewed, it is easier to update and most important, its performance is similar to commercial software. Efficiency of chosen Elmer software was previously verified for magnetodynamics studies up to 10 kHz [1].

In our study we were using newly developed Elmer [2] module – *Vectorial Helmholtz for* electromagnetic *waves* [3] – for simulation of microwave waveguides and cavities.

2 Materials and Methods

Elmer module for electromagnetic waves was examined in three chosen cases, proving its usability and correctness in simple applications. First one is just rectangular waveguide. Second and third one are models of coaxial to rectangular waveguide couplers for 2.45 GHz and 5.8 GHz, respectively. Carried out modeling allowed verification whether new software has any potential for real-life use.

Presented 3D geometry on Figs. 2, 4 and 7 visualize internal volume of waveguides without surrounding walls. Those models have to be discretized by defining mesh – determining matrix for finite elements. 3D models are generated using CAD program and exported to *.stp* format. Those files are compatible with Netgen [4], which is also open-source meshing software. Figure 1 presents generated 3D mesh on one of the investigated geometry.

It is essential, to determine the right mesh size. By making cells smaller, it is possible to achieve more accurate results but it significantly increases the number of elements, which heavily affects memory and CPU usage, so the simulation takes more time.

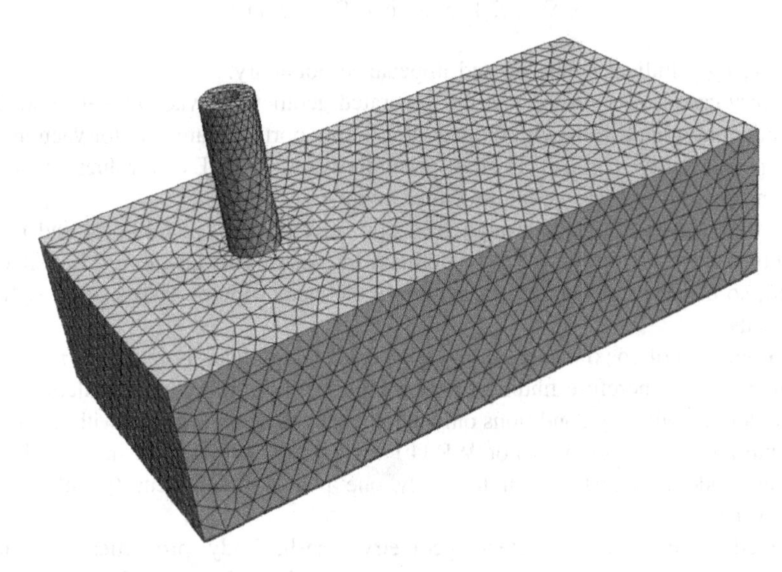

Fig. 1. View of sample surface mesh, generated using Netgen

On the other hand big cells will calculate faster, but the results might not reflects real phenomenon.

Next, appropriate solver (Elmer Vectorial Helmholtz in this case), boundary conditions (BC) and other model properties, are set using Elmer GUI [5]. This preprocessor handles all parameters and writes them to start file. The start file defines model equations, location of mesh and simulation properties. Chosen values, i.e. preconditioning, can affect simulation convergence.

3 Theory

In Elmer Vectorial Helmholtz module propagation of electromagnetic waves in space is described by Maxwell's equations. From time-harmonic form (1) and (2), simplified curl-curl equation (3) is used in software module:

$$\nabla \times \boldsymbol{E} = -\mathrm{j}\omega\mu \boldsymbol{H} \tag{1}$$

$$\nabla \times \boldsymbol{H} = \mathrm{j}\omega\varepsilon \boldsymbol{E} + \boldsymbol{J} \tag{2}$$

$$\nabla \times \mu^{-1} \nabla \times \boldsymbol{E} - \omega^2 \varepsilon \boldsymbol{E} = \mathrm{j}\omega \boldsymbol{J} \tag{3}$$

where: $\boldsymbol{E}$ – electric field intensity, $\boldsymbol{J}$ – electric current, $\boldsymbol{H}$ – magnetic field intensity, permeability: $\mu = \mu_0 \mu_r$, permittivity: $\varepsilon = \varepsilon_0 \varepsilon_r$ $\omega = 2\pi f$.

Dirichlet (4) and Robin (5) boundary conditions for Eq. (3) applied in module are [2]:

$$\vec{n} \times E = f \text{ on } \Gamma_E \tag{4}$$

$$\vec{n} \times \nabla \times E + \alpha \vec{n} \times (\vec{n} \times E) = g \text{ on } \Gamma_Z \tag{5}$$

where: Γ_E, Γ_Z – indicates electric and impedance boundary.

In every case it is assumed that the generated geometry is vacuum surrounded by well conducting metal walls with input and/or output ports. Parameters for vacuum are: relative permeability $\mu_r = 1.0$, relative permittivity $\varepsilon_r = 1.0$. Exact values of permeability and permittivity are coded in Elmer module.

For microwaves entering through feed port, electric field at this port and propagation constant of medium have to be defined. In case of coaxial cable, TEM mode is assumed, so electric field in radial direction is defined, which consists of two phasor components.

Incorporation of coaxial cable in model geometry significantly increases number of elements in mesh, therefore finding solution requires more time and calculations. It is easier to define boundary conditions only at the end wall of waveguide, without coaxial cable entering the waveguide. For WR340 waveguide, as modeled in cases below, dominant mode is TE10, which has only one phasor component (parallel to the narrow wall).

Defined simulation parameters: geometry, mesh, body properties, equations, boundary conditions, preconditioning, etc. are used by solver to calculate solution.

Calculations are carried out by consecutive iterations, during which convergence is indicated.

Elmersolver saves calculated results in *.vtu* format. File contains electric and magnetic field distribution across defined geometry. To carry out data-analysis and visualization of simulations results ParaView [6] open-source software was used. Visualized simulation results are presented on Figs. 9, 10, 11, 12, 13, 14, 15, 16, 17, 18, 19 and 20.

Summarizing, workflow is following:

1. Defining 3D model.
2. Mesh generation (Netgen).
3. Setting boundary conditions and simulation properties.
4. Solving (Elmersolver).
5. Analyzing results (ParaView).

Computer used for the simulations: CPU Intel i7-4770 (4 cores), 16 GB RAM, 1 TB HDD, GPU GTX 760.

4 Calculation

4.1 Simple Geometry

First test case set up to prove accuracy of new software, was model of straight rectangular waveguide WR340 with internal dimensions 43.18 mm × 86.36 mm. It was assumed as open on both ends, with wave entering through one and leaving at the other. Incoming wave is at TE10 mode, so electric field is parallel to narrow walls and perpendicular to direction of propagation. CAD model of described setup is shown on Fig. 2. Wave is entering through port no. 1, walls 2 and 3 are assumed as metal. Walls 4, 5 are not visible (also assumed as metal) and output port 6 (no constraints).

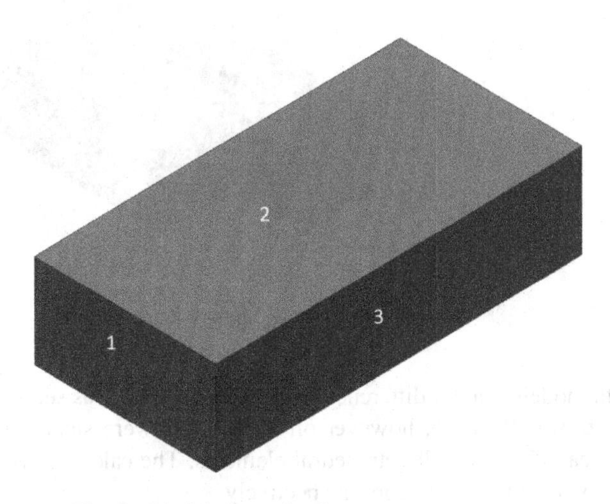

Fig. 2. Model view, simple geometry

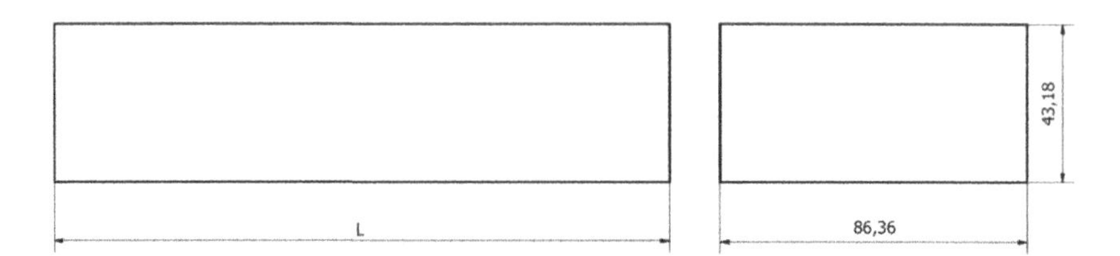

Fig. 3. Model dimensions in mm, simple geometry

Figure 3 shows dimensions of established model. Two different values of L2 was chosen due to wavelength in waveguide. For 2.45 GHz and WR340, $\lambda_w \approx 173.4$ mm [7]. Overall length L equals $1 \cdot \lambda_w \approx 173.4$ mm or $5/4 \cdot \lambda_w \approx 216.75$ mm.

In both models mesh size was set to be fine, maximum element size was 0.004 m, therefore there were ca. 41k elements. The calculations took about ~30 s on single core.

4.2 Coaxial to Rectangular Waveguide Coupler at 2.45 GHz

Example of more sophisticated geometry is the coaxial to rectangular WR340 waveguide coupler (Fig. 4) similar to real waveguide component shown on Fig. 6. Dimensions of this section are presented on Fig. 5. Modeling of microwave field distribution was carried out with two different overall waveguide lengths 173.4 mm and 303.45 mm. As exercised with simple geometry it is associated with microwave wavelength in waveguide. Microwave input antenna was placed in a quarter of λ_w from ending wall, and L2 in this case equals $3/4 \cdot \lambda_w \approx 130.05$ mm or $6/4 \cdot \lambda_w \approx 260.1$ mm.

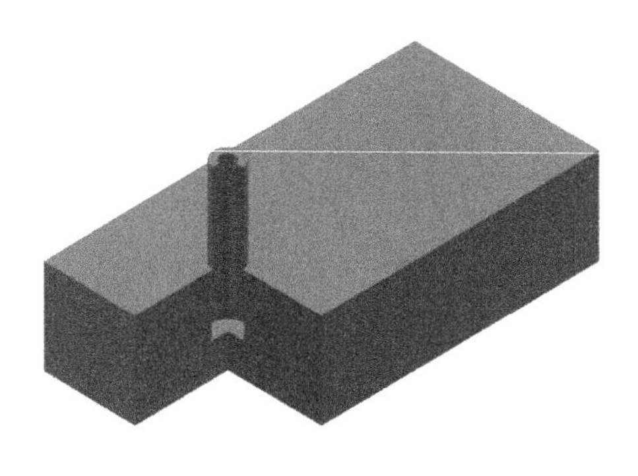

Fig. 4. Model view, coupler coax-WR340

In both models (length difference) most of mesh size was set to be fine, maximum element size was 0.004 m, however on some faces were smaller 0.002 m, therefore there were ca. 55k and 107k tetrahedral elements. The calculations took about ~63 s and ~130 s on single CPU core, respectively.

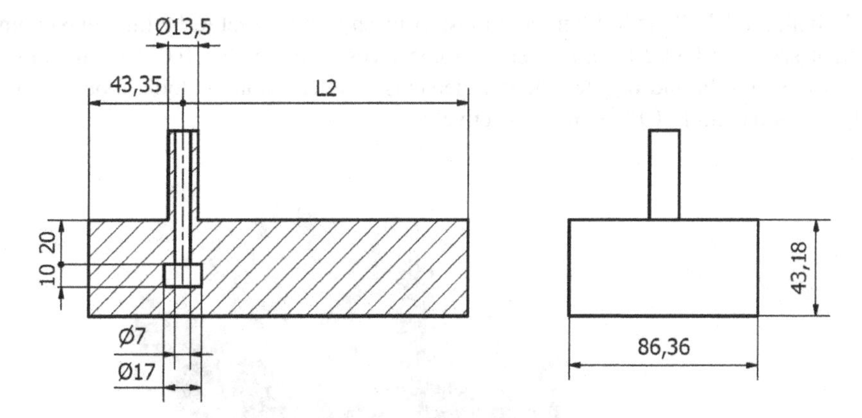

Fig. 5. Model dimensions in mm, coupler coax-WR340

Fig. 6. Exemplary coaxial to rectangular waveguide coupling devices, left WR340, right WR159

4.3 Coaxial to Rectangular Waveguide Coupler at 5.8 GHz

In order to fully determine the usability of the software, equipment for frequency other than 2.45 GHz was modeled. For this purpose magnetron source of 5.8 GHz and WR159 waveguide line was chosen. As in the previous case, coaxial to rectangular waveguide coupler was modeled, similar to one shown on Fig. 6 (right). It's CAD model is presented on Fig. 7.

Model dimensions are shown on Fig. 8. In WR159 waveguide the wavelength $\lambda_w \approx 67.72$ mm. Microwave input antenna was placed in a quarter of λ_w from ending wall and L2 equals $\frac{3}{4} \cdot \lambda_w \approx 50.79$ mm or $1 \cdot \lambda_w \approx 67.72$ mm.

In both models (length difference) most of mesh size was set to be fine, maximum element size was 0.002 m, however on some faces were smaller 0.001 m, therefore there were ca. 42k and 62k tetrahedral elements. The calculations took about ~16 s and ~73 s on single CPU core, respectively.

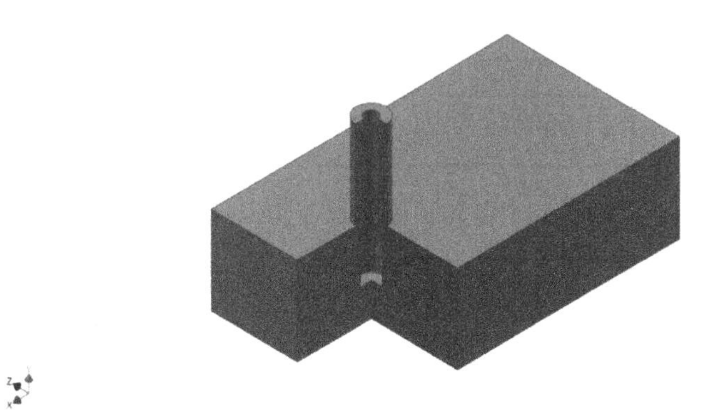

Fig. 7. Model view, coax-WR159 coupler

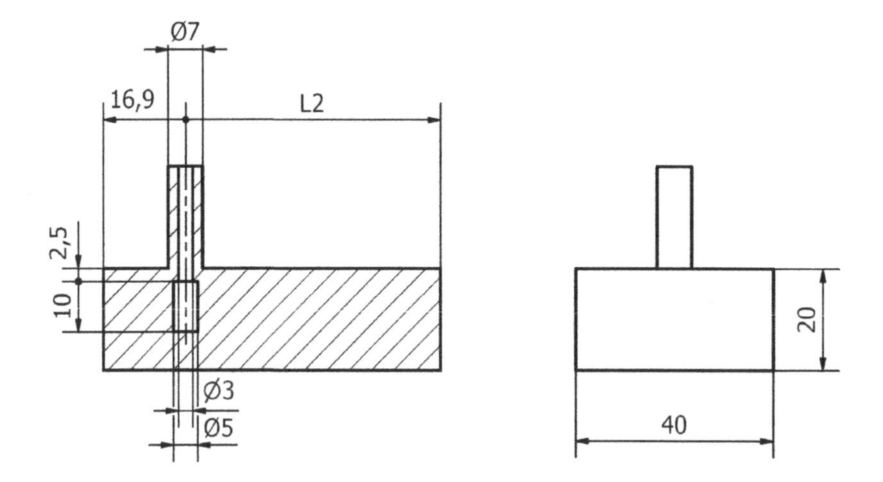

Fig. 8. Model dimensions in mm, coax-WR159 coupler

5 Results

For presented cases visualization of electric and magnetic field calculated by Elmer software, was conducted using ParaView software. Below results of this investigation is presented, consisting visualization of vectors for both fields and electric field distribution along central plane.

5.1 Simple Geometry

It is clearly noticeable that results of modeling and simulations conform with analytical calculations of wave in waveguide. On Figs. 7 and 8 vectors of magnetic and electric fields are presented. Their directions indicate that TE10 mode is present through entire length of waveguide.

Software correctly calculates the electromagnetic field in given geometry, however there are so called "artifacts" at the edges of the geometry, where vectors are not aimed in the right direction. Nerveless this issue doesn't affect correctness of simulation.

Figures 11 and 12 present electric field distributions in waveguides of different lengths (field magnitude doesn't have same scale on both figures). It can be noticed that there is exactly 1 or 1¼ of wave inside the waveguide, accordingly.

Fig. 9. Top view, M field vector, simple geometry, L = 173.4 mm

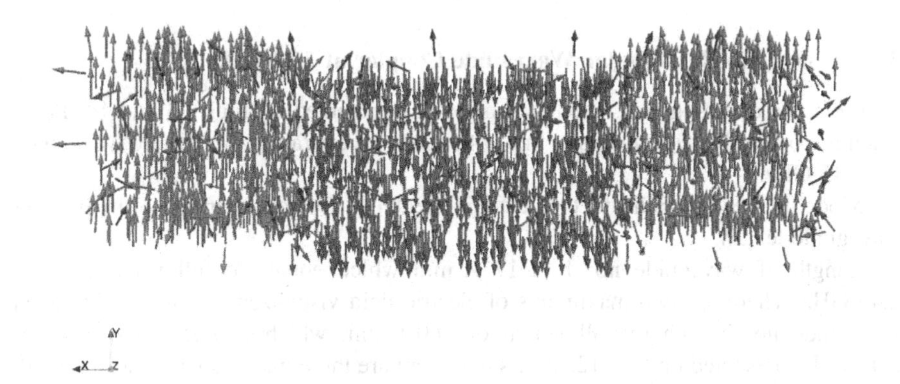

Fig. 10. Side view, E field vectors, simple geometry, L = 173.4 mm

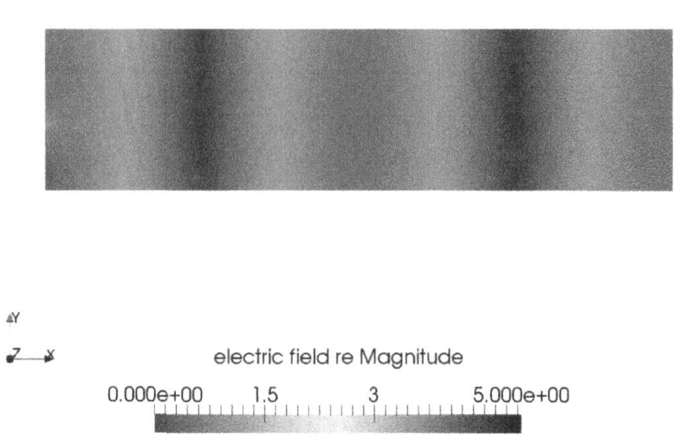

Fig. 11. Electric field distribution, clip on XY plane, simple geometry, L = 173.4 mm

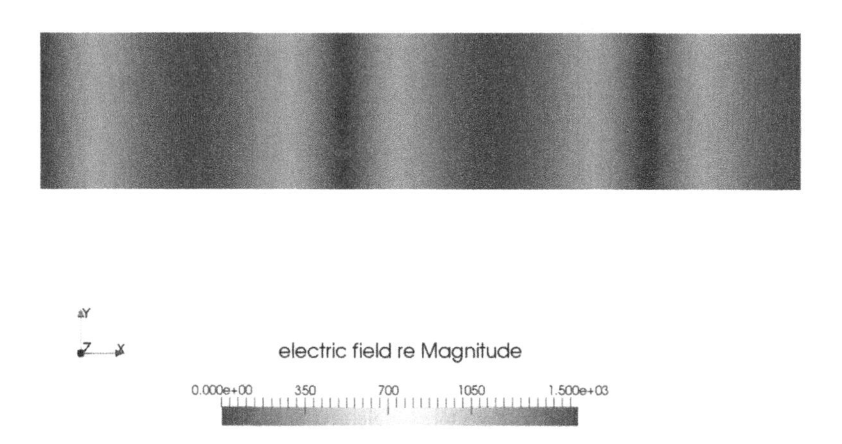

Fig. 12. Electric field distribution, clip on XY plane, simple geometry, L = 216.75 mm

5.2 Coaxial to Rectangular Waveguide Coupler at 2.45 GHz

Both electric and magnetic fields are correct, as presented on Figs. 13 and 14. Propagating mode is TE10, which is in line with theory of operation for WR340 waveguide for 2.45 GHz.

Modeling results indicate that electric field distribution changes with different waveguide length.

Length of waveguide no. 1 is 173.4 mm which equals to full wavelength of 2.45 GHz. There are two maximums of electric field visualized on Fig. 15. Modeled waveguide no. 2 with overall length of 260.1 mm, which is 1.75 wavelength of 2.45 GHz, presented on Fig. 12, shows that there are three maximums of electric field. In this case electric field magnitude in waveguide no. 2 is almost 2 times higher than in waveguide no. 1. On the presented visualizations one can observe standing waves inside rectangular waveguide. This occurs because both ends are shorted.

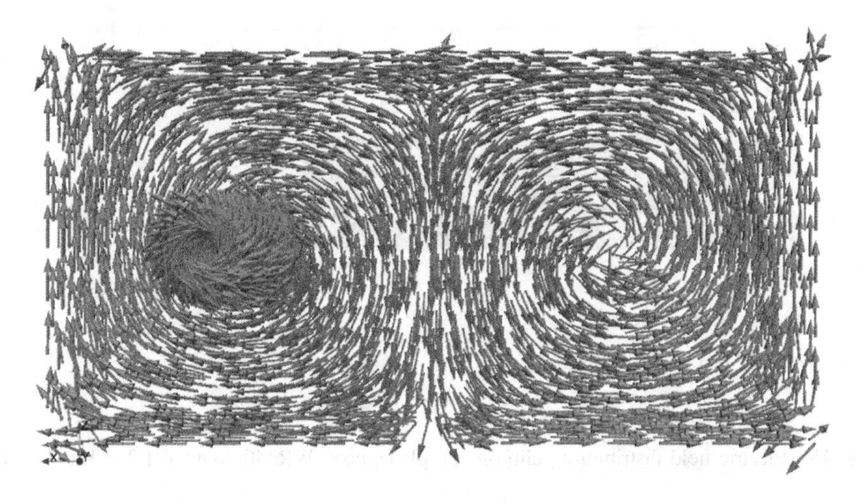

Fig. 13. Top view, M field vector, coax-WR340 coupler, L2 = 130.5 mm

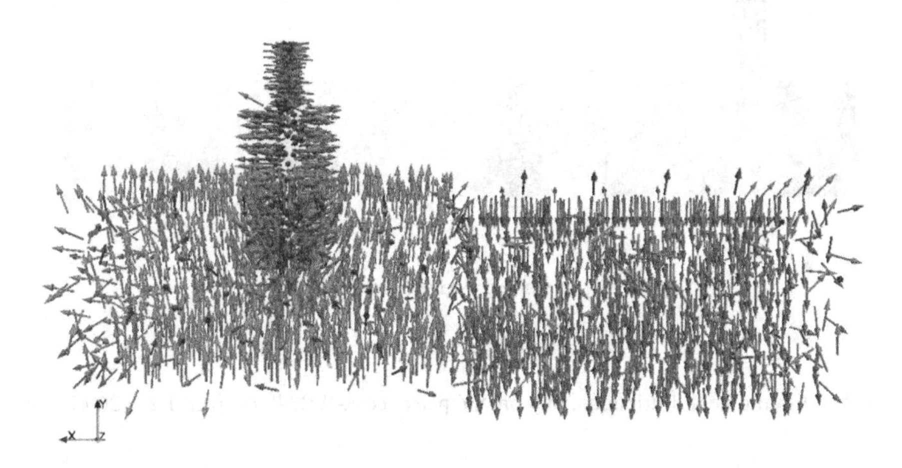

Fig. 14. Side view, E field vectors, coax-WR340 coupler, L2 = 130.5 mm

5.3 Coaxial to Rectangular Waveguide Coupler at 5.8 GHz

At frequency of 5.8 GHz results are very similar to examined WR340 case for 2.5 GHz, and are also consistent with theory of operation for rectangular waveguides. Vectors of magnetic and electric fields shown on Figs. 17 and 18 are correct. For better illustration of the data, Figs. 17 and 18 of WR159 waveguide are enlarged, so the dimensional scale is not as on figures for WR340.

Electric field distribution is presented on Figs. 19 and 20. There are two peaks of full wave on first model, because ends are shorted and standing wave is formed.

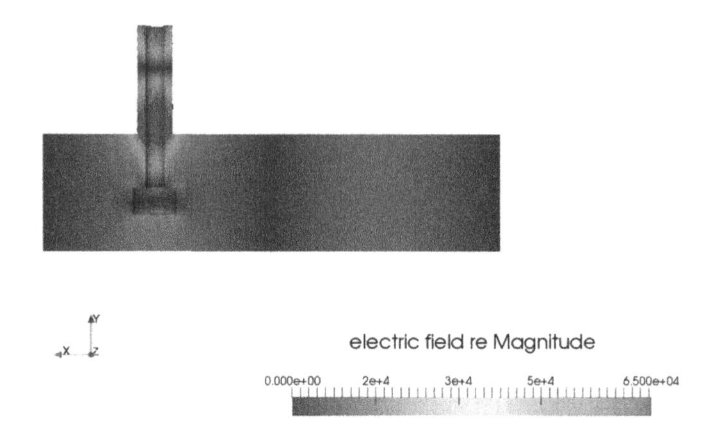

Fig. 15. Electric field distribution, clip on XY plane, coax-WR340 coupler, L2 = 130.5 mm

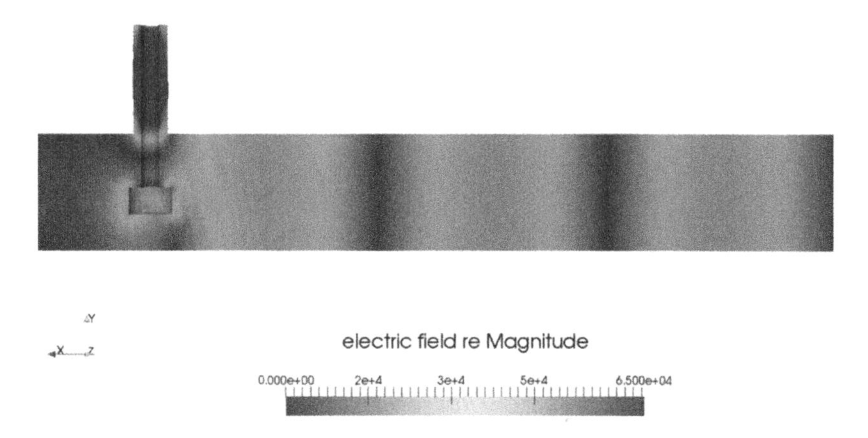

Fig. 16. Electric field distribution, clip on XY plane, coax-WR340 coupler, L2 = 260.1 mm

6 Discussion

Application of new Elmer Vectorial Helmholtz module in modeling of electromagnetic field distribution in waveguides was very successful. Results are as expected, and consistent across different configurations, i.e. TE10 mode inside WR340 and WR159 waveguides.

Successful verification of this new open-source tool allows its application in design of microwave waveguide systems with resonant cavities.

Modeling of simple geometries presented on Figs. 2, 4 and 7, allows determination of key waveguide parameters to be correctly matched to right frequency and operation. This includes correct mode and field distribution in modeled waveguide.

Next, modeling of resonant cavities with waveguide launcher allows precise matching and optimization of developed microwave apparatus. The most important

Fig. 17. Top view, M field vector, coax-WR159 coupler, L2 = 50.79 mm

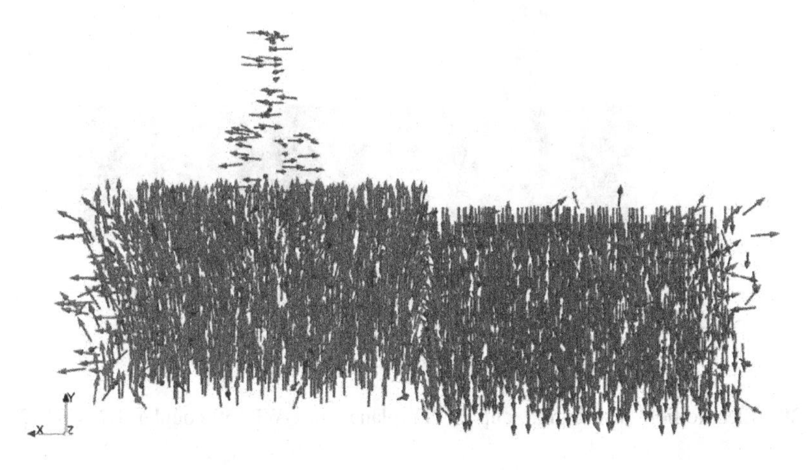

Fig. 18. Top view, E field vector, coax-WR159 coupler, L2 = 50.79 mm

benefit of FEM modeling with use of new Elmer module is the possibility to easily change the parameters of modeled object and to obtain calculations results relatively fast. It is especially important in modeling of resonant cavities.

Preparation of simulation require finding optimal configuration of mesh size, preconditioning, solver etc., that allows overcoming problems with convergence. With right simulation values, the calculations are fast and accurate. The important issue in preparation of model properties is the mesh size around small and complex geometries, i.e. connection of coaxial cable to waveguide. When the mesh is coarse, there are problems with convergence and results. In this case ElmerSolver did not calculate the

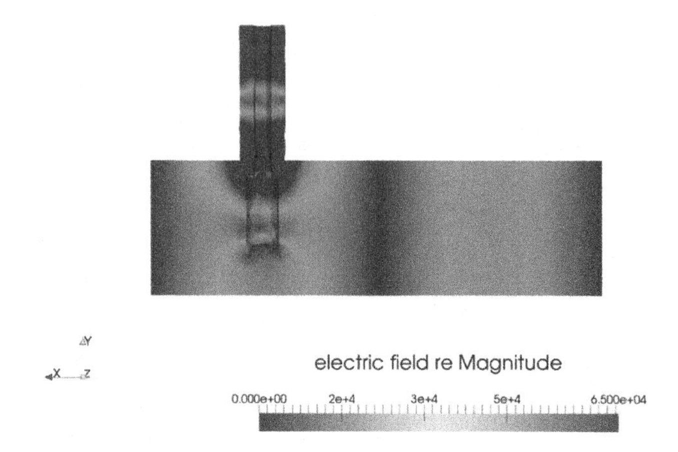

Fig. 19. Electric field distribution, clip on XY plane, coax-WR159 coupler, L2 = 50.79 mm

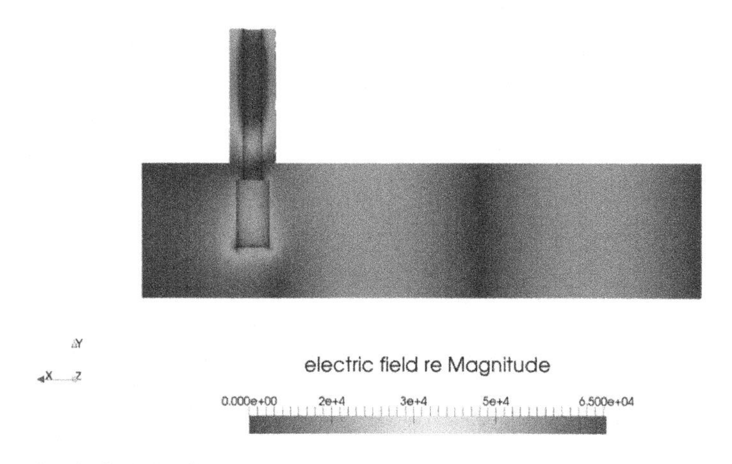

Fig. 20. Electric field distribution, clip on XY plane, coax-WR159 coupler, L2 = 67.72 mm

microwaves entering the waveguide. It is necessary to make the mesh finer, which will result in bigger mesh size and longer computation time, but the results will be correct.

In presented models irregularities occur, theirs origin is explained by numerical noise which comes from discretization of model.

In Industrial Research Institute for Automation and Measurements research project is carried out to develop microwave moisture analyzer for specialized applications. The aim of this project is to develop microwave drying system based on waveguide with resonant cavity. The key task to achieve this goal is to develop and optimize microwave waveguide line with resonant cavity. Works are carried out to apply 2.45 GHz and 5.8 GHz frequency for this application. Project includes also development of measurement and control system of microwave magnetron sources that was carried out.

Carried out modeling, part of which is presented in hereby article, allowed identification and optimization of key system dimensions and development of test stand.

The test stand consists of magnetron microwave source, waveguide magnetron head launcher, waveguide isolator, measurement ports and resonant cavity. The system is equipped with measurement system allowing continuous measurements of: electric power to the magnetron transformer and from transformer to magnetron, microwave power forwarded to the cavity, microwave power reflected from the cavity, sample temperature and mass.

The test stand control system consists of manual and automatic microwave power control connected to computer SCADA system, allowing continuous data acquisition and control.

Developed test stand allowed carrying out research over drying parameters of chosen samples. Preliminary research data indicates that results of modeling using Elmer and based on own calculations correspond each other.

7 Conclusions

The new Elmer module *Vectorial Helmholtz for electromagnetic waves* investigated in this study proved its potential in application of modeling microwave waveguide systems. This open-source noncommercial software equipped with true electromagnetic waves simulations in 3D is effective tool for broad spectrum of applications.

Investigated cases show that this software proves itself in R&D applications and in educational use, which is also a breakthrough in small and medium scale simulations. With some effort, it is possible to make fully open-source environment for CAD/CAE and use it for developing advanced RF equipment. This software is opportunity especially for small high-tech companies.

Together with other Elmer modules it is possible to carry out complex multiphysics simulations, similar to currently undergoing modeling of electromagnetic waves, heat transfer and fluid flow to extensively test out chosen cavity models.

Acknowledgments. This work was carried out within a research project no. PBS2/B3/18/2013 financed by The National Center for Research and Development (Narodowe Centrum Badań i Rozwoju).

References

1. Szewczyk, R., Salach, J., Ruokolainen, J., Råback, P., Stefko, K., Nowicki, M.: Noise assessment in Whitney elements based forward transformation for high resolution eddy current tomography. In: Szewczyk, R., Zieliński, C., Kaliczyńska, M. (eds.) Progress in Automation, Robotics and Measuring Techniques. AISC, vol. 352, pp. 219–224. Springer, Heidelberg (2015). doi:10.1007/978-3-319-15835-8_24
2. (2015). https://www.csc.fi/web/elmer
3. Råback, P., Malinen, M., Ruokolainen, J., Pursula, A., Zwinger, T. (eds.): Elmer Module Manual, CSC – IT Center for Science (2015)

4. (2015). http://sourceforge.net/projects/netgen-mesher/
5. Ruokolainen, J., Malinen, M., Råback, P., Zwinger, T., Pursula, A., Byckling, M.: Elmer Solver Manual, CSC - IT Center for Science (2014)
6. (2015). http://www.paraview.org/
7. Das, A., Das, S.: Microwave Engineering, pp. 71–72. Tata McGraw-Hill Education (2000)

Industrial Automation

How to Increase Efficiency of Automatic Control of Complex Plants by Development and Implementation of Coordination Control System

Igor Korobiichuk[1(✉)], Anatoliy Ladanyuk[2], Dmytro Shumyhai[2], Regina Boyko[2], Volodymyr Reshetiuk[3], and Marcin Kamiński[4]

[1] Institute of Automatic Control and Robotics, Warsaw University of Technology, Warsaw, Poland
kiv_igor@list.ru

[2] National University of Food Technologies, Kiev, Ukraine
{ladanyuk, rela}@ukr.net, shumygai@gmail.com

[3] National University of Life and Environmental Sciences of Ukraine, Kiev, Ukraine
volodymyr.reshetiuk@hotmail.com

[4] Industrial Research Institute for Automation and Measurements PIAP, Warsaw, Poland
mkaminski@piap.pl

Abstract. Complex technological plants work in different technological modes during its exploitation that causes changes of plants parameters. Changeable parameters of the plant may lead to changeable efficiency of automatic control systems or even to the abnormal functioning of the plant if such automatic control systems have static value of controller parameters. Controller's all-mode tuning is usually used to avoid abnormal functioning of the automatic control system in different modes, but such strategy reduces profits. There are many ways to improve the quality of automatic control. The most common is to use adaptive tuning that allows to find optimum controller's parameters of the system for every technological mode of the plant. Another approach to increase efficiency is to solve coordination task. Such approach can be used for plants with significant nonlinear connections. As example, the article describes developed coordination control system with decision support system for sugar refinery.

Keywords: Coordination · Decision support system · Fuzzy logic · Sugar refinery

1 Introduction

Quality of the automatic control is the basis of economic effect of industrial control systems [1]. Controller's "low (all-mode) tuning" is the main disadvantage of existing control systems, which reduces profits. Frequent changes of plant parameters are the prime cause of low quality tuning and reduced efficiency of control systems.

R. Szewczyk and M. Kaliczyńska (eds.), *Recent Advances in Systems, Control and Information Technology*, Advances in Intelligent Systems and Computing 543, DOI 10.1007/978-3-319-48923-0_23

These changes are caused by changeable mode of plants. The quality of work with such tuning is obviously worse [2].

There are many ways to improve the quality of automatic control: to use various kinds of disturbances compensators; to use models for the prediction of the controlled variable; to change controller parameters depending on the values of controlled variable and controlled device [3]; to use adaptive control.

For systems with significantly changeable parameters of the plant there are two possible strategies that can lead to the normal automatic control functioning. Those strategies allow to exclude permanent shutdown and transition to manual control:

1. To use “low” but all-mode tuning.
2. To provide adaptive tuning for changing parameters of the plant (automatic tuning).

All types of automatic tuning use three fundamentally important stages: identification, calculation of controller parameters, controller tuning [4]. In some cases despite the auto-tuning, the controller cannot provide quality control for reasons that do not depend on controller’s algorithms. For example, the plant may be poorly designed (dependable loops); sensors can be located in the wrong place and have poor contact; installation may be done with some inaccuracy, etc. [5].

Those are general methods for increasing the efficiency for either control systems with significant nonlinear connections or systems with lack of significant nonlinear connections [6]. Also there is completely different approach to increase efficiency – to solve coordination task.

2 Coordination Task

Approach to increase efficiency by solving the coordination task works only with plants that can be considered as bunch of subsystems and some of these subsystems must have complex nonlinear connections. The main goal is to find the optimal set point for controllers without changing parameters Kp, Ki, Kd. So controller parameters could stay low and efficiency of automatic control will be increased not by decreasing the amplitude of oscillatory process but by moving plot of process to its optimal set point. Usually we know limits of set point, but every time optimal value is different. According to this method the coordination task must be solved. Coordination task means consideration of complex nonlinear connections between selected subsystems [7].

For example the sugar refinery will be considered as complex nonlinear plant. At first the considered plant should be analyzed and decomposed. Always there is optimal number of subsystems for each TC. Further we will be considering following subsystems: extraction, purification and evaporation, as they are the most complex and have significant nonlinear connections.

The nonlinearity is explained the next way: the increasing value of raw juice consumption reduces the loss of sugar in the pulp, and that is improving efficiency of TC, but at the same time it is rising costs of evaporation, and that is reducing efficiency. So our goal is to find the optimal value of raw juice consumption considering values of sugar refinery parameters [7].

As a result of researches the structure of coordination control system was designed as shown in Fig. 1 [8]. In Fig. 1 selected subsystems, PLC and PC with SCADA for each subsystem are displayed on the bottom level, and the structure of developed control system is displayed on the top. Developed control system consists of a database that contains current values of parameters of each subsystem and decisions that are made by coordination control system, blocks of coordination (COORD) and Decision Support System (DSS). Coordination block is using mathematical models to predict the optimal value of parameters to increase efficiency of TC, so that profit would be maximum. Coordination solution should minimize probability of emergency situations or abnormal mode. DSS block consists of two components: extraction and evaporation.

Block "extraction" makes decision using fuzzy logic methods: using a "fuzzification" to transform value of parameter into fuzzy values; "analysis" to find appropriate situation in "rules base"; "defuzzification" to convert values in clear form. Block "extraction" consists of most common 35 rules that contains 8 technological variables. This block makes a clear decision about set point for 5 technological variables. Block "evaporation" of Decision Support System consists of 10 formed the most common abnormal situations. In case of emergency situations or abnormal mode this block displays possible causes and suggestions for resolving the current problems. Block "evaporation" includes 27 selected possible causes and appropriate for them actions. Decision is made by comparing the current situation with the situations in "situations base". If the automatic mode is selected, decisions made by the "coordination block" and "DSS" will be written to the database. With SCADA optimal values of variables are read from the database and transferred to the PLC and then set by actuators.

Coordination is solved only if the system is running in normal mode. If the situation is abnormal it is inappropriate to solve the coordination task, and the main task is to return to normal mode by situational control.

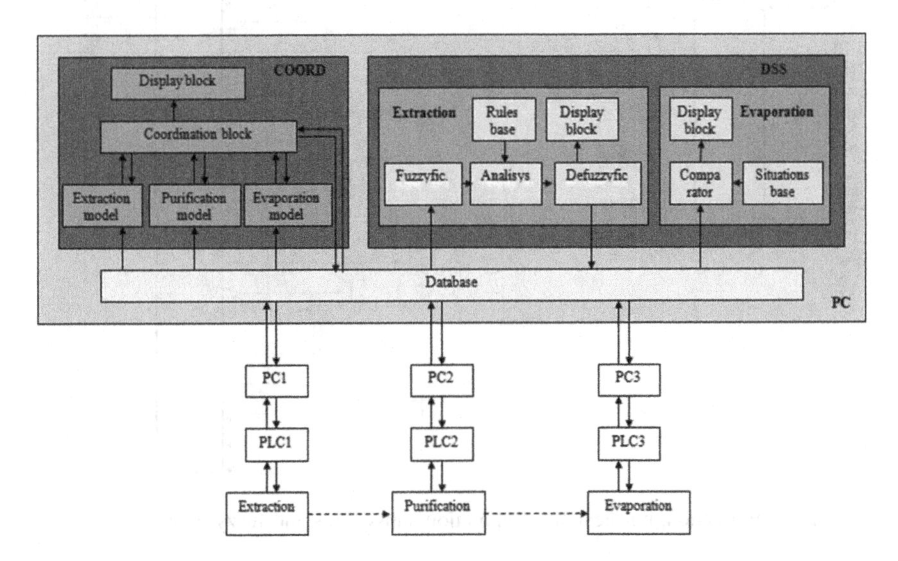

Fig. 1. The structure of coordination control system with decision support system

The developed Coordination control system with DSS for sugar refinery is additional software to existing Industrial control systems. In real-time system reads from the database 22 value of technological variables that are required for the task solution. The control decision is made to change the value of 5 process variables: raw juice consumption, beet consumption, rotation speed in extraction machine, level and temperature in extraction machine.

Coordination control system with DSS of sugar refinery is developed in MATLAB [5]. Extraction, purification, evaporation subsystems and fuzzy control block for DSS are set in Simulink (Figs. 2 and 3).

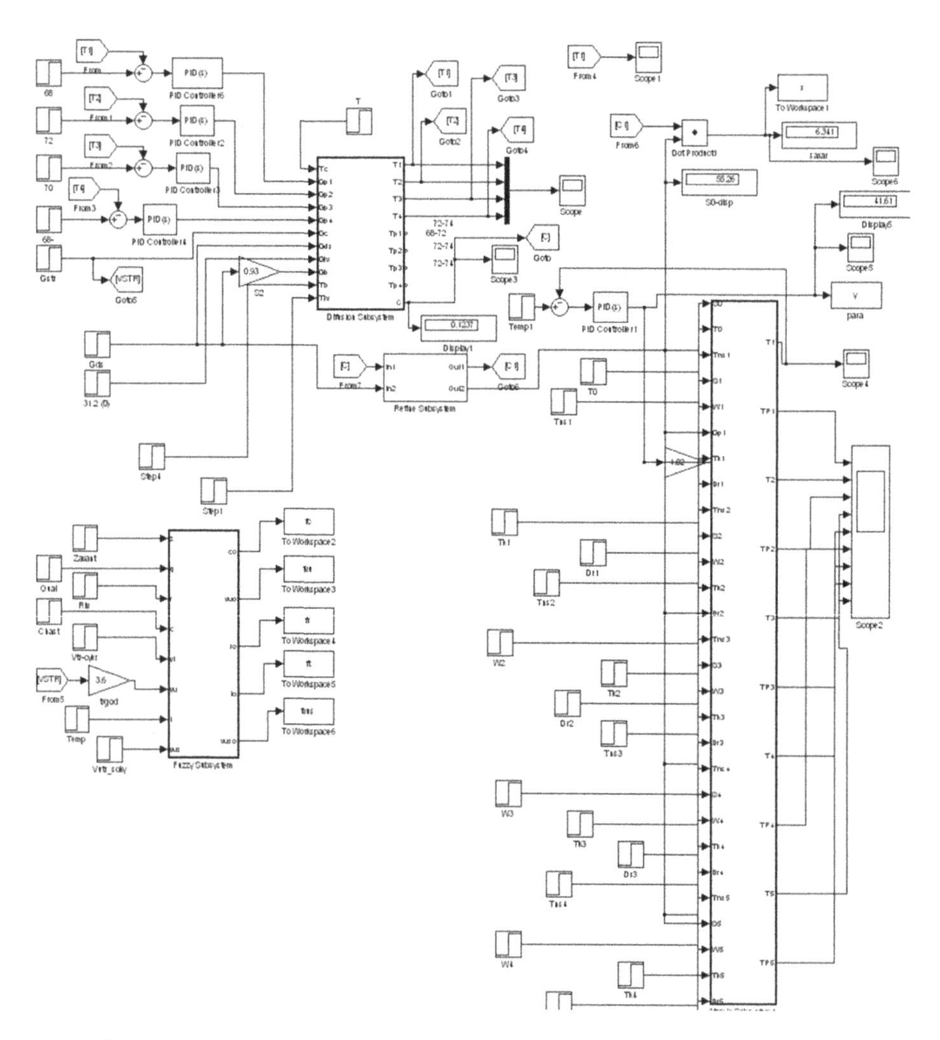

Fig. 2. Extraction, purification, evaporation subsystems and fuzzy control block

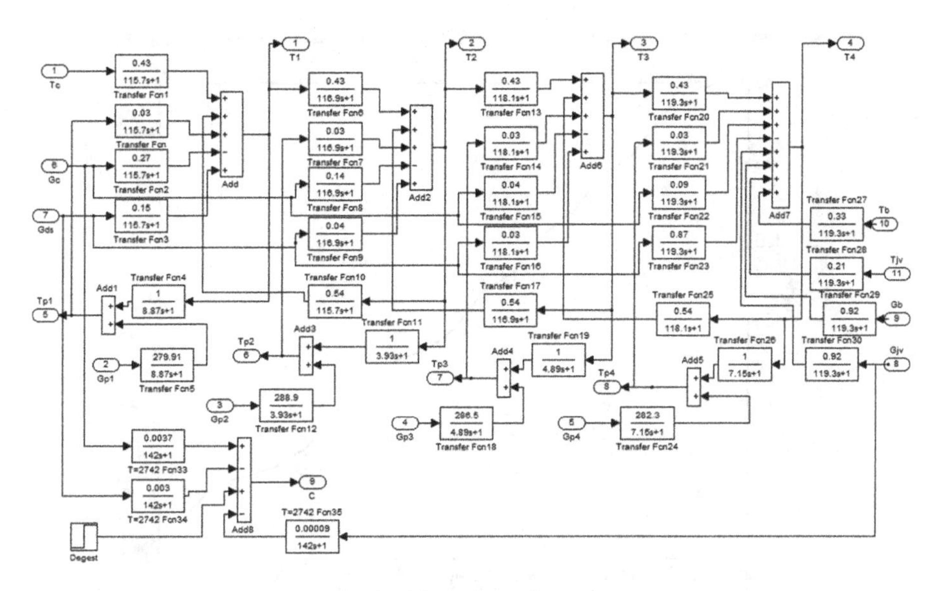

Fig. 3. Extraction subsystem

Coordination task is to find the optimum value of raw juice consumption, so that profit would be maximum. So global criteria is the following:

$$P = 4,6grn/kgGsugar\text{-}0,1984grn/kgGsteam\text{-}0.338grn/kgGbeet\text{-}Const$$

where *Gsugar* – amount of sugar output; *Gsteam, Gbeet* – steam consumption and beet consumption are variables that nonlineary depend on raw juice consumption; *Const* – fixed costs that are not dependent on process mode: salary, rent, etc.

Fixed costs are not calculated, because when potential profits with different values of raw juice consumption are compared, the fixed costs are self-destructed.

We got the next plot of optimal value of raw juice consumption versus beet consumption (Fig. 4).

The main window of created program consists of two blocks (Fig. 5). In coordination block the values of beet consumption, current raw juice consumption, optimal raw juice consumption, economy effect and the plot of global criteria (income minus variable costs) versus raw juice consumption are displayed.

The second block (DSS) consists of extraction block and evaporation block. The extraction block displays decisions in fuzzy form that are made for 5 process variables. By pressing details the current value and the calculated optimal value of shown variables are displayed in clear form. The evaporation block displays ten most common emergency situations. When such situation occurs it becomes red. By pressing this situation the list of possible causes and suggestions for resolving the current problems is displayed.

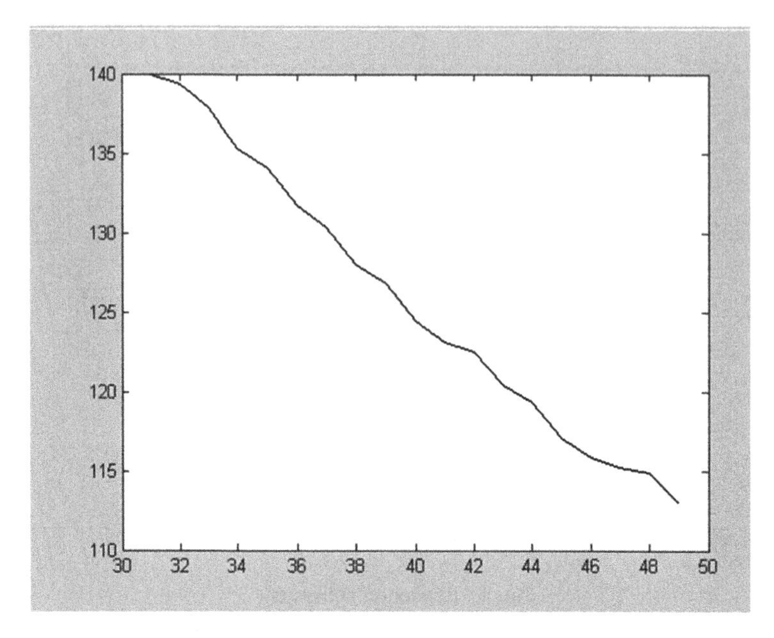

Fig. 4. Optimal value of raw juice consumption (%) versus beet consumption (kg/s)

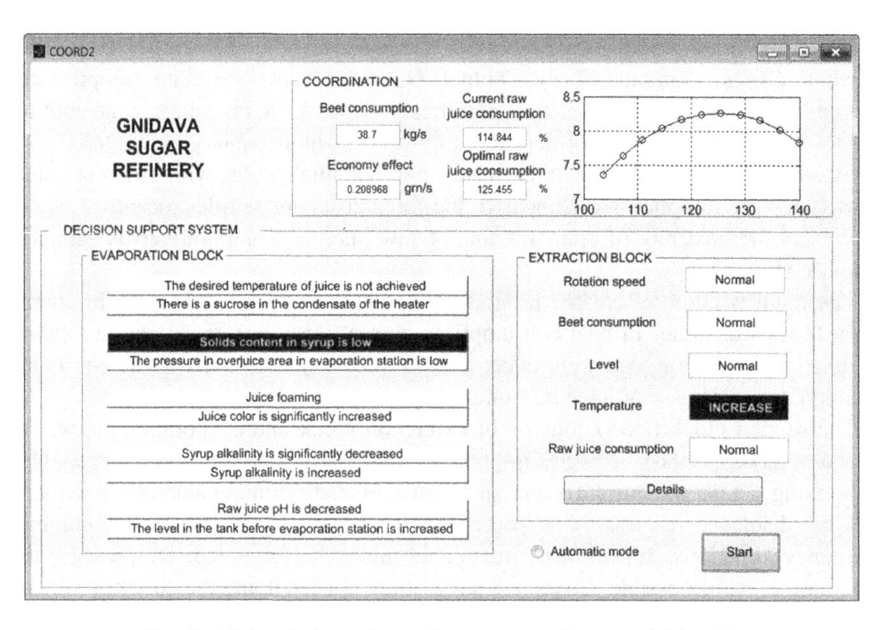

Fig. 5. Main window of coordination control system with DSS

3 Conclusions

One of the approaches to increase efficiency of automatic control system of complex nonlinear plant (for example sugar refinery) is to solve coordination task. Nowadays at the sugar refinery the profit at current technological mode is not predicted and optimal values of technological parameters are not calculated. As optimal values of technological parameters differ every time we developed program of coordination control system with decision support system that increases efficiency of sugar refinery by searching the optimal value of raw juice consumption in real time. That system also forecast economy effect from changing current values of technological parameters to optimal values.

References

1. Kuo, B.C., Golnaraghi, F.: Automatic Control Systems, 609 p. John Wiley & Sons Inc., USA (2002)
2. Shteinberg, S., Serejun, L., Zaluckiy, I., Varlamov, I.: Problems of creation and operation of effective control systems. Ind. ACS Controllers **7**(7), 1–7 (2004)
3. Korobiichuk, I.: Mathematical model of precision sensor for an automatic weapons stabilizer system. Measurement **89**, 151–158 (2016). doi:10.1016/j.measurement.2016.04.017
4. Izerman, R.: Digital control systems, Mir, 541 p. (1984)
5. Korobiichuk, I., Podchashinskiy, Y., Shapovalova, O., Shadura, V., Nowicki, M., Szewczyk, R.: Precision increase in automated digital image measurement systems of geometric values. In: Jabłoński, R., Brezina, T. (eds.) Advanced Mechatronics Solutions. AISC, vol. 393, pp. 335–340. Springer, Heidelberg (2016). doi:10.1007/978-3-319-23923-1_51
6. Korobiichuk, I., Bezvesilna, O., Ilchenko, A., Shadura, V., Nowicki, M., Szewczyk, R.: A mathematical model of the thermo-anemometric flowmeter. Sensors **15**, 22899–22913 (2015). doi:10.3390/s150922899
7. Ladanyuk, A., Shumygai, D., Boiko, R.: Situational coordination of continuous technological complexes subsystems. J. Autom. Inf. Sci. **45**(8), 68–74 (2013). Begell House, New York
8. Shumygai, D.: Automation of coordination of subsystems of technological complex of sugar refinery. Ph.D. thesis, 212 p. (2014)

Formalization of Energy Efficiency Control Procedures of Public Water-Supply Facilities

Liudmyla Davydenko[1(✉)], Viktor Rozen[2], Volodymyr Davydenko[3], and Nina Davydenko[1]

[1] Lutsk National Technical University, Lutsk, Ukraine
L.Davydenko@mail.ru
[2] National Technical University of Ukraine "Kyiv Polytechnic Institute", Kyiv, Ukraine
auek@ukr.net
[3] National University of Water Management and Nature Resources Use, Rivne, Ukraine

Abstract. The article considers the issue of information support of the energy efficiency control system of public water-supply facilities. The aim of the article is to develop principles of forming a field-specific information space and formalizing its procedures to improve the efficiency of control. Formalized description of subjects and selection of procedures-algorithms is suggested to be made taking into account the type of subject and the assignment of research task. Submission of input and output data is expected in the form of information blocks. Energy efficiency control procedures have been formalized on the basis of object-oriented technology. Formed classes cover attributes of a subject, environment, energy consumption parameters, as well as planning methods and energy consumption control. Classes with common procedures are incorporated into the appropriate class categories. WEB-focused technology is suggested to be used to share information. This will create a single information space, provide the ability to process information on energy efficiency of water-supply facilities in real time and facilitate integration of control procedures into energy management of company.

Keywords: Public water supply system · Energy efficiency control · Information space

1 Introduction

Energy efficiency is one of the priorities of the EU. In the context of energy reserves reduction and energy market value increase, the issue of improving energy efficiency refers to the strategic objectives of states and is a priority of individual organizations and businesses. Technical approaches are used to reduce energy consumption. They rely on investment in energy saving technologies. Today, approaches aimed at creating an appropriate management system [1] are a key element of energy efficiency. Energy Management Standard ISO 50001:2011 [2] provides management approach that involves the use of the best management practices and gives almost unlimited possibilities in the field of energy efficiency to any company. This requires improvements to

R. Szewczyk and M. Kaliczyńska (eds.), *Recent Advances in Systems, Control and Information Technology*, Advances in Intelligent Systems and Computing 543, DOI 10.1007/978-3-319-48923-0_24

existing and development of new control functions and procedures, as well as their integration into energy management system. Systematic energy management necessitates operational rather than periodic energy efficiency control [3], for both production system as a whole and its structural elements.

Currently, experts and scientists conduct studies on efficient energy use in various fields of social production [4]. The study has resulted in a significant number of publications on implementation of energy management systems [2, 5, 6], performance monitoring and energy efficiency level [3, 5, 7], etc. However, the issues related to organization of energy efficiency integrated control in public water-supply systems, which would take into account the particularity of functioning of the whole system and its structural elements, have been insufficiently considered.

The objective of the study is to improve the effectiveness of the energy integrated control of public water-supply facilities by creating a field-specific information space based on formalized description of its subjects and elements-entities given the particularity of operation of water-supply facilities and specification of control procedure tasks.

2 Materials and Methods of Study

Public water-supply system consists of many elements that consume certain types of energy for implementation of technological process, are characterized by certain initial conditions, are at different hierarchical levels and have their own performance features. It is a complex system with ordered hierarchical structure and extensive network of relationships between its elements. The analysis of functioning of a complex production system requires examining the operation conditions of its facilities and evaluating the efficiency of available types of resources. The system is represented as a set of components necessary for its existence and functioning [8]:

$$S \underset{def}{\equiv} \langle Z, STR, TECH, COND \rangle, \tag{1}$$

where $Z = \{z\}$ is the set of functioning objectives;
$STR = \{STR_{vyr}, STR_{org}, \ldots\}$ is the set of structures (production, organizational) that implement the goal;
$TECH = \{\text{meth, means, alg}, \ldots\}$ is the set of technologies (methods, tools, algorithms);
$COND = \{\varphi_{ext}, \varphi_{in}\}$ is conditions of the system: factors that affect its functioning.

The fact that solution to the problem of energy efficiency is aimed at stabilization of functioning and improvement of systems with hierarchical structure causes hierarchy of the problem. The problem for each hierarchical level is a combination of sub-problems and at the same time can be considered as a part of a more complex problem. The problem of energy efficiency control of public water-supply facilities should be considered as one being composed of sub-problems of various ranks: energy efficiency of production, of technological processes, of structural elements and component units, as well as management effectiveness and organization of technological process.

Each level of energy control task implies its own goals and objectives (necessarily subordinated to a single ultimate goal), as well as its own structure of sub-problems [9].

Operation of a complex industrial system is determined by technological processes that implemented by its subjects pursuant to the target within the specific field. Management impacts on field-specific subjects are applied to ensure their effective functioning. One of the structure-forming elements distinguished functionally (pumping unit, pumping station, production process, hierarchical level, etc.) is a field-specific subject. The subject should be clearly described by a set of parameters that reflect the system of quantitative parameters of its original state, operation efficiency and energy consumption efficiency.

In general, each field-specific subject should be presented as a set (tuple) [10]:

$$Sub \underset{def}{\equiv} \langle name, St, Fn \rangle, \tag{2}$$

where *name* is the name of a field-specific subject; *St* is the set of original state parameters; *Fn* is the set of performance indexes.

Each of the energy efficiency parameters has a factor identity. They constitute the following factors: climatic, technical, technological, power and operational. This allows for representation of a set in the form of a tuple of relevant factors:

$$St = \langle Cl, Thn, Thl, Pw \rangle; \tag{3}$$

$$Fn = \langle Thn, Thl, Pw, Op \rangle, \tag{4}$$

where *Cl* is the set of parameters *P*, pertaining to a particular field-specific subject, which constitute a climatic factor; similarly, *Thn* – technical factor; *Thl* – technological factor; *Pw* – power factor; *Op* – operational factor.

Formation of a set of parameters of energy efficiency requires consideration of hierarchical identity of the selected research object (field-specific subject) and identification of research task class. This will give a structured set of parameters that provide a sufficient degree of detail for the selected level of problem statement and don't require a detailed description of subjects of lower levels [9].

Information space of energy efficiency control system of public water sector is a specially organized collection of attributes (parameters) of its components which along with quantitative and qualitative values (parameters of these attributes) distinguishes one subject from another. Construction of information space implies a formalized description of subjects by a complex of parameters inherent to them by using certain description techniques [11]. Each subject is classified by its type; identified by a list of its own properties; has a list of subjects of which it consists or to which it is included according to a certain subordination and with which it has relations. Functional status of the subject is defined by list of indicators that reflect parameters of its operation. Thus, quantitative attributes-characteristics of efficiency control subject are technical and technological parameters, energetic characteristics and energy efficiency indicators that describe the efficiency of its initial state and regime organization.

Besides, the main criterion determining the efficiency of water supply system is provision of consumers with water in the required volume. Water consumption is uneven and is influenced by many factors. Incompleteness and unreliability of initial information leads to errors of water supply planning, which causes irrational operation of pumping stations, energy overconsumption, excessive pressures and greater water loss due to leakage in the water supply network [7]. Organization of water supply regime should closely correspond to water consumption regime. Therefore, monitoring of environmental factors that affect should be one of the constituent elements of energy efficiency control system. Consideration of the factors is necessary for implementation of procedures-algorithms of water-supply and energy consumption planning regimes.

Functional elements-entities of information space of integrated energy efficiency control include: schemes, rules, algorithms and alternate calculations; results of calculation execution and control; reporting forms. Each subject may have its own (not typical for other subjects) procedures-algorithms for calculation, the results of which are used in the algorithms of higher level subjects. At the same time, there are calculation procedures-algorithms typical for all field-specific subjects, regardless of their hierarchical identity.

Selection of elements-entities and procedures-algorithms is not only defined due to functional features of the field-specific subject, but also depends on the specification of research task.

Control system should provide regular recording of energy consumption and its fluctuation, as well as possibility to detect, by analyzing energy efficiency performance, certain energetic aspects and processes that need to be improved [12]. Changes in energy efficiency indicators should be measured relative to the baseline energy consumption ("standard"), recorded in the output energy profile, as well as given the best examples of efficient energy use [7]. Energy efficiency control requires constant analysis of energy efficiency dynamics, identification of trends to deterioration (improvement), operational determination of non-random decrease (increase) in energy efficiency levels and validated assessment of reasons for such changes [3, 13].

Thus, energy efficiency control system in water-supply facilities should include [7]:

(1) operational energy efficiency control subsystem, which provides current control of water-supply dynamics as a factor that determines efficient energy consumption regime; current control of dynamics of energy efficiency indicators in terms of their compliance with certain ranges of energy efficiency levels; control of compliance with energy consumption "standard";
(2) energy efficiency benchmarking subsystem that contains procedures for comparison of dynamics of energy efficiency indicators with analogical indicators of the best facilities of the same type; procedures for comparative analysis of compliance of actual energy consumption regime with the "standard" regime of the best facilities of the same type.

Correct setting of objective variables, description of field-specific subjects and selection of necessary calculation procedures-algorithms is an important moment for further application of energy control procedures and obtaining adequate results for making decisions as to improvement of energy efficiency.

All field-specific input and output data should be presented in the form of information blocks [11]:

- Block information on facilities is to describe the state of subjects; contains information on subjects of energy efficiency control process;
- Block information on facility parameters is to describe indicators which are characterized by sets of numerical values and reflect the state of subjects;
- Block information on calculation algorithms is to describe calculation patterns for generating values of facility calculated parameters, which are indicators of energy efficiency in the performance calculations according to various methods depending on the task assigned and hierarchical identity of facility;
- Block information on calculation results is to describe alternate calculations for field-specific facilities and to store obtained data for further use in reporting documents;
- Block information on reporting forms documents is to prepare and generating standard forms of documents which contain calculation results.

The set of structural and functional dependences available in the system would act as the element of information search algorithm, i.e. would indicate sequence of data selection, order of necessary calculations and control and reporting procedures.

3 Research Results

Energy efficiency integrated control procedures in public water-supply system is formalized using object-oriented technology. Environmental facilities are simulated by classes with associated properties and existential rules, i.e. populations that share common characteristics and qualities. The class contains properties of a facility (defines facility data structure, rules observed by facilities) and methods (functions) that have access to the data of a facility, process them and perform certain operations and tasks. Architecture of procedure for energy efficiency integrated control is shown in Fig. 1.

There are three categories of classes: WEB-service is a set of classes that are combined by the procedure for obtaining initial information about the object of study;

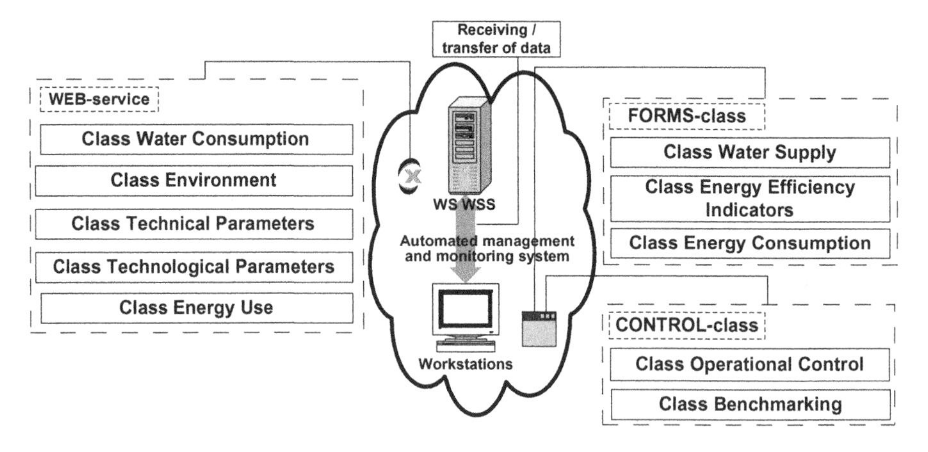

Fig. 1. Architecture of energy efficiency integrated control in public water supply system

FORMS-class is a set of classes combined by computational procedures and models; CONTROL-class is a set of classes combined by procedures for direct energy efficiency control. Properties of class are represented by quantitative characteristics of the object of study, and methods are represented by calculation algorithms, procedures, communications, actions, functions, etc., which ensure functionality of class (or its model). Description of properties and methods of classes is defined by the type of research object (specific-entity subject) subject to energy efficiency control and the specification of the research task.

To provide operation of the architecture proposed, it is necessary to adjust procedures for input and output data exchange according to research task assigned. In accordance with ISO 50001 current global trends in energy efficiency management provide widespread implementation of automated utility metering systems, process management systems (they are a part of the overall monitoring system) and energy management information systems. Availability of automated control systems facilitates collection and processing of information on each of the field-specific facilities. Their connection to the INTERNET provides communication between control points of management entities and central server of a company. This will allow for consolidation of data in the unified system. WEB-service (WS) Water Supply System (WSS) should be installed to implement energy efficiency control procedures of public water-supply facilities in a central server. This will provide for real-time processing of necessary output information, tracking of the temporal evolution of events, assessing of controlled quantities mutual dependence, etc.

4 Conclusions

The procedure for generation of information space to control energy efficiency of water-supply facilities requires formalization of description of facilities and functional field-specific elements-entities and consideration of links between technological processes and their components, as well as environmental impact. Formalized description of subjects involves the formation of the set of energy efficiency indicators, covering technical and technological parameters which characterize the efficiency of initial state and regime organization. The set of indicators should be structured, i.e. consider the hierarchical identity of the selected research object and identification of research task class. Selection of elements-entities and energy efficiency control procedures-algorithms is determined by functional features of a field-specific facility and depends on specification of research tasks, whereas sequence of data selection and order of calculations is determined by structural and functional links.

Transmission of data flows between field-specific subjects and central server of a company on the basis of Web-oriented systems will create a single information space and provide the ability to process information on regime parameters and energy efficiency indicators of the structural elements of water-supply system as a whole in real time.

References

1. European Commission (2009) Integrated Pollution Prevention and Control (IPPC). Reference Document on Best Available Techniques for Energy Efficiency [Электронный ресурс], 430 p. Institute for Prospective Technological Studies, European IPPC Bureau, Seville (2008). http://eippcb.jrc.ec.europa.eu/reference/ene.html, http://www.russian-city-climate.ru/
2. ISO 50001:2011. Committee draft. Energy management systems — Requirements with guidance for use
3. Nahodov, V.F., Petskova, O.O., Ivanko, D.O.: Monitoring of energy consumption indicators in the energy management system. Power Industry, Ecology, Human, Scientific Works of NTUU “KPI”, IEE. - Kyiv, NTUU “KPI”, 480 p., pp. 210–217. IEEE (2015)
4. Korobiichuk, I., Bezvesilna, O., Ilchenko, A., Shadura, V., Nowicki, M., Szewczyk, R.: A mathematical model of the thermo-anemometric flowmeter. Sensors **15**, 22899–22913 (2015). doi:10.3390/s150922899
5. Kovalko, O.M., Novoseltsev, O.V., Evtukhova, T.O.: Introduction to the Theory of Energy Efficiency of Multilevel Systems: Methods and Models of Energy Management in Housing and Public Utility Sector: NAS of Ukraine, Institute of Engineering Thermophysics, 252 p. (2014)
6. Inshekov, E.N.: ISO 50001 “Energy Management System” from origin to commercialization/EN inshekov. Energy Saving, Power Industry, Energy Audit **10**(116), 53–55 (2013)
7. Davydenko, L.V.: Principles of formation of integrated energy monitoring system for the water-supply and sanitation enterprise. Power Ind. Econ. Technol. Ecol. **3**, 107–115 (2015)
8. Volkova, V.N., Emelyanova, A.A. (eds.): System Theory and Analysis in Business Management: Directory, 848 p. (2006). M.: Finance and Statistics
9. Davydenko, L.V.: Indicators system creation for the energy efficiency benchmarking of municipal power system facilities. Problemele energeticii regionale **1**(27), 58–70 (2015)
10. Rozen, V.P., Davydenko, L.V., Davydenko, V.A.: Formation of information field for evaluating the energy efficiency level in public water-supply system. Bull. M. Ostrogradskyi KSPU **4**(63), 50–53. KSPU, Kremenchug
11. Borukaev, Z.H., Ostapchenko, K.B., Grytsiuk, L.I.: Computer model of energy efficiency monitoring: aspects of information modeling. Power Ind. Electrification **1**, P.3–P.7 (2007)
12. Korobiichuk, I., Shavursky, Yu., Nowicki, M., Szewczyk, R.: Research of the thermal parameters and the accuracy of flow measurement of the biological fuel. J. Mech. Eng. Autom. **5**, 415–419 (2015). doi:10.17265/2159-5275/2015.07.006
13. Korobiichuk, I., Shostachuk, A., Shostachuk, D., Shadura, V., Nowicki, M., Szewczyk, R.: Development of the operation algorithm for a automated system assessing the high-rise building. Solid State Phenom. **251**, 230–236 (2016). doi:10.4028/www.scientific.net/SSP.251.230

The Measuring System for Bee Hives Environmental Monitoring

Jerzy Niewiatowski[1(✉)] and Paweł Nowak[2]

[1] Industrial Research Institute for Automation and Measurements PIAP, Al. Jerozolimskie 202, 02-486 Warsaw, Poland
jniewiatowski@piap.pl

[2] Faculty of Mechatronics, Warsaw University of Technology, sw. A. Boboli 8, 02-525 Warsaw, Poland

Abstract. The research on living organisms require high standards for non-invasiveness of the measurement setups both in operation principle as well as in the utilized materials. Adding an energy to the measured object may lead to disturbance of body functions and measurement interference. Also wrongly selected materials for construction of measurement setup may impact on organisms response. The specific of the beekeepers work is monitoring of bees family growth in conjunction with the season. Beekeeper makes decisions which are a result of expected results of breeding, state of bee family and a season. Intensity of conducted measurements is labour-intensive for the worker and overloading for the bees. In order to minimize the frequency of hive inspections a measurement setup for environmental parameters of a bee hive was developed in Industrial Research Institute for Automation and Measurements PIAP. It was anticipated, that, basing on proper measurements, the state of bee family may be determined without the need of hive opening. The main task was development of measurement setup in a way which will not interfere with bees normal life and even be invisible to them. Also an attempt was made to develop a tools for tests for possible (presented in literature [4]) reduction of parasite *varroa destructor*. This reduction presumably may be obtained by the optimization of a size of honeycomb cell. With the utilization of possibilities of 3D printing workshop, a method for creating stencils for wax foundation was developed. This method allows to create wax foundations with the varied size of honeycomb cell, which may also change monotonically in a single wax foundation.

Keywords: Measurement setup · Bee hive environment · Matrix · Mites

1 Introduction

With the growth of information systems, monitoring systems with real time data analysis are becoming more common [8–10]. Modern methods of data processing require proper measurement setup mounted on the monitored object, containing control system which provides communication via external interfaces as well as well-designed and properly displaced sensors for measured parameters. Nowadays, with actual state of technology development, biggest issues during measurement systems construction concerns the sensors. The main difficulty is to provide proper power supply for sensors

R. Szewczyk and M. Kaliczyńska (eds.), *Recent Advances in Systems, Control and Information Technology*, Advances in Intelligent Systems and Computing 543, DOI 10.1007/978-3-319-48923-0_25

mounted on distant objects. Also the design of the sensor may present a challenge for the constructor if the measurements are conducted in unusual conditions. Example of such non-standard are measurements conducted inside a bee hive. Object of the measurement is typically far in a field with no access to electricity. Also the operating temperature range of the sensors has to be properly wide, or shelters for the measurement systems have to be additionally built. Conducting measurements of an inside of a bee hive is even more challenging due to the contact with a living organism – bee family. Measurement system shall not interfere with the bees live and provoke them to attack on sensors which are foreign bodies to the bees. Additionally such attack would be a disruption in the bees life and obtained measurement results would not be reliable.

In order to fulfill above requirements, in developed measurement system sensors were deluged in the bee wax and control electronic setup was placed in a drilled side bar made of metal. Top bar was modified by adding cable outlets which allowed to move the measurement data outside of the bee hive. In order to obtain proper thermal isolation and remove thermal bridges (which may lead to increase of the nest moisture) top bar remained wooden. Measurement frames are connected through the signal outlets into the RS-485 network and with RS-485/USB converter with PC, which provides data distribution and power supply for the system as presented in Fig. 1. This setup was used for temperature and sound measurements.

Fig. 1. Bee hive with the measurement setup; 1 – measurement frames, 2 – RS-485/USB converter, 3 – laptop providing data distribution and power supply

2 Temperature Measurement

Measurement system for temperature distribution in a bee hive was previously reported in [7] and was the first stage of development of series of measuring instruments for metering the bee hive. Construction of the temperature measurement frame is presented

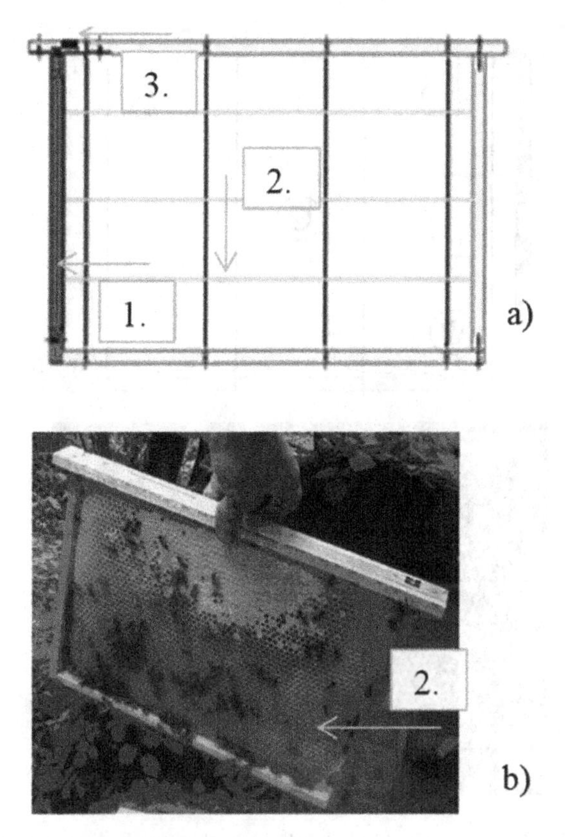

Fig. 2. Construction (a) and result of bees work (b) on the temperature measurement frame: 1 – metal beam with electronic control setup, 2 – NTC stripes deluged in the wax foundation sheet, 3 – cable outlet in the top beam

in Fig. 2. During its construction important data about possibilities of casing of wax foundation sheet with embedded measurement elements were obtained. 3D visualization of temperature distribution inside the bee hive allows:

- identification of shape and movement of winter cluster,
- determination of too big/too small nest relative to the winter supplies,
- determination of bee family's multitude,
- conducting of veterinary examinations – determination of bees reaction to nursing.

3 Sound Measurement

Construction of the sound measurement frame based on previously mentioned setup, is presented in Fig. 3.

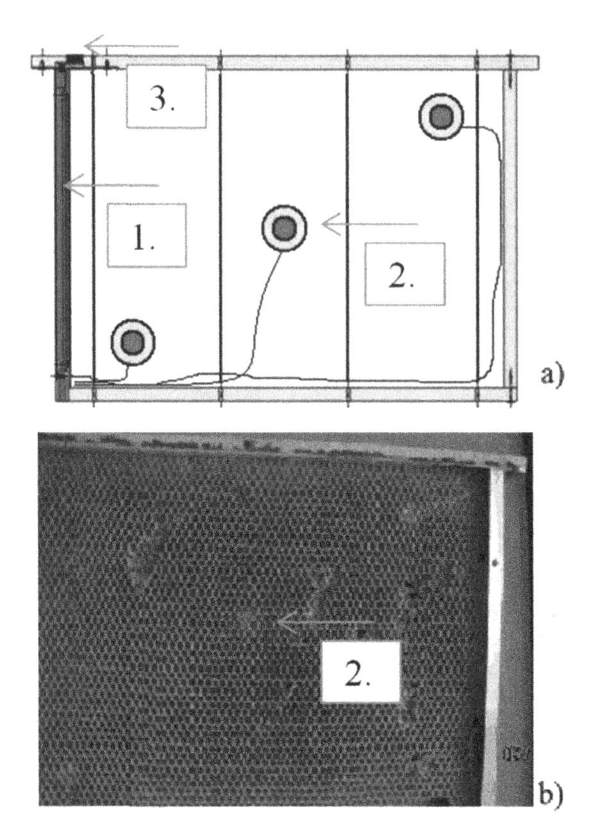

Fig. 3. Construction (a) and result of bees work (b) on the sound measurement frame; 1 – metal beam with electronic control setup, 2 – piezoelectric element deluged in the wax foundation sheet, 3 – cable outlet in the top beam

3.1 Listening Watch

Motivation for development of the noise measurement setup were previously reported observations [1, 2] as well as promising results of a recent studies [5, 6]. Those studies connected the noise in the hive with the bee family state, as presented in Tables 1, 2 and 3.

In this paper we present an idea for determination of the frame filling.

3.2 New Idea of Automatic Identification of Bee Family State and Frame Filling Determination

Results of the previous measurements [7] on bees tolerance of the objects deluged in the wax foundation, data about bees biology and positive results of studies reported in [5, 6] present high probability of successful development of setup for identification of bee family's state and frame filling type. This identification contains both qualitative (whether frame contains maggots, honey or beebread) as well as quantitative

Table 1. Listening sounds (time domain)

Sound of wintering [1]	Bees family state
Smooth, quiet hum	Normal course of wintering
Smooth, loud hum	Superheated, suffocated family
Smooth, loud and sharp sound	Too big aeration of the hive
Loud hum with sharply, violently fly out	Too tight nest, thirst
Quiet hum with the sound of the rustle of leaves	Hunger
Restless, whining hum with the sound of single bees	Bee interregnum
Uneven, very loud and fidgety hum	Filled rectum, possible danger of purging

Table 2. Listening and automatic analysis of sounds via FFT (frequency domain)

Sound of queen [2, 5]	Bees family state
Toot	Queen walking on the honeycombs
Quack	Young queens in swarm cells answer to the queen bee's tooting
Singing near the young queen	Bees are preparing for queens replacement

Table 3. Modelled sounds

Sounds, for which managed to find mathematical models [5, 6]
Robbing behavior
Swarm mood
Medical condition
Drone's presence
Stress

differentiation (amount of maggots, honey, etc.). With proper arrangement of the sensors, determination of the filling distribution may be possible as well. This data, combined with temperature distribution measurement, would give an overall picture of the bee nest without the need of hive opening.

In order to conduct the identification, well know source of the sound is required. According to [5, 6] such source of sound may be bees activity, which causes honeycombs vibrations. For the identification procedure proper fragments of sound records, describing uniform background, periodic or aperiodic oscillations or short vibrations with increased amplitude, can be used.

Developed sound measurement system ensures full consistency overt time – measurement is synchronized in all circuits with accuracy up to 250 ns. For acoustic frequency range this ensures simultaneity of the measurements, which is crucial for the correlation methods.

Figure 4 presents exemplary signal obtained from the sound sensor which was deluged in the wax foundation. This is a sensors response to the external scrapes on the wax foundation surface, which confirms the capabilities of measurement setup before implementation in a bee hive. Further research confirmed bees willingness to the wax foundation rebuild.

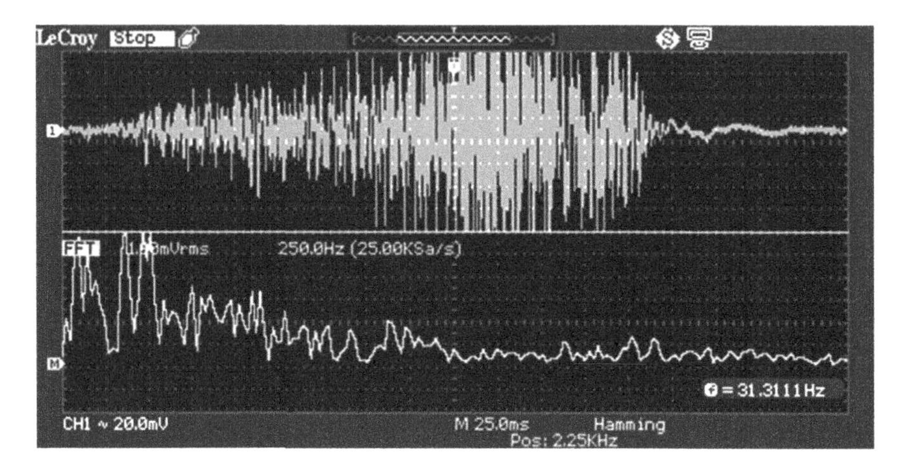

Fig. 4. Amplitude response (in the time and frequency function) of the piezoelectric membrane, deluged in the wax foundation sheet, to the scrapes

4 Wax Foundation Matrixes

In order to deluge measurement elements in the wax foundation sheet, proper matrix had to be designed. Construction presented in Fig. 5 allows sheets thickness adjustment as well as precise positioning of deluged measurement stripes.

Patter for wax foundation extrusion was designed in a CAD software, created with additive manufacturing on PolyJet 3D printer and cast in silicone. This method is multistep, but provides proper cohesion of extruded cells on both sides of the sheets. Due to this method wax foundation sheet of arbitrary cell size can be created.

Based on the above method, wax foundation sheet with monotonically transition of honeycomb cell size was designed and created. The center of the sheet was formed by small cells (e.g. 4.7 mm size) whereas the edges were formed by a larger cell (e.g. 6.2 mm size).

Motivation for development of such wax foundation sheet were data presented in [3, 4], which describe possibility of increase of bees vigour by its size reduction. Bees return to their normal size which increases its ability to resist risk of mites infection such as *varroa destructor* and *acarapis wood.*

Natural honeycomb has alternating size of cell, which differs due to the:

- cells for the workers larvae,
- cells for the drones larvae,

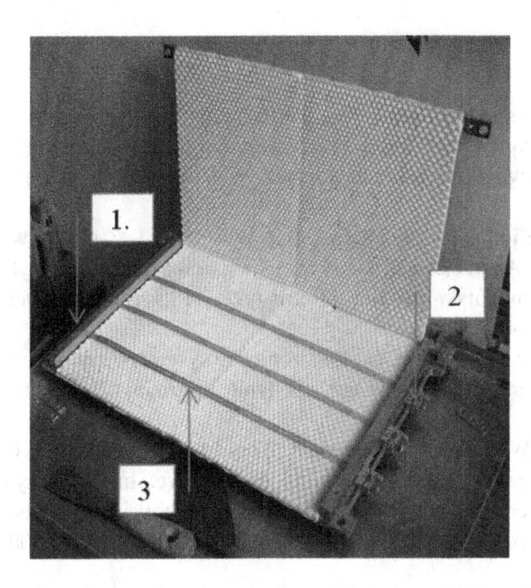

Fig. 5. Matrix for wax foundation sheet production with the possibility of measurement objects deluging. 1 – beam fixing the thickness, 2 – slide positioners of measurement stripes, 3 – delunged measurement stripe

- queens cells,
- honey cells.

Also different bees breed, which characterize with different size, produce honeycomb with different cell size.

Presented method of relatively easy creation of wax foundation sheets with different patterns provides useful tool for further research on the ways of the cells variance for the best projection of the natural honeycombs shape. This research may provide data about necessity of different size bees in a single family, as well as prove the possibility of bees development orientation by proper wax foundation structure.

5 Conclusions

Assumption of development of a system for bee hive parameters measurements, which does not interfere with the bees, was verified and confirmed in practice. Measurements of a temperature distribution confirmed usefulness of the method for both research objectives and in typical work in bee-garden. Initiated research with the sound measurements confirmed non-invasiveness of the method.

Obtainment of possibility of building of a matrix for wax foundation sheet with monotonically transition of honeycomb cell size is an incentive for further research [3, 4] on bees auto-sanitization based on precise regulation of their size and composure.

References

1. Ostrowska, W.: Gospodarka pasieczna. PWRiL (2013)
2. Pszczoły, G.W.: Poradnik hodowcy, Wydawnictwo RM, Warszawa (2014)
3. Thomas, K.: (www.imkereikober.de). http://www.miodpodkarpacki.pl/produkty/produkty_weza.php
4. Baudoux, U.: The influence of cell size. Bee World, 1/33 & 1/34. www.beesource.com
5. Krzywoszyja, G., Andrzejewski, G.: System wspomagania diagnostyki rodzin pszczelich, Uniwersytet Zielonogórski Instytut Informatyki i Elektroniki, Materiały X Konferencji Naukowej SP (2014)
6. Krzywoszyja, G., Andrzejewski, G.: Elektroniczny system oceny stanu biologicznego rodzin pszczelich, Uniwersytet Zielonogórski Instytut Informatyki i Elektroniki, PAK **59**(11) (2013)
7. Niewiatowski, J., Winiarski, W., Nowak, P., Topolska, G., Grzęda, U., Gałek, M.: Issues and problems with measuring the temperature in the hive. In: Szewczyk, R., Zieliński, C., Kaliczyńska, M. (eds.) Challenges in Automation, Robotics and Measurement Techniques. AISC, vol. 440, pp. 797–805. Springer, Heidelberg (2016). doi:10.1007/978-3-319-29357-8_70
8. Internet monitoring of beehives. www.e-ruche.fr
9. Open Energy Monitor-Bee Hive Monitor. http://openenergymonitor.org/emon/beemonitor
10. Nova Labs Helps Local Beekeepers Collect Hive Data. www.nova-labs.org/blog/

A Formal Model in Control Systems Design

Małgorzata Kaliczyńska[1(✉)], Stanisław Lis[2], Marcin Tomasik[2], and Tomasz Dróżdż[2]

[1] Industrial Research Institute of Automation and Measurements PIAP, Al. Jerozolimskie 202, 02-486 Warsaw, Poland
mkaliczynska@piap.pl
[2] Faculty of Production and Power Engineering, University of Agriculture in Krakow, Balicka St. 116b, 30-149 Krakow, Poland

Abstract. The article describes a method for compiling the control algorithm using computer simulation in MATLAB/Simulink software. The process of the closed control system design is the implementation of the methodology using a formal model. The essence of the control system operation is presented in the block diagrams. Logical verification (off-line simulation) of the correct functioning of the control system based on the formal model was carried out. Simulation has not revealed operation errors – the model calculated the correct output signal values for the simulated values of the input signal. The laboratory stand for hardware verification (on-line simulation) has been developed. Some elements of a real control object were included in the feedback loop of the virtual control system, stored in the computer memory. The course of the hardware verification of the proposed solution has been presented. With a specific signal from the real measuring element, the system calculated the control signal for the real actuator correctly.

Keywords: Control · Formal model · MATLAB · Driver

1 Introduction

An essential component of a properly functioning technical system is the properly configured control system. If during the control system designing, there is a possibility of identifying the dynamic properties of the experimental control of the object, then the tuning of the created system, is carried out based on the characteristics obtained during the experiment [1]. The design process calls for a different approach to the problem when the designer does cannot determine the characteristics of the control object experimentally. In this case, it is necessary to apply an alternative methodology. Analysis of a formal model of the technical system may provide a lot of relevant information about the control system. Such a model is understood as an abstract description of some features of the system, expressed using formal tools in the form of block diagrams and mathematical formulae [2]. The proposal of the methodology assuming the use of a formal model in the design process of the control system has been presented in the article. A set of guidelines is based on the recommendations from the literature [3–5]. Block diagram in Fig. 1 presents the general assumptions of the project cycle phases, according to the proposed methodology.

R. Szewczyk and M. Kaliczyńska (eds.), *Recent Advances in Systems, Control and Information Technology*, Advances in Intelligent Systems and Computing 543, DOI 10.1007/978-3-319-48923-0_26

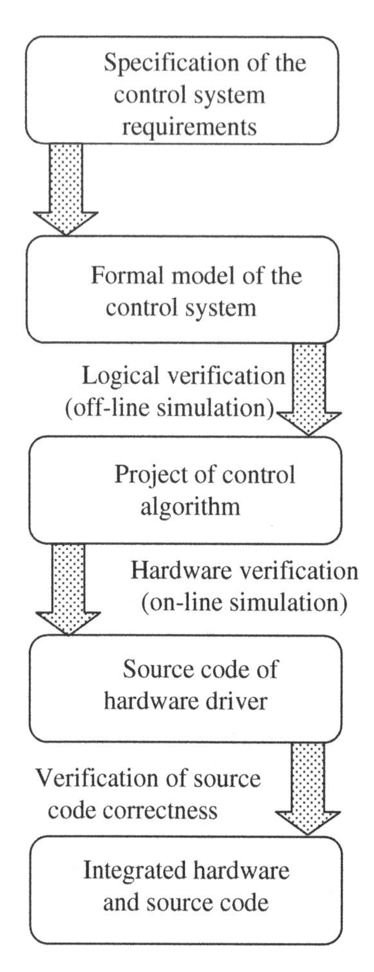

Fig. 1. Process of application development using the design methodology based on a formal model

The cycle begins with determining the requirements for the designed control system, then a formal model of the control system is prepared. Its correctness is verified with the off-line simulation. In case of the unsatisfactory result, it is necessary to tune the model parameters or redesign it. Activities related to the modification of the model and the simulation of its operation are carried out repeatedly, until the desired results are obtained and comply with the design assumptions. The control algorithm design resulting from these steps is verified in real time using the physical object – it is the on-line hardware simulation. At this stage, the real elements representing the controlled object are entered into the virtual loop control system. Coupling between the real elements, and the designed controller, is made using the I/O cards. Based on the positively verified control algorithm, the source code is developed which is implemented in the real hardware system. For the system prepared like that, the verification of the correctness of code operation is carried out. The outcome is the integrated and calibrated control system.

During the evolution of the application according to the methodology discussed, cycle phases may be partly skipped and their sequence, presented in the diagram, does not have to be followed closely.

Project cycle according to the presented methodology is discussed on the example of the temperature control system design. The goal was to develop the control algorithm for a hardware driver, controlling the temperature in a system.

2 Specification of Requirements

In this case, the objective function of the control system is to keep the temperature in the object at a pre-set level. The controlling element in a closed control system can be the two-state or PID regulator. Currently, it is also possible to use the more advanced technical solutions based on algorithms from the field of artificial intelligence (fuzzy logic and artificial neural networks). The considered example, however, does not require any sophisticated solutions that would feature an unnecessary computing burden of the hardware driver, therefore, the project assumes the use of a two-state regulator.

3 Control System Analysis

The general assumptions for the control system, which ensure the implementation of the conditions presented in Sect. 2 are illustrated by the diagram in Fig. 2.

The control system illustrated in the diagram above consists of the following functional elements: control object, an actuator in the form of a fan, measuring element (temperature sensor), summing node and setting device.

The operation of the system is as follows: the signal generated by the sensing element with the controlled quantity measurement information is compared with the set

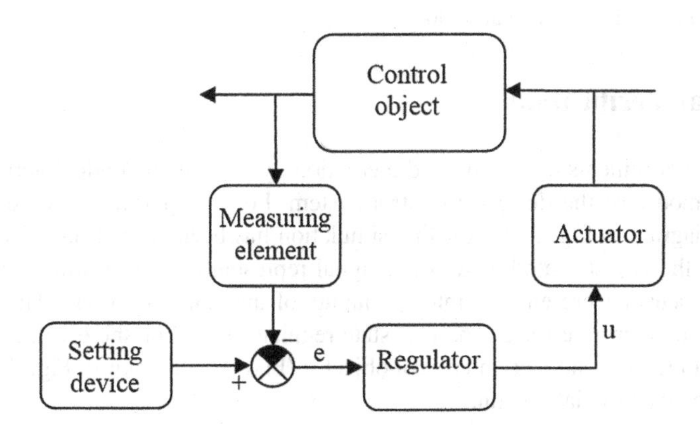

Fig. 2. Block diagram of a closed-loop temperature control system in the object

point in the summing node. The calculated difference is the control error **e**. The signal which represents the difference is entered to the regulator input, which calculates the feedback value **u** to be performed by the actuator on the control object.

4 Formal Model of the Control – Logical Verification

The analysis carried out based on a formal model provides the essential technical information on the system. Thus, at the next stage of the project cycle, a formal model of a closed-loop control system has been developed using MATLAB/Simulink software. Literature recommendations [2, 6–9] were followed in analysis of computer modelling and calculations methods for modern automation systems. The assumptions concerning the control system were written in the form of mathematical relationships which, in accordance with the nomenclature of Simulink software, took the form of functional system blocks (Fig. 3).

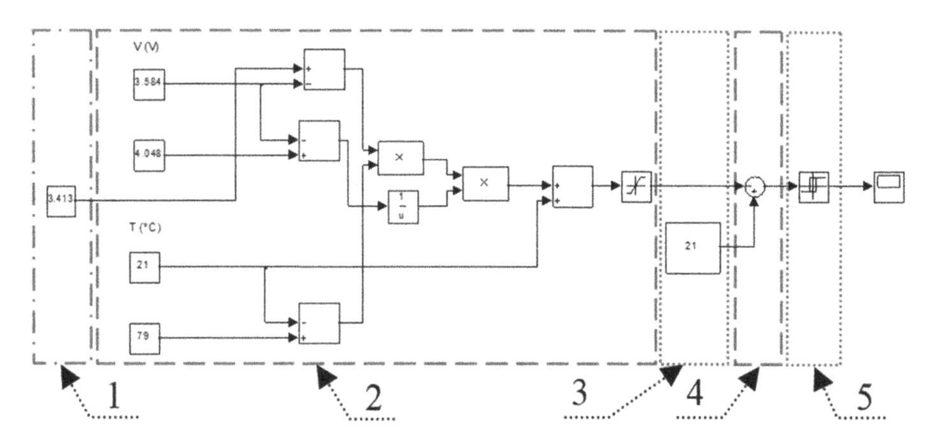

Fig. 3. Formal model of closed-loop temperature control system in the object – off-line simulation: 1 – block representing the sensor signal, 2 – scaling module, 3 – adjuster block, 4 – the summing node, 5 – regulator block

5 Logical Verification

One of the requirements of the adopted methodology [2] was the logical verification of the formal model of the designed control system. For this purpose, according to the presented diagram in Fig. 3, the off-line simulation has been carried out. The set point signal from the adjuster module and the signal representing the information from the temperature sensor were entered into the inputs of the summing node. The summing node generated control error **e**. The two-state regulator based on the **e** value calculated the signal of feedback impact on control object **u** (Fig. 3). Diagram in Fig. 4 illustrates the course of the simulation run.

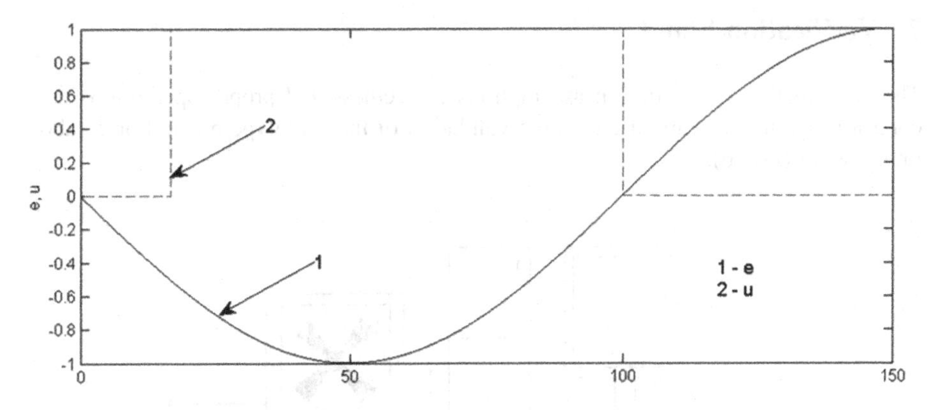

Fig. 4. The course of the off-line simulation

When the control error signal **e** = 0 occurs, the two-state regulator was giving back the control signal **u** with the value of 0, and when **e** $\neq$ 0 occurs, signal **u** equals 1. It should be noted that the simulation process showed no functional errors of the virtual control system.

6 Formal Model of the Control System – Hardware Verification

The next step was to verify the hardware. The hardware verification was carried out by on-line simulation, which consisted in merging the actually used objects of the control system into a virtual feedback loop control system. The system was fitted with the analogue input and digital output blocks so that the formal model could work with real elements (Fig. 5).

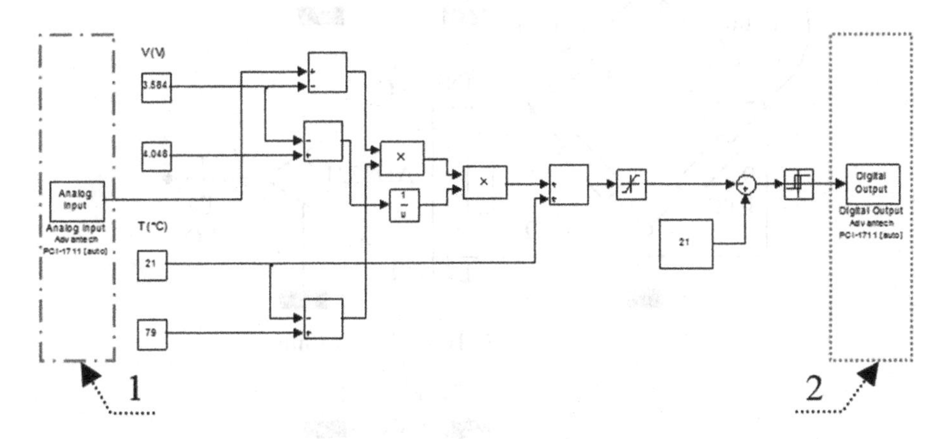

Fig. 5. Formal model of closed-loop temperature control in the object – on-line simulation: 1 – analogue input block, 2 – digital output block

7 Verification Stand

The confirmation of the design assumptions correctness and proper operation of the control program was carried out as the validation of the prototype method on a laboratory stand (Fig. 6).

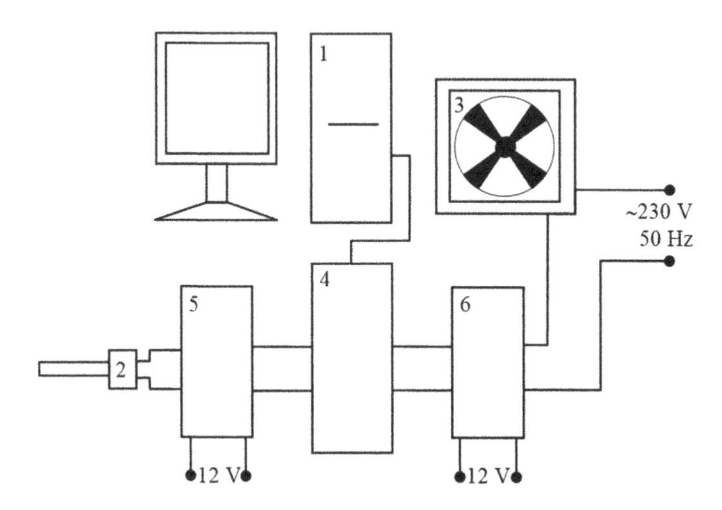

Fig. 6. Diagram of the test stand: 1 – PC computer with I/O PCI–1711 card, 2 – measuring element, 3 – actuator, 4 – PCLD–8710 clamp terminal of I/O card, 5 – amplifier

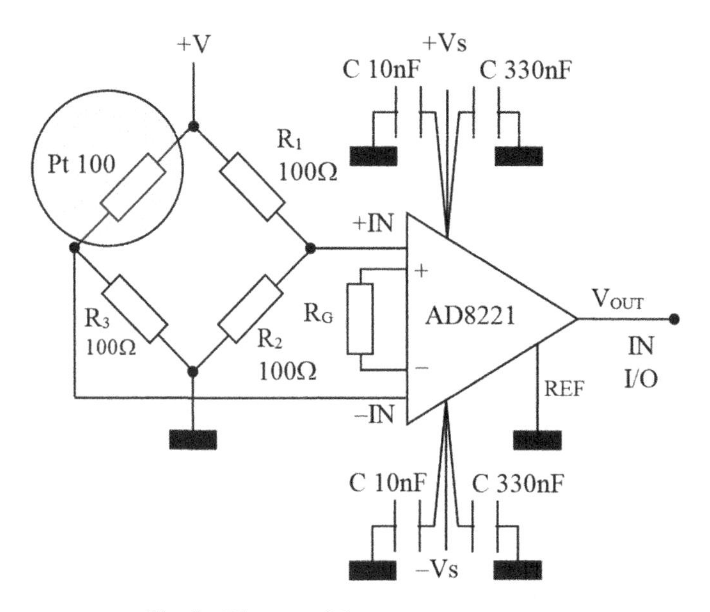

Fig. 7. Diagram of the measurement system

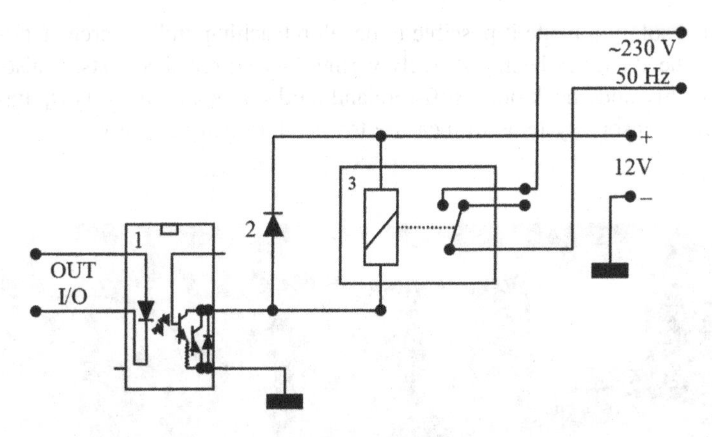

Fig. 8. Diagram of overload protection module along with electromagnetic relay: 1 – optocoupler H11G1, 2 – rectifier diode, 3 – electromagnetic relay

Connection between the formal model for closed-loop control stored in computer memory and physical elements representing the control object was implemented with the I/O PCI–1711 card. The measuring system consisting of a Pt 100 temperature sensor and a signal amplifier, was connected through the PCLD–8710 terminal clamp to the analogue input of the I/O card. Actuators of the control system (fans, overload protection module with electromagnetic relay) were connected to the digital outputs of the I/O card.

The measurement system made with the AD 8221 amplifier, controlled by measuring bridge is presented in Fig. 7 [10]. Figure 8 presents a diagram of the overload protection module along with the electromagnetic relay.

The temperature control stand is shown in Fig. 9.

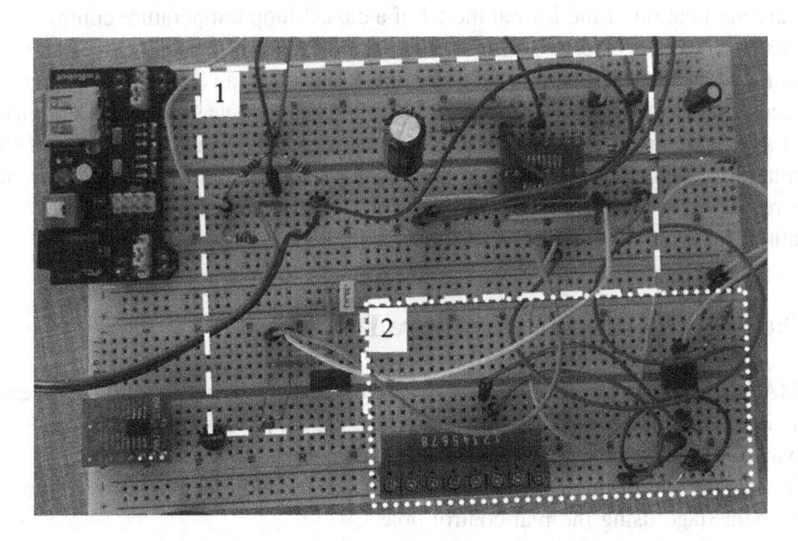

Fig. 9. The test stand made in the Laboratory of Automation in the Faculty of Production and Power Engineering, University of Agriculture: amplifying circuit – 1 and protection circuit – 2

Presented solution made it possible to develop teaching and research stands shown in Fig. 10. The stand has been positively verified during repeated tests. It also has the ability to modify and check other software and hardware solutions. It is equipped with the additional sensory systems to measure temperature and humidity.

Fig. 10. Stand for hardware verification

8 Hardware Verification

Hardware verification of the formal model of a closed-loop temperature control system was carried out on a laboratory stand (Fig. 9). The control system was designed to maintain a constant temperature. During the simulation an interfering signal was introduced. The control system response to the interference was correct – it turned on actuators (fans), causing loss of control of the temperature to the pre-set level. During the simulation, there were no errors in the operation of the control system. The course of the report of a control error **e** and control signal **u** was the same as for logical simulation, for which the graph of the analysed values is presented in Fig. 4.

9 The Source Code for Hardware Driver

The MATLAB/Simulink environment used in the design work allows to export automatically the developed control algorithm into C programming language code. This way, it can be used as a hardware driver source code.

The integration of hardware system and source code is possible at the final development stage using the real control object.

10 Conclusions

1. Logical verification (off-line) of the formal model of temperature control system developed in MATLAB/Simulink, proved that it performed properly. The model calculated the correct value of the output signal with the simulated values of the input signal.
2. On-line hardware simulation carried out on the test stand confirmed the adopted assumptions. The control system reacted in a correct way in response to the introduced disturbance – turned on the actuator, resulting in maintaining the controlled value at a pre-set level.
3. The algorithm of the two-state regulator controlling the temperature in the object, guarantees maintaining its value at a pre-set level.
4. The presented design methodology based on a formal model allows for tuning the control system, where there is no possibility of experimental identification of the object dynamic properties.

References

1. Tarnowski, W.: Projektowanie układów regulacji automatycznej. Ciągłych z liniowymi korektorami ze wspomaganiem za pomocą Matlab'a. Wydawnictwo Uczelniane Politechniki Koszalińskiej, Koszalin (2008). Chap. 6
2. Tadeusiewicz, R.: Biocybernetyka. Metodyczne podstawy dla inżynierii biomedycznej. PWN, Warszawa (2014). Chap. 1
3. Kurytnik, I.P., Lis, S., Dróżdż, T., Telega, A.: The method of rapid prototyping in the control algorithm development for a solar installation. Measur. Autom. Monit. **2014**(5), 325–328 (2014)
4. Mrozek, B., Mrozek, Z.: MATLAB i Simulink. Helion, Gliwice (2004). Chap. 11
5. Real-Time Testing Using Simulink: Continuous Verification and Validation. Speedgoat GmbH, Liebefeld (2014). http://www.speedgoat.ch/Applications.aspx
6. Klempka, R., Stankiewicz, A.: Modelowanie i symulacja układów dynamicznych. AGH, Kraków (2006). Chap. 1–3
7. Osowski, S., Stankiewicz, A.: Modelowanie i symulacja układów i procesów dynamicznych. Oficyna Wydawnicza Politechniki Warszawskiej, Warszawa (2006). Chap. 1, 2, 8, 9
8. Tadeusiewicz, R., Piwniak, G.G., Tkaczow, W.W., Szaruda, W.G., Oprzędkiewicz, K.: Modelowanie komputerowe i obliczenia współczesnych układów automatyzacji. AGH, Kraków (2004). Chap. 1
9. Tarnowski, W.: Modelowanie systemów technicznych. WUPK, Koszalin (2004). Chap. 1
10. Kitchen, C., Counts, L.: A Designer's Guide to Instrumentation Amplifiers. 3rd Analog Devices, Inc., U.S.A. (2006). Chap. 1. http://www.analog.com/static/imported-files/design_handbooks/5812756674312778737Complete_In_Amp.pdf#page=25&zoom=auto,-204,140

Automation of Evaporation Plants Using Energy-Saving Technologies

Anatoliy Ladanyuk, Olena Shkolna(✉), and Vasil Kyshenko

National University of Food Technologies, Kiev, Ukraine
{ladanyuk, evlens}@ukr.net, kvdl948@gmail.com

Abstract. The article deals with methods and ways of energy saving automatic control of sugar factory evaporators based on modern control theory, including a number of approaches that can provide both basic tasks of finished-product output and saving of energy and material resources. A great importance of an evaporator plant is admitted not only in a technological complex, but also in a heating complex of a sugar factory as a supplier of secondary steam for other subsystems of technological complex, such as vacuum devices and various heaters, to ensure their proper functioning.

Keywords: Evaporator · Technological processes · Energy saving technologies · Energy saving control · Food production

1 Introduction

The problem of energy-saving control in the sphere of production is being paid much attention to in recent years. The methods, means and algorithms of the efficient control are applied to provide the solutions to the basic tasks of final product manufacturing with some special limitations related to energy and material resources use [1]. It takes place at automation of different technological objects (technological processes, units and complexes). The modern theory of control allows applying some approaches to the systems of complex objects. These approaches are directly used to determine high technical and economic indexes of functioning by means of using smart technologies [2], adaptive and robust-optimal control, diagnostic methods and prognostication, situational and precedential control, etc. [3, 4]. All these methods help to save either material, or energy resources at product cost decrease and provision of its quality.

The given article considers the modern methods of evaporator (E) control at sugar factory.

2 Formulation of the Problem and Investigation Methodology Selection

Considering the problem of evaporator (E) energy saving control, it is recommended to determine the main aspects of (E) functioning from the point of view of system analysis, its place in technological and heat power complexes, its impact on general

R. Szewczyk and M. Kaliczyńska (eds.), *Recent Advances in Systems, Control and Information Technology*, Advances in Intelligent Systems and Computing 543, DOI 10.1007/978-3-319-48923-0_27

technical and economic indexes of sugar factory work. This means (E) features and work indexes have to be estimated in complex way.

Technological and thermal technical features of evaporator functioning in regard to energy saving is presented in diverse by their completeness sources [5–11]. It is showed, that at the conditions of fuel and energy resource (FER) cost increase, their fraction in the processing cost of 1 t of sugar beets is 30–32% and even more and in sugar cost is up to 20% and more. Fuel consumption amounts nearly 6%. Its decrease causes the enterprise profitability increase correspondingly and is considered by heat reduction consumed by technological needs. It exceeds 80% of the overall FER balance cost and reduces heat losses (technological vapor from power plants). So, it makes approximately 32% beets mass (specific losses of heat energy 160 Mcal/t of beets). But the enterprises in Western Europe have the indexes 17–23% and 87–110 Mcal.

The evaporator at a sugar factory is the main consumer of technological vapor. It performs the technological function by thickening diffusive juice, thermal heating function by providing customers with the secondary vapor (vacuum apparatuses and heaters) and power plants with condensate.

It is necessary to take into account some complex interconnected factors while estimating the indexes of E functioning. For example, in order to reduce vapor losses, it is advisable to decrease the total amount of venting from E units. At the same time it is recommended to support the required syrup concentration by the correct venting distribution. It can result in evaporation of the necessary amount of water from juice otherwise; it can decrease the amount of diffusive juice (pumping). Possible compression of the secondary vapor is considered, as well as the decrease of diffusive juice pumping. It is achieved due to the introduction of deep pulp squeezing, redirecting the pulp pressing water to the diffusive process, beet chip and feed water quality increase, etc.

One more problem is the application of film apparatuses at the final units of evaporator which leads to the concentration of 70–72% for syrup and provides the continuous E functioning at optimal mode. It is achieved because of the qualitative juice purification, silt forming inhibitor application, etc. So, the modern automation system design is becoming of the highest priority and is considered further in the given article.

It is also important to mention that the decrease of diffusive juice pumping can cause, firstly, syrup thickening increase. Evaporator lacks juice and water has to be supplied into purified juice or E units. Secondly, evaporation multiplicity increase and venting shift to rear units lead to E efficiency exceed compared to the required one.

Some articles, for example [8], can show the cases of the Harrington function and pinch analysis method application in order to estimate heat energy systems at sugar factories. The Harrington function (desirability function) is applied to solve multi criterion tasks, for instance, to optimize multidimensional objects. It is used to find solution for energy saving tasks by selecting one of the variants for efficiency estimation. Pinch analysis methodology is focused on the minimization of chemical process energy losses by calculating thermal dynamically implemented target energy. It is an integration process (heat and heat power energy). As a result, it has been calculated, that 60–70 kg of returned vapor is taken to process 100 kg of sugar beets, the factories of Western Europe take 40 kg, but the application of film vapor apparatuses can reduce it up to 25 kg. The total implementation of technological and heat energy measures allows decrease of energy consumption to 20–45%. The effective modern automation systems will facilitate this as well.

Evaporators are considered as the components (of subsystem) of heat power complex. It is connected to other subsystems with numerous material and energy flows and the processes of operation information exchange. Any of these systems is characterized by specific modes and functioning indexes, optimality criteria and mathematical models. The terms of energy saving control for complex dynamic objects are presented in [12]. It highlights the necessity to use special the automatic and automated systems.

3 Methods of Energy Saving Control

In order to formalize the task of energy saving control the mathematical model of object is presented in form within the coordinates of state:

$$\begin{cases} \dot{X} = Ax + Bu + D_1 w \\ Y = Cx + D_2 w \end{cases} \tag{1}$$

where: $X(t) \in R^n$ – the coordinate state vector, $U(t) \in R^m$ – the vector of control actions; $Y(t) \in R^l$ – the vector of output variables, $W \in R^{m_1}$ – disturbances or task signals, A, B, C, D_1, D_2 – the matrixes, here matrix A has always $n \times n$, where n – the system order.

Optimality criteria:

$$I_1 = \int_{t_0}^{t_k} f_0(U(t))dt \tag{2}$$

or

$$I_2 = \int_{t_0}^{t_k} U^2(t)dt \tag{3}$$

to minimize energy consumption, or

$$I_3 = \int_{t_0}^{t_k} |U(t)|dt \tag{4}$$

to minimize fuel consumption.

Tasks (1)–(4) consider function $f_0(U(t))$ which determines the form of functionality, where t_0, t_k – the beginning and end of control time interval. The value intervals are set as the values for the state coordinates, output variables and control actions.

Evaporator functioning can be characterized by situational uncertainty which corresponds to the multiplicity of states of functioning H:

$$\dot{X} = f_n(X, U, t, \gamma_h), h \in H, \quad (5)$$

where γ_h – the massif of model parameters in state h.

Special attention has to be paid to the formulation and the methods of multi criteria task solutions [14].

The task of optimal control system development has been of current interest up to now. This system is able to provide synthesis of energy saving control actions in real time at changes of object functioning state. The principle of maximum is used to determine the optimal programs by applying optimality criteria (2), (3), (4). But the object range here is restricted when it is necessary to count the programs again at changes of initial data within the time interval of control, for instance, for multistage processes. The method of dynamic development is applied to such objects, although, some difficulties can appear because of large dimension of the task. The problems of control limit consideration, term fulfillment and weight coefficient selection can also take place when applying the methods of analytical design of optimal regulators (ADOR). The mentioned above methods are not able to provide the obtaining of effective control actions at situational control. One of the methods to solve the mentioned problems is the method of synthesized variables [10] application. It allows the fast determining of form and parameters of optimal control directly for the determined massif of initial data, i.e. object model parameters, control limitations, etc.

Taking into account the importance of evaporator in heat energy complex at sugar factory, it is necessary to select the methods of energy saving control. The connections of different subsystems, their stage, possibilities to plan the losses of energy sources within a single integrated structure of automation system has to be considered as well. The methods of proactive and energy saving control [12–15] can be used for this purpose. The generalized assessment of technological and heat power complexes is the criterion of energy efficiency. This criterion considers either production volumes or resources consumption needed for this production. The prediction models can also be used. These models are based on the statistical data related to: output-resource consumption. However, such models do not possess high precision and are not able to consider constant changes of production situations. Here, the principle of proactive control is the most efficient when the technical and economic indexes of production are used before the resources are over. The simple methods of operative estimation of technical and economic indexes are, as a rule, incorrect and statistically shifted. For example, the direct operative estimation of the specific indicator of product energy consumption is determined as:

$$a(t) = \frac{P(t)}{E(t)} \quad (6)$$

where $P(t)$ – the current volume of production made, $E(t)$ – the current volume of the energy consumed. Task (6) is incorrect because the current energy volume $E(t)$ is directed not to the current output of product $P(t)$, but to future product $P(t + \tau_3)$ due to inertia and delays in technological objects. It does not show the current power consumption. Far more reasonable and correct estimation is:

$$a(t) = \frac{P(t)}{E(t - \tau_3)} \tag{7}$$

where τ_3 is the technological process delay, but in this case estimation $E(t - \tau_3)$ is retrospective one, the resource is over. The prediction estimation can be used here:

$$a(t) = \frac{P(t + \tau_3)}{E(t)} \tag{8}$$

but marginally approximated model of the process (e.g., the delay link) is applied here. The statistic estimation

$$a_{cp} = \frac{P_{cp}}{E_{cp}} \tag{9}$$

is correct, but not operative one. It does not show the process of object dynamics.

The constructive approach is one that is based on the inverse models of object dynamics. These models are formed on the method of exponential filtering method. Formally, the object operator is presented in the fraction rational form with the delay link:

$$W(p) = \frac{\sum_{j=0}^{m} b_j p^j}{\sum_{i=0}^{n} a_i p^i} e^{-p\tau_3}, m < n, \tag{10}$$

where: a_j, b_j – the coefficients, $p = \frac{d}{dt}$ – the differentiation operator. Then, the inverse operator $W^{-1}(p)$ cannot be applied and it requires the implementation of extra mathematical methods, e.g., output signal analysis in polynomial basis. It is done to design the differential part with prognosis, which is a component of the inverse operator.

4 Discussion

Proactive estimations of dynamics can be applied to technical and economic indexes. For example, the current values of technical and economic indexes as well as their proactive values and cumulative estimation of resource saving are showed on the screen on system level.

Evaporator function in non-stationary modes and its evaporative capacity are described by the equation:

$$S_0(1 - \frac{CP_1}{CP_2}) = \frac{KF\Delta t}{\eta}, \tag{11}$$

where: S_0 – the amount of juice at input, [t/h]; CP_1, CP_2 – are correspondingly the content of dry matters in juice and syrup, [%]; K – the heat transfer coefficient, [W/m^{2}k]; F – is the heating surface, [m^2]; Δt – is useful temperature difference, [K]; η – is the vaporization heat.

In order to achieve the energy saving of evaporator control it is recommended to apply the well-known dependency of heat transfer coefficient on juice (syrup) level. This level is supported in different units ranging 35–60% from the heating surface length of tubes. It corresponds to the maximum value K. The methods of robust and extreme control are efficient in this case. The heat transfer coefficient change in 1, 5–2 times in the process of E functioning has also to be taken into account.

5 Conclusions

1. Energy saving control using evaporator implies considering its significance for the functioning in both technological and power heat complexes.
2. Heating vapor saving is possible provided that the levels of juice (syrup) in the units are supported when the heat transfer coefficient gets its highest value.
3. The system approaches and methods based on the existing technological and thermo technical requirements are recommended to use. The connections with other subsystems have to be taken into account as well: by losses (pumping) of juice with diffusive unit, by losses of secondary vapor with product unit (vacuum apparatuses).
4. The functional structure of evaporator automated system is developed on the methods of modern theory of control, information technologies which are implemented to optimization task solutions, situational, precedential and adaptive control. They are provided with information from subsystems of technological and energy monitoring.
5. The implementation of proactive control using time series, prognostic models, coordination algorithms, data base and knowledge creation for subsystems of decision making in terms of uncertainty and risks is a promising direction to increase the efficiency of sources of energy.

References

1. Korobiichuk, I., Bezvesilna, O., Ilchenko, A., Shadura, V., Nowicki, M., Szewczyk, R.: A mathematical model of the thermo-anemometric flowmeter. Sensors **15**, 22899–22913 (2015)
2. Korobiichuk, I., Podchashinskiy, Y., Shapovalova, O., Shadura, V., Nowicki, M., Szewczyk, R.: Precision increase in automated digital image measurement systems of geometric values. Adv. Intell. Syst. Comput. **393**(1), 335–340 (2016)
3. Karachun, V., Mel'nick, V., Korobiichuk, I., Nowicki, M., Szewczyk, R., Kobzar, S.: The additional error of inertial sensor induced by hypersonic flight condition. Sensors **16**(3) (2016). doi:10.3390/s1603029
4. Korobiichuk, I.: Mathematical model of precision sensor for an automatic weapons stabilizer system. Measurement **89**, 151–158 (2016). doi:10.1016/j.measurement.2016.04.017147

5. Stangeyev, K.O., keHrystenko, V.U.: The ways of energy saving for sugar production, Tutorial, p. 32 (2002)
6. Vasylenko, S.M., Shtangeyev, K.O.: Sources of energy saving at sugar factories. Sugar of Ukraine **1**(57), 40–44 (2010)
7. Stangeyev, K.O., Hrystenko, V.I., Vasylenko, T.P., Vasylenko, S.M.: Energy saving at sugar factories. Sugar of Ukraine **2**(98), 14–17 (2014)
8. Boldarev, S.A.: The methods of energy saving in sugar production (review). Integrated technologies and energy saving, pp. 28–33 (2007)
9. Ladanyuk, A.P., Kravchuk, A.F., Kurylenko, O.D.: Heat complex reconstruction: basic parameters and process control tasks. Sugar of Ukraine **4–5**(34), 26–32 (2003)
10. Kravchuk, A.F., Eremenko, B.A.: Technical and economic estimation of evaporator at sugar factory. Sugar of Ukraine **4–5**(34), 33–35 (2003)
11. Golybin, V.A., Fedorchuk, V.A., Lavrenova, M.A., Denisova, E.A.: The ways energy efficiency increase of sugar plant. Messenger of VGUIT **1**, 185–188 (2014)
12. Matveykin, V.G., Muromtsev, D.Y.: Theoretical Background of Energy Saving Control for Dynamic Modes of Production and Technical Use: Monograph, p. 128. «Publishing house Mache building- 1» (2007)
13. Shnaider, D.A.: Proactive control based on the criterion of energy efficiency of heat power complexes in metallurgy production. Messenger of UrGU, Big System Control (25), 215–230 (2010)
14. Ladanyuk, A.P., Kyshenko, V.D., Shkolna, O.V.: Evaporator control in conditions of uncertainty: intellectualization of applied functions. Sci. Works NUHT **21**(6), 7–14 (2015)
15. Prokopenko, Y.V., Ladanyuk, A.P., Sokol, R.M.: The detection of emergency situations for vacuum apparatus of batch action. Technol. Audit Prod. Reserves **6/3**(26), 22–27 (2015)

Adaptive Control of Dynamic Load in Blooming Mill with Online Estimation of Process Parameters Based on the Modified Kaczmarz Algorithm

Vadim Kharlamenko[1], Sergii Ruban[1], Igor Korobiichuk[2(✉)], and Oleg Petruk[3]

[1] State Institution of Higher Education "Kryvyi Rih National University", Kryvyi Rih, Ukraine
vadim-kharlamenko@ukr.net, ruban_sa@i.ua

[2] Institute of Automatic Control and Robotics, Warsaw University of Technology, Warsaw, Poland
kiv_igor@list.ru

[3] Industrial Research Institute for Automation and Measurements PIAP, Warsaw, Poland
opetruk@piap.pl

Abstract. This paper considers approaches of solving actual task of reducing dynamic load of lines in the blooming mill during rolling, which occurs due to a number of uncontrollable factors. An approach that combines modified Kaczmarz algorithm and robust algorithm of speed gradient is employed to estimate past and current values of the state, provide dynamic compensation of uncertainties and changes of control object parameters during metal compression on blooming, as well as to form adaptive control law for solving the problem of reducing the dynamic load in working rolls of blooming mill. Analysis of the results of simulation of proposed methods has shown high efficiency of transient processes with provision of opportunity to compensate dynamic changes in parameters of the research object during exploitation.

Keywords: Blooming mill · Adaptive control · Online parameter estimation · Speed gradient algorithm · Modified kaczmarz algorithm

1 Introduction

The main disadvantage of metal compression on blooming is uncontrollable dynamic load of lines in the state, which occurs due to a number of uncontrollable factors. Sharp change of the dynamic load on working rolls contributes to phenomena such as slipping.

Despite the significant amount of research and development, the existing process automation system of metal compression does not always meet modern requirements

R. Szewczyk and M. Kaliczyńska (eds.), *Recent Advances in Systems, Control and Information Technology*, Advances in Intelligent Systems and Computing 543, DOI 10.1007/978-3-319-48923-0_28

and are not effectively solve the problems in the real production conditions, which are caused by the presence of defects in the casting, constructive peculiarities of equipment, and features of technological methods in the operators work. These objectives include reducing the dynamic load on the electromechanical equipment of main drive of working rolls of blooming mill, timely determination of slipping and reduction in specific consumption of energy caused by fluctuations in energy-power parameters. The general lack of existing methods of control of compression on blooming is to save for the operator of a leading role, and the operator reaction time to a slipping occurrence depends on his experience and state. Insufficient efficiency of solving of these tasks leads to a reduction technical and economic indicators of metal compression process.

Existing methods of process control of compression on blooming does not allow take full account of random factors causing occurrence of slipping of work rolls. Therefore, the solution to this problem should be sought in the use of modern control techniques based on adaptive methods, using online estimation of control object parameters.

The analysis of existing methods of control of dynamic load in the blooming mill led to the conclusion that robust algorithm of speed gradient should be used for forming adaptive control law, and modified Kaczmarz algorithm should be used for online estimation of control object parameters.

Speed gradient scheme is based on the idea of setting parameters in the opposite direction to the speed of target functional changing along the trajectory of generalized object, which is adjusted [1–4].

2 Adaptive Control of Dynamic Load in Blooming Mill with Using of Robust Algorithm of Speed Gradient

The use of robust algorithms of speed gradient allows providing capacity for work of synthesized systems under stochastic disturbances and nonstationarity of control object parameters [1]. Speed gradient algorithm has a two-level structure. Control algorithm (on the first level) depends on the vector of controller parameters Θ. For each $\chi \in \Xi$ it must provide the achievement of the target condition. The second level algorithm provides adjustment of vector Θ for achievement of target conditions at the unknown value of $\chi \in \Xi$ [1].

Speed gradient algorithm (SGA) sets changing rule for vector Θ, which is given by adapter equation

$$\frac{d(\Theta(t) + \Psi(X(t), \Theta(t), t))}{dt} = -\Gamma \cdot \nabla_{\Theta}\omega(X(t), \Theta(t), t), \quad (1)$$

where $\Gamma = \Gamma^T > 0$ – ($m\Theta \times m\Theta$) – gain coefficient matrix,

$$\omega(X(t), \Theta(t), t) = \bar{q}(X(t), \Theta(t), t) \quad (2)$$

for the integral functional represents total derivative with respect to time by virtue of system trajectory, $\Psi(X, \Theta, t)$ – some vector function that satisfies pseudo gradient condition

$$\Psi(X(t), \Theta(t), t)^T \cdot \nabla_\Theta \omega(X(t), \Theta(t), t) \geq 0. \tag{3}$$

Condition (3) is met when

$$\Psi(X(t), \Theta(t), t)^T = \Gamma \cdot \nabla_\Theta \omega(X(t), \Theta(t), t). \tag{4}$$

SGA, represented as (2), is a generalized form. In this case, due to several advantages [1], is advisable to use differentiated form

$$\frac{d\Theta(t)}{dt} = -\Gamma \cdot \nabla_\Theta \omega(X(t), \Theta(t), t). \tag{5}$$

Algorithm robustness is ensured by entering a dead zone for the target function [3].

The components of the adaptive control law will be determined by the following expressions

$$\begin{aligned} U_S(t) &= -\gamma |E(t)| sign(B^T HE(t)), \\ \frac{dK_x(t)}{dt} &= -\gamma_1 B^T HE(t) X(t)^T, \\ \frac{dK_G(t)}{dt} &= -\gamma_2 B^T HE(t) G(t)^T, \end{aligned} \tag{6}$$

where $\gamma_1 > 0$, $\gamma_2 > 0$, $\gamma_S > 0$ – gain coefficients; $U_s(t)$ – signal component of control law; $K_x(t)$, $K_y(t)$ – regulator coefficients; B – control matrix; H – matrix obtained as a result of solving the Lyapunov equation; $E(t)$ – vector of control errors; $G(t)$ – vector of input actions; $X(t)$ – state vector.

3 Online Estimation of Parameters of Metal Compression on Blooming Based on the Modified Kaczmarz Algorithm

Operating conditions of blooming mill require the use of methods consistent refinement of model parameters used to form the vector of control actions in real time. For this purpose it is advisable to use recursive algorithm [5, 6]. Among the wide variety of recursive algorithms it is expedient to use the Kaczmarz projection algorithm.

To solve this problem it is advisable to present Kaczmarz algorithm in the iteration form [7]

$$\theta(n) = \theta(n-1) + \frac{y(n) - \theta^T(n-1) \cdot x(n)}{x^T(n) \cdot x(n)} \cdot x(n), \tag{7}$$

where $n = 1, 2, 3, \ldots, m$ – algorithm iterations; $x(n)$ – vector of input signals; $y(n)$ – vector of output signals; $\theta(n)$ – vector of parameters to be estimated.

Analysis of convergence of calculation results with using of Kaczmarz algorithm showed that the major shortcoming of the proposed algorithm is the low convergence when there is a high degree of correlation between the input $x(n)$ and $x(n-1)$. In addition, when the components of the vector $x(n)$ tends to zero, the value of $x^T(n) \cdot x(n)$ may turn out equal to zero, i.e., parameter estimation may be unstable [8].

To improve the efficiency of the online estimation algorithm is expedient to use the average values of results of measurement of input actions and output values of the control object that will allow to refine the model parameters even in the conditions of insignificant changes of the measurement results. It will also allow to implement the correction of algorithm parameters to reduce the sensitivity of estimation $\theta(n)$ of the algorithm parameters to the measurement errors [9–11].

To solve this problem we used approach described in [9], which provides correction of estimation $\theta^2(n)$ according to the following expression

$$\hat{\theta}^2(n) = \theta^2(n) + \alpha(n) \cdot \left(\hat{\theta}^2(n-1) - \theta^2(n)\right), \tag{8}$$

which is a first-order difference equation with variable coefficient $\alpha(n)$, and the law of its variation is given by expression

$$\alpha(n) = \begin{cases} 1, \; if \; \rho^2\left(\hat{\theta}^2(n-1), \tilde{\theta}(n)\right) > \rho^2\left(\hat{\theta}^2(n-1), \hat{\theta}^2(n)\right) \\ \qquad 0, \; otherwise \end{cases} \tag{9}$$

$$\rho^2\left(\hat{\theta}^2(m), \theta(k)\right) = \frac{\left(y(k) - \hat{\theta}^{2T}(m) \cdot x(k)\right)^2}{x^T(k) \cdot x(k)}, \tag{10}$$

where $\rho^2\left(\hat{\theta}^2(m), \theta(k)\right)$ – square of the distance from the point $\hat{\theta}^2(m)$ to the k-th hyperplane [9].

Given the modifications (8)–(10) generalized algorithm of model parameters correction used to determine the static moment load of electric drives of working rolls based on the modified Kaczmarz algorithm takes the form

$$\theta(n) = \hat{\theta}^2(n) + \frac{\Delta\bar{y}(n) - \hat{\theta}^{2T}(n) \cdot \Delta\bar{x}(n)}{\Delta\bar{x}^T(n) \cdot \Delta x(n)} \cdot x(n). \tag{11}$$

Despite the modification of the basic algorithm (7), proposed algorithm (11) also belongs to the class of projection algorithms. According algorithm (11), estimates $\frown{\theta}^2(n)$ are an updated orthogonal projections of estimates $\hat{\theta}^2(n-1)$ on the hyperplane formed by corresponding vectors $x(n)$ and $x(n-1)$.

4 Simulation Results

In order to determine effectiveness of the proposed method in Figs. 1, 2 below compares of transient processes for angular velocity of work rolls, armature current and the static moment load, obtained with using of the proposed automated control system of dynamic load based on adaptive control with online parameter estimation on the basis of modified Kaczmarz algorithm, and classic dual-circuit system of a subordinate control based on typical PI regulators.

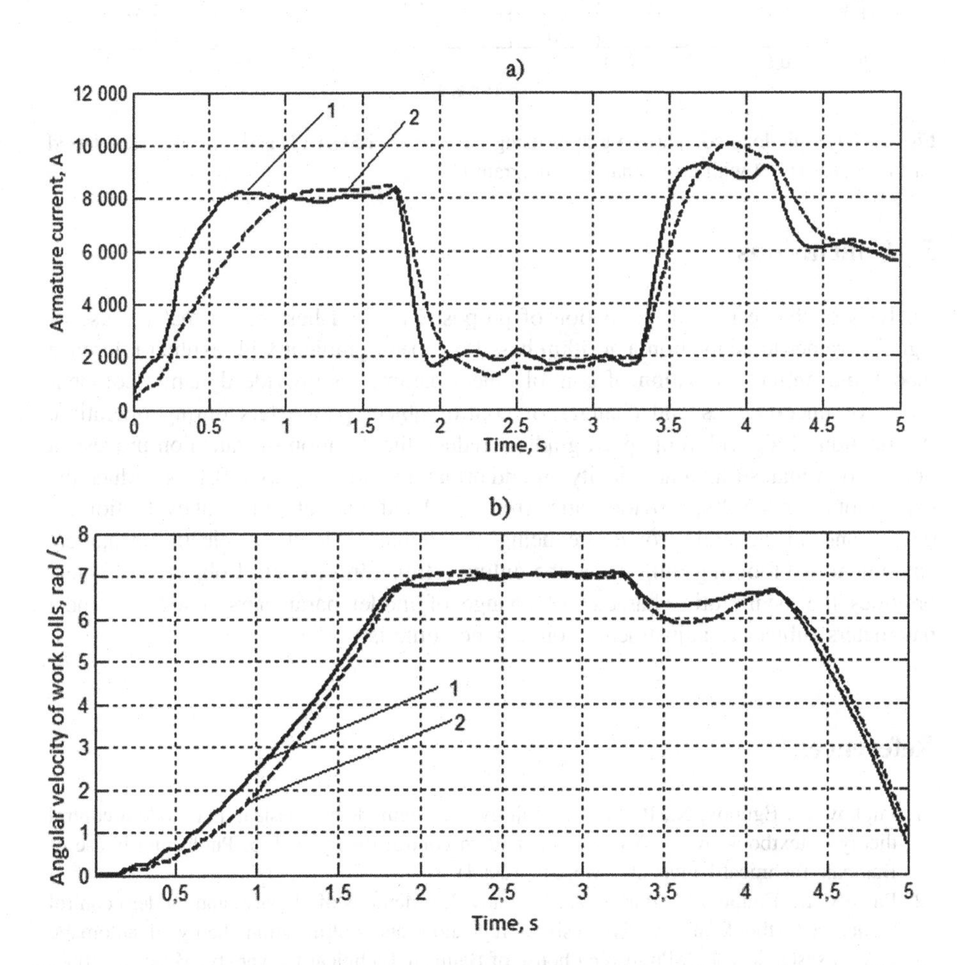

Fig. 1. Graphs of transient processes when using the proposed automatic control system (1) and typical system based on PI regulator (2) for the speed of the work rolls (a) and armature current (b)

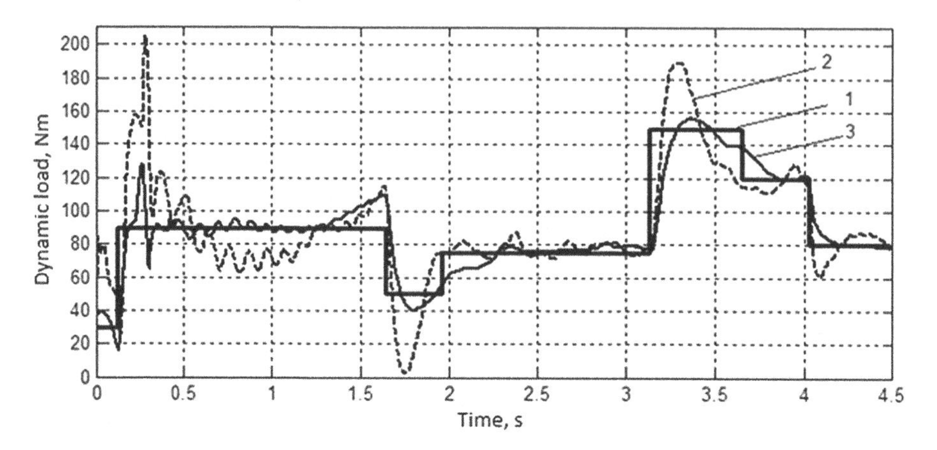

Fig. 2. Dynamic load comparison graphs using the proposed SGA (3) and typical system based on PI-regulators (2) with a nominal load diagram (1)

5 Conclusions

Analysis of the results of simulation of proposed method has shown that the use of signal-parametric adaptation algorithm based on speed gradient with explicit reference model and online estimation of control object parameters provide dynamic compensation of uncertainties and changes of control object parameters during operation. Application of algorithm of speed gradient reduce the duration of transition process at output coordinates (angular velocity ω_1 and armature current I_{arm}) to 0.17 s, reduce the overshoot on $\sigma = 4{,}2\%$, provide static error $\varepsilon_s = 0$ and quadratic integral evaluation $I = 0.647$, and in general provide reducing the dynamic load on electromechanical equipment of blooming mill. Thus, the online estimation of control object parameters provides the asymptotic character of change of model parameters based on object parameters subject to adjustment. Convergence time is 0.01 s.

References

1. Pupkov, K., Egupov, N., Barkin, A., Zaitsev, A.: Methods of classical and modern control theory: a textbook in 5 v. Methods of modern control theory, vol. 5. Publishing house of Bauman Technical University, Moscow (2004)
2. Pupkov, K., Egupov, N., Barkin, A., Voronov, E.: Methods of classical and modern control theory: a textbook in 5 v. Synthesis of regulators and optimization theory of automatic control systems, vol. 2. Publishing house of Bauman Technical University, Moscow (2000)
3. Kim, D.: The theory of automatic control. T. 2. Multi-dimensional, nonlinear, optimal and adaptive systems. FIZMATLIT, Moscow (2004)
4. Korobiichuk, I., Podchashinskiy, Y., Shapovalova, O., Shadura, V., Nowicki, M., Szewczyk, R.: Precision increase in automated digital image measurement systems of geometric values. In: Jabłoński, R., Brezina, T. (eds.) Advanced Mechatronics Solutions. AISC, vol. 393, pp. 335–340. Springer, Heidelberg (2016). doi:10.1007/978-3-319-23923-1_51

5. Bukreev, V.G., Paraev, Ju.I., Shamin, A.M., Chashhin, A.K.: The algorithm of parameter identification of electromechanical object on the basis of the theory of sensitivity. Bull. Tomsk Polytech. Univ. **3**, 143–146 (2000)
6. Kharlamenko, V.Y.: Adaptive identification of load moment on the working rolls of break-down mill based on inverse dynamic problem. Metall. Mining Ind. **2**, 49–51 (2014)
7. Il'in, V.P.: The iteration method of Kaczmarz and its generalizations. Siberian J. Ind. Math. **3**(27), 39–49 (2006)
8. Karelin, A.E., Svetlakov, A.A.: Synthesis and some results of studies projection recurrent algorithms of estimating the parameters of linear models of static objects. In: Automated Systems of Information Processing, Management and Design, vol. 9, pp. 152–164. Publishing House of Tomsk State University of Control Systems and Electronics, Tomsk (2004)
9. Avedyan, E.D., Zipkin, Ya.Z.: Generalized Kaczmarz algorithm. Autom. Telemech. **5**, 195–197 (1978)
10. Korobiichuk, I., Bezvesilna, O., Ilchenko, A., Shadura, V., Nowicki, M., Szewczyk, R.: A mathematical model of the thermo-anemometric flowmeter. Sensors **15**, 22899–22913 (2015). doi:10.3390/s150922899
11. Korobiichuk, I.: Mathematical model of precision sensor for an automatic weapons stabilizer system. Measurement **89**, 151–158 (2016). doi:10.1016/j.measurement.2016.04.017

A Programmable Logic Controller (PLC); Programming Language Structural Analysis

Yulia Kovalenko(✉)

National Aviation University, Kiev, Ukraine
yleejulee22@gmail.com

Abstract. Application of Programmable Logic Controller in the sphere of industrial automation substantially simplified technological processes control. New data exchange systems and new algorithms are being developed. It leads to enormous variety of comptrollers. We consider a programmable logic controller (PLC), a device that performs control of the physical processes of the algorithm written in it, oriented to work with devices developed through the input sensor signals and output signals to the actuators. PLCs are designed to work in real-time systems. Note that one of the advantages of the PLC system is modular. That is, the ability to combine and mix of types of input and output devices in a manner that best suits the application. Each of them differs by specific set of functions, unique construction and certain control language. In this article we describe classification of PLC, that can help to choose one, and also present the structural analysis of PLC programming languages.

Keywords: A programmable logic controller (PLC) · Programming languages · Real-time system · Diagram type language SFC

1 Introduction

Controllers may vary in several aspects, namely the manufacturing country, capacity and scope of use. The first factor lost its relevance some time ago, since the quality of controllers is increasing rapidly worldwide. In terms of capacity, PLCs are subdivided into nanocontrollers, small, medium, large and extra-large controllers. They differ by the number of I/O, network interfaces, memory, word length and main CPU speed. However, the most important factor to consider when selecting a programmable controller is, generally, the scope of its application:

– specialized controller with built-in functions. They feature a low capacity. These controllers already have an action program embedded, and you can only change its settings. This type of controllers is often able to implement various functions, and this determines a set of I/O modules. These PLC are commonly used in a small mechanical installation or a small gear;
– a controller to implement logical dependencies. This type of controllers is also called smart relay, and even the language of controlling such devices is similar to the relay circuits in many respects. It contains a large library of ready logic functions. It is also engaged in blocking of standard actuators. It is used mainly in the

R. Szewczyk and M. Kaliczyńska (eds.), *Recent Advances in Systems, Control and Information Technology*, Advances in Intelligent Systems and Computing 543, DOI 10.1007/978-3-319-48923-0_29

engineering and tank construction. A set of I/O modules is designed for digital channels;
- a controller for implementation of computing and logical functions. It is used almost everywhere thanks to a special generic nature. CPU can handle both logical and computational problems. Typically, such devices are not limited to a single control language, and another coprocessor is added for a better computational efficiency;
- crash protection controller. They are remarkable for fault tolerance, high availability and reliability, which is achieved by different methods of diagnostics and redundancy;
- telemechanical automation system controller. These controllers are generic too, but the scope of their application is quite narrow, namely supervisory systems of monitoring the objects distributed about the terrain. Here the most important thing is components for information data transmission at a distance via a wireless network.

Thus, we consider a programmable logic controller (PLC), a device controlling the physical processes according to an algorithm recorded therein, oriented to work with devices through a branched input of sensor signals and signal outputs to the actuators [1]. PLCs are designed to work in real-time systems. Let's note that one of the PLC advantages is modularity, i.e. an ability to combine the types of input and output devices in a way, which best suits the application [2, 3].

PLC structure can be subdivided into four parts, i.e. input-output modules, CPU, memory and terminal (Fig. 1).

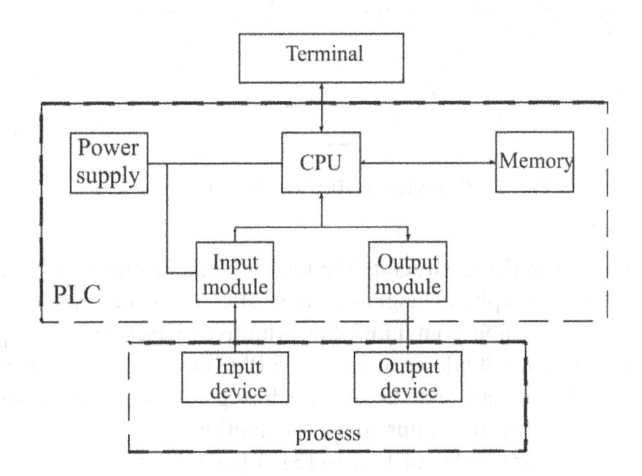

Fig. 1. Programmable logic controller (PLC) structure

2 Materials and Methods

Focusing on IEC 61131-3 [4–6], we determine the syntax and, for a smaller part, semantics of four programming languages for PLCs, as well as an aid for application structuring (SFC diagram-type language).

The first four languages include the LD ladder diagram language, FBD functional diagram language, ST textual high-level language and IL textual low-level language. LD, based on the ladder diagram, allows describing the logic functions. In FBD language, the functionality is presented as graphic blocks. ST textual language is close to the Pascal language, and operates the procedural, conditional and cycle operators. IL is a device-independent assembler-like language. While ST and IL are textual languages, and LD and FBD are graphic ones, SFC, in addition to its own syntax, can be used in conjunction with any textual or graphic language (Fig. 2).

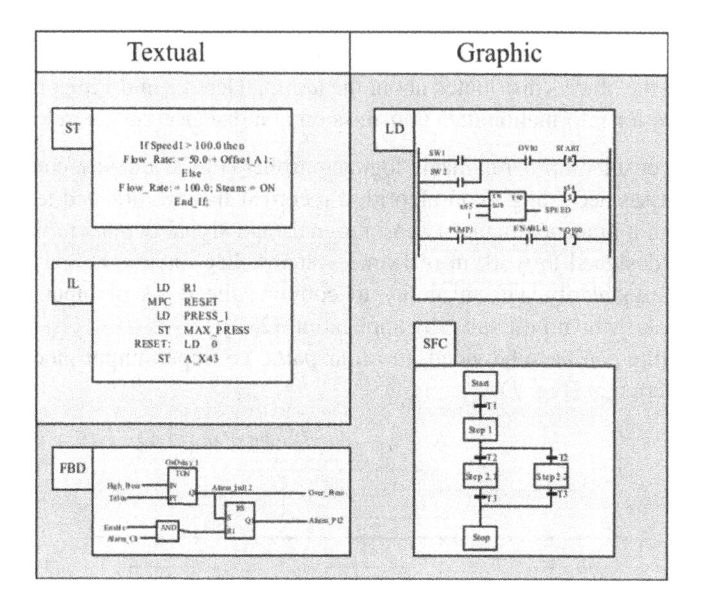

Fig. 2. Overview of IEC 61131-3 languages

SFC was created in order to structure the internal organization of PLC applications or function blocks. This graphical language called SFC (sequential function charts) is a generalized terminal machine containing the primitives describe serial, parallel and alternative lines of conduct. It allows dividing the PLC application (or function blocks) by the sets of interrelated steps and statuses. Each step corresponds to a set of actions, and each status is related to the status transfer condition.

Let's single out the principles of IEC 61131-3 languages:

- the entire application is subdivided into a set of functional elements, Program Organization Units (POU). These elements may be implemented using standard languages and consist of three categories [7]:
  - POU functions while running return only one value, defined by one of the standard types, function result and an arbitrary set of additional output data. These functional elements do not have a certain status (do not contain any information about the status), i.e. a function call with the same arguments always produces the same result;

  - POU functional units return one or more values as the result. The functional unit status is maintained run after run. It features a probabilistic aftereffect, so the call with the same arguments can return different results;
  - POU applications can be defined as a "logical connection of all elements and structures of the programming language required to process the selected signal to control the mechanism or process via a programmable controller" (IEC, 2003). Their definition and use is equivalent to functional units. They also can use the previous two POU categories as additional items.
- standard requires strict data typing. Specifying the data types can detect most errors in the application prior to its running;
- there are means to run various program fragments at different times, at different rates, and in parallel. For example, one program fragment can scan the terminal sensor 100 times per second, while the second fragment will scan the sensor with a rate of once every 10 s;
- to perform operations in a specific sequence triggered by time or events, a special Sequential Function Chart (SFC) language is used;
- the standard supports structures for description of heterogeneous data. For example, the pump bearing temperature, pressure and the "on-off" status can be described by a single "Pomp" structure, which can be transfer within the application as a single data element;
- standard provides for sharing of all five languages, so the most convenient language can be selected for each piece of the problem;
- the application written for a single controller can be transferred to any controller compliant with IEC 61131-3 [6].

So, let's analyze five languages of the IEC 61131-3 standard for PLCs. The use of each language in the PLC application involves two steps [8]:

- define the IEC 61131-3-based semantics;
- define the transition systems representing the language elements, and define the general rules of element layout in the application for ST and LD languages. This leads to definition of operational semantics.

The operational semantics consists of a transition system modeling the PLC running PLC cycle, plus one transition system for each language, representing the behavioral rules of the selected language, and several transition systems modeling the application elements. Modelling of behavioral rules depends on the chosen language.

SFC language structure is converted in the transition system calculating the application change algorithm. The approach of LD and ST languages is more successful when compiling the application in the transition system.

IL (Instruction List) is a textual low-level language. It is generic and often used as a common intermediate language into which other languages are translated. IL is a linearly-oriented language, its main advantage is the ease of study. IL can be programmed using any text editor. It is commonly used to solve small problems with a small number of branches and to write the most critical places in the application, as it allows creating the highly efficient and optimized functions.

At the moment, the third edition of IEC 61131-3 [9] declared the IL obsolete and undesirable for use.

The ST language uses a different approach. ST (Structured Text) is a high-level language, comprising a plurality of designs to assign the values to variables, call the functions and function blocks, write the expressions, conditional branches, choose the operators, and construct the iterative processes. This language is mainly intended to perform complex mathematical calculations, describe the complex functions, function blocks and applications. Its advantages over IL is a very concise formulation of programming tasks, clear program structure in the blocks of operators and major structures to control the running progress. ST algorithm is subdivided into several steps (operators) used to calculate and assign the values in order to control the run, call or exit from POU. In contrast to IL, ST can be defined by several lines or several operators, which may be written in one line.

LD (Ladder Diagram) language is based or the ladder flowcharts with horizontal links between the vertical power rails, executed sequentially. This programming language is developed mainly to process the Boolean signals (true/false). The buses link the LD network on the left and on the right. From the left bus, the current, controlled by "1" signal status, reaches all connected elements. Depending on their status, the elements either allow the current to proceed to the following elements, or interrupt the flow.

LD-network is described by the vertical and horizontal lines, as well as intersection points. The contact performs a logical operation based on the value from the incoming line and the required variable value. The logical operation type depends on the contact type. The value derived from the right connected line is the desired result.

However, it is hard to use the LD to implement complex algorithms, since it does not support routines, functions, encapsulation and other application structuring agents to improve the programming quality. These shortcomings make it difficult to reuse the software components, making the application long and complicated for maintenance. Another downside is that only a small part of the application fits the computer screen or operator panel during programming.

Despite these shortcomings, the LD language is one of the most common in the world [3], although it is used to program simple tasks only (Fig. 3).

FBD (Functional Block Diagram) is a graphical programming language using the functional block diagrams to submit an application to the PLC. It is based on the basic

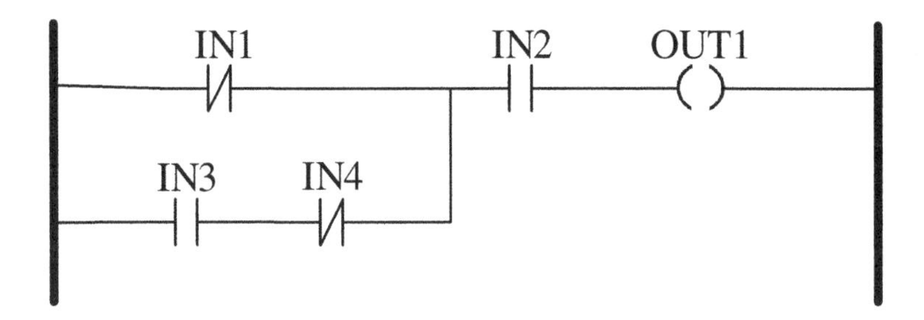

Fig. 3. Example of application in the Ladder Diagram language

block diagrams caused by a graphical representation of electrical circuits and basic language elements, i.e. blocks. Block is a subroutine, a function, or a function block. The functional blocks encapsulate the data and methods, like the object-oriented programming languages, but do not support inheritance and polymorphism. Another block feature is parameterization of inputs and outputs, which can be represented by various types. For example, the ADD block function may have from two to any admissible number of inputs, and a summation of all inputs will be fed to the output.

FBD borrows the symbols of Boolean algebra and, as Boolean symbols have inputs and outputs, which can be connected to each other, FBD is more efficient to represent the structural information than the ladder diagram language. Also, in addition to the basic elements of graphic languages, FBD has some unique elements (Fig. 4).

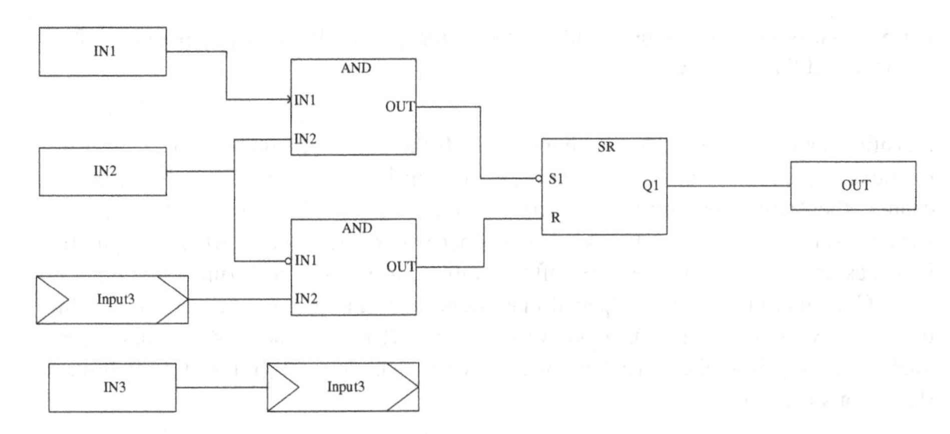

Fig. 4. Example of an application in the Functional Block Diagram language

The diagram-type SFC (Sequential Function Chart) language allows representing a POU application or a PLC functional block using the graphical and textual notation system. SFC is essentially an aid to structure the applications by breaking the main control application branch into smaller components and monitoring of their implementation. Its basis is the mathematical formalism of Petri nets (PNs), which allows describing the processes in the form of bipartite directed graphs [10] (Fig. 5). The graphical representation allows clearly defining the application running progress, and allows the design of serial and parallel application processes.

The operation of any T_j transition within the marked network leads to a change in the marking. Running of the smaller program components (e.g., processes or branches) depends both on conditions determined by the application, and on behavior of the input and output data. The components are programmed directly in one of the other IEC 61131-3 languages. The processes with a stepper behavior are particularly suitable for programming in SFC.

The first level of structuring in SFC is a network made up of elements called steps and statuses. Step can be active or inactive. When it is active, the required commands are run until the step becomes inactive. The step status substitution is determined by the

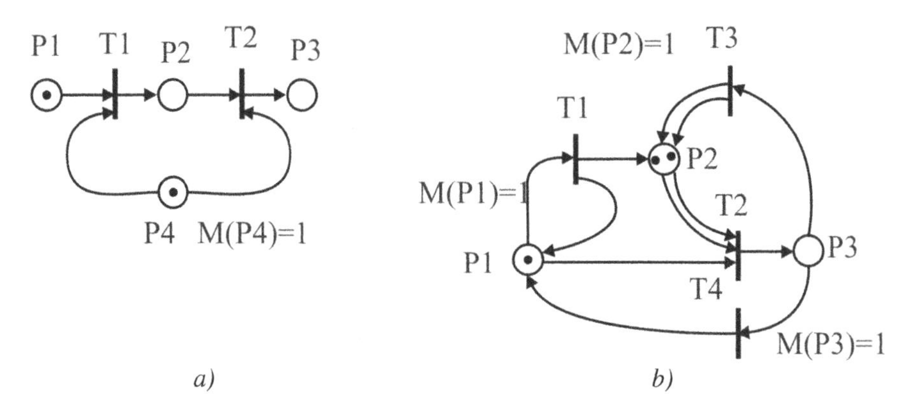

Fig. 5. Process presentation in the form of bipartite graphs. Where: P_i is positions, T_j is transitions, $M(P_i)$ - markings

transition status, which is a logical expression. If the transition condition becomes true, the next step becomes active and the previous step is deactivated. With the transition change, the "active" property moves from an active step to its successor or successors; such movement forms a network. This property may be divided when the parallel branches are run, and then restored after completion of the branch run.

IEC-actions in steps have special classifiers determining how they are run within the step: cyclic running (N), a one-time running (P), etc. There is a total of nine qualifiers, including the classifiers with saving (S), delayed (D) and time-limited (L) actions (Fig. 6).

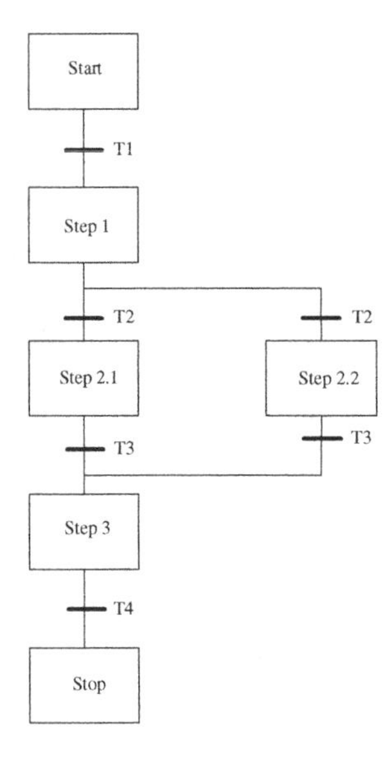

Fig. 6. A set of Sequential Functional Chart network steps combined by transitions

3 Results and Discussion

After analysis of the programming languages, we conclude that the controlling systems operating in real-time logic processes can be programmed on SFC. LD, FBD and IL are the programming languages suitable for formulation of the main action and for operating systems, which can be described by simple logical operations or logic signals. ST language can be used mainly to create the software modules with a mathematical context, for example, to describe the control algorithms. Perhaps, the most powerful and versatile tool from the IEC 61131-3 languages is a SFC + ST combination. SFC language graphics facilitates the language learning. The presence of common roots with Petri nets partially eliminated the synchronization and parallelism issues. Programming of operations with analog and logic variables is comfortable enough due to the use of the textual Pascal-like ST language. The command flow control does not cause problems. A significant advantage of the approach is event-based nature, naturally supported through a "step-transition" mechanism. The application debugging can be facilitated by visual tracing of the control flow. The weaknesses of SFC language (like Petri nets) is an abstraction and structuring [11], which, as in previous cases, adversely affects the difficulties of programmable algorithms and their quality (reliability and maintainability). SFC + ST approach is inferior to FBD and LD in terms of convenient programming of the algorithm parallelism and thus is more suitable to program the linear algorithmic sequences.

4 Conclusions

A typical example of use is combined algorithms of the logic and analog control. The programming experts combining the knowledge of programming languages and algorithmic peculiarities of the automated process are proposed to be the users [9].

Where the PLC or PLC software permits, all created applications or application parts have to be simulated before launch. This allows detecting and eliminating the errors at an early stage, which would then reduce the complexity and cost of such applications.

References

1. Korobiichuk, I.: Mathematical model of precision sensor for an automatic weapons stabilizer system. Measurement **89**, 151–158 (2016)
2. Korobiichuk, I., Podchashinskiy, Y., Shapovalova, O., Shadura, V., Nowicki, M., Szewczyk, R.: Precision increase in automated digital image measurement systems of geometric values. In: Jabłoński, R., Brezina, T. (eds.) Advanced Mechatronics Solutions. AISC, vol. 393, pp. 335–340. Springer, Heidelberg (2016). doi:10.1007/978-3-319-23923-1_51
3. Korobiichuk, I., Shostachuk, A., Shostachuk, D., Shadura, V., Nowicki, M., Szewczyk, R.: Development of the operation algorithm for a automated system assessing the high-rise building. Solid State Phenomena **251**, 230–236

4. Bonfatti, F., Monari, P.D., Sampieri, U.: IEC 1131-3 programming methodology. Software engineering methods for industrial automated systems. CJ International Editions (1997). ISBN 2-9511585-0-5
5. Ohman, M., Johansson, S., Arzén, K.E.: Implementation aspects of the PLC standard IEC 1131-3. IFAC Control Engineering Practice 123 **6**(4), 547–555 (1998)
6. Lewis, R.W.: Programming industrial control systems using IEC 113-3 Revised edition, 329 p. The Institution of Electrical Engineers, London, UK (1998)
7. Barbosa, H., Déharbe, D.: Formal verification of PLC programs using the B method. In: Proceedings of the Third International Conference on Abstract State Machines, Alloy, B, VDM, and Z, pp. 353–356 (2012)
8. De Smet, O., Couffin, S., Rossi, O., Canet, G., Lesage, J.-J., Schnoebelen, Ph., Papini, H.: Safe Programming of PLC Using Formal Verification Methods. Ecole Normale Suprieure, Chaire De Fabrications, France (2000)
9. IEC 61131-3:2013 Programmable controllers - Part 3: Programming languages
10. Anisimov, N.A., Golenkov, E.A., Kharitonov, D.I.: Composition approach to development of parallel and distributed systems based on Petri nets. Programming (6) (2001)
11. Ziubin, V.E.: PLC programming: IEC 61131-3 languages and possible alternatives. Ind. ACSs Control. (11), 31–35 (2005)

Energy-Efficient Electrotechnical Complex of Greenhouses with Regard to Quality of Vegetable Production

Igor Korobiichuk[1(✉)], Vitaliy Lysenko[2], Volodymyr Reshetiuk[2], Taras Lendiel[2], and Marcin Kamiński[3]

[1] Institute of Automatic Control and Robotics, Warsaw University of Technology, Warsaw, Poland
kiv_igor@list.ru

[2] National University of Life and Environmental Sciences of Ukraine, Kiev, Ukraine
lysenko@nubip.edu.ua,
volodymyr.reshetiuk@hotmail.com,
taraslendel@rambler.ru

[3] Industrial Research Institute for Automation and Measurements PIAP, Warsaw, Poland
mkaminski@piap.pl

Abstract. Vegetable growing in greenhouses is an important sector of agriculture, providing the population with fresh vegetables throughout the year. In greenhouses that are typical representatives of hothouse plants, technological parameters of growing plants microclimate support systems, and their functioning is accompanied by costly energy resources, which in turn affects the cost of production. Modern systems of providing growing plants process usually do not track their states in response to the action of natural disturbances, which makes it impossible to accumulate knowledge about the plant but do not to predict its productivity. Therefore, as a result, there is a need to develop energy-efficient method of managing electrical complex in the greenhouse, which takes into account the biological component of the facility during its operation to maximize profits.

Keywords: Electrotechnical complex · Greenhouses · Mathematical model · Energy-efficient method · Quality of vegetable production

1 Introduction

The main objective of any enterprise, including greenhouse plants, is to maximize profits. At the same time the increase in mass manufactured products, does not always lead to profit maximization as a significant place takes quality of the products. Control systems which are used today in greenhouse structures, able in some cases to observe the growth and development of the main indicators of one plant, using stationary sensors photometric quantities but do not consider the quality of products.

R. Szewczyk and M. Kaliczyńska (eds.), *Recent Advances in Systems, Control and Information Technology*, Advances in Intelligent Systems and Computing 543, DOI 10.1007/978-3-319-48923-0_30

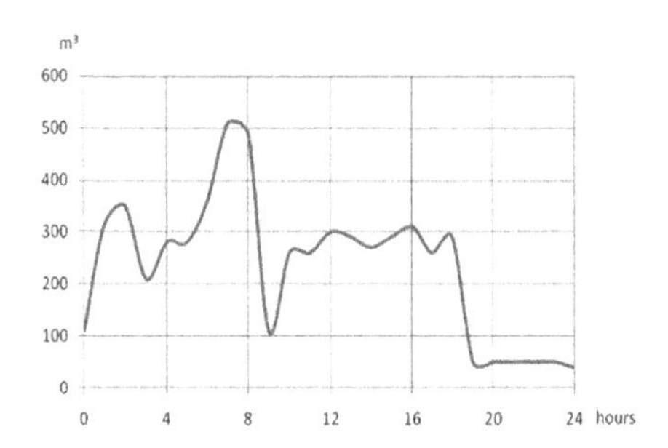

Fig. 1. The daily consumption of natural gas control systems in greenhouses №9 PJSC «Combinat «Teplychny»

The analysis of energy costs for growing vegetable plants in greenhouses the peculiarities processes of crop plants under glass. It was determined that one of the most important components in the greenhouse heating system is supporting the necessary conditions for growing plants, and for its operation consumes the largest amount of energy. Thus, measurements revealed that for the spring period daily gas consumption for hothouse №9 PJSC «Combinat «Teplychny» is 6000 m^3 (Fig. 1). Besides this greenhouse complex electrical overnight consumes about 6000 kW•h of electricity (Fig. 2), which largely determines the cost of production (energy share in the cost of tomatoes is close to 70 %).

The main influence on the amount of energy consumed perform system management strategies form the object of a biotechnological greenhouse.

For these greenhouses feature a biotech facility proposed for the formation of maximizing profit management strategies electrical engineering complexes that accompany the production of vegetable production in greenhouses, to not only the parameters of the microclimate and the quality of products [1].

Formation of the management strategies is given by the analysis of process parameters and their actual values [2–4]. But in such strategies ignores the biological component of the greenhouse – plants – and its own response to natural disturbances such as: solar radiation, temperature and humidity of the environment (Fig. 3).

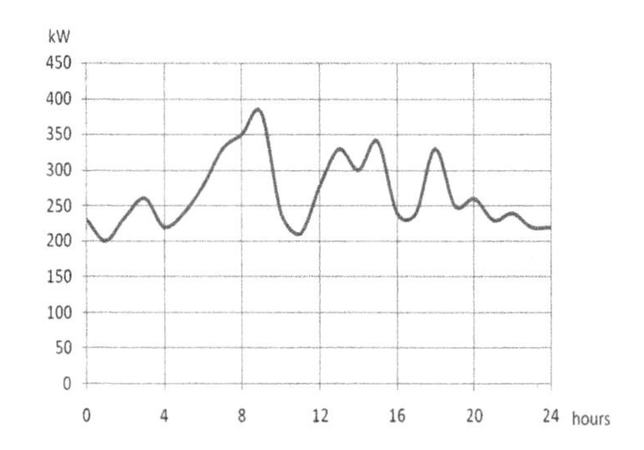

Fig. 2. Daily power consumption of electrotechnical complexes of greenhouses №9 PJSC «Combinat «Teplychny»

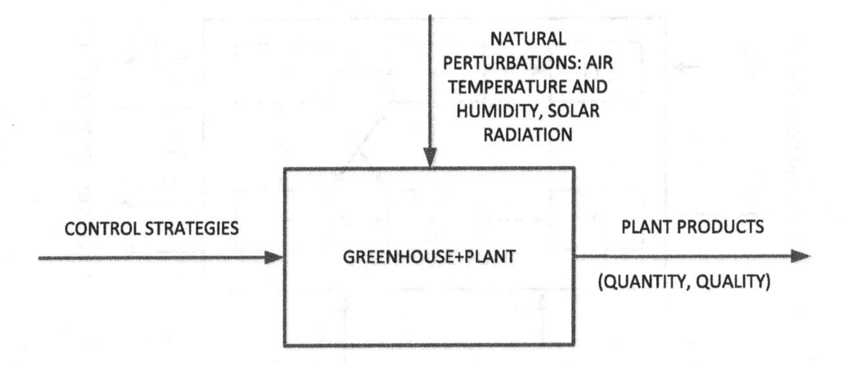

Fig. 3. Input and Output action greenhouses with plants like biotech facility

2 Materials and Methods

The condition in which the plant is located in the space of one of the greenhouses in PJSC «Combinat Teplychny», which is influenced by internal and external disturbances (temperature plants, temperature and humidity, changes in solar radiation) made it possible to use the results of measurements [5, 6] to create management strategies of electrotechnical complex. Defined that the major factors in the plant development is the temperature of the plant and the environment temperature around it [7].

The conducted measurements in greenhouses revealed significant irregularity in temperature, due to a large number of heat transfer surfaces, which contribute to heat loss, and hence the uneven development of plants. So we must consider that the greenhouse is an object with distributed parameters, to assess plant conditions and formation control actions.

For this purpose has been developed mathematical model that takes into account the spatial coordinate in width sections and allowed to calculate the temperature based on the influence of external perturbations (Fig. 4). In this space one section of the greenhouse conditionally divided into 8 temperature zones to the width of the greenhouse into account design features section [7, 8].

It was decided that each section affects the balance of the greenhouse temperature and the amount of heat for *i*-th areas Q_i depend on the amount of heat, Q_{i+1} and Q_{i-1}, which is given to or received from neighboring areas (i + 1, i-1). Thus, we can write:

$$Q_i = Q_{t,i} + Q_{s,i} - Q_{sr,i} - Q_{v,i} + Q_{i+1} + Q_{i-1} \tag{1}$$

where i = 1, …, 8, Q_t – the amount of heat coming from the heating system; Q_s – the amount of heat received from the sun, J; Q_v – heat loss through the roof greenhouses and end walls, *J*; Q_{i+1}, Q_{i-1} – the amount of heat coming from the adjacent temperature zones, *J*; $Q_{sr,i}$ – the amount of heat absorbed by plants *i*-th zone, J.

Due to the geometric dimensions of temperature zones greenhouse, its thermal characteristics (coefficient of heat transfer from the water to the wall of the pipe, the pipe wall to the air greenhouses on air greenhouse to the glass wall of the greenhouse, the glass to the outside air) equation of thermal balance for the *i*-th zone will look like:

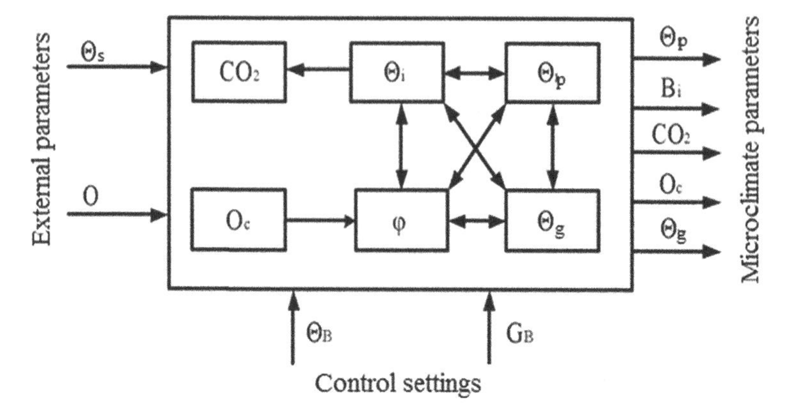

Fig. 4. Block diagram of the climate models used in greenhouses, taking into account external parameters and control parameters: Θ_g – soil temperature; Θ_P – average air temperature; CO_2 – concentration of carbon dioxide in the section; O_c – the intensity of solar radiation in the section; Θ_B – hot water temperature;Θ_c – ambient air;G_B – the cost of hot water; φ – humidity; O – the intensity of solar radiation; Θ_s – the temperature outside; Θ_i – the temperature in the section

$$\begin{aligned} 6,443 \cdot 10^5 \cdot \Theta_i = 1,658 \cdot 10^3 \cdot \left(\Theta_{w,i} - \Theta_i\right) + 3 \cdot S_{k,i} \\ + 4 \cdot \left(S_{b,i} - S_{k,i}\right)(\Theta_i - 20) + 0,026 \cdot S_{i-1,i}(\Theta_{i-1} - \Theta_i) \\ + 0,026 \cdot S_{i+1,i}(\Theta_{i+1} - \Theta_i) - 3,3 \cdot S_{k,i}. \end{aligned} \quad (2)$$

where Θ_i – the temperature in i-th area $\Theta_{w,I}$ – the coolant temperature in i-th zone; $S_{b,i}$ – area; the area of the lateral surface in i-th area; $S_{k,i}$ – in the area of the roof in i-th zone.

The model includes the effect of each zone surrounding temperature zones, namely in the first section and eight areas considered heat exchange with the environment through the end surface. Solar energy is passed through the roof of each area considering its inclination. This in mind final balance for greenhouse will be recorded in the form of equations:

$$\begin{cases} 6,443 \cdot 10^5 \cdot \Theta_i = & 11 \cdot S_t(\Theta_{w,1} - \Theta_1) + 1,181 \cdot 10^3 \cdot (\Theta_1 - 20) \\ & + 8,19 \cdot (\Theta_2 - \Theta_1) - 25,8; \\ 6,443 \cdot 10^5 \cdot \Theta_i = & 11 \cdot S_t(\Theta_{w,i} - \Theta_i) + 3 \cdot S_{k,i} + 4 \cdot S_{b,i} + S_{k,i})(\Theta_i - \Theta_3) \\ & + 0,026 \cdot S_{i-1,i}(\Theta_{i-1} - \Theta_i) + 0,026 \cdot S_{i+1,i}(\Theta_{i+1} - \Theta_i) - 3,3 \cdot S_{k,i}, \quad i = 2\ldots7; \\ 6,443 \cdot 10^5 \cdot \Theta_i = & 11 \cdot S_t(\Theta_{w,8} - \Theta_8) + 3 \cdot S_{k,8} + 1,181 \cdot 10^3 \cdot (\Theta_8 - \Theta_{i+1}) \\ & + 8,19 \cdot (\Theta_7 - \Theta_8) - 25,8. \end{cases} \quad (3)$$

The mathematical model (3) was investigated using MATLAB/Simulink software environment for these input parameters O_s = 300 W/m^2 and Θ_c = 20 °C. The air temperature in different areas after the transition (Fig. 5) stabilized in the range of 23.5 to 26 °C, and the acceleration in this case is from 2000 to 6000 s. This discrepancy is

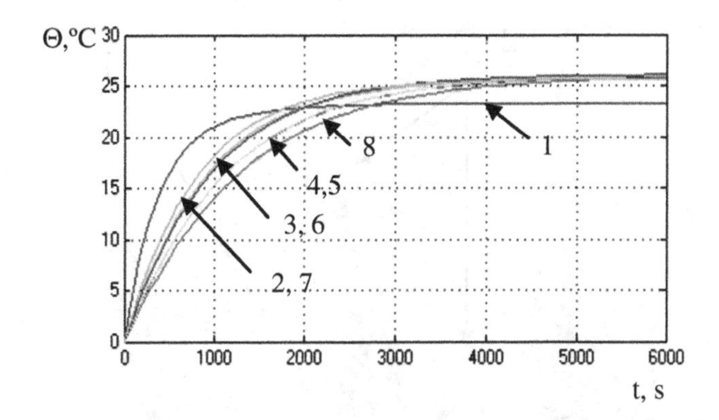

Fig. 5. Change the temperature in the greenhouse: 1…8 – zone section

explained by the different amounts of each zone and the influence of external disturbances in the extreme zone, leading to uneven heat transfer medium from the greenhouse heating. The results of calculation according to mathematical models and experimental measurements of temperatures in cross-section (relief temperature) section of the greenhouse from 1 to 8 zone shown in Fig. 6. This standard deviation is 1.05 °C, confirming the adequacy of the model.

To monitor the development of the plants, the definition of the relevant coefficients of mathematical models of plants was created artificial microclimate chamber (fitotron), consisting of: a vegetative cover utensils for growing tomatoes; reservoir of nutrient solution; pump; humidifier; device for feeding CO_2; lamp lighting chamber; fans; air heaters; temperature sensors; ambient light sensor; CO analyzer; microprocessor unit of measurement and control.

To create a mathematical model depending on the quality of tomatoes from the effects of process parameters in artificial microclimate chamber (Fig. 7) conducted active multifactorial experiment. This input factors are: temperature plants, temperature, humidity, gas composition of air, light, and output – quality tomatoes [1].

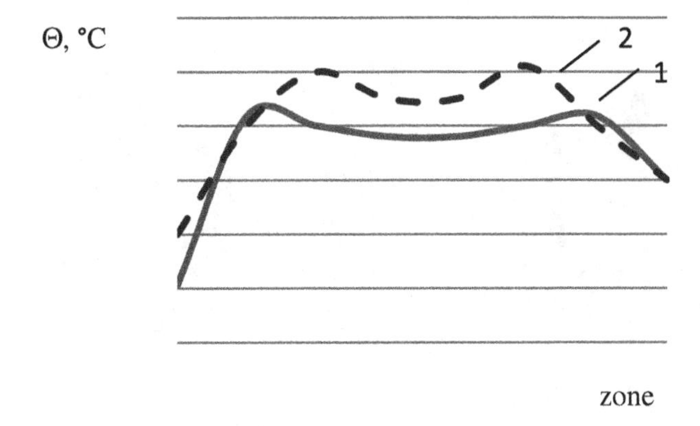

Fig. 6. Average temperature: 1 – calculated temperature; 2 – measured temperature value used for greenhouses (static mode)

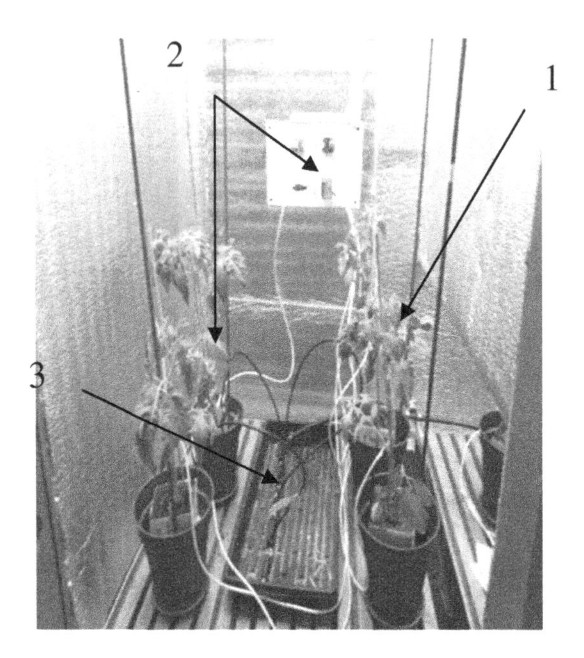

Fig. 7. Camera artificial microclimate: 1 – tomato plants; 2 – sensors; 3 – drip irrigation

3 Result and Discussion

To assess the impact of process parameters, quality and quantity of plant products for profit facilities were closed ground studies using neural networks (multilayer perceptron) when input parameters are air temperature and plant humidity and the quantity and quality of plant production, and output – profit [1, 9].

We studied the neural networks of different structures where the highest accuracy showed the network with the structure "multilayer perceptron" (Fig. 8a, b). To reduce the magnitude of the error using the "Network Builder" in the software package

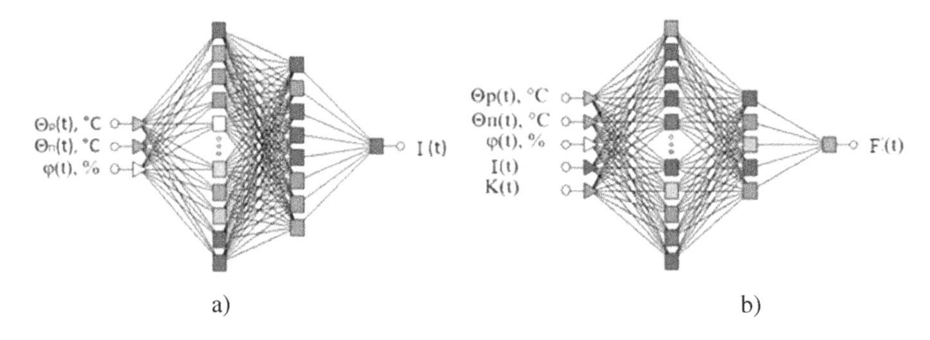

Fig. 8. The architecture of neural networks: (a) NM1 (for quality), (b) NM2 (for assessing profit)

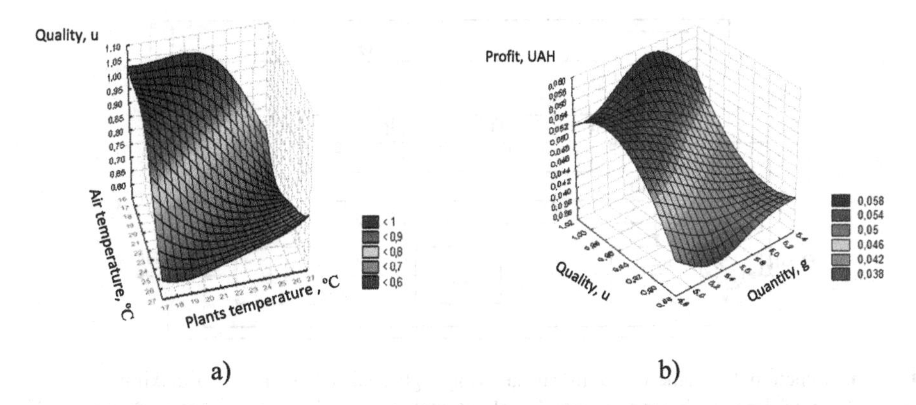

Fig. 9. The results of neural networks: (a) the dependence of product quality of air temperature and plant; (b) dependence of profit on the quantity and quality of production

Statistica 6.0 conducted additional training, through which neural network architecture that provided the training error of 2.4 %, and control – 1.9 %.

The results of the neural network dependence of quality of tomato plants and air temperature (Fig. 9) and profit dependent on the number of products and quality (Fig. 9b) taking into account the costs of technological requirements.

Analysis of the results (Fig. 9), we conclude that in the temperature within 20… 24 °C maximum possible profit is 0.06 UAH per one bush at night, with the increase of quality production of 6 g/h.

In order to determine the factors that maximize profit, Harrington used the desirability function. Generalized optimization criterion F (function desirability Harrington in dimensionless form) delivers current to maximize profits production [6] and is defined as:

$$F\left(I\left(\Theta_p,\Theta_{P,}\varphi\right),B\left(\Theta_p,\Theta_{P,}\varphi\right)\right)=\left(I\left(\Theta_{p,}\Theta\varphi\right)\right)0.5\cdot p,\left(B\left(\Theta_{p,}\Theta\varphi\right)\right)0.5\rightarrow max,\mathrm{p},$$

$$I\left(\Theta_{p,}\Theta_{p,}\varphi\right)\rightarrow max, B\left(\Theta_p,\Theta_{,p}\,\varphi\right)\rightarrow min,$$

where I – function product quality; B – growing cost function, Θ_p – temperature plants, °C; Θ_n – air temperature, °C, φ – air humidity %; figure 0.5 is used as a weighting factor (made equal, as I and B are equivalent to revenue production).

The maximum profit is achieved by manufacturing conditions, temperature plants should be 21.3 °C, 22.5 °C air temperature and humidity – 60 %. This ensures the production of the index I within the "very good" at minimal energy costs.

The results made it possible to create a management system biotechnological object, which is building greenhouses [10] (Fig. 10), consisting of: subsystem decision 1 (includes filtering the input block 2, block neural network forecasting product quality 3, neural network evaluation unit profit 4, block 5 decision, the control unit 6); mobile robotic unit monitoring process parameters 7; local control system 8 (consisting of local automatic control device 9, actuators 10, control object 11).

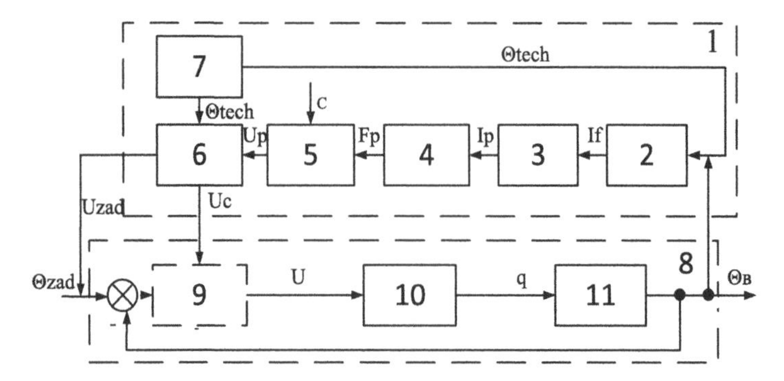

Fig. 10. Functional scheme of control in growing high-quality vegetables to maximize profits: Θ_{tech} – signal from perceiving elements; I_f – about qualitative biological parameter component; I_p – predicted value product quality; F_P – estimated value of income; C – these cost components of profit; U_p – choice of strategy management; U_{zad} – given management action; U_c – established management strategy; U – handed management strategy; q – performance management; Θ_{ZAD} – set value signal; Θ_B – output signal

4 Conclusions

The energy efficient method of managing complex electrical greenhouses for the cultivation of plant products, which based on the analysis of natural disturbances can consider the parameters of the microclimate in the greenhouse, plant conditions, product quality control creates effects that maximize revenue production. Specified mathematical model of the spatial distribution of temperature in the greenhouse as an object with distributed parameters. Adequacy of the mathematical model of convergence confirmed by the results of field measurements and simulation (standard deviation do not exceed 1.05 °C).

References

1. Lysenko, V., Lendiel, T.: Modeli dlya formuvannya optimalnix strategij keruvannya u sporudax zakritogo gruntu. Visnik Agrarnoï Nauki **10**, 45–48 (2015)
2. Korobiichuk, I., Bezvesilna, O., Ilchenko, A., Shadura, V., Nowicki, M., Szewczyk, R.: A mathematical model of the thermo-anemometric flowmeter. Sensors **15**, 22899–22913 (2015)
3. Korobiichuk, I.: Mathematical model of precision sensor for an automatic weapons stabilizer system. Measurement **89**, 151–158 (2016)
4. Karachun, V., Mel'nick, V., Korobiichuk, I., Nowicki, M., Szewczyk, R., Kobzar, S.: The additional error of inertial sensor induced by hypersonic flight condition. Sensors **16**(3) (2016)
5. Lysenko, V., Bolbot, I., Lendiel, T.: Vejvlet-analiz v fitometrii rastenij. Sbornik nauchnyx trudov «aktualnye voprosy sovremennoj nauki», 163–173 (2014)
6. Lysenko, V., Bolbot, I., Miroshnik, V., Lendiel, T.: Temperatura roslin yak parametr dlya regulyuvannya. Naukovij visnik NUBIPU **209**(1), 64 –72 (2015)

7. Lysenko, V., Bolbot, I., Lendiel, T.: Fitotemperaturnij kriterij ocinki rozvitku roslini. Naukovij zhurnal «Energetika i avtomatika» **3**(11), 122–128 (2013)
8. Lysenko, V., Miroshnik, V., Lendiel, T.: Modeling the spatial distribution of temperature zones in the greenhouse. Naukovij zhurnal «bioresursi i prirodokoristuvannya» **7**(1, 2), 159–165 (2015)
9. Lysenko, V., Miroshnik, V., Lendiel, T.: Optimizaciya viroshhuvannya tomativ v teplici z vikoristannyam funkciï bazhanosti xarringtona. Zhurnal «Avtomatizaciya texnologichnix ta biznes-procesiv» **7**(4), 33–39 (2015)
10. Lysenko, V., Bolbot, I., Lendiel, T., Chernov, I.: Pat. 103274 ua, MPK G05b 13/00 (2015.01): Sistema upravlinnya biotexnichnimi obektami. Patent opublikovano 10.12.2015, Byul. № 23

Method of Control for Gasification Reactor in Ecological Technology of Biomass Waste Utilization

Jakub Szałatkiewicz[1,2(✉)], Grzegorz Zielono[1], Łukasz Obrzut[1], and Marzena Szałatkiewicz[1]

[1] Phoenix Technologies Sp. z o.o., ul. Trębacka 4 lok. 336, 00-074 Warsaw, Poland
jakub.szalatkiewicz@phoenixtechnologies.com.pl
[2] Przemysłowy Instytut Automatyki i Pomiarów, Al. Jerozolimskie 202, 02-486 Warsaw, Poland
jszalatkiewicz@piap.pl

Abstract. This article presents method of control of new throughput type gasification reactor for gasification of biomass. Reactor allows precise control of gasification process via adjustment of main variables which are: temperature, biomass flow, oxidants mass and temperature. Pointed variables allow control of reactor gas production, efficiency, throughput, gas composition and energy consumption. Control parameters are varied via SCADA system that provides process visualization, adjusting variables and data acquisition. In this work, experimental results are presented and pointed correlations affecting reactor work parameters with preliminary optimization directions and results.

Keywords: Gasification · Reactor · Control · Biomass · Waste

1 Introduction

Company Phoenix Technologies Sp. z o.o., carried out research project to develop new innovative technology allowing waste utilization for energy generation. The project focused on biomass utilization technology including development of new reactor construction. The aim of the work was to allow energy generation in gas fired cogeneration units on gas generated from biomass waste. To achieve this goal the biomass feedstock needs to be gasified, in other words converted into gaseous form of flammable gas. There are known processes (pyrolysis, gasification, plasma gasification, and other) that allows gasification of biomass and other feedstocks, but the generated gas is low quality as its heating value is only 5–6 MJ/m^3 [1–6] Also higher heating values of the synthesis gas was reported up to 10 MJ/m^3 [7, 8]. Those processes mainly include shaft gasification and partially incineration of the feedstock inside the reactor as air is used as oxidant. Those solutions allows generation of heat inside the reactor, which increases the heat transfer, but the gas quality is significantly lower. Also the construction of those reactors (vertical shaft) limits theirs control parameters. Feedstock moves inside by gravity from the top to the bottom as pyrolysis/gasification/combustion takes place, and

R. Szewczyk and M. Kaliczyńska (eds.), *Recent Advances in Systems, Control and Information Technology*, Advances in Intelligent Systems and Computing 543, DOI 10.1007/978-3-319-48923-0_31

ash is being removed. There are no available data of range of operation in those reactors, but it is hard to expect they could be operated below 50% of load. Also other types of reactors constructions are under investigation however they are not so popular [9].

To overcome existing limitations of gasification technologies new technology was developed basing on different assumptions and targets. The gas quality in terms of heating value was expected to be above 12 MJ/m^3. The process should allow broad range of operation in terms of throughput control, reaching technical minimum of about 30% maximum load. Also process should allow high (above 95%) utilization of chemical energy available in processed feedstock.

Above assumptions were the goals that carried out research project allowed to reach, and the new technology was developed.

2 Materials and Methods

The new developed technology is based on different than currently exercised by other technologies assumptions.

High heating value (above 12 MJ/m^3) of gas generated from the process needs to be fee of nitrogen, which lowers the heating value. This requires airtight reactor construction and airtight method of feedstock loading.

The broad range of reactor throughput operation 30%–100% requires precise and stable control over feedstock loading and processing. Also temperature control is required to operate in such broad range of throughput.

High utilization of chemical energy in processed waste requires good and uniform contact of processed feedstock whit oxidant. This requires special reactor construction, and control of oxidant input, parameters, and type.

Those goals on one hand limit the possible solutions, and on the other hand allows precise selection of methods necessary to reach the right result.

Also in every reactor construction there are important limitations that needs to be taken into account [10]. On one hand we know the best materials that are capable to operate in very high temperatures, but the cost of the final solution will be unacceptable for production. Tungsten is capable to work in very high temperatures, but instead of them the high temperature steels are applied which are much cheaper and easier to form and apply. For example commercially available AISI 309S is good quality high temperature resistant steel that can operate in temperature of 1050/950 °C [11]. Also price of this steel is in good relation to its parameters so it is one of the materials that can be used in reactor or other hot working parts.

Basing on listed above assumptions test stand was developed including gas tight feeding system, gastight reactor, gas cooling and cleaning system, oxidant preparation system, control and measurement system, and gas utilization system.

The designed reactor has 6.3 m^2 of external surface, and the internal volume of 0.35 m^3. It allows processing of biomass and other feedstocks in controlled throughput from 10% to 100% speed. For wood biomass example it is capable to investigate up to about 80 kg/h of biomass "sawdust" throughput. This limitation is the result of feeding system handling of biomass sawdust.

Reactor control allows control over process parameters and its products. In case of this work the key goals of control are:

- Keep the process parameters in given (optimal) range,
- Repeatability (stability) of process products, over time, and in fact of variable feedstock parameters,
- Range of operation capability, throughput minimum and maximum,
- Safety for users, environment and equipment,
- Automation, allowing automatic operation without human supervision,
- Certainty of operation.

There are many ways of reactor control, however each one needs to be analyzed in scope of given exemplary reactor variables and limitations. In case of developed reactor analysis was carried out that allowed identification of variables and constants. There are two stages for reactor control degrees of freedom. First is in the early phase of development where the geometry and materials are also the variables. And the second is, when the design is given and some variables became constant. Below variables and constants are listed in regard of developed test stand.

The variables are:

- Speed of feedstock feeding – the feedstock mass stream in the reactor, available for the process,
- Type and parameters of feedstocks (assumed as constant),
- Process substrates, additions i.e. oxidant, catalysts, etc. (reduced to substrates parameters, mass, temperature, pressure),
- Temperature,
- Pressure,
- Turbulence,
- Loses, for example heat loss.

The constants are:

- Geometry, surface, volume,
- Materials used for construction,
- The way of substrates feeding to the reactor.

The above variables and constants can be reduced to key reactor and process parameters.

Variables:

- Feedstock feeding speed/mass stream 10–80 kg/h.
- Oxidant parameters: steam 0–50 kg/h temp up to 250 °C.
- Temperature: up to 1050 °C.
- Pressure –100 mBar to +100 mBar.

Constants:

- Reactor surface: 6.3 m^2, and volume 0.35 m^3.
- Reactor construction material: AISI 309S (15–27 W/mK) [11]

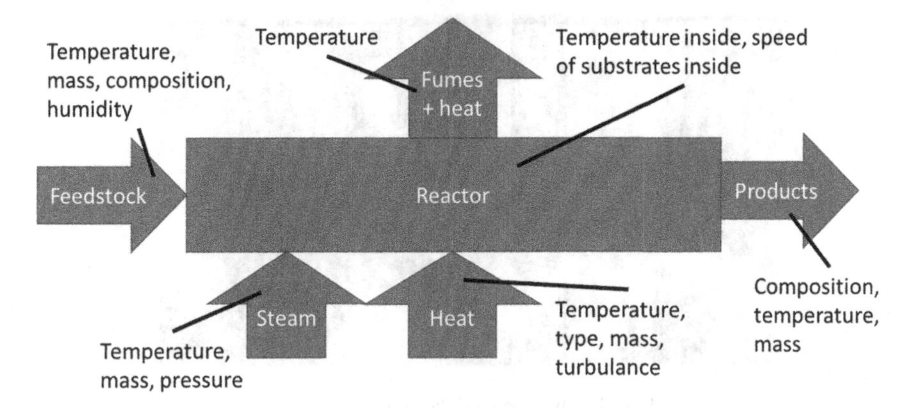

Fig. 1. Block diagram of the reactor

- Heat loss, is dependent on temperature, but in given reactor construction it is a function of temperature, and throughput.
- Feedstock, assumed to be uniform – wood biomass.

Figure 1. Presents block diagram of the developed reactor.

Also the reactor and its auxiliary systems were controlled by PLC controller with SCADA software. The SCADA main screen is presented in Fig. 2. The SCADA and PLC controller were communicating together allowing data acquisition via SCADA system, and devices control via PLC directly. Such organization provides high safety in case of PLC – SCADA communication failure. Exemplary PLC control protocols were: PID controller allowing automated gas blower speed adjustment depending on syngas gas pressure on one hand measured in the reactor (–10 mbar target), and (40 mbar) target measured on gas supply to the gas engine, and the gas burner. Another one was

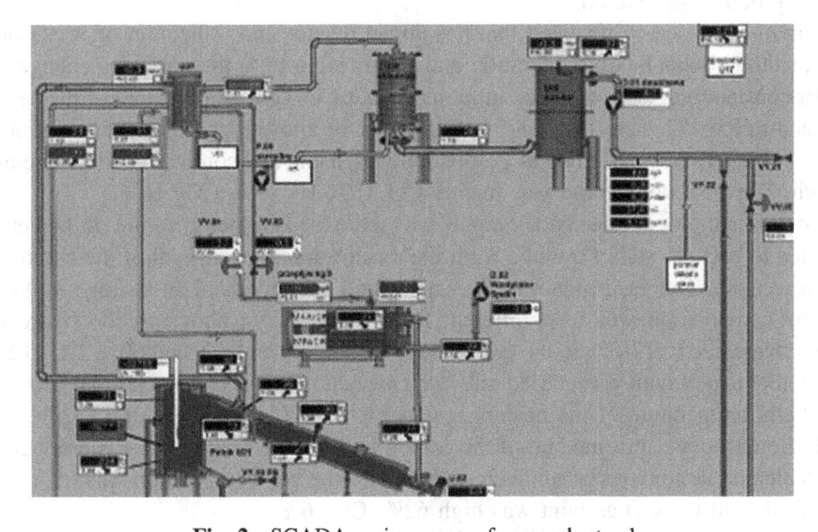

Fig. 2. SCADA main screen of research stand

Fig. 3. Overview of the research stand

PID to control temperature in heat exchanger and simultaneously limiting the pressure in steam generator, basing on steam pressure, and the temperature in the heat exchanger through which the steam was blown away. Also other control protocols were implemented for safety reasons, and automation of the test stand. For example pumping of the scrubber was applied when the level in it was reached. Another is filling of the water tanks and steam generator basing on water levels, and so on.

The PLC and SCADA control of test stand allowed to automatize and handle many process variables to allow the research and safety reactor operation.

In the Fig. 3 the overview of test stand is presented.

3 Results

The investigated reactor construction allows broad range of process control essential for best products generation.

Carried out research revealed that developed reactor and utilization of waste biomass method allows flexible operation and good response to the variables change.

For constant biomass feedstock input mass set to 35 kg/h, temperature of the heater regulated to 850 °C, stem flow of 5 kg/h the reactor allowed complete gasification of carbon in the leftover ash. The gas production was stable and continuous of an average 31 kg/h. The gas composition was mainly CO 51%, H_2 20%, CO_2 16%.

For constant biomass feedstock input mass set to 50 kg/h, temperature of the heater regulated to 850 °C, stem flow of 5 kg/h the reactor did not gasify all of the carbon in the leftover ash. The gas production was stable and continuous of an average 39 kg/h. The gas composition was also different from the example above as the hydrogen amount decreased to 17%, and the CO amount was higher and reached 60%, CO_2 12%.

Another experiment was carried out with constant biomass feedstock input mass set to 50 kg/h, temperature of the heater regulated to 950 °C, stem flow of 15 kg/h. The gasification process consumed all of the carbon in the leftover ash. The gas production was stable and continuous of an average 56 kg/h. The gas composition was hydrogen reach 27%, and the CO amount was high 62%, CO_2 6%.

Next experiment was carried out with constant biomass feedstock input mass set to 25 kg/h, temperature of the heater regulated to 700 °C, stem flow of 5 kg/h. The gasification process did not consumed all of the carbon as it was still in the leftover ash. The gas production was stable and continuous of an average 20 kg/h. The gas composition was hydrogen 14%, and the CO amount was high 47%, CO_2 19%.

4 Discussion

Presented analysis and results for control for gasification reactor in ecological technology of biomass waste utilization method provides key functional data and directions for further works and initial optimization.

Analyzed data indicates that key correlations in presented method of control for gasification reactor are as expected and in line with theory of operation, and process control demands.

The throughput – feedstock mass affects the gasification efficiency. Increasing feeding rate the gas amount rises, but the chemical energy utilization decreases and in the leftover ash also solid carbon is still present. On the other hand increasing the feeding rate to a certain point affects positively the gas composition as the amount of hydrocarbons rises. Also the throughput of the reactor and the produced gas mass increases. The main limitation of this variable is the heat and temperature available for the process in the reactor. Decreasing of the temperature in the reactor also decreases the gasification process which results in lower chemical energy utilization, and shifting the reaction from CO reach gas into CO_2 reach gas that affects its heating value.

The temperature – increasing the reactor temperature allows increasing its throughput and generation of CO reach gas and CO_2 amount reduction. Higher the temperature the faster the process and the throughput can be increased. Also the higher the temperature the higher speed of the gasification reactions. However the main limitation of this variable is the temperature resistance of the reactor materials, and on the other hand the heat transfer through the reactor heated wall (thermal conductivity of the material). The drawback of the temperature increment is the higher heat consumption and higher heat loses.

Substrates parameters – in experiments stem was used, and its mass and temperature was regulated. The higher the steam temperature the less energy is consumed in the reactor for the gasification/dissociation reactions. The lower the oxidant temperature the higher the amount of heat that needs to be consumed from the reactor. If the oxidant temperature is too low and its mass is to high the process can be stopped due to temperature drop inside the reactor. The oxidant have also impact on the produced gas composition, as steam reacts with carbon and forms CO and H_2, and next another reaction of CO and steam can be observed that forms CO_2 and another H_2. So the higher the steam amount the more H_2 is produced in the gas.

For maximization of throughput and usage of chemical energy in processed biomass, the reactor shout be run on maximum temperature, excess of oxidant should be added, and the throughput should be regulated to the point where no leftover carbon is present in the ash. It is essential for this operation to keep in mind, or apply in control software, the operation limits and safety procedures, as the increase of the temperature may result in reactor destruction. Also in some observed cases the temperature inside the reactor tend to rise above the heater temperature. Such situations are dangerous and commonly are the cause of research stand components damage or rapid and increased corrosion that shortens its usage time.

To operate the reactor on minimal parameters, it is possible to decrease its parameters to technical minimum. To sustain its operation for a given period of time without shutting it down. For example such situation is a result of the need, to allow quick incensement of reactor parameters to its nominal levels, faster than from cold start. To do so the maximization of usage/utilization of chemical energy in processed feedstock is one of the parameters. Another are decrease of the throughput, and oxidant amount and the temperature reduction. This leads also to heat loss decrease. Dependably on feedstock parameters the throughput is limited to 10–30%, and the temperature to about 700 °C, which is necessary for gasification process but also limits the heat usage.

5 Conclusions

Presented in this paper method of control for gasification reactor in ecological technology of biomass waste utilization allows precise operation of the developed reactor. The main variables and correlations were identified, and verified in experiments. Those variables are essential for further adjustment works and full automatic control development and implementation.

The new knowledge gathered during research project allows control and automation of developed technology. Also depending on the optimization target it is possible to eater way maximize throughput, maximize chemical energy utilization in processed feedstock, or when necessary to run the reactor on minimal throughput without its shutting down.

Acknowledgments. This work was carried out within a research project co-financed by the European Regional Development Fund under the Innovative Economy Operational Programme for years 2007–2013, Measure 1.4 Support for goal-oriented projects, Priority I Research and development of modern technologies. Project no. UDA-POIG.01.04.00-14-170/10-00.

References

1. EKOD gasifier Zamer company. http://www.zamer.com.pl/main.php?fid=359&pg=10&id_lang=0
2. ICHPW gasifier. http://www.imp.gda.pl/projekty/ps4/prototypy/zgazowarka-ichpw/
3. CRB Energia company gasifier. crbenergia.pl
4. Grochal, B.: Technologiczne i ekonomiczne aspekty zgazowania biomasy, Warsztaty wykorzystanie biomasy w inwestycja miejskich. http://six6.region-stuttgart.de/sixcms/media.php/773/7.%20B.%20Grochal,%20RIMAMI,%20Poland,%20biomass%20gasification.pdf
5. Zaporowski, B., Szczerbowski, R., Wróblewski, R.: Analiza efektywności energetycznej i ekonomicznej elektrociepłowni małej mocy opalanych biomasą. Polityka Energetyczna Tom **10**(2), 367–378 (2007)

6. Suna, Y., Zhanga, Z., Liua, L., Wanga, X.: Integration of biomass/steam gasification with heat recovery from hot slags: Thermodynamic characteristics. Int. J. Hydrogen Energy **41**(14), 5916–5926 (2016). doi:10.1016/j.ijhydene.2016.02.110
7. Duan, W., Yu, Q., Wu, T., Yang, F., Qin, Q.: Experimental study on steam gasification of coal using molten blast furnace slag as heat carrier for producing hydrogen-enriched syngas. Energy Convers. Manag. **117**, 513–519 (2016). doi:10.1016/j.enconman.2016.03.051
8. Balu, E., Lee, U., Chung, J.: High temperature steam gasification of woody biomass – a combined experimental and mathematical modeling approach. Int. J. Hydrogen Energy **40**(41), 14104–14115 (2015). doi:10.1016/j.ijhydene.2015.08.085
9. Wang, K., Yu, Q., Qin, Q., Hou, L., Duan, W.: Thermodynamic analysis of syngas generation from biomass using chemical looping gasification method. Int. J. Hydrogen Energy **41**(24), 10346–10353 (2016). doi:10.1016/j.ijhydene.2015.09.155
10. Szałatkiewicz, J., Szewczyk, R., Budny, E., Missala, T., Winiarski, W.: Identification of thermal response, of plasmatron plasma reactor. In: Szewczyk, R., Zieliński, C., Kaliczyńska, M. (eds.) Recent Advances in Automation, Robotics and Measuring Techniques. AISC, vol. 267, pp. 265–274. Springer, Heidelberg (2014)
11. AISI 309S steel specification: Acerinox, "Refractory Austenitic stainless steel ACX 340". http://www.acerinox.com/opencms901/export/sites/acerinox/.content/galerias/galeria-descargas/galeria-documentos-producto/ACX340-low.pdf

Robots and Robotic Systems

Coordination and Cooperation Mechanisms of the Distributed Robotic Systems

Igor Korobiichuk[1(✉)], Yuriy Danik[2], Pavlo Pozdniakov[2], and Dorota Jackiewicz[3]

[1] Institute of Automatic Control and Robotics, Warsaw University of Technology, Warsaw, Poland
kiv_igor@list.ru

[2] Zhytomyr Military Institute n.a. S.P. Korolyov, Zhytomyr, Ukraine
zhvinau@ukr.net, pozdner86@gmail.com

[3] Industrial Research Institute for Automation and Measurements PIAP, Warsaw, Poland
djackiewicz@piap.pl

Abstract. The success of robotics and artificial intelligence over the past decades led to increased interest in theoretical research and practical application of distributed robotic systems. The advantages of robots are improvement of reliability, reduction of time and cost of resolving such problems as reconnaissance, surveillance, search and rescue missions, patrolling, etc. At the same time, collective use of robots requires solving a number of specific problems arising in collective management, including distribution and redistribution of tasks, coordination, cooperation and collective decision-making. This article provides an overview of existing methods, algorithms and principles of robotic system coordination and cooperation, known from the theory of multi-agent systems, swarm intelligence and evolutionary modeling. It is established that most of them only partially solve the problem of coordination and cooperation of distributed robotic systems under real-life conditions. The paper defines basic methods of solving tasks and problems of coordination and cooperation of distributed robotic systems under conditions of incomplete a priori information, changes in operational environment, goals and objectives.

Keywords: Robotic systems · Application · Coordination · Multi-agent systems · Incomplete a priori information

1 Introduction

The successes in the development of robotics and artificial intelligence have caused the great interest to the theoretical research and practical application of the distributed robotic systems. The group robot application allows finding the effective solutions to such tasks as: exploring, observation, mapping, search and rescue operations, patrolling, remote object diagnostics. Having started in the late 80-s of the previous century, the researches of multi robotic systems resulted in the number of successful projects. There are ACTRESS [1], COFER [2], CEBOT [3], MARTHA [4], ALLIANCE [5], AMADEUS [6], MURDOCH [7], ASyMTRe [8] and 4D-RCS [9].

R. Szewczyk and M. Kaliczyńska (eds.), *Recent Advances in Systems, Control and Information Technology*, Advances in Intelligent Systems and Computing 543, DOI 10.1007/978-3-319-48923-0_32

The main advantages of the group robot application compared to a single one are the greater reach, the increased reliability, the flexibility, the lower cost, the potential possibility for the development and the complexity of the tasks to be solved. This can be achieved due to the effect of behavior emergence. It means that the simple rules of an individual behavior form the complex organized behavior of the group. However, the group robot application requires the distinct interaction of the robots and their interaction with the environment. There is also a demand for planning and group task solutions. In more general case at ask consists of the problem of coordination and the distributed robotic system cooperation.

There are a lot of methods, algorithms, and principles of coordination and the distributed robotic system cooperation in the theory of the distributed artificial intelligence and multi-agent systems [10, 11]. However, the application of such mechanisms to the distributed groups of robots is limited. It is explained by the fact that the theory of multi-agent systems aims to investigate the distributed computer systems. These systems have the separate nods and the environment of functioning is well-structured. But the methods of the distributed artificial intelligence mostly use the program agents. They are not enough efficient in the conditions of real physical environment and can be applied to solve the limited number of tasks only.

Therefore, the purpose of the given article is to cover well-known methods, algorithms and principles of group control, and also to determine the effective mechanisms of coordination and cooperation of the distributed robotic systems.

2 The Review of Well-Known Mechanisms of Coordination and Cooperation of the Distributed Robotic Systems

In order to determine the efficient variants of coordination and cooperation of the distributed robotic systems let us consider the general strategies which are used for the control in technical, social and natural groups. As a rule, these strategies are the strategies of centralized, decentralized or combined control.

The project "MARTHA" [4] is an example of the centralized approach. Its concept is as following. The system of a robot group control consists of two main parts. They are the central station and the board computers with the connection controller. The central station is installed stationary at the control point. It resolves the task of planned actions for the whole group of robots. Besides, the central station provides the interaction between a human-operator and every single robot in case of unpredictable situations and it is also used to formulate the target task. The board computer is used to solve the task of a robot's rout. The robots exchange information with the central station and other robots from the group by radio links.

The advantages of the centralized strategy of group control are the simplicity of the organization and the algorithmization. On the other hand, this strategy possesses a lot of essential disadvantages. First of all, it is the fact, that the central control unit (the leader) has a task to optimize the whole group member actions. At the same time, the complexity of this task increases exponentially with the increase in number of group objects. This causes the increase of time for the decision-making. Therefore, the task solution of the group control by applying such strategy is recommended to obtain beforehand.

The mentioned disadvantage can be eliminated to some extend by applying the strategy of decentralized control. The similar approach was implemented by the scientists of Nagoya University. They developed the system DARS (Distributed Autonomous Robotic System) [12]. It allowed improving the algorithms, the planning methods and controlling the coordinated actions of the robot group. The methods of the action robot group control in the mode "driving-driven" [13] are developed in the University of Pennsylvania, the USA. Another example of group actions of the robot team is the organization of antimine protection of waters. Its concept is developed within the project Office Naval Research's Very Shallow Water / Surf Zone Mine Countermeasure [14].

Some of the researches are conducted to implement the models of biological object interactions [15–17]. These models are based on the algorithms used by swarms of bees, ants and others. The methods applied research to the outer phenomena of living organism behavior aspects. One of the most spread approach of robot group application is "foraging" task solution. It is based on the model of food behavior of gregarious living organisms: birds, bees, ants, etc.

One of the possible variants of the distributed robotic system coordination is the application of the principles of situation control [18–21]. The idea of the approach is based on the sets of various types of robot behavior. Every robot obtains several types of behavior to be used depending on a situation.

One more example of the solution to the problem of multi agent interaction organization in robot groups is the approach offered at the Robotics Institute at Carnegie Mellon University. The methods of free market economy [22, 23] are recommended to be applied to coordinate the actions in multi robotic systems which solve the tasks of exploring and mapping.

The market architecture of robot group behavior control is the base of this approach. The market economy, as a rule, is not oriented on the centralized control. The participants are free to sell and to purchase the goods and services; they sign contracts and get profit. The same approach was applied to the robot group which performed the distributed task of information collection in the environment with the known infrastructure [24]. The robots in the group interact providing high total efficiency by maximizing their own contributions.

Therefore, it is possible to conclude, that the decentralized strategy is recommended to be applied to the large groups of robots. This strategy will enable every robot to make its own decision by exchanging information with other members of the group in order to optimize the group decision. The decentralized approach provides the linear increase of time to make group decision at the increase of group members. Unlike the decentralized approach, the centralized strategy causes the exponential increase of time for decision-making at the increase of group members.

The advantages of the decentralized strategy are its high durability and the relative simplicity of the task which is solved by every member of the distributed robotic system. But, at the same time, the strategy of decentralized control in the distributed robotic systems possesses the complexity of coordination process and algorithmization.

In order to find the effective solution to the coordination task for the distributed robotic system let us apply the state space method [25].

3 The Coordination of the Distributed Robotic Systems

In general, the task of coordination of the distributed robotic systems can be formulated as following.

Let some group $\mathfrak{R}$, which consists of N robots $R_j(j = \overline{1,N})$ function in the environment E. The state of every single robot $R_j \in \mathfrak{R}(j=\overline{1,N})$ at the moment of time $t$is described by the vector-function $\boldsymbol{r}_j(t) = [r_{j,1}(t), r_{j,2}(t), \ldots r_{j,h}(t)]^{\mathrm{T}}$. The state of robot group $\mathfrak{R}$ is made by the vector $\mathfrak{R}(t) = [\boldsymbol{r}_1(t), \boldsymbol{r}_2(t), \ldots, \boldsymbol{r}_N(t)]^{\mathrm{T}}$. The state of the environment, where the j-robot functions – $\boldsymbol{e}_j$ at the moment of time t is described by the vector $\boldsymbol{e}_j(t) = [e_{1,j}(t), e_{2,j}(t), \ldots e_{w,j}(t)]^{\mathrm{T}}$. Then the state of the environment, where the robots of the group function, provided that the environment is stationary, at the moment of time t is described by the vector $\boldsymbol{e}(t) = [\boldsymbol{e}_1(t), \boldsymbol{e}_2(t), \ldots \boldsymbol{e}_N(t)]^{\mathrm{T}}$.

The robots and the environment, interacting with each other, form the system "group of robots – environment". The state of this system at the moment of time t is described by the couple $\boldsymbol{s}_c = \langle \mathfrak{R}, \boldsymbol{e} \rangle$. The multiplicity of various states of the system "group of robots – environment" is described by the points $N{\cdot}(h + w)$ –dimensional space states $\{s_c\}$. As the initial and final (target) states of the system "group of robots – environment" the following states are used

$$\boldsymbol{s}_c^0 =< \mathfrak{R}^0, \boldsymbol{e}^0 >,\ \boldsymbol{s}_c^f =< \mathfrak{R}^f, \boldsymbol{e}^f > \tag{1}$$

correspondingly.

The state of the system "group of robots–environment " $\boldsymbol{s}_c^{\hat{t}} =< \mathfrak{R}^{\hat{t}}, \boldsymbol{e}^{\hat{t}} >$ at current moment is called the current one.

Every robot $R_j \in \mathfrak{R}\ (j=\overline{1,N})$ is able to perform actions which are described by the vector $a_j(t) = [a_{1,j}(t), a_{2,j}(t), \ldots, a_{m,j}(t)]^{\mathrm{T}}$. Here the multiplicity of actions which can be performed by robot $\boldsymbol{R}_j \in \mathfrak{R} - \{\boldsymbol{a}\}_j$. The multiplicity of actions which can be performed by the group of robots is represented by the united multiplicity of actions of some robots in the group: $\{\boldsymbol{a}_c\} = \{\boldsymbol{a}\}_1 \cup \{\boldsymbol{a}\}_2 \cup \ldots \cup \{\boldsymbol{a}\}_N$.

The actions, which are performed by the group of robot sat the moment of time t, can be described by the vector function $\boldsymbol{a}_c(t) = [\boldsymbol{a}_1(t), \boldsymbol{a}_2(t), \ldots, \boldsymbol{a}_N(t)]^{\mathrm{T}}$. The changes in the state "group of robots – environment" are described by the system of differential equations of the type

$$\dot{s}_c = \mathbf{f}_c(\boldsymbol{s}_c(t), \boldsymbol{a}_c(t)). \tag{2}$$

Here, both the situations and the robots can have some limitations:

$$s_c(t) \in \{\boldsymbol{s}_c^p(t)\} \subset \{\boldsymbol{s}_c\}, \quad \boldsymbol{a}_c(t) \in \{\boldsymbol{a}_c^p(t)\} \subset \{\boldsymbol{a}_c\}, \tag{3}$$

where: $\{s_c^p(t)\}$ – the multiplicity of the permissible at the moment of time t states of the system "group of robots – environment";

$\{\boldsymbol{a}_c^p(t)\}$ – the multiplicity of the permissible at the moment of time t actions of the robot group.

Taking into consideration the introduced above indexes, it is possible to state the task of the distributed robotic system coordination which is in the determination on the interval $[t_0, t_f]$ such optimal actions $\bar{a}_j(t)$ for every robot $\boldsymbol{R}_j \in \mathfrak{R}$, which can transform the system "group of robots – environment" from the initial state into final (target). Here, the equation system (2) and the limitation (3) have to be performed and the functional extremum has to be provided

$$\begin{aligned} \boldsymbol{Y}_c &= \int_{t_0}^{t_f} \boldsymbol{F}(\boldsymbol{s}_c(t), \boldsymbol{a}_c(t), t)dt \\ &= \int_{t_0}^{t_f} \boldsymbol{F}(\boldsymbol{r}_1(t), \boldsymbol{r}_2(t), \ldots, \boldsymbol{r}_N(t), \boldsymbol{e}(t), \boldsymbol{a}_1(t), \boldsymbol{a}_2(t), \ldots, \boldsymbol{a}_N(t), t)dt, \end{aligned} \tag{4}$$

It can help to estimate the quality of the control process and the target of the robot group functioning is stated.

The coordination of robots can be static or dynamic one. The static coordination is also known as deliberative [26] or off-line coordination [27]. The dynamic coordination (reactive [26] or on-line coordination [27]) of the distributed robotic system occurs during the whole performance of the target task. Depending on the conditions of the distributed robotic system functioning, it is possible to highlight three classes of the coordination tasks of robot group. They are: the tasks of coordination in the stationary structured environments, the tasks of coordination in the dynamic non-determined situations and the tasks of coordination in the conditions of counterpart resistance.

The task of coordination of the distributed robotic systems in the dynamic non-determined situations is the most complex one from the algorithmic point of view. This task is becoming even more difficult to solve when the active organized resistance is present.

The task of finding the optimal solution to coordinate robots in dynamic non-determined environments is becoming complex due to the necessity to perform control during the time when the state $s_c(t)$ of the system "group of robots – environment" does not dramatically change.

Let us consider the method of group control to find the effective group solution of the distributed robotic system.

4 The Cooperation of the Distributed Robotic Systems

The offered approach is based on the principles of group interaction. These principles are used by the groups of people who perform the common work without a manager or a leader. This method is based on the idea, that every single robot of the group determines the next actions by itself. These actions are targeted to achieve the group aim in the current situation the best. The decision on the next actions is made by every robot on the basis of information about its own state, the state of the environment around the robot and the selected for it actions of other robots in the group.

In general, the processes in the system "group of robots – environment" are continuous, but the vector functions of control (actions) $\boldsymbol{a}_c(t)$ does not possess continuity. This is caused by such factors as:

- the discontinuity of data representation in digital calculation devices of robot control systems;
- the resolution (by time) of the sensor devices;
- the recurrence of calculations.

Due to these reasons the group actions of robots $\boldsymbol{a}_c(t)$ are the discontinuous function of time which can extremely change at some definite moments of time $t\Delta t$ (t – the discontinuous time, i.e. t = 0, 1, 2, …).

Therefore, the discontinuous task formulation is limited to the task of robot selection and performance at current moment of time such group actions $\boldsymbol{a}_c(t)$, which are able to provide the extremum (maximum, if the benefit of robot group is estimated or minimum if the expenses are estimated) of the target functional.

$$Y_c(\hat{t}) = \sum_{t=\hat{t}}^{t_f-1} \boldsymbol{F}(\mathfrak{R}(t), \boldsymbol{e}(t), \boldsymbol{a}_c(t), \boldsymbol{g}(t))\Delta t \tag{5}$$

where: $\boldsymbol{g}(t)$ – the vector of resistance.

In this case the task of every robot in the group $\boldsymbol{R}_j \in \mathfrak{R}$ $(j=\overline{1,N})$ is to perform such actions $\boldsymbol{a}_c(t), t = 0, 1, 2, \ldots t_{f-1}$, which can lead to achieve the aim of the robot R_j, and to maximize the functional at the same time

$$\boldsymbol{Y}_j(\hat{t}) = \sum_{t=\hat{t}}^{t_f-1} \boldsymbol{F}_j(\boldsymbol{r}_j(t), \boldsymbol{e}_j(t), \boldsymbol{a}_j(t), \boldsymbol{g}_j(t))\Delta t \rightarrow \boldsymbol{Y}_j^{\max}(\boldsymbol{Y}_j^{\min}), \tag{6}$$

at the initial conditions

$$\begin{aligned} \boldsymbol{r}_j^0 = \boldsymbol{r}_j(t_0) = \boldsymbol{r}_j(\hat{t}), \quad \boldsymbol{e}_j^0 = \boldsymbol{e}_j(t_0) = \boldsymbol{e}_j(\hat{t}), \\ \boldsymbol{a}_j^0 = \boldsymbol{a}_j(t_0) = \boldsymbol{a}_j(\hat{t}), \quad \boldsymbol{g}_j^0 = \boldsymbol{g}_j(t_0) = \boldsymbol{g}_j(\hat{t}), \end{aligned} \tag{7}$$

at the equations of connection

$$\begin{aligned} &s_j(\hat{t}+1) = \mathbf{f}_j(s_j(\hat{t}), \boldsymbol{a}_1(\hat{t}), \boldsymbol{a}_2(\hat{t}), \ldots, \boldsymbol{a}_j(\hat{t}), \ldots, \boldsymbol{a}_N(\hat{t}), \boldsymbol{g}_j(\hat{t})), \\ &j = \overline{1, N}, \quad \hat{t} = \overline{0, t_f - 1}. \end{aligned} \tag{8}$$

at the limitation

$$\boldsymbol{a}_j(t) \in \{\boldsymbol{a}_j^p(t)\}, \quad \boldsymbol{r}_j(t) \in \{\boldsymbol{r}_j^p(t)\}, \quad \boldsymbol{e}_j(t) \in \{\boldsymbol{e}_j^p(t)\}. \tag{9}$$

The given approach to the solution of discontinuous task of robot group cooperation is based on the application of the iteration procedure, when the robots select their next actions successively. In order to implement the iteration procedure of the optimization task solution (6)–(9), it is recommended to number the robots in the appropriate way. The idea of the iteration method of the optimization task solution (6)–(9) is as following. Every robot R_j in the groups elects its next action on the basis of information about the group aim, its current state, the current state of the environment, the current state of an obstacle (if there is any) and the selected by other robots actions for it. This corresponds to the iteration formula

$$a_j^{k+1} = f_j\left(a_1^{k+1}, \ldots, a_{j-1}^{k+1}, a_j^k, a_{j+1}^k, \ldots, a_N^k, r_j^0, e_j^0, g_j^0\right), \quad j = \overline{1, N},$$

here $k = 0, 1, 2, 3, \ldots$ – is the iteration number.

Here, the limitations (9) have to be taken into consideration,

It is important to mention, that the permissible actions $\{a_j^p\}^0$ performed by the robot $\boldsymbol{R}_j \in \mathfrak{R}$ at current moment of time are only those actions among the multiplicity of permissible ones $\{a_j^p\}$, at the condition of $\Delta \mathbf{Y}_c^j \geq 0$.This in equality can be considered as a condition for the robots not to bring any loss to the group in whole.

The action a_j^{k+1} of the robot R_j is taken into account by all other robots in the group when selecting their next current actions. That is why, when the robot R_j selects its next current action a_j^{k+1} by means of solving the discontinuous task (6)–(9), all the rest robots in the group have to make their next selection of the optimal actions. It has to be done in order to achieve the aim of the group, because the previous selection has not taken into account the next actions of robot R_j. The iteration cycles of optimization are repeated until the target functional growth (5) stops or the values of this growth reduce to the value which can be ignored.

The procedure of robot group cooperation by applying the method of group control is considered to be completed, if at the completion K of the iteration cycles there are no changes in the actions of robots in the group which can cause the growth of the target functional. Here, the total time t_{ga} to solve the task of optimization for the group actions is

$$t_{ga} = (t_{opga} + t_{it})N \cdot K \tag{10}$$

Where

t_{opga} – is the time of optimization procedure for a single robot action implementation,

t_{it} – is the time to implement the procedure of the information transfer. This information is about the selected action by any robot in relation to the rest of the robots in the group,

N – is a number of robots in the group,

K – is a number of iteration cycles

On the other hand, to achieve the correspondence of the robot actions to the current situation (in real time mode) by applying the method of group control, it is necessary to obtain the fulfillment of the following condition

$$t_{ga} \leq \tau_c, \tag{11}$$

where τ_c – is the time when the situation in the system "group of robots – environment" changes (i.e., the time when the robot parameter state change in such way, that they can be measured by robot sensors and taken into account while selecting the next actions).

Equations (10) and (11) show that the permissible number of robots in the group, when the condition (11) is fulfilled, has to be limited by the value

$$N_{\max} = \frac{\tau_c}{K\left(t_{opga} + t_{it}\right)} \tag{12}$$

where $N_{\max}$ – the maximum permissible number of robots in the group, when the condition $t_{ga} \leq \tau_c$ is fulfilled.

This split can be implemented by applying the method of hierarchical clustering. In this case every cluster contains the robots which constantly interact between each other and their number should not exceed the maximum number of robots which arable to provide the controllability. The selection of actions for every single robot in such case is performed by taking into account the current action of other robots of the given cluster. The clusters interact between each other only at the stage of the target subtask distribution.

The multiplicity of targets in the hierarchically organized group of robots is also distributed by hierarchy principle. If the group of robots is split into the clusters of h levels, then the multiplicity of targets is split into the multiplicity of target subtasks of h levels. The cluster leaders participate in the process of the target subtask distribution. These tasks are given beforehand or during the clustering process.

The split of a large robot group into sub clusters of the constant, but not the same number of robots is performed as following. Initially, the first cluster with the number $N_1 < N_{\max}$, is formed. Its robots determine their actions by means of the iteration procedure of group action optimization (6)–(9). After the given procedure is completed, some number of robots N_0 stays in the first cluster. Their actions are the most effective. The rest $N_1 - N_0$ form the basis for the second cluster formation. At the second stage the number N_2 of new robots joins $N_1 - N_0$ robots. The procedure is repeated similar to the first stage. The next cluster is formed the same way as the previous one. The clustering procedure occurs until the group task is totally distributed.

5 Conclusions

Therefore, the given article presents the review of some coordination and cooperation methods of the distributed robotic systems. It is found, that the most effective distributed robotic systems are those, which implement the strategy of decentralized control. Due to the fact, that the time for decision-making of group actions in such systems is significantly less, they possess higher durability.

The article presents the formulation of the task of coordination of the distributed robotic system in state space. The mechanism of robot group cooperation is proposed. This method is based on the method of group control. The mechanism looks like the systematic decision-making as to next actions of robots, taking into consideration the group aim and changes in current situation. The actions performed by robots have to be of the current interest within the current situation. This means, that they have to be directed to achieve the aim by the optimal (or close to optimal) way. In order to solve the task of cooperation in large group of robots, it is proposed to split the group into sub clusters by hierarchical principle. It can provide the group controllability and allow reducing the total time to make decisions.

At the same time, it is possible to conclude about the lack of the efficient approaches to solve the task of group control with the implementation of the decentralized strategy in the condition of dynamically changing situations nowadays. Extra researches have to be conducted to solve the specific problems such as the determination of the optimal structure of the robot group to solve the definite tasks, the group resource distribution, localization, planning of shift and group decision-making by the members of the group, the provision of the reliable communication.

References

1. Asama, H., Matsumoto, A., Ishida, Y.: Design of an autonomous and distributed robot system: ACTRESS. In: Proceedings of IROS 1989, Tsukuba, Japan, pp. 283–290, September 1989
2. Caloud, P., Choi, W., Latombe, J.-C., Le Pape, C., Yim, M.: Indoor automation with many mobile robots. In: Proceedings of IROS 1990, Ibaraki, Japan, pp. 67–72, July 1990
3. Fukuda, T., Ueyama, T., Kawauchi, Y., Arai, F.: Concept of cellular robotic system (CEBOT) and basic strategies for its realization. Comput. Electr. Eng. **18**(1), 11–39 (1992)
4. Alami, R., Fleury, S., Herrb, M., Ingred, F., Robert, F.: Multi-robot cooperation in the MARTHA project. IEEE Robot. Autom. Mag. **5**(1), 36–47 (1998)
5. Parker, L.E.: ALLIANCE: an architecture for fault tolerant, cooperative control of heterogeneous mobile robots. In: Proceedings of IROS 1994, Munich, Germany, pp. 776–783, September 1994
6. Karada, T., Oikawa, K.: AMADEUS: a mobile, autonomous decentralized utility system for indoor transportation. In: Proceedings of the International Conference on Robotics and Automation, vol. 4, pp. 2229–2236. IEEE (1998)
7. Gerkey, B.P., Matarić, M.J.: Murdoch: publish/subscribe task allocation for heterogeneous agents. In: Proceedings of Agents 2000, Barcelona, Spain, pp. 203–204, June 2000
8. Tang, F., Parker, L.E.: ASyMTRe: automated synthesis of multi-robot task solutions through software reconfiguration. In: Proceedings of ICRA 2005, Barcelona, Spain, pp. 1501–1508, April 2005. doi:10.1109/ROBOT.2005.1570327
9. Meystel, A.M., Albus, J.S.: Intelligent Systems: Architecture, Design, and Control, 716 pages. Wiley Interscience (2002)
10. Weiss, G.: Multiagent Systems: A Modern Approach to Distributed Artificial Intelligence. The MIT Press, Cambridge (1999)
11. Wooldridge, M.: An Introduction to MultiAgent Systems. John Wiley & Sons, Chichester (2002)
12. Kaga, T., Fukuda, T.: An oscillation analysis on distributed automations robotic system. In: Proceedings of the International Conference on Robotics and Automation, vol. 4, pp. 2846–2851. IEEE (1998)

13. Desai, J.P., Osstrowski, J., Kumar, V.: Controlling formations of multiple mobile robots. In: Proceedings of the International Conference on Robotics and Automation, vol. 4, pp. 2864–2869. IEEE (1998)
14. Stokey, R., Freitag, L., Grund, M.: A compact control language for AUV acoustic communication. In: Proceedings of the IEEE Oceans 2005 Europe, vol. 2, pp. 1133–1137 (2005)
15. Shtovba, S.D.: Ant Algorithms. Exponenta Pro. Math. Appl. (4), 70–75 (2003)
16. Bonavear, E., Dorigo, M.: Swarm Intelligence: From Natural to Artificial Systems, 307 p. Oxford University Press (1999)
17. Dorigo, M.: Swarm intelligence, ant algorithms and ant colony optimization. In: Reader for CEU Summer University Course «Complex System», pp. 1–38. Central European University, Budapest (2001)
18. Balch, T., Arkin, R.C.: Motor schema-based formation control for multiagent robot team. In: Proceedings of First International Conference on Multiagent Systems, San Francisco, pp. 10–16, June 1995
19. Ali, K.S., Arkin, R.C.: Multiagent Teleautonomous Behavioral Control. Mach. Intell. Rob. Control **1**(2), 3–10 (2000)
20. Pospelov, D.A.: Situational Control: The Theory and Practice. Science, Moscow (1986)
21. Korobiichuk, I., Bezvesilna, O., Ilchenko, A., Shadura, V., Nowicki, M., Szewczyk, R.: A Mathematical Model of the Thermo-Anemometric Flowmeter. Sensors **15**, 22899–22913 (2015). doi:10.3390/s150922899
22. Dias, M.B., Stentz, A.: A free market architecture for distributed control of a multirobot system. In: Proceedings of the 6th International Conference on Intelligent Autonomous Systems (IAS), Venice, Italy, July 2000
23. Stentz, A., Dias, M.B.: A Free Market Architecture for Coordinating Multiple Robots tech. report CMU-RI-TR-99-42, Robitics Institute, Carnegie Mellon University, December 1999
24. Zlot, R., Stentz, A., Dias, M.B., Thayer, S.: Multi-robot exploration controlled by a market economy. In: IEEE International Conference on Robotics and Automation (ICRA), May 2002
25. Kaliaev, I.A., Gayduk, A.R., Kapustjan, S.G.: The distributed systems of actions planning of robots collectives, 292 p. Yanus, M. (2002)
26. Iocchi, L., Nardi, D., Salerno, M.: Reactivity and deliberation: a survey on multi-robot systems. In: Hannebauer, M., Wendler, J., Pagello, E. (eds.) BRSDMAS 2000. LNCS (LNAI), vol. 2103, pp. 9–32. Springer, Heidelberg (2001). doi:10.1007/3-540-44568-4_2
27. Todt, E., Raush, G., Suárez, R.: Analysis and classification of multiple robot coordination methods. In: Proceedings of ICRA 2000, San Francisco, CA, USA, pp. 3158–3163, April 2000

Selected Aspects of Implementation of "EU Occupational Safety and Health (OSH) Strategic Framework 2014–2020" Connected with Automation & Robotics

Marcin Słowikowski(✉), Jacek Zieliński, Zbigniew Pilat, Michał Smater, and Wojciech Klimasara

Przemysłowy Instytut Automatyki i Pomiarów PIAP, Industrial Research Institute for Automation and Measurements, Warsaw, Poland
{mslowikowski, jzielinski, zpilat, msmater, wklimasara}@piap.pl

Abstract. The European Commission has adopted a Strategic Framework on Health and Safety at Work 2014–2020. This document identifies three major challenges for health and safety at work. The first is improvement of implementation of existing health and safety rules, in particular by enhancing the capacity of Micro and Small Enterprises (SME) to put in place effective and efficient risk prevention strategies. Next is improvement of the prevention of work-related diseases by tackling new and emerging risks without neglecting existing risks. The third challenge is tackling demographic change. This article shows selected aspects and tools connected with introduce the Strategic Framework in the industrial environment, with high level of automation & robotics. The Online Interactive Risk Assessment (OiRA), a web platform providing sectoral risk assessment tools, is presented. Authors propose an approach for creation of OHS trainings, dedicated for A&R manufacturing installations. There are discussed training curriculum and dedicated ICT training solution.

Keywords: Information technologies · Online Interactive Risk Assessment · Occupational health and safety · Training

1 Introduction

Nowadays safe and healthy work environment for over 217 million workers in the EU is a strategic goal for the European Commission, Member States and the other EU institutions and bodies. Risk prevention and the promotion of safer and healthier conditions in the workplace are key not just to improving job quality and working conditions, but also to promoting competitiveness. Keeping workers healthy has a direct and measurable positive impact on productivity, and contributes to improving the sustainability of social security systems.

Already done research show investment in OSH as to the well-being of workers and is cost-effective. According to International Social Security Association (ISSA) and

R. Szewczyk and M. Kaliczyńska (eds.), *Recent Advances in Systems, Control and Information Technology*, Advances in Intelligent Systems and Computing 543, DOI 10.1007/978-3-319-48923-0_33

BenOSH (Benefits of Occupational Safety and Health) study investments in this area can produce high ratios of return, averaging 2.2, and in a range between 1.29 and 2.89. The major challenges, outstanding problems and key strategic objectives have been published in the special communication from the Commission to the European Parliament, the Council, the European Economic and Social Committee and the Committee of the Regions [1].

2 Strategic Framework on Health and Safety at Work 2014–2020

This Framework aims at ensuring that the EU continues to play a leading role in the promotion of high standards for working conditions both within Europe and internationally, in line with the Europe 2020 Strategy.

The Strategic Framework identifies three major health and safety at work challenges:

- To improve implementation of existing health and safety rules, in particular by enhancing the capacity of micro and small enterprises to put in place effective and efficient risk prevention strategies.
- To improve the prevention of work-related diseases by tackling new and emerging risks without neglecting existing risks.
- To take account of the ageing of the EU's workforce.

The Strategic Framework proposes to address these challenges with a range of actions under seven key strategic objectives:

- Further consolidating national health and safety strategies through, for example, policy coordination and mutual learning.
- Providing practical support to small and micro enterprises to help them to better comply with health and safety rules. Businesses would benefit from technical assistance and practical tools, such as the Online Interactive Risk Assessment (OiRA), a web platform providing sectoral risk assessment tools.
- Improving enforcement by Member States for example by evaluating the performance of national labour inspectorates.
- Simplifying existing legislation where appropriate to eliminate unnecessary administrative burdens, while preserving a high level of protection for workers' health and safety.
- Addressing the ageing of the European workforce and improving prevention of work-related diseases to tackle existing and new risks such as nanomaterials, green technology and biotechnologies.
- Improving statistical data collection to have better evidence and developing monitoring tools.
- Reinforcing coordination with international organisations, such as the International Labour Organisation (ILO), the World Health Organisation (WHO) and the Organisation for Economic Co-operation and Development (OECD) and partners to contribute to reducing work accidents and occupational diseases and to improving working conditions worldwide.

3 Online Interactive Risk Assessment

One of objectives of Strategic Framework on Health and Safety at Work 2014–2020 is to provide practical support to small and micro enterprises to help them to better comply with health and safety rules. Businesses would benefit from technical assistance and practical tools, such as the Online Interactive Risk Assessment (OiRA), a web platform providing sectoral risk assessment tools. OiRA system is available on web page http://www.oiraproject.eu/oira-tools (Fig. 1).

Fig. 1. Online interactive Assessment

OiRA – is a web platform that enables the creation of sectoral risk assessment tools in different languages with in an easy and standardized way. At this moment tools are available on English, Spanish, French, Bulgarian, Dutch, Greek, Finnish, Portuguese, Slovenian, Lithuanian and Latvian. Tools are dedicated to different sectors.

From the point of view of enterprises using automated and robotized solutions in their daily operations the most interesting part is this dedicated directly for production (Fig. 2).

Assessment online tools can help to identify of OHS risks and operates on a basis of users responses for questions grouped in following sections covering most important aspects of safety at work issues:

- General OHS management
- Communication and hosting venue(s)
- Venue environment considerations
- Stage- related operations
- Staging special elements
- Props, Costumes, Make-Up and Wings
- Performers
- Audience
- Touring
- Added risks

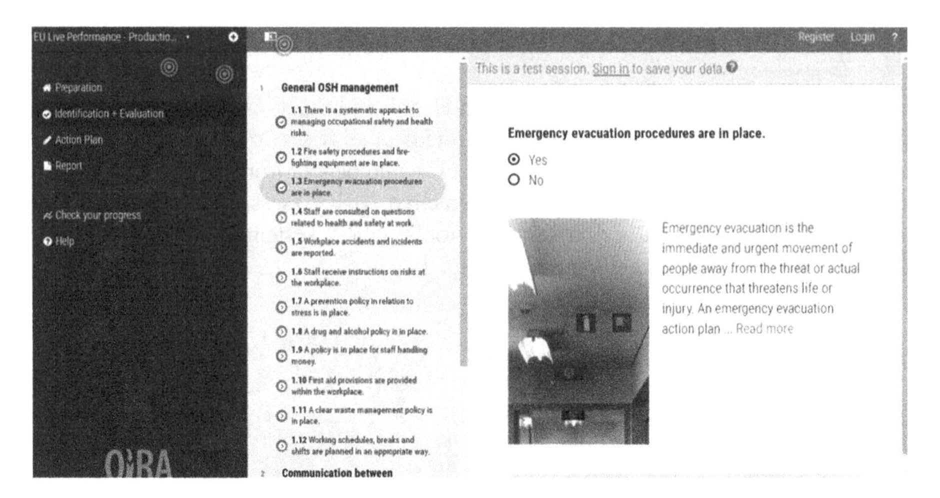

Fig. 2. Example Online interactive Assessment question

Proposed systematized approach allows quick identification of OHS risks in production company. These tools can also be used as supporting material for OHS trainings.

4 Needs Analysis on Health and Safety at Work

During the initial phase of realization of ERASMUS+project related to OHS "Quality Management in Occupational Health and Safety Training and Practice" consortium partners conducted in depth analysis of end-users needs.

Study of sample group was realized in three countries taking part in the project. Studies were conducted by 6 different institutions from which four were from Turkey, one from Poland and one from Italy. Partners collected 936 questionnaires which gives good representation of study samples, the figure below shows study participation rates (Fig. 3).

The results of conducted research shows that the general awareness about various concepts regarding the OHS aspect is rather low in surveyed companies and there are significant market needs for dedicated trainings which of course have to be adapted to specific risks and possible technical safety solutions. Trainings must be customized to end-user work place OHS risks as well as students style of learning. Especially important is to provide information about effective and efficient risk prevention strategies and prevention of work-related diseases. Results of survey were used as a base for the preparation of the structure and contents of training adapted to real training needs of target groups.

The main conclusions from the survey showed that target groups main benefits from the developed training course shall be designed with the goal of enabling managers,

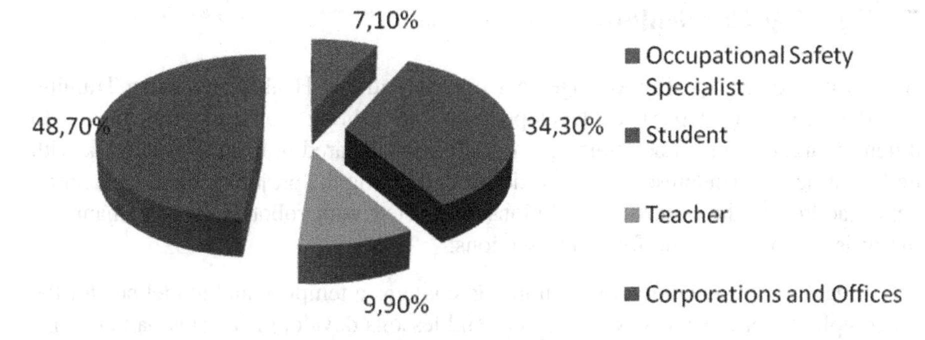

Fig. 3. Survey participation rates

supervisors, employee and workers in general to understand health and safety issues, laws, rights and responsibilities and provide the employees with essential knowledge needed to create a safe work place environment for themselves and those around them. Final aim is to avoid dangerous behaviours and workplace injuries. As a result of the course the trainee should:

- understand the general industry safety and health standards;
- understand specific risks and hazards on the workplace;
- understand how to address possible standard violations;
- be able to implement preventative measures to decrease safety risks.
- learn how to meet job responsibilities in a safe manner and respect company's standards.

To meet those objectives each course has been split into following sections:

- The first part, a general section covers general matters applicable to all the workplaces: it includes competences about current legislation at a European and National level, general elements and definitions, understanding the OHS legislation applicable to the work environment, management of the OHS in the office environment and is applicable to all the workplaces.
- The second part is specific for selected working environment: it provides the specific target group students the basics concepts of elements, roles and responsibilities on occupational health and safety, listing a series of subjects and elements that employees have to deal with on their job.
- At the end of each module, students have to pass an exam to enter the next module. An evaluation system is based on mandatory multi-choice questionnaire is be designed and delivered at the end of each module to assess achieved outcomes: only giving the correct answers (or a high percentage of them) will permit the student to pass from a module to the other. Each questionnaire will include at least ten questions regarding the topics just studied, with the aim to test the knowledge of students.

5 Training Curriculum

As part of project "Quality Management in Occupational Health and Safety Training and Practice" project partners developed sets of Training Curriculums dedicated to different target groups. Each training curriculum is prepared in form of handbook with unified structure. The presented below developed during the project training curriculum which addresses the issues and solutions connected with robotization and manufacturing is composed of the following sections:

- Guidelines for generation of the training, contains a template and guidelines for the completion of the courseware material and lessons development. This part contains a number of guidelines to help the developers to structure and complete courses in a consistent and coherent manner.
- Guidelines for organization of training, contains a template and guidelines for the organization of the training course related to the technology for different target audiences.
- Guidelines for execution of training, contains: material for introductory classroom training – this includes several sets of viewgraphs for different target groups, and instructions how to carry out introductory training; guidelines for learners how to use eLearning training courses, guidelines for collaborative learning and guidelines on how to support learners (mentor role).
- List of all lessons for target groups and all functionalities, which were specified and developed during the project.
- The qualifications offered by courses according to EQF, contains chosen EQF as the main basis for the development of procedures for unified qualification assessment in the area of A&RM systems selection, specification, introduction and usage.
- Services, contains the learning services framework.

6 Dedicated ICT Training Solution

The learning services framework was adapted from the work of Britain and Liber (2004), who have prepared a report for JISC on the services that a Virtual Learning Environment (VLE) should support. The framework uses the widely recognised seven principles of effective teaching as the basis for examining the potential of VLEs to enhance learning. Furthermore it combines the conversational model which is a well-known model of effective teaching practice for academic learning and the viable system model (VSM) which is a model for the design and diagnosis of effective organisational structures drawn from management cybernetics, to create an evaluation questionnaire. These two models have been used complementary, with the first providing a model for incorporating effective teaching and learning practice into an e-learning environment and the second providing a number of criteria from an organizational perspective which influence whether the system will facilitate or inhibit the ease with which a pedagogic model such as the conversational model can be used within that system (Fig. 4).

Fig. 4. Quality Management in Occupational Health and Safety Training and Practice, learning system starting page

This framework includes the module, the programme and the student components and has been selected based on the following criteria:

- It supports evaluation of learning system characteristics – The authors have evaluated the most important VLEs that are available either as commercial or open software environments, including ATutor, Forma LMS, ILIAS. and Moodle. The last one is particularly interesting because it has been used as the basis of the final learning system.
- It enables evaluation of key VLE pedagogical aspects – These include increased programme level support, a greater level of flexibility, support of pedagogical innovation, support of content reusability, conformance to standards, variety of student tools (added collaboration features).
- It is based on widely accepted models – As mentioned above, the framework combines the conversational and viable system models while it is based on an earlier version, which has received good reviews.

7 Conclusion

The presented paper shows parts of topics and works connected with creation of OHS training materials. Authors have used materials, which are result of two running OHS training oriented projects: “Quality Management in Occupational Health and Safety Training and Practice” and “Strategic Partnership for Occupational Safety and Health”, both co-funded by the Erasmus+programme of the European Union. Results of these projects will be available as Open Educational Resources OER.

Disclaimer. The European Commission support for the production of this publication does not constitute an endorsement of the contents which reflects the views only of the authors, and the Commission cannot be held responsible for any use which may be made of the information contained therein.

References

1. Communication from the Commission to the European Parliament, the Council, the European Economic and Social Committee and the Committee of the Regions on an EU Strategic Framework on Health and Safety at Work 2014–2020. http://eur-lex.europa.eu/legal-content/EN/TXT/?uri=CELEX%3A52014DC0332
2. International Social Security Association (ISSA), The return on prevention: Calculating the costs and benefits of investments in occupational safety and health in companies. http://www.issa.int
3. Klimasara, W.J., Zieliński, J., Słowikowski, M., Pilat, Z.: Modern trainings in field of safety in automatized and robotized installations using advanced information technology solutions. Pomiary Automatyka Robotyka **16**(2), 177–182 (2012)
4. International Social Security Association (ISSA). https://www.issa.int/
5. BenOSH (Benefits of Occupational Safety and Health) study. http://ec.europa.eu/social/BlobServlet?docId=7417&langId=en
6. Project Quality Management in Occupational Health and Safety Training and Practice web information page. http://www.e-ohstrainingschool.com/
7. Project Strategic Partnership for Occupational Safety and Health information page. http://sposh.piap.pl/
8. Britain, S., Liber, O.: A Framework for the Pedagogical Evaluation of eLearning Environments (2004)
9. Chickering, A.W., Gamson, Z.F.: Seven principles for good practice in undergraduate education. AAHE Bull. **39**(7), 3–7 (1987)
10. Laurillard, D.: Rethinking University Teaching – A Framework for the Effective Use of Educational Technology. Routledge, London (1993)
11. Beer, S.: Diagnosing the System for Organizations. Wiley, Chichester (1985)

Diagnostic Systems of Mobile Robot Technical State

Svitlana Marchenkova(✉)

National Aviation University, Kiev, Ukraine
s.marchenkova90@gmail.com

Abstract. Consider the definition of the technical state of the mobile robot using test and diagnostic tools. The analysis determined the interaction between the object of diagnosis, control and diagnostic tools. Since this interaction is the process of applying for facility diagnosis of multiple actions (outputs) and multiple-shift analysis and responses (outputs) to these actions. Actions for mobile robot may come from control and diagnostic agents or external (on the system of diagnosis) signals defined working algorithm of the object. Depending on the mode of operation of differentiated functional system test and diagnosis. Summarizes the functional diagrams of these systems. The use of functional diagnosis system to verify proper operation and troubleshooting the most critical equipment, assemblies and systems mobile robot that violate normal functioning. From this we can conclude that these systems work when the mobile robot is used for other purposes. They can be used in simulation mode operation work. In this case, should be provided simulation workflows. Such use of functional diagnosis should be used during debugging and repair system. Revealed that the development of diagnostic systems for interaction between object and vehicle diagnostics to be resolved following tasks: feasibility selecting the type and purpose of the system diagnostics; analysis of the physical processes occurring in the facility diagnosis.

Keywords: Mobile robot · Automated system management · Monitoring · Quality assessment · Designing system diagnostics

1 Introduction

In today's world, robots occupy an increasingly significant position. They are used in a variety of fields, such as military service, industry, science, medicine and even education. In recent years, the interest in construction of artificial intelligence system in the so-called multi-agent approach increased greatly [1, 2]. In robotics, this paradigm is typically used only for implementation of functions such as learning, purposeful functioning or decision-making in various management systems (MS). Modern robotic systems must integrate a whole set of sophisticated hardware and algorithms for statistical processing and analysis of multimodal data.

After analysis of the mobile robot operation and thanks to the findings of numerous studies, it was found that the annual productivity of robots by the end of their service life is reduced 1.5 to 2 times compared to the original productivity, and the safety

R. Szewczyk and M. Kaliczyńska (eds.), *Recent Advances in Systems, Control and Information Technology*,
Advances in Intelligent Systems and Computing 543, DOI 10.1007/978-3-319-48923-0_34

design is reduced too. Over the service life of a mobile robot, the costs of its maintenance and repair exceed the initial ones 5 to 7 times. An integrated approach to solving of technical economic and social problems associated with introduction of robots allowed to release the workers from monotonous labor [3].

Robot effectiveness factors are:

(a) mobility in extreme conditions, which do not allow participation of human personnel;
(b) increase in the speed, accuracy and stability of key machinery indicators [4];
(c) reduction of the number of people and their removal from life-threatening areas;
(d) exclusion of operators' error.

Technical diagnosis is a part of the engineering processes of acceptance, maintenance and repair of robots; it is a process of determination of the diagnosed object technical condition with certain accuracy without disassembly and dismantling.

Reliability of diagnostics providing complete information on the status and faults of a mobile robot largely determines its performance.

Due to the growing requirements for reliability and quality, the problem of identifying hidden defects occurring, in particular, through various equipment failures, including those of sensors [5], becomes highly relevant. The principal feature of the proposed concept of a sharp improvement of equipment reliability due to elimination of the impact of failures thereon is that, unlike all used approaches to solution of this problem, not failure locations, but failure sources are identified and registered. The robot considered as an object of diagnostics is a complex system consisting of a large number of heterogeneous components and subsystems (traffic, communication, surveillance, management). Autonomous robots as complex technical objects (CTO) require the status diagnostics. The objective of the robot status diagnostics is to identify and report the abnormal situations and malfunctions in operation of CTO main elements and subsystems.

2 Materials and Methods

Traditional **structure of the diagnostic system** of autonomous robots (Fig. 1) consists of a multichannel measurement system, telemetry transmitter placed on the robot and telemetry receiver with a conversion unit, data display and operator (located at the command post) [6]. The large amounts of information are processed at the command post, and the robot operation is assessed by the operator.

The recording parameters of mobile robots include signals from location sensor, controls, and CCTV in IR and visible range. The calculated parameters of the object are the coordinates of location with reference marks on the area map, movement direction and speed, as well as condition of the major components and subsystems. The comparison of recorded and calculated parameters determines the need for a widespread use of the new information technologies in the processing and analysis of measurement results [7].

Implement of neural network measuring stations into CTO diagnostics is a relevant task [1] for real-time processing of the measured indirect signals and subsequent data analysis.

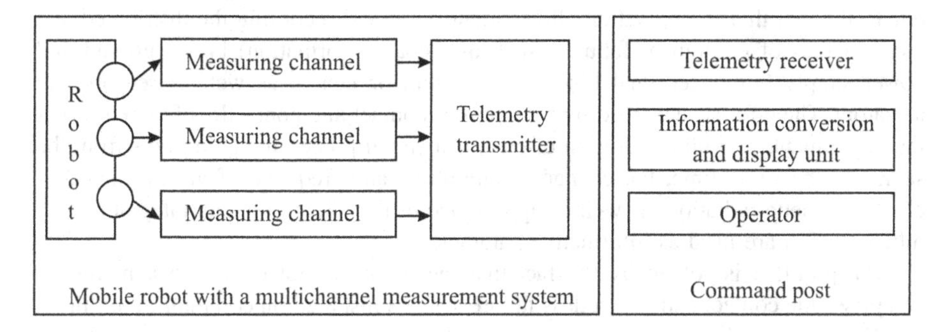

Fig. 1. Traditional structure of the mobile robot diagnostic system

The concept of the proposed class of measuring systems differs from the known implementation environments; consideration of a neural network [8] as a super-multichannel measuring system; approach to diagnostics of CTO status as systemic measurements of an integral parameter (subject to formation of a hierarchical structures of CTOs, construction of integrated parameters of each hierarchy level, common registration and analysis of diagnostic parameter change at all hierarchy levels) [9].

The principal feature of the new concept of a sharp increase in equipment reliability by elimination of the impact of failures thereon it is that, unlike all previously used approaches to solution of this problem, not the failure locations but the failure sources are identified and recorded. Depending on the principles of formation and obtaining of informative features, the aggregate of which estimates the failing condition of equipment elements and failure sources, various methods of identification and registration of the sources of failures with a significant priority are suggested [9]. The emergence of new properties of the equipment passive elements in the intermediate state between "non-faulty" and "faulty", in particular such as differentiation and signal integration, allows addressing the issue of duration of failures occurring due to hidden defects.

Let's consider the example of using the hybrid techniques in the anthropomorphic robot management system [10]. For this purpose, the hardware includes the contact and contactless sensors of failures by informative signs of signal "differentiation", "integration" and "electromagnetic radiation", implementing the hybrid methods for diagnostics of latent defects. Meanwhile, the connections (connectors), interface buses, management buses, as well as internal and external electromagnetic interference are identified as sources of failures. Technical effect implies the ability to design, develop, create and operate the failure-free equipment (similar to the trouble-free equipment).

3 Results and Discussion

It is proposed to increase the reliability and quality of service, ensured by identification and registration of hidden defects found due to failures in equipment operation, as well as identification and registration of internal and external electromagnetic interferences due to the fact that the operation allows identifying and recording the distributed and local sources of equipment failures, such as signal (information) buses, ground and power supply buses, terminal blocks (connectors or jacks), as well as sensors and actuators. The effect is achieved by inclusion of contact and contactless failure sensors into equipment, as well as addition of electrical signal processing algorithms from the same. At the same time, the change in amplitude and frequency features, increased electromagnetic radiation, as well as appearance of the effect of signal integration and differentiation are used as informative parameters.

The problem is solved by the fact that the mobile robot control system further comprises the contact and contactless failure sensors installed on communication lines, interface buses or in the vicinity (up to 2 cm) to the communication line or interface bus to identify the internal and external electromagnetic interference from the power failures in the form of connectors/jacks, front-end buses, control, earthing and power buses, while the system is adapted to the algorithmic signal processing from the said failure sensors.

The solution to the problem of informative parameter of electrical signal integration is based on detection of hidden defects of the device in the form of increased (tens or hundreds of times) conductor, which represents an integrating link with the subsequently included micro capacity (e.g., hundredths of picofarads).

Previously, the operator (human) teaches the neural networks to "touch" and "cover" different objects using the vision systems. The robot is managed from the computer through the microcontroller by means of signaling for actuator enabling. Meanwhile the quality of the object coverage is ensured by the microcontroller through internal and external sensing elements of tactile sensors.

In parallel with online operation of units and devices, the sensors installed on communication lines between units and devices monitor the data lines for hidden defects therein, manifested in the form of crashes or presence of internal or external electromagnetic noise.

The failure sensors are installed, for example, using clips. A simultaneous response of sensors on different communication lines and non-response indicates the source of disruption in the form of external electromagnetic interference. Simultaneous response indicated an internal electromagnetic interference. The main difference made by sensors included in equipment is the fixed signal volume, depending on the distance to the failure source.

A promising direction for further research in the theory and practice of failure, in my opinion, is the study of hidden defects in active elements, in particular, formed by defect-impurity structure [9].

Upon diagnostics for evaluation of the robot technical status (unit), the so-called output processes of the operating mechanism are used. There are working output processes (e.g., capacity consumption or output, power consumption, heat exchange with environment) and related output processes (e.g., noises, vibration, light phenomena, etc.). Each output process is quantitatively assessed using the relevant parameters (e.g., the output power can be evaluated by appropriate value and its increase rate). There is a functional connection between structural parameters and parameters of the initial process, so the values of the latter allow assessing the technical condition of the mobile robot and its operation quality adequately. The boundary value of the output process parameter indicates a defective condition of the mobile robot and determines the need for maintenance or repair. Knowing the nature, rate of change of the output process setting and its boundary value, the operation resource until the next maintenance or repair can be defined.

Depending on the amount of information contained in the output process parameters, they can be generalized or private. The first describe the technical condition of the unit as a whole (e.g., path and time of the robot acceleration to a set speed, power consumption at a given distance, etc.), private, i.e. technical condition of a specific mechanism, system (e.g., steering wheel backlash, knocking in the engine, etc.).

Not all output processes can serve as diagnostic features. In order to be able to use the initial process setting as a diagnostic one, it should meet the following requirements:

- to be functionally important to evaluate the system technical condition;
- to be unequivocal, i.e. its transition from a growing to descending function (or vice versa) should be missing, depending on the time between the mobile robot or change of its structural setting from the original to the boundary value;
- to be sensitive (informative)
- to feature repeatability at multiple measurements, characterized by a degree of value dispersion in respect of the average value of the parameter under constant measurement conditions;
- to allow separating and localizing failures of various elements of the object at their place of origin;
- to ensure manufacturability and cost-effectiveness due to the easy determination of the parameter upon diagnostics, relevant labor and material costs.

The reliability of diagnostics results largely depend on the load, speed and heat modes of operation. Therefore, to obtain high-quality diagnostic information, the appropriate devices setting and maintaining the optimum load, speed and heat modes are used.

The methods of diagnosing the technical condition of the robots and units are characterized by the physical nature and the way of measuring the diagnostic parameters most suitable for use depending on the diagnostics task. Currently, there are three main groups of diagnostic methods: by the output parameters of performance properties, by geometric parameters, by parameters of related processes.

Currently, the studies to develop the new and improve existing methods of diagnostics regarding design of mobile robots become more complex every year. The microelectronics components and microprocessor technology is changed. The same diagnostic feature can often be installed using several methods of diagnostics.

Diagnostics **reliability**, providing complete information about failure, determines its operability. Improvement of diagnostics accuracy will allow reducing the operating costs and labor-intensity of one robot diagnostics.

Analysis of the research performed for diagnostics of a mobile robot showed that setting and solving of research problems for the purpose of improving the accuracy of system diagnostics are new and relevant for robotics. The problem solution will improve the robot efficiency while performing operations.

These methods will facilitate the forecasting of the mobile robot condition and enhance the diagnostics efficiency and accuracy.

The accuracy of mobile robot diagnostics upon short-term and long-acting failures is determined by the time of their occurrence and detection upon performance of diagnostic operations.

4 Conclusions

The approach based on integration of existing components allows a significant acceleration of development and obtaining of a complete operating system relatively easy, as well as enables the reuse of the previously created components or further integration into third-party MSs.

The analysis determined the interaction between the object of diagnosis and control diagnostic tools. The use of the functional diagnosis system to verify the proper operation and troubleshoot the most critical equipment, assemblies and systems of a mobile robot disrupting the normal operation were considered. The simulated workflows of the functional diagnosis should be used during system debugging and repair. It was revealed that the development of a diagnostic system for interaction between the object and diagnostic tools required solving the following tasks: feasibility of selecting the type and purpose of the diagnostics system and analysis of the physical processes occurring in the diagnosed object.

References

1. Wooldridge, M.J., Jennings, N.R.: Intelligent agents: theory and practice. Knowl. Eng. Rev. **10**(2), 115–152 (1995)
2. Zambonelli, F., Jennings, N.R., Wooldridge, M.J.: Developing multiagent systems: the gaia methodology. ACM Trans. Softw. Eng. Method. **12**(3), 317–370 (2003)
3. Zambonelli, F., Omicini, A.: Challenges and research directions in agent-oriented software engineering. J. Auton. Agents Multiagent Syst. **9**(3), 253–283 (2004)
4. Korobiichuk, I.: Mathematical model of precision sensor for an automatic weapons stabilizer system. Measurement **89**, 151–158 (2016). doi:10.1016/j.measurement.2016.04.017
5. Korobiichuk, I., Bezvesilna, O., Ilchenko, A., Shadura, V., Nowicki, M., Szewczyk, R.: A mathematical model of the thermo-anemometric flowmeter. Sensors **15**, 22899–22913 (2015). doi:10.3390/s150922899

6. Korobiichuk, I., Podchashinskiy, Y., Shapovalova, O., Shadura, V., Nowicki, M., Szewczyk, R.: Precision increase in automated digital image measurement systems of geometric values. In: Jabłoński, R., Brezina, T. (eds.) Advanced Mechatronics Solutions. AISC, vol. 393, pp. 335–340. Springer, Heidelberg (2016). doi:10.1007/978-3-319-23923-1_51
7. Shushlyapin, S.V.: Measure the accuracy and reliability of the diagnosis units tractor transmission. Tractor power in crop production, – Coll, scientific. tr. HGTUSKH, Vol. 5, 135–140 (2002)
8. Bishop, C.M.: Neural Networks for Pattern Recognition. Oxford University Press (1995)
9. Dianov, V.N.: Avtomatika i telemekhanika [Automation and telemechanics], no. 7, pp. 119–138 (2012)
10. Talanchuk, P.M., Golubkov, S.P., Maslov, V.P., et al.: Sensory v kontrol'no izmeritel'noy tekhnike [Sensors in inspection technologies], 173 p. Tekhnika, Kiev (1991)

Positioning of Industrial Robot Using External Smart Camera Vision

Jacek Dunaj(✉)

Industrial Research Institute for Automation and Measurements PIAP,
Aleje Jerozolimskie 202, 02-486 Warsaw, Poland
jdunaj@piap.pl

Abstract. This paper contains information about the positioning of industrial robot based on data obtained from the smart camera vision. The camera identifies the element on the pallet and sends information about its position to the robot controller. The application program of robot analyzes the information received, performs appropriate calculations and moves the detected items. The presented method was used in the implementation of the two operating installations which use robots from ABB and Kuka. The paper describes what were the foundation for the construction of these installations and what tasks each of them had to perform robot. The article contains a description of the communication between the robot and the vision system that uses digital IO signals as well as standard interfaces: Ethernet and RS-232. Also application programs of a robot ABB IRb-140 and vision camera Sick IVC-2D used in the construction of one of these installations are presented.

Keywords: Industrial robot · Vision system · Robot positioning · Ethernet · RS-232

1 Introduction

Movement of the robot is a shift TCP point. TCP point is selected by the programmer characteristic point of tool mounted on the manipulator flange. Changing the position of the TCP during automatic or manually moving the manipulator may occur as a result of:

- Rotation axes robot
- Move along the axis of the Cartesian coordinate system associated with the base of manipulator,
- Move along the axis of the Cartesian coordinate system associated with the tool tip,
- Move along the axis of one of the few Cartesian coordinate systems defined by the programmer.

Programming of robot motions carried out by the teaching. It involves manual feeding point TCP to the successive trajectory's points and storing their coordinates. Every move instruction of robot program has information about coordinates, which tool is selected and in which coordinates movement is carried out. In advanced

R. Szewczyk and M. Kaliczyńska (eds.), *Recent Advances in Systems, Control and Information Technology*, Advances in Intelligent Systems and Computing 543, DOI 10.1007/978-3-319-48923-0_35

programming, described by the Kuka company as "expert programming", move instructions do not have to refer only to the points defined during the teaching. User can perform the movement to the point to which the software assigned coordinates of the point defined during the teaching, and then they are modified, e.g., by the addition to the X offset of 100 mm.

One of the many tasks you can perform application robot is to take an element and its movement. This point may be fixed in position in the chosen coordinate system or a location changes to a predetermined extent. In the first case the robot program implements the motion of the same point in space and that it was possible to take the element, it must always be in the same position. In the second case, the robot program must get information, what is the current position of the element in the selected coordinate system, and therefore can eliminate the need for its mechanical, accurate positioning.

Information about the location undertaken element can be obtained by various methods. One of the cheapest, but not necessarily the easiest is mounting a proximity sensor on the robot manipulator and connect it to one of the binary inputs. Then, moving the robot manipulator in several directions determined the coordinates of the points at which the sensor detects an element and performing the appropriate conversion can accurately determine the location of the object. Another way is to use smart camera vision, the application identifies the object on the resulting picture, determine its position, and then send the relevant information to the robot controller.

The idea of moving the robot manipulator to point taking elements whose coordinates robot application received from the external equipment for the first time used at Industrial Research Institute for Automation and Measurements PIAP during the implementation of the model of the technological line for the Poznan University of Technology (Fig. 1). Used here ABB IRb-140 robot coupled with intelligent camera switcher IVC-2D Sick (Fig. 2).

Fig. 1. Laboratory of modelling flexible manufacturing line Poznan university of technology

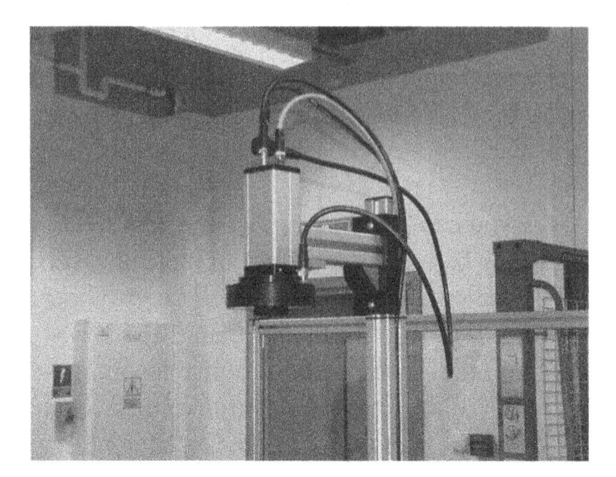

Fig. 2. The camera Sick IVC-2D with a resolution of 1600 × 1200 pixels

2 Task of Vision System and Robot

The task of illustrating the control of the robot ABB IRb-140 by the vision system Sick IVC-2D defined as follows. On the conveyor belt are moved black pallets with dimensions approx. 40 cm × 40 cm (Fig. 3). On the pallet there are two randomly arranged elements:

- an element in the shape of a cylinder 6 cm tall and 3 cm in diameter laid on the base pallet,
- a ring-shaped element with a height of 3 cm and diameter: 5.5 cm outer and inner 3 cm

Fig. 3. A pallet with elements, on the left side is robot manipulator IRb-140, on the right side is the stand of the Sick camera

Pallet with both elements moves in the field of view Sick IVC-2D camera with a resolution of 1600 × 1200 pixels, which is stopped. After stopping the pallet camera's application program has to identify the position of the two elements and send information about the position to the robot. The task of the robot is to take up the cylinder, insert it into the ring, and then move the manipulator to the home position. An additional difficulty is that the robot and the vision system for reasons of didactic could not be placed in a light-housing, and so is exposed to different lighting conditions.

3 Selection Tool and Robot Coordinate System

The tool used to perform the measurements and the task is shown in Fig. 4. It consists of two parts: a fixed gripper mounted on the robot flange and the movable element in the shape of a cylinder ending with a cone on the one hand, and with flange on the other hand. The tool itself is "created" by closing gripper clamps on cylinder in such a way that the lower part of the flange touches to these clamps. As a point TCP so built tools adopted apex of the cone, and the robot application program assigned it as *nameCylinderGripper*.

Fig. 4. Robot tool *CylinderGripper*

The mount point of camera were chosen in such a way that its optical axis is perpendicular to the plane of the pallet and crossed its center. The axis does not necessarily perfectly hit the center of the pallet, because the position is calibrated by software. In carrying out the task it was assumed that the two coordinate systems, i.e. The camera and the user coordinate system of robot, described in its application program as *CameraCoordinateSystem*, they overlap, i.e., both X and Y axes are common and common point is defining the beginning of both of these systems.

4 Calibration Position

In order to system robot–camera work correctly, after assembly and immobilization of some regulated settings must be software-calibrated. To do this, using "live" camera marked on the pallet three points defining three of the four vertexes of rectangle bounding the field of camera view. Then measure the distance in millimeters between the points:

- point (0, 0) (the origin of coordinate) and a point (X, 0) which is the extreme "right" point of sight of the camera. The measurement result is divided by 1600.
- point (0, 0) (the origin) and the point (0, Y) which is the extreme "low" point of sight of the camera. The measurement result is divided by 1200.

Numbers 1600 and 1200 define the resolution of the camera in pixels. When properly selected points, both ratios should be equal. The result determines the conversion factor pixels per millimeter and calculate value is entered into the robot application program as a constant named *PixelOnMmFactor*.

The next step is to define the robot's coordinate system associated with camera's coordinate system. It uses tool *CylinderGripper* (Fig. 4) and the points defining the bounding box of the camera. After defining the system, using a tool *CylinderGripper* and coordinate system *CameraCoordinateSystem*, must be set manipulator robot in such a position that the plane of the terminal gripper is parallel to the plane of the pallet. From this point, any further movement of the robot is performed by changing only the X, Y and Z coordinates without changing the orientation of the tool. Point TCP (end cone) should move to the origin *CameraCoordinateSystem*. Then, using the teaching function, make correction of position of the point named *OriginCoordinatesPoint*. This point will be the point of reference for positioning the robot manipulator when making the cylinder and inserting it into the ring.

The camera's application program uses the stored patterns to determine the position of the geometric center of cylinder and ring. These patterns (picture elements), like the field of view of the camera, change depending on the height of the camera above the pallet. To determine the current patterns should be placed cylinder base on a pallet as close to the optical axis of the camera, and then using the preview feature software Sick IVC Studio precisely focus the image at the top edge and secure adjustment of the lens. After determining the lighting, cylinder pattern and then the ring pattern is introduced using the Setup function of two instructions Shape Locator of camera application program.

5 Execution of Task

Both elements, which camera identifies the position are axially symmetrical. This makes it easier task, because in order to take the cylinder and put it into the ring enough to know the position of geometric centers of both elements.

Camera's application program analyzes the image and tries to recognize geometric centers of two elements. It stores them as two pairs of numbers $[X_{1k}, Y_{1k}]$ (cylinder) and $[X_{2k}, Y_{2k}]$ (ring). Both pairs are the coordinates of points in the coordinate system of the camera. Since the beginning and the directions of the axis of the robot coordinate system coincide with the coordinate system camera, the information obtained from the camera can be used for positioning the manipulator. Each of these coordinates is expressed in pixels, so there is a natural number in the range of 0 to 1600 (coordinates X_{1k}, X_{2k}), or in the range of 0 to 1200 (coordinates Y_{1k}, Y_{2k}) (1600 × 1200 a resolution of the camera in pixels, respectively, in the X and the Y axis). Camera's program are not converted pixels to millimeters, because robot's application due to the possibility of programming easier and more convenient can do this. The camera's program transmits four integers to the robot's controller. Using constant *PixelOnMmFactor* defined during the calibration, robot's application program calculates the coordinates received from pixels to millimeters as follows:

X_{1r} [mm] = PixelOnMmFactor * X_{1k} [pixels] →X coordinate of the cylinder
Y_{1r} [mm] = PixelOnMmFactor * Y_{1k} [pixels] →Y coordinate of the cylinder
X_{2r} [mm] = PixelOnMmFactor * X_{2k} [pixels] →X coordinate of the ring
Y_{2r} [mm] = PixelOnMmFactor * Y_{2k} [pixels] →Y coordinate of the ring
where the subscript "r" means the robot, the subscript "k" means the camera.

The values appearing on the left side are the coordinates of the geometric centers of cylinder and ring in a working robot coordinate system referred to in the application as *CameraCoordinateSystem*.

During testing, it turned out that the error of settings robot manipulator in points with coordinates calculated in accordance with the above formulas, increases as a function of the distance of the object identified from the optical axis of the camera. With high probability the cause of this error could be:

- too little accurate positioning of the camera relative to the plane of the pallet – the optical axis of the camera does not cut exactly at 90° plane discus,
- too small precision marking points defining the field of camera's view. Lack of precision is the result of use "live view" of the camera. Consequently, this leads to the fact that the camera coordinate system and the robot coordinate system are moved to be slightly offset from one another,
- the plane pallet is located at a different distance from the camera lens than the upper surfaces of the identified elements. In addition, the height of the cylinder and the ring are different.

Looking for a way to position correction is noted that:

- error setting the robot manipulator increases linearly as a function of distance from the optical axis of the camera, so from a point [800, 600] in the camera coordinate system,
- maximum values of error setting the manipulator are different depending on whether they concern the cylinder or the ring (there are different heights of these elements in relation to the camera lens),
- the maximum values of error setting manipulator for the selected element placed near the line bounding the field of camera view is different for the X-axis and Y-axis,
- maximum value of error setting manipulator for the selected element in the direction of the axis X varies depending on whether the point is to the left or to the right from the point of coordinate X = 800 [pixel],
- maximum value of error setting manipulator for the selected element in the direction of the axis Y varies depending on whether the point is above or below from the point of coordinate Y = 600 [pixel].

These errors are repetitive for calibrated system, so they can be corrected by software. To do this, set the robot in step-by-step mode and put the ring in the vicinity of the optical axis of the camera (point [800, 600], see Fig. 5). Then sequentially restart the robot application without the adjustment position, each time putting the cylinder in the vicinity of the next four points shown in Fig. 5. As soon as the robot program moves gripper to take the cylinder without correction of position it is necessary to measure [mm] unless that position differs from the ideal position. Subsequent measurements determine the maximum misalignment of the gripper in the X direction to the left (X < 800) and right (X > 800) from the optical axis of the camera (positions marked left and right in Fig. 5), the maximum misalignment of the gripper in the Y direction above (Y < 600) and below (Y > 600), the optical axis of the camera (position marked top and bottom in Fig. 5). Repeat this procedure for the ring.

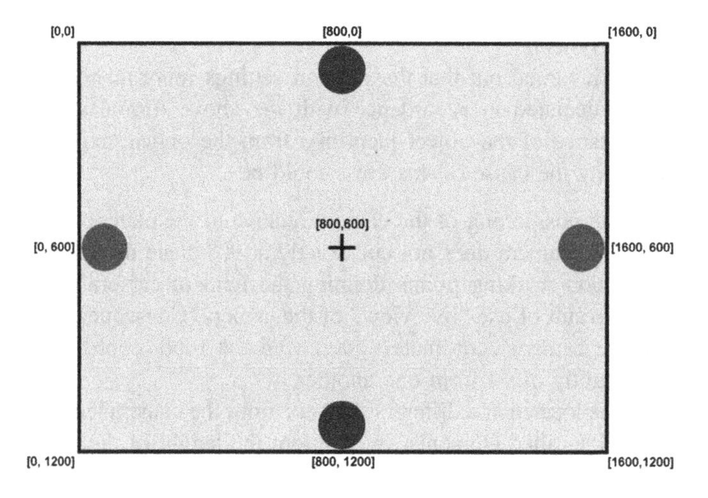

Fig. 5. Measurement points adjustments of gripper's position

Having measured the maximum errors positions should introduce them to the robot program by assigning them the following variable:

- MaxCorrectionX1 [mm] – it refers to the X coordinate of the cylinder,
- MaxCorrectionY1 [mm] – it refers to the Y coordinate of the cylinder,
- MaxCorrectionX2 [mm] – it refers to the X coordinate of the ring,
- MaxCorrectionY2 [mm] – it refers to the Y coordinate of the ring

Since the robot manipulator positioning errors grow linearly with distance from the optical axis of the camera, to correct the geometric center coordinates of the cylinder and the ring are determined from the following relationships:

X_{1r} [mm] = PixelOnMmFactor * X_{1k} [pixels] + [MaxCorrectionX1 * (800 - X_{1k})/ 800]

Y_{1r} [mm] = PixelOnMmFactor * Y_{1k} [pixels] + [MaxCorrectionY1 * (600 - Y_{1k})/ 600]

X_{2r} [mm] = PixelOnMmFactor * X_{2k} [pixels] + [MaxCorrectionX2 * (800 - X_{2k})/ 800]
Y_{2r} [mm] = PixelOnMmFactor * Y_{2k} [pixels] + [MaxCorrectionY2 * (600 - Y_{2k})/ 600]
where the subscript "r" means the robot, the subscript "k" means the camera.

The coordinates of all points of the trajectory robot program calculates relative to *OriginCoordinatesPoint* point defined during software calibration of system. During the movement of the manipulator robot does not change the orientation of the tool, so the plane of gripper clamps still remains parallel to the plane of the pallet.

6 Controlling a Robot by the Master PLC Controller and Sick IVC-2D Camera

Industrial robot IRb-140 cooperates with the master PLC controller and Sick IVC-2D camera with the digital inputs and outputs. Additionally, robot application receives information about position of two elements from the camera by Ethernet network.

6.1 Digital Inputs and Outputs from/to IRb-140 Robot

Table 1 shows the digital input and output signals to/from a robot. This signals is used to synchronize operations of manipulator with other devices lines. The individual columns have the following meanings:

- column named Name specifies the name of the signal which has been described in robot software system and under which occurs in the robot application program,
- column named Type of Signal specifies the type of signal: digital input, digital output,

Table 1. Digital inputs and outputs to/from IRb-140 robot

Name	Type of signal
DI_000_OpRequest	Digital Input
DI10_01_GripperClosed	Digital Input
DI10_02_GripperOpened	Digital Input
DI10_10_IdentificationAck	Digital Input
DI10_11_IdentificationOK	Digital Input
DO_000_Ready	Digital Output
DO_001_OpPending	Digital Output
DO_002_OpComplete	Digital Output
DO_003_OpError	Digital Output
DO_004_OpFault	Digital Output
DO_005_CycleONr	Digital Output
DO10_01_CloseGripper	Digital Output
DO10_02_OpenGripper	Digital Output
DO10_09_StartIdentification	Digital Output

6.1.1 Robot Digital Signals to Operate Sick IVC-2D Camera

Robot digital signals to operate Sick IVC-2D camera are:

DI10_10_IdentificationAck
- This signal is coupled to the output No. 0 of camera (Output number = 0),

DI10_11_IdentificationOK
- This signal is coupled to the output No. 1 of camera (Output number = 1),

DO10_09_StartIdentification
- This signal is coupled to the input No. 0 of camera (Input number = 0),

Significance of these signals was described in section "Digital input/output signals to/from the camera Sick IVC-2D"

6.1.2 Robot Digital Signals to Operate the Gripper

Robot digital signals to operate the gripper are:

DI10_01_GripperClosed
DI10_02_GripperOpened
- These robot input signals are coupled with the gripper's sensors. Reading the status:
  - DI10_01_GripperClosed = 1
  - DI10_02_GripperOpened = 0

  It means that the gripper clamps are in the "closed" state. Reading the status:
  - DI10_01_GripperClosed = 0
  - DI10_02_GripperOpened = 1

  It means that the gripper clamps are in the "open" state.

DO10_01_CloseGripper
DO10_02_OpenGripper
- These robot output signals are used to open/close the gripper clamps. These signals operate alternately, i.e. sets:
  - DO10_01_CloseGripper = 1
  - DO10_02_OpenGripper = 0

  causes tightening gripper clamps, activation:
  - DO10_01_CloseGripper = 0
  - DO10_02_OpenGripper = 1

  frees the gripper clamps. Any other combination of [0, 0] or [1, 1] does not change the state of the gripper clamps.

6.1.3 Robot Digital Signals to Cooperate with the Master PLC Controller

To cooperate with the master PLC controller robot uses a single digital input and five digital output signals.

DI_000_OpRequest
- This is the input signal to the robot. Master PLC controller set this signal (DI_000_OpRequest = 1) to start the task associated with the recognition of the position and movement of elements on the pallet.

DO_000_Ready

- This is output signal from the robot supported by the system software. The state of this signal is determined by a combination of the two logic signals: DO_005_CycleON and DO_001_OpPending:
  DO_000_Ready = (DO_005_CycleON).AND.(NOT DO_001_OpPending)

DO_001_OpPending

- This is the output signal from the robot supported by the application. If the robot is in a state of waiting for the task to start (waiting until the input signal DI_000_OpRequest will have a value of "1"), then the output will be maintained state "0". At the time of maintenance moving elements on the pallet, robot program sets the value of this signal (DO_001_OpPending = 1) and maintains the state until they complete the task (regardless of outcome).

DO_002_OpComplete

DO_003_OpError

- These output signals from the robot are supported by the application. At the launch of the task both of these signals are reset. Once the task is completed, which is signaled DO_001_OpPending = 0, the robot program sets the value of both signals according to the following algorithm. Combination:
  - DO_001_OpPending = 0
  - DO_002_OpComplete = 1
  - DO_003_OpError = 0

  It means that the vision system recognized both of the elements, robot moved the elements and manipulator returned to the home position. Combination:
  - DO_001_OpPending = 0
  - DO_002_OpComplete = 0
  - DO_003_OpError = 1

  means that the vision system did not recognized at least element, and manipulator returned to the home position. Combination:
  - DO_002_OpComplete = 1
  - DO_003_OpError = 1

  It is not permitted and should not occur.

DO_004_OpFault

- This is output signal from the robot supported by the system software. This signal determines whether attached to the robot emergency stop.

DO_005_CycleON

- This is output signal from the robot supported by the system software. This signal determines whether the robot application program is running (Cycle ON).

6.2 Digital Input/Output Signals to/from the Camera Sick IVC-2D

To synchronize operations between Sick IVC-2D camera and robot it uses one digital input No. 0 (Input number = 0) and two digital outputs No. 0 and No. 1 (Output number = 0 and Output number = 1). They operate as follows:

Input number No 0

- This is the input signal to the camera coupled with the robot output DO10_09_StartIdentification. If the robot sets this signal "1", this means that the pallet with elements for identification is situated in the field of view of camera and does not move. In response, the camera's application sets to "0" both outputs: Output number = 0 and Output number = 1, identifies the position of the two elements on the pallet, sends information about their location to the robot. Finally, properly values of outputs Output number = 0 and Output number = 1 are set.

Output number No 0

- This is the output signal from the camera coupled to the robot input DI10_10_IdentificationAck. "0" logic (low status) on the output means, that the camera application program is in progress to identify the position of elements on the pallet. "1" logic (high state) means that the identification of the position of elements on the pallet ended, position information was sent to the robot by Ethernet, and whether the identification was successful or not determines the Output number No 1.

Output number No 1

- This is the output signal from the camera coupled to the robot input DI10_11_IdentificationOK. The robot application program should interpret this input only when the input signal DI10_10_IdentificationAck is equal to "1". Combination:
  - DI10_10_IdentificationAck = 1
  - DI10_11_IdentificationOK = 1

  It means that the identification of the position of both elements on the pallet was successful. Combination:
  - DI10_10_IdentificationAck = 1
  - DI10_11_IdentificationOK = 0

  It means that the identification of the position of at least one of the pallet failed.

6.3 Information Sent from the Camera Sick IVC-2D to IRB-140 by Ethernet

Camera application program analyzes the image to recognize the cylinder and the ring. It determines the position of the geometrical centers and stores them as two pairs of coordinates $[X_{1k}, Y_{1k}]$ and $[X_{2k}, Y_{2k}]$. Each of these coordinates is expressed in pixels ("units" of the image) so there is a natural number from 0 to 1600 (coordinates X_{1k}, X_{2k}), or from 0 to 1200 (coordinates Y_{1k}, Y_{2k}). Therefore, there are four natural numbers to transfer from the camera to the robot controller. For this purpose Ethernet interface is used as a transmission channel. Each of the four coordinates is not transmitted as a value only as a string of four decimal digits. So the frame from camera to the robot has the following format:

#XXXX YYYY VVVV ZZZZ\n\r

where:

#
– is a sign of the beginning of frame,
#XXXX
– four decimal digits specifying coordinate X_{1k} [in pixels] of the cylinder's geometric center,
#YYYY
– four decimal digits specifying coordinate Y_{1k} [in pixels] of the cylinder's geometric center,
#VVVV
– four decimal digits specifying coordinate X_{2k} [in pixels] of the ring's geometric centers,
#ZZZZ
– four decimal digits specifying coordinate Y_{2k} [in pixels] of the ring's geometric center,
\n\r
– two characters, respectively new line and carriage return, signifying the end of the frame.

Single space characters are inserted between strings: XXXX, YYYY, VVVV, ZZZZ. These spaces, new line sign and carriage return sign allow without additional transformations to display for testing purposes the entire frame on the screen of robot programming panel using the library function TPWrite(). The frame are not included in any checksums.

The transmission of information to an Ethernet network is done using TCP protocol. The camera acts as a client (the camera initiates the connection) and the robot acts as a server. Addresses:

- TCP address of each device is set using the system software of each,
- TCP robot's address to which the camera transmits package sets up the camera application,
- TCP camera's address that a robot uses to listen whether the frame was sent sets the robot's application program.

7 Application Program of IRb-140 Robot

Program of IRb-140 robot, which implements the task described in the previous sections, is stored in robot's mass store as a file *Poznan.prd*. It contains references to four modules:

MainModule.mod
– this module provides a definition of the main program **main()** performing the task,
GripperService.mod
– this module contains a definition of two procedures:
  - OpenGripper ()
  - CloseGripper ()

performing opening and closing the gripper. Each of these procedures includes a pair of instructions controlling gripper's valves, a pair of instructions waiting for confirmation from the sensors state of the gripper and the instructions of the additional delay.

TCPZ_CylinderGripper.mod

– this module provides a definition of the tools shown in Fig. 4.

Wobj_CameraCoordinateSystem.mod

– this module provides a definition of the robot's coordinate system named *CameraCoordinateSystem* in which the robot performs the task. This system coincides coordinate system of camera.

Tool *CylinderGripper* and coordinate system *CameraCoordinateSystem* are defined by teaching.

This article will concentrate on the module *MainModule.mod* containing the definition of robot source program. Module *MainModule.mod* starts declarations/definitions of constants and variables used in the program:

– Definitions of the variables used to Ethernet communicate with Sick IVC-2D camera by Ethernet:

```
VAR socketdev ClientSocket;
VAR socketdev TmpSocket;
```

– Definition of a variable used as a frame buffer received from the camera:

```
VAR string ReceivedString
```

– Definitions of string variables are used as character arrays to store separate parts of the buffer *ReceivedString* (separate parts of the frame). Each such array contains 4 digits consisting of 4 numbers corresponding to the coordinates X_1, Y_1, X_2, Y_2. the positioning of the cylinder and the ring on the pallet:

```
VAR string StringX1
VAR string StringY1
VAR string StringX2
VAR string StringY2
```

– Definitions of *num* type variables in which they are stored the coordinates X_1, Y_1, X_2, Y_2. after they have been interpreted based on the content of the relevant character tables:

```
VAR num OffsetX1   !value calculated on the ground of StringX1 table,
VAR num OffsetY1   !value calculated on the ground of StringY1 table,
VAR num OffsetX2   !value calculated on the ground of StringX2 table,
VAR num OffsetY2   !value calculated on the ground of StringY2 table.
```

– Definition of auxiliary program flag:

```
VAR bool OK_Flag
```

- Declaration of the robot's work coordinate system associated with the coordinate system of the camera:
  `PERS wobjdata CameraCoordinateSystem`
- Definition of a point [0, 0, 0] defines the beginning of the camera's work coordinate system (the origin of the camera). This point is determined by "teaching" program during the calibration system:
  `CONST robtarget OriginCoordinatesPoint`
- Declaration of the pallet's point from which the robot takes a cylinder and declaration of the pallet's point where the captured cylinder puts into the ring. The coordinates of these points are calculated on the basis of information from the vision system, adjustments resulting from the conversion of pixels to millimeters and additional adjustments related to the distance of the object from the optical axis of the camera:
  ```
  VAR robtarget CylinderPoint
  VAR robtarget RingPoint
  ```
- Tool declaration `CylinderGripper:`
  `PERS tooldata CylinderGripper`
- The definition of a constant defining the coefficient of conversion pixels per millimeter. If you change the height of the camera above the pallet must modify the value of this constant:
  `CONST num PixelOnMmFactor`
- Declaration of auxiliary variable used to determine the following points that define the trajectory of the robot during the task:
  `VAR robtarget AuxPoint`
- Declarations of variables to store the maximum values of the corrections (in millimeters). These variables are used for correcting coordinates X_1, Y_1, X_2, Y_2 respectively, of the cylinder and the ring. The values of these variables are determined experimentally:
  ```
  VAR num MaxCorrectionX1
  VAR num MaxCorrectionY1
  VAR num MaxCorrectionX2
  VAR num MaxCorrectionY2
  ```
- Declarations of variables to store the calculated values of the corrections (in millimeters). These variables are used for correcting coordinates X_1, Y_1, X_2, Y_2 respectively, of the cylinder and the ring:
  ```
  VAR num AdditionalCorrectionX1
  VAR num AdditionalCorrectionY1
  VAR num AdditionalCorrectionX2
  VAR num AdditionalCorrectionY2
  ```

After the declarations/definitions of constants and variables MainModule.mod contains instructions of robot's program. The program repeatedly uses TPWrite() library function to display messages (print control) on the robot's programming panel:

– The beginning of the program: open the gripper, the calculation of the rest position of the manipulator (so that the manipulator with gripper was out of sight of the camera), moving the manipulator to this point, reset DO_001_OpPending signal (see description):

```
OpenGripper;
AuxPoint := Offs(OriginCoordinatesPoint,
   0,-100,-100);
MoveL AuxPoint, v100, z50,
   CylinderGripper\WObj:=CameraCoordinateSystem;
TPWrite "   ";
TPWrite "Robot at START point";
SetDO DO_001_OpPending, 0;
```

– Waiting for the start signal of the program (waiting until the digital input DI_000_OpRequest shows "1"):

```
TPWrite "Wait for START signal DI_000=1...";
WaitDI DI_000_OpRequest, 1;
```

– Sets digital outputs from the robot to the master PLC controller:

```
SetDO DO_001_OpPending,  1;
SetDO DO_002_OpComplete, 0;
SetDO DO_003_OpError,    0;
```

– Initiate a vision control by switching on approx. 0.3 s a digital output signal DO10_09_StartIdentification to the camera:

```
SetDO DO10_09_StartIdentification, 1;
WaitTime 0.3;
SetDO DO10_09_StartIdentification, 0;
TPWrite "Start vision identification...";
```

– Setting the transmission channel supported by Ethernet in a state of waiting for the frame from Sick IVC-2D camera. The network can simultaneously operates several devices, but which is meant for robot decide the parameters of SocketBind instruction. The waiting time for delivery amounts a few seconds (the time can be changed by robot controller settings). In the absence of shipping software system reports an error, which must be manually confirmed. A received frame is inserted into the buffer `ReceivedString`:

```
SocketCreate TmpSocket;
SocketBind   TmpSocket, "10.1.6.84", 65000;
SocketListen TmpSocket;
SocketAccept TmpSocket, ClientSocket;
SocketReceive ClientSocket\Str:=ReceivedString;
SocketClose ClientSocket;
SocketClose TmpSocket;
```

- Convert a string contained in the ReceivedString buffer into four strings. These strings contain four decimal numbers in each sequence (sequence of the coordinates X_1, Y_1, X_2, Y_2). The conversion involves the separation using the function **StrPart()**. Then transforming these strings to the corresponding values (function **StrPart()**

```
StringX1 := StrPart(ReceivedString,2,4);
StringY1 := StrPart(ReceivedString,7,4);
StringX2 := StrPart(ReceivedString,12,4);
StringY2 := StrPart(ReceivedString,17,4);
!
OK_Flag := StrToVal(StringX1,OffsetX1);
OK_Flag := StrToVal(StringY1,OffsetY1);
OK_Flag := StrToVal(StringX2,OffsetX2);
OK_Flag := StrToVal(StringY2,OffsetY2);
```

Functions **StrPart()** and **StrToVal()** are library functions of Rapid language used to program ABB robots.

- Determining the initial value of additional corrections (in pixels) of coordinates X_1, Y_1, X_2, Y_2. The need for additional adjustments due to the fact that the farther element is detected from the optical axis of the camera (that is, from a point [800, 600] expressed in pixels) that the greater mistake vision system gives its coordinates. It was assumed that this error increases linearly as a function of distance from the optical axis:

```
AdditionalCorrectionX1 := 800 - OffsetX1;
AdditionalCorrectionX2 := 800 - OffsetX2;
AdditionalCorrectionY1 := 600 - OffsetY1;
AdditionalCorrectionY2 := 600 - OffsetY2;
```

The numbers 800 and 600 occurring in the above conversions are half of the Sick IVC-2D camera resolution 1600 × 1200 pixels.

- Determination of the maximum value (in millimeters) additional adjustments coordinates X_1, Y_1, X_2, Y_2 of geometric centers of cylinder and ring. The values of these adjustments have been selected experimentally by measurements performed during the robot and vision system with the exception of correcting the position (for these measurements should be "uncomment" next four instructions):

```
!AdditionalCorrectionX1 := 0;
!AdditionalCorrectionX2 := 0;
!AdditionalCorrectionY1 := 0;
!AdditionalCorrectionY2 := 0;
!
MaxCorrectionX1 := 17;
IF (OffsetX1 > 800) THEN
      MaxCorrectionX1 := 16;
ENDIF
!
MaxCorrectionX2 := 8;
IF (OffsetX2 > 800) THEN
      MaxCorrectionX2 := 10;
ENDIF
!
MaxCorrectionY1 := 9;
IF (OffsetY1 > 600) THEN
      MaxCorrectionY1 := 10;
ENDIF
!
MaxCorrectionY2 := 5;
IF (OffsetY2 > 600) THEN
      MaxCorrectionY2 := 8;
ENDIF
```

- Calculation of the final value (in millimeters) additional adjustments coordinates X_1, Y_1, X_2, Y_2 of geometric centers of cylinder and ring. The following conversion based on the assumption that the position error given by the vision system increases linearly as a function of distance from the optical axis of the camera:

```
AdditionalCorrectionX1 := (MaxCorrectionX1 *
AdditionalCorrectionX1) / 800;
AdditionalCorrectionX2 := (MaxCorrectionX2 *
AdditionalCorrectionX2) / 800;
AdditionalCorrectionY1 := (MaxCorrectionY1 *
AdditionalCorrectionY1) / 600;
AdditionalCorrectionY2 := (MaxCorrectionY2 *
AdditionalCorrectionY2) / 600;
```

- Determination of adjusted coordinates (in millimeters), X_1, Y_1, X_2, Y_2 of geometric centers of cylinder and ring after the amendments described in the preceding paragraphs. Then designate the point of taking up the cylinder and the point of insertion of the cylinder to the ring (more precisely: calculated coordinates X and Y, coordinate Z is inserted explicitly in the parameters of movement instructions):

```
OffsetX1 := (PixelOnMmFactor * OffsetX1) +
            AdditionalCorrectionX1;
OffsetX2 := (PixelOnMmFactor * OffsetX2) +
            AdditionalCorrectionX2;
OffsetY1 := (PixelOnMmFactor * OffsetY1) +
AdditionalCorrectionY1;
OffsetY2 := (PixelOnMmFactor * OffsetY2) +
            AdditionalCorrectionY2;
!
TPWrite "Received: " + ReceivedString;
TPWrite "X1=" + StringX1 + " Y1=" + StringY1 + "
X2=" + StringX2 + " Y2=" + StringY2;
!
CylinderPoint := Offs (OriginCoordinatesPoint,
OffsetX1,OffsetY1,0);
RingPoint := Offs (OriginCoordinatesPoint,
OffsetX2,OffsetY2,0);
```

- Wait until the camera application program confirms the termination of the vision control ("1" logic on the input DI10_10_IdenificationAck to the robot) and thus confirm the information sent to an Ethernet network:

```
WaitDI DI10_10_IdentificationAck, 1;
```

- If the digital signal DI10_11_IdentificationOK from the camera to the robot is "0" it means that the camera's application program does not detect at least one element on the pallet. In this case, the robot does not move manipulator, only opens and closes the gripper, and then sets the three signals to the master controller line:

```
IF DI10_11_IdentificationOK = 0 THEN
       CloseGripper;
       OpenGripper;
       TPWrite "Element(s) not founded";
       SetDO DO_001_OpPending,  0;
       SetDO DO_002_OpComplete, 0;
       SetDO DO_003_OpError,    1;
       ExitCycle;
ENDIF
```

- Frame containing information about the position of the cylinder and the ring is transmitted as in the case of non-detection of the element(s) by the vision system. In this case, in the right place the frame instead of the actual coordinates will be inserted strings 9999 9999.

```
#XXXX YYYY VVVV ZZZZ\n\r
```

- Implementation of the movement of the robot manipulator to the point of taking up the cylinder:

```
TPWrite "Move to cylinder's take up point...";
AuxPoint := Offs(CylinderPoint,0,0,-100);
MoveL AuxPoint, v300, z50,
   CylinderGripper\WObj:=CameraCoordinateSystem;
WaitTime\InPos, 0.0;
AuxPoint := Offs(CylinderPoint,0,0,25);
MoveL AuxPoint, v30, z50,
   CylinderGripper\WObj:=CameraCoordinateSystem;
WaitTime\InPos, 0.0;
CloseGripper;
AuxPoint := Offs(CylinderPoint,0,0,-50);
MoveL AuxPoint, v50, z50,
   CylinderGripper\WObj:=CameraCoordinateSystem;
WaitTime\InPos, 0.0;
```

– Implementation of the movement of the robot manipulator to the point of insertion the cylinder into the ring:

```
TPWrite "Move to insertion point...";
AuxPoint := Offs(RingPoint,0,0,-50);
MoveL AuxPoint, v50, z50,
   CylinderGripper\WObj:=CameraCoordinateSystem;
WaitTime\InPos, 0.0;
AuxPoint := Offs(RingPoint,0,0,25);
MoveL AuxPoint, v30, z50,
   CylinderGripper\WObj:=CameraCoordinateSystem;
WaitTime\InPos, 0.0;
OpenGripper;
AuxPoint := Offs(RingPoint,0,0,-100);
MoveL AuxPoint, v50, z50,
   CylinderGripper\WObj:=CameraCoordinateSystem;
WaitTime\InPos, 0.0;
```

– Implementation of the movement of the robot manipulator to the rest point:

```
TPWrite "Move robot to rest point...";
AuxPoint := Offs(RingPoint,0,-50,-100);
MoveL AuxPoint, v300, z50,
   CylinderGripper\WObj:=CameraCoordinateSystem;
WaitTime\InPos, 0.0;
!
AuxPoint:= Offs (OriginCoordinatesPoint,
   0,-100,-100);
```

```
MoveL AuxPoint, v300, z50,
   CylinderGripper\WObj:=CameraCoordinateSystem;
```

- Set properly values of digital signals to the master PLC controller with information about the successful completion of the movement of the detected elements:

```
SetDO DO_001_OpPending,  0;
SetDO DO_002_OpComplete, 1;
SetDO DO_003_OpError,    0;
```

8 Application Program of Sick IVC-2D Camera

8.1 The Algorithm Implemented by the Program

Program of Sick IVC-2D camera performs the following algorithm:

1. At a signal from the IRB-140 camera takes pictures of the pallet with elements. Then, the image is subjected to binarization, i.e. converting to zero-one (brighter areas of the picture will be filled with white color, the darker areas of the picture – will be filled black color).
2. The camera identifies both elements by comparing the subsequent parts of the image with the stored patterns of the cylinder and the ring images. The result of the identification are two pairs of integers $[X_{1k}, Y_{1k}]$ and $[X_{2k}, Y_{2k}]$ defining coordinates of centers of both elements.
3. Each of the four coordinates $[X_{1k}, Y_{1k}]$, $[X_{2k}, Y_{2k}]$ is converted into four strings consisting of four digits each. As a result of this conversion is obtained 16 decimal digits of which program the camera completes the frame to the robot,
4. The camera opens a transmission channel supported by Ethernet network to the robot controller, and then sends a ready frame,
5. After sending frame, camera's program sets appropriate digital signals to the robot.

8.2 Implementation of the Program

The camera program was developed in Sick IVC Studio 3.2_CR52a. The application program executing the tasks consists of the following elements:

1. Main program performing the tasks described in the previous section
2. Subroutine converting natural number into the numeric 4-digit string. This routine is called four times by the main program for the implementation of conversion of each of the four coordinates X_{1k}, Y_{1k}, X_{2k}, Y_{2k} defining geometric centers of the cylinder and the ring.
3. Image memory banks:
   - No. 0 bank to store the actual image of pallet with elements to identify,
   - No. 1 bank for storing the image of the bank No. 0 after the binarization. In this image the program will look for the elements.
   - No. 2 bank for storing the image of the identified cylinder,
   - No. 3 bank for storing the image of the identified ring.

   Images stored in banks 2 and 3 are not used in the program for further analysis. If the camera is attached PC with installed Sick IVC Studio 3.2_CR52a that these images can be helpful in observing the operation of the application program.
4. Data block (fragments of camera's RAM) to which from the flash memory are copied patterns of cylinder and images of the ring. Identifying the elements camera's program uses:
   - Data block #0 of the reference image cylinder,
   - Data block #0 of the reference image ring.
5. The additional array to store constants and variables used by the program. The array also includes the intermediate results of calculations generated by different instructions. It is used as a place to provide the parameters and result of execution of the subroutine conversion of natural number on the 4-digit string.

The exact description of all program instructions and information about the organization used memory contains [1].

During several hours of testing robot-camera station it confirmed, that the correct operation of the vision system is dependent on the lighting conditions. For reasons of didactic the station could not be placed in a light-housing, so the only way to adopt the camera application to changing lighting conditions were patches with software. They consisted of stepwise parameter modification Lower Threshold of instruction Binarize, which converts the actual image with many degrees of gray on the image with only white and black colours. Depending on the parameter Lower Threshold brighter parts of the image will be filled with white color, and darker parts of the image will be filled with black color. This parameter can be modified within the loop of the program, as follows:

- The program takes a picture of pallet with elements
- A small value of Lower Threshold parameter suitable for low light conditions is set,
- The beginning of the loop: A program using the instructions Binarize modifies the pallet's image with both elements. The original image remains unchanged.

- A program attempts to identify the position of the cylinder on a modified image. If the cylinder is detected, it will be stored coordinates of its center, and the program switches off the search for the cylinder in subsequent cycles software loop,
- A program attempts to identify the position of the ring on a modified image. If the ring is detected, it will be stored coordinates of its center, and the program switch off the search for the ring in subsequent cycles software loop,
- The program increases the value of Lower Threshold parameter, checks conditions for the implementation of the next cycle of the program's loop and if they are satisfied that jumps to the beginning of the loop,
- Jump from a loop is possible only when identified position of both elements, or if the parameter value Lower Threshold exceeded the allowable upper limit.

Multiple trials have confirmed that the vision system is more resistant to changing lighting conditions. At certain locations elements on the pallet, the cylinder and the ring may be detected at different values of Lower Threshold parameter. The disadvantage of this solution is that the loop software significantly extends the analysis of vision.

9 Station Demonstration Shows at Automaticon 2014

The PIAP Institute completed another station that uses an external vision system for robot positioning. It was showed at Automaticon 2014 – The Trade Fair for Automation, Control, Measurement and Robotics. In this case, used a robot Kuka KR16 with the KRC2 control equipment and Cognex vision system.

The task of illustrating cooperation Kuka KR16 robot, vision system from Cognex and weight Radwag is as follows: On the table placed black (matte) metal plate measuring approx. 59 cm × 45 cm. Above it on a stand at a height of 114 cm suspended Cognex camera with a resolution of 640 × 480 pixels. Point mount of the camera selected in such a way that its optical axis intersects the center of the plate. Plate size and the height of camera's suspension are chosen in such a way that the image plate approximately fill the area of field of view of the camera.

On the plate operator sets in any order three identical cans. Weight cans are different from each other and are chosen so that one has the correct value, the second is lighter and a third heavier than normal weight (underweight and overweight). The vision system determines position of the cans and submit summary information about their location to the robot. Robot transfers cans onto the weight's conveyor. After weighing, depending on the measurement result robot conveys each can at one of the three locations described labeled "Correct weight", "Underweight" and "Overweight".

As in the case of station with ABB robot and Sick's camera, in this application robot Kuka KR16 cooperates with Cognex camera using two types of connections:

- Digital inputs and digital outputs to synchronize operations performed by the application software of vision system and robot
- RS-232 interface by which the vision system transmits to the robot information about the location of cans. The choice of RS-232 interface here was not accidental. Both robot Kuka KR16 control system KRC2 and Cognex camera interfaces include RS-232 and Ethernet. But in the case of robot Kuka with KRC2 control

system it is not possible to use Ethernet interface without additional software system to support Ethernet.

Cooperation begins with set ("1") digital signal by the robot:

```
SIGNAL WyzwalanieKamery $OUT[3]
```

to the camera, which initiates photo of plate with elements, and image analysis to detect the position of the cans. The result of this analysis, the camera transmits to the robot using two digital signals:

```
SIGNAL ACK_KontrolaWizyjna     $IN[7]
SIGNAL KontrolaWizyjnaOK$IN[8]
```

The value of "1" of the first signal, it means that the camera has completed the analysis of the image, the value of "1" the second signal indicates that correctly identified the location of all three cans. In both cases, the camera's application transmits to the robot position information, in the absence of identification frame contains random values.

The position information is transmitted in the form of three pairs of numbers: $[X_{1k}, Y_{1k}]$, $[X_{2k}, Y_{2k}]$ $[X_{3k}, Y_{3k}]$. Each of the six coordinates, expressed in pixels, so it can be a natural number of 0 to 640 (coordinates Y_{1k}, Y_{2k}, Y_{3k}) or in the range of 0 to 480 (X_{1k}, X_{2k}, X_{3k}). The numbers 640 and 480 is the camera resolution – respectively in the Y-axis and X axis to send the camera to the robot controller is six natural numbers. Each of them is converted to a string of four digits. The frame contains sign indicating the start of the frame, 24 digits specifying the coordinates and two characters defining the end of the frame. Frame are not included any checksums.

This station with Kuka robot and Cognex camera works like a station made for the Poznan University of Technology. In the same way you have to correlate with each work coordinate system of the Kuka robot and coordinate system of the camera. The calibration, the measurements and adjustments to the position in which the manipulator is to take the cans are similar.

References

1. Dunaj, J., Pachuta, M.: Opis programu robota ABB IRb-140 i programu kamery SICK IVC-2D wykonanych na potrzeby Laboratorium Modelowania Elastycznych Linii Produkcyjnych Politechniki Poznańskiej. Przemysłowy Instytut Automatyki i Pomiarów (2013)
2. Dunaj, J.: Opis programu demonstracyjnego robota Kuka KR16 na targi Automaticon 2014. Przemysłowy Instytut Automatyki i Pomiarów (2014)
3. Jarzembski, B.: Modułowy system zautomatyzowanej linii produkcyjnej. Pomiary Automatyka Robotyka, nr 10/2013
4. ABB Robotics: User's Guide 3HAC 7793-1 For BaseWare OS 4.0.60
5. ABB Robotics: Instrukcja obsługi IRC5 z panelem FlexPendant. Copyright 2004–2011 ABB
6. ABB Robotics: Operating manual: Robot Studio. Copyright 2008-2011 ABB
7. ABB Robotics: Product specification: Robot Studio. Copyright 2008-2011 ABB
8. SICK AG: Application Programming IVC-2D Reference Manual Copyright SICK AG 2011-07-01
9. SICK AG: Industrial Vision Camera IVC-2D – A high performance 2D smart camera. Copyright SICK AG 2011-07-01

10. SICK AG: Industrial Vision Camera IVC-2D Copyright SICK AG 2011-07-01
11. KUKA Roboter GmbH: KUKA System Software (KSS) – KR C2/KR C3 Expert Programming. Issued: 26 Sep 2003
12. KUKA Roboter GmbH: KUKA System Software (KSS) – KR C2/KR C3 Configuration (External Automatic). Issued: 20 Jan 2004
13. KUKA Roboter GmbH: KR C... System Variables. Issued: 30 Mar 2004
14. KUKA Roboter GmbH: Programming CREAD/CWRITE and related statements for KUKA System Software 5.4, 5.5, 5.7. Issued: 19.06.2007 Version 1.3

Application of Direct Metal Laser Sintering for Manufacturing of Robotic Parts

Maciej Cader[1(✉)] and Dominik Wyszyński[2]

[1] Industrial Research Institute for Automation and Measurements PIAP, Warsaw, Poland
mcader@piap.pl
[2] Cracow University of Technology, Kraków, Poland
wyszynski@mech.pk.edu.pl
http://www.m6.mech.pk.edu.pl
http://www.piap.pl

Abstract. Application of Rapid Tooling technology for manufacturing of robotic parts became possible due to development of Direct Metal Laser Sintering method. The intensive progress in this method improvement was an effect of interdisciplinary cooperation of material science, physics and production engineers. DMLS opened new horizons for many fields of e.g., industrial and medical applications. Production of customized demanding robotic parts requires effective method and reliable materials. The article presents results of materials analysis used in DMLS technology, in purpose of use robot parts manufacturing. It shows the main properties of the main materials and the most important properties of DMLS technology in the context of the use for the manufacture of robot's parts. In the article authors show the famous machines and the cost simulation of production of selected parts.

Keywords: Rapid tooling · DMLS · 3D printing · Robotics · Materials

1 Introduction

Rapid development of additive manufacturing methods in recent 20 years enabled possibility of application of almost any engineering solution [1–4]. Diversity of methods and availability of materials places additive manufacturing in top 21st century production technologies [5–8]. Industrial design, aviation, automotive, tooling, medical and pharmaceutical industries use additive methods to improve our life quality and comfort. Daily life of modern human is supported by discrete presence of technical solutions which lion's share is automation and robotics.

Industrial Research Institute for Automation and Measurements PIAP as a national institute for decades has been engaged in research and development in area of preparing and implementing new technologies, automation systems, production plant and specialist measuring equipment in various branches of industry. Forty years of close cooperation with industry have brought a series of new designs and significant implementations. Institute PIAP also prepared and started up the production of mobile robots supporting antiterrorist missions [9].

R. Szewczyk and M. Kaliczyńska (eds.), *Recent Advances in Systems, Control and Information Technology*, Advances in Intelligent Systems and Computing 543, DOI 10.1007/978-3-319-48923-0_36

2 Manufacturing of Robotic Parts

Industrial robots mean great precision of executed operation, realization of tasks that cannot be executed by man, such as: operation in chemically hazardous environment, transport of heavy elements and optimum usage of space. Institute PIAP has many years of experience in the field of robotization, which enables offer of realization of complex projects, starting with analysis of opportunities of automation in technical and economic aspects, drawing up of design, selection of the best elements and components guaranteeing high level of reliability of the final product, execution and start-up of the system, training of personnel operating installed equipment, ending with guarantee and post-guarantee service.

Various technologies are realized on robotized work stations including welding and joining, chamfering of sheets, palletization and depalletization, trimming and polishing, coating, operation of machines in production sockets, transport between operation facilities [9].

Complication of tailored robotic solutions demands flexible, reliable, time and cost effective manufacturing potential. Especially if selection of the best elements and components guaranteeing high level of reliability of the final product are required. Additive manufacturing methods seem to be perfect compromise between flexibility, versatility and cost and time efficiency due to their reliability, repeatability and wide range of engineering materials.

2.1 Additive Manufacturing – Direct Metal Laser Sintering by EOS

Additive Manufacturing relates to a method where CAD 3D data is applied to create a part in sequence of layers by depositing and consolidation of used material. Term 3D printing becomes common equivalent for Additive Manufacturing. Nevertheless, the second one is the most adequate because describes a unconventional but still professional production technique which is clearly distinguished from conventional methods of material removal. Subtractive machining of a part from the solid block, for example, can be replaced by Additive Manufacturing that builds up components layer by layer using materials which are applicable in fine powder form. A range of different alloys, metals, plastics and composite materials can be applied.

The technology that enables growth manufacturing of metallic fully functional parts by means of laser beam sintering or melting of powder bed is called Direct Metal Laser Sintering. It can be also used as Rapid Prototyping technique for construction of illustrative and functional prototypes. Additive Manufacturing becomes one of the most intense developing technologies and now is used increasingly in Series Production. It enables manufactures in the most varied sectors of industry to create individual profile for themselves based on new customer's benefits, cost-saving potential and flexibility to meet sustainability goals.

The advantage of Additive Manufacturing can be clearly seen in areas where conventional manufacturing reaches its limits. This technology is attractive where a new approach to design and manufacturing is required. It enables a design-driven manufacturing process – where design determines production and not the other way around. Additive Manufacturing enables achieving extremely complicated light and stable structures, design free form, the optimisation and integration of functional features, as well as the manufacturing of small batch sizes at reasonable unit costs and a high degree of product customisation even in serial production.

The process starts by filling the working chamber with protective gas (N, Ar) and applying a thin layer of powder material on the building platform. A powerful (100–400 W) laser beam then melts the powder at selective area defined by the computer-generated component design data. The building platform is then lowered and another layer of powder is spread. Once again the material is melt so as to bond with the previously built layer below at the defined area (part cross-section) [2]. The scheme of DMLS manufacturing process is presented below on Fig. 1 (Figs. 2, 3, 4, 5, 6, 7, 8, 9, 10, 11, 12, 13, 14, 15, 16 and 17).

Building strategies in DMLS:
Direct Part® – standalone parts.
Direct tool®:
Parts integrated with building platform.

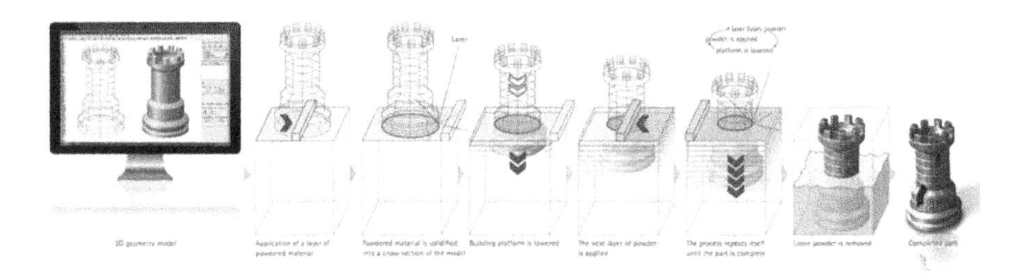

Fig. 1. General functional principle of laser sintering [10]

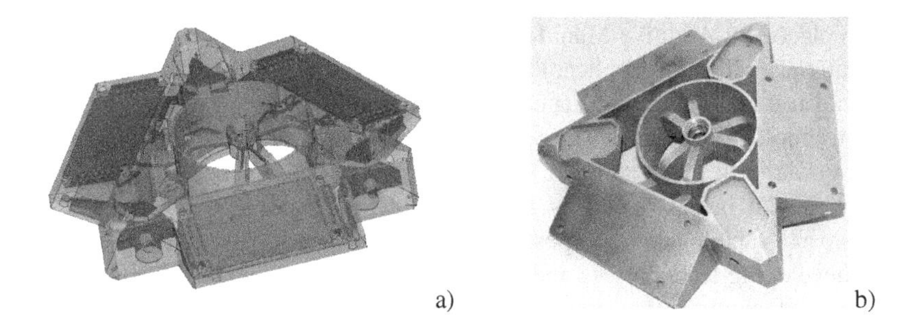

Fig. 2. The transformer body with internal cooling channels: (a) 3D CAD model, (b) manufactured part [10]

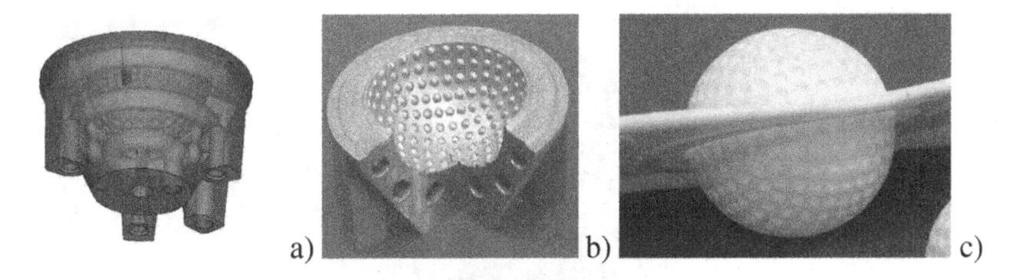

Fig. 3. Molding tool insert with optimized cooling channels: (a) 3D CAD model, (b) cross-section of tool insert, (c) final product [10]

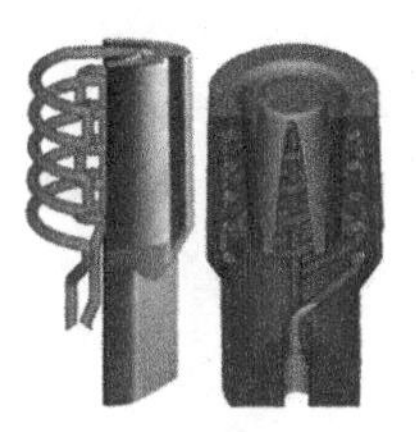

Fig. 4. Examples of moulding inserts designed and manufactured by DMLS from Maraging steel [10].

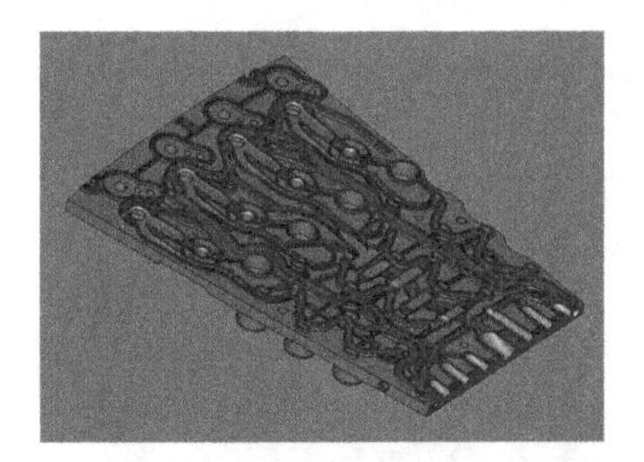

Fig. 5. Complicated shape internal cooling channels [10]

Fig. 6. The EOS StainlessSteel GP1cross section [10]

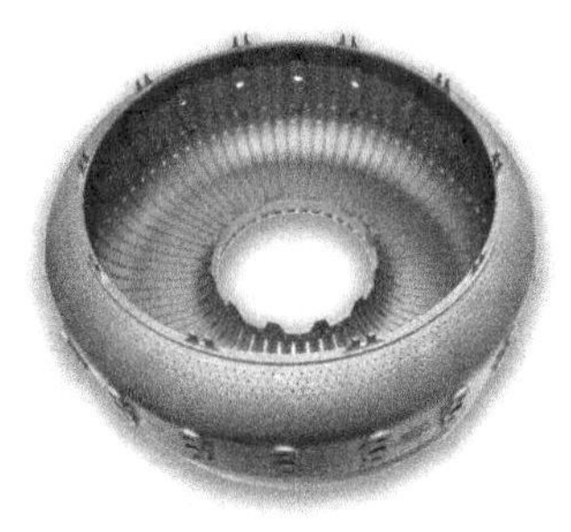

Fig. 7. Turbine combustion chamber, built in EOS StainlessSteel PH1 [10]

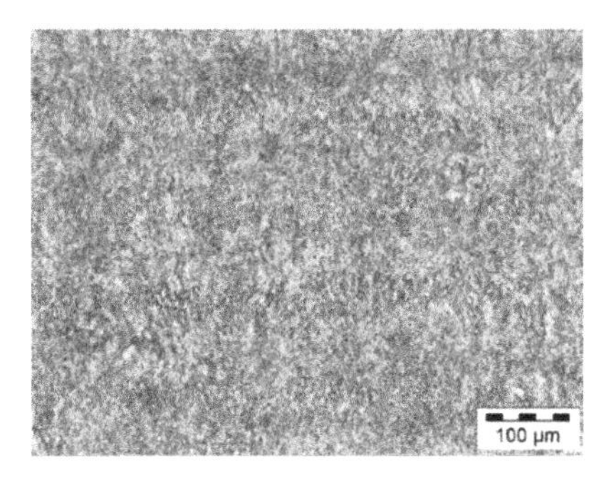

Fig. 8. The EOS StainlessSteel PH1cross section [10]

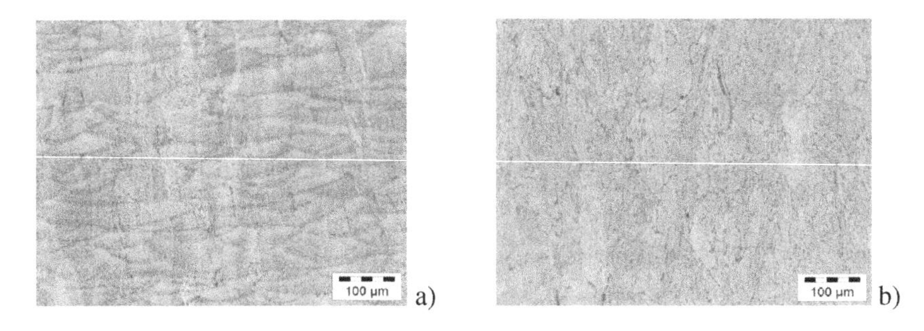

Fig. 9. Photograph of IN718 structure: (a) before heat treatment, (b) after heat treatment [10]

Typical dimensional accuracy achieved in standard DMLS process is ± 20 to 50 μm for small parts or 0.2 % for bigger ones. The minimum wall thickness available depends on applied material but is not less than 0.3 mm. Roughness after shot peening Ra ~ 3–8 μm (depends on material type) and Rz < ~ 0.5 μm after polishing. Applied powder layer thickness vary between 20 to 50 μm and determines part detail definition. The volume build ratio depends on power of applied laser source, layer thickness, part geometry and material type. Laser beam scanning speed is up to 7 m/s (Tables 1, 2, 3, 4, 5, 6, 7, 8, 9, 10 and 11).

Fig. 10. Stator ring built in EOS Nickel Alloy IN718 11]

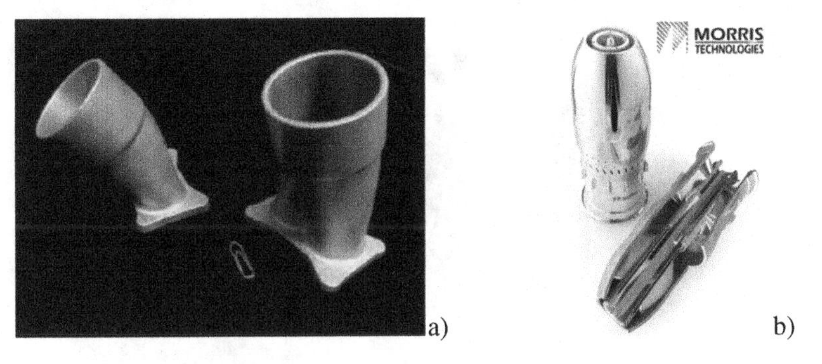

Fig. 11. Examples of application: (a) exhaust system [10], (b) fuel injection – swirlers [11]

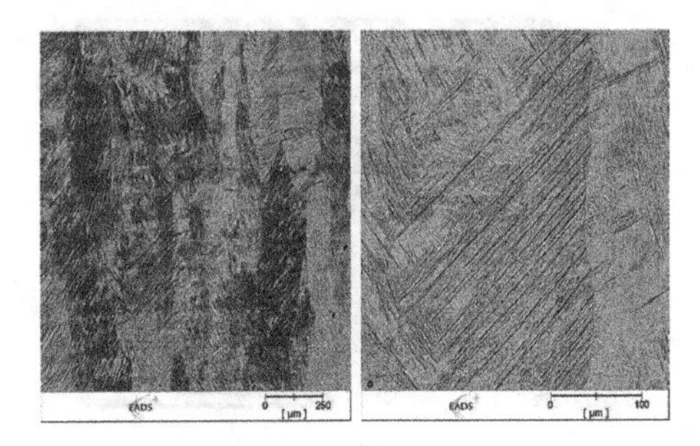

Fig. 12. Tipical martensite structure with grains growing from layer to layer, preferable z axis orientation, grain size > > layer thickness [10]

2.2 Materials Used in DMLS

Direct Laser Metal Sintering method offers wide variety of engineering materials. It enables engineers to design freeform functional parts and tools. The most interesting from the robotics' demands point of view are those materials which comply rigorous standards. The material list presented in the table below comprise material types, standards compatibility and main field of application (the greyed one are the most suitable for application in robotic industry). All of them are material's science recent

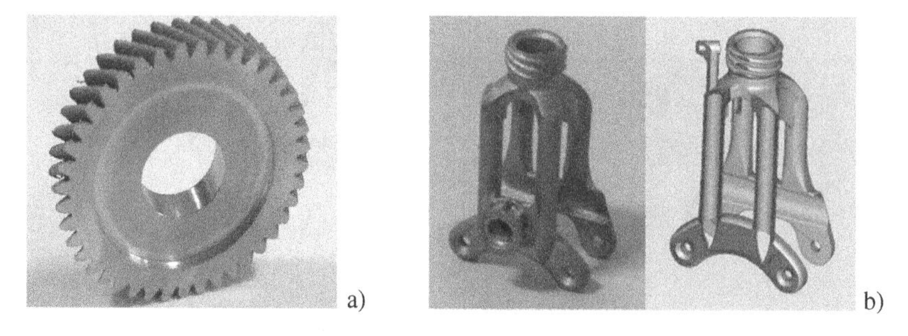

Fig. 13. Gear and thin wall brake part [10]

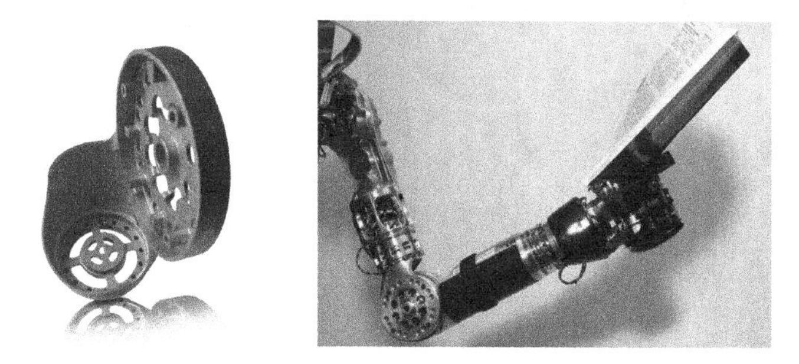

Fig. 14. Humeral mount for a fully integrated prosthetic arm [12]

Fig. 15. The robotic arm [12]

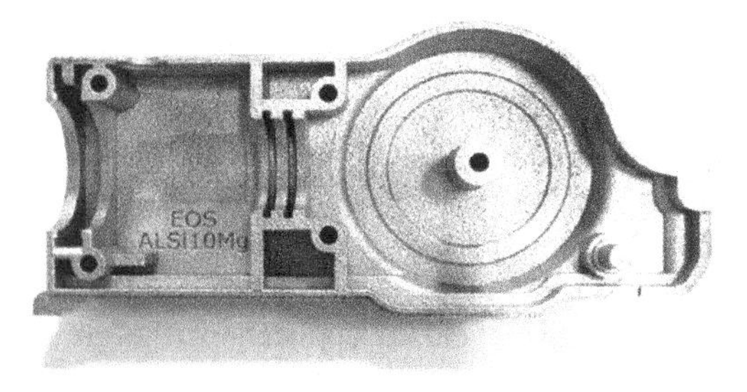

Fig. 16. Tailored body made of AlSi10 Mg [10]

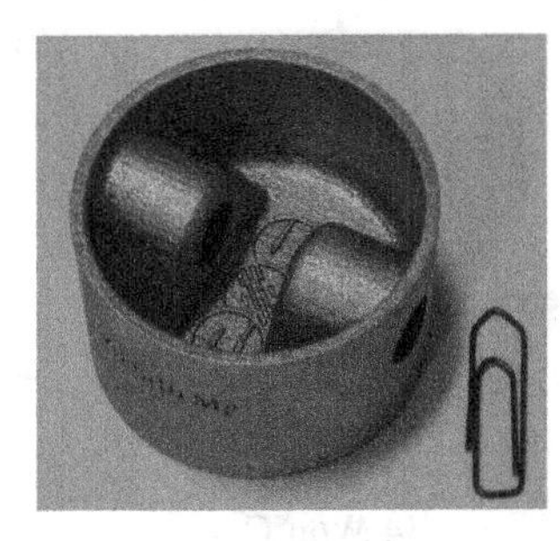

Fig. 17. Piston and shaft made of AlSi10 Mg [10]

Table 1. Metallic materials for DLMS [10]

Product class	Commercial name	Compatibility	Typical application
Maraging steel	EOS MaragingSteel MS1	18 Mar 300 / 1.2709	Series injection moulding tools; mechanical parts
Stainless steel	EOS StainlessSteel GP1	Stainless steel 17-4 / 1.4542	Functional prototypes and series-production parts; mechanical engineering and medical technology
	EOS StainlessSteel PH1	Hardenable stainless steel 15-5 / 1.4540	Functional prototypes and series-production parts; mechanical engineering and medical technology
	EOS stainlessSteel 316L	1.4404 / UNS S31673	Lifestyle: jewellery, functional elements in yachts, spectacle frames, etc. Aerospace: supports, brackets, etc. Medical: functional prototypes and series-production parts in e.g. endoscopy and orthopaedics
	EOS StainlessSteel CX	Tooling grade steel	Manufacturing of injection moulding tools for medical products or products from corrosive plastics
Nickel alloy	EOS NickelAlloy	IN718 Inconel™ 718, UNS N07718, AMS 5662, mat. # 2.4668	Functional prototypes and series-production parts;mhigh-temperature turbine components
	EOS NickelAlloy	IN625 Inconel™ 625, UNS N06625,AMS 5666F, mat. # 2.4856 etc.	Functional prototypes and series-production parts; high-temperature turbine components
	EOS NickelAlloy HX	UNS N06002	Components with severe thermal conditions and high risk of oxidation, e.g. combustion chambers, burner components, fans, roller hearths and support members in industrial furnaces
Cobalt chrome	EOS CobaltChrome MP1	CoCrMo super alloy, UNS R31538, ASTM F75	Functional prototypes, series-production parts, mechanical engineering, medical technology, dental
	EOS CobaltChrome SP2	CoCrMo super alloy	Dental restorations (series-production)
	EOS CobaltChrome RPD	CoCrMo super alloy	Removable partial dentures
Titanium	EOS Titanium Ti64	Ti6Al4V light metal	Functional prototypes and series-production parts; aerospace, motorsports etc.
	EOS Titanium Ti64ELI	Ti6Al4V ELI	Functional prototypes and series-production parts in medical technology
Aluminium	EOS Aluminium AlSi10Mg	AlSi10Mg light metal	Functional prototypes and series-production parts; mechanical engineering, motorsports etc.

Table 2. Mechanical properties of EOS MaragingSteel MS1 [10]

Properties	As built	After age hardening (6 h @ 490°C)
Ultimate tensile strength	1100 ± 100 MPa	1950 ± 100 MPa
Yield strength (R_p 0.2 %)	1000 ± 100 MPa	1900 ± 100 MPa
Elongation break	8 ± 3 %	2 ± 1 %
Young's modulus	180 ± 20GPa	
Hardness	33–37 HRC	50–54 HRC
Ductility	45 ± 10 J	11 ± 4 J
Relative density	~ 100 %	

Table 3. Properties of EOS StainlessSteel GP1

Properties	Value
Relative density	~ 100 %
Density	7,8 g/cm^3
Maximal operating temperature	550°C
Melting range	1350–1430°C
Thermal conductivity	
@ 20°C	13 W/m°C
@ 100°C	14 W/m°C
@ 200°C	15 W/m°C
@ 3000°C	16 W/m°C
Coefficient of thermal expansion	
20–600°C	$14{\cdot}10^{-6}$ m/m°C

Table 4. Properties of EOS StainlessSteel PH1 [10]

Property	As built	After age hardening (6 h @ 490°C)
Ultimate tensile strength		
Horizontal (XY)	1150 ± 50 MPa	Min. 1310 MPa (typ. 1450 ± 100 MPa)
Vertical (Z)	1050 ± 50 MPa	Min. 1310 MPa (typ. 1450 ± 100 MPa
Yield strength (R_p 0.2 %)		
Horizontal (XY)	1050 ± 50 MPa	Min. 1170 MPa (typ. 1300 ± 100 MPa)
Vertical (Z)	1000 ± 50 MPa	Min. 1170 MPa (typ. 1300 ± 100 MPa)
Elongation break		
Horizontal (XY)	16 ± 4 %	Min. 10 % (typ. 12 ± 2 %)
Vertical (Z)	17 ± 4 %	Min. 10 % (typ. 12 ± 2 %)
Hardness	30–35 HRC	Min. 40 HRC
Relative density	~ 100 %	
Density	7.8 g/cm^3	
Thermal conductivity		
As manufactured	13.8 ± 0.8 W/m°C	15.7 ± 0.8 W/m°C
Specific heat		
As manufactured	460 ± 20 J/kg°C	470 ± 20 J/kg°C

achievement, their composition was carefully designed and mechanical and thermal properties duly checked. The detailed description of some of them will be given later on in the text.

Table 5. Properties of EOS Nickel Alloy IN718@20°C

Property	As built	Heat treated per AMS 5662	Heat treated per AMS 5664
Tensile strength			
Horizontal (XY)	1060 ± 50 MPa		
Vertical (Z)	980 ± 50 MPa	Min. 1241 MPa (typ. 1400 ± 100 MPa)	Min. 1241 MPa (typ. 1380 ± 100 MPa)
Yield strength (R_p0,2 %)			
Horizontal (XY)	780 ± 50 MPa		
Vertical (Z)	634 ± 50 MPa	Min. 1241 MPa (typ. 1400 ± 100 MPa)	Min. 1241 MPa (typ. 1380 ± 100 MPa)
Elongation break			
Horizontal (XY)	27 ± 5 %		
Vertical (Z)	31 ± 5 %	Min. 12 % (typ. 15 ± 3 %)	Min. 12 % (typ. 18 ± 5 %)
Hardness	30 HRC	47 HRC	43HRC
Young modulus			
Horizontal (XY)	160 ± 20 GPa		
Vertical (Z)		170 ± 20 GPa	

Maraging Steel EOS MaragingSteel MS1 is a martensite-hardenable steel. Its chemical composition corresponds to European 1.2709. This material distinguish high strength combined with high toughness. The parts made of MS1 can be easily machined after the building process and post-hardened up to 54 HRC. The material has

Table 6. Properties of EOS Nickel Alloy IN718@695°C [10]

Property	Heat treated per AMS 5662	Heat treated per AMS 5664
Tensile strength R_m		
Vertical (Z)	Min. 965 MPa typ. 1170 ± 50 MPa	typ. 1210 ± 50 MPa
Yield strength (R_p 0.2 %)		
Vertical (Z)	Min. 862 MPa typ. 970 ± 50 MPa	Typ. 1010 ± 50 MPa
Elongation break		
Vertical (Z)	Min 6 % typ. 16 ± 3 %	20 ± 3 %
Stress-Rupture Properties		
Vertical (Z)	Min. 23 h@689 MPa	
	51 ± 5 h (final applied stress to rupture 792.5 MPa)	81 ± 10 h (final applied stress to rupture 861.5 MPa)

Table 7. Thermal properties of parts made of IN718 [10]

Property	Heat treated per AMS 5662
Coefficient of thermal expansion	
over 25–200°C	approx. 12,5–13,0 · 10^{-6} m/m°C
over 25–750°C	approx. 16,6–17,2 · 10^{-6} m/m°C
Maximum operating temperature for parts under load	approx. 650°C
Oxidation resistance up to	approx. 980°C

Table 8. Mechanical properties of parts at 20°C [10]

Properties	As built	Stress relieved (6 h@ 1150°C)
Tensile strength		
Horizontal direction (XY)	1350 ± 100 MPa	1100 ± 100 MPa
Vertical direction (Z)	1200 ± 150 MPa	1100 ± 100 MPa
Yield strength (Rp 0.2 %)		
Horizontal direction (XY)	1060 ± 100 MPa	600 ± 50 MPa
Vertical direction (Z)	800 ± 100 MPa	600 ± 50 MPa
Elongation at break		
Horizontal direction (XY)	(11 ± 3) %	min. 20 %
Vertical direction (Z)	(24 ± 4) %	min. 20 %
Young modulus		
Horizontal direction (XY)	200 ± 20 GPa	200 ± 20 GPa
Vertical direction (Z)	190 ± 20 GPa	200 ± 20 GPa
Fatigue life		
max. stress to reach 10 milions cycles	approx. 560 MPa	
max. stress to reach 1 milion cycles	approx. 660 MPa	
Hardnes	approx. 35–45 HRC	

Table 9. Thermal properties of parts [10]

Properties	As built
Coefficient of thermal expansion	
over 20–500°C	typ. 13.6 × 10^{-6} m/m °C
over 500–1000°C	typ. 15.1 × 10^{-6} m/m °C
Thermal conductivity	
@20°C	typ. 13 W/m°C
@300°C	typ. 18 W/m°C
@500°C	typ. 22 W/m°C
@1000°C	typ. 33 W/m°C
Maximum operating temperature	approx. 1150°C
Melting range	1350–1430°C

Table 10. Mechanical properties of parts

Properties	As built	Heat treated (at 800°C for 4 h in argon inert atmosphere)
Tensile strength		
Horizontal direction (XY)	typ. 1290 ± 50 MPa	min. 930 MPa typ. 1100 ± 40 MPa
Vertical direction (Z)	typ. 1240 ± 50 MPa	min. 930 MPa typ. 1100 ± 40 MPa
Yield strength (Rp0,2)		
Horizontal direction (XY)	typ. 1140 ± 50 MPa	min. 860 MPa typ. 1000 ± 50 MPa
Vertical direction (Z)	typ. 1120 ± 80 MPa	min. 860 MPa typ. 1000 ± 60 MPa
Elongation at break		
Horizontal direction (XY)	typ. (7 ± 3)%	min. 10 %
Vertical direction (Z)	typ. (10 ± 3)%	min. 10 %
Young modulus (XYZ)	typ. 110 ± 15 GPa	
Hardness	typ. 320 ± 12 HV5	

Table 11. Mechanical properties of the AlSi10 Mg 200°C made of parts [10]

Property	As built
Tensile strength	
Horizontal direction (XY)	typ. 360 MPa
Vertical direction (Z)	typ. 390 MPa
Yield strength (Rp 0.2 %)	
Horizontal direction (XY)	typ. 220 MPa
Vertical direction (Z)	typ. 210 MPa
Young's modulus (isotropic)	typ. 70 GPa
Elongation at break	
Horizontal direction (XY)	typ. 8 %
Vertical direction (Z)	typ. 6 %

also excellent polishability and good thermal conductivity. That predispose it for manufacturing moulding tool inserts. Other common applications are small series-production parts, tooling (e.g. aluminium die casting), mechanical engineering and aerospace [10].

EOS StainlessSteel GP1 Parts made of powder that chemical composition corresponds to European 1.4542 have good mechanical properties, especially unique ductility in laser processed state. Could be widely used in a variety of engineering applications like functional metal prototypes, small series products, individualised products or spare parts. Typically is used for manufacturing of functional prototypes, series-production parts, mechanical and medical engineering.

EOS StainlessSteel PH1 This precipitation hardening steel chemical composition conforms to DIN 1.4540. Parts made of PH1 have excellent mechanical properties due to the precipitation hardening. Normally is applied in a variety of medical, aerospace and other engineering applications that require high hardness and strength. This kind of steel can be used for part-building applications like functional metal prototypes, small series products, tailored products or spare parts. Typically can be used for manufacturing of functional prototypes, series-production parts, mechanical engineering, medical technology.

EOS NickelAlloy IN718 Nickel based heat resistant alloy that composition corresponds to 2.4668. This type of precipitation-hardening nickel-chromium alloy has good tensile, fatigue, creep and rupture strength at temperatures up to 700°C. The material has also excellent corrosion resistance in variety of corrosive environments. This alloy is perfect for high temperature applications like gas turbine parts, instrumentation parts, power and process industry parts. IN718 has also outstanding excellent cryogenic properties and potential for cryogenic applications.

EOS CobaltChrome MP1 This super alloy powder enables manufacturing parts of a cobalt-chrome-molybdenum-based material. MP1 is characterized by excellent mechanical properties (strength, hard-ness etc.), resistance against corrosion and temperature. Cobalt-chrome alloys are widely used in biomedical applications like dental and medical implants but also for high-temperature engineering applications such as in aero engines. Parts built from this super alloy are nickel-free (< 0.1 % nickel content) and are characterized by a fine, uniform crystal grain structure. As built material meets the chemical and mechanical specifications of ISO 5832-4 for cast CoCrMo implant alloys, as well as the specifications of ISO 5832-12 for wrought CoCrMo implants alloys except remaining elongation. The remaining elongation can be increased to fulfil even these standards by high temperature stress relieving or hot isostatic pressing (HIP). Parts made of CobaltChrome MP1 can be machined, spark-eroded, welded, micro shot peened, polished and coated if required. Material is suitable for biomedical applications and for parts requiring high mechanical properties in elevated temperatures (500–1000°C) and with good corrosion resistance. Due to the layer wise building method, the parts have a certain anisotropy, which can be reduced or removed by appropriate heat treatment [10].

The goal of presented above fuel injection swirlers was to optimise airflow and create cooling with integrated fuel channels. Application of DLMS in MP1 enabled highly complex design built as 'one piece', dramatic reduction in time to create (2 weeks vs. 6 weeks), significant cost reduction – min. 50 % less and increased robustness.

EOS Titanium Ti64 This well-known Ti6Al4 V light alloy has outstanding mechanical properties and corrosion resistance combined with low specific weight (density 4.41 g/cm^3). The ELI version (extra-low interstitials) has particularly high purity. Parts built in EOS Titanium Ti64 have a chemical composition corresponding to

ISO 5832-3. This material is ideal for many high-performance engineering applications, for example in aerospace and motor racing, and also for the production of biomedical implants. Due to the layer-wise building method, the parts have a certain anisotropy, which can be reduced or removed by appropriate heat treatment [10].

The humeral mount for a fully integrated prosthetic arm was developed by Deka Research for the Defence Advanced Research Projects Agency (DARPA) for Revolutionizing Prosthetics Solution. This complex part was produced quickly and cost-effectively, with high quality, single step, fully automatic build process with the part accuracy, surface quality and mechanical properties at least as good as with casting.

AlSi10 Mg 200°C Typical casting alloy of good casting properties is typically used for cast parts with thin walls and complex geometry. Offers good strength, hardness and dynamic properties and is therefore also used for parts subject to high loads. Parts made of this material are ideal for applications which require a combination of good thermal properties and low weight (2.67 g/cm^3). They can be machined, spark-eroded, welded, micro shot-peened, polished and coated if required. Processing of AlSi10 Mg 200°C is realized at elevated building platform temperature of 200°C (normally 80°C) which minimises internal stresses typically generated in DMLS parts. The laser sintering/melting process is characterized by extremely rapid melting and re-solidification. This produces a metallurgy and corresponding mechanical properties in the as built condition which is similar to T6 heat treated cast parts. Due to the layer-wise building method, the parts have a certain anisotropy. Suitable heat treatment can be used for further improvement of part properties and reduction of anisotropy. Conventionally cast components in this type of aluminum alloy are often heat treated to improve the mechanical properties, for example using the T6 cycle of solution annealing, quenching and age hardening [10].

References

1. Bagsik, A., Josupeit, S., Schoeppner, V., Klemp, E.: Mechanical analysis of lightweight constructions manufactured with fused deposition modelin. In: AIP Conference Proceedings, p. 696 (2014). doi:10.1063/1.4873874
2. Ning, F., Cong, W., Qiu, J., Wei, J., Wang, S.: Additive manufacturing of carbon fiber reinforced thermoplastic composites using fused deposition modeling. Compos. Part B **80**, 369–378 (2015). doi:10.1016/j.compositesb.2015.06.013
3. Pranjal, J., Kuthe, A.M.: Feasibility study of manufacturing using rapid prototyping: FDM approach. Procedia Eng. **64**, 4–11 (2013). doi:10.1016/j.proeng.2013.08.275
4. Kumar, S., Kruth, J.-P.: Composites by rapid prototyping technology. Mater. Des. **31**(2), 850–856 (2010). doi:10.1016/j.matdes.2009.07.045
5. Wei, G., Yunbo, Z., Devarajan, R., Karthik, R., Yong, C., Williams, C.B., Wang, C.C.L., Shin, Y.C., Zhang, S., Zavattieri, P.D.: The status, challenges, and future of additive manufacturing in engineering. Comput. Aided Des. **69**, 65–89 (2015). doi:10.1016/j.cad.2015.04.001

6. Moilenen, J., Vaden, T.: Manufacturing in motion: first survey on the 3D printing community. Statistical Studies of Peer Production, 15 March 2016. http://firstmonday.org/ojs/index.php/fm/article/view/4271/3738
7. Pham, D.T., Gault, R.S.: A comparison of rapid prototyping technologies. Int. J. Mach. Tools Manuf **38**(10–11), 1257–1287 (1998). doi:10.1016/S0890-6955(97)00137-5
8. Wohlers Report: 3D Printing and Additive Manufacturing State of the Industry, Annual Worldwide Progress Report, Wholers Associates, USA (2015)
9. http://www.piap.pl
10. http://eos.info
11. http://www.morristech.com (GE Aviation)
12. http://www.dekaresearch.com

Inspection Robots in Hard Coal Mines

Maciej Cader[1(✉)] and Leszek Kasprzyczak[2]

[1] Industrial Research Institute for Automation and Measurements PIAP, Warsaw, Poland
mcader@piap.pl
[2] Institute of Innovative Technologies EMAG, Katowice, Poland
l.kasprzyczak@ibemag.pl
http://www.ibemag.pl
http://www.piap.pl

Abstract. The review of mining inspection robots designed for use in coal mines developed in different countries was performed. The functionality of the robots and – if available – their technical parameters were described. The designed was performed based on scientific articles and media coverage accompanying the most frequently mining disasters. Subsequently, the results were discussed the results of two projects accomplished in a consortium consisting of the Institute of Innovative Technologies EMAG and the Industrial Research Institute for Automation and Measurements PIAP, where two prototypes of inspection robots named Mobile Inspection Platform and Mining Mobile Inspection Robot designed for work in hazardous of explosion of methane and/or carbonaceous dust (group I of EN 60079-0) were created. The functionality of both robots was presented and the implemented legislative explosion proof techniques were given. Functional and traction testing carried out by mine rescuers were also discussed in Introduction.

Keywords: Mining inspection robots · Mining robotization · Explosion proof techniques in robotization

1 Introduction

The robotization of modern mines is currently a strategic issue due to the exploitation of more and more deeper loads and as a result of Hazards existing in mines, in particular climatic hazards (higher temperature and humidity), bursts and emissions of hazardous gases (methane, sulphide), endogenous fires and related to them emissions of carbon monoxide and carbon dioxide, as well as bursts of water and rocks [1, 2]. Due to these reasons, man should be removed as far as possible from a hazardous zone (mining area), and the most favorably would be for him to control the mining process and the transport from the surface of a mine. The robots equipped with a sensor suitable for measuring concentration of hazardous gases and climate conditions should also participate in rescue missions ahead of rescuers' groups, providing them with information on conditions existing in a mining excavation, what should guarantee greater safety. Such necessity is being recognized all over the world as evidenced by the multitude of construction solutions of mining robots in different countries [3–5]. This issue is also

R. Szewczyk and M. Kaliczyńska (eds.), *Recent Advances in Systems, Control and Information Technology*, Advances in Intelligent Systems and Computing 543, DOI 10.1007/978-3-319-48923-0_37

being raised by mines rescue stations, which sometimes are forced to cease a rescue operation due to dangerous conditions in an operation area and unacceptable risk for rescuers [6].

In this article the constructions of mining robots were described based on scientific articles and media coverage accompanying mainly mining disasters and miners trapped underground [6–30]. In their context, there were presented achievements related to the implementation by a consortium of the Institute of Innovative Technologies EMAG and the Industrial Research Institute for Automation and Measurements PIAP of the project of designinig the Mobile Inspection Platforms and the Mining Mobile Inspection Robot, which were financed by the national budget of the National Centre for Research and Development.

2 Mining Inspection Robots – Review of Literature

2.1 Numbat

On the basis of a review of literature it can been stated that the first who made an effort of building a robot designed for the coal industry were the construction engineers of the Commonwealth Scientific and Industrial Research Organization CSIRO from Australia in the 1990s. A mining reconnaissance robot NUMBAT (Fig. 1) is characterized by the following technical parameters [8–10]:

- 8 wheels with tyres (there is an engine 750 W powering the wheels on each side of the robot),
- operating time: 5–8 h,
- speed: 2 km/h,
- 40 accumulators Ni-Cd, 140 Ah,
- size: 2.5 m × 1.65 m,
- sealed body filled with nitrogen and overpressure,
- measurement of CO, CO_2, O_2, H_2, velocity airflow, pressure and temperature,

Fig. 1. Numbat from CSIRO [3]

- cylinders filled with reference gases are on the vehicle allowing periodic calibration of sensors,
- 4 monochrome cameras and optionally an infrared camera,
- communication through the optical fiber unwinding/winding by the unwinder.

According to information found on a website of the organization www.csiro.au, an entrepreneur interested in the implementation of the robot's prototype into production was being sought. This robot has never been used in any rescue operation [3].

2.2 Groundhog

In the Carnegie Mellon University in the USA in 2003–2005 the robot named Groundhog (Fig. 2) was built as an autonomous vehicle for scanning abandoned mining excavations [3, 11, 12]. The robot is characterized by the following technical parameters:

Fig. 2. Groundhog [13]

- hydraulic drive system with a pomp placed on the vehicle,
- velocity: 0.15 m/s,
- range: 3 km,
- four-wheel drive,
- turn front and rear driving axles,
- power demand while driving: 1 kW,
- power demand for measurement and processing: < 100 W,
- equipped with 8 lead-acid accumulators,
- equipped with a laser scanner.

2.3 Wolverine V-2

In the robot Wolverine V-2 (Fig. 3) of Remotec company is being exploited by the Mine Health and Safety Administration of America in the USA.

Fig. 3. Wolverine V-2 [15]

The robot was used in several operations, though unsuccessfully [3, 5, 14]. The robot is 127 cm high, it weighs 544 kg and it is drived by explosion proof motors. It comprises 3 cameras with lighting, sensors of atmosphere (for measuring concentration of toxic and explosive gases) and it is characterized by night vision, voice communication in two directions and a manipulator arm. The robot's range is 1.5 km. The communication occurs through the optical fiber [15]. The cost is $265,000.

2.4 Gemini-Scout

In the Sandia National Laboratories in the USA the robot named GEMINI-SCOUT Mine Rescue Robot (Fig. 4) was developed. The robot consists of two trucks moving on caterpillar treads. The trucks are connected by a link of two levels of flexibility enabling them to lean sideways towards each other and allowing the vehicle to bend in the middle while the front truck is climbing or driving down an obstacle. The robot can drive through a water obstacle 46 cm deep, as well as manage to climb − stairs. It enables voice communication between staff participating in an operation and victims [16].

Gemini-Scout is characterized by the following technical parameters:

- PTZ head with a color and infrared camera,
- sensors for measuring concentrations of gases and temperature,
- inputs/outputs available for additional equipment,

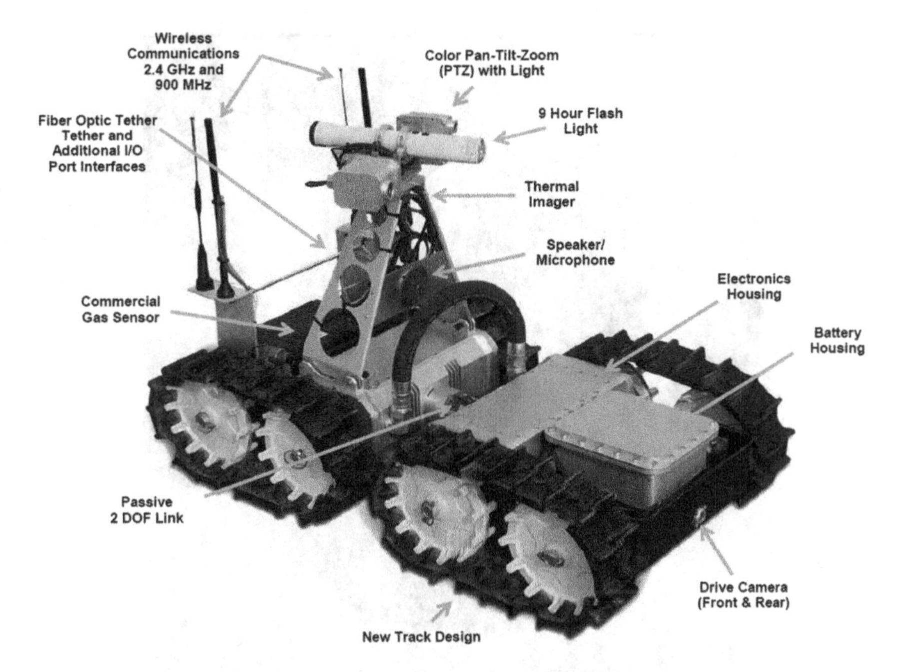

Fig. 4. GEMINI-Scout z Sandia National Laboratories [16]

- velocity: 5.63 km/h,
- weight: 86 kg,
- accumulators Ni-MH of a minimum guaranteed lifetime of 4 h,
- size: 122 cm × 61 cm × 61 cm (length × width × height),
- wireless communication 2.4 GHz and 900 MHz.

Convincing proofs evidencing excellent traction abilities of this robot are published on websites [17–19].

2.5 Chinese Mining Robots

Furthermore, in literature [20–23] there can be found several Chinese constructions:

- CMU-1 developed in the China University of Mining and Technology,
- LURKER-1 and LURKER-3 developed by the Robotics Research Centre of Shandong University.

Unfortunately, the authors of these articles focus primarily on a description of the robot's kinematics rather than presenting evidence of explosion proof protections, which allow to use the robots in the coal mining industry.

It seems that the promising constructions are the robots of Tangshan Kaicheng Electronic company seated in Hebei Province, presented during the mining fair in Pekin (Fig. 5), The representatives of the company claim that their annual production capacity amounts to 1800 items as of 2012 [24, 25].

Fig. 5. Mining robots of Tangshan Kaicheng Electronics company

2.6 Mining Mobile Inspection Robot (GMRI)

Mining Mobile Inspection Robot (in Polish: Górniczy Mobilny Robot Inspekcyjny, GMRI) (Fig. 6) was developed by the institutes EMAG and PIAP in a project financed by the national Polish budget in 2008–2010.

Fig. 6. Mining Mobile Inspection Robot on testing in a mine and during EMC testing

It was designed with an intention to meet the requirements of category M1 of EN 50303 standard, which permits the robot to be continuously used in methane and/or coal dust explosion zones. It was achieved due to the use of a pneumatic actuators with compressed nitrogen developed by PIAP Institute and the limitation of velocity of movable elements. After recompression, nitrogen cools the interior of actuators, what constitutes an additional protection against heating of the friction elements (a cylinder and a piston). The electronic circuits of GMRI robot created by EMAG Institute were designed as intrinsically safe of the highest category "ia" of EN 60079-11 norm. A notified body under the ATEX directive issued an EC certificate for GMRI robot confirming its performance in the explosion proof techniques applied and category M1.

GMRI robot passed the electromagnetic compatibility tests with a positive result in the accredited laboratory of EMAG Institute, as described in the study [26].

Functionality and traction characteristics of GMRI were studied by the rescuers of Central Mines Rescue Station in Bytom (Centralna Stacja Ratownictwa Górniczego S. A.) in an operating mine at the level of 726 m. The experts of CSRG prepared the research report [31].

In Table 1 the parameters of GMRI robot were presented. In addition to these, the robot was equipped with a machine vision containing two monochrome cameras (one directed to the front of the robot and the other to its rear) with lighting. Transmission of the control signals and the measured data, as well as the video streams occurred through two pairs of cooper cables (one pair of the cables is designed for recharging of the intrinsically safe accumulator Ni-Cd and for the two-way alternate transmission of data full-duplex in a standard of modem V.23, whereas the other pair of the cables is designed for the transmission of video streams from the cameras in RS-485 standard). Electronic modules of GMRI robot and flow of signals were presented in Fig. 7.

There is a rich literature in English available on the subject of construction specifications and technical assumptions of GMRI [4, 27–29].

Table 1. Basic technical parameters of GMRI robot

Explosion-proof construction category	Ex I M1
Pneumatic power supply	200 bar
Electric power supply (of a stationary part)	36–250 V AC
Operation temperature	0–60 °C
Mass	approx. 200 kg
Measuring ranges	
Methane	0–100%
Carbon monoxide	0–1000 ppm
Oxygen	0–25%
Temperature	0–60 °C
Humidity	0–100%

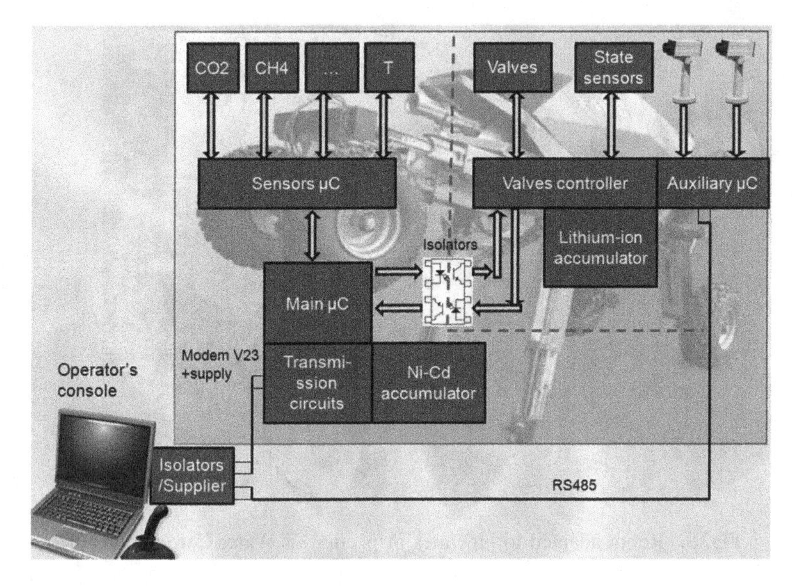

Fig. 7. Data transmission between the programmable components of GMRI

2.7 Robot Adapted to Pipelines Penetration

The robot adapted to pipelines inspection is a product of Water Corporation from Australia. It was sent to assist in the rescue operation in New Zealand, where in the Pike River coal mine 29 miners were trapped in November 2010 [6]. Unfortunately the media reported on the robot's damage.

The robot is equipped with four driving wheels with tyres and one auxiliary wheel placed on the back (Fig. 8). The robot is remotely controlled, equipped with cameras, lighting and equipment suitable for testing atmosphere. The communication medium is the optical fiber and the robot's range is 6 km. From the figure 8 it can be deduced that the robot's body in the front and in the rear is largely oblique, which facilitates clearing the way through loose obstacles.

2.8 Sewer Robot

After 33 miners were buried in Chile in August 2010, the media [7] reported on the increasing need of mobile robots for the Chilean mine industry. In Fig. 9 the robot developed in Canada in the research center Penguin Automated Systems Inc. was presented. It can be seen that the body's weight is distributed between four wheels on each side held in pairs by the control arms. On a website of the center www.penguinasi.com the videos showing the produced robots were presented. They lead to the conclusions that the robots move very slowly and the wheels suspension enables them to pass over obstacles such as railway tracks. The value of the purchase contract of the robots concluded by the Chilean company CODELCO was estimated at 3 million dollars [7].

Fig. 8. Robot adapted to pipelines inspection of Water Corporation [6]

Fig. 9. The robot of Penguin Automated Systems Inc. [7]

2.9 CAESAR

Figure 10 illustrates the construction of CAESAR (Contractible Arms Elevating Search And Rescue) developed in the University of KwaZulu-Natal in the Republic of South Africa as a fire-fighting robot. Artificial intelligence enables it to detect dangers and victims with use of communication systems, gas detection systems and the analysis of the views of unstable environment [3].

Fig. 10. CAESAR from the University of KwaZulu-Natal [30]

2.10 Mobile Inspection Platform (MPI)

The Mobile Inspection Platform (MPI) is a technology demonstrator developed by the Institute of Innovative Technologies EMAG and the Industrial Research Institute for Automation and Measurements PIAP.

Division of the work between the two institutes resulted from the complementarity of each other's competences and it constituted a model example of an excellent cooperation. PIAP Institute developed i.a. the mechanical components of the platform (flame-proof enclosures constituting the body, pressurized enlosure, unwinder of the optical fiber [and it selected the optical fiber], supple pipe, enclosure for the intrinsically safe equipment) and it selected the adequate wheels – rims and tyres, as well as the drive units containing the brushless DC motors, the bevel gears and the retarders. The scope of works performed by EMAG Institute included creating of the electronic and the programmable electronic components of the robot (i.a. measuring systems, machine vision, transmission units, communication controllers and motor controllers, electric accumulators and enclosures for the particular components, as well as measuring and control software where necessary). Detailed technical specification is included in the unpublished documentation prepared by each institute with regard to the respective phases of the project. A photo of MPI with the most important components was presented in Fig. 11.

Subsequently, the project results were presented and the standards whose requirements are met by the construction were specified.

MPI has four wheels with tyres, each powered by an independent drive unit. PIAP Institute selected the steel rims and the electrically conductive tyres in order to avoid accumulation of electrostatic charge. The drive unit consists of the brushless DC motor (BLDC), the bevel gear, the retarder and the motor controller. Each drive has its own lithium polymer accumulator at a nominal voltage of 37 V. The motor is made with the type of explosion protection increased safety "e" of EN 60079-7, whereas the gear and the retarder are secured by an oil immersion "o" of EN 60079-6. Printed electronic

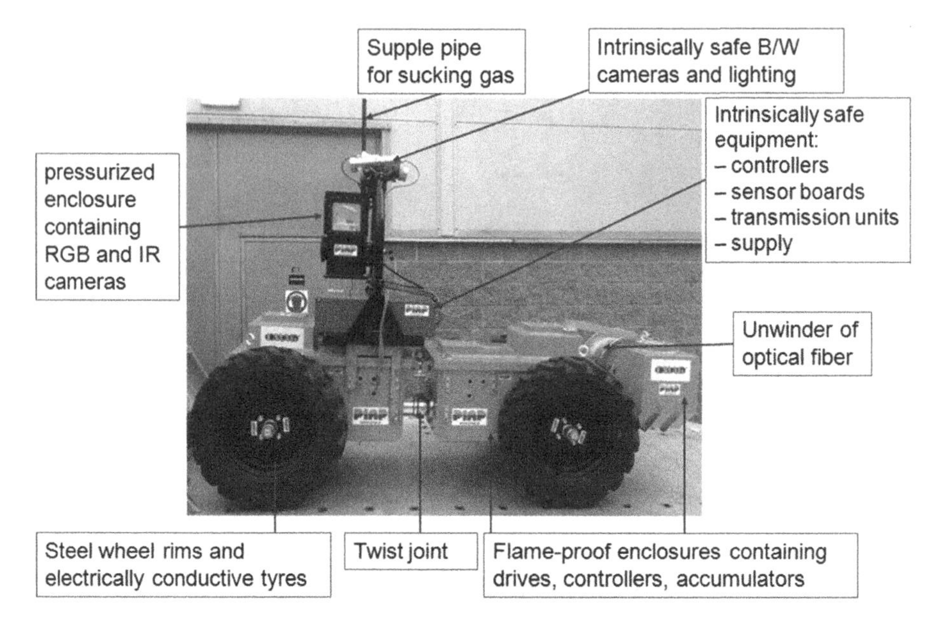

Fig. 11. Mobile Inspection Platform MPI

circuits of the motor controllers and the accumulators are encepsuled according to EN 60079-18, and the connections of the motor controllers and the cells of the accumulators meet the requirements of the increased safety "e". The drives were covered by flame-proof enclosures "d" of EN 60079-1 constituting the vehicle's body. Such solutions ensure double explosion proof protection, which in the light of EN 50303 standard requirements allows to obtain category M1, that is a continuous use in environment filled with explosive gases of group I and/or coal dust.

There are two emergency stop switches (the "kill switches") for the purpose of the emergency stop of the machine, placed on the front and the rear truck of MPI. Together with the relays and the contactors they fulfill *the emergency stop function* at the level of Performance Level PLc according to EN ISO 13849.

The Mobile Platform Inspection is equipped with two identical modules of the gas meters. Infrared sensors were applied for measuring concentration of carbon dioxide and methane at a low and high range and electrochemical sensors were applied for measuring concentration of oxygen and carbon monoxide. Methane is measured at a range of 0…100%, carbon dioxide at a range of 0…5%, carbon monoxide 0…1%, oxygen at a range of 0…25%, temperature –40…+120 °C and humidity at a range of 0…100%. Due to the fact that usually there is a high humidity (generally around 100%) within the collapsed area, a system drying gas mixture before putting it on the gas sensors was developed. Examined gas is taken through the supple pipe and dried by the dryer, and then it is placed in the measurement enclosure containing sensors of gas concentration and sensors controlling the proper functioning of the system. The internal sensor of temperature has a diagnos function as regards functioning of the sensors at a permissible temperature range (up to +60 °C). Similarly, the sensor of humidity has a diagnostic function and it informs on quality of gas samples drying. The flow sensor provides information on whether a pomp is effectively sucking air. While the external sensors of temperature and humidity located outside the measurement enclosure provide information on climate conditions around the robot.

In order to increase safety of the rescuers, there was applied a redundant solution redundancy of the gas meters in the robot. It is insufficient to rely on the results of a single sensor due to its limited reliability. Especially important in this matter are the measurements of poisonous carbon monoxide and explosive methane. In functional safety set out in EN 61508 standard the voting principles were provided. In metrology system of MPI the voting principle 2oo2 (2 out of 2) was applied. The measurement results in both channels need to be compatible as regards the accepted tolerance in order to consider the measurement as reliable. Thus, for the headquarters of the operation to take a decision on letting the rescuers in the hazardous zone, the measurement results in both channels must indicate permissible concentration of hazardous gases.

Machine vision consists of four cameras and lightning. Two monochromatic cameras (black and white) of EMAG production are build as intrinsically safe of the category "ia" according to EN 60079-11 standard. These cameras do not have any drive and one of them is directed to the front of the robot, while the other is directed to its rear. A high resolution RGB camera is placed in the pressurized enclosure type "px" according to EN 60079-2. It has a drive enabling its control in the vertical and horizontal plane. Below the RGB camera there is an infrared camera with no drive.

Pressure inside of the television tower is higher so that it prevents a gas mixture from reaching the interior of the tower. In the enclosure an intrinsically safe protective device monitoring external and internal pressure difference was located. If unsealing of the enclosure occurs, the cameras are disconnected from the power supply. The protective device has safety integrity level SIL2 according to EN 50495, which together with the explosion proof construction "px" allows obtaining category M1 for this unit of the robot.

Four illuminating lamps were placed on the supple pipe – two of them directed to the front of the robot and the other two directed to its rear. They are remotely activated by an operator of a measuring and control console.

Control commands are sent by an operator from the measuring and control console (a computer of increased environmental factors resistance – IP54, shock and vibration resistant) equipped with an advanced joystick. There are two possible methods of remote communication with MPI: wireless communication (via Wi-Fi) and communication thought the optical fiber. During MPI's drive to the final destination where an operation is to be carried out, wireless communication is being used, whereas after putting the robot outside an insulation dam, the optical fiber (with special fire performance) is being used. The optical fiber is unwound from the unwinder attached to the rear truck of MPI.

Control commands are received by the robot in TCP computer and then distributed to the respective microprocessor controllers. Commands ensuring the robot's movement are sent from TCP computer to the respective motor controllers BLDC. Commands starting measurements are sent from TCP computer to two sets of sensors (each has its own microcontroller). Data streams from RGB camera and IR camera are sent directly via TCP computer and the optical fiber to the operator's console.

Fig. 12. MPI in CSRG's mining excavations adopted to training

The abovementioned transmission units were designed according to the intrinsically safe explosion proof technique of the category "ia".

2.10.1 Functional and Traction Testing of MPI

Functional testing was assigned to experts of Central Mines Rescue Station in Bytom (CSRG – Centralna Stacja Ratownictwa Górniczego S.A.). The tests were performed in mining excavations adapted to training within the area of CSRG (Fig. 12). The screen presenting measuring and control software of the operator's console was presented in Fig. 13.

Below the quotation of point 5 "Final assessment of testing results" of the report [32] prepared by CSRG:

"Functional testing of the Mobile Inspection Platform model MPI performed lead to the following conclusions:

(1) ***the selected characteristics and properties** would enable to replace a man – a rescuer by MPI platform in the identified situations during a performance of underground mining*

Fig. 13. Screen presenting measuring and control software of MPI [32]

Fig. 14. MPI driving in a mining excavation [33]

excavation inspections and measurement of environmental parameters, in particular in explosive environments or air unfit for breathing.

(2) ***the construction of the robot's mobile part*** *sufficiently meet the selected characteristics and properties (…). The view transmitted from the cameras in the absence of any permanent lightning in the mining excavation was good quality, what given the necessity of watching the route by a remote operator enables to effectively steer MPI platform.*

(3) ***the functioning of the operator's station*** *(the console and the power supply) was correct. Wireless communication (over a distance of approx. 25 m) in the device steering application is reliable. Steering of MPI platform by an operator via joystick used in the prototype version allowed effective steering of the platform movement direction and of the installed RGB camera. Under the assumed level of functional testing assessment, the same*

Fig. 15. Overcoming obstacles constituting a mine infrastructure by MPI [33]

can be stated as regards the functioning of the metrology application (the research did not provide for metrology assessment)."[1]

Traction testing was carried out by mine rescuers in mining excavations, where the robot covered a distance of approx. 250 m, overcoming a ramp of 30° incline, a relatively powder surface of the tunnel (loose rocks on which the vehicle's wheels were sometimes spinning), a mine railway tracks, switches and other typical obstacles for a mine infrastructure (passing by equipment and devices, a material stored, wagons standing on tracks, stands, enclosures, etc. – Figs. 14 and 15). Traction testing was successfully completed and a film documenting a course of the tests and the field tests can be found on [34].

3 Conclusions

Annually increasing number of mining robots constructions is reflecting the fact that there is a great niche in the market as regards demand for such goods. The designers of the first mining robot Numbat of the Australian CSIRO were seeking an entrepreneur interested in the implementation of their construction into production. While, the robot Gemini-Scout of the Sandia National Laboratories became an object of interest of the company Black-I Robotics from Boston, who finally bought the robot's license [3]. Also the abovementioned Chinese company Tangshan Kaicheng Electronic evaluates its production capacity to 1800 robots per year [24, 25]. The sale market of mining inspection robots should be accessible due to the fact that the headquarters of a rescue operation do not want to unnecessarily put rescuers at high risk related to entering into zones threatened by an explosion and explosive zones. Since police and military sappers have adequate equipment suitable for detonation of explosives for years,

[1] The authors' own translation.

it might be expected that similar solutions will be introduced in the mining industry in the following years.

However, in comparison to sapper robots, mining robots apart from meeting functional requirements need to fulfill wide range of requirements provided for by law, namely the EU directives, in particular the most important ATEX directive (94/9/EC), EMC directive (2004/108/EC) and MD directive (2006/42/EC). Designers must prove that the given construction is in compliance with the relevant directives. Common explosion proof techniques such as flame-proof enclosures, pressurized enclosures and encepsulation result in the increase of mass and size of devices, and consequently, in the deterioration of their functionality. Therefore, it is substantial to apply absolutely optimal design, which on the one hand meets the necessary explosion proof requirements and on the other keep the highest possible functionality of a robot.

References

1. Cioca, I.L., Moraru, R.I.: Explosion and/or fire risk assessment methodology: a common approach, structured for underground coalmine environments. Arch. Min. Sci. **57**(1), 53–60 (2012)
2. Trenczek, S.: Levels of possible self-heating of coal against current research. Arch. Min. Sci. **53**(2), 293–317 (2008)
3. Green, J.: Mine rescue robots requirements. In: Outcomes from an Industry Workshop, Robotics and Mechatronics Conference (RobMech), pp. 111–116 (2013)
4. Kasprzyczak, L., Trenczek, S., Cader, M.: Robot for monitoring hazardous environments as a mechatronic product. J. Autom. Mob. Robot. Intell. Syst. (JAMRIS) **6**(4), 57–64 (2012)
5. Murphy, R., Kravitz, J., et al.: Mobile robots in mine rescue and recovery. IEEE Robot. Autom. Mag. **16**, 91–103 (2009)
6. WA robot called in to help NZ mine rescue, 23 November 2010. http://www.watoday.com.au/wa-news/wa-robot-called-in-to-help-nz-mine-rescue-20101123-184vo.html. Accessed 16 March 2016
7. Sudbury robot goes to Chile, Robot designed with recent mining disasters in mind, 25 October 2011. http://www.cbc.ca/news/canada/sudbury/sudbury-robot-goes-to-chile-1.1091254. Accessed 16 March 2016
8. Hainsworth, D.W., et al.: Teleoperation user interfaces for mining robotics. Auton. Robot. **11**, 19–28 (2011)
9. Ralston, J.C., Hainsworth, D., et al.: Recent advances in remote coal mining machine sensing, guidance, and teleoperation. Robotica **19**, 513–526 (2001)
10. Ralston, J.C., Hainsworth, D.: The numbat: a remotely controlled mine emergency response vehicle. In: Zelinsky, A. (ed.) Field and Service Robotics, pp. 53–59. Springer, London (1998)
11. http://www.cs.cmu.edu/afs/cs/Web/People/groundhog/robots_ghog.html. Accessed 16 March 2016
12. Thruny, S., Hähnel, D., et al.: A system for volumetric robotic mapping of abandoned mines. In: Proceedings of the ICRA 2003 (2003)
13. http://www.cs.cmu.edu/~3D/mines/groundhog/web/indexall.html. Accessed 16 June 2015

14. Murphy, R., Kravitz, J.: Preliminary report: rescue robot at Crandall Canyon, Utah, mine disaster. In: IEEE International Conference on Robotics & Automation, Pasadena, pp. 19–23 (2008)
15. http://www.msha.gov/SagoMine/robotdetails.asp. Accessed 16 June 2015
16. http://www.sandia.gov/index.html Gemini_Scout_Handout_Final.pdf. Accessed 16 March 2016
17. https://share.sandia.gov/news/resources/news_releases/miner-scou/#.VYA55_mvFpg. Accessed 16 March 2016
18. https://www.youtube.com/watch?v=gLjwfUh1_1w. Accessed 16 March 2016
19. Hambling, D.: Next-Gen Coal Mining Rescue Robot, 23 August 2010. http://www.popularmechanics.com/science/energy/a6095/next-gen-coal-mining-rescue-robot/. Accessed 16 March 2016
20. Rongb, X., et al.: Mechanism and explosion-proof design for a coal mine detection robot. Adv. Control Eng. Inf. Sci. **5**, 100–104 (2011)
21. Li, Y., et al.: Explosion-proof design for coal mine rescue robots. Adv. Mater. Res. **211–212**, 1194–1198 (2011)
22. Wang, W., et al.: Kinematics analysis for obstacle-climbing performance of a rescue robot. In: IEEE International Conference on Robotics and Biomimetics, pp. 1612–1617 (2007)
23. Gao, J., et al.: Coal mine detect and rescue robot technique research. In: IEEE International Conference on Information and Automation, pp. 1068–1073 (2009)
24. Tangshan Kaicheng Electronic to produce coal mine rescue robots, 03 November 2010. http://www.whatsonxiamen.com/tech475.html. Accessed 16 March 2016
25. N China plant ready to produce coal mine rescue robots, 02 November 2010. http://www.chinadaily.com.cn/business/2010-11/02/content_11490552.htm. Accessed 16 March 2016
26. Kasprzyczak, L., Pietrzak, R.: Electromagnetic compatibility tests of mining mobile inspection robot. Arch. Min. Sci. **59**(2), 427–439 (2014)
27. Kasprzyczak, L., Trenczek, S., Cader, M.: Pneumatic Robot for Monitoring Hazardous Environments of Coal Mines. Solid State Phenomena, Mechatronic Systems and Materials IV, pp. 120–125 (2013)
28. Kasprzyczak, L., Trenczek, S.: Mobile inspection robot made in the M1 category for monitoring explosive environments of coal mines. In: Proceedings of 22nd World Mining Congress & Expo, vol. 3, Istanbul, pp. 457–462 (2011)
29. Kasprzyczak, L., Nowak, D., Szwejkowski, P., et al.: Sensors for measurement mining atmosphere parameters of mobile inspective robot. Mechanizacja i Automatyzacja Górnictwa **12**(466), 19–30 (2009)
30. http://mecheng.ukzn.ac.za/Research-Areas/MR2GSearchandRescueDivision.aspx. Accessed 16 March 2016
31. Raport z badań trakcyjnych Prototypu Górniczego Mobilnego Robota Inspekcyjnego GMRI, Centralna Stacja Ratownictwa Górniczego S.A., Laboratorium Badania i Opiniowania Sprzętu, Bytom (September 2010, unpublished)
32. Raport z badań funkcjonalnych Mobilnej Platformy Inspekcyjnej MPI, Centralna Stacja Ratownictwa Górniczego S.A., Laboratorium Badania i Opiniowania Sprzętu, Bytom, p. 8 (December 2014, unpublished)
33. Raport z badań trakcyjnych Mobilnej Platformy Inspekcyjnej MPI, Centralna Stacja Ratownictwa Górniczego S.A., Laboratorium Badania i Opiniowania Sprzętu, Bytom, p. 8 (June 2015, unpublished)
34. https://www.youtube.com/watch?v=zA6pysXa9-E. Accessed 16 March 2016

Drop Test of Tactical Mobile Robot

Bartosz Blicharz(✉) and Maciej Cader

Industrial Research Institute for Automation and Measurements PIAP,
Warsaw, Poland
bblicharz@piap.pl

Abstract. The aim of the research presented in this article was to obtain boundary conditions necessary to run a computer simulation of the robot's drop on the basis of the actual drop test. The Tactical Throwing Robot TRM developed by the Industrial Research Institute for Automation and Measurements PIAP was used in testing. The moment of the robot's contact with the ground was recorded by a high-speed camera. The video and adequate math calculations were the grounds for obtaining the boundary conditions and several additional values, which subsequently allowed to verify convergence of the results obtained for the real object and the simulation model. The verified simulation model was then used in the analysis consisting of the reduction of the robot's tyre weight.

Keywords: FEM simulation · Geometry simplifying · Tactical mobile robot · Boundary conditions

1 Introduction

1.1 Tactical Mobile Robots

Tactical mobile robots are adapted to reconnaissance, in particular audio-visual. These constructions are used by the special forces during their missions if an inspection of a hard-to-reach area is required. The robots are usually sent to zones hazardous to human life, health and safety [1] by being thrown to a contaminated area or dropped into a building through a window. This specific use of the tactical mobile robots compared to other mobile robots requires their special construction. Firstly, the robot need to have an adequate mechanical strength so that it is not damaged as a result of the impact loading arising during the robot's contact with the ground after a drop. And secondly, the robot's breaking acceleration must comply with the resistance of all electronic components [2, 3]. Due to the fact that the Tactical Throwing Robot TRM produced by the Industrial Research Institute for Automation and Measurements PIAP meets all of the abovementioned conditions, it was used in testing.

1.2 Tactical Throwing Robot TRM

The construction of the first generation Tactical Throwing Robot TRM was presented in Fig. 1. The robot consists of two wheels (B) with an independent motor, hub caps (A)

R. Szewczyk and M. Kaliczyńska (eds.), *Recent Advances in Systems, Control and Information Technology*, Advances in Intelligent Systems and Computing 543, DOI 10.1007/978-3-319-48923-0_38

protecting it against damage during a fall to the one side, a main body (C) with sensors and internal electronics, and a tail (D) made of a metal rope with a stabilizer in a ball shape. TRM movement is a result of independent steering of the rotational speed of the wheels. The robot's movement direction is selected via joystick placed on a control console.

Fig. 1. Construction of the first generation TRM

The basic parameters of the robot were presented in Table 1 below based on its data sheet available at [4].

Table 1. Basic parameters of TRM

Parameter	Value
Weight	1.4 kg
Maximum speed	3.1 km/h
Drop height	< 9 m
Basic equipment	Camera, microphone, LED lightning
Lifetime	Approx. 1 h
Size	205 mm × 167 mm × 190 mm

1.3 Finite Element Method

The Finite Element Method (FEM) is a numerical technique by which a physical object can be depicted as a simplified computer model [5]. It allows to obtain an approximate reaction of an object on different external factors. There are many complex FEM systems designed for computer simulations, in which the actual determinants are replaced by the respective simulation environment. Boundary conditions are used for this purpose. The most popular simulation programs are: ANSYS, MD ADAMS, COMSOL and MSC MARC [6–9]. Depending on their complexity, it is possible to perform analyses of different types of parameters: thermal, resistance, electric, magnetic, as well as analyses of vibrations, fluids and dynamics. Furthermore, multiple physical models and multiple physical phenomena may be used. However, it should be noted that correctness and accuracy of results obtained from FEM simulations depend on accuracy of modelled boundary conditions including geometry, materials properties, loads, supports, connections, mesh, and other analysis settings such as: solver type, number of steps and convergence criteria.

1.4 High-Speed Cameras

High-speed cameras enable to record hundreds and thousands rates per second. It need to be noted though that the higher frame rate the smaller image size and consequently the lower image quality. This relation was presented in Table 2 below with respect to Mega Speed HHC X4 camera used in testing [10].

Table 2. Image size and frame rates relation of Mega Speed HHC X4

Mega Speed HHC X4	
Image size [px]	Frame rates [fps]
1280 × 1020	600
1280 × 720	850
800 × 600	1025
640 × 480	1280
320 × 240	2500

High-speed cameras are used in visual analyses of fast-changing phenomena difficult to be perceived by a human's eye. For example, the cameras are being installed in production halls for a detailed verification of production process.

1.5 Research

The aim of the research presented in this article was to obtain boundary conditions necessary to run a computer simulation of the robot's drop on the basis of the actual drop test, as well several additional values, which subsequently allowed to verify convergence of the results obtained for the real object and the simulation model. The verified simulation model was then used in the analysis consisting of the reduction of weight of TRM robot's tyre.

2 Laboratory Testing

The first stage of the research was the Tactical Throwing Robot drop test performed in laboratory testing. For this purpose a special test bench presented in Fig. 2 was construed.

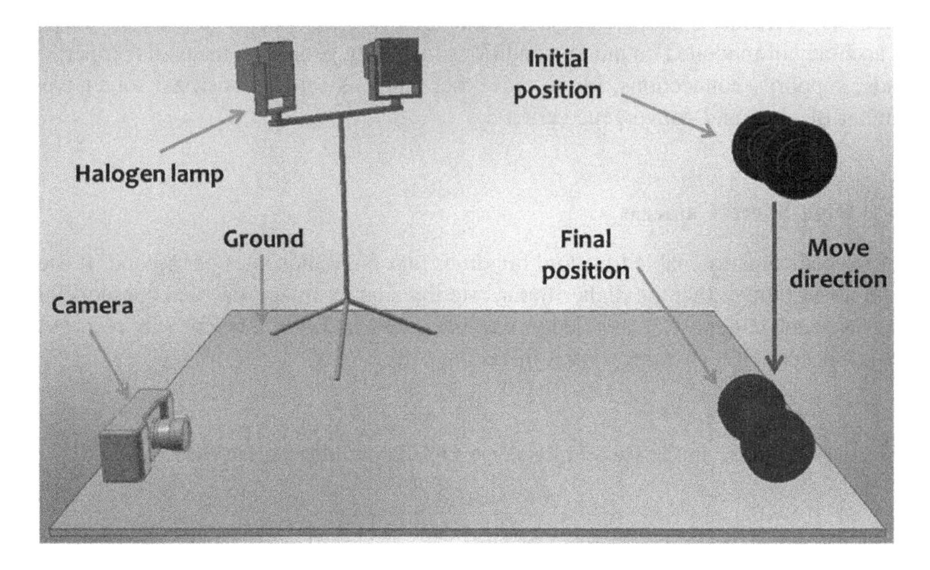

Fig. 2. Test bench

TRM was dropped from a height of 8.3 m (the second floor of the building). The moment of the robot's contact with the ground was recorded by a high-speed camera – Mega Speed HHC X4, while the ground was illuminated by a double 500 W halogen lamp in order to reduce exposure time and consequently increase clarity of the camera view. The experiment was repeated 10 times. The selected settings of the camera were presented in Table 3 below.

Table 3. Selected settings of the camera

Setting	Value
Frame rates	800 fps
Image size	1280 × 720 px
Focal length	25 mm
Diaphragm	2.8

Laboratory testing aimed to provide the following **boundary conditions**: drop reaction force and an area affected by drop reaction force. In order to obtain these values, many additional values needed to be measured.

The maximum tyre deflection value was calculated on the basis of two frames of the video presented in Fig. 3.

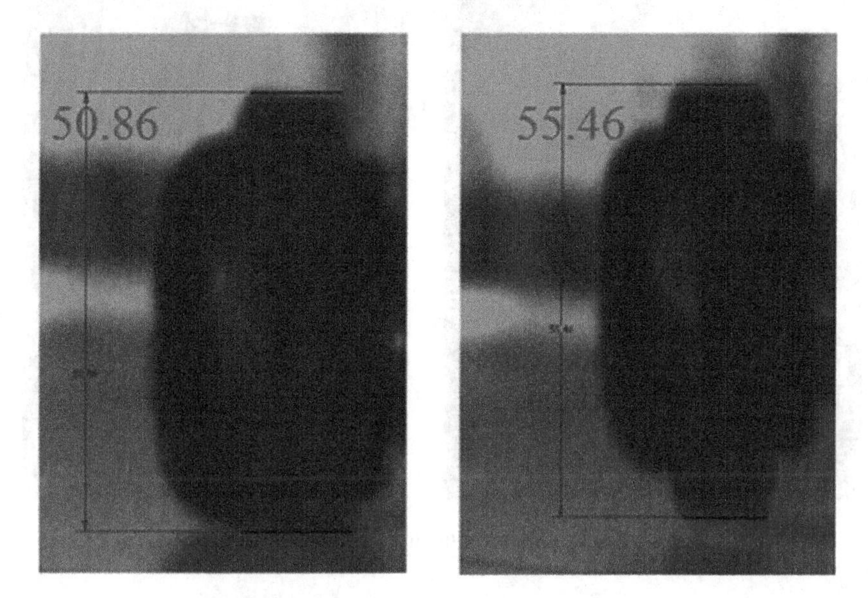

Fig. 3. Frames relevant for the maximum tyre deflection value calculation

The left frame shows the robot's tyre before contacting the ground. Its height was market by h_p. The right frame presents the robot's contact with the ground while the tyre was of maximum deformation (this is one of four frames capturing the contact with the ground). Its height was market by h_u. Finally, a height of the robot's tyre in reality (h_r) was measured. All these values were used in the following calculation of the maximum tyre deflection value (f):

$$f = \frac{h_u}{h_p} \cdot h_r \tag{1}$$

The maximum tyre deflection value is equal to breaking distance during the fall. Breaking time (t_h) is the product of time elapsing between two frames (Δt) and numbers of frames between the last frame before contacting the ground and the frame presenting the tyre's maximum deformation (q).

$$t_h = \Delta \mathrm{t} \cdot \mathrm{q} \tag{2}$$

The maximum speed of the robot during the fall (v_m) was calculated on the basis of two frames of the video presented in Fig. 4.

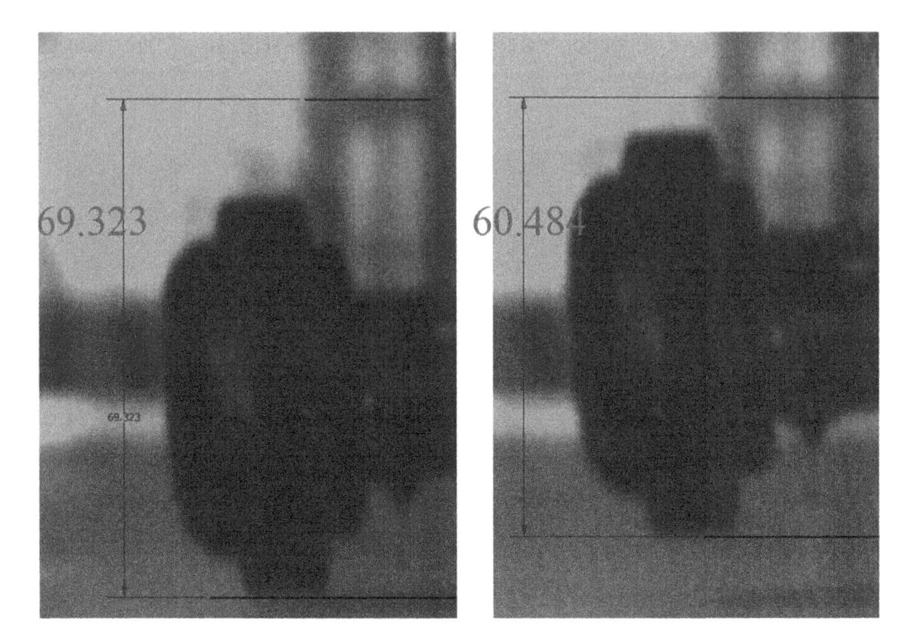

Fig. 4. Frames relevant for the maximum speed calculation

The left frame is the last frame before the robot's contact with the ground. The right frame is the frame directly preceding the left frame. A distance between the tyre's bottom edge and the fixed point of reference (Δs) was divided by time elapsing between two frames (Δt), and multiplied by the quotient of the tyre's heights h_r and h_p.

$$v_m = \frac{\Delta s}{\Delta t} \cdot \frac{h_r}{h_p} \tag{3}$$

The calculated value of the maximum speed was verified by a theoretical speed v_t on the following grounds: the law of conservation of energy, kinetic energy (E_k) and potential energy (E_p).

$$E_k + E_p = const \tag{4}$$

$$E_k = \frac{mv^2}{2} \tag{5}$$

$$E_p = mgh \tag{6}$$

$$v_t = \sqrt{2gh} \tag{7}$$

The maximum breaking acceleration of the robot during the fall (a) was calculated with the assumption that uniform deceleration motion during the break occurred. Due to that fact the braking acceleration was equal to the quotient of the maximum speed (v_m) and breaking time (t_h).

$$a = \frac{v_m}{t_h} \tag{8}$$

The first boundary condition – drop reaction force (F) was calculated as the product of the robot's mass (m) and the maximum breaking acceleration (a).

$$F = m \cdot a \tag{9}$$

The second boundary condition – the area affected by drop reaction force (the deflected area) is determined by the maximum tyre deflection value. As was shown in Fig. 5 below, 8 sprockets of the tyre were deflected.

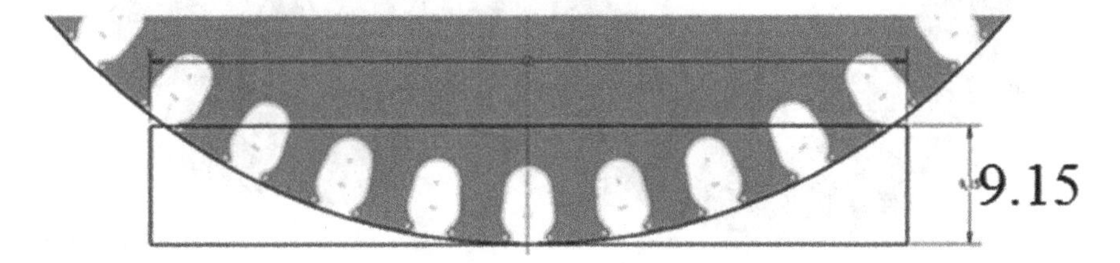

Fig. 5. Area affected by drop reaction force

The results of the calculations made in laboratory testing were presented in Table 4 below.

Table 4. The results of calculations in laboratory testing

Maximum tyre deflection	f	9.15 mm
Breaking time	t_h	33.34 ms
Maximum speed	v_m	12.75 m/s
Theoretical speed	v_t	12.76 m/s
Breaking acceleration	a	5,100 m/s^2
Robot's mass	m	1.4 kg
Drop reaction force	F	7.140 N

3 Simulation Testing

3.1 Part I

The second stage of the research was performed in FEM program: ANSYS 13.0. However, the simulation was preceded by simplifying the robot's geometry in CAD program: Autodesk Inventor Professional 2014. Due to load symmetry, half of TRM model was removed. Furthermore, structure of the main body and the tyre was simplified by removing unnecessary elements such as bearings and screws. Additionally, the tyre was divided into two parts – internal and external, which was presented in Fig. 6.

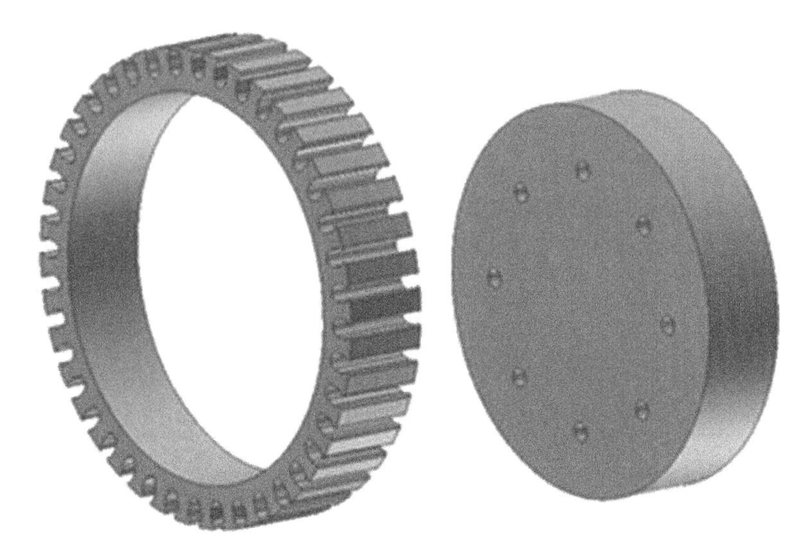

Fig. 6. Internal and external part of the robot's tyre

The robot's modifications were presented in Fig. 7.

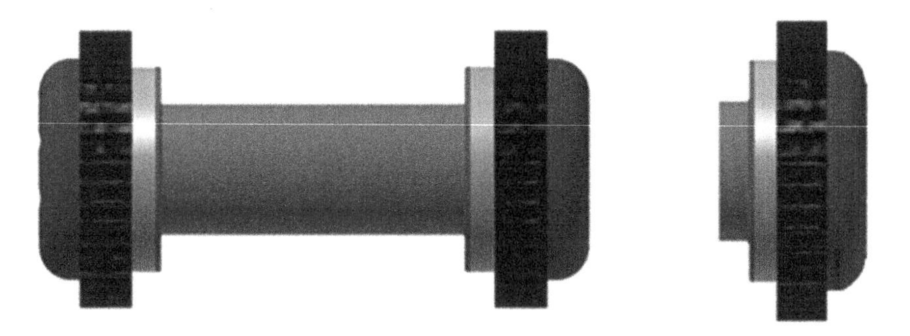

Fig. 7. Simplification of the robot's geometry in Autodesk Inventor Professional 2014

The simplified model of TRM was imported to ANSYS 13.0. Contacts between the robot's elements were modelled as rigid. The mesh density on the external part of the sprockets was presented in Fig. 8.

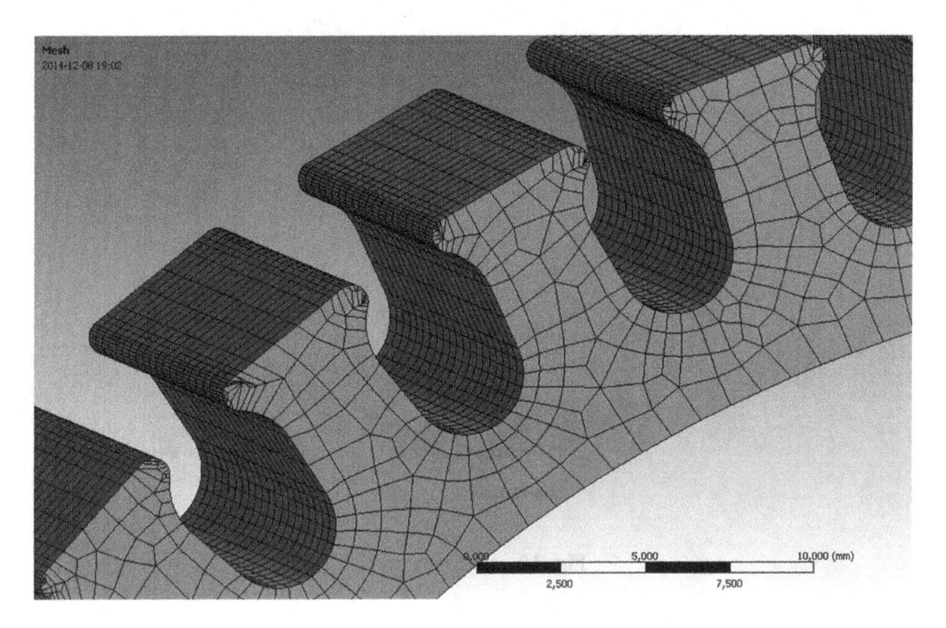

Fig. 8. Mesh density

Half of the drop reaction force was applied to the deflected area, which was presented in Fig. 9.

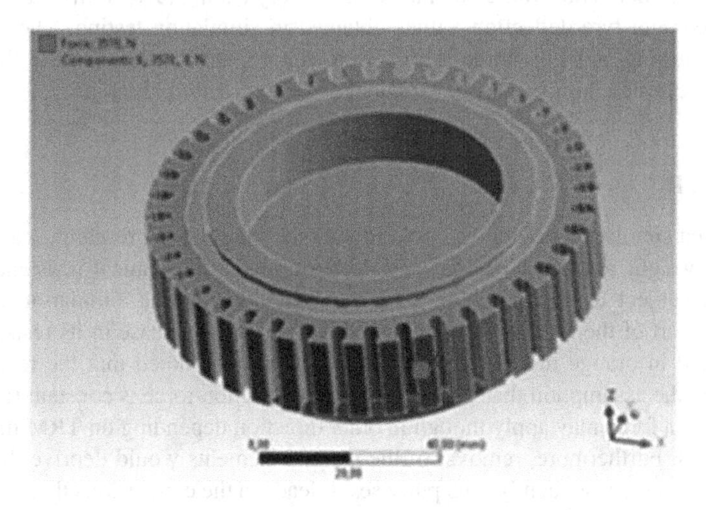

Fig. 9. Force applied to the deflected area

Fixed support shown in Fig. 10 was applied due to much higher stiffness of body elements than stiffness of the tyre.

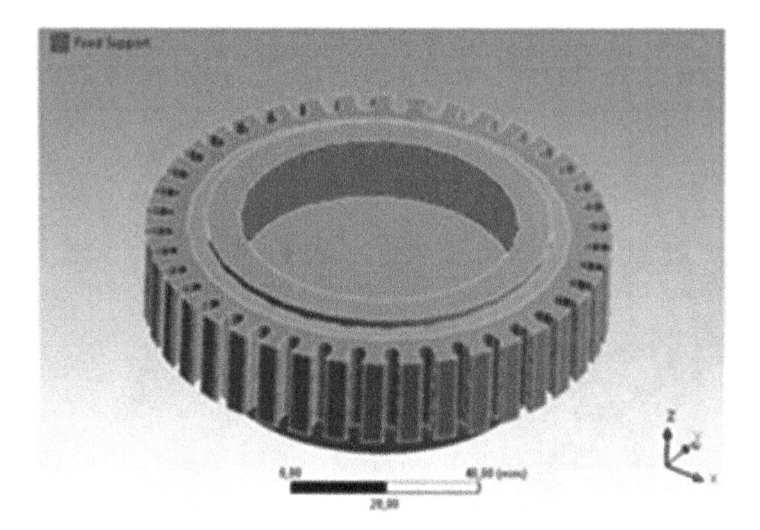

Fig. 10. Fixed support

The simulation was calculated. The precise reading of **the results** was allowed due to the probe function. The maximum tyre deflection was measured in the direction of TRM movement based on the two middle sprockets of the deflected area. Deflection of the tyre's sprockets were shown in Fig. 11.

Two different deflection values were obtained – 9.27 mm and 9.16 mm. The maximum tyre deflection value obtained in laboratory testing (9.15 mm) was compared to the maximum tyre deflection values obtained in simulation testing. The results of laboratory testing and simulation testing were convergent at a level of 1.3 %. Due to that fact it can be claimed that the simulation was correctly prepared.

3.2 Part II

The verified simulation model was used in the third stage of the research – analysis of the tyre's weight reduction without any decrease in the mechanical resistance of the robot and subject to the initial drop height. According to the simulation results, a significant part of the tyre could be removed without any decrease in its resistance (an area marked in orange in Fig. 12b). However, it must be noted that the results were obtained in the assumption that a direction of drop reaction force is constant (Fig. 12a). Drop reaction force may apply though in other direction depending on TRM movement after the fall. Furthermore, removal of the marked elements would deprive the tyre its round shape and consequently – its purposes. It leads to the conclusions that any area of the tyre should be removed and the tyre was designed optimally for the purpose of a drop up to a height of 9 m.

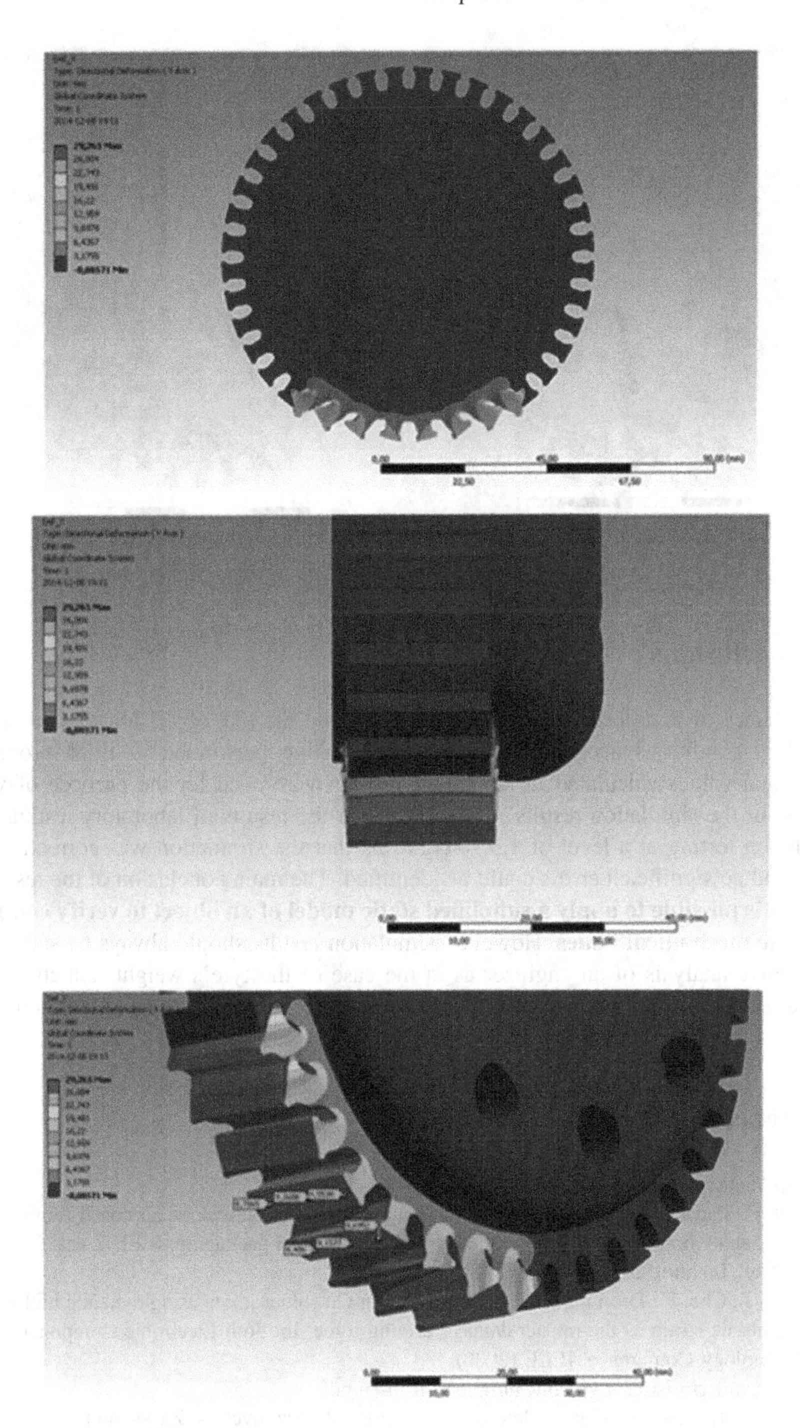

Fig. 11. Deflection of the tyre's sprockets

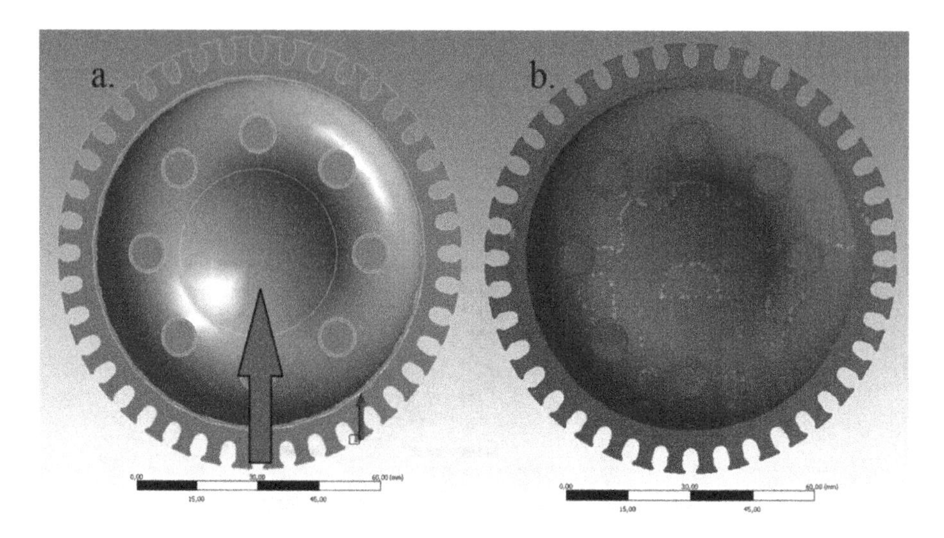

Fig. 12. Analysis of the tyre's weight reduction (Color figure online)

4 Conclusions

Performance of non-destructive laboratory testing on the real object allowed to obtain boundary conditions necessary for simulation testing performed in FEM program. Additional values calculated in laboratory testing were used for the purpose of verification of the simulation results. Convergence of the results of laboratory testing and simulation testing at a level of 1.3 % confirms that the simulation was correctly prepared and no significant errors could be identified. The main conclusion of the research is that **it is possible to apply a simplified static model of an object to verify complex dynamic mechanical issues**. However, simulation results should always be subject to substantive analysis of an engineer as in the case of the tyre's weight reduction presented in this article. The results themselves may suggest impracticable solutions.

References

1. http://antiterrorism.eu/en/portfolio-posts/trm/
2. Lall, P., Panchagade, D., Choudhary, P., Gupte, S.: Failure-envelope approach to modeling shock and vibration survivability of electronic and MEMS packaging. IEEE Trans. Compon. Packag. Technol. **31**, 104–113 (2008)
3. Pang, J., Che, F.: Drop impact analysis of Sn-Ag-Cu solder joints using dynamic high-strain rate plastic strain as the impact damage driving force. In: 56th Electronic Components and Technology Conference. IEEE (2006)
4. http://antiterrorism.eu/wp-content/uploads/trm-en.pdf
5. MESco Tarnowskie Góry: Metoda Elementów Skończonych – Zarys podstaw teoretycznych, R1
6. http://ansys.com/

7. http://www.mscsoftware.com/product/adams/
8. http://www.comsol.com
9. http://www.mscsoftware.com/product/marc/
10. http://www.avicon.pl/sites/default/files/other/speed_chart_v30.pdf

Human-Robot Interaction in the Rehabilitation Robot Renus-1

Jacek Dunaj, Wojciech J. Klimasara, and Zbigniew Pilat(✉)

Industrial Research Institute for Automation and Measurements PIAP, Warsaw, Poland
{jdunaj,wklimasara,zpilat}@piap.pl

Abstract. Renus-1 is a mechatronic system for support of rehabilitation of the patients with dysfunctions of the upper arm. It is the End-effector based device. It contacts the patient's limb only at its most distant part, the handle. The mechanical structure of the manipulator is a SCARA like, with 3 DOF. The robot control system uses commercial components of Mitsubishi Electric, typically used in industrial automation. Each axis is moved by synchronous motor with permanent magnets, controlled by individual servo cooperating with the PLC CPU. As the user interface the standard PC station was used. It is working under control of specialized application software. PC is equipped with 2 screens: the first for the patient and the second one for the therapist. Two main modes of work are available: teaching and training. In the training mode it is possible to perform active as well as passive rehabilitation. In both cases patient has current information about the course of the exercise. These data are stored in the PC mass memory, with possibility to send via LAN to the external computer station for further processing. The handle is equipped with the six-axis sensor (forces and torques). Additional for safety reasons patient has special permission button. It must be pressed, so that the robot can perform a translational motion. Also Emergency Stop button is mounted.

Keywords: Rehabilitation robotics · Human-Robot interaction

1 Introduction

In the initial period of the robotics development the effort of research groups have been focused primarily on the industrial applications of robots. Those were also the first implementation of robotics technology [1]. It was only in the 80s, that the concept of introduction of robots in human environment [2] have become real. A new, separate class of devices, called service robots, appeared. Feature, which from the beginning distinguishes these two groups: industrial and service, is a way of communication between human and robot (Human Robot Communication HRC). In case of industrial robots most important task of the HRC system, was allowing the operator to teach the robot its job, start/stop the work in automatic mode and then provide information about the status of device and realized task. In general, engineers were engaged in the installation, programming and maintenance. Skilled workers, after appropriate trainings, were employed as operators on the direct manufacturing positions. In the case of

R. Szewczyk and M. Kaliczyńska (eds.), *Recent Advances in Systems, Control and Information Technology*, Advances in Intelligent Systems and Computing 543, DOI 10.1007/978-3-319-48923-0_39

service robots, it was assumed from the beginning that they will be used by ordinary people, who do not have the technical preparation. At the same time the tasks expected to be performed by service robots, imposed the requirement of transfer between them and their operators far greater variety of information, in much larger quantities. Essentially, these robots are working close to the people, so they must receive and transmit in return from/to them information up to date. Therefore, the problem of communication is still one of the most serious obstacles to the wide application of service robots. It can be assumed, that equipping robots with an effective, yet simple communication systems, will largely determine the acceptance of these devices by future users. This applies in its entirety to assisted robots, especially these designed to support motor rehabilitation of patients.

Three types of communication are used in the exchange of information between human and service robot [3]. In the initial design of these devices, efforts have been made to use voice communication. The aim was to equip the robot with the ability to use natural human language. The work on the synthesis and speech recognition have been carried out intensively in many centers around the world for several years. Still, the effects of these actions, concerning the use of natural language, do not go beyond research laboratories. Subsequent projects and programs do not offer the possibility of practical use of the results in the form of speech and hearing apparatus for a robot. At conferences specifically dedicated to communication of human with robots [9, 10], networks [11] and in journals there are still a lot of information about the possibility of using other methods of communication, both visual and tactile. These solutions are more frequently implemented in practical applications in models of service robots offered on the market [4]. This refers to both commercial, as well as personal robots, including these intended to help in home and in direct assist of human, for the elderly and disabled, performing tasks related to health. An example of such a device is Renus-1, the mechatronic active system supporting the physical rehabilitation of patients after stroke or orthopedic disorders.

2 Renus-1 Mechanical Structure

Renus-1 is a system assisting rehabilitation of upper limbs. Its mechanical structure is the end-effector-based type [6]. It means, that device contact the patient's limb only at its most distal part, which simplifies the structure of the device. However, this feature can reduce the control of the position of particular joints because of many possible degrees of freedom.

Manipulator. The basic component of the system's mechanical structure is the manipulator (Fig. 1) that leads the patient's upper limb along the defined multi-plane trajectory. The spatial motion range of the manipulator corresponds to the motion range of a healthy human arm. The arm of the manipulator consists of rigid components connected with joints and mounted through a joint to the electrical drive units and carriage, that can move on vertical guides fixed to a motionless column (pos. 4). The axes of the joints of the manipulator's arm are vertical (similar as in the SCARA type robot manipulator). The components of the manipulator are driven by the electrical drive units (pos. 3 and 6) that enable the horizontal movement of the manipulator's arm. The vertical movement of the carriage is driven by electrical drive unit (pos. 2) with the chain gear (pos. 1) and counterweight (pos. 5). From the kinematic point of view, the mechanical structure of this manipulator is a mechanism with three degrees of freedom. The arm end with the manipulator's mechanical interface is equipped with the force and torque sensor (pos. 7). The manipulator and the control cabinet of the robot are mounted on a movable supporting frame.

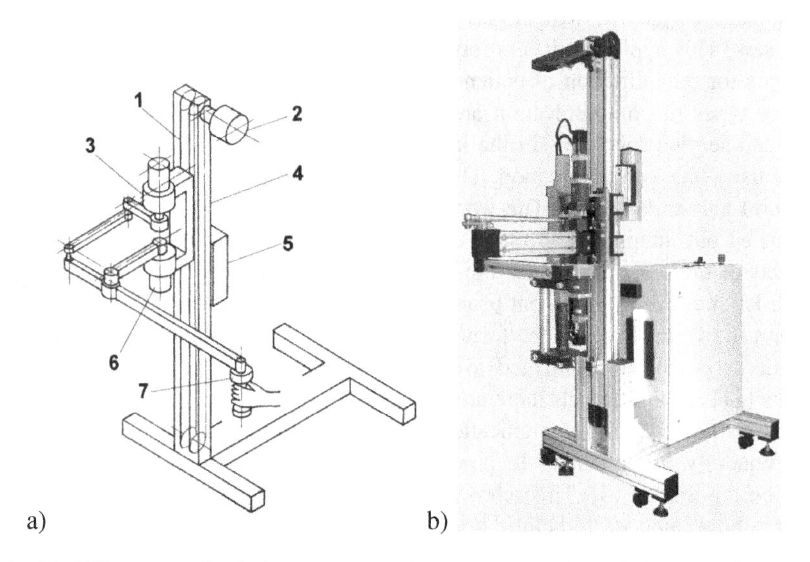

Fig. 1. RENUS-1 (a) mechanical structure: 1 – chain gear, 2 – drive unit for vertical movement, 3 – drive unit for horizontal movement, 4 – column with vertical guides for carriage, 5 – counterweight, 6 – drive unit for horizontal movement, 7 – multi-axis force/torque sensor; (b) general view of the model device

Multi-axis force/torque sensor. The multi-axis force sensor is located at the end of the manipulator. One part of this sensor is attached to the end of the arm of the manipulator, and the second part is attached to the gripper for the hand of trained person. During exercises the signals from this sensor are transmitted to the control system of the Renus-1 robot. Therefore value of forces between the robot ram and upper limb of the patient are monitored by the control system of the robot. It is important for the safety of patient (see p. 6) and makes possible the exercises with the controlled resistance to motion.

3 Renus-1 Structure of the Control System

In the first model of the Renus-1 [7], to build the control system, standard components, well proved in the industrial environment, have been used. The block diagram of the control system is shown in Fig. 2. As the CPU the PLC model Q02HCPU from

Mitsubishi Electric has been selected. It manages the work of the whole manipulation system. CPU cooperates with positioning module QD75M4 and 3 axis controllers, AC servo amplifiers MR-J2S series (MELSERVO-J2-Super), each for one axis. Digital Inputs/Outputs are responsible for controlling lights and read the status of the buttons, which are used for operation of the robot. CPU communicates through the USB channel with the external PC. On this computer the application software Renus.exe is installed (see p. 4). It realizes the Human-Robot communication interface. It is possible to connect 2 monitors to the PC: the first one for the patient and the second one for the therapist. The PC can also be connected to the LAN of the rehabilitation center or to the router in the patient home. In this way it is possible to assure remote access to the Renus-1 from any place in the World, through the Internet.

The control system is installed in the control cabinet, situated on the carrying frame of the whole device. Robot Renus-1 is powered from a standard electrical network, 1-phase, 230 V.

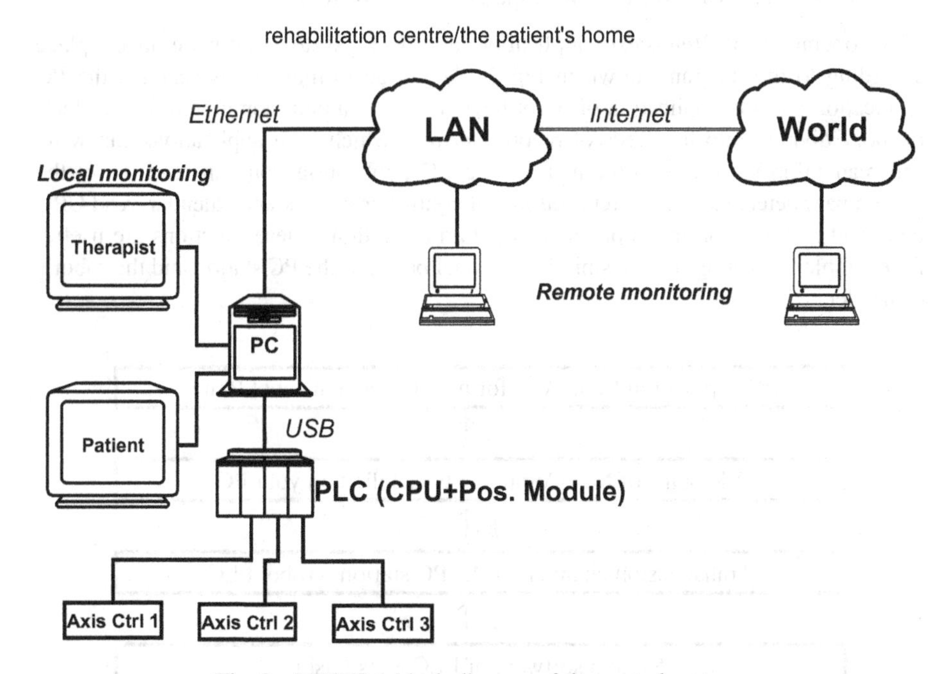

Fig. 2. RENUS-1 block diagram of the control system

4 User Interface

Rehabilitation robot Renus-1 device is not fully autonomous, because for its operation PC station with Windows operating system and management software is required [3]. This software consists of the following components:

1. Mitsubishi MX Components software, which allows the user to define a transmission channel between the PC and PLC robot. The information between the two devices is exchanged through this channel. Software MX Components provides Active-X controls, which are used by the application Renus.exe to directly support robot.
2. Application Renus.exe uses MX Components package, which directly controls the robot, manages the databases associated with the exercise of rehabilitation and acts as the operating panel. The application also uses additional text configuration file **Renus.ini** in which the information needed to modify its operation are contained.
3. Database of trajectories, created in Access format and stored on the PC in the file RenusBazaDanychTrajektorii.mdb. The database contains information about the defined trajectories of robot Renus-1.
4. Database of patients, created in Access format and stored on the PC in the file RenusBazaDanychPacjentow.mdb. The database contains information about patients' exercises performed with the help of robot Renus-1.

Cooperation of Renus.exe application and PLC system software takes place according to the diagram shown in Fig. 3. Acting according to this scheme, the PC application and the application of robot have access to a common area of controller's memory divided into hundreds of records to/from which both applications can write and read information. In addition, from the PC application you can read or set the system parameters such as current settings of controller's clock and calendar, read CPU type, and start or stop the application program execution. These functions are useful, for example for testing the transmission channel between the PC station and the robot's controller.

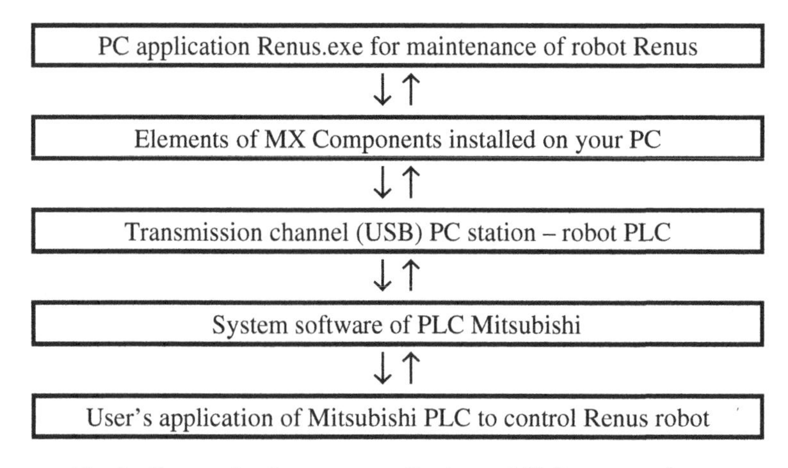

Fig. 3. Cooperation Renus.exe application and PLC system software

PC application for maintenance of robot Renus-1 is contained in the file Renus.exe. The main dialog box of this application (Fig. 4) consists of twelve boxes of information and nine function keys:

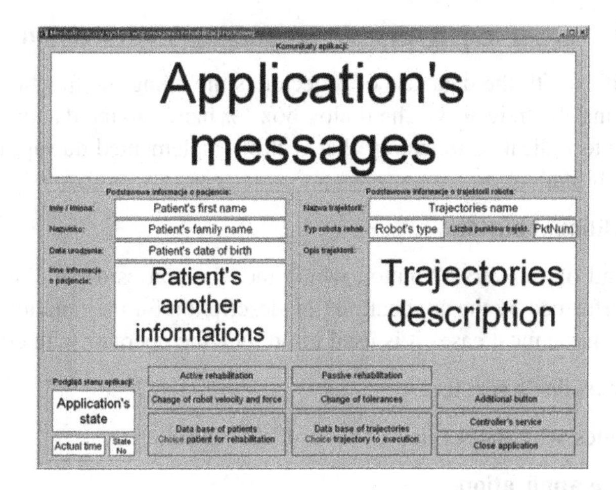

Fig. 4. The main dialog box of **Renus.exe** application

In its upper part there is a window in which the application displays messages for the operator of the rehabilitation robot. Below, on the left side there is a group of windows which displays some of the information read from the database of patients and on the right side - part of the information read from the database of trajectories. In the lower left corner the application shows the clock of the computer, the description and number of stage activities currently performed by the application. Function buttons have the following meanings:

Button **Active rehabilitation**

- Starts the activities related to the handling of active rehabilitation.

Button **Passive rehabilitation**

- Starts the activities related to the handling of passive rehabilitation.

Button **Change of robot velocity and force**

- Starts the activities related to changes in the setting of power (active rehabilitation) with which manipulator affects the patient and velocity (passive rehabilitation) with which manipulator performs trajectory.

Button **Change of tolerances**

- Starts operations related to changes in the setting of tolerance, bringing each engine to the next positions defining the trajectory realized during active rehabilitation.

Button **Data base of patients/Choice patient for rehabilitation**

- Starts the activities with the database of patients. The dialog box for handling the database also allows the operator to select the patient who will perform rehabilitation exercises with the help of robot Renus-1.

Button **Data base of trajectories/ Choice trajectory to execution**

- Starts activities with the database of trajectories including the function of teaching, that is defining the trajectory. The dialog box for handling the database also allows the operator to indicate a trajectory that will be implemented during the active and passive rehabilitation.

Button **Additional button**

- This is a kind of "universal" button which meaning varies depending on activities currently performed by the application (the description for the function displayed on the button). In the most cases it is used to provide the operator with error messages.

Button **Controller's service**

- Starts activities with the service of robot PLC controller.

Button **Close application**

- Ends the execution of Renus.exe application.

5 Basic Modes of Operation

Defining trajectory involves manual moving of the handle by the operator. During this movement the robot's controller reads the position of each servo motor shaft at regular intervals and then stores the read positions in its memory.

Playing trajectory is the reversed action, i.e. the defined trajectory is read into the controller from robot's memory, and then all three engines are brought to subsequent saved positions. Renus-1 robot does not play linear or circular motion, such as industrial robots, moving the manipulator "from position to position" of the three shafts of synchronous motors instead.

As far as control is concerned, the system can operate in four modes:

- recording the trajectory by the manipulator (teaching mode) by leading the robot's arm along the required trajectory by the therapist;
- phase I – passive (manipulator reproduces the preset trajectory);
- phase II – active (the patient reproduces the trajectory along the model displayed on the screen;
- Z axis calibration, i.e. balancing the weight of the patient's arm positioned limply in the nest of the robot's joint.

The servo drive control system is integrated with the computer by the station management software. As far as the software is concerned, the system can operate in the following modes:

- performing basic tasks (creating databases, selecting operating modes);
- exercise configuration (selection of the model exercise and its parameters);
- performing the exercise (reproduction of the preset trajectory);
- recording the results (analysis and recording of the exercise results);
- browsing through the history of results (analysis of the recorded exercise results).

In the current version of the application software teaching is the only way to define the trajectory of the robot Renus-1. It involves manual displacement of manipulator in space and storing the read position in the PLC memory every 1 s. Each position of the manipulator defines the position of $\mathbf{A_1}$, $\mathbf{A_2}$, $\mathbf{A_3}$ axes. Three servo motors work related to their base position and the underlying values of three forces $\mathbf{F_{A1}}$, $\mathbf{F_{A2}}$ and $\mathbf{F_{A3}}$ exerted by the operator on the axes of these servo motors when the trajectory is defined.

The function of teaching robot can run in the input mode of new record (new trajectory) or modify an existing database record. To make teaching of the robot possible, the PC and the robot controller has been coupled with an efficient communication channel, which allows transferring commands, reading robot's status registers and transmitting trajectory of the robot.

Dialog box window of teaching the robot contains information about the type of the robot, the number of points defining the trajectory and the current value of the positions (angles $\mathbf{A_1}$, $\mathbf{A_2}$, $\mathbf{A_3}$) axis of the three engines of the robot manipulator, and the current values of forces ($\mathbf{F_{A1}}$, $\mathbf{F_{A2}}$ and $\mathbf{F_{A3}}$), with which the operator interacts with the axes of each of the motors. These information are read from the status registers of the robot's controller at 0.1 s interval and updated on the screen.

In addition, the six windows of the application display the positions of the waveforms $\mathbf{A_1}$, $\mathbf{A_2}$, $\mathbf{A_3}$ and forces $\mathbf{F_{A1}}$, $\mathbf{F_{A2}}$ and $\mathbf{F_{A3}}$, versus time so that the next pixel on the graph corresponds to reading the next position or force. If the number of points exceeds the width of the window (in pixels) graphs are shifted by one pixel to the left, freeing up the space for the next reading. In the course of teaching positions consecutive points of the trajectory are stored in the registers of PLC, their transmission and recording in the PC's database takes place after the teaching is completed.

Robot Renus-1 enables two types of rehabilitation:

- active rehabilitation which consists of single or multiple playback of the selected trajectory by the patient; during this operation the patient moves robot's manipulator along the selected trajectory,
- passive rehabilitation which consists of single or multiple playback of the selected trajectory by robot's manipulator; during this operation the robot moves the patient's limb along the selected trajectory.

Performing both types of rehabilitation requires selecting and verifying the trajectory using the dialog box to choose the trajectory form the database.

The manipulator has a distinguished position called the base position. This is the location the readings of the positions $\mathbf{A_1}$, $\mathbf{A_2}$, $\mathbf{A_3}$ of three axis of robot's motors are 0. Base position of the manipulator is independent from the selected trajectory. Performing rehabilitation always starts from the base position, and after the rehabilitation manipulator is automatically moved to this position. Both active and passive rehabilitation can be carried out in three playback trajectory modes:

- **Single playback of the selected trajectory:** The starting point is the base position, from which the patient moves manipulator (active rehabilitation) or the robot moves the patient's limb (passive rehabilitation) through the following trajectory points to the last defined point. After reaching this point, manipulator is automatically moved to the base position.

- **Multiple playback of the selected trajectory with back to the base position:** The starting point is the base position, from which the patient moves manipulator (active rehabilitation) or the robot moves the patient's limb (passive rehabilitation) through the following trajectory points to the last defined point. When this point is reached, the manipulator is automatically moved to the base position and the cycle is repeated.
- **Multiple playback of the selected trajectory without back to the base position:** The starting point is the base position, from which the patient moves manipulator (active rehabilitation) or the robot moves the patient's limb (passive rehabilitation) through the following trajectory points to the last defined point. After reaching this point, the manipulator does not move automatically to the base position, moving to the first point of the trajectory instead. Then the cycle is repeated.

Playback mode selection can be done either before the start of the exercise or while it is being implemented. In each of the three described modes, the operator can manually stop the execution of the exercise.

6 Safety Solutions

Assurance of safety of the patient, who use the robotic device for rehabilitation exercises, is the crucial requirement, which fulfillment is demanded from designers. In the Renus-1 robot some chosen solutions were implemented.

The special enabling switch, in the form of button, was introduced. It is connected directly to the PLC, using the cable. During execution of the rehabilitation exercise (playing the trajectory), no movement of the manipulator is possible if the switch is not pressed. The switch can be activated by patient or by therapist, and must be pressed throughout the exercise.

On the bottom wall of the control cabinet the Emergency Stop button is installed. Pressing this button stops all drives of the robot immediately.

A particularly important component of the RENUS-1 rehabilitation system is the multi-axis force and torque sensor mounted at the end of the manipulator's arm on one side and to the patient's limb on the other side. This sensor measures the forces and torques between the manipulator's arm and the limb of the patient. It is used in a feedback loop to control the movements of the manipulator's arm during the exercises. Especially it can issue a warning if the resistance of the patient is too large. This could mean blocking the patient's arm or attempt to move beyond scope of its motion. In both cases continuing the exercise could be dangerous for the patient. Then the move of the manipulator is stopped immediately.

If any of the safety devices is activated appropriate message appears on the PC screen. Continuation of the exercise is possible only when the cause of emergency is removed and the therapist confirms it.

7 Summary

The presented HRI system was developed for the first robot model Renus-1. Its primary purpose was to enable the testing of the model, primarily functional and then in terms of therapeutic usefulness. This order of priorities is reflected in the solutions used. The proposed HRI is not very attractive visually and not easy to use. Its usage requires some practice. On the other hand, applied imaging of waveforms may facilitate the assessment of the quality of the exercise performed. It seems, however, that this method of presentation should be available only for the therapist. The patient should be provided with simpler tools that graphically show specific exercise course and the rehabilitation progress. Promising in this area are modern solutions in the field of virtual reality (VR) or brain-computer interface (BCI) [8].

Acknowledgements. The paper is a result of the project: RoboReha – Robotics in Rehabilitation, LdV – TOI no. 13310 0530. This project is partially funded by the European Commission. This paper reflects the authors' opinion only. Neither the European Commission nor the National Agency takes any responsibility for any information contained herein.

References

1. Siciliano, B., Khatib, O. (eds.): Springer Handbook of Robotics. Springer-Verlag, Heidelberg (2008)
2. Engelberger, J.F.: Robotics in Service. MIT Press, Cambridge (1989)
3. Dunaj, J., Klimasara, W.J., Pilat, Z., Rycerski, W.: Human-robot communication in rehabilitation devices. JAMRIS J. Autom. Mob. Rob. Intell. Syst. **9**(2), 9–19 (2015). ISSN 1897-8649
4. Maciejasz, P., et al.: A survey on robotic devices for upper limb rehabilitation. J. Neuroeng. Rehabil. **11**(3), 29 (2014). doi:10.1186/1743-0003-11-3
5. Goodrich, M.A., Schultz, A.C.: Human–robot interaction: a survey. Found. Trends Hum. Comput. Interact. **1**(3), 203–275 (2007). doi:10.1561/1100000005
6. Koukolová, L.: Overview of the robotic rehabilitation systems for upper limb rehabilitation. Transfer inovácií **30**(2014), 146–149 (2014)
7. Pilat, Z., Klimasara, W.J., Juszyński, Ł., Michnik, A.: Research and development of rehabilitation robotics in Poland. Appl. Mech. Mater. **613**, 196–207 (2014). Trans Tech Publications, Switzerland, ROBTEP 2014
8. Gomez-Rodriguez, M., Grosse-Wentrup, M., Hill, J., Gharabaghi, A., Schölkopf, B., Peters, J.: Towards brain–robot interfaces in stroke rehabilitation. In: Proceedings of IEEE International Conference on Rehabilitation Robotics, pp. 1–6 (2011)
9. The International Conference on Social Robotics. http://www.icsr2013.org.uk/
10. The International Conference on Human-Robot Interaction. http://humanrobotinteraction.org/2013/
11. A Research Portal for the HRI Community. http://humanrobotinteraction.org/

Soft Flexible Gripper Design, Characterization and Application

Jan Fraś(✉), Mateusz Maciaś, Filip Czubaczyński, Paweł Sałek, and Jakub Główka

Industrial Research Institute for Automation and Measurements, al. Jerozolimskie 202, 02-486 Warsaw, Poland
{jfras,mmacias,fczubaczynski,psalek,jglowka}@piap.pl
http://www.piap.pl

Abstract. This paper presents a design concept of soft flexible gripper dedicated for delicate objects manipulation. Traditional grippers are composed of rigid components and consist of finite number of discrete joints. The more joints they have the better they can adapt for the specific object. However, manipulating with fragile objects still requires precise control and some kind of measurement as well. In this paper the Authors propose soft flexible gripper that is able to adapt for the manipulated object shape without any additional computational effort nor any sensors. The gripper is made only of flexible materials such as rubber and silicone. Since the gripper lacks of any discrete joints and is actuated through smooth deformation of its body, it can take very complex shapes and thus easily adapt to the surface of grasped object. The mechanism is based on soft pneumatic actuators developed by the authors and its configuration can be easily redesigned in order to extend its application for special purposes. In this paper the design and experimental characterization of the gripper prototypes is presented. Possible applications are discussed.

Keywords: Robotics · Soft manipulator · Gripper · Grasping · Artificial muscle

1 Introduction

One of the most fundamental challenges in robotics is object grasping. Conventional grippers and graspers are expensive and sometimes unsuitable for tasks where the manipulated object weight, its shape or the environment are uncertain. Moreover, manipulating delicate objects with a traditional rigid gripper requires sophisticated sensing and high precision that might not always be easily provided. Rigid grippers designed to handle fragile objects are often complex or can handle only specified type of items. One of the possible solutions is replacement of the gripper by a soft device that can passively adapt to the object and to the environment. For that reason, soft robotics has been extensively investigated by

R. Szewczyk and M. Kaliczyńska (eds.), *Recent Advances in Systems, Control and Information Technology*, Advances in Intelligent Systems and Computing 543, DOI 10.1007/978-3-319-48923-0_40

scientists, and many different kinds of soft mechanical structures has been proposed [1–3]. Soft robots offer high flexibility, adapt to the external conditions and interact more safely with human than any rigid machine. In this paper we present a soft flexible gripper that is able to manipulate with fragile objects and adapt to unstructured shapes. In the following sections design, manufacturing process and the gripper capabilities are presented. Then, possible applications are discussed. In general two issues are discussed when characterizing robotic grippers: ability to fixture object (often called grasping), and task of manipulating object with fingers [4]. In this paper we will focus on grasping only.

2 Design of the Gripper

2.1 Finger Construction

A crucial element of the gripper are fingers. The fingers are derived from artificial muscles technology developed first in late 1950 [5,6], principle of their operation is, however, significantly different from the initial muscles technology. An actuator based on the similar principle our fingers are, was first described in 1991 and called flexible micro-actuator [2]. The most important difference between artificial muscle and the actuator described in this paper is that the force exerted by actuator and the one generated by muscle have opposite directions. In our design each finger is a cylinder made of silicone with a pressure actuated chamber aligned with its central axis, Fig. 1. Two kinds of silicone are used - EcoFlex 0050 for softer and SmoothSil 950 for stiffer parts. The actuation chamber (c) is made of the softer material and reinforced with tight helix (e) (made of polyester thread) in order to limit radial expansion of the finger [7]. The inner fingers' side (f) is made of the stiffer silicone and formed in a bellow shape (g) in order to increase grasping capabilities. For the outer side of the finger (a) the softer silicone is used, so that the pressure application results in bending (j) as the inner part of the finger elongates less than the outer one. The tip (d) and the bottom (h) of the finger are sealed with the stiffer silicone. An actuation fluid is provided via small rubber pipes connected to the actuation chamber at the base of each finger.

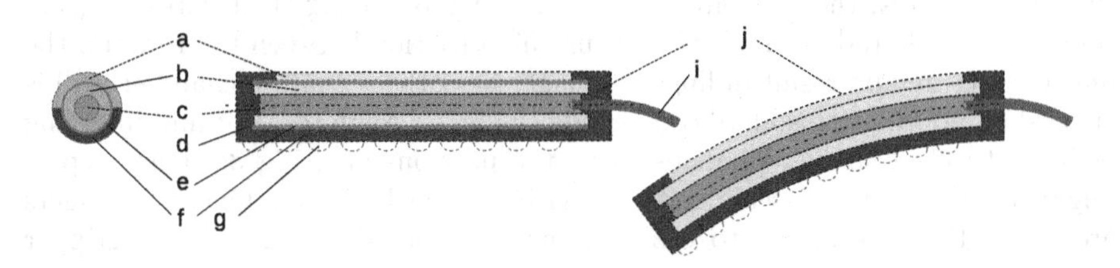

Fig. 1. Single finger design. a - outer silicone layer, b - actuation chamber wall, c - actuation chamber, d - finger's tip, e - reinforcement, f - inner layer, g - bellow shaped surface, h - finger's base, i - pressure cable, j - actuated finger. Light and dark grey colour corresponds with soft and stiff silicone, respectively.

2.2 Gripper Construction

The gripper consists of a number of fingers. In particular, every single finger can be considered as a gripper as it can coil around the object and grasp it. The fingers can be arranged symmetrically or asymmetrically, actuated in groups or working independently each other. Fingers are bound to the gripper body with the same silicon type that is used for the inner finger side. In this paper we present few patterns of the fingers arrangements, however, due to simple design the setup can be very easily rearranged and adapted for a specific application. Exemplary arrangements are presented in Fig. 2.

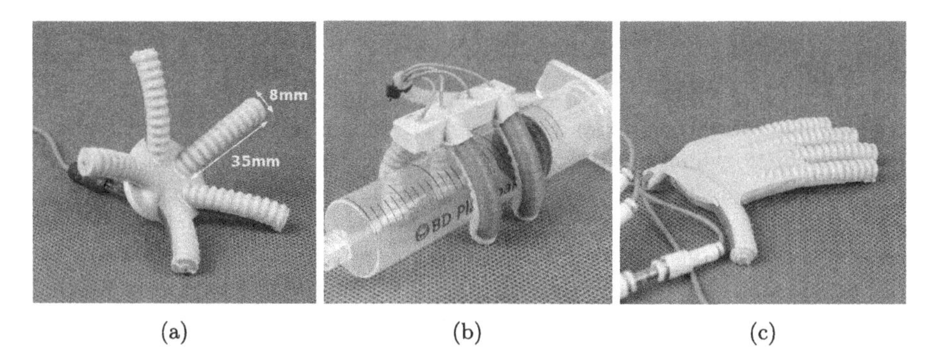

(a) (b) (c)

Fig. 2. Example grippers with different fingers arrangements: (a) with axial and (b) linear symmetry, (c) human hand inspired.

3 Manufacturing

The manufacturing process consists of several steps. The fingers are prepared first. For each finger special cylindrical chamber rod with removable core is prepared. Next, the rods are tightly braided using a low-diameter polyester thread, Fig. 3a. Then, the chamber reinforcement prepared in this way is covered with silicone layer separately - the soft outer part and the inner harder one. After the silicone cures, the rods are removed by first removing their cores, Fig. 3b. Removing whole rod at once would cause high friction between the rod and the finger structure and result in finger damage. Next the actuation chamber wall is created by pouring the soft silicone inside the finger reinforcement and inserting dedicated rod of smaller diameter than the previous one, Fig. 3c. Last step of finger manufacturing is sealing its tip with the hard silicone. Once the fingers are ready, they are bound together in the final mould that forms the gripper body, Fig. 3d.

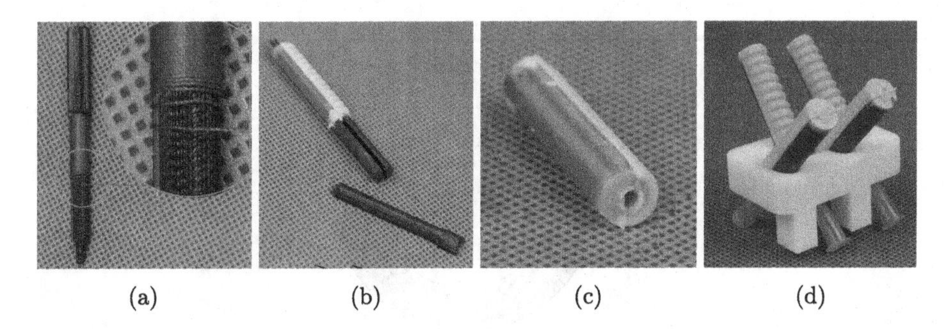

(a) (b) (c) (d)

Fig. 3. Manufacturing steps: (a) chamber reinforcement preparation, (b) rod removal (core goes first), (c) inner chamber wall moulding, (d) final moulding.

4 Characterization

The described gripper has been tested in terms of geometry and generated forces, partially covering possible characterization [8,9]. The force generated by the single finger depending on its length and the overall gripper capabilities were examined. The bending angle of the finger as a function of pressure was determined and compared with proposed mathematical model.

4.1 Single Finger Bending

Mathematical Description. The pressure applied into the actuation chamber results in a force acting in a cross-section of the finger. Considering the cross-section perpendicular to the finger neutral axis, the force is perpendicular to the cross-section plane. Its value is proportional to the pressure and the chamber cross-section area. As the Young modulus of the materials used in the finger construction differs at the cross-section, the bending neutral axis is shifted towards the inner finger part, Fig. 4.

Since the activation camber is aligned with the finger geometrical centre axis, the pressure applied results in the force that is not aligned with the bending neutral axis. That leads to a bending moment in the cross-section. The cross-section geometry does not change along the finger and due to the Pascal's law, force acting in all the cross-sections has the same value. Thus the bending moment along the finger is constant. The bending moment around neutral axis resulting from pressure can be expressed as (1). The P states for pressure value, d corresponds with the distance of the neutral bending axis from cross-section's geometrical centre that is a point of internal force acting, A states for the actuation chamber cross-section area and equals $\pi(r_3)^2$.

$$M = PdA \tag{1}$$

The neutral axis position can be obtained from the assumption that tensions for pure bending compensates on both sides of that axis. In such case the tension at the point (x, y) is expressed by (2),

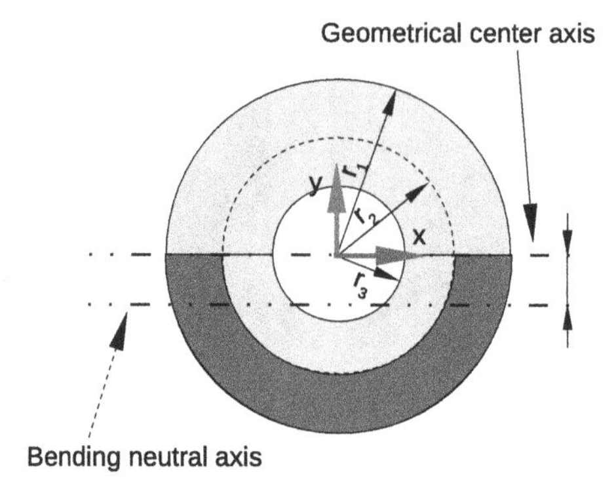

Fig. 4. Finger cross-section. Light grey and dark grey colours correspond with soft and stiff silicone, respectively.

$$\sigma(x,y) = E(x,y)\frac{x-d}{\rho} \tag{2}$$

where $E(x, y)$ states for the Young's modulus of the material used at (x, y) and ρ states for the curvature radius of the neutral plane at the cross-section. Hence, the neutral axis position d can be obtained from equilibrium condition (3) [10].

$$\iint_{xy} \sigma(x,y)dxdy = 0 \tag{3}$$

Since the ρ value is a constant the equation can be rewritten as (4).

$$\iint_{xy} E(x,y)(x-d)dxdy = 0 \tag{4}$$

Then the curvature radius ρ for a certain bending moment M can be derived from (5),

$$\rho_M = \frac{\sum^n E_n I_n}{M} \tag{5}$$

where E_n and I_n states for Young modulus and moment of inertia in respect to neutral bending axis of the cross-section n^th component, respectively. Once the position of the neutral bending axis is determined the moments of inertia can be obtained using parallel axis theorem.

The pressure applied into the actuation chamber causes not only bending, but elongation as well. The change of length can be derived from Hooke's law (6),

$$\Delta l = \frac{l_0 F}{\sum^n E_n A_n} \tag{6}$$

where l_0 represents rest finger length, E_n and A_n states for the area and Young modulus of the relevant cross-section component, respectively. F is stretching force that equals to pressure value multiplied by an activation chamber cross-section area. Hence, the overall bending angle can be expressed by the (7).

$$\alpha = \frac{l_0 + \Delta l}{\rho_M} \tag{7}$$

Thus the bending angle for a specific finger length is a quadratic function of the pressure.

Experiment. The finger bending angle as a function of pressure has been measured. Due to limited data regarding material's Young modulus and not very precise manufacturing process (3D printed moulds, manual moulding, etc.) obtaining parameter values to compare the results with theoretical model is not easy. However, the second order polynomial fit presents good approximation of the gathered data, thus the principle of the model seems to be in force. The length of an active finger part was 20 mm and the pressure range was 0–0.6 bars. The empirical data is presented in Fig. 5.

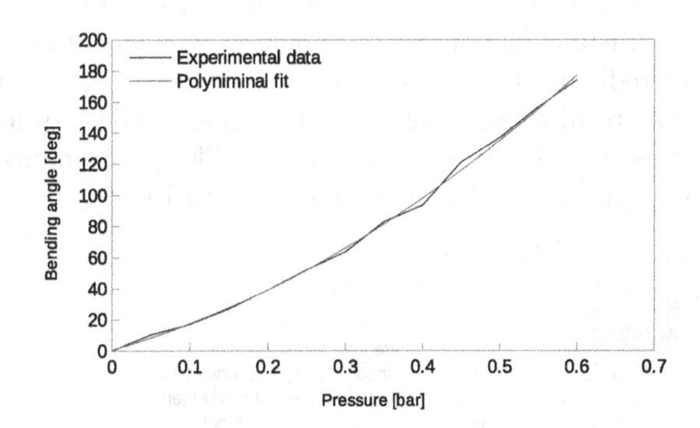

Fig. 5. Bending as a function of pressure.

4.2 Force Generated by a Single Finger

The finger capabilities in terms of generated force values has been examined. The finger was mounted horizontally and the force has been measured as a function of pressure. The point at which the force has been measured was constant in space. For the measurement precise scale was used. The weight exceeded the finger capabilities, thus its position on the scales was stable during the experiment. The measurement setup and the results of the experiment are presented in Fig. 6

4.3 Grasping Capabilities

Single finger gripper would have very limited application capabilities. In this section the performance of a grasper composed of five fingers is presented.

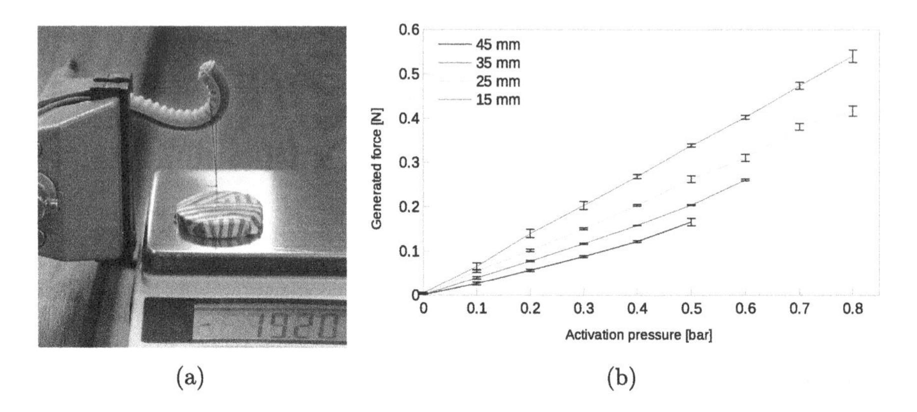

Fig. 6. Finger capabilities examination: (a) experimental setup, (b) force generated by fingers measured for different length fingers.

Two manipulated objects of different kinds were tested. The force of the grasp was obtained for each object for various actuation pressures. The experiment was performed by grasping the object of known weight with the maximal available pressure and reducing the pressure until the object was dropped. Due to high variation of the results the experiment was repeated 6 times for each weight value. The tested use case is presented in Fig. 8a. The performance for irregular soft object and for spherical rigid one is presented in Fig. 8b

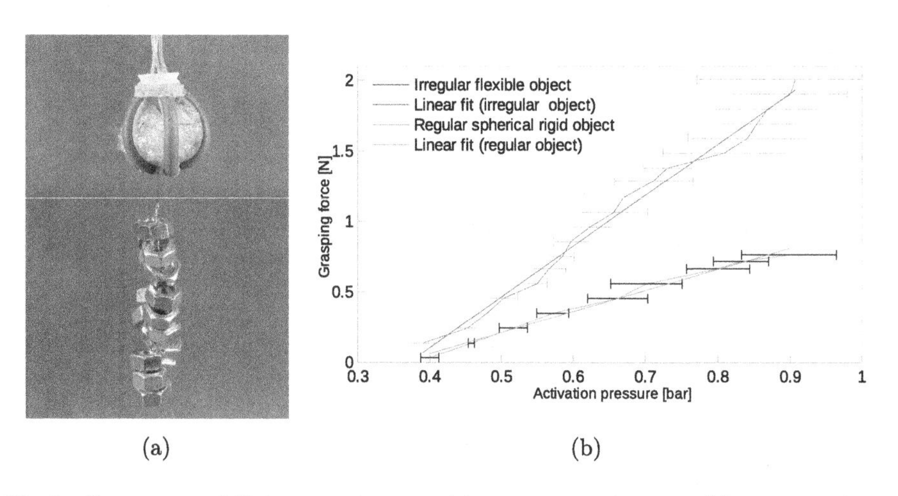

Fig. 7. Grasper capabilities examination: (a) experimental setup, (b) grasping force comparison.

5 Usage Scenario Discussion

There are many potential applications of flexible graspers in various scenarios. One of the examples is dealing with soft, flexible or delicate materials, in industries like apparel or shoe manufacturing [11]. Soft robots can enable us to make exoskeletons or wearable protetics, and provide more unobtrusive way to interface with human body [12]. Another intensively investigated application in recent years is fruit picking that require caution due to objects fragility [13,14]. Highly demanding potential usage scenario is forensic evidences collection [15]. Since the evidences may be fragile and of irregular or unexpected shapes they may be difficult to grasp using conventional gripper. Experience of PIAP's experts gained from working in this field in cooperation with crime scene investigators

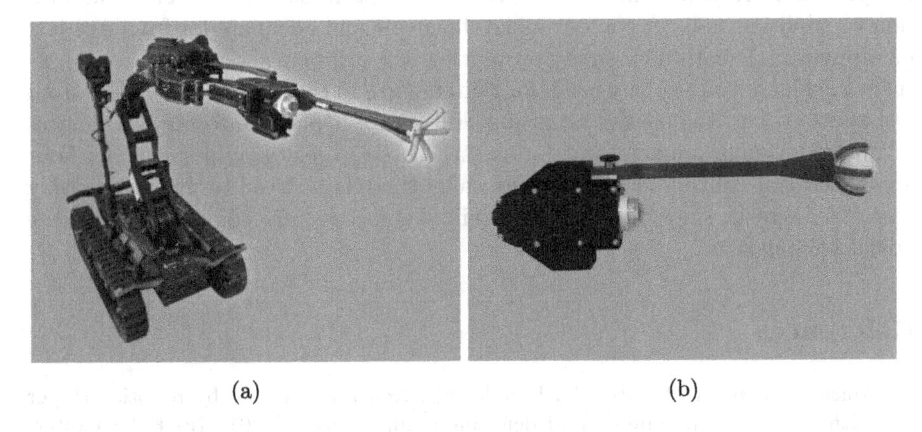

(a) (b)

Fig. 8. Mobile security robot equipped with the soft gripper: (a) the robot, (b) the device designed to actuate the soft gripper and to be carried by robot.

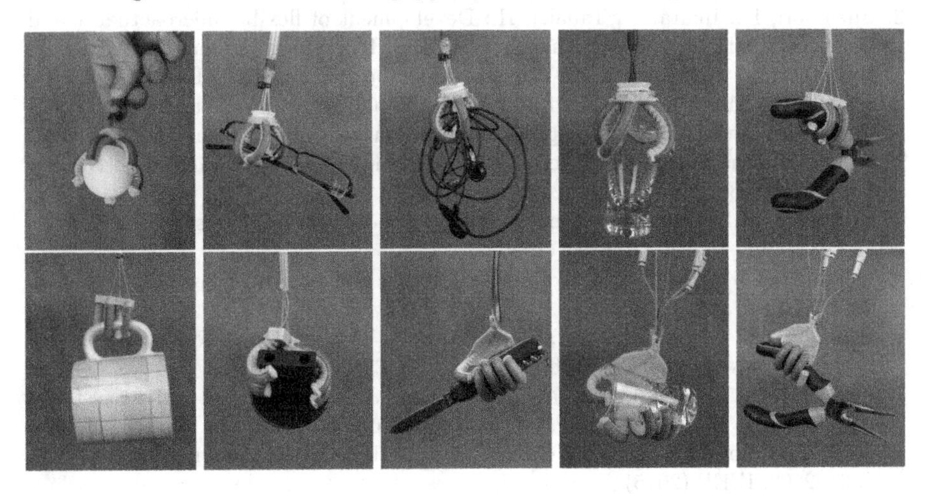

Fig. 9. Examples of grasped objects.

concludes that such a soft grippers may be very useful for remote physical crime evidence gathering. One of the PIAP's robots equipped with the soft gripper is presented in Fig. 8. Generally handling fragile objects (fruits, human body, forensic evidences), in very different situations and usages can become domain dominated by soft robots [16], especially soft grippers such as the one described in this paper.

In the Fig. 9 few example handled objects are presented.

6 Summary

Traditional grippers are very widely utilized devices. However in some cases, the usage of rigid tools is problematic due to their stiffness. Soft and flexible graspers and grippers demonstrate great potential in such cases. Their adaptation capabilities and soft contact make them safe and adequate for fragile object manipulation. Due to their safety properties, soft grippers and grasper are extensively considered as a substitution for rigid devices in the human-present environments. Moreover, the passive adapting skills make it possible to use soft devices in uncertain environments and for unstructured object manipulation without any additional control effort. In many cases there is no need to use any sensing to realise a task, since the gripper flexibility successfully compensates all the object irregularities.

References

1. Maruyama, R., Watanabe, T., Uchida, M.: Delicate grasping by robotic gripper with incompressible fluid-based deformable fingertips. In: 2013 IEEE/RSJ International Conference on Intelligent Robots and Systems, pp. 5469–5474, November 2013
2. Suzumori, K., Iikura, S., Tanaka, H.: Development of flexible microactuator and its applications to robotic mechanisms. In: 1991 IEEE International Conference on Robotics and Automation, 1991, Proceedings, pp. 1622–1627. IEEE (1991)
3. Hassan, T., Manti, M., Passetti, G., d'Elia, N., Cianchetti, M., Laschi, C.: Design and development of a bio-inspired, under-actuated soft gripper. In: 2015 37th Annual International Conference of the IEEE Engineering in Medicine and Biology Society (EMBC), pp. 3619–3622, August 2015
4. Cutkosky, M.R.: Human grasp choice and robotic grasp analysis. In: Dextrous Robot Hands, vol. 1, pp. 5–31 (1990)
5. Gaylord, R.H.: Fluid actuated motor system and stroking device, US Patent 2,844,126, 22 July 1958
6. Klute, G.K., Czerniecki, J.M., Hannaford, B.: Artificial muscles: actuators for biorobotic systems. Intl. J. Robot. Res. **21**(4), 295–309 (2002)
7. Fraś, J., Czarnowski, J., Maciaś, M., Główka, J., Cianchetti, M., Menciassi, A.: New stiff-flop module construction idea for improved actuation and sensing. In: 2015 IEEE International Conference on Robotics and Automation (ICRA), pp. 2901–2906. IEEE (2015)

8. Falco, J., Marvel, J., Messina, E.: A roadmap to progress measurement science in robot dexterity and manipulation. US Dept. Commer., Nat. Inst. Std. Technol., Gaithersburg, MD, USA. Technical report NISTIR, 7993 (2014)
9. Klute, G.K., Czerniecki, J.M., Hannaford, B.: Mckibben artificial muscles: pneumatic actuators with biomechanical intelligence. In: 1999 IEEE/ASME International Conference on Advanced Intelligent Mechatronics, 1999, Proceedings, pp. 221–226. IEEE (1999)
10. Dyląg, Z., Jakubowicz, A.S., Orłoś, Z., Naukowo-Techniczne, W.: Wytrzymałość materiałów. Wydawnictwa Naukowo-Techniczne (2013)
11. Robotics 2020 multi-annual roadmap for robotics in europe call 2 ict24 (2015) – horizon 2020. 2nd ed. Report, euRobotics, 2015
12. Wood, R.J., Walsh, C.J.: Smaller, softer, safer, smarter robots. Sci. Transl. Med. **5**(210), 210ed19 (2013)
13. Qian, S., Zhang, L., Yang, Q., Bao, G., Wang, Z., Qi, L.: Research on adaptive multi-contract grasping model of flexible pneumatic finger. In: 2009 International Conference on Mechatronics and Automation, pp. 2823–2828, August 2009
14. Bao, G., Yao, P., Cai, S., Ying, S., Yang, Q.: Flexible pneumatic end-effector for agricultural robot: design and experiment. In 2015 IEEE International Conference on Robotics and Biomimetics (ROBIO), pp. 2175–2180, December 2015
15. Kowalski, G., Maciaś, M., Wołoszczuk, A.: Recent advances in automation, robotics and measuring techniques. In: Selected Issues of Collecting Forensic Evidence with a Mobile Robot, pp. 431–440. Springer International Publishing, Cham (2014)
16. Filip, I., Mazzeo, A.D., Shepherd, R.F., Chen, X., Whitesides, G.M.: Soft robotics for chemists. Angewandte Chemie Intl. Edn. **50**(8), 1890–1895 (2011)

Dynamics Model of a Three-Wheeled Mobile Robot Taking into Account Slip of Wheels

Maciej Trojnacki(✉)

Industrial Research Institute for Automation and Measurements PIAP, Warsaw, Poland
mtrojnacki@piap.pl

Abstract. The problem of modeling of dynamics of a three-wheeled mobile robot is analyzed in this paper. The robot has two non-steered driven wheels and a caster. Kinematic structure of the robot and its kinematics are described. Dynamics model of the robot dedicated for control applications is derived. It takes into account tire-ground contact conditions and slip of wheels. The tire-ground contact conditions are characterized by coefficients of friction and rolling resistance. The tire model in a simple form, which considers only the most important effects of tire-ground interaction, is applied. Electromechanical model of a servomotor used for driving the robot is also included.

Keywords: Wheeled mobile robot · Wheel slip · Dynamics model · Tire model · Drive unit model

1 Introduction

Ground mobile robots are vehicles whose task is realization of desired motion in indoor and outdoor environments. Robots of this kind are more and more widely used in industrial applications, e.g. for inspection [3], in special missions, e.g. connected with neutralization of improvised explosive devices [10], and in order to help people in their everyday life. They include teleoperated robots, examples of which are products of PIAP Institute [14], and robots with various levels of autonomy of movement, whose examples are autonomous robotic vacuum cleaners and lawn mowers. Among various locomotion systems the solutions involving wheels and tracks are predominant.

Wheeled mobile robots (WMRs) are particularly attractive because of the simplicity of design of driving system, and at the same time they can achieve similar mobility to tracked vehicles. Common are also hybrid solutions, that is, combining features of wheeled and tracked locomotion systems, example of which is PIAP SCOUT robot [14]. Much less common are commercial solutions involving walking, crawling or reconfigurable robots.

WMRs are equipped with various types of wheels, which include: non-steered wheels, steered wheels, casters and mecanum (Swedish) wheels. In addition, in WMRs are used: wheels with pneumatic tires, wheels with non-pneumatic tires with various types of fillings (e.g. in form of a foam) and wheels made of metal only (e.g. in case of planetary rovers [5] and robots dedicated for removing the effects of the nuclear disaster).

R. Szewczyk and M. Kaliczyńska (eds.), *Recent Advances in Systems, Control and Information Technology*, Advances in Intelligent Systems and Computing 543, DOI 10.1007/978-3-319-48923-0_41

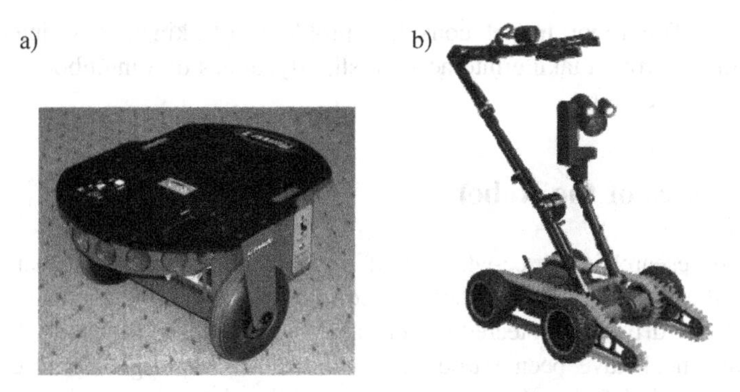

Fig. 1. Examples of wheeled mobile robots: (a) differentially driven Pioneer 2-DX [13], (b) skid-steered PIAP SCOUT [14]

Based on above mentioned types of wheels various kinematic structures of WMRs can be created. They include the following structures: differentially driven (e.g. Pioneer 2-DX [13] – see Fig. 1a), skid-steered (e.g. PIAP SCOUT [14] – see Fig. 1b), car-like [1], omnidirectional [9] and rover-type [5].

The differentially driven WMRs are typically used indoors. They are characterized by simplicity of design and simultaneously good maneuverability, in particular they can rotate in place (so-called pivot turning). This group includes also the two-wheeled robots operating on the principle of an inverted pendulum (e.g. WRUT [6]).

The choice of type of wheels and kinematic structure of the robot determines the conditions of interaction of wheels with the ground. For most of previously mentioned kinematic structures of robots excluding the skid-steered, if their movement is realized with low velocity and acceleration then slip of wheels may be negligible. Therefore, during modeling the movement of such robots one often assumes that the wheels roll without slip. This approach is reflected, among others, in the works [4, 8].

In a general case, however, movement of WMRs can also be carried out with higher velocities and accelerations, which will be associated with occurrence of slip of wheels. The problem of modeling of dynamics of differentially driven WMRs taking into account slip of wheels is rarely discussed in the literature. One of exceptions is the work [12]. Significant here is the question in which cases or conditions robot motion without slip of wheels can be assumed, and when slippage should be taken into consideration as it affects the accuracy of robot motion.

Analyzing the conditions of interaction of wheels with the ground in case of WMRs one can use the results of research in the automotive field as a first approximation (representative examples of works can be [7, 11]). However, wheels of WMRs, especially of the lightweight WMRs, usually have different geometry and properties in comparison to car wheels. Robotic wheels often have non-pneumatic tires, which demands realization of dedicated investigations. As a result, for this type of vehicles different requirements on tire models should be taken into consideration, as highlighted in the work [2].

Therefore, the objective of this paper is the development of dynamics model of a three-wheeled differentially driven robot, which incorporates the slip of wheels phenomenon. The robot model considers problems of: kinematics, interaction of wheels with the ground taking into account slip, dynamics of a multibody system and properties of drive units.

2 Kinematics of the Robot

The object of research is a three-wheeled differentially driven robot, kinematic structure of which is illustrated in Fig. 2a. The robot consists of the following main components: 0 – body, 1–2 – driven non-steered wheels, and 3 – caster. The following designations for the i-th wheel have been introduced in the model: A_i – geometric center, r_i – geometric (unloaded) radius, θ_i – spin angle. In addition, in case of the caster the turning angle ψ_3 is introduced. Distance between the front wheels is W ($A_1R = A_2R = W/2$) whilst distance form turning axis of a caster to the point R of the robot body is L. In the robot model, shift of caster spin axis with respect to the axis of its turning is neglected.

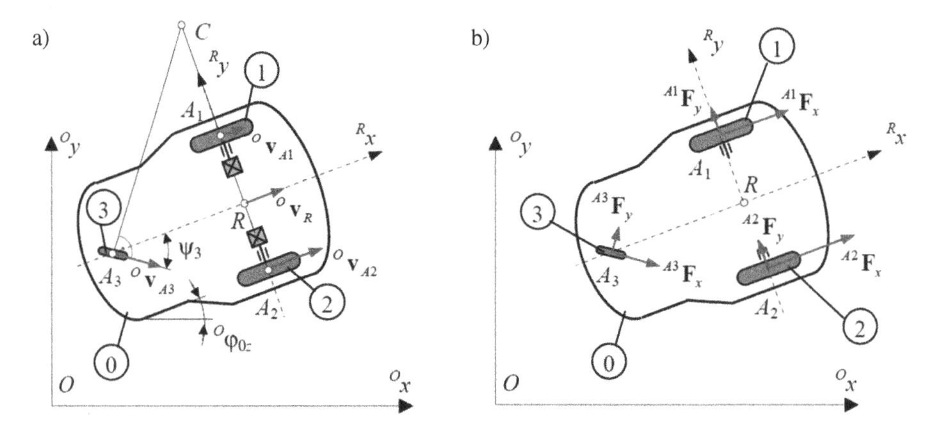

Fig. 2. Three-wheeled differentially driven mobile robot: (a) kinematic structure, (b) diagram of reaction forces acting on the robot in the wheel-ground contact plane

It is assumed that robot motion is realized in ${}^Ox{}^Oy$ plane of the fixed coordinate system $\{O\}$. The moving coordinate system, considered as rigidly connected to the robot body, is denoted $\{R\}$. Position and orientation of the mobile platform are described by the vector of generalized coordinates:

$$ {}^O\mathbf{q} = \left[{}^Ox_R,\ {}^Oy_R,\ {}^O\varphi_{0z}\right]^T, \tag{1}$$

where: ${}^Ox_R$, ${}^Oy_R$ are coordinates of the point R belonging to the mobile platform, and ${}^O\varphi_{0z}$ denotes angle of rotation of mobile platform about ${}^Oz$ axis with respect to the fixed coordinate system $\{O\}$.

In case of plane motion of the mobile platform, point R velocity vector depends on angular velocity ${}^R\dot{\varphi}_{0z}$ and radius of curvature R_z of the path according to the formula:

$$ {}^{R}v_{R} = {}^{R}\dot{\varphi}_{0z}R_{z}. \tag{2} $$

For tracking control of robot motion, it is necessary to solve the inverse kinematics problem, that is, to determine desired angular velocities of spin of wheels from desired motion of the robot body. One of possible approaches to tracking control is to make assumption that robot motion is given in the form of vector of generalized velocities ${}^{R}\mathbf{v}_{d} = \left[{}^{R}v_{Rd}, {}^{R}\dot{\varphi}_{0zd}\right]^{T}$, which corresponds to desired trajectory of robot motion in the form of vector of desired generalized coordinates ${}^{O}\mathbf{q}_{d} = \left[{}^{O}x_{Rd}, {}^{O}y_{Rd}, {}^{O}\varphi_{0zd}\right]^{T}$.

If during robot movement slip of wheels does not occur, the inverse kinematics problem can be solved on the basis of the relationship:

$$ \dot{\theta}_{d} = \begin{bmatrix} \dot{\theta}_{1d} \\ \dot{\theta}_{2d} \end{bmatrix} = \frac{1}{r}\begin{bmatrix} 1 & -W/2 \\ 1 & W/2 \end{bmatrix} {}^{R}\mathbf{v}_{d}. \tag{3} $$

In addition, it is assumed that the angular velocity of spin of the caster and its turning angle can be determined based on relationships:

$$ \dot{\theta}_{3} = \sqrt{({}^{R}v_{Rx})^{2} + (L^{R}\dot{\varphi}_{0z})^{2}/r_{3}}, \tag{4} $$

$$ \mathrm{tg}(\Psi_{3}) = -L^{R}\dot{\varphi}_{0z}/{}^{R}v_{Rx} \quad => \quad \Psi_{3} = \mathrm{atan2}(-L^{R}\dot{\varphi}_{0z}, {}^{R}v_{Rx}). \tag{5} $$

3 Dynamics of the Robot

The goal of this work is to develop a simplified dynamics model of the robot, taking into account slip of wheels.

It is assumed that robot is under action of the following external forces:

- ground reaction forces ${}^{R}\mathbf{F}_{Ai} = \left[{}^{R}F_{Aix}, {}^{R}F_{Aiy}, {}^{R}F_{Aiz}\right]^{T}$ acting on each wheel,
- gravity force ${}^{R}\mathbf{G} = m_{R}\,{}^{R}\mathbf{g}$, where m_{R} denotes total mass of the robot.

Components ${}^{R}\mathbf{F}_{Aix}$ and ${}^{R}\mathbf{F}_{Aiy}$ ($i = \{1, 2, 3\}$) of ground reaction forces acting in the plane of wheel-ground contact are shown in Fig. 2b. Gravity force vector ${}^{R}\mathbf{G}$ is a function of gravitational acceleration vector ${}^{R}\mathbf{g} = \left[{}^{R}g_{x}, {}^{R}g_{y}, {}^{R}g_{z}\right]^{T}$ and it is applied at the robot mass center, whose position is described by the vector ${}^{R}\mathbf{r}_{CM} = [{}^{R}x_{CM}, {}^{R}y_{CM}, {}^{R}z_{CM}]^{T}$

Under action of the mentioned forces, according to the Newton's 2nd law, robot body moves with linear acceleration ${}^{R}\mathbf{a}_{CM} = \left[{}^{R}a_{CMx}, {}^{R}a_{CMy}, {}^{R}a_{CMz}\right]^{T}$.

The following dynamic equations of motion of the robot mass center are valid:

$$ m_{R}{}^{R}a_{CMw} = \sum_{i=1}^{3} {}^{R}F_{Aiw} + m_{R}{}^{R}g_{w}, \tag{6} $$

where $w = \{x, y, z\}$.

As far as caster is concerned one can calculate projections of a vector of ground reaction force ${}^{A3}\mathbf{F} = [{}^{A3}F_x, {}^{A3}F_y, {}^{A3}F_z]^T$ on the axes of robot coordinate system using formulas:

$$ {}^{R}F_{A3x} = {}^{A3}F_x \cos\Psi_3 - {}^{A3}F_y \sin\Psi_3,\ {}^{R}F_{A3y} = {}^{A3}F_x \sin\Psi_3 + {}^{A3}F_y \cos\Psi_3,\ {}^{R}F_{A3z} = {}^{A3}F_z. \tag{7} $$

On the assumption that the robot does not rotate about ${}^{R}x$ and ${}^{R}y$ axes, ${}^{R}y_{CM} = 0$, it is possible to write the following two equations of equilibrium of moments of force about ${}^{R'}x$ and ${}^{R'}y$ axes, which are situated on the wheel-ground contact plane:

$$ \sum M_{R'x} = m_R {}^{R}a_{CMy}(r + {}^{R}z_{CM}) + ({}^{R}F_{A1z} - {}^{R}F_{A2z})W/2 = 0, \tag{8} $$

$$ \sum M_{R'y} = -m_R {}^{R}a_{CMx}(r + {}^{R}z_{CM}) + m_R g {}^{R}x_{CM} + {}^{R}F_{A3z}L = 0. \tag{9} $$

On the assumption that the ground surface is horizontal and even as well as all wheels have contact with this surface, the gravitational acceleration vector expressed in the robot coordinate system $\{R\}$ is equal to: ${}^{R}\mathbf{g} = [0, 0, -g]^T$, where $g = 9.81$ (m/s^2).

In this case, based on previous equations it is possible to determine normal components of the ground reactions forces (on the assumption that accelerations of the robot mass center are known ${}^{R}a_{CMx}$, ${}^{R}a_{CMy}$, and ${}^{R}a_{CMz} = 0$) in the form:

$$ {}^{R}F_{A1z} = m_R\Big((-{}^{R}a_{CMx}/(2L) - {}^{R}a_{CMy}/W)(r + {}^{R}z_{CM}) + g(1/2 + {}^{R}x_{CM}/(2L))\Big), \tag{10} $$

$$ {}^{R}F_{A2z} = m_R\Big((-{}^{R}a_{CMx}/(2L) + {}^{R}a_{CMy}/W)(r + {}^{R}z_{CM}) + g(1/2 + {}^{R}x_{CM}/(2L))\Big), \tag{11} $$

$$ {}^{R}F_{A3z} = m_R\big({}^{R}a_{CMx}(r + {}^{R}z_{CM})/L - g{}^{R}x_{CM}/L\big). \tag{12} $$

For calculation of values of normal components of the ground reaction forces, knowledge of motion parameters of the robot mass center *CM* and of geometric centers of wheels, that is, characteristic points A_i, is necessary. In the calculations, values of those parameters from the previous calculation step, that is, from time instant $t - \Delta t$ are used, where Δt is the adopted time step. In the first calculation step, known initial conditions are taken into account.

The longitudinal slip ratio for the i-th wheel is determined from the formula:

$$ \lambda_i = \begin{cases} 0 \quad \text{for} \quad {}^{Ai}v_x = 0 \text{ and } v_{oi} = 0, \\ (v_{oi} - {}^{Ai}v_x)/\max({}^{Ai}v_x,\ v_{oi}) \quad \text{for other} \quad {}^{Ai}v_x \text{ and } v_{oi}, \end{cases} \tag{13} $$

where $v_{oi} = \dot{\theta}_i r_i$ and ${}^{Ai}v_x$ are respectively velocity at the wheel circumference and longitudinal component of velocity of wheel geometric center.

The tire adhesion coefficient on longitudinal direction for this wheel is calculated using Kiencke tire model [9] modified by the author, that is, from the formula:

$$\mu_{ix} = \begin{cases} \frac{2\mu_p \lambda_p \lambda_i}{\lambda_p^2 + \lambda_i^2} & \text{for} \quad |\lambda_i| \leq \lambda_p, \\ a_{\lambda x} \lambda_i + b_{\lambda x} \mathrm{sgn}(\lambda_i) & \text{for} \quad |\lambda_i| > \lambda_p, \end{cases} \tag{14}$$

where λ_p denotes the value of longitudinal slip corresponding to the value of maximum tire adhesion coefficient μ_p.

Modification of the original Kiencke dependency is connected with calculation of the tire adhesion coefficient for $|\lambda_i| > \lambda_p$. In this case the $\mu_{ix}(\lambda_i)$ characteristics is approximated by straight lines, so that, for longitudinal slip equal to $\pm 100\%$, the adhesion coefficient attains value corresponding to the sliding tire adhesion coefficient μ_k.

Coefficients $a_{\lambda x}$ and $b_{\lambda x}$ in (15) are described by formulas:

$$a_{\lambda x} = \frac{\mu_p - \mu_k}{\lambda_p - \lambda_{max}}, b_{\lambda x} = \mu_p - a_x \lambda_p, \tag{15}$$

where μ_k denotes sliding tire adhesion coefficient (coefficient of kinetic friction), and λ_{max} = 100% is a maximum value of the longitudinal slip analyzed in the present work.

Longitudinal component of ground reaction force for the i-th wheel in the wheel coordinate system $\{A_i\}$ depends on current value of the adhesion coefficient on longitudinal direction, according to relationship:

$$^{Ai}F_x = \mu_{ix} {}^{Ai}F_z, \tag{16}$$

where $^{Ai}F_z$ = $^{R}F_{Aiz}$ because values of the normal components of the ground reaction forces do not depend on angle of wheel turning.

In turn, lateral slip angle for i-th wheel is determined from the formula:

$$\alpha_i = \begin{cases} 0 & \text{for} \quad {}^{Ai}v_y = 0, \\ \arctan2({}^{Ai}v_y, |{}^{Ai}v_x|) & \text{for} \quad |{}^{Ai}v_y| > 0, \end{cases} \tag{17}$$

where $^{Ai}v_x$ and $^{Ai}v_y$ are respectively longitudinal and transversal velocity of the geometric center of the i-th wheel.

Knowing the lateral slip angle it is possible to calculate adhesion coefficient on lateral direction for i-th wheel. To this end the following approximate (as compared to H.B. Pacejka model [7]) relationship is introduced:

$$\mu_{iy} = -\mu_{ymax} \tanh(\beta \alpha_i), \tag{18}$$

where μ_{ymax} denotes maximum value of adhesion coefficient on lateral direction (it is assumed that $\mu_p \geq \mu_{ymax} \geq \mu_k$), and β is certain positive coefficient.

Hence, lateral component of the ground reaction force for i-th wheel is calculated from the formula:

$$^{Ai}F_y = \mu_{iy} {}^{Ai}F_z. \tag{19}$$

For known values of longitudinal and lateral components of the ground reaction forces for particular wheels, it is possible to calculate components of acceleration of point R of robot body on longitudinal and lateral directions using equations:

$$^{R}a_{CMx} = \sum_{i=1}^{3} {}^{R}F_{Aix}/m_R, \; {}^{R}a_{CMy} = \sum_{i=1}^{3} {}^{R}F_{Aiy}/m_R. \tag{20}$$

For the robot mobile platform, it is also possible to write the following dynamic equation of motion resulting from its rotation about ^{R}z axis with angular acceleration $^{R}\ddot{\varphi}_{0z}$:

$$I_{Rz}{}^{R}\ddot{\varphi}_{0z} = \sum_{i=1}^{3} {}^{R}T_{Aiz} + ({}^{R}F_{A2x} - {}^{R}F_{A1x})(W/2) - ({}^{R}F_{A1y} + {}^{R}F_{A2y}){}^{R}x_{CM} - {}^{R}F_{A3y}(L + {}^{R}x_{CM}), \tag{21}$$

where: I_{Rz} – mass moment of inertia of the robot about the ^{R}z axis, $^{R}T_{Aiz}$ – the so-called self-aligning torque associated with rotation of the i-th wheel about ^{R}z axis, resulting from friction forces acting in the tire-ground contact.

After making assumption that $^{R}T_{Aiz} \approx 0$, that is, the self-aligning torque is negligibly small in comparison to the remaining moments of force acting about the ^{R}z axis, it is possible to determine value of the angular acceleration associated with robot rotation about ^{R}z axis in the form:

$$^{R}\ddot{\varphi}_{0z} = (({}^{R}F_{A2x} - {}^{R}F_{A1x})(W/2) - ({}^{R}F_{A1y} + {}^{R}F_{A2y}){}^{R}x_{CM} - {}^{R}F_{A3y}(L + {}^{R}x_{CM}))/I_{Rz}, \tag{22}$$

and then calculate, by integration, the value of angular velocity of the robot mobile platform about this axis, that is $^{R}\dot{\varphi}_{0z}$.

For each of the robot wheels, after neglecting the friction in kinematic pairs, it is possible to write dynamic equation of motion associated with wheel spin:

$$I_{iy}\,\ddot{\theta}_i = \tau_i - {}^{R}F_{Aix}r - {}^{R}F_{Aiz}rf_r\mathrm{sgn}(\dot{\theta}_i), \; I_{3y}\,\ddot{\theta}_3 = -{}^{A3}F_x\,r_3 - {}^{A3}F_z\,r_3f_r\mathrm{sgn}(\dot{\theta}_3), \tag{23}$$

where: $i = \{1, 2\}$, I_{iy}, I_{3y} – mass moments of inertia of the driven wheel and caster about their spin axes, τ_i – driving torque, f_r – coefficient of rolling resistance, $\dot{\theta}_i$, $\dot{\theta}_3$ and $\ddot{\theta}_i$, $\ddot{\theta}_3$ – angular velocity and acceleration of spin of those wheels, respectively.

After taking into account the above relationships, it is possible to determine driving torques (in case of solving the inverse dynamics problem) or angular accelerations (for forward dynamics problem) from the following equations:

$$\tau_i = I_{Wy}\,\ddot{\theta}_i + {}^{R}F_{Aix}r + {}^{R}F_{Aiz}rf_r\mathrm{sgn}(\dot{\theta}_i), \tag{24}$$

$$\ddot{\theta}_i = \left(\tau_i - {}^{R}F_{Aix}r - {}^{R}F_{Aiz}rf_r\mathrm{sgn}(\dot{\theta}_i)\right)/I_{iy}, \; \ddot{\theta}_3 = \left(-{}^{A3}F_x\,r_3 - {}^{A3}F_z\,r_3f_r\mathrm{sgn}(\dot{\theta}_3)\right)/I_{3y}. \tag{25}$$

The described simplified model of robot dynamics can be also enhanced with the model of its drive units. For this purpose it is assumed that:

- each drive unit consists of identical DC motor, encoder, and transmission system,
- drive units are not self-locking, that is, they can freely turn under the influence of external moments of force,
- mass moments of inertia of the rotating elements of the DC motor, encoder and gear unit are small in comparison to mass moments of inertia of the wheels, that is why they are neglected.

The DC motor model of the *i*-th drive unit is described by the following dependences:

$$\frac{di_i}{dt} = \left(u_i - k_e n_d \dot{\theta}_i - R_d i_i\right) / L_d, \ \tau_i = \eta_d n_d k_m i_i, \tag{26}$$

where: u_i – motor input voltage, i_i – rotor current, L_d, R_d – respectively inductance and resistance of the rotor, k_e – electromotive force constant, k_m – motor torque coefficient, n_d, η_d – gear ratio and efficiency factor of the transmission system, respectively.

Above described model of drive units of the robot can be used in case of solving the forward dynamics problem.

After transforming the above equations to the form:

$$i_i = \tau_i / (\eta_d n_d k_m), \ u_i = k_e n_d \dot{\theta}_i + L_d \frac{di_i}{dt} + R_d i_i, \tag{27}$$

one can also determine current flowing in the armature winding and the value of motor input voltage necessary for realization of motion of the wheel with desired angular velocity and with desired driving torque exerted by the wheel.

This form of model of drive units can be used when one solves inverse dynamics problem.

4 Conclusions and Future Works

In the present work the dynamics model of a three-wheeled mobile robot was described. Developed model can be used in computer simulations of robot motion with high velocities and accelerations. In this case the slip of wheels will affect robot motion, therefore it should be taken into account during synthesis of control system. For this reason, robot motion in these circumstances cannot be realized only on the basis of the wheels control. It is necessary to introduce the mobile platform velocity controller (based on measurement of linear and angular velocities of the robot) and/or robot position and course controller (using e.g. GPS for position measurement).

Future works will include application of the robot dynamics model in synthesis and investigations of control systems, including those based on robot dynamics model.

References

1. Baturone, I., Gersnoviez, A.A.: A simple neuro-fuzzy controller for car-like robot navigation avoiding obstacles. In: IEEE International Fuzzy Systems Conference (2007)
2. Dąbek, P., Trojnacki, M.: Requirements for tire models of the lightweight wheeled mobile robots. In: Awrejcewicz, J., Kaliński, Krzysztof, J., Szewczyk, R., Kaliczyńska, M. (eds.) Mechatronics: Ideas, Challenges, Solutions and Applications. AISC, vol. 414, pp. 33–51. Springer, Heidelberg (2016). doi:10.1007/978-3-319-26886-6_3
3. Giergiel, M., Kurc, K., Małka, P., Buratowski, T., Szybicki, D.: Dynamics of underwater inspection robot. Pomiary Automatyka Robotyka **17**, 76–79 (2013)
4. Hendzel, Z.: An adaptive critic neural network for motion control of a wheeled mobile robot. Nonlinear Dyn. **50**, 849–855 (2007). Springer
5. Iagnemma, K., Dubowsky, S.: Mobile Robots in Rough Terrain. Estimation, Motion Planning, and Control with Application to Planetary Rovers, Springer Tracts in Advanced Robotics, Vol. 12 (2004)
6. Kędzierski, J., Tchoń, K.: Feedback control of a balancing robot. In: Proceedings of 14th IFAC International Conference on Methods and Models in Automation and Robotics (2009)
7. Pacejka, H.B.: Tire and Vehicle Dynamics, 2nd edn. SAE International and Elsevier (2005)
8. Padhy, P.K., et al.: Modeling and position control of mobile robot. In: The 11th IEEE International Workshop on Advanced Motion Control, Nagaoka, Japan (2010)
9. Ping, L.Y.: Slip Modelling, estimation and control of omnidirectional wheeled mobile robots with powered caster wheels. Doctorial Thesis, National University of Singapore, Singapore (2009)
10. Szynkarczyk, P., Czupryniak, R., Trojnacki, M., Andrzejuk, A.: Current state and development tendency in mobile robots for special applications. In: Proceedings of the International Conference WEISIC, Vol. 8, pp. 30–41 (2008)
11. Wong, J.Y.: Theory of Ground Vehicles, 3rd edn. Wiley-Interscience (2001)
12. Zadarnowska, K., Oleksy, A.: Motion planning of wheeled mobile robots subject to slipping. J. Autom. Mobile Robot. Intell. Syst. **5**, 49–58 (2011)
13. ActivMEDIA_ROBOTICS, Pioneer 2 Mobile Robot – Operations Manual, ActivMEDIA ROBOTICS Peterborough, USA (2000)
14. Mobile robots for counter-terrorism (PIAP). http://www.antiterrorism.eu

Development of the Human-Robot Communication in Welding Technology Manufacturing Cells

Zbigniew Pilat[1(✉)], Jozef Varga[2], and Mikulas Hajduk[2]

[1] Industrial Research Institute for Automation and Measurements PIAP, Warsaw, Poland
zpilat@piap.pl
[2] Technical University of Košice, Kosice, Slovakia
{josef.varga,mikulas.hajduk}@tuke.sk

Abstract. Currently, approx. 29 % of all industrial robots are installed working on the welding processes, the vast majority of them in the automotive industry. They are mostly mounted in the lines of arc welding or spot welding, which are joined with the automated system of handling and transport. In such a configuration, HRC is less important, because the entire manufacturing installation manages the master control system of the line, the hall or the whole plant. In the near future industrial welding robotics await big changes. Clearly it increases interest in robotics among SMEs. The allow for operation of robotic installations through workers with limited physical efficiencies and cognitive abilities is inevitable. It refer elderly and disabled people. It becomes more and more real to put a man in the working area of robots – the idea called collaborative robots. The implementation of these concepts is often dependent on the development of effective, efficient, "user friendly" Human-Robot-Communication system. The paper presents the development and current state of HRC for welding robots. On this basis the trends and anticipated developments of HRC solutions in the robotic welding installations are described.

Keywords: Robotics welding · Human-Robot communication

1 Introduction

In the development of the robotics different phases can be indicated. Modern robotics, during its formation in the 50 s of the XX century, had an industrial face. The general aim of designers of the first robots, was to replace the human when working hard, in difficult or dangerous conditions [1]. Therefore, the greatest progress during that phase has been made in the area of mechanics, drives and motion control. In the 80 s the new branch, service robotics has started [2]. In the first commercial models, many solutions well proved in the industrial environment were used. Then specific field was developed especially intensive and reached higher level. Constructors of the next industrial robots are willing to use these best practices from service robots. One of such areas was Human-Robot Communication.

R. Szewczyk and M. Kaliczyńska (eds.), *Recent Advances in Systems, Control and Information Technology*, Advances in Intelligent Systems and Computing 543, DOI 10.1007/978-3-319-48923-0_42

According to recent data from International Federation of Robotics IFR approx. 29 % of all robots installed in industry today operates in the welding processes [3]. There is no indication that the situation is changing in the near future. Among the robots launched in 2013, 28 % were also used in welding. These are primarily the position of lines and arc welding and spot welding, working mainly in the automotive industry.

This situation is especially observable in the Central and Eastern Europe [4]. During last 20 years many manufacturers of cars and subassemblies for them, have moved their plants to the countries like Poland, Slovakia, Czech Republic, Hungary, Romania. These factories are generally heavily saturated by robotics & automation technique, mainly in the welding technologies. Robotic cells and lines are often operated by unskilled workers, trained for a new job on the short, intensive courses. Many of these people don't know foreign languages very well, especially English. In this situation good, effective Human-Robot Communication is very important from efficiency of the manufacturing plant, quality of production and safety of employees point of view.

2 Communication of Human with Environment

Communication is the process of organizing messages in order to create meaning [5]. The message, that is information, is stored and encoded using different types of signals. Organizing involves converting messages, according to the used transmission channel and the physical transfer between at least two objects. Such a participant of the communication process can be in general a living creature or device. When one of them is the human it is referred to as personal communication. In general, six possible variations, when it comes to the type of communication participating sites, can be recognized:

- man ↔ man
- man ↔ animal
- man ↔ machine
- machine ↔ animal
- machine ↔ machine
- animal ↔ animal

Man-machine communication includes also communication between human and robot. Important feature of this communication process is the number of participants. In general, there are four cases [6]:

- one man ↔ one robot
- one man ↔ many robots
- many people ↔ one robot
- many people ↔ many robots

Way to communicate depends on the type of the signal that is used to transmit information. One of possible classifications of signals, used in communication of the human with the environment considerations, is presented below:

- mechanical signals,
- chemical signals,
- sound signals
- optical signals,
- electrical signals,
- electromagnetic signals (radio waves).

In addition to the character of the signal to describe the communication it is necessary to specify a method for producing the signal (generator) and its reception (receiver). People use their senses to receive communication signals. In relations with other people or animals in a natural way a man uses a voice (sound signals), for which the sense of hearing is needed.

Very often a man uses optical signals, using the sense of sight. This happens when a person has a weakened hearing apparatus or effective voice communication is not possible, because of the distance between the interlocutors or existing distortions. This method of information exchange can be called visual communication.

The men use also the mechanical signals to communicate, using their sense of touch. Usually the scope and complexity of the content, in this case is much less than with the voice or visual communication. Typical message, like suitable nudge, pat or stroking usually means assessment (acceptance, opposition) or the feelings that one wants to convey to another (praise, reprimand). However, it cannot be forgotten, that the sense of touch is used when reading the Braille. Individual characters (letters are encoded by dots arranged in the correct order) are recognized by the touch of one's fingers. This way of exchanging information, can be called touch communication. The communication using chemical signals is also possible. It involves the sense of smell (olfactory communication), and the sense of taste (flavor communication). This type of communication very common in the animal world, but is rarely used by people.

3 Development of HRC in Industrial Robots

The control system of the robot is responsible for the external communication, especially with the human/operator. Since one of the basic tasks of communication is to create and enhance application (utility) program of the robot, the way (organization and technical realization) of programming and then memorizing and storing these programs has a significant impact on the organization and technical implementation of the communication.

The control systems of the first industrial robots cooperated with an external memory block, in which the program that specifies the robot action had been stored. In the beginning different solutions were used. These included storage drum and perforated tapes, in which a change of the program was followed by the replacement of media. There were also systems programmed by changing the electrical connections.

This generation of robots was characterized by a very limited ability to communicate with the operator. In general, there were the buttons (e.g. "START/STOP") and lamps (e.g. "OPERATION/PAUSE"). So it was a communication through the exchange of digital (two-state) signals. For their handling, a special communication block was introduced in the control systems. Its task was also to control the cooperating devices (e.g. open/closure of the gripper).

In subsequent years, robot control systems were equipped with special, separate devices to communicate with the operator: teach pendant (TP) or programming unit. They were used to operate the robot and to create/change of the application programs. This involved further development of the robot control systems. In the early 70 s of last century the microprocessor controllers were introduced. They were equipped with the integrated internal electronic memory, in which main (system) control program of the robot, as well as current utility program have been stored. In addition, control systems of the robots have had external memory, which served as a storage of utility programs. These programs can be exchanged by uploading them to the internal memory of the control system. The use of external memory, from which users required durability and the ability to make and store the copies, forced introducing new solutions. The tape memory was very well fitted for this purpose. Other external storage solutions, also called mass storage, were based on those used in computers, consecutively EEPROM memory card, battery-backed RAM and different types of floppy disks (FD).

Next breakthrough in storage solutions for robots, was associated with the introduction of computer control system structures, which included hard drives (HD). It meant the end of the problems with lack of memory. The capacity of HD has enabled storing everything: system, current utility program, as well as entire libraries of other procedures and programs on a single device. An external data store functions were taken over by standard computers. Today they are usually working in the structures of local area networks (LAN) of the manufacturing plants. Then a USB Flash memory is used to transfer the program from/to the robot as a standard. At the same time, today's control computers of the industrial robots have the ability to directly connect to the network through digital communication channels used in office computers (RS, USB, Ethernet) or in industrial networks (Profibus, CAN, DeviceNet, etc.).

Like the storage solutions, devices for communication with the operator have gone through the long evolutionary path. In the first models of robots only digital signals were used. Even the first teach pendants with digital displays were still controlled using 2-state (digital) signals. The manual movement of the robot was realized using buttons. Usually a pair of buttons was assigned to each axis, realizing the movement to " + " and "–". The TP was equipped with a number of light-emitting diodes, which informed about the selected robot statuses. Information on displays were coded. For example, the programming unit of IRB-6/60 robots (ASEA) had a display, which consisted of four digital, 7-segment modules, which practically allowed displaying only digits and some letters. An operating panel (OP: buttons and lamps on the control cabinet door) was also used to operate this robot. Such a human-robot communication system, that uses OP on the cabinet and TP, which enabled to reach the vicinity of the mechanical part of the robot, has become a standard used by all manufacturers of industrial robots for many years (Fig. 1).

The next phase of communication between robot and human development, was associated primarily with the development of hardware and software of robot controllers. A major step was to equip TP with the alphanumeric displays. This allowed presenting the information in natural language. Further improvement was to extend the screens, giving two lines of information. This in turn helped to implement a

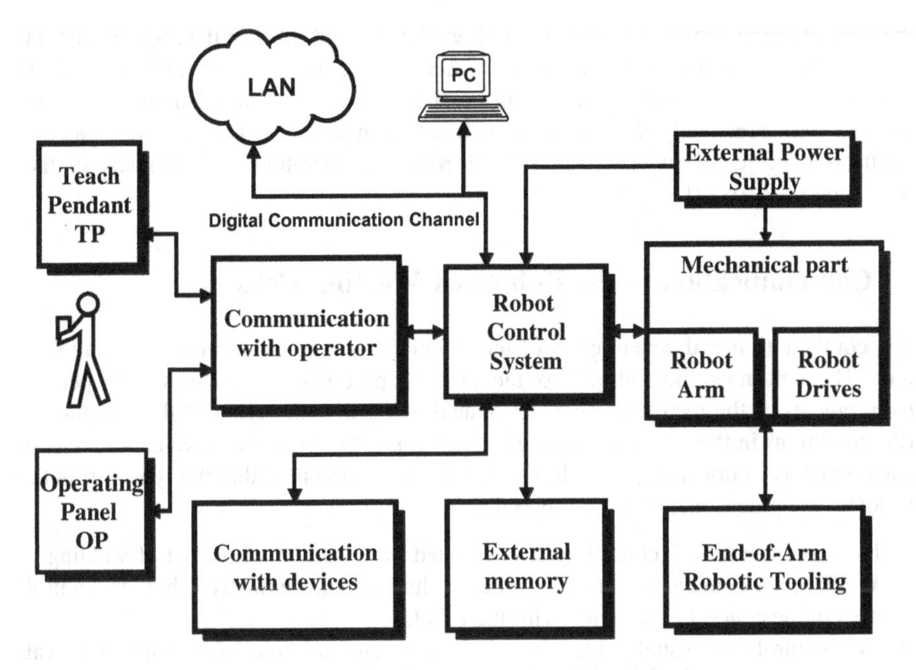

Fig. 1. General block diagram of an industrial robot

conversational system of communication between operator and robot. One line was used as the so-called command line, and the second as a menu line. At the same time the way to manually control the movement of the robot was changed. Joystick control was used with increasing frequency. The first such solution has been presented by ABB in TP of IRB robots, working with S3 control system. Today, both approaches are in use: the buttons and the joystick.

Subsequent changes in the HRC devices were related to the display of TP. The larger screen was introduced, which contained several lines of text. It turned out to be very comfortable, especially when viewing and editing the robot's program. The next step was the introduction of a graphical display, which enabled the implementation of environment modeled on Windows operating system, popular in the world of PCs. The first TP of this type implemented in the mass-produced robot, has been introduced in the KUKA robots, equipped with control systems of KRC family. For manual control of movement, the set of keys, placed next to the screen or a special joystick, called "space mouse" can be used. To activate the manual movement of the robot, the operator must press so-called authorization button (located on the rear panel). This is one of the solutions which improve the safety of the robot. In general, in recent years safety issues, in addition to the requirements of technological processes, are the most common cause of changes and modifications of robots, their controls and interfaces including communication with the operator.

It seems that only the introduction of the possibility of joint work by a man with the robot, and therefore operation of man in the workspace of robot, will force the introduction of new methods and solutions for human-robot communication. Discussions on this type of organization of interaction between human and robot in the implementation of tasks in manufacturing processes, have been taking place for several years. It is getting closer to a modification of current restrictive safety regulations, which in existing form practically make the common action impossible. The first step was the publication of technical specifications for robots cooperating with humans in the beginning of 2016 [7].

4 Communication in the Robotized Welding Cells

The configuration and construction of the robotic welding cell depends on many factors. The cell must be tailored to the task of production and meet the specific requirements of the recipient. They are related to the overall layout of the installation (the conditions in the hall, the organization of transport), the size and weight of welded parts, machines cooperating, etc. In general, it can be assumed that the typical robotic welding equipment includes the following:

- Industrial robot with control system, adapted to cooperate with a set of welding,
- Welding set containing current source, cooling system, wire feed, burner, control system, designed to cooperate with the robot,
- Cell controller – usually PLC in the proper configuration, often with graphical display as an operator panel,
- Positioner in configuration and with parameters suitable for the executed task,
- Fixing device, often equipped with sensors for recognize the presence of welded elements,
- Safety fences, equipped with sensors for access control,
- The security system (light curtains, door locked, the system of "emergency stop" safety controller),
- Fields of storage/warehouses for details in/out,
- Auxiliary equipment, e.g. a system to verify the tool definition, sensory systems, burner cleaning machine, filtering unit, etc.

Today most of these devices are equipped with an intelligent control, providing communication with other machines or industrial local area networks (LAN) via transmission channels. This enables the exchange of content-rich information, e.g. diagnosis of the state of the devices themselves, as well as the process as a whole. These solutions also enable connection of the robotized cell controller to the global Internet network, either directly or through a network of the factory. Thereby access to the robotic installation is available from any place in the world. It is big advantage for maintenance and service. Then, supervision, monitoring and sometimes the service of such an installation in remote mode can be performed from anywhere at any time, by the integrators, consultants, suppliers or high-level control systems. The figure below shows a schematic flow of information in the typical robotic welding cell (Fig. 2).

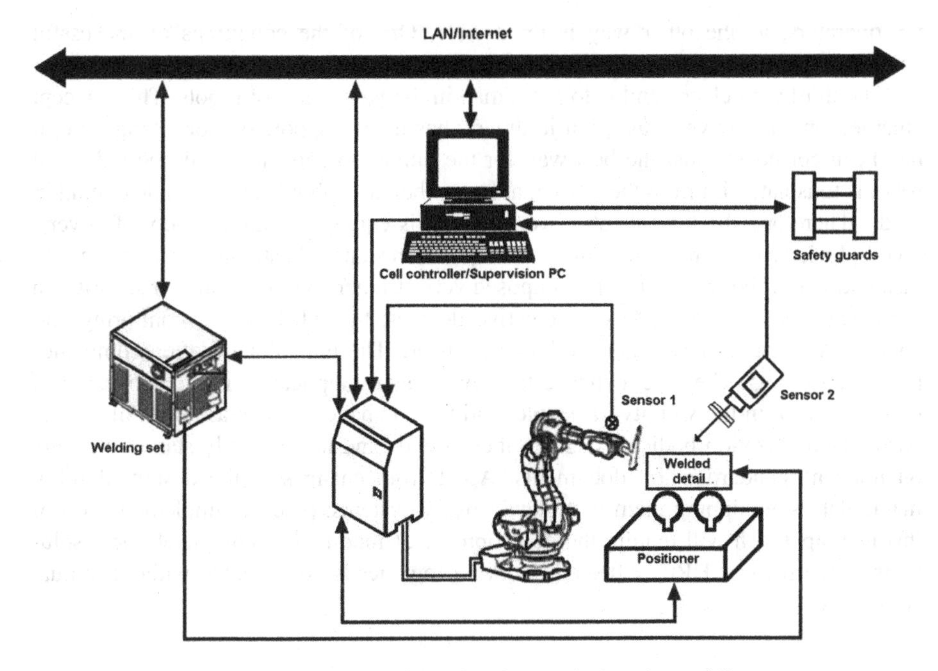

Fig. 2. The flow of information in a typical socket welding

5 Challenges of the Near Future and Consequences for HRC in Welding Applications

Until recently, robotic welding systems worked mostly in large factories that could provide funds for investment and adequate technical staff for operation and maintenance. Today an increase in interest in robotics among SMEs is obvious. Such companies generally do not have, and are unlikely to hire specialists with extensive knowledge and experience in the field of robotics. Maintaining such a specialist for several robots will not be profitable. Therefore, the communication system must be extremely 'user-friendly', so the person with the general technical competence and a very superficial knowledge of the robot can effectively supervise its work, modify the program and carry out basic maintenance and checks.

Another clear trend is the release for operation of robotic installations through workers with limited physical efficiencies and cognitive abilities. It refers to the elderly and disabled people. As the average life span becomes longer, prolongation of the professional activity is inevitable. It can be expected that people older than now will be employed to operate the robotised welding installations. These people will often have limited locomotion, manipulation or cognitive possibilities. Also, for many years there has been a strong trend to activate the people with disabilities, congenital or acquired as a result of illness or accident. Such individuals may also be employed at workplaces, where cooperation with the robot will be required. Robots working with elderly and disabled people must be able to adapt to the specific characteristics of its partners i.e. operators, as the other way is impossible. One of the conditions of successful cooperation is using the proper HRC.

The third very clear trend is to put a man in the workspace of robots. This concept emerged only a few years ago, but it already has many supporters. For a long time, it had been considered that the best way for the human to cooperate with the industrial robot is to isolate him from the device, at least when the robot is in automatic operation mode. Therefore, the typical robot working cell is equipped with a variety of covers, fences, locks, etc. As the result, there are separation standardization documents on safe installation of robotic [8–10]. They imposed very stringent requirements for security on the design, both on the side of protective devices, as well as their monitoring and control. At present, these rules are being changed. The initiators are often companies producing robots. They see a chance for new fields of application in the admission of robots to work in the vicinity of people, and thus a chance to increase sales of robots. Also, standardization bodies recognize these trends and have already started working on relevant standardization documents. Apart from changes in the design of robot manipulators, equipping them with new sensory systems, practical implementation of this new approach will require the development of modified or completely new solutions in the field of HRC. After all, working together is not possible without mutual communication.

6 Summary and Final Conclusions

Considering the future development trend of HRC in robotic welding systems, the direct interaction and remote access should be treated separately.

Currently, HRC which is the most widely used in industrial robots for direct interactions, is the type using a graphical user interface. It meets today's requirements of robotic production systems. A way to communicate human directly with industrial robots will become increasingly similar to the communication with service robots. Ultimately, both variants will aim to the human-to-human type of communication. The first step in the realization of this vision is to equip robots working in industrial installations, in particular welding, with more than one sense and additional communication channels. Because of development of the collaborative robots' concept, it can be the sense of touch.

The reasons for the remote access development can be presented in two groups. The first one includes efforts for purpose of assuring effective maintenance and servicing. The second is directed at remote programming and training. In both cases the technical development of robotic welding cells will be required. It refers to installation of advanced vision systems, with possibility to observe the cell from different points. Also new solutions connected with Internet of Things will be introduced (direct access to the welding equipment, sensors). On the remote stands, especially for programming and training, the virtual reality looks very promising. Systems which will enable remote training in virtual environment will be useful not only for workers but also for vocational schools and technical universities.

Acknowledgements. The paper presents the results of researches supported by EU within the project Rusos "Robotics for teachers of secondary vocational schools", 2015-1-SK1-KA202-008970, under the ERASMUS + Program. This publication represents only author's opinion and neither the European Commission nor the National Agency is not responsible for any of the information contained in it.

References

1. Siciliano, B., Khatib, O.: Springer Handbook of Robotics. Springer, Heidelberg (2008)
2. Engelberger, J.F.: Robotics in Service. MIT Press, Cambridge (1989)
3. World Robotics 2014 Industrial Robots, published by the IFR Statistical Department, hosted by VDMA Robotics + Automation, Germany (2014)
4. Vagaš, M., Hajduk, M., Semjon, J., Páchniková, L., Jánoš, R.: The view to the current state of robotics. In: Advanced Materials Research, vol. 463–464, pp. 1711–1714 (2012). ISSN: 1022-6680
5. Morreale, S.P., Spitzberg, B.H., Barge, J.K.: Human Communication: Motivation, Knowledge and skills. Wadsworth, a division of Thomas Learning Inc., Canada (2001)
6. Burke, J.L., Murphy, R.R.: Final report for the DARPA/NSF interdisciplinary study on human–robot interaction. IEEE Trans. Syst. Man Cybern. PART C: Appl. Rev. **34**(2), 103–112 (2004)
7. ISO/TS 15066:2016, Robots and robotic devices – Collaborative robots
8. EN ISO 10218-1:2011 Robots and robotic devices – Safety requirements for industrial robots – Part 1: Robots (ISO 10218-1:2011)
9. EN ISO 10218-2:2011 Robots and robotic devices – Safety requirements for industrial robots – Part 2: Robot systems and integration (ISO 10218-2:2011)
10. Machinery Directive 2006/42/EC

Modular Robotic Toolbox for Counter-CBRN Support

Grzegorz Kowalski(✉), Adam Wołoszczuk, Agnieszka Sprońska, Damian Buliński, and Filip Czubaczyński

Przemysłowy Instytut Automatyki i Pomiarów PIAP, Warsaw, Poland
{gkowalski, awoloszczuk, aspronska, dbulinski, fczubaczynski}@piap.pl

Abstract. The aim of this paper is to present an approach to design of the robotic payloads toolbox based on addressing end-user requirements. First, a state of the art in robotic counter-CBRN technologies is presented and the problem statement is defined. The idea of payloads toolbox is then described and set in a global architecture context, related to first-responder systems, other robotised solutions and IT systems. Next, the robot subsystem along with operator's console and payload are presented. Subsequently, example payload solutions for toolbox are described as the effects of certain research and development activities carried out within national and international R&D projects, in which PIAP has participated. The example solutions are presented to show approach to the problem of remote sampling and identification of the CBRN threats. A summary of the article and overall research conclusions are presented at the end of the paper.

Keywords: Mobile robot · Counter-CBRN · Interoperability · Sampler · Sensor · Payload

1 Problem Statement and Market Analysis

CBRN[1] incidents pose a lot of threats for first responders, especially those resulting from intentional dispersion. During ROTA[2] events the type of released substances is usually known, because their source can be identified. Intentionally dispersed or hidden residual threats of unknown origin can cause severe health damages or death of the exposed persons. These kind of threats can occur either in outdoor (e.g. mass crisis in public space) or indoor environment (e.g. police inspection of suspicious facilities). Responders unaware of presence of such substances are directly exposed or can accidentally put themselves under exposure (e.g. during inspection of canisters) [3]. One of the possible ways of addressing this kind of situations is through using remote tools capable of:

- checking for the presence of the threats, identifying them and measuring their level,

[1] CBRN – Chemical, Biological, Radiological, Nuclear [2].
[2] ROTA – Release Other Than Attack [4].

R. Szewczyk and M. Kaliczyńska (eds.), *Recent Advances in Systems, Control and Information Technology*, Advances in Intelligent Systems and Computing 543, DOI 10.1007/978-3-319-48923-0_43

- sampling small amounts of suspicious substances for laboratory analysis when on-site identification is not possible.

However, such tools still have to be transported into hot zone. A person transporting the tools to the place of interest would be put in danger of exposure, therefore, tools need to be transported by an unmanned means such as mobile robots.

Nowadays, multiple robotic solutions are commercially available, with majority designed mainly for IEDD/EOD purposes. These robots are usually a part of the antiterrorist squads' equipment and are used in the role of remote eyes and arms. Equipped with multiple cameras and strong manipulators IEDD/EOD robots perform inspection tasks, disposal of suspicious packages and controlled destruction of dangerous materials. A teleoperated mobile robot, equipped with a set of accessories for detection and sampling of unknown substances is able to carry out CBRN inspection and sampling activities. Mobile robot providers, in response to an increasing CBRN threat, are adapting their solutions towards dedicated counter-CBRN platforms, by adding the CBRN sets of accessories, which the robot can be equipped with. A brief overview of the example counter-CBRNe equipment configuration from various robot providers is presented below.

Talon robot from Qinetiq, equipped with CBRNe kit is a Qinetiq standard mobile robot based solution (Talon) adapted for CBRNe operations. The CBRNe equipment comes as modular configuration with optional manipulator arrangements [6] and it consists of a range of CBRNe sensors such as: Smiths Detection LCD3.3 for TICs[3] detection and CWAs[4] identification, Canberra AN/UDR-14 for measuring of gamma radiation dose, RAE Systems MultiRAE for volatile gases detection, Flir Fido XT for explosives tracing, Thermo Scientific FirstDefender RMX for precursors, explosives and chemicals identification and Thermo Scientific RadEye GN+ for gamma and neutron radiation detection. Robot control unit provides display of sensor data and audio alarms for detection of hazards. The set does not contain any sampling devices. Figure 1 presents Talon robot with modular CBRNe/HAZMAT kit.

Fig. 1. Qinetiq Talon robot equipped with modular CBRNe/HAZMAT Kit (Source: https://www.qinetiq-na.com/wp-content/uploads/datasheet_THAZ-15-10.pdf)

[3] TICs – Toxic Industrial Chemicals [2].

[4] CWAs – Chemical Warfare Agents [2].

The 510 PackBot CBRNe robot from iRobot is equipped with payloads for supporting CBRNe operations, containing the same set of sensors as Qinetiq's Talon, only with addition of Drager X-am 5600 sensor for detection of gases. Operator's console provides sensor data display and audio alarms, but also image capture, terrain map and points of interests marking as well as registration of sensor data in text table format with date and time stamp [5]. The robot, as shown in Fig. 2, does not provide sampling capabilities.

Fig. 2. 510 PackBot CBRNe from iRobot (Source: http://photos.prnewswire.com/prnvar/20140905/143601)

Another examples of robotic solutions dedicated for CBRNe response are Cameleon C and Cobra Mk2 from ECA. Cameleon C, as shown on Fig. 3, is a medium class robot featuring a set of 3 interfaces for CBRN sensors, explosimeter and dosimeter, as well as sampling capability for gases (up to 8 Tenax and Draeger tubes) and liquids (up to 6 liquid samples) with automatic sampling process [7]. Cobra Mk2, presented in Fig. 4, is a small, portable mobile robot with integrated chemical and radiological detection module for quick reconnaissance in the field [8].

Fig. 3. Cameleon C from ECA (Source: http://www.ecagroup.com/en/solutions/cameleon-c)

Fig. 4. Cobra Mk2 C from ECA (Source: http://www.ecagroup.com/en/solutions/mini-ugv-cbrn-operations)

Cobham NBCmax service robot, presented in Fig. 5, is a universal mobile sensor platform that can be equipped with various sensors to detect and investigate chemical, biological, explosive and toxic substances. The collected data about recognized substances is transmitted online to the control panel of the robot and processed for the operator. Additionally, the robot is equipped with a set of sampling payloads to be used only with NBCmax robot, which has a Cartesian manipulator control that allows full automation of certain robot operations (such as picking up or putting back the sampler to its housing) [9]. Figure 6 presents the set of NBCmax samplers: shovel sampler for picking granular substances, cotton bud sampler for liquids sampling, swab sampler for usage in ion-scanners and mini gripper collection of various solid objects.

Fig. 5. Cobham NBCmax service robot (Source: http://www.cobham.com/media/853508/NBC%20CBRNE%20Service%20Robot%20-%20banner.jpg)

Review of the above presented solutions leads to a conclusion that the approach presented by robotic providers seems to be targeted towards the users already having mobile platform of a certain producer or those planning to acquire the whole new robot with integrated CBRN equipment. In practice, however, entities such as military counter-CBRN units, firefighters or any other first-response services are often already equipped with mobile robots of various producers, which are in service for years. In case the users wish to extend the functionalities of those robots and their manufacturer

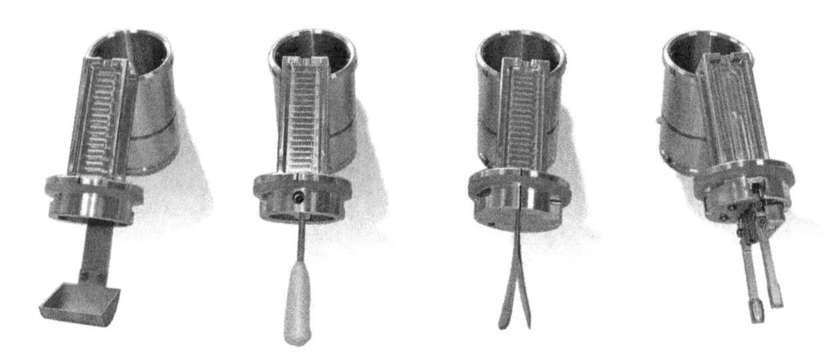

Fig. 6. Set of samplers for NBCmax developed by Cobham: (a) shovel sampler (b) cotton bud sampler (c) swab sampler (d) mini gripper (Source: http://www.cobham.com/mission-systems/unmanned-systems/remote-controlled-robotic-solutions/nbcmax-cbrne-service-robot/nbcmax-cbrne-service-robot-brochure-extract/)

does not provide dedicated CBRN payloads, the users are forced to buy a new platform with the given payload or apply some ad-hoc solutions using the equipment at hand.

A good approach to address this issue would be to introduce a system, which components may, but not need to, operate as whole integrated system, as it was done in the PROTEUS project coordinated by PIAP Institute. The project provided a complete crisis-response robotic system consisting of three different Unmanned Ground Vehicles (UGVs) including counter-CBRN robot), Unmanned Aerial Vehicle (UAV), Operator Control Units (OCUs), mobile command centre, mobile robot operators centre, wearable sensors and training simulator. System components have been integrated within IT component, allowing high-level system management [10]. However, each of the main system components can also be used separately as a stand-alone solution: UGVs with portable console can successfully operate outside PROTEUS system, mobile robot operators centre can be used for robot transportation and control and wearable sensors can be carried either by human or by robot. This approach does not coerce the application of the whole system, thus lowers the entry barrier for implementation into users practice thanks to modularity and scalability of the solution.

Experience and conclusions from the development of the PROTEUS system, encouraged PIAP researchers and engineers to continue the introduction of the platform independent payloads, that can be used not only with UGVs produced by PIAP Institute, but also with other robotic platforms with none or minimal adaptation. This activities resulted in the concept of modular, platform-independent solutions, bringing flexibility to the users assembling the necessary, custom configuration of the counter-CBRN robotic subsystem.

2 Modular Toolbox Concept

Robot subsystems consist of the mobile robot, operator control unit and a set of robot accessories (toolbox). Information made available by the providers of the commercially available solutions of CBRN robotic systems, as described in previous paragraph, indicates that all of them are based on a closed architecture, allowing to work only within their own subsystems, which makes host platform dependent and giving very limited configuration flexibility. To address this issue a modular CBRN toolbox concept was developed, with a high level of interoperability, enabling the toolbox to work with different host platforms and external IT systems. Within this concept, the toolbox shall be configured from a set of tools which meet wide range of requirements for variety of missions, allowing fast reconfiguration of accessories mounted on the host platform and being easy to use and maintain. The toolbox shall also enable alternative, platform independent usage (e.g. handheld) of the accessories and have the ability to undergo a decontamination process. All those requirements had to be fulfilled by the design of every toolbox' component, starting from the mechanical design of the sampling and sensing tools up to the architecture of the control software.

Mechanical integration of the tools with the robotic platform can be done twofold – part of the items can be fixed-mounted either on the robot's base or manipulator, the other part is lodged in the accessory bank, that is attached to the robot's base. First option is designated for the ambient measurement tools, such us most of the environmental sensors (e.g. weather station, gamma radiometer). The tools are mounted basing on quick fasteners (Fig. 7) which fix the tool to a section of NATO Accessory Rail (NAR) shown in Fig. 8. Tools that need to operate strictly in the spot of interest (e.g. liquid sampler, chemical sensors for substances identification or concentration measurement) are equipped with a custom-standard rail, that slides into one of the corresponding accessories bank's holder and they are grasped by the robot using gripper-slots on the tool's case. The accessories bank is also fixed to NAR with quick fasteners, so it can be easily and fast detached, when not in use. This kind of mounting and transporting of the tools allows fast unaided manual dismantling of current on-robot accessory set. Combination of mounting via NAR with quick fasteners and accessories bank gives a wide variety of configuration and interchangeability of the whole set of accessories, depending on the mission requirements.

Fig. 7. Quick fastener for NAR (Source: http://img-cdn.redwolfairsoft.com/upload/product/img/QD006-1L.jpg)

Software architecture of the toolbox bases on pairing of the radio-controlled accessories with the robot or directly with the separate OCU (depending on the available robotic system). Pairing is performed on a similar basis as in the Bluetooth

Fig. 8. Section of NATO Accessory Rail (Source: http://www.brownells.com/userdocs/products/p_093000030_1.jpg)

technology, where the user can choose the devices from the list of available resources, as presented in Fig. 9. One device can be paired with one OCU/robot at once. When the pairing is completed, the OCU displays the list of paired accessories by the set of icons. Clicking the selected icon opens accessory's individual GUI window (Fig. 10), which enables control of the corresponding device (sensors and electro-mechanical samplers) and displays the relevant information, such as battery level, communication link strength, measurements data or over-threshold concentration alerts.

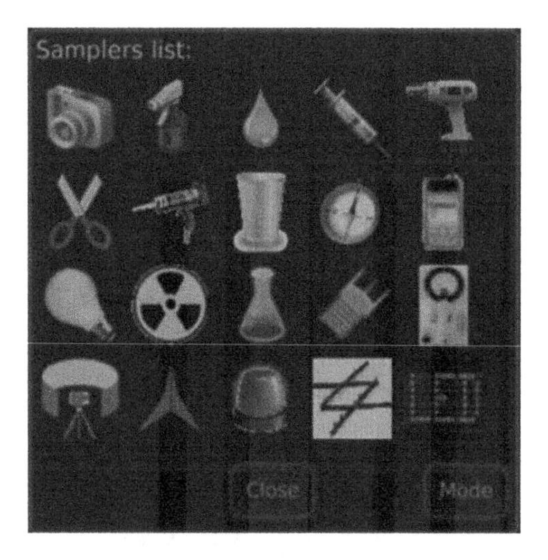

Fig. 9. List of payloads available for pairing

Platform independent usage of the majority of accessories is also provided. Simple mechanical devices (such as ground sampler) can easily be used manually or with the use of a manipulator. Advanced accessories (electromechanical) can be either controlled by the dedicated OCU (direct pairing) or by user interface (UI) located on the cover of the accessory.

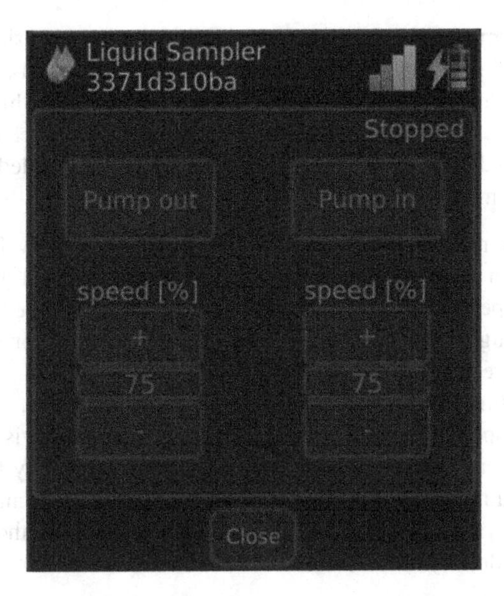

Fig. 10. GUI window for liquid sampler

This kind of approach gives the users much more flexibility than the closed architecture systems. If toolbox components are used together with PIAP robot, the robot and the payloads share the same OCU. Each tool connects wirelessly with the robot's OCU and displays the payload information and statuses via integrated GUI of the OCU. If the payloads are used with a platform not originating from PIAP, they are controlled by a dedicated OCU while the robot is operated by its original control panel.

Handheld usage of the toolbox items is also possible via the inbuilt UI which features two capacitive buttons hidden under the cover of the device which are marked with convex dots. Basic device functions, such as: activation and deactivation, starting/stopping operation sequence, checking battery level or starting device's identification process in the OCU, are executed using a pattern of button hits and finger swipes over the dots. Device's statuses are communicated by the corresponding illumination combination of three LEDs placed over the dot buttons on the casing. Successful finger swipe pattern recognition is also confirmed by the sound signal. When the tool is a sensor with the original (embedded) user interface (of the sensor provider) the results of measurements can be read directly from the device and the dot-and-led interface is used only to check battery level.

3 Information Distribution Levels

Integration level of the robot and its accessories varies from the lowest one – providing only mechanical interfaces for sensors while the readings are viewed on the original device's display with robot's on-board cameras, to fully integrated solutions supporting both visual and sensory inspection, when the information about residual threats is sent to the robot's control panel. The components may operate as stand-alone items or they may share the information in a structured manner. Authors of this paper defined a framework under the term of the Information Distribution Level (IDL) in order to describe the scope and range of information feedback that is provided to the user by the remote robotic system[5]:

- **IDL1** – the user receives no telemetric or visual data directly from the payloads only by looking at them using the robot's default cameras. IDL1 gives a high degree of platform independence and is robust, but a big disadvantage are good visibility conditions and high-resolution image transmission required for the operator to be able to read the results,
- **IDL2** – information from payloads is sent to a dedicated OCU, which is additional to default robot operator console. This IDL covers telemetric, visual and audio data gathered by payloads and sent to the payloads OCU carried by the user,
- **IDL3** – information from payloads carried by the robot is integrated with the communication system of the robot operator console and both the payloads and the robot are controlled by the same OCU enhanced with additional payloads' GUI windows,
- **IDL4** – information from payloads is distributed within the IT architecture of the whole robotic system and can be shared by any component within this system. This level of information distribution also encompasses open interfaces to existing online infrastructures, data bases and communication systems.

This terminology will enable better positioning of robotic security systems and payloads in the context of the information distribution.

4 Exemplary Toolbox' Items

PIAP's ground sampler (Fig. 11) is designated for acquiring soil/ice/sand samples up to 300 cm^3. The device can be configured from several different components (end effectors and gripper pads) depending on the size of the robotic gripper and type of ground to be collected. The ground sampler is an example of IDL1, as the device is fully mechanical, the user receives no feedback from the device and must rely only on visual transmission from robot cameras. While on the robot, the sampler is transported in dedicated holder, mounted directly to NAR.

SPME adsorber (Fig. 12) is an example of electromechanical device, controlled from OCU, which may be treated either as IDL2 or IDL3, depending on the platform

[5] Full elaboration of the Information Distribution Level framework concept will be a subject to a separate research paper.

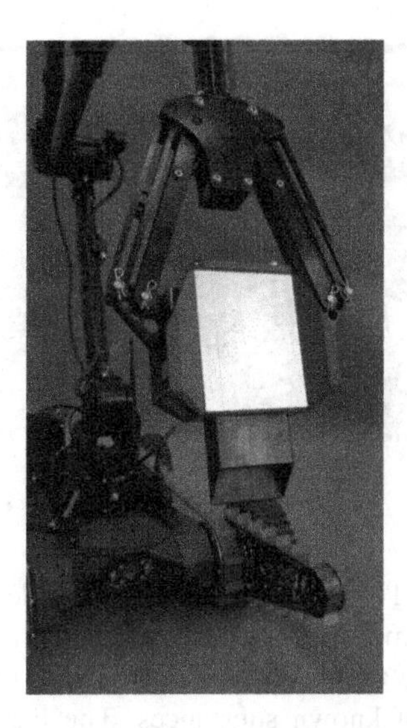

Fig. 11. Ground sampler in robot's gripper

used. The sampler was designed for the acquisition of chemical particles from air or liquid onto disposable SPME fiber. After the exposure, the substances, that residue on the extracted fiber, can be further analysed in the gas chromatograph. This tool is supplementary to the direct on-site measurements carried out with the use of chemical sensors.

Fig. 12. SPME adsorber in robot's gripper

PIAP's modular toolbox contains also a set of two radiological and chemical sensors. First one is designated for alfa, beta, gamma and X radiation detection and measurement (Fig. 13). Measured data is sent to the OCU in real time, so the operator is able to seek local differences (e.g. radiation source seeking). Real time data analysis is used to provide alarms for the user, stating excess of the set threshold of the measured radiation. Although, the sensor lacks of radionuclide identification, it provides display of the radiation power in various unit. The device can be operated both from OCU's GUI and genuine interface on its housing.

Fig. 13. Radiological sensor in robot's gripper

The second sensor (Fig. 14) is mentioned for TICs detection and CWAs identification. It provides real time measurement data of the gas concentration and over-threshold alarms via the OCU. The device's is able to identify and name CWAs according to a default list of known substances. The list of TICs is larger than the CWA's list, but for the TICs the device provides only indication levels, without the identification details. Remote usage via the OCU additionally provides data of gas concentration measurements.

Fig. 14. Chemical sensor

Both above mentioned sensors were developed by integration of commercially available standalone sensors and enhancing them with links for bidirectional communication with the OCU, mechanical interfaces for mounting on the robot (transportation in the accessories' bank) and cross-toolbox common wireless charging. Similarly to SPME adsorber, they may be considered either as IDL2 or IDL3, depending on the platform used. PIAP's operator's console supports every IDL3 tool within the toolbox. IDL2 and IDL3 in the context of PIAP's payloads are presented in Fig. 15.

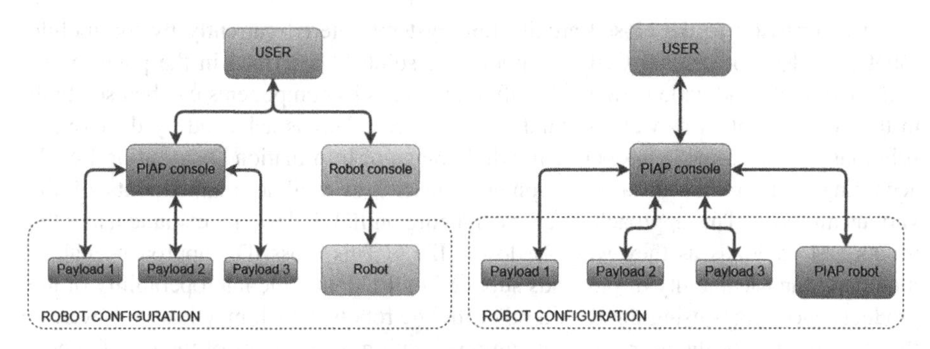

Fig. 15. Idea of IDL2 (left) and IDL3 (right) payload configuration

IDL4, as the highest level of information distribution and system integration, is supported by the example of PROTEUS system or EDEN toolbox [1, 11], which were created in the course of R&D projects. On IDL4 the tools and the robots are integrated within a common framework, sharing modules and data through standardised interfaces. Figure 16 presents the PROTEUS system general architecture as an example of IDL4.

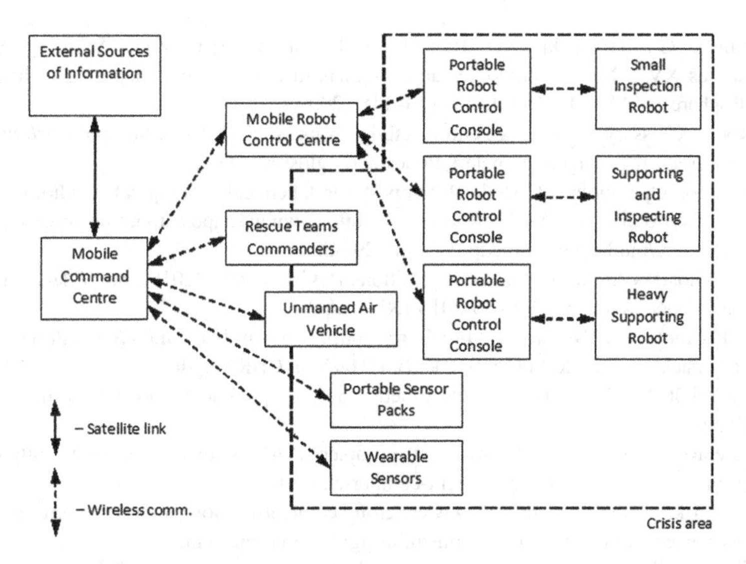

Fig. 16. PROTEUS system architecture as example of IDL4 [1]

5 Summary and Conclusions

This paper encompasses considerations over the optimal application and data distribution in robotic solutions for countering and mitigating CBRN incidents, based on the brief analysis of the solutions currently available on the market. It presents the concept of the modular toolbox for supporting counter-CBRN operations in relation to the framework of Information Distribution Levels, as defined by the authors, followed by a short overview of example toolbox items provided by PIAP.

In comparison to the closed architecture systems offered currently by the mobile robot providers for counter-CBRN support, the solution described in the paper gives high modularity and wide variety of configurations, as its components can be used both in the system context as well as stand-alone devices. This is achieved by developing solutions that are capable to operate in whole range of Information Distribution Levels according to users' needs and equipment capacity as well as requirements of the various missions. This approach, even when being on IDL4, does not exclude using the robots and payloads as they were on lower IDLs. This cross-IDL approach enables modularity and scalability of payloads supplied by PIAP and the interoperability of the products encourages using them with other mobile robots, which may lead to lowering the entry barrier for the users already equipped with mobile robots of their preference.

Acknowledgements. Part of the research leading to these results has received funding from the European Community's Seventh Framework Programme (FP7/2007-2013) under grant agreement no. 313077.

References

1. Kowalski, G., Trojnacki, M.: Proteus as Industrial Crisis Zone Securing System, Tehnomus XV – New technologies and products in machine manufacturing technologies, Ed. Bucharest: MATRIX ROM, pp. 113–118 (2009)
2. CBRN Glossary. http://ec.europa.eu/dgs/home-affairs/what-we-do/policies/crisis-and-terrorism/securing-dangerous-material/docs/cbrn_glossary_en.pdf
3. Concept of Operations of Medical Support for Chemical, Biological, Radiological and Nuclear Environments, AMedP-7(D) (2007). https://www.shape.nato.int/resources/site6362/medica-secure/publications/amedp-7%28d%29.pdf
4. EUMC Glossary of Acronyms and Definitions Revision 2015 (2016). http://data.consilium.europa.eu/doc/document/ST-6186-2016-INIT/en/pdf
5. 510 Packbot CBRNe robot brochure. http://www.irobot.com/~/media/Files/Robots/Defense/PackBot/%20iRobot_510_PackBot_HazMat_CBRNe.pdf
6. Talon robot brochure. https://www.qinetiq-na.com/wp-content/uploads/datasheet_THAZ-15-10.pdf
7. Cameleon C website. http://www.ecagroup.com/en/solutions/cameleon-c, http://www.ecagroup.com/en/solutions/lightweight-counter-cbrn-ugv
8. Cobra Mk2 website. http://www.ecagroup.com/en/solutions/mini-ugv-cbrn-operations, http://www.ecagroup.com/en/solutions/mini-ugv-hazmat-incidents
9. NBCmax robot website. http://www.cobham.com/mission-systems/unmanned-systems/remote-controlled-robotic-solutions/nbcmax-cbrne-service-robot/
10. PROTEUS project website. http://www.projektproteus.pl/en/
11. EDEN project website. www.eden-security-fp7.eu

Tire Models for Studies of Wheeled Mobile Robot Dynamics on Rigid Grounds – A Quantitative Analysis for Longitudinal Motion

Przemysław Dąbek[1(✉)] and Maciej Trojnacki[2]

[1] Industrial Research Institute for Automation and Measurements PIAP, Al. Jerozolimskie 202, 02-486 Warsaw, Poland
pdabek@piap.pl

[2] Institute of Vehicles, Warsaw University of Technology, Narbutta 84 Street, 02-524 Warsaw, Poland
mtrojnacki@simr.pw.edu.pl

Abstract. The work is concerned with quantitative analysis of tire models from the point of view of their use in studies of dynamics of lightweight wheeled mobile robots. The analyses are carried out for the four-wheeled skid-steered mobile robot dedicated for investigations of robot kinematics and dynamics. Robot kinematics and definitions of parameters describing wheel slips are discussed. Two tire models selected for the quantitative analysis are presented. The model of robot dynamics used during studies which allows simulation of longitudinal motion enhanced with drive unit model is described. Within work, simulation investigations of the full vehicle with successively connected different tire models are conducted. Results of simulations are benchmarked against data obtained during experiments with the real mobile robot.

Keywords: Wheeled mobile robot · Rigid ground · Tire-ground model · Dynamics model · Drive unit model · Longitudinal motion · Wheel slip · Simulation research · Experimental investigations

1 Introduction

Modern methods of computer aided engineering analysis (CAE) increasingly become more widespread and more important in the process of design of wheeled mobile robots. The approach to design involving CAE is promoted in the 2014–2020 Strategic Research Agenda proposed by the public private partnership for robotics in Europe – SPARC [1].

The computer aided engineering methods enable development of virtual prototypes whose common applications include: (1) process of development of machine mechanical structure, where they enable multiple tests to be carried out to choose optimal values of parameters from the point of view of the given requirements and (2) synthesis of machine control algorithms, where virtual prototypes are used as plants (controlled objects).

R. Szewczyk and M. Kaliczyńska (eds.), *Recent Advances in Systems, Control and Information Technology*, Advances in Intelligent Systems and Computing 543, DOI 10.1007/978-3-319-48923-0_44

The virtual prototypes of mobile robots are based on mathematical models of dynamics, in which the key element is the model describing the effector interaction with environment. The purpose of this model is determination of effector forces and moments of force, which besides the gravity force, virtually alone determine motion of a robot. In case of the wheeled mobile robots, the role of effectors is usually played by wheels, most often equipped with rubber tires which interact with grounds of various properties.

So far, the problems of modeling of the tire-ground interaction were tackled mainly from the point of view of requirements of automotive vehicles, which over the years resulted in development of many tire models with diverse capabilities [2]. However, it turns out that the wheeled mobile robots, especially lightweight robots, differ significantly from automotive vehicles. Significant differences can be pointed out in at least the following areas: applications, maneuvers performed, vehicle design, types of ground and tire parameters [3]. It seems that there is little work done so far in the field of modeling of tire-road interaction for lightweight wheeled robots. This gap is sometimes noticed by researchers like in [4], where authors observe that tire-terrain models for lightweight robots on rigid terrain are not as readily available as models for heavier vehicles. On the other hand, the models must be highly accurate in order to yield results useful in the engineering practice. Several studies concern wheel-terrain interaction of planetary rovers, where a rigid wheel is considered [5] or a flexible wheel of special design made primarily of metal [6].

The aim of the work is to evaluate two commonly used tire models from the point of view of their use in conditions typical for the wheeled mobile robots. The analysis focus is on modeling the tire-ground contact patch and the case of tire interaction with rigid grounds.

The work extends current knowledge about capabilities of the existing tire models in application to wheeled robot dynamics studies. In this paper longitudinal motion of the robot is analyzed.

2 Mathematical Model

2.1 Main Modeling Assumptions

In the present work mathematical model of the vehicle comprises full-vehicle model and the tire model, which is a separate component of the full-vehicle model.

Full vehicle model. Scope of the study is limited to longitudinal motion, therefore plane model of vehicle is assumed. The object of modeling is the existing PIAP GRANITE lightweight four-wheeled research platform with non-steered wheels.

Tire model. Existing tire models for rigid ground surfaces can be divided into tire-ground interaction models for generation of shear forces in the tire-ground contact patch only and complete tire models which include additionally model of radial forces generation. In the present work it is assumed that models from the first previously mentioned category will be used and investigated. The tire-ground interaction model will be complemented with simple linear force-deflection dependency to determine tire normal force as a function of radial deflection.

The first category of tire models includes the following models (the list not exhausitive): Pacejka [2], Fiala [7], Dugoff [8], Brush [9] and Kiencke [10]. In the present work Pacejka and Dugoff models will be analyzed, because both are widely used by the vehicle dynamics researchers community.

2.2 Robot Kinematics

For the purpose of model development, the following coordinate systems will be introduced:

- the fixed coordinate system $\{O\}$,
- the moving coordinate systems with axes parallel to the fixed coordinate system: $\{CM\}$ – origin at the robot center of mass, $\{A_i\}$ – origins at centers of wheels, $\{C_i\}$ – origins at the intersections of ${}^{Ai}z$ axes with tire tread surfaces,
- the moving coordinate systems with axes fixed to the robot body 0 (rotating with body 0): $\{CM'\}$ – origin at the robot centre of mass, $\{A'_i\}$ – origins at centers of wheels,
- the moving coordinate systems with axes spinning with robot wheels: $\{A''_i\}$ – origins at centers of wheels,

and in the above $i = \{f, b\}$, where f denotes pair of front wheels, and b, pair of rear wheels (Fig. 1a).

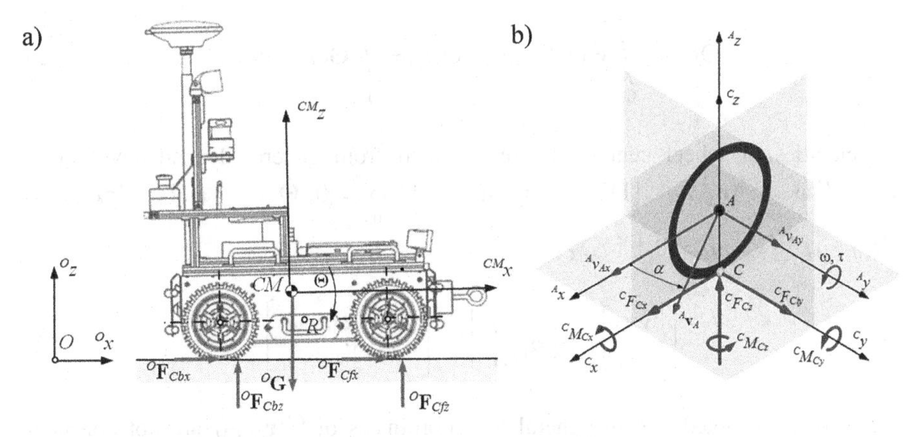

Fig. 1. Forces and moments of force: (a) acting on the mobile robot, (b) acting on the tire (according to ISO convention)

It is assumed that robot motion is realized in ${}^{O}x{}^{O}z$ plane of the fixed coordinate system $\{O\}$.

Positions of the characteristic points CM, A_i, C_i of the vehicle in the fixed coordinate system $\{O\}$ are given respectively by position vectors ${}^{O}\mathbf{r}_{CM}$, ${}^{O}\mathbf{r}_{Ai}$, ${}^{O}\mathbf{r}_{Ci,}$ where coordinates of those vectors have the form ${}^{O}\mathbf{r}_j = [{}^{O}x_{jx}, {}^{O}y_{jy}, {}^{O}z_{jz}]^T$, and $j = \{CM, A_i, C_i\}$.

Angle of robot body rotation Θ about ${}^{CM'}y$ axis of the $\{CM'\}$ coordinate system is measured from ${}^{CM}x$ (${}^{CM}z$) axis to ${}^{CM'}x({}^{CM'}z)$ axis, with positive sense of the angle marked in Fig. 1a.

Angle of wheel rotation θ_i is measured in a similar way to the Θ angle, from ${}^{Ai'}x({}^{Ai'}z)$ axis to ${}^{Ai''}x({}^{Ai''}z)$ axis ($\{A_i''\}$ coordinate system spins with the wheel).

If acceleration ${}^{O}\mathbf{a}_{CM} = \left[{}^{O}a_{CMx}, {}^{O}a_{CMy} = 0, {}^{O}a_{CMz}\right]^T$ of the point CM is known, then velocity ${}^{O}\mathbf{v}_{CM} = \left[{}^{O}v_{CMx}, {}^{O}v_{CMy} = 0, {}^{O}_{CMz}\right]^T$ and position of the point CM are obtained by successive integrations:

$$ {}^{O}\mathbf{v}_{CM}(t) = \int_0^t {}^{O}\mathbf{a}_{CM}\mathrm{d}t + {}^{O}\mathbf{v}_{CM}(0), \quad {}^{O}\mathbf{r}_{CM}(t) = \int_0^t {}^{O}\mathbf{v}_{CM}\mathrm{d}t + {}^{O}\mathbf{r}_{CM}(0). \tag{1} $$

In the above equation it is assumed that point CM is in rectilinear motion during experiment, and therefore total acceleration of the point CM is assumed to be identical with the tangential acceleration of the point CM (normal acceleration neglected). In fact, the path of the CM point only slightly differs from straight line during acceleration and deceleration phases of motion in the performed simulation experiments.

Similarly, if angular acceleration of the robot body $\ddot{\Theta}$ is known, then Θ angle and its rate of change are given by:

$$ \dot{\Theta}(t) = \int_0^t \ddot{\Theta}\mathrm{d}t + \dot{\Theta}(0), \quad \Theta(t) = \int_0^t \dot{\Theta}\mathrm{d}t + \Theta(0). \tag{2} $$

Velocities of wheel centers A_i are obtained from general dependency ${}^{O}\mathbf{v}_{Ai} = {}^{O}\mathbf{v}_{CM} + {}^{O}\dot{\varphi} \times {}^{CM}\mathbf{r}_{Ai/CM}$ [11], after setting ${}^{O}\underset{0}{\dot{\varphi}} = [0, \dot{\Theta}, 0]^T$ and ${}^{CM}\mathbf{r}_{Ai/CM} = \left[{}^{CM}x_{Ai/CM}, {}^{CM}y_{Ai/CM} = 0, {}^{CM}z_{Ai/CM}\right]^T$:

$$ \begin{bmatrix} {}^{O}v_{Aix} \\ {}^{O}v_{Aiz} \end{bmatrix} = \begin{bmatrix} {}^{O}v_{CMx} \\ {}^{O}v_{CMz} \end{bmatrix} + \begin{bmatrix} \dot{\Theta}\, {}^{CM}z_{Ai/CM} \\ -\dot{\Theta}\, {}^{CM}x_{Ai/CM} \end{bmatrix} \tag{3} $$

It should be noted that in general the coordinates of ${}^{CM}\mathbf{r}_{Ai/CM}$ are not constant, because of robot body rotating through angle Θ. On the other hand, the vector ${}^{CM'}\mathbf{r}_{Ai/CM} = {}^{CM'}\mathbf{r}_{Ai}$ has constant coordinates, because it is the position vector of the point A_i of robot body in the $\{CM'\}$ coordinate system, and the robot body is considered rigid. The vector ${}^{CM}\mathbf{r}_{Ai/CM}$ is obtained from vector ${}^{CM'}\mathbf{r}_{Ai}$ by applying rotation through angle Θ:

$$^{CM}\mathbf{r}_{Ai/CM} = {}^{CM}_{CM'}\mathbf{R}\,{}^{CM'}\mathbf{r}_{Ai} \tag{4}$$

where ${}^{CM}_{CM'}\mathbf{R}$ is the matrix of rotation from coordinate system $\{CM'\}$ to $\{CM\}$ given by:

$${}^{CM}_{CM'}\mathbf{R} = \begin{bmatrix} \cos\Theta & 0 & \sin\Theta \\ 0 & 1 & 0 \\ -\sin\Theta & 0 & \cos\Theta \end{bmatrix}. \tag{5}$$

The vector $^{CM'}\mathbf{r}_{Ai}$ can be expressed as vector sum $^{CM'}\mathbf{r}_{Ai} = {}^{CM'}\mathbf{r}_R + {}^{CM'}\mathbf{r}_{Ai/R}$ where the coordinates of the vectors are as follows:

$$^{CM'}\mathbf{r}_R = \left[{}^{CM'}x_R,\ 0,\ {}^{CM'}z_R\right], \tag{6}$$

$$^{CM'}\mathbf{r}_{Af/R} = [L/2,\ 0,\ 0], \tag{7}$$

$$^{CM'}\mathbf{r}_{Ab/R} = [-L/2,\ 0,\ 0]. \tag{8}$$

Position vectors of points A_i in the fixed coordinate system $\{O\}$ are given by the following relationship:

$$^{O}\mathbf{r}_{Ai} = {}^{O}\mathbf{r}_{CM} + {}^{CM}\mathbf{r}_{Ai/CM}. \tag{9}$$

Finally, the points of contact of tires with the ground C_i have their positions in the fixed coordinate system {O} given by:

$$^{O}\mathbf{r}_{Ci} = {}^{O}\mathbf{r}_{Ai} + {}^{Ai}\mathbf{r}_{Ci}, \tag{10}$$

where $^{Ai}\mathbf{r}_{Ci}$ is the vector of position of point C_i in the moving coordinate system $\{A_i\}$ of axes parallel to the fixed system $\{O\}$. It is assumed that the point C_i lies on the ^{Ai}z axis, thus it has only one non-zero coordinate, that is, $^{Ai}\mathbf{r}_{Ci} = [0, 0, {}^{Ai}z_{Ci}]^T$. Because norm $\|{}^{Ai}\mathbf{r}_{Ci}\|$ describes the tire radius and it is assumed that tires can deform in radial direction, the following constraint holds for $^{Ai}z_{Ci}$

$$0 \geq {}^{Ai}z_{Ci} \geq -r_{0i}, \tag{11}$$

where r_{0i} is the radius of undeformed tire for i-th wheel. The value of $^{Ai}z_{Ci}$ coordinate depends on position of wheel center A_i with respect to ground, where ground is assumed to be Oxy plane, that is on $^{O}z_{Ai}$, and can be calculated from the following relationship:

$$^{Ai}z_{Ci} = \begin{cases} -r_{0i} & \text{for} \quad {}^{O}z_{Ai} > r_{0i}, \\ -{}^{O}z_{Ai} & \text{for} \quad r_{0i} \geq {}^{O}z_{Ai} \geq 0, \\ 0 & \text{for} \quad {}^{O}z_{Ai} < 0. \end{cases} \tag{12}$$

Finally, if the deformed radius of the tire is equal to

$$r_i = \left\| {}^{Ai}\mathbf{r}_{Ci} \right\| = \left| {}^{Ai}z_{Ci} \right|, \tag{13}$$

the radial deformation of the i-th tire Δr_i is given by:

$$\Delta r_i = r_{0i} - r_i. \tag{14}$$

2.3 Wheel Slip Definitions

Let the longitudinal wheel slip velocity be given by:

$${}^{O}v_{Six} = {}^{O}v_{Aix} - v_{pi} \tag{15}$$

where v_{pi} is wheel circumferential velocity given by:

$$v_{pi} = \dot{\theta}_i \, r_{si} \tag{16}$$

and r_{si} is called the slip radius [2].

Let the lateral wheel slip velocity be given by:

$${}^{O}v_{Siy} = {}^{O}v_{Aiy}. \tag{17}$$

Then, let the longitudinal κ_i and lateral $\tan\alpha_i$ slip ratios be defined by the following formulas (signs compliant with ISO convention [2]):

$$\kappa_i = \begin{cases} 0 \quad \text{for} \quad \max(|v_{pi}|, \, |{}^{O}v_{Aix}|) = 0, \\ -{}^{O}v_{Six}/\max(|v_{pi}|, \, |{}^{O}v_{Aix}|) \quad \text{for} \quad \max(|v_{pi}|, \, |{}^{O}v_{Aix}|) \neq 0, \end{cases} \tag{18}$$

$$\tan\alpha_i = \begin{cases} 0 \quad \text{for} \quad \max(|v_{pi}|, \, |{}^{O}v_{Aix}|) = 0, \\ -{}^{O}v_{Siy}/\max(|v_{pi}|, \, |{}^{O}v_{Aix}|) \quad \text{for} \quad \max(|v_{pi}|, \, |{}^{O}v_{Aix}|) \neq 0, \end{cases} \tag{19}$$

2.4 Tire-ground Interaction Models

The components of reaction forces and moments of force generated by a rubber tire in the tire-ground contact patch and studied in the vehicle dynamics, are listed below and shown in Fig. 1b:

- ${}^{O}F_{Cix}$ – longitudinal force,
- ${}^{O}F_{Ciy}$ – lateral force,
- ${}^{O}F_{Ciz}$ – normal force,
- ${}^{O}M_{Cix}$ – overturning moment,
- ${}^{O}M_{Ciy}$ – rolling resistance moment,
- ${}^{O}M_{Ciz}$ – (self-)aligning moment.

In the present work, two models are subject to analysis, namely Pacejka model in accordance with MF-Tire/MF-Swift 6.1.2 Equation Manual [2] and Dugoff model in a modified form proposed by Lo Bianco and Gerelli [12]. Below equations of both models are given.

Pacejka model. The Pacejka model is given by the following equation for longitudinal tire-ground contact force:

$$^{O}F_{Cix} = D_{ix}\sin[C_{ix}\arctan(B_{ix}\kappa_i - E_{ix}(B_{ix}\kappa_i - \arctan(B_{ix}\kappa_i)))], \tag{20}$$

where: B_{ix} is the stiffness factor, C_{ix} the shape factor, D_{ix} the peak value and E_{ix} the curvature factor, and in general all coefficients are given individually for each i-th tire.

The B_{ix} through E_{ix} coefficients can be expressed in terms of tire structural and operational parameters, for example according to equations given below, which have the simplest physically sensible form (for full set of equations refer to work [2]):

$$B_{ix} = {}^{O}F_{Ciz}p_{Kx1}/(C_{ix}D_{ix} + \varepsilon_{ix}), \tag{21}$$

$$C_{ix} = p_{Cx1}, \tag{22}$$

$$D_{ix} = \mu_i {}^{O}F_{Ciz}, \tag{23}$$

$$E_{ix} = p_{Ex1}, \tag{24}$$

where: μ_i is the tire-ground friction coefficient, ε_{ix} is a certain small value to handle division by zero problem, and p_{Cx1}, p_{Ex1}, p_{Kx1} are model parameters.

One may also note that the tire longitudinal slip stiffness $k_{i\kappa}$ can be expressed as:

$$k_{i\kappa} = B_{ix}C_{ix}D_{ix}. \tag{25}$$

Dugoff model (modified by Lo Bianco and Gerelli). For longitudinal tire-ground contact force, the modified Dugoff model is given by:

$$^{O}F_{Cix} = -(k_{i\kappa}/k_i)\ \|\mathbf{F}_i\|\ \cos(\psi_i), \tag{26}$$

where the norm of the tire total shear force $\|\mathbf{F}_i\|$ is given by:

$$\|\mathbf{F}_i\| = \mu_i {}^{O}F_{Ciz}\begin{cases} p_i & \text{for} \quad p_i \leq 0.5 \\ 1 - 1/(4p_i) & \text{for} \quad p_i > 0.5 \end{cases}, \tag{27}$$

where

$$p_i = \frac{\lambda_i k_i}{2\mu_i {}^{O}F_{Ciz}}, \tag{28}$$

and

$$\lambda_i = \begin{cases} 0 & \text{for} \quad \max(|v_{pi}|,\ |^{O}v_{Aix}|) = 0, \\ \sqrt{\kappa_i^2 + \tan^2 \alpha_i} & \text{for} \quad \max(|v_{pi}|,\ |^{O}v_{Aix}|) \neq 0, \end{cases} \tag{29}$$

$$k_i = \sqrt{(k_{i\kappa} \cos(\psi_i))^2 + (k_{i\alpha} \sin(\psi_i))^2}, \tag{30}$$

and in Eq. (30) $k_{i\alpha}$ is the tire lateral slip stiffness.

The angle ψ_i between slip speed vector and the ^{Ai}x axis of the wheel reference frame is given by:

$$\psi_i = \begin{cases} 0 & \text{for} \quad ^{O}v_{Six} = 0, \\ \arctan 2(^{\mathrm{O}}\mathrm{v}_{\mathrm{Siy}}, {}^{\mathrm{O}}\mathrm{v}_{\mathrm{Six}}) & \text{for} \quad ^{O}v_{Six} \neq 0, \end{cases} \tag{31}$$

where $\psi_i \in \langle -\pi,\ +\pi \rangle$ thanks to using MATLAB four-quadrant inverse tangent function.

The form of Eq. (29) for the case different than zero deserves a comment, because in [12] it is given originally in the following form:

$$\lambda_i = \|\tilde{\mathbf{v}}_i\| / \max(|v_{pi}|,\ |^{O}v_{Aix}|). \tag{32}$$

After recalling from [12] that

$$\tilde{\mathbf{v}}_i = \begin{bmatrix} ^{O}v_{Six} \\ ^{O}v_{Siy} \end{bmatrix}, \tag{33}$$

calculating norm $\|\tilde{\mathbf{v}}_i\|$ and inserting into (32), we obtain:

$$\lambda_i = \sqrt{^{O}v_{Six}{}^2 + {}^{O}v_{Siy}{}^2} \Big/ \max(|v_{pi}|,\ |^{O}v_{Aix}|), \tag{34}$$

after simple transformation we arrive at:

$$\lambda_i = \sqrt{\frac{^{O}v_{Six}{}^2}{\max^2(|v_{pi}|,\ |^{O}v_{Aix}|)} + \frac{^{O}v_{Siy}{}^2}{\max^2(|v_{pi}|,\ |^{O}v_{Aix}|)}} \tag{35}$$

which is equivalent to:

$$\lambda_i = \sqrt{\kappa_i^2 + \tan^2 \alpha_i}. \tag{36}$$

Tire-ground friction coefficient. Both models include tire-ground friction coefficient μ_i in the equations. Value of this coefficient depends on several factors, one of which is wheel slip velocity ${}^{O}v_{Six}$. In the present work this dependency will be expressed following [12] as:

$$\mu_i = \mu_s/(1 + A_s\,|{}^{O}v_{Six}|), \tag{37}$$

where A_s is a constant parameter and μ_s is coefficient of static friction between tire and ground surface materials.

2.5 Normal Ground Reaction

Normal ground reaction on the tire ${}^{O}F_{Ciz}$ is a function of kinematics of tire deformation when the wheel axle presses tire against ground surface. In the present model, the following formula will be used to evaluate ${}^{O}F_{Ciz}$:

$${}^{O}F_{Ciz} = k_{ti}\Delta r_i + c_{ti}\,\Delta\dot{r}_i\,\mathrm{sgn}(\Delta r_i), \tag{38}$$

where: k_{ti} is tire radial stiffness, c_{ti} tire radial damping coefficient, and sgn() is the signum function.

2.6 Robot Dynamics Model

It is assumed that the robot is under action of the external forces: ground reaction forces ${}^{O}\mathbf{F}_{Ci} = [{}^{O}F_{Cix}, {}^{O}F_{Ciy}, {}^{O}F_{Ciz}]^T$ acting on each wheel and gravity force ${}^{O}\mathbf{G}_{CM}$. Additionally, wheels are subject to moment of rolling resistance ${}^{O}\mathbf{M}_{Ci} = [0, {}^{O}M_{Ciy}, 0]^T$.

The moment of rolling resistance is given by:

$${}^{O}M_{Ciy} = {}^{O}F_{Ciz}f_{ir}\mathrm{sgn}(\dot{\theta}_i)r_i. \tag{39}$$

Knowing the moment of rolling resistance it is possible to determine the coefficient of rolling resistance with dimension of length:

$$e_i = {}^{O}M_{Ciy}/{}^{O}F_{Ciz}. \tag{40}$$

Gravity force vector ${}^{O}\mathbf{G}_{CM}$ is a function of gravitational acceleration vector ${}^{O}\mathbf{g}$ = $[0, 0, {}^{O}g_z]^T$ and its application point is at the robot mass centre:

$${}^{O}\mathbf{G}_{CM} = m_R^O\mathbf{g}, \tag{41}$$

where m_R denotes robot total mass.

Above mentioned forces and moments of force acting on the robot and on the wheel are illustrated in Fig. 1a.

Under action of the mentioned forces, according to the Newton's 2nd law, the robot moves with translational acceleration ${}^{O}\mathbf{a}_{CM}$ and rotational acceleration $\ddot{\Theta}$.

Dynamic equations of motion in ${}^{O}x{}^{O}z$ plane have the form:

$$m_R\,{}^{O}a_{CMx} = \sum_{i=1}^{2} {}^{O}F_{Cix}, \tag{42}$$

$$m_R\,{}^{O}a_{CMz} = \sum_{i=1}^{2} {}^{O}F_{Ciz} - m_R\,{}^{O}g_z, \tag{43}$$

$$I_{CM'y}\,\ddot{\Theta} = {}^{CM}z_{Ci/CM}\sum_{i=1}^{2} {}^{O}F_{Cix} - ({}^{CM}x_{Cf/CM} + e_f){}^{O}F_{Cfz} - ({}^{CM}x_{Cb/CM} - e_b){}^{O}F_{Cbz} \tag{44}$$

where: ${}^{CM}z_{Ci/CM}$, ${}^{CM}x_{Ci/CM}$ are coordinates of the vector ${}^{CM}\mathbf{r}_{Ci/CM} = {}^{CM}\mathbf{r}_{Ai/CM} + {}^{Ai}\mathbf{r}_{Ci}$, $I_{CM'_y}$ – mass moment of inertia of robot about the axis ${}^{CM'}y$ passing through the mass centre of the robot.

Moreover, for each of the robot wheels it is then possible to write dynamic equation of motion associated with wheel spin:

$$I_{Ai''y}\,\ddot{\theta}_i = \tau_i - {}^{O}F_{Cix}r_i - {}^{O}F_{Ciz}e_i, s \tag{45}$$

where: $I_{Ai''y}$ – mass moment of inertia of the wheel about its spin axis ${}^{Ai''}y$, τ_i – driving/braking torque applied to wheel axle.

2.7 Drive Model and Controller

The described model of robot dynamics includes also the model of drive units. It is assumed that: each of the robot drive units consists of identical DC motor, encoder, and transmission system. The DC motor model of the i-th drive unit is described by the following relationships:

$$\frac{\mathrm{d}i_i}{\mathrm{d}t} = \left(u_i - k_e n_d\dot{\theta}_i - R_d i_i\right)\Big/L_d, \quad \tau_i = \eta_d n_d k_m i_i, \tag{46}$$

where: L_d, R_d – respectively inductance and resistance of the rotor, k_e – electromotive force constant, k_m – motor torque coefficient, n_d – gear ratio of the transmission system, u_i – motor voltage input, i_i – rotor winding current, η_d – efficiency of transmission.

Motion of robot drive units is governed by linear drive controller. Based on the desired and actual angles and angular velocities of wheel spin, that is, respectively vectors: $\boldsymbol{\theta}_d = [\theta_{fd}, \theta_{bd}]^T$ and $\dot{\boldsymbol{\theta}}_d = [\dot{\theta}_{fd}, \dot{\theta}_{bd}]^T$ as well as $\boldsymbol{\theta} = [\theta_f, \theta_b]^T$ and $\dot{\boldsymbol{\theta}} = [\dot{\theta}_f, \dot{\theta}_b]^T$, the controller determines the control signal vector $\mathbf{u}$ in (V) for drives of particular wheels.

After assuming that errors of angles and angular velocities of spin for driven wheels are defined as:

$$\mathbf{e}_\theta = \boldsymbol{\theta}_d - \boldsymbol{\theta},\ \mathbf{e}_\omega = \dot{\boldsymbol{\theta}}_d - \dot{\boldsymbol{\theta}}, \tag{47}$$

the control signal $\mathbf{u}_c = [u_{cf}, u_{cb}]^T$ for motors of driven wheels can be determined using the control law:

$$\mathbf{u}_c = k_P \mathbf{e}_\theta + k_D \mathbf{e}_\omega, \tag{48}$$

where: k_P, k_D – controller gains, and $\mathbf{e}_\theta = [e_{\theta f}, e_{\theta b}]^T$, $\mathbf{e}_\omega = [e_{\omega f}, e_{\omega b}]^T$.

The current $\mathbf{i} = [i_f, i_b]^T$ flowing through the rotor windings is limited using the saturation function of the form:

$$\mathbf{i} = \mathrm{sat}(\mathbf{i},\ \mathbf{i}_{min},\ \mathbf{i}_{max}), \tag{49}$$

where: $\mathbf{i}_{max} = [i_{lim}, i_{lim}]^T$, $\mathbf{i}_{min} = [-i_{lim}, -i_{lim}]^T$.

Additionally, limitation on the power of drives is taken into account, because power of each drive unit is 150 W for the configuration of robot used during experiments.

3 Research

3.1 Desired Motion

It is assumed that robot motion consists of three phases: accelerating with maximum acceleration a_{CMmax} on the distance of l_r, steady motion with constant velocity ${}^{O}v_{CMd} = v_{CMu}$ and braking with maximum deceleration – a_{CMmax} on the distance of l_h. The desired path of motion is a straight line of length L_p.

The maximum value of desired linear velocity is assumed equal to v_{CMu} = 1.5 m/s, maximum acceleration during accelerating and braking phases a_{CMmax} = 150 m/s^2 and length L_p = 3.8 m. The maximum acceleration was assumed to reflect step change in velocity taking into account discrete time intervals at which desired control signals are fed into real robot controller (0.01 s).

Lengths of acceleration and braking distances, l_r and l_h respectively, are determined for the given velocity profile based on v_{CMu} velocity and maximum acceleration (deceleration) a_{CMmax}. Next, based on the knowledge of v_{CMu} velocity, a_{CMmax} acceleration and the L_p length, characteristic time instants are determined.

3.2 Measurement and Control System. Data Post Processing

The heart of the control and data acquisition system is the on-board laptop PC running Ubuntu Xenomai OS and dedicated software. This system can be qualified as a hard-real time system, which means that the processing loop is supposed to meet the real-time criteria at every iteration. The PC is connected with the robot CAN bus using USB-CAN adapter.

Control variables are desired velocities of spin of wheels $\dot{\theta}_d = [\dot{\theta}_{fd}, \dot{\theta}_{bd}]^T$. Those velocities are sent directly to drive velocity controllers for the left and right-hand side of the robot. Measured quantities include: actual velocities of spin of wheels

$\dot{\theta} = [\dot{\theta}_f, \dot{\theta}_b]^T$, linear acceleration of the characteristic point I of the robot $\mathbf{L}^{Ia}$ from accelerometers.

Measurements of acceleration and angular velocity are provided by tactical grade Inertial Measurement Unit, Analog Devices ADIS 16488 [13] and by 4 pieces of commercial grade STM iNEMO M1 units [14].

Raw empirical data are post-processed using the algorithm described in [15], which is enhanced to perform data fusion from 5 IMU sensors (averaging of angular velocities from selected gyros). Additionally, manual change in the time history of pitch angle is made by adding a small constant correction at the level of 0.1 degree in order to obtain linear velocity time history which is equal to zero when the robot stops.

3.3 Simulation Environment

The simulation studies were carried out in the MATLAB/Simulink environment. For simulations Runge-Kutta fixed-step solver was chosen with fixed-step size Δt.

3.4 Robot, Tire and Ground Parameters

The object of the study is a lightweight four-wheeled mobile robot called PIAP GRANITE (Ground Robot for ANalyzes of wheels Interaction with various TErrain). This robot has been developed for the purpose of performing vehicle dynamics research. In the configuration used during empirical investigations for the present work it had all wheels driven independently by DC servomotors with gear units and encoders. The real look of the robot is shown in Fig. 2a.

Fig. 2. Four-wheeled skid-steered mobile robot PIAP GRANITE: (a) current configuration, (b) robot tire

For the simulation studies the values of parameters of tire models and of the PIAP GRANITE robot shown in Tables 1, 2 and 3 are assumed. It is assumed that parameters for all robot wheels are the same, so subscripts i in the symbols in tables are omitted where not necessary.

Table 1. Parameters of the tire models used during simulations

Symbol	Name, unit	Value
p_{Cx1}	Pacejka model basic parameter for C, –	1.6
p_{Ex1}	Pacejka model basic parameter for E, –	0.975
p_{Kx1}	Pacejka model basic parameter for B, –	18
k_{α}	Tire lateral slip stiffness, N/rad	0
k_{κ}	Tire longitudinal slip stiffness, N/unit slip	1926
A_s	Friction reduction factor, s/m	0.8
μ_s	Static friction coefficient, -	1.05

Table 2. Auxiliary parameters of tires

Symbol	Name, unit	Value
c_t	Radial damping coefficient, Ns/m	500
k_t	Radial stiffness, N/m	20000
r_s	Slip radius, m	0.0946 ($0.98\ r_0$)
r_0	Undeformed radius, m	0.0965
f_r	Rolling resistance coefficient, –	0.03

Table 3. Parameters of the vehicle

Symbol	Name, unit	Value
m_R	Robot total mass, kg	43.4
${}^{CM}\mathbf{r}_R$	Position of point R in $\{CM\}$, m	[0.063, 0, –0.081]
$I_{CM'y}$	Mass moment of inertia, pitch, kg.m^2	2.80
${}^{O}g_z$	Gravitational acceleration, m/s2	9.81
L	Wheelbase, m	0.425
$I_{A''y}$	Mass moment of intertia wheel, kg.m^2	0.01
n_d	Wheel-motor gear ratio, –	53
L_d	Inductance of rotor winding, H	0.0000823
R_d	Resistance of rotor winding, Ohm	0.317
k_e	Electromotive force constant, Vs/rad	0.0301
k_m	Motor torque coefficient, Nm/A	0.0302
η_d	Efficiency of transmission, –	0.5
k_P	PD controller P constant, V/rad	14.28
k_D	PD controller D constant, Vs/rad	142.8
i_{lim}	Allowable current for rotor windings, A	15

The following initial conditions were assumed: ${}^{O}\mathbf{v}_{CM}(0) = [0,\ 0,\ 0]^T$ (m/s), ${}^{O}\mathbf{r}_{CM}(0) = [0, 0, r_0 + {}^{CM}z_R]^{\ T}$ (m), $\dot{\Theta}(0) = 0$ rad/s, $\Theta(0) = 0$ rad.

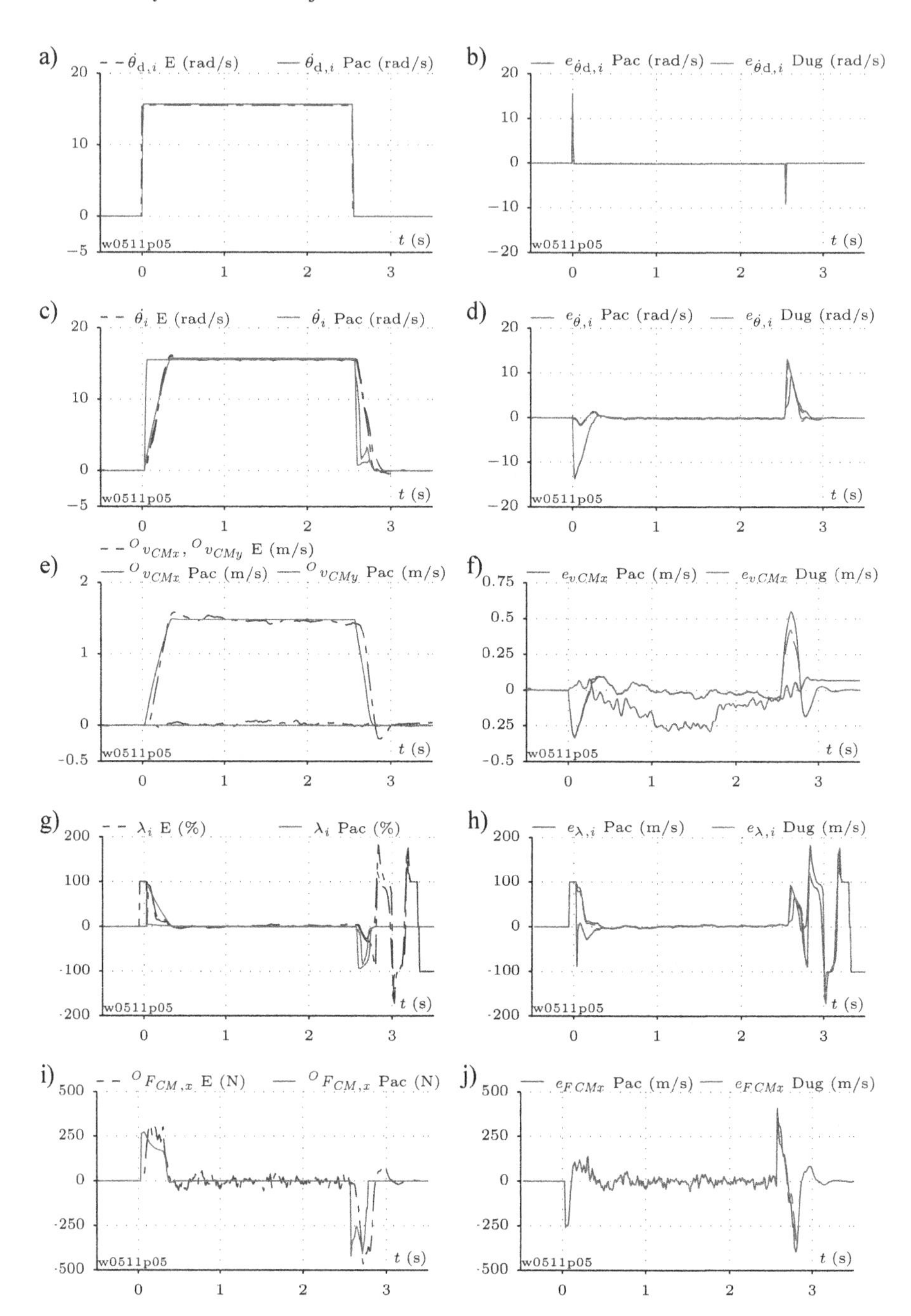

Fig. 3. Comparison of simulation results obtained using two tire models: a, b – time-histories of desired angular velocities of wheels, c, d – time histories of actual angular velocities of wheels, e, f – time histories of linear velocities of robot mass center, g, h – longitudinal wheel slip ratios, i, j – net longitudinal force applied at robot mass center

4 Results

In Fig. 3 results of the study are presented. Time histories of reference parameters from empirical experiment are shown using blue lines, of Pacejka model using red lines, and of Dugoff model using green lines. In the left column of the figure, are shown examples of obtained time histories, whereas in the right column, the errors of time histories of particular tire models against empirical data. The errors for both models are shown on a single plot to enable better comparison of results.

5 Summary and Future Works

The most important conclusions from the study are as follows:

- Results of simulations from both tire models are practically indistinguishable during period of robot motion in the steady state.
- Results of simulations based on any of the tire models differ significantly from the empirical results during initial acceleration phase, and during final braking phase. The errors during those phases can be attributed to both the model of the vehicle (independent from tire models used) and to modeling of the tire (one possible hypothesis is that tire rotational stiffness should be taken into account).
- There are differences between both tire models, although their parametrization was equivalent, and the differences are most pronounced during braking phase.

Directions of future works will include:

- Development of the tire model to take into account rotational stiffness and generally aiming at development of the high fidelity of lightweight robot simulation.

Acknowledgements. The work has been realized as a part of the project entitled "Dynamics modeling of four-wheeled mobile robot and tracking control of its motion with limitation of wheels slip". The project was financed from the means of National Science Centre of Poland granted on the basis of decision number DEC-2011/03/B/ST7/02532.

References

1. SPARC: Strategic Research Agenda For Robotics in Europe (2014–2020). http://www.eu-robotics.net/cms/upload/PPP/SRA2020_SPARC.pdf
2. Pacejka, H.B.: Tire and Vehicle Dynamics. Elsevier, New York (2012)
3. Dąbek, P., Trojnacki, M.: Requirements for tire models of the lightweight wheeled mobile robots. In: Awrejcewicz, J., Kaliński, K.J., Szewczyk, R., Kaliczyńska, M. (eds.) Mechatronics: Ideas, Challenges, Solutions and Applications. AISC, vol. 414, pp. 33–51. Springer, Heidelberg (2016). doi:10.1007/978-3-319-26886-6_3
4. Ray, L.R., Brande, D.C., Lever, J.H.: Estimation of net traction for differential-steered wheeled robots. J. Terramech. **46**, 75–87 (2009)
5. Kobayashi, T., Fujiwara, Y., Yamakawa, J., Yasufuku, N., Omine, K.: Mobility performance of a rigid wheel in low gravity environments. J. Terramech. **47**, 261–274 (2010)

6. Heverly, M., Matthews, J., Lin, J., Fuller, D., Maimone, M., Biesiadecki, J., Leichty, J.: Traverse performance characterization for the mars science laboratory rover. J. Field Robot. **30**, 835–846 (2013)
7. Fiala, E.: Seitenkraefte am rollenden Luftreifen (1954)
8. Dugoff, H., Fancher, P.S., Segel, L.: An analysis of tire traction properties and their influence on vehicle dynamic performance, 1 February 1970
9. Svendenius, J., Gäfvert, M.: A semi-empirical dynamic tire model for combined-slip forces. Veh. Syst. Dyn. **44**, 189–208 (2006)
10. Kiencke, U.W., Daiss, A.: Estimation of tyre friction for enhanced ABS-systems. JSAE Rev. **2**, 221 (1995)
11. Trojnacki, M.: Modelowanie dynamiki mobilnych robotów kołowych. Oficyna Wydawnicza PIAP, Warszawa (2013)
12. Bianco, C.G.L., Gerelli, O.: An alternative model for the evaluation of tyre shear forces under steady-state conditions. In: IEEE/RSJ International Conference on Intelligent Robots and Systems, IROS 2008, pp. 1983–1989. IEEE (2008)
13. ADIS16488A (Rev. A) – ADIS16488A.pdf. http://www.analog.com/static/imported-files/data_sheets/ADIS16488A.pdf
14. STEVAL-MKI121V1: Discovery kit for INEMO-M1 – DM00075064.pdf. http://www.st.com/st-web-ui/static/active/en/resource/technical/document/user_manual/DM00075064.pdf
15. Trojnacki, M., Dąbek, P.: Determination of motion parameters with inertial measurement units – Part 1: mathematical formulation of the algorithm. In: Awrejcewicz, J., Szewczyk, R., Trojnacki, M., Kaliczyńska, M. (eds.) Mechatronics: Ideas for Industrial Applications. AISC, vol. 317, pp. 239–251. Springer, Heidelberg (2015)

Sensors and Metrology

New Method for Calculation the Error of Temperature Difference Measurement with Platinum Resistance Thermometers (PRT)

Tadeusz Goszczyński(✉)

Industrial Institute of Automation and Measurement PIAP, Warsaw, Poland
tgoszczynski@piap.pl

Abstract. PRTs are often used for precise measurements of temperature difference in laboratories and industrial installations. For the range between 0 °C to 661 °C Standard EN 60751 defines the equation for PRT resistance vs temperature relation as quadratic equation. Therefore calculations characteristics of the temperature difference measured as two resistances of the two thermometers implies calculation the roots of many quadratic equations. This paper presents method for calculations allowing engineer to calculate in the first step, with a short simple program, crucial points on the error characteristics of every selected pair of PRTs. User can than choose a best application range for temperature and for temperature difference range and calculate values of errors in those ranges.

Keywords: Temperature sensors · Difference measurement systems · Error sources

1 Introduction

For precise definition of temperature difference measured in two points most often Platinum Resistance Thermometers are used. PRTs are used in laboratories and in industrial installations. Standard EN 60751 defines the equation for PRT resistance/temperature relation as quadratic equation. For the range between 0 °C to 661 °C the equation is:

$$r(t) = R_0\left(1 + At + Bt^2\right) \tag{1}$$

Here, $r(t)$ is the resistance at temperature t, R_0 is the resistance at 0 °C, and the constants (for an alpha = 0.00385 platinum RTD) are:

$$A = 3.9083 \cdot 10^{-3} \tag{2}$$

$$B = -5.7750 \cdot 10^{-7} \tag{3}$$

R. Szewczyk and M. Kaliczyńska (eds.), *Recent Advances in Systems, Control and Information Technology*, Advances in Intelligent Systems and Computing 543, DOI 10.1007/978-3-319-48923-0_45

Therefore calculations characteristics of the temperature difference measured as two resistances of the two thermometers implies calculation the roots of many quadratic equations.

2 Temperature Difference Calculations

Having defined by producer or in our laboratory parameters R_{01}, A_1, B_1, and R_{02}, A_2, B_2 of both thermometers of selected pair we can next calculate for every temperature difference in full range of t_1 and t_2 resistances of this thermometers:

$$R_{t1} = R_{01}\left(1 + A_1 t_1 + B_1 t_1^2\right) \tag{4}$$

$$R_{t2} = R_{02}\left(1 + A_2 t_2 + B_2 t_2^2\right) \tag{5}$$

After substituting the results into two quadratic equations with parameters R_0, A and B having values according to Standard EN 60751:

$$R_{t1} = R_0\left(1 + At_{1cal} + Bt_{1cal}^2\right) \tag{6}$$

$$R_{t2} = R_0\left(1 + At_{2cal} + Bt_{2cal}^2\right) \tag{7}$$

equations roots $t_{1\mathrm{cal}}$ and $t_{2\mathrm{cal}}$ are calculated and the relative error is determined from equation:

$$e_t = \frac{(t_{1cal} - t_{2cal}) - (t_1 - t_2)}{(t_1 - t_2)} \tag{8}$$

In order to calculate errors in full range of temperature user needs to do many thousands calculations of quadratic equations. Since values of resistance difference error are close to values of calculated temperature difference error we propose use those calculations for: first step – estimation of the resistance error characteristic values, and then in second step – standard calculation in few points of temperature characteristic in the limited temperature range where calculation of resistance difference error was satisfying for planned application.

Korytkowski et al. [2] proposed using resistance values as an output signal of the thermometers. In this procedure define first measured resistance difference:

$$\Delta r_{sp} = R_{01}\left(1 + A_1 t_1 + B_1 t_1^2\right) - R_{02}\left(1 + A_2 t_2 + B_2 t_2^2\right) \tag{9}$$

and the standard resistance difference in the same point of characteristics:

$$\Delta r_{st} = R_0\left(1 + At_1 + Bt_1^2\right) - R_0\left(1 + At_2 + Bt_2^2\right) \tag{10}$$

the relative error will be:

$$e_r = \frac{\Delta r_{sp} - \Delta r_{st}}{\Delta r_{st}} \tag{11}$$

3 Methods of Calculations for the Temperature Range Definition

In commonly known procedures – error calculations are proposed for many points laying on limit lines of the planned measurements area. (Then maximum values of error on these lines and in the area inside it are calculated).

In proposed here procedures maximum error value is predefined by application and the calculations should give as a result values for temperature range with the predefined values of error for every completed pair of thermometers. Than the user can choose easily the pair of thermometers for used range of temperatures.

The presented method of calculation can also be used by users or by manufactures and other laboratories to define the range where measurement of the temperature difference is assured with an error value less than maximum permissible error specified by user.

3.1 Extremum Points on Error Characteristics

Goszczyński in [3] proposed using the resistance signals for localisation of an extremum point on heat meters error characteristics in heat meters validation stand. In this article the method is used in looking for maximum point in temperature difference measurement error characteristic of the pair.

$$\Delta r_{sp} - \Delta r_{st} = (R_{01} - R_{02}) + (R_{01}A_1 - R_{02}A_2)t + (R_{01}A_1 - R_0A)\Delta t + (R_{01}B_1 - R_{02}B_2)t^2 + (R_{01}B_1 - R_0B)2t\Delta t + (R_{01}B_1 - R_0B)\Delta t^2 \tag{12}$$

where:

$$\Delta t = t_1 - t_2 \text{ and } t = t_2 \tag{13}$$

New variables are used to simplify equation: w_1, w_2, w_3, w_4, w_5

$$\begin{aligned} R_{01} - R_{02} = w_1; \quad R_{01}A_1 - R_{02}A_2 = w_2; \quad R_{01}A_1 - R_0A = w_3; \\ R_{01}B_1 - R_{02}B_2 = w_4; \quad R_{01}B_1 - R_0B = w_5 \end{aligned} \tag{14}$$

and new form of the equation is:

$$\Delta r_{sp} - \Delta r_{st} = w_1 + w_2t + w_3\Delta t + w_4t^2 + 2w_5t\Delta t + w_5\Delta t^2 \tag{15}$$

$$e_r = \frac{w_1 + w_2t + w_3\Delta t + w_4t^2 + 2w_5t\Delta t + w_5\Delta t^2}{R_0\left(1 + At_1 + Bt_1^2\right) - R_0\left(1 + At_2 + Bt_2^2\right)} \tag{16}$$

In order to find a place of extremum on the characteristic we define equation for derivative over temperature:

$$e'_{r(t)} = \frac{w_2 + 2w_4 t + 2w_5 \Delta t}{R_0 A \Delta t + R_0 B \Delta t^2 + 2R_0 B t \Delta t} - \frac{2R_0 B \Delta t (2w_4 t^2 + 2(w_2 + 2w_5 \Delta t)t + 2(w_1 + w_3 \Delta t + w_5 \Delta t^2))}{(R_0 A \Delta t + R_0 B \Delta t^2 + 2R_0 B t \Delta t)^2} \tag{17}$$

and find roots of the equation for zero value of the derivative:

$$t = \frac{-2w_4(R_0 A \Delta t + R_0 B \Delta t^2) \pm \sqrt{del}}{4w_4 R_0 B \Delta t} \tag{18}$$

where:

$$del = 4w_4^2(R_0 A \Delta t + R_0 B \Delta t^2)^2 - 4w_4 R_0 B \Delta t\big((w_2 + 2w_5 \Delta t)(R_0 A \Delta t + R_0 B \Delta t^2) - 4R_0 B \Delta t(w_1 + w_3 \Delta t + w_5 \Delta t^2)\big) \tag{19}$$

We will not define derivative over temperature difference since author in [3] had prove that there are no extremum places on it but it's value always continuously decrease or increase with temperature difference change. Calculations of many pairs of the thermometers with error figures produced by MATLAB shows the shape of ridge or rows, descending or ascending in continuous way (Figs. 1 and 2).

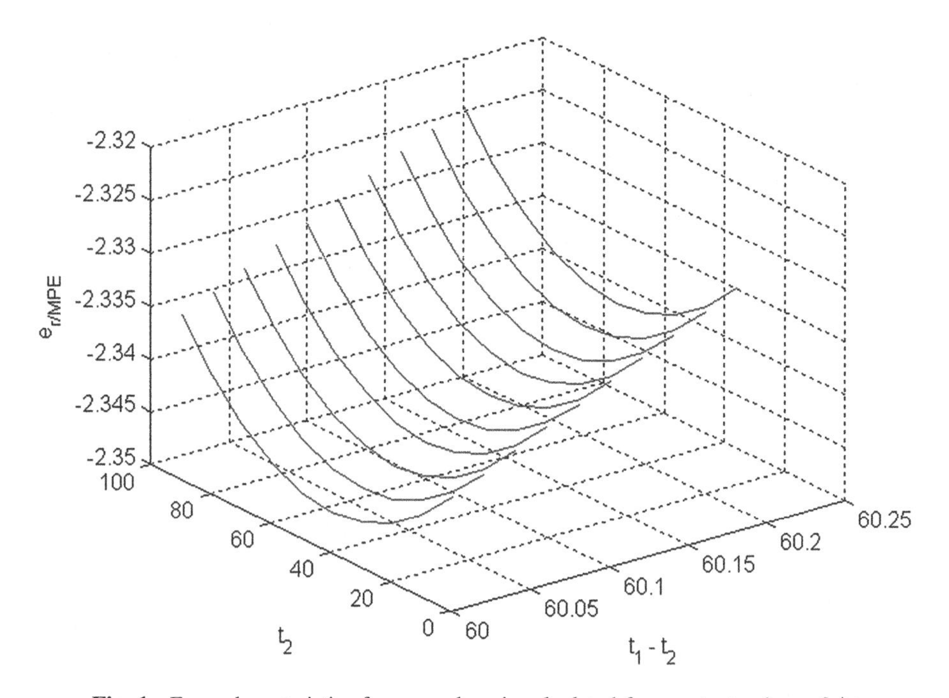

Fig. 1. Error characteristics for example pair calculated for constant values of Δt

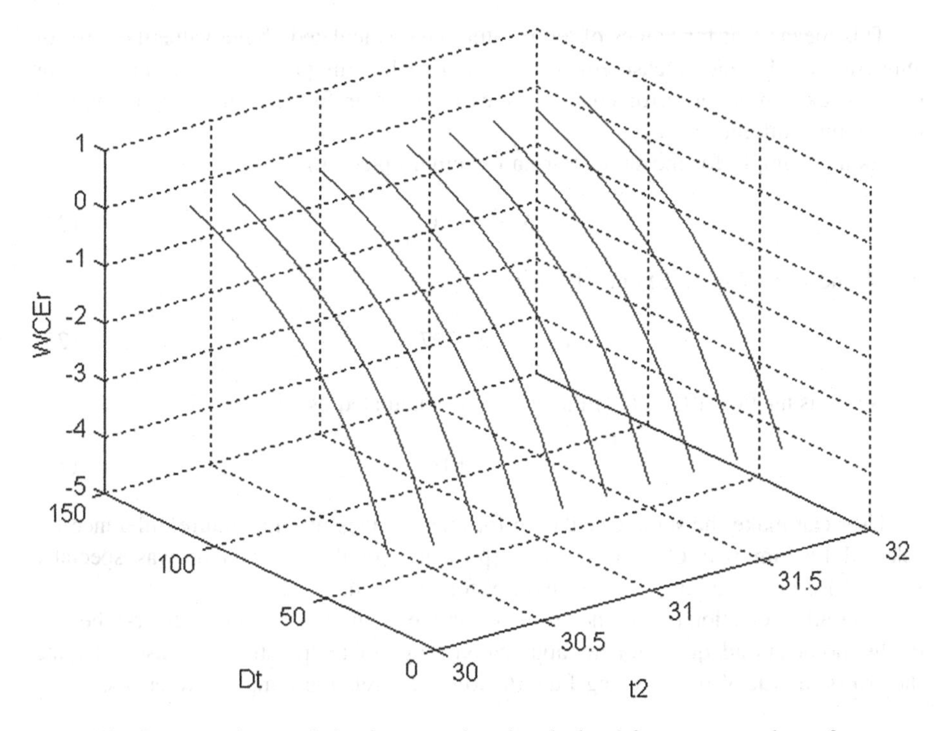

Fig. 2. Error characteristics for example pair calculated for constant values of t

4 Methods of Determination of Zero Error Places and Maximum Error Values on Characteristics

Using Eq. (9) with zero value as a result we can find the temperature of bath value when the difference in measurements will be zero:

$$0 = R_{01}\left(1 + A_1 t_1 + B_1 t_1^2\right) - R_{02}\left(1 + A_2 t_1 + B_2 t_1^2\right) \tag{20}$$

$$(R_{01}B_1 - R_{02}B_2)t_1^2 - (R_{01}A_1 - R_{02}A_2)t_1 + (R_{01} - R_{02}) = 0 \tag{21}$$

The pair was chosen with the special characteristics for showing in the best way extremum point and zero error point on its characteristics:

$$\begin{aligned} R_{01} &= 100.0;\ A_1 = 3.9663 \cdot 10^{-3};\ B_1 = -5.7985 \cdot 10^{-7}; \\ R_{02} &= 100.7, A_2 = 3.9383 \cdot 10^{-3}, B_2 = -5.8250 \cdot 10^{-7} \end{aligned} \tag{22}$$

Calculating the root t_1 of this quadratic equation for these parameters gives:

$$t_1 = t_2 = \ +10.509\,^{\circ}\mathrm{C} \tag{23}$$

This means that for values of temperature near calculated above value the error of temperature difference measurement is the lowest for this pair. Now we can look for place of extremum on error characteristics of the pair for chosen by user range of temperature difference.

As an example for the same pair an extremum place for:

$$\Delta t = 10\,^{\circ}\mathrm{C} \tag{24}$$

was found using Eqs. (18) and (19) as:

$$t_2 = +22.297\,^{\circ}\mathrm{C} \tag{25}$$

Next, using Eqs. (9) to (11) the error was calculated as:

$$e_r = 0.18 \tag{26}$$

User can make the same calculation for many values of temperature difference for selected by him pair of thermometers (presented in the article pair was specially selected for presentation purposes and not for low values of errors).

User after selection (using the presented above method of resistance errors) the pair of thermometers adequate for his application range of temperature can also calculate the errors in crucial points using Eqs. (6) to (8) for relative temperature errors.

5 Other Sources of Errors and Interchangeability

PRT interchangeability, referred to as resistance tolerance, is the "closeness of agreement" in the R vs T relationship of a PRT to a predefined nominal R vs T relationship. IEC 60751 offer several tolerance options to choose from, and manufacturers may offer other options. The interchangeability of a PRT may be a significant source of the total error when measuring temperature as stated by Tegeler et al. [4]. Two ways to reduce this error are to select a PRT with a tight interchangeability specification, and/or to use a transmitter that can be matched to a specific PRT R vs T relationship which can virtually eliminate this error.

5.1 Causes of Interchangeability Error

Many factors contribute to the interchangeability of any given PRT. One factor to consider is that the nominal R vs T relationship contained in the standards is only an approximation of how PRTs actually behave, it is based on the best fit that could be obtained using a second order equation for temperatures above 0 °C, and a third order equation for temperatures below 0 °C. Another factor, and arguably the largest contributor, has to do with the specific design, materials, and manufacturing details used to construct a PRT. Variations in the composition and purity of the platinum used, differences and variation between insulating materials, the amount of precision in adjusting the resistance during manufacture, and assembly process resistance shifts all affect the specific R vs T performance for any given PRT. Other factors to consider are the effect that external influences have on the resistance. PRTs must withstand exposure to basic levels of pressure, vibration, mechanical shock, and temperature extremes while remaining within the resistance tolerance band.

6 Other Possibilities of Measurements

The relation between temperature and resistance for the range: –200 °C to 0 °C is given by the Callendar-Van Dusen equation:

$$r(t) = R_0\left(1 + At + Bt^2 + (t-100)Ct^3\right) \tag{27}$$

where:

$$C = -4.1830 \cdot 10^{-12} \tag{28}$$

In cases where the special requirements for precision are demanded and user has got to use his own thermometers the proposed calculation program can define the change in length of platinum wire in one thermometer to get its resistance value needed for reaching the zero on characteristics of error in demanded range of temperature value.

Using the presented here method and computer with access to data base with measurement data of many platinum thermometers users could collect the pair for demanded precision of measurement in given range of temperature.

7 Conclusions

In present paper solutions that are emerging in using platinum thermometers used for temperature difference are presented. Understanding these issues and their mathematical derivations and algorithms are essential for later works on this matter. The author proposed methods of analysis for calculation of errors needed for using pairs of PRT in varied applications. The author have undertaken the experimental trials, which preliminarily prove the presented calculations very useful. The method should also be used by laboratories offering pairs of PRTs for sale. Specifying values on temperature characteristics where errors have minimal and maximal values would be very useful for their clients. The next step is planned to widen the analysis on measurements in temperature range from –200 °C to 0 °C and for calculating values of corrections of platinum wire resistance (length in one thermometer) needed for better accuracy in the range of measurements.

References

1. European Standard EN 60751 (IEC Publication 751) Industrial platinum resistance thermometer sensors
2. Korytkowski, J., Goszczynski, T., Jachczyk, E.: Methods of computer testing system for examining the accuracy of temperature probes of heat meter. In: Automation 1998 Conference, Warsaw, Poland, pp. 335–342, 11–12 March 1998. (in Polish)
3. Goszczynski, T.: Error determination for heat meter validation. Heat Transf. Eng. **31**(1), 83–89 (2010)
4. Tegeler, E., Heyer, D., Siebert, B.: Uncertainty of the calibration of paired temperature sensors for heat meters. In: Tempmeko 2007 Conference, Lake Louise, Canada (2007). www.tempmeko2007.org
5. Goszczynski, T.: Measuring Instrument for errors of platinum resistance thermometers pairs P-389244 – Patent Office RP

Measurement of the Number Servings of Milk and Control of Water Content in Milk on Stall Milking Machines

Volodymyr Kucheruk(✉), Pavel Kulakov, and Natalia Storozhuk

Department of Metrology and Industrial Automation,
Vinnytsia National Technical University, Vinnytsia, Ukraine
{kucheruk, kulakov, storozhuk}@vntu.edu.ua

Abstract. In this article the dependence of the output voltage of photo-receiver based on photodiode-operational amplifier pair from relative mass fraction of milk in water-milk solution, during infrared radiation is passaging through it, is received. The method for determining the optimum wavelength for providing maximum sensitivity of measuring the relative mass fraction of milk is created. The results of theoretical research are confirmed by an experiment.

Keywords: Stall milking machine · Water-milk solution · Infrared radiation · Photo-receiver

1 Introduction

Stall milking machines provide machine milking of cows in stalls, transporting milk by means of milk line in the milk container, filtration of milk and its pumping from milk containers to the refrigerator or mixing machine using milk pump, measuring of milk yield, which is received by each milker. In such milking machines animals are divided into groups and refined in line along the milk line, there are four or eight such lines in a barn, each line is served by one milker, milking takes place using between two and four milking machines. Milk, obtained from each animal in the line, accumulates in the dispenser, located at the end of each line. After the milk of determined volume was accumulated in the dispenser, automatic drain valve triggeres and a portion is absorbed into the milk container, from which it enters the cooling or mixing machine [1]. Milker wages are calculated according to the number of portions of milk, which is formed by the dispenser, and which are calculated by special milk portion meters.

Each milker has a bowl of water on stall milking machines, which is necessary to prepare the animals for milking. There are cases when milker, using machine, absorbs this water in the milk line to improve milk yield indices, which influences on his salary. Identifying these cases visually is very difficult, because human eye can't distinguish the water-milk solution of pure milk. The same problem occurs when purchasing milk in small private farms. Dishonest providers dilute milk with water and pass it to the receiving station. Existing measuring parameters of milk are expensive and do not allow operational control of the availability of water in milk, and it is not possible to detect milker, which provides milk falsification with their help. Therefore, creating a

R. Szewczyk and M. Kaliczyńska (eds.), *Recent Advances in Systems, Control and Information Technology*,
Advances in Intelligent Systems and Computing 543, DOI 10.1007/978-3-319-48923-0_46

method for rapid measurement of relative mass fraction of milk in water-milk solution is an important and urgent task.

2 Analisys of Studies

Nowadays, to identify the facts of falsification of milk in the lab the measuring control of its freezing temperature is used [2]. This method can't be used to stall milking machines during milking.

At stall milking installations with milk line, milk portions meters are used with the sensor, the principle of which follows [3]. After passing a portion of milk from the dispenser through the drain valve, the milk enters the flexible hose. Inside of the hose there is a tube with mounted electrodes. With the passage of portions of milk through a tube between the electrodes the electrical conductivity changes, which is a sign for dispenser to operate. If you change the conductivity, sensor generates a voltage signal that is directly proportional to the value of conductivity. If the milk contains the water, and the temperature changes, and milk foams, the initial conductivity conversion error significantly increases. The water in the milk leads to a decrease of its conductivity, but, due to the presence of the above factors, revealing a slight dilution of milk is impossible.

Also to count the portions of milk they use optical sensor [4], the principle of which follows. A portion of the milk passes through the tube, which is integrated with infrared LED and an infrared phototransistor, which operates in key mode. With the passage of portions of milk the light flow interrupts, therefore an impulse of the determined duration forms in the output of phototransistor. Basing on the measurement of output signal of phototransistor, the fact of passing of portions of milk is established. When there is a large amount of water in the milk, the luminous flow passes through it with little loss of power. As a result, a portion of milk is not counted, but finding a small amount of water in the milk using the aforementioned sensor is also impossible.

3 Problem Formulation

To further creation of a mean of measuring operational control of presence of water in the milk, we need to obtain the dependence of the output voltage of photo-receiver based on photodiode-operational amplifier pair, from the relative mass fraction of milk in water-milk solution, when the infrared radiation passes through solution. To maximize the sensitivity of an aforementioned mean of measuring operational control, we must create a method for determining the optimum wavelength of infrared radiation. The results of theoretical research are necessary to confirm by the results of the experiment.

4 Main Part

At the stall facilities obtained from each animal in the milk line, milk accumulates in the dispenser, located at the end of each line. After serving of milk in the dispenser accumulates of a certain volume, automatic drain valve triggers, and a portion is transported to an automatic mixer or refrigerator. The capacity of the dispenser is P_{DM} = 10 l/min. In the presence of high-performance animals in the stall milking machines milk dispenser with drain solenoid valve is used, the capacity of such dispenser is P_{DM} = 12 l/min. At present, at the stall milking machines with milk pipes, milk meters of milk portions with the sensor are used, the principle of which follows. After a portion of milk passes from the dispenser through the drain valve, the milk enters the flexible hose. In the intersection of the hose the tube with mounted electrodes is situated. With the passage of servings of milk through a tube, between the electrodes the electrical conductivity changes, this is a sign of triggering dispenser. If you change the conductivity, sensor generates a voltage signal, that is directly proportional to its value and duration of which is equal to the duration of the passage of the portion of milk. On the basis of the amplitude parameters measuring and duration of this signal, the fact of the portion passages is established. Using this sensor, in the presence of water in the milk, changes in temperature, foaming milk, the error of conductivity initial conversion significantly increases. This leads to the fact that the indicator of portions starts to behave unpredictably – skip some portion or counts portions in their absence. Also to count servings of milk optical sensor is used, the principle of which follows. A portion of the milk passes through the tube, in which is integrated infrared LED and phototransistor which operates in key mode. With the passage of servings of milk the flow of light is interrupting, and as a result in the output of phototransistor a pulse of certain duration is formed. On the basis of the measurement of the signal duration in phototransistor output, the fact of the portion of milk passages is established. In this case, in presence of the foaming milk, at the phototransistor output appears random sequence of short pulses, which is actually a noise component added to the signal. Consequently, in many cases, error of the measurement of the signal duration in phototransistor output increases, and accordingly, the counting the servings error too. Also, the duration and waveform of the signal from above sensors depends on the drain pollution and other components of the dispenser, configuration of milk pipes, the degree of foaming milk, the duration of the reverse drain, contamination of the sensor elements, which also leads to increased counting error. The process of measuring transformation of the intensity of light flow to voltage considered in [4].

The authors proposed and discussed servings of milk counter, formed by dispenser, with the function of detecting water content in milk. Let`s consider the principle of its operation on the basis of functional circuit that is shown in Fig. 1 and the time charts that are shown in Fig. 2.

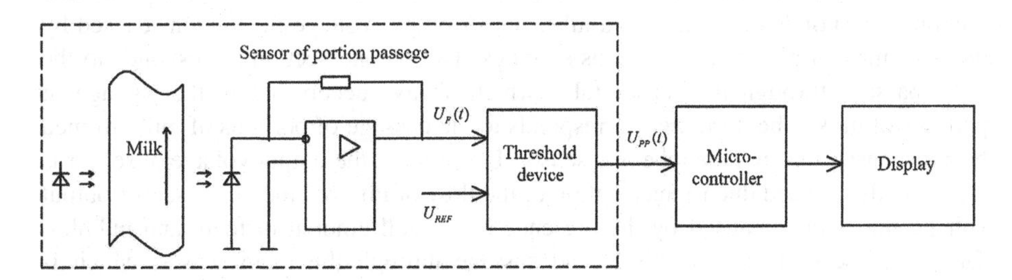

Fig. 1. Functional diagram of the counter of portions of milk, formed by dispenser of stall milking machine, with the function of detecting water in milk

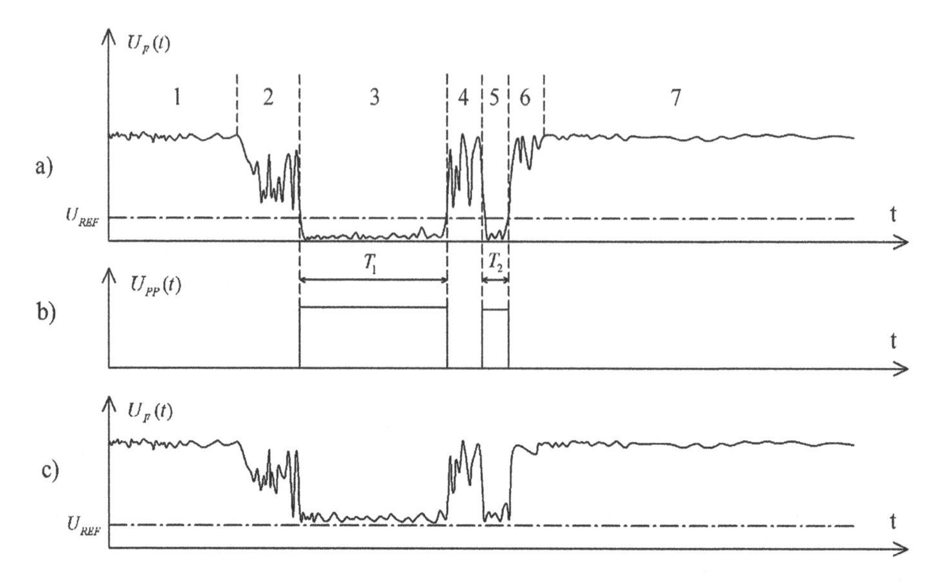

Fig. 2. Time diagrams of the counter of portions of milk, formed by dispenser of stall milking machine, with the function of detecting water in milk: (a) the timing diagram of the output voltage of photodetector, during portions of milk passage, free of water; (b) timing diagram of the output voltage of threshold device, during portions of milk passage, free of water; (c) the timing diagram of the output voltage of photodetector, during portions of milk with water passage

The structure of the sensor of passing a portion includes a tube through which is carried the draining portions of milk formed by dispenser. In the tube there is a source of infrared radiation, opposite to him is a photodiode, which is part of the photodetector based on a pair of photodiode-operational amplifier. The output signal of the photodetector reaches the input of threshold device with which it is compared with a certain reference voltage U_{REF}. Output signal of threshold device $U_{PP}(t)$ comes to a discrete entrance of microcontroller, which measures the duration of the signal. On the basis of measuring the duration the fact of the portion of milk passages is established, information about the number of counted servings visualizes by using the display.

Consider Fig. 2a, where are the timing diagram of the output voltage of photodetector based on a pair of photodiode-operational amplifier during passing portions of milk, in which there is no water. On this graph seven areas is highlighted. The first section corresponds to no milk in a tube, sensor output voltage fluctuations caused by drops of milk drip from the previous servings. The second section corresponds to the foam passage through the sensor tube, which always occurs before the passage of portions of milk. The third area corresponds to the passage of portions of milk, formed by a dispenser, through the tube of a sensor. In this case, the output voltage of sensor is significantly reduced due to interruption of the flow of infrared radiation, sensor output voltage fluctuations caused by the presence of a small amount of foam and bubbles. The fourth section corresponds to the passage through the foam sensor, which is formed at the end passage of the portion. The emergence of the fifth area is caused by the fact, that the small portion of milk is not absorbed in milk pipes and merges in the opposite direction. The sixth area is due to the presence of a small amount of foam that follows part of the portion that merges back. The seventh section of the graph as the first, corresponds to no milk in a sensor tube. Let`s consider Fig. 2b, which shows the timing diagram of the output voltage of threshold device when portions of milk passage with no water. This signal consists of two rectangular pulses with duration T_1 and T_2 which are formed when the sensor output voltage based on a pair of photodiode-operational amplifier $U_F(t)$ is less than the reference voltage U_{REF}. Impulse with duration T_1 corresponds to the passage of portions of milk from dispenser through the sensor, second pulse with duration T_2 corresponds to reverse drain of milk. The value T_1 and T_2 depend on the contamination of the dispenser and its elements, the height of the location and configuration milk pipes, milk pipes vacuum mode, volume of portions of milk. Informative in terms of identifying the passage of portion is only the first impulse of threshold device output. The duration of the pulse T_1, with the volume of portions 1 L, ranges from $T_{1MIN} = 3.2$ s to $T_{1MAX} = 4.6$ s. Accordingly, the criterion for identifying the portion passage is achieving the duration of the first pulse of threshold device output, value T_{1MIN}, namely

$$T_1 = T_{1MIN}. \tag{1}$$

After the identification of passing portion of milk and its enrollment, the time interval T_Z is formed by program, during the interval the output signal of threshold device is not analyzed. This is required to back drain triggered not the enrollment of additional portions. It is clear that the sum of the duration of the minimum time interval of portions passing T_{1MIN} and time interval T_Z shall not exceed the duration of minimum time interval between dispenser formed portions T_{DMPMIN}, which is given by

$$T_{DMP\,MIN} = 1/P_{DM}. \tag{2}$$

Based on this

$$T_Z = T_{DMP\,MIN} - T_{1MIN} = 1/P_{DM} - T_{1MIN}. \tag{3}$$

In the presence of water content in the milk, the optical density of the passage of infrared radiation is reduced, accordingly increases the output voltage of photodetector based on photodiode-operational amplifier pair. A small amount of water in the milk, based on measuring the output voltage of photodetector, it is impossible to detect, because of the possibility of passing a certain amount of infrared radiation through a layer of milk without additives, reflection radiation, radiation diffraction and it's scattering by fat balls, non-monochromatism of radiation source. The value of the reference voltage U_{REF}, which corresponds to a value of mass fraction of milk in water and milk solution, depends on a large number of random factors and determined experimentally. Let`s consider Fig. 2c, in which there is timing diagram showing the sensor output voltage while portion passing, in which water is present. In the case of presence of certain water content in milk, output signal of linear sensor is larger than the reference voltage U_{REF}. As a result, at the output of threshold device a pulse will form which corresponds to the passage portions of milk, which are formed by dispenser. Consequently, the relevant portion of milk with a high content of water will not count.

Filtering of the measuring signal to reduce random measurement errors may be carried out by using an amplitude detector [5].

The weakening of monochromatic radiation substance is determined by Beer-Lambert-Bouguer law, that binds the input and output intensity I of the optical radiation as it passes through substance [6]

$$I = I_0 \cdot e^{-\alpha cd}, \tag{4}$$

where I_0 – the intensity of radiation, that falls on substance; d – thickness of a layer; c – concentration of the substance; α – attenuation coefficient.

Beer-Lambert-Bouguer law can be written using decimal logarithms

$$I = I_0 \cdot 10^{-k(\lambda)cd}, \tag{5}$$

where $k(\lambda)$ – extinction coefficient, which is a function of the wavelength of optical radiation λ.

Skipping by substance the radiation is characterized by transmittance K_{PR}, defined as

$$K_{PR}(\lambda) = I/I_0 = 10^{-k(\lambda)cd}. \tag{6}$$

An important characteristic of the substance is its optical density $D(\lambda)$, determined by the expression

$$D(\lambda) = \log(I/I_0) = k(\lambda)cd. \tag{7}$$

Passing through a solution of the components, radiation by each of them is absorbed in different ways. The resulting absorption is derived by additive superposition of individual components. Accordingly, the optical density n – component mix is determined by the expression

$$D(\lambda) = \sum_{i=1}^{n} k_i(\lambda) c_i d, \tag{8}$$

where $k_i(\lambda)$ – extinction coefficient of i-component of the mix; c_i – concentration; i – component of the mix.

For the water-milk solution $n = 2$, accordingly, its optical density $D_{VM}(\lambda)$ is determined by the expression

$$D_{VM}(\lambda) = d(k_M(\lambda) \cdot c_M + k_V(\lambda) \cdot c_V), \tag{9}$$

where $k_M(\lambda)$ – milk extinction coefficient; c_M – concentration of the milk in water-milk solution; $k_V(\lambda)$ – water extinction coefficient; c_V – concentration of the water in water-milk solution.

The volume of water-milk solution, that absorbs radiation, is given by

$$V_K = V_M + V_V = m_M/\rho_M + m_V/\rho_V, \quad (10)$$

where V_M – volume of milk; V_V – volume of water; m_M – mass of the milk in water-milk solution; ρ_M – density of milk; m_V – mass of the water in water-milk solution; ρ_V – density of water.

Concentration of the water c_V in water-milk solution defined as

$$c_V = \frac{m_V}{V_K} = \frac{m_V}{m_M/\rho_M + m_V/\rho_V} = m_V \cdot \frac{\rho_M \cdot \rho_V}{m_M\rho_V + m_V\rho_M}. \quad (11)$$

Concentration of the milk c_M in water-milk solution

$$c_M = \frac{m_M}{V_K} = \frac{m_M}{m_M/\rho_M + m_V/\rho_V} = m_M \cdot \frac{\rho_M\rho_V}{m_M\rho_V + m_V\rho_m}. \quad (12)$$

The relative mass fraction of milk in water-milk solution is defined as

$$\eta = \frac{m_M}{m_M + m_V}. \quad (13)$$

From the expression (13), after simple transformations, we get

$$m_V = m_M\left(\frac{1}{\eta} - 1\right). \quad (14)$$

Substituting (14) to (11) and (12), after transformations we obtain expressions that link the concentration of milk and water in the milk-water solution with relative mass fraction of milk

$$c_V = \frac{\rho_M\rho_V\left(\frac{1}{\eta} - 1\right)}{\rho_V + \rho_M\left(\frac{1}{\eta} - 1\right)}, \quad (15)$$

$$c_M = \frac{\rho_M\rho_V}{\rho_V + \rho_M\left(\frac{1}{\eta} - 1\right)}. \quad (16)$$

Substituting (15) and (16) to (9), and after transformations we obtain the dependence of the optical density of the water-milk solution from the relative mass fraction of milk

$$D_{VM}(\lambda) = \frac{d\rho_M\rho_V\left(k_M(\lambda) + k_V(\lambda)\left(\frac{1}{\eta} - 1\right)\right)}{\rho_V + \rho_M\left(\frac{1}{\eta} - 1\right)} \quad (17)$$

In view of (8), after transformations we obtain

$$I = I_0 \cdot 10^{-\frac{d\rho_M \rho_V \left(k_M(\lambda) + k_V(\lambda)\left(\frac{1}{\eta} - 1\right)\right)}{\rho_V + \rho_M\left(\frac{1}{\eta} - 1\right)}}. \quad (18)$$

Expression (18) is a mathematical model of optical radiation absorption by water and milk solution. This expression links the intensity of optical radiation passing through the solution with relative mass fraction of milk in solution. According to [7], the output voltage of photo-receiver U_F, based on photodiode-operational amplifier pair, in a first approximation is given by

$$U_F = IS_{I0}(\lambda)R_{ZZ}S_{VD}, \quad (19)$$

where S_{VD} – square of photosensitive layer of photodiode; $S_{I0}(\lambda)$ – spectral sensitivity of photodiode; R_{ZZ} – resistance in the feedback loop of the operational amplifier.

Substituting (18) in (19) and obtain the dependence of the output voltage of photo-receiver, based on photodiode-operational amplifier pair, from the relative mass fraction of milk in water-milk solution.

$$U_F(\eta) = I_0 S_{I0}(\lambda) R_{ZZ} S_{VD} 10^{-\frac{d\rho_M \rho_V \left(k_M(\lambda) + k_V(\lambda)\left(\frac{1}{\eta} - 1\right)\right)}{\rho_V + \rho_M\left(\frac{1}{\eta} - 1\right)}}, \quad (20)$$

Solving the Eq. (20) as concerns η, obtain an expression that links the relative mass fraction of milk in water-milk solution with an output voltage of photo-receiver, based on photodiode-operational amplifier pair

$$\eta = \frac{\rho_M \log \frac{U_F(\eta)}{I_0 S_{I0}(\lambda) R_{ZZ} S_{VD}} + d\rho_M \rho_V k_V(\lambda)}{(\rho_M - \rho_V) \cdot \log \frac{U_F(\eta)}{I_0 S_{I0}(\lambda) R_{ZZ} S_{VD}} + d\rho_M \rho_V (k_V(\lambda) - k_M(\lambda))}. \quad (21)$$

Infrared spectrometry is widely used for quality control of foodstuffs [6]. Milk and milk-water solution refer to substances with a significant degree of absorption of infrared radiation in the near region of the spectrum, characterized by wavelengths of 0.75 μm to 2.5 μm [6]. This fact leads to high sensitivity of means of measuring the mass fraction of milk in the water-milk solution, based on infrared spectrometry methods.

Figure 3 shows typical experimental spectral transmission characteristics of infrared radiation of near areas of spectrum for milk and water, which thickness is $d = 10\,\text{mm}$ [6].

The spectral transmittance characteristics of substance is the dependence of transmittance, defined by (6), from the wavelength of the optical radiation. Extinction coefficient and transmittance are connected by the dependence

$$k(\lambda) = -\frac{1}{cd} \cdot \log K_{PR}(\lambda). \quad (22)$$

It should be noted, that the shape of the spectral transmission characteristics is highly dependent on the thickness of matter d. The mathematical expression, that

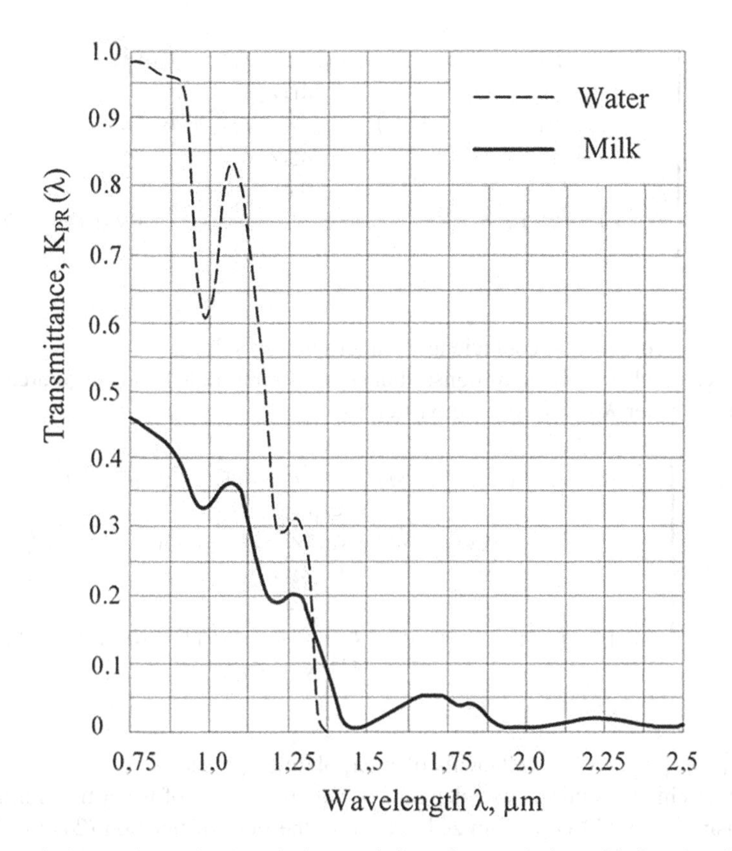

Fig. 3. Typical experimental spectral characteristics of transmission of infrared radiation for milk and water, thickness of which is d = 10 mm

describes a spectral transmission characteristic, can be obtained only by interpolation results of experimental data.

Let`s find the optimum wavelength of optical radiation, when the sensitivity of mean of the measurement will be maximum. It is clear, that sensitivity will be maximum at the wavelength, at which ratio of transmittance of water to milk transmission ratio will be maximized. In this case the maximum reduction of the absorption of infrared radiation by milk is provided, due to the presence of water.

To determine the optimum value of wavelength of infrared radiation, let's hold interpolation of spectral transmission characteristics of infrared radiation near spectrum areas for milk and water. We know that high quality results of interpolation are achieved with cubic spline function, which consists of pieces of cubic polynomials, with using which pulsations of interpolating function are not possible [8]. Cubic splines are cubic parabola, which, in each case, are just passing through two control points, which are determined from experimental spectral characteristics. Thus, the spectral transmittance characteristics of infrared radiation of near areas of the spectrum for milk $K_{PRM}(\lambda)$ can be represented as an interpolating function:

$$K_{PRM}(\lambda) = \begin{cases} a_{1M} + b_{1M}(\lambda - \lambda_1) + c_{1M}(\lambda - \lambda_1)^2 + d_{1M}(\lambda - \lambda_1)^3, \\ \lambda \in [\lambda_1, \lambda_2]; \\ a_{2M} + b_{2M}(\lambda - \lambda_2) + c_{2M}(\lambda - \lambda_2)^2 + d_{2M}(\lambda - \lambda_2)^3, \\ \lambda \in [\lambda_2, \lambda_3]; \\ \ldots \\ a_{N-1M} + b_{N-1M}(\lambda - \lambda_{N-1}) + c_{N-1M}(\lambda - \lambda_{N-1})^2 + d_{N-1M}(\lambda - \lambda_{N-1})^3, \\ \lambda \in [\lambda_{N-1}, \lambda_N]; \end{cases} \tag{23}$$

where $a_{iM}, b_{iM}, c_{iM}, d_{iM}$ – coefficients of interpolating splines.

The spectral characteristic of transmittance of infrared radiation of near areas of the spectrum for water $K_{PRV}(\lambda)$ can be written as

$$K_{PRV}(\lambda) = \begin{cases} a_{1V} + b_{1V}(\lambda - \lambda_1) + c_{1V}(\lambda - \lambda_1)^2 + d_{1V}(\lambda - \lambda_1)^3, \\ \lambda \in [\lambda_1, \lambda_2]; \\ a_{2V} + b_{2V}(\lambda - \lambda_2) + c_{2V}(\lambda - \lambda_2)^2 + d_{2V}(\lambda - \lambda_2)^3, \\ \lambda \in [\lambda_2, \lambda_3]; \\ \ldots \\ a_{N-1V} + b_{N-1V}(\lambda - \lambda_{N-1}) + c_{N-1V}(\lambda - \lambda_{N-1})^2 + d_{N-1V}(\lambda - \lambda_{N-1})^3, \\ \lambda \in [\lambda_{N-1}, \lambda_N]; \end{cases} \tag{24}$$

where $a_{iV}, b_{iV}, c_{iV}, d_{iV}$ – coefficients of interpolating splines.

To determine the optimal wavelength, for which the ratio of water transmittance to milk transmittance will be maximized, let`s find the ratio of function (24) to (23)

$$K_{PRMV}(\lambda) = \frac{K_{PRV}(\lambda)}{K_{PRM}(\lambda)} = \begin{cases} \frac{a_{1V} + b_{1V}(\lambda-\lambda_1) + c_{1V}(\lambda-\lambda_1)^2 + d_{1V}(\lambda-\lambda_1)^3}{a_{1M} + b_{1M}(\lambda-\lambda_1) + c_{1M}(\lambda-\lambda_1)^2 + d_{1M}(\lambda-\lambda_1)^3}, \\ \lambda \in [\lambda_1, \lambda_2]; \\ \frac{a_{2V} + b_{2V}(\lambda-\lambda_2) + c_{2V}(\lambda-\lambda_2)^2 + d_{2V}(\lambda-\lambda_2)^3}{a_{2M} + b_{2M}(\lambda-\lambda_2) + c_{2M}(\lambda-\lambda_2)^2 + d_{2M}(\lambda-\lambda_2)^3}, \\ \lambda \in [\lambda_2, \lambda_3]; \\ \ldots \\ \frac{a_{N-1V} + b_{N-1V}(\lambda-\lambda_{N-1}) + c_{N-1V}(\lambda-\lambda_{N-1})^2 + d_{N-1V}(\lambda-\lambda_{N-1})^3}{a_{N-1M} + b_{N-1M}(\lambda-\lambda_{N-1}) + c_{N-1M}(\lambda-\lambda_{N-1})^2 + d_{N-1M}(\lambda-\lambda_{N-1})^3}, \\ \lambda \in [\lambda_{N-1}, \lambda_N]; \end{cases} \tag{25}$$

At the optimum wavelength λ_{OPT} of the infrared radiation there is the relation

$$K_{PRMV}(\lambda_{OPT}) = \max_{[\lambda_1, \lambda_N]} K_{PRMV}(\lambda). \tag{26}$$

Thus, finding the optimum wavelength is reduced to solving the standard problem of determining the maximum of function (25). In Fig. 4a, there are the results of interpolation using cubic splines of spectral transmission characteristics of water and milk, which are on 1, and in Fig. 4b – a graph of the function described by (25).

As follows from Fig. 4, the optimum wavelength of infrared radiation for determining the relative mass fraction of milk in water-milk solution at 10 mm is about

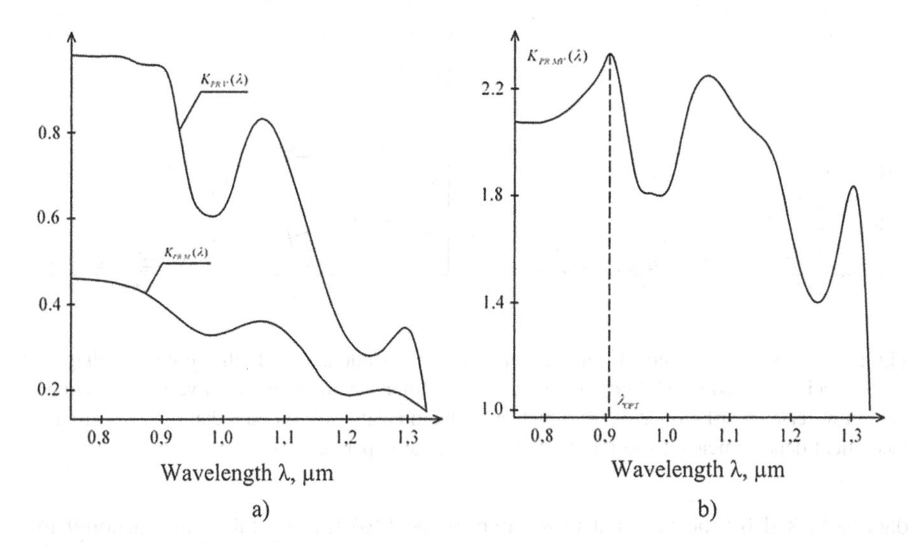

Fig. 4. Results of interpolation of spectral transmission characteristics of water and milk and graph of the ratio of their interpolating functions: (a) interpolating functions of spectral transmission characteristics of water and milk; (b) the ratio of interpolating functions of spectral transmission characteristics of water and milk

0.91 μm. In accordance with (20), to ensure a minimum threshold of sensitivity, it is desirable to maximize approximation wavelength, which corresponds to the maximum of spectral characteristics of the photodiode, to the optimum wavelength of infrared radiation. In other words, ideally, there must be performed the ratio

$$\lambda_0 = \lambda_{OPT}. \tag{27}$$

In the Fig. 5a there is theoretical dependencies family $U_F(\eta)$ of output voltage of photo-receiver, based on photodiode-operational amplifier pair, from the relative mass fraction of milk in water-milk solution at different values of the intensity of infrared radiation. The above dependencies are defined by the expression (20), the value of I_0 during their construction chosen so as to ensure equality of theoretical and experimental values at zero relative mass fraction of milk in water-milk solution. Figure 5b shows a graph of experimental dependence $U_{FE}(\eta)$ of the output voltage of photo-receiver, based on photodiode-operational amplifier pair, from the relative mass fraction of milk in water-milk solution at different values of current of infrared LEDs, which correspond to different values of the intensity of infrared radiation.

In experimental studies, as the radiator is used IR LED EL IR11-21C manufactured by Everlight Americas Inc., which has a nominal wavelength of the infrared radiation of 0.94 μm and a maximum current of 100 mA. Photo-receiver, based on photodiode-operational amplifier pair was implemented on photodiode S1336-18BQ manufactured by Hamamatsu Photonics, whose spectral characteristics has a maximum at a wavelength of radiation of 0.96 microns, and which at this wavelength has integrated current sensitivity of 0.5 A/W. A certain difference between the experimental and theoretical

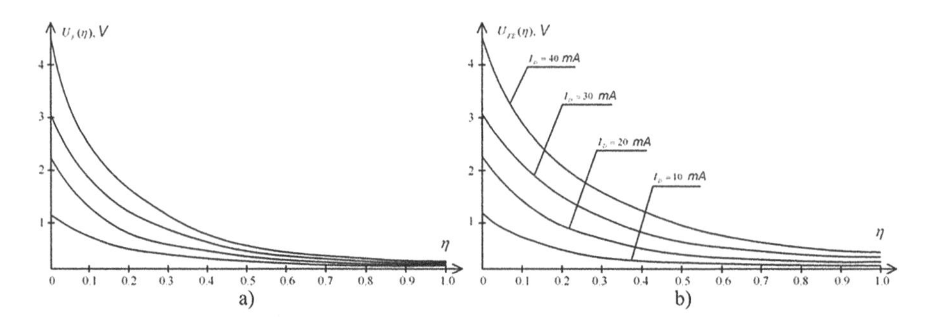

Fig. 5. Family of theoretical and experimental dependencies of the output voltage of photo-receiver, based on photodiode-operating amplifier pair from the relative mass fraction of milk in water-milk solution at different values of the intensity of infrared radiation: (a) a family of theoretical dependencies; (b) a family of experimental dependencies

data is caused by the fact that in the expression (20) it's not taken into account that there are the reflection of infrared radiation from the water-milk solution, radiation diffraction and its scattering in fat balls in solution, not a monochromatism of sources of radiation.

5 Conclusions

In the process of theoretical studies, we obtained the dependence of photo-receiver output voltage, based on photodiode – operational amplifier pair, from the relative mass fraction of milk in water-milk solution. Based on the above dependence it's possible to implement the mean of measuring control of the presence of water in the milk during the milking at stall milking facilities in order to reveal the falsification of milk by milkers. We proposed the method of determining the optimum, for measuring the relative mass fraction of milk, wavelength of infrared radiation, which passes through the water-milk solution. The adequacy of the obtained theoretical results is confirmed by experimental research.

References

1. Hand book of milk processing, dairy products and packaging technology, Engineers India Research Institute, p. 510 (2007)
2. Topel, A.: Chemistry and Physics of Milk, p. 472. VEB Fachbuchverlag, Germany (1976)
3. Milking machine UDM-200 [Text]. Technical description and operating instructions, Bratslav, p. 165 (2002)
4. Podzharenko, V., Kulakov, P.: Photoelectric angle converter. In: Proceedings of SPIE 4425, Selected Papers from the International Conference on Optoelectronic Information Technologies, p. 452, 12 June 2001

5. Kucheruk, V., Ovchynnykov, K., Molchaniuk, M., Kurytnik I.P.: The usage of the linear interpolating filter for an accurate fluctuation fading time measuring activated in LC-circuit. Przegląd Elektrotechniczny, pp. 68–70 (2013). ISSN 0033-2097, R. 89 NR 8/2013
6. Stesler, H.W., Ozaki, Y., Kawata, S., Heise, H.M.: Near-Infrared Spectroscopy: Principles, Instruments, Applications, p. 361. Wiley-VCH, New York (2002)
7. Kucheruk, V.Y.: Photoelectrical measuring conversion of area-voltage [Text]. In: Kucheruk, V.Y., Palamarchuk, Y.A., Kulakov, P.I., Gnes, T.V., Blohin, Y.Y. (eds.) Optoelectronic Information and Energy Technologies, vol. 27(1), pp. 139–145 (2014)
8. Singh, S., Burry, J., Watson, B.: Approximation theory and spline functions. In: Proceeding of the NATO Advanced Study Institute on Approximation Theory and Spline Functions, Canada Newfoundland (1983)

Investigation of the Appropriate Method of Mounting Tested Elements in the Test Stand for Temperature Characteristics of Ultra-Precise Resistors

Andrzej Juś[1(✉)], Paweł Nowak[1], Roman Szewczyk[2], and Weronika Radzikowska-Juś[3]

[1] Industrial Research Institute for Automation and Measurements PIAP, Al. Jerozolimskie 202, 02-486 Warsaw, Poland
ajus@piap.pl
[2] Faculty of Mechatronics, Warsaw University of Technology, sw. A. Boboli 8, 02-525 Warsaw, Poland
[3] Faculty of Civil Engineering and Geodesygen, Military University of Technology, Sylwestra Kaliskiego 2, 00-908 Warsaw, Poland

Abstract. Paper presents problems and results of research of the method of mounting the test objects in the test stand. This test stand has been developed to measure thermal characteristics of ultra-precise resistors such as UPR 0.5 D10 distinguished by very low temperature coefficient of resistance (TCR). Main advantage of the utilized test stand is high accuracy (reaching up to 0.05 ppm/°C) of TCR measurement. Obtaining such high measurement accuracy would be not possible without appropriate method of mounting resistors in the test stand. Paper presents two possible methods of mounting resistors and compares their pros and cons, both in function of measurement accuracy as well as ease of use.

Keywords: Resistance measurement · Stable resistors · Mounting of resistors · Temperature coefficients · TCR

1 Introduction

Due to the need of research of temperature parameters of the precise reference resistors the test stand was developed [1]. The reason of that need was the desire to reduce high costs of the utilized elements. Precise reference resistors are very expensive, and their cost can be a significant part of a total device cost. So it has proved advantageous to select those elements from the series of cheaper components.

The precision of determination the temperature characteristic of the researched resistors is very high – its reliability (during research of very stable resistors) is on the level of 0.05 ppm/°C and has been previously reported in [1–3]. However to achieve such accuracy a number of problems had to be solved during the development of the test stand. One of the most important was the selection of the method of mounting resistors into the test stand, in the way which does not influence on the results.

R. Szewczyk and M. Kaliczyńska (eds.), *Recent Advances in Systems, Control and Information Technology*, Advances in Intelligent Systems and Computing 543, DOI 10.1007/978-3-319-48923-0_47

2 Test Stand and Measurement Method

Developed test stand is designed to measure a variation of the resistance of the tested resistors. This determinate the measurement method – the measurements may be carried out as differential. Such approach resulted with increased accuracy of the tests and, same time, facilitated the development of the test stand [4–6].

Considering the necessity of an automation of the test process and high measurement accuracy, method based on DC compensator was used. The principles of the method operation is presented in Fig. 1.

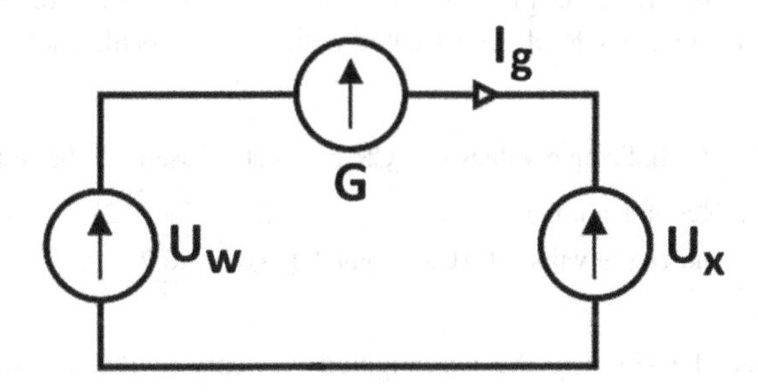

Fig. 1. Principle of DC compensation measurements. U_w – reference voltage source, U_x – measured voltage, G – galvanometer [7–9]

In the measurement loop of the developed test stand voltage sources U_w and U_x have been replaced by the resistors R_x and R_w with the same nominal resistance value and connected in series. The resistors are powered from ultra-stable current source with current of value I_g. In such situation the galvanometer indicates the difference of voltage drops on the resistors, which can be expressed by the formula:

$$R_x - R_w = \frac{U_x - U_w}{I_g} \tag{1}$$

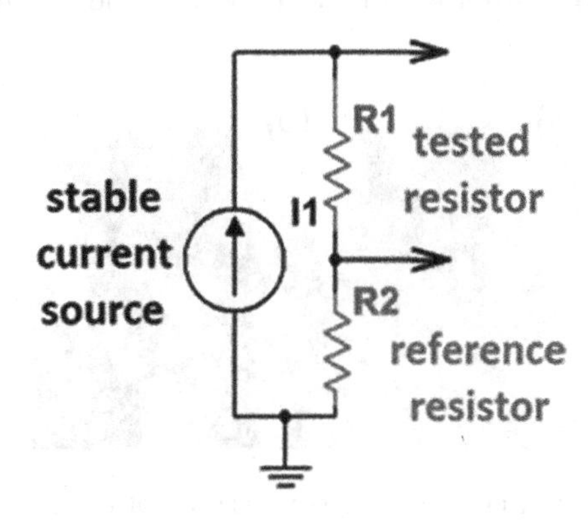

Fig. 2. Principle of operation of the measurement loop [1–3]

Diagram showing the principle of the measurement loop operation is presented in Fig. 2.

From the measurement loop the signals of voltage drops are outputted to differential amplifier circuit. In the output of circuit the amplified proportional to the difference of resistors R1 and R2 values signal is obtained. A detailed description of the system was previously reported in [1].

3 Results of Investigation

The chapter presents the description of an influences of: wires, resistors leads and connector on the value of TCR of investigated resistors. It has also selection of optimal

Table 1. Indicative values of TCR of resistors used for the tests

Resistor number	21	33	27
Indicative value of TCR (ppm/°C)	+0.3	–0.2	–0.5

solutions in terms of mounting the investigated resistors in the test stand. Indicative values of the TCRs of resistors used during tests are presented in the Table 1.

3.1 Influence of Lead of Signal to Resistors

Firstly, an influence of the soldering of the resistors leads on the tests results was studied. The research relied on the measurements of the elements TCR while the wires were soldered to the legs at different distances (as far as possible (a) and as close as possible (b)) from the body of the resistors. Both cases are presented in Fig. 3.

The purpose of the tests was not to obtain the quantitative assessment of impact (distance of wires soldering from resistors body) but only to determine if any influence exists, so the study was carried out for only one resistor. For each method of the wires soldering, two measurements were performed (with re-soldering between them). During the tests TCR of resistor no. 21 was measured, and as a reference, resistor no.

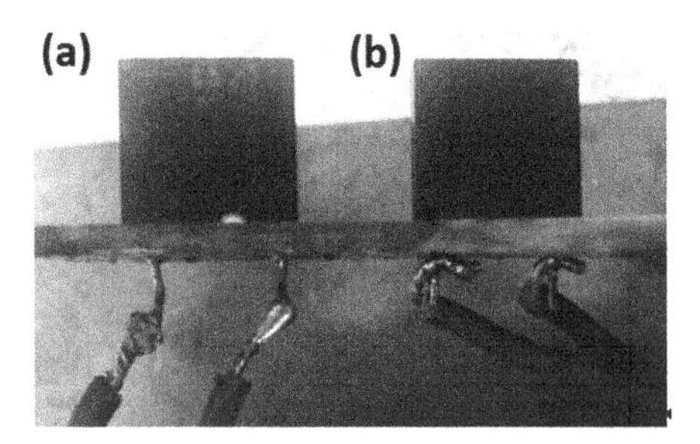

Fig. 3. The method of soldering wires to resistors legs – at different distances (possibly far (a) and close (b)) to its body

Table 2. Value of the TCR in the function of distance of soldering wires from the body of resistors

Nr	Distance from body of resistors	Mean value of TCR (ppm/°C)
1	close	0.28–0.29
2	close	0.28
3	far	0.37
4	far	0.38

33 was selected (only resistor 21 was heated). Results of investigation are summarized in Table 2.

As a result of the studies similar effect was observed twice – resulting temperature coefficient value was approx. 0.09 ppm/°C higher when the wires were soldered to the end of the tested resistor legs. Because the diameter of the legs (0.36 mm^2), is almost identical to wire (0.35 mm^2), the observed effect was not caused by their different diameters, but by its own, substantial value, of the TCR of resistors legs.

Accordingly, the wires should be soldered to the legs of resistor possibly close to its body. An additional advantage of such solution is the most faithful mapping of a connection of the resistor in working conditions (thru-hole mounting into PCB and soldered).

3.2 Influence of Wires

The measurement signal of the resistors disposed between the heating blocks is derived from test stand using copper wires (Fig. 3). Those wires are soldered to the resistors leads and taken outside through the gap in the screening slab.

To determine whether the heating of wires has a significant impact on the obtained results the following calculations were carried out [5, 6, 10]:

$$R_{Cu} = \rho_{Cu} \frac{l}{S} \tag{2}$$

where:

- R_{Cu} – resistance of copper wire [Ω],
- ρ_{Cu} – copper resistivity $\rho_{Cu} = 0.01725(\frac{\Omega * mm^2}{m})$,
- l – wire length [m],
- S – cross sectional area of wire (mm^2),

$$R_2 = R_1[1 + \alpha(T_2 - T_1)] \tag{3}$$

where:

- R_2 – resistance after temperature change [Ω],
- R_1 – resistance before temperature change [Ω],
- α_{Cu} – copper temperature coefficient of resistance $\alpha_{Cu} = 0.004[\frac{1}{°C}]$,

- T_1 – initial resistors temperature [°C],
- T_2 – resistors temperature after heating [°C]

$$\Delta R = R_2 - R_1 \tag{4}$$

where:

- ΔR – wire resistance change due to heating

For the calculation it was assumed that the test resistor is heated by 12 °C and in such situation the wire with a length of 0.2 m is heated by 2°C. Wires of different thickness were considered. Calculation results were converted to the TCR values related to the resistors of values 100 Ω, 200 Ω and 2.4 kΩ.

The estimated impact of the wires on the obtained values of resistors temperature coefficients can be large, especially for small resistor values – in this case, to disregard an acceptable result was obtained only for a cable of 1 mm^2 and 2.5 mm^2 cross section (required accuracy is 0.05 ppm/°C).

To verify the results of calculations (Table 3) two cycles of tests for 200 Ω resistor were conducted. The difference between cycles was cross-section area of connecting wires (0.35 mm^2 and 2.5 mm^2). The tests were conducted on a pair of resistors 21 and

Table 3. Summarized results of the calcualtion of the impact of changes in the resistance of conncecting wire on the value of tested resistors TCR

Parameter					
Diameter of wire (mm^2)	0.25	0.35	0.5	1	2.5
Resistance of wire (mΩ)	13.8	9.86	6.9	3.45	1.38
Absolute resistance increment (mΩ)	0.1104	0.0788	0.0552	0.0276	0.0104
TCR increment (ppm/°C) (for 100 Ω resistor)	0.092	0.066	0.046	0.023	0.0092
TCR increment (ppm/°C) (for 200 Ω resistor)	0.046	0.033	0.023	0.012	0.0046
TCR increment (ppm/°C) (for 2.4 kΩ resistor)	0.004	0.003	0.002	0.001	0.0004

33 (21 as object of test and 33 as a reference). Example characteristics of the resistance variation in the function of temperature obtained during tests (for both thickness of wires) are presented in Fig. 4. While the obtained values of TCR are summarized in Table 4.

Comparing the results of the calculations (Table 3) for 200Ω resistor for wires of thickness 0.35 and 2.5 mm^2 (column 2 and 5) the expected value of TCR for 0.35 mm^2 wire is about 0.028 ppm/°C higher than value of TCR for 2.5 mm^2 wire. While the difference obtained during the tests (Table 4) is 0.06 ppm/°C.

Such result confirm a significant impact of cross-sectional area of wires on the results of TCR of researched resistors. Empirical result is even higher than calculated. The cause may be the higher than assumed level of heating of wires during experiment

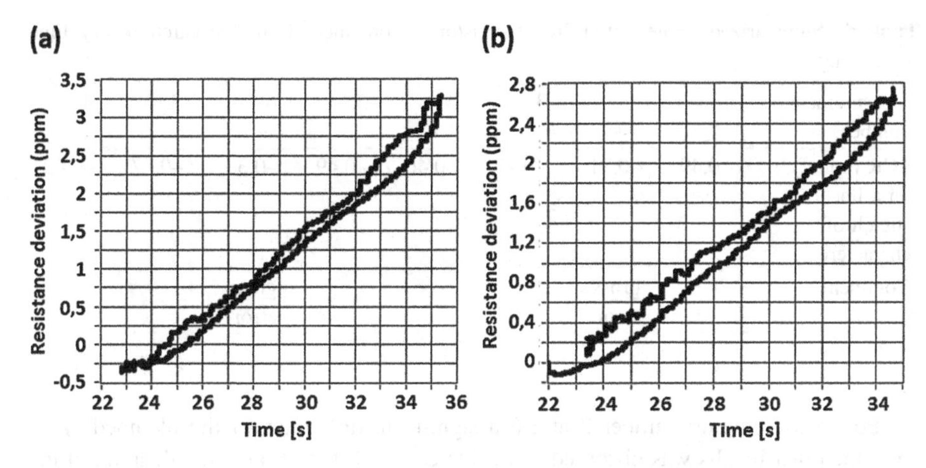

Fig. 4. Example characteristics of the resistance variation in the function of temperature obtained during tests for resistors 21 as object and 33 as reference for different thickness of wires: (a) 0.35 mm, (b) 2.5 mm

Table 4. Summarized results of the tests of resistance variation in the function of temperature for different thickness of wires

Nr	Cross-sectional area of wires (mm^2)	Average value of TCR (ppm/°C)	Fig.
1	0.35	0.28	4 (a)
2	2.5	0.22	4 (b)

or thermocouple effect [11, 12]. Due to the difficulties in the determination of level of wires heating during experiment this influence is hard to compensate. So it is recommended to use wires of possibly high cross-sectional area (especially for a low value resistors like 100 Ω or 200 Ω).

3.3 Influence of Soldering

During the search of the optimal method of mounting resistors in the test stand each test considered every time soldering of resistors. Advantage of that solution is the most reliable reproduction of the working conditions. However chosen method of mounting resistors must provide high stability and repeatability of indicated values during all tests. Moreover very important issue is the repeatability of soldering procedure and immutability of resistors parameters despite possible heat transfer on resistors during this operation. In order to check that matter multiple tests of resistor number 27 were performed.

The first characteristic was obtained during tests is presented in Fig. 4a. Then next 7 analogical tests were conducted. Before each TCR measurement tested resistor (no. 27) was unsoldered and again soldered into a THT board that are commonly used of FR4 PCB [13]. The values of resistor TCR obtained in each test are summarized in Table 5 (characteristics are analogical to Fig. 4a so they are not presented here).

Table 5. Summarized values of TCR of resistor 27 obtained during research (every time resoldering)

Measurement number	1	2	3	4	5	6	7	8
TCR (ppm/°C) on a linear stretch of characteristic	–0.49	–0.71	–0.47	–0.46	–0.49	–0.36	–0.47	–0.42
Comments		gross error				gross error		

For measurements number 2 and 6 a significant difference of the obtained value form the other results was observed. Probable cause of that was the overheating of the solder or leaving a significant amount of a rosin in the solder site. In each experiment solder used was the same type PbSn type, as also lead free solders are available with addition of silver [14]. So that results (2 and 6) have not been taken into account in the calculation of standard deviation.

Standard deviation allows to quantify the amount of variation of the results. For the calculations the following relation were used (5) [15]:

$$s = \sqrt{\frac{\sum_{i=1}^{n} (x_i - \overline{x})^2}{n - 1}} \quad (5)$$

where:

- s – standard deviation of the set,
- x_i –results,
- $\overline{x}$ – arithmetic average,
- n – number of results.

To calculations MS Excel software package was used. Following results were obtained:

- average value: $\overline{x}$ = –0.467 ppm/°C
- standard deviation of the set: s = 0.0236 ppm/°C

Table 6. Summarized values of TCR of resistor 27 obtained during research (without re-soldering)

Number of soldering	1	2	3	4	5	6
TCR (ppm/°C) on a linear stretch of characteristic	–0.42	–0.47	–0.46	–0.44	–0.46	–0.45

For comparison a series of tests without every time re-soldering of resistors were made. Results are summarized in Table 6.

For measurements, without every time re-soldering, value of standard deviation were calculated as well (according to the relations (5)). Following results were obtained:

- average value: $\overline{x} = -0.45$ ppm/°C,
- standard deviation of the set: s = 0.0163 ppm/°C

Values obtained in both cases (with every time re-soldering and without re-soldering) allow to stat that the action of soldering have slight influence on TCR determination. Average value is close to each other in both cases. As well as value of standard deviation, which is a little higher in the situation of re-soldering. Only in the case of errant soldering (Table 5 columns 2 and 6) may have been significant deviation of results. However, these mistakes can be eliminated by good reproducibility of soldering.

Table 7. Summarized values of TCR of resistor 27 obtained during research (every-time re-srewing of resistor)

Number of screwing	1	2	3	4	5	6
TCR (ppm/°C) on a linear stretch of characteristic	–0.62	–0.67	–0.59	–0.56	–0.59	–0.62
Number of screwing	7	8	9	10	11	12
TCR (ppm/°C) on a linear stretch of characteristic	–0.61	–0.59	–0.62	–0.52	–0.54	–0.61

3.4 Influence of Screw Connector

Study were conducted in order to determine the influence of placing the resistor 27 into screw connector on the value of resistor TCR. Tests were conducted analogical to the tests presented in previous points (3.1–3.3). Results are summarized in Table 7.

For results presented in Table 7 standard deviation (according to the relations (5)) and mean value were calculated. Following results were obtained:

- average value: $\overline{x} = -0.595$ ppm/°C,
- standard deviation of the set: s = 0.0386 ppm/°C

Dispersion of the results is greater than in cases of every time re-soldering or no re-soldering and equals 0.15 ppm/°C (Table 7). Also the value of results standard deviation is greater than in previously analyzed cases. Same time average value of results have offset about –0.13 ppm/°C comparing to the results obtained in case of soldering resistors. All observation indicated major influence of screw connector on the results.

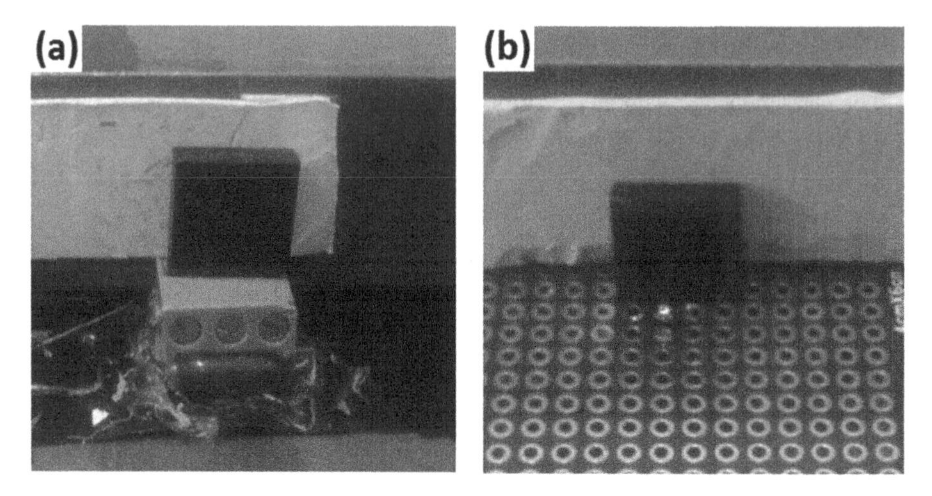

Fig. 5. Considered way of mounting resistors into the test stand: (a) –resistor in screw connector, (b) – resistor soldered into THT board

Table 8. Advantage and disadvantages of considered solutions of mounting resistors

Screw connector		Soldering	
+	Easy replacement of components	–	Necessity of every-time resoldering
+	Short time required to stabilize the indication before measurement (time for elements self-heating)	–	Significant time required for completely cooling of resistor before measurement – strong influence of resistor heating during its soldering
–	Offset caused by the screw connector – probably different value of offset for different units of screw connectors	+	No offset
–	The method of mounting other than in the target system of resistor	+	Measurements conditions most similar to the working conditions in the target system of resistor
–	Probability of wearing screw connector out during utilization of test stand	+	No wearing

4 Advantages and Disadvantages of Considered Ways of Fixing Resistors and the Selection of the Optimal Solution

Two ways of mounting resistors into test stand were took into account: in screw connector and every time resistors resoldering into THT board (Fig. 5). Both solutions have its advantages and disadvantages (Table 8). However the main determinant of the solution selection was to obtain the highest accuracy of TCR determination.

Table 9. Average values and standard deviation of the sets obtained with different methods of mounting resistors

Resistor no 27	Every-time resoldering	Without resoldering	Every-time screw driving into screw connector
Average value of TCR (ppm/°C)	–0.467	–0.45	–0.595
Standard deviation of TCR (ppm/°C)	0.0236	0.0163	0.0386

Based on the summarization presented in (Table 8) it can be observed that each method of resistors mounting has some advantages and disadvantages. Main disadvantages of resoldering resistors in a THT board are workload of resoldering and possibility of defective soldering, which causes significant errors. While main advantages of that method of mounting are the best simulation of the working conditions in the target system and high repeatability of the results (Table 9) – standard deviation is just a little higher from the standard deviation obtained in the set of tests without resoldering.

Main advantage of mounting resistors in the screw connectors is easy replacement of tested resistors. However that method of mounting have several significant disadvantages too:

- connector caused offset (Table 9 – average value of TCR is about 0.13 ppm/°C lower than with resoldering resistors case),
- in time screw connector can wear out – what may cause additional errors,
- value of standard deviation (Table 9) is noticeably higher in this case.

Given the required high accuracy of measurements presented arguments enjoin the use of every time resoldering method of mounting resistors in the test stand.

5 Conclusion

The optimal way of mounting resistors in the test stand is its every time resoldering into THT board. Such method ensure the best simulation of the working conditions in the target system. This avoids offset of the TCR value caused by screw connector and errors caused by its wearing. Moreover ensures high repeatability of the results (provided that there is no overheating of the solder).

References

1. Nowak, P., Juś, A., Szewczyk, R., Pijarski, R., Nowicki, M., Winiarski, W.: Test stand for temperature characteristics of ultra-precise resistors. In: Awrejcewicz, J., Szewczyk, R., Trojnacki, M., Kaliczyńska, M. (eds.) Mechatronics: Ideas for Industrial Applications, vol. 317. Springer, Heidelberg (2015). doi:10.1007/978-3-319-10990-9_32

2. Juś, A., Nowak, P., Szewczyk, R., Nowicki, M., Winiarski, W., Winiarski, W., Radzikowska, W.: Assesment of temperature coefficient of extremely stable resistors for industrial applications. In: Awrejcewicz, J., Szewczyk, R., Trojnacki, M., Kaliczyńska, M. (eds.) Mechatronics: Ideas for Industrial Applications, vol. 317, pp. 297–306. Springer, Heidelberg (2015). doi:10.1007/978-3-319-10990-9_27
3. Juś, A., Nowak, P., Szewczyk, R.: Automatic system for identification of temperature parameters of resistors based on self-heating phenomena. In: Szewczyk, R., Zielinski, C., Kaliczyńska, M. (eds.) Progress in Automation, Robotics and Measuring Techniques, vol. 352, pp. 91–100. Springer, Heidelberg (2015). doi:10.1007/978-3-319-15835-8_11
4. Horowitz, P., Hill, P.: The Art of Electronics, 2nd edn. Cambridge University Press, Cambridge (1989)
5. Korytkowski, J.: Digital controlled resistance synthesis electronic circuit for precise simulation of thermo-resistance sensors (Układ elektroniczny cyfrowej syntezy rezystancji do dokładnej symulacji rezystancyjnych czujników temperatury). Pomiary Automatyka Robotyka **17**(5), 86–92 (2013)
6. Warsza, Z.L.: New approach to the accuracy description of unbalanced bridge circuits with the example of pt sensor resistance bridges. J. Autom. Mob. Robot. Intell. Syst. **4**(2), 8–15 (2010)
7. Buckingham, H., Price, E.N.: Principles of Electrical Measurements. English Universities, London (1966)
8. Golding, E.W.: Electrical Measurement and Measuring Instruments, 3rd edn. Sir Issac Pitman and Sons, London (1960)
9. Chwaleba, A., Poniński, M., Siedlecki, A.: Metrologia elektryczna. Wyd. 8 zmienione. Wyd. Naukowo-Techniczne, Warszawa 1979 (2003)
10. Stepowicz, W.J., Górecki, K.: Materiały i elementy elektroniczne. Wyd. 2. Wydawnictwo Akademii Morskiej w Gdyni, Gdynia (2008)
11. Rowe, D.M.: Thermoelectrics Handbook: Macro to Nano. CRC Press, Boca Raton (2005)
12. Turkowski, M.: Przemysłowe sensory i przetworniki pomiarowe. Wyd. PW, Warszawa (2002)
13. Szałatkiewicz, J.: Energy recovery from waste of printed circuit boards in plasmatron plasma reactor. Pol. J. Environ. Stud. **23**(1), 277–281 (2014)
14. Szałatkiewicz, J.: Metals content in printed circuit board waste. Pol. J. Environ. Stud. **23**(6), 2365–2369 (2014)
15. Cowan, G.: Statistical Data Analysis. Oxford University Press, Oxford (1988)

Acoustic Radiation Energy Focus in a Shell with Liquid

Volodimir Karachun and Viktorij Mel'nick(✉)

National Technical University of Ukraine, Kiev Polytechnic Institute,
Kiev, Ukraine
karachun11@i.ua, bti@fbt.ntu-kpi.kiev.ua

Abstract. When the geometric dimensions of the cylindrical housing are much greater than the sound wavelength, the excitation zones. There is thus a focusing of penetrating acoustic radiation energy, which takes the form of a caustic surface. This results in a sharp concentration of the sound wave energy inside a liquid static part of the suspension, accompanied by a significant increase in the sound pressure level. Upon housing irradiation with a wide enough sound beam, which, by the way, is observed under operating conditions in the form of a reverberant space, the coincidence resonance may occur for both transverse and circumferential waves. Furthermore, the liquid static parts may also have the radial bands of sound radiation energy concentration as a result of interference phenomena.

Keywords: Acoustic radiation energy · Cylindrical body · Zone kaustikos · Liquid

1 Introduction

When the geometric dimensions of the cylindrical housing are much greater than the sound wavelength, the excitation zones, i.e. areas where the condition of coincidence of the incident of the wave track speed with the wave speed excited in the shell can be observed with a sufficient degree of accuracy in the liquid static part [1]. There is thus a focusing of penetrating acoustic radiation energy, which takes the form of a caustic surface (from the Greek kaustikós – burning, flaming). This surface is the bending of acoustic beams emanating from the same point of internal surface of the cylindrical body of the floating device as a consequence of its longitudinal and transverse vibrations. This results in a sharp concentration of the sound wave energy inside a liquid static part of the suspension, accompanied by a significant increase in the sound pressure level [2].

2 Materials and Methods

Let's disclose some aspects of the nature of emerging energy focusing of the past acoustic radiation. For the purpose of clarity, let's study the action mechanism on the middle frame (Fig. 1). Thus, the sound wave R affects the cylindrical body from the

R. Szewczyk and M. Kaliczyńska (eds.), *Recent Advances in Systems, Control and Information Technology*, Advances in Intelligent Systems and Computing 543, DOI 10.1007/978-3-319-48923-0_48

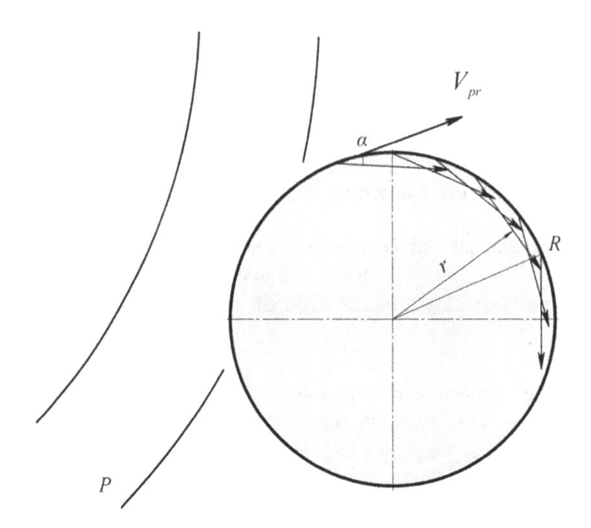

Fig. 1. Penetrating acoustic radiation energy focusing

outside. It generates *circumferential* vibrations $U_\varphi(t, z, \varphi)$ in the material, which propagate in the parallel direction with the speed of V_{pr}, that is on the side surface of the housing, and bending, *radial* vibrations $W(t, z, \varphi)$ in the section plane with the speed of V_{pp}.

First of all, let's analyze the action mechanism of the *circumferential* wave. Considering the side of the hull to be a shell of a large wave size, i.e., having properties $1 < kR$ (k is the wave number), we can consider a single element of the frame inner surface as a plate of zero curvature, the velocity of longitudinal waves in which coincides with circumferential speed of the shell V_{pr}.

If the longitudinal wave speed V_{pr} is greater than sound speed c_0 in the liquid, i.e.

$$V_{pr} > c_0, \tag{1}$$

then the wave running along the parallel will radiate a sound wave into the liquid, and its direction will have angle α versus the velocity vector V_{pr}, defined by expression

$$\sin \alpha = \frac{c_0}{V_{pr}}. \tag{2}$$

Consequently, much of the sound wave energy will be concentrated near the r_1 radius circumference (Fig. 1)

$$r_1 = R \cos \alpha \tag{3}$$

For example, if we assume the housing inner cavity radius equal to R = 1.5 sm for clarity purposes, the housing material can be considered aluminum (V_{pr} = 6400 m/s, V_{pp} = 3080 m/s), the liquid static suspension can be considered, for example, glycerol

(c_0 = 1923 m/s at t = 20 °C), and the frequency of the ultrasonic beam f = 42 kHz. Then, the wave size of the housing inner cavity will be determined by the following formula

$$kR = \frac{\omega}{c_0}R = \frac{2\pi f}{c_0}R \approx 3.43. \tag{4}$$

Therefore we can write

$$\sin\,\alpha = \frac{c_0}{V_{pr}} \approx 0.3004;\ \alpha = 17°30'. \tag{5}$$

Whereas:

$$r_1 = R\,\cos\,\alpha = 1.43\,\text{sm}, \tag{6}$$

i.e. *kaustikos* surface is located 0.7 mm away from the housing inner surface.

Obviously, if $V_{pr} = c_0$, then $\alpha = \frac{\pi}{2}\mathit{rad}$ and the wave running along the parallel will emit a sound wave into the liquid, which will cross the housing axis. Thus, the *aberration* will disappear and caustic surface will become a locus of the points of energy concentration located on the longitudinal axis of the housing.

Similarly, when $V_{pp} = c_0$, then $\beta = \frac{\pi}{2}\mathit{rad}$ and caustic surface is not formed by the housing transverse wave.

3 Results and Discussion

It is interesting to estimate the temperature effect on the degree of energy concentration in the liquid static part of the suspension. The speed of sound in the liquid upon its temperature variation is determined as follows

$$c(t) = c_0 + \alpha(t - t_0). \tag{7}$$

For glycerol, the temperature coefficient is $\alpha = -1.8\frac{m}{s\cdot\deg}$ at t_0 = 20 °C. Then, when t = 60 °C, the speed of sound in the liquid static suspension will be reduced to 1851 m/s. When t = 0 °C, conversely, the speed of sound will increase to 1959 m/s. So when t = 200 °C, the speed of sound in the liquid will be equal to 1899 m/s. It follows that maintaining a constant temperature of the liquid static part (and zero gradient throughout its volume) creates conditions for management of the process of "*zone kaustikos*" occurrence.

For the same reason, *shear wave* (flexural wave in the radial direction) will result in energy concentration near the circumference of radius r_2:

$$\sin\beta = \frac{c_0}{V_{pp}} \approx 0.67; \tag{8}$$

$$\beta = 42°; r_2 = R\,\sin\,\beta = 1.68\,\text{sm}.$$

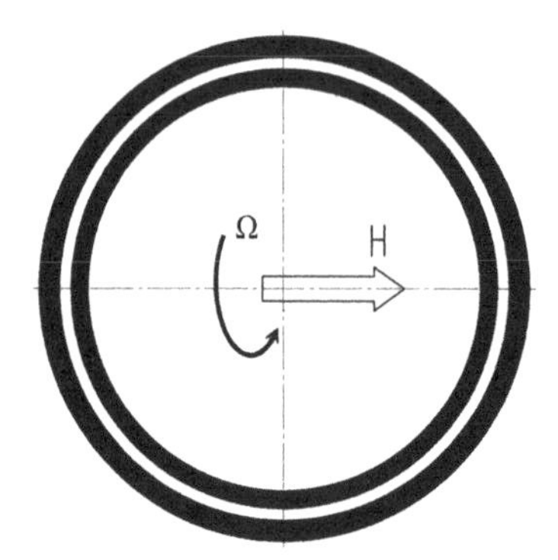

Fig. 2. Penetrating radiation energy focusing in a liquid static part of the housing

Since the housing radius is assumed to be 2 cm, it is obvious that caustic surface of radius r_2 will disappear, and the surface of radius r_1 will remain (Fig. 2).

Caustic surface of radius r_1 will clearly demarcate the regions of acoustic shadow in the liquid static part.

It is obvious that selection of an appropriate housing material and liquid can influence the nature of the surface of *zone kaustikos* (from the Greek *zone*, a surface between any boundaries, characterized by certain features) or, for example, make them discrete-continuous [3].

If we use the methods of radiation acoustics, we can classify the phenomenon under consideration as a kind of aberration (from the Greek aberration, deviation from normal) of the sound waves. Aberration-free structures are known to have the caustic surface transformed into a point, and thus, in this case, it will be located on the longitudinal axis of the housing.

It can be argued that the presence of excitation zones inside the shells is caused by coincidence of the incident wave trace speed with speed of the wave generated in the shell [4, 5].

Upon housing irradiation with a wide enough sound beam, which, by the way, is observed under operating conditions in the form of a reverberant space, the coincidence resonance may occur for both transverse and circumferential waves (Fig. 1).

4 Conclusions

Furthermore, the liquid static parts may also have the radial bands of sound radiation energy concentration as a result of interference phenomena. They take the form of alternating dark (acoustic shadow) and light bands, though not intensive as surfaces of radius r_1 and r_2 [2].

As shown by experimental studies, zone kaustikos surfaces in a section plane are confocal to the housing interior. Therefore, circumferential caustic in this case is confocal to the inner circumference of the device housing section.

References

1. Karachun, V., Mel'nick, V., Korobiichuk, I., Nowicki, M., Szewczyk, R., Kobzar, S.: The additional error of inertial sensor induced by hypersonic flight condition. Sensors **16**(3) (2016). doi:10.3390/s1603029. ISSN: 1424-8220
2. Makarov, V.I., Fadeev, N.A.: On wave radiation by shells in the sound field. Acoust. Mag. **IV** (2), 261–263 (1960)
3. Karachun, V.V., Tryvailo, M.S., Melnyk, V.M., Rudenko, O.S.: Utility model patent No. 66311, Ukraine, IPC (2011) G01S19 / 20. Float gyroscope. U 201108294. – Application dated July 01, 2011. – Published on December 26, 2011. Bulletin 24, 1 p. (2011)
4. Shenderov, E.L.: Hydroacoustic Wave Problems, 352 p. Sudostroyeniye (1972)
5. Zaborov, V.I.: Walling Sound Insulation Theory, 116 p. Stroyizdat (1962). References: 113–114

Torque and Capacity Measurement in Rotating Transmission

Ivan Grabar[1], Igor Korobiichuk[2(✉)], and Oleg Petruk[2]

[1] Zhytomyr National University of Agriculture and Ecology, Zhytomyr, Ukraine
ivan-grabar@ukr.net
[2] Industrial Research Institute for Automation and Measurements PIAP, Warsaw, Poland
{ikorobiichuk, opetruk}@piap.pl

Abstract. The task of measuring, monitoring and recording of the torque, mechanical stress and capacity is relevant for most transmissions of the process and transport machines. The importance of this problem is increased by an order in terms of unsteady loads (random, impulsive, alternating, etc.). The disadvantage of its correct solution by experimental methods is substantially limited by the need for expensive and unreliable, environmentally dangerous current collectors. To measure the dynamic and kinematic parameters of the rotating transmission by modern information computer technologies, we propose equipping the mechanical transmission with two additional half-couplings with point magnets fixed thereon, the scope of each includes one or more Hall sensors. The application of the developed technique allows determining the torsional rigidity of the transmission experimentally, and allows recording the actual torque and capacity per turn (or a part of the turn) of the transmission of a process or transport machine and record the same online in the computer electronic memory. This, in turn, allows an online assessment of the residual transmission resource, machine operation per shift and generate its operation history, which provides an invaluable information both to the machine developer and owner.

Keywords: Mechanical transmission · Tachometer · Measurement systems · Torque · Capacity · Rotating transmission

1 Introduction

The task of measurement, control and registration of torque, mechanical stresses and power occurs in most cases of transmissions of technological machine and transporters. The importance of this problem is increasing in the conditions of non-stationary loads (random, impulse, alternating, etc.). The tensometric methods are often applied to increase the accuracy of the correct solution to this task. It will be shown later, that their application is significantly reduced with the necessity to use the expensive and fault-tolerant methods of the transmission of the signal from rotor to stator. Moreover, these signals are not reliable. At the same time, the application of the perfect means of technological process parameter control is the main task of the production automation. The control of the angular speed, and especially, the torque and capacity, allows

R. Szewczyk and M. Kaliczyńska (eds.), *Recent Advances in Systems, Control and Information Technology*, Advances in Intelligent Systems and Computing 543, DOI 10.1007/978-3-319-48923-0_49

preventing the technological machinery overloading. It also makes possible to register the real history of their operation. The given systems obtain the special significance at drive unit and transmission tests, at their maximum load bearing capacity research, at the investigation of the dynamic parameters and transmission efficiency, gear boxes, reducers, fans, pumps. The bolted connection assemley automation, the crank connection rod and crank rocker mechanisms and transporters operation also require the experimental measurement (Ω M N).

2 The Current State of the Problem

Nowadays the market offers a great variety of experimental means for measuring the angular velocity, torque and the capacity on the transmissions which rotate [1–17]. Such methods as: the method of measuring of the centrifugal force which is proportional to the square of the angular velocity; the induction and magnet induction methods, stroboscopic method and others have gained a good reputation among those which have been studied. The laser portative fototachometers of the types AKIP-9201 and AKIP-9202 [14] demonstrated their bests at the application. They allow determining the angular velocities by non-contact method; possess the comfortable system of the laser sighting on the object which is being investigated. The given tachometers are capable of providing the effective measurements of rotating shafts of energy for manufactory equipment, as well as for household appliances. Besides, they are comfortable to be applied to the scientific researches. It is important to mention, that the given fototachometers function in the range of 100000 RPM.

The modern means of the angular velocity measurement are also offered by automobile manufacturers, especially for ABS system [15]. Figure 1 shows the example of such tachometer, where 1 – is the permanent magnet, 2 – is the trunk; 3 – is the winding; 4 – is the iron circuit; 5 – is the pulse drive.

The given system is characterized by the high precision for non-stationary processes. It is also able to measure the modes with great values of angular accelerations for both signs.

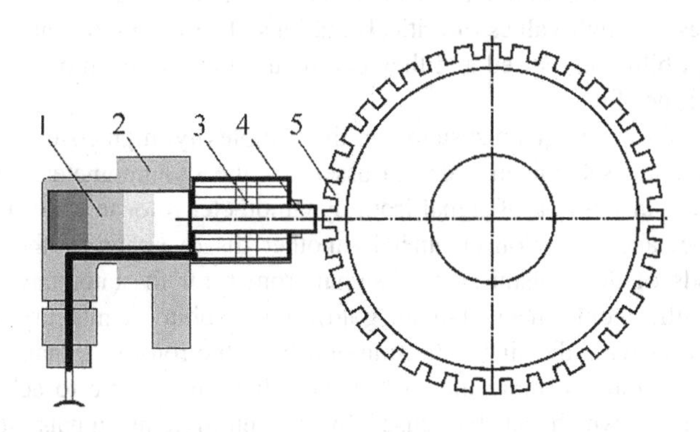

Fig. 1. Construction of tachometer

The modern systems of torque measurement are offered by the famous German companies such as HBM, Burster, Japanese company Kyowa and Korean firm Dacell, and also by other world leaders in mechatronics and automation. The market offers a great variety of dynamometers of torque which are based on the tensometry application, i.e. the converting shaft mechanical deformations into electric signal and signaling the computer by contact or non-contact way. Figure 2 demonstrates the samples of torque dynamometer.

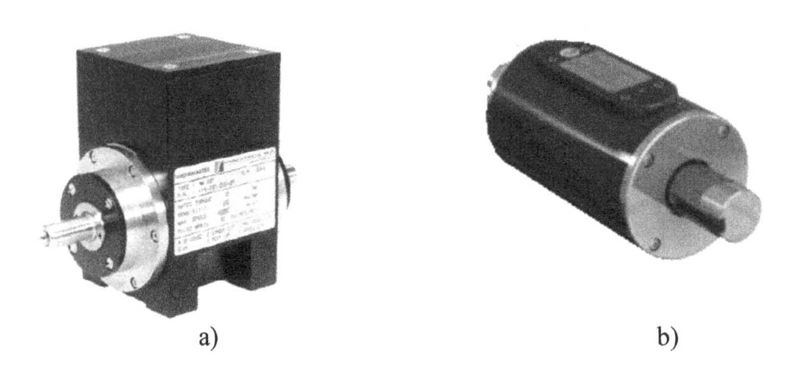

a) b)

Fig. 2. The torque dynamometers [16, 17]

3 Task Formulation

[10] describes the system of non-contact sensor of angular velocity and torque with the transmission of information from dynamometer rotor to stator. The tenso-system is attached to the sensor of the dynamometer. The system includes four tenso-sensors linked by bridge circuit and it is connected to the power supply system. The strain gauge bridge supply is misbalanced by torque. Here the misbalance signal is proportional to the value of torque. The misbalance signal is transmitted to the transmitter signal by the tenso-amplifier and is sent to stator receiver. The receiver is placed on the rotor. The given system is widely used in the modern systems of the measurement of torque on the shaft of the rotating transmission. It possesses huge rigity and it means that it possesses the high values of critical rotations. These rotations are characterized by high susceptibility and speed and they can be used at any form of a torque feature depending on time (Fig. 3).

The drawbacks of the given system are: the complexity, high cost, the necessity to break the kinematic scheme in order to assembly the dynamometer. The low fault tolerance at the transmission of signal from dynamometer rotor to stator using optical, radio or magnet and induction channel is another disadvantage of the system. The contact methods of the transmission of signal from rotor to stator have many more limitations. Neither noble metals (silver, gold), nor graphite or mercury moving contacts are able to provide the signal transmission from the rotor of dynamometer to the stator at high revolutions (more than 100 RPM). It is impossible to achieve without some modifications which can be caused by the intermittent signals, sparking, and

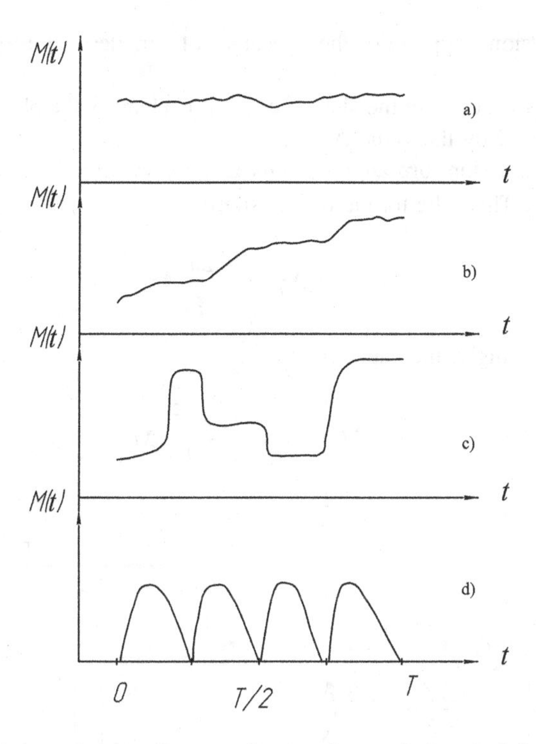

Fig. 3. The typical dependencies of torque change of transmission shaft for its single rotation

mercury viscosity. The application of the supersensitive percolation-fractal sensors which can replace the tenso-amplifiers [3, 12] does not change the situation. It has been shown [1], that the great breakthrough in tensometry elimination (both contact and non-contact) is possible to achieve by denying using it at all. The signal about the value of torque is formed by the system of the direct measurement of the deformation of the shaft sensor without converting it into electric signal of tensometric bridge. The given system has demonstrated a lot of advantages at torque measurement, angular velocity and capacity for transmissions which are characterized by great values of twist angle (long shafts) and little revolutions (up to 300 RPM). The optical systems (laser radiator and optrone receivers), induction systems and Hall sensors were used to obtain the non-contact measurement of the twist angle.

4 The Research Results

In order to find the partial solution to the problem of rotating transmission functioning in the wide range of operating modes, it is recommended to apply the Dynamometer-Tachometer-Watt-meter by Professor Grabar [2].

We suggest equipping the mechanical transmission by two half-couplings with the fixed spot magnets on them. A Hall sensor D1–D2 (Fig. 4) is placed on the coverage of each magnet. It is performed to measure the dynamic and kinematic parameters of the rotating transmission applying the means of modern information-computer technologies.

If the torque M is active in the shaft section, this changes the phase angle between spot magnets 1 and 2 by the value $\Delta\varphi$.

Then, the transmission turn angle $\varphi = \omega t$ and $\Delta\varphi = \omega\Delta\tau$.

Hence: $\omega = \frac{2\pi}{\tau}$. Then, the torque on the shaft:

$$M = C \cdot \Delta\phi = C\frac{2\pi}{\tau}\,\Delta\tau \tag{1}$$

And, correspondingly, the capacity:

$$N = M \cdot \omega = C\left(\frac{2\pi}{\tau}\right)^2\,\Delta\tau \tag{2}$$

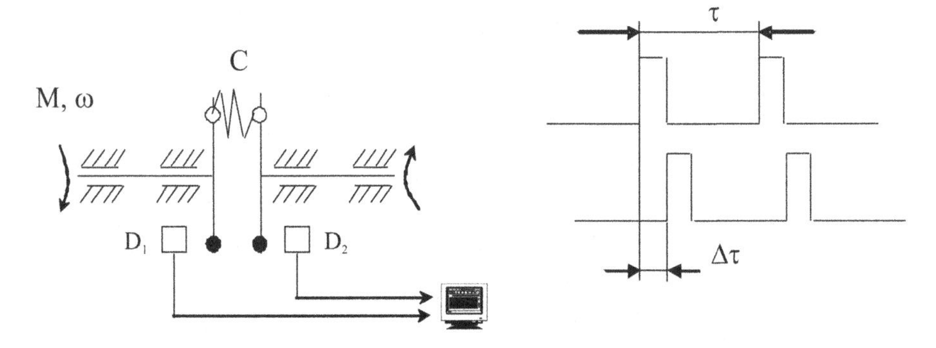

Fig. 4. Construction of measurement systems

Therefore, we obtain the opportunity to register the real torque and the capacity of every single revolution (or a part of a revolution) of any technological machine or transporter. It is done by measuring τ and $\Delta\tau$ only, and by the experimental determination of the transmission torsional rigity **C** in the section D1–D2. It allows online registering of these processes and storing them in the computer's memory. In its turn, it makes possible to provide online estimation of the remaining transmission resource, the operation of machines per shift and to form the history of its operation. This provides both the developer and the owner of the machine with the valuable information.

5 The Experimental Research

The testing of the offered system efficiency to measure the angular velocity, torque and the rotation transmission efficiency was experimentally conducted with two-rotor timber grinder (Fig. 5).

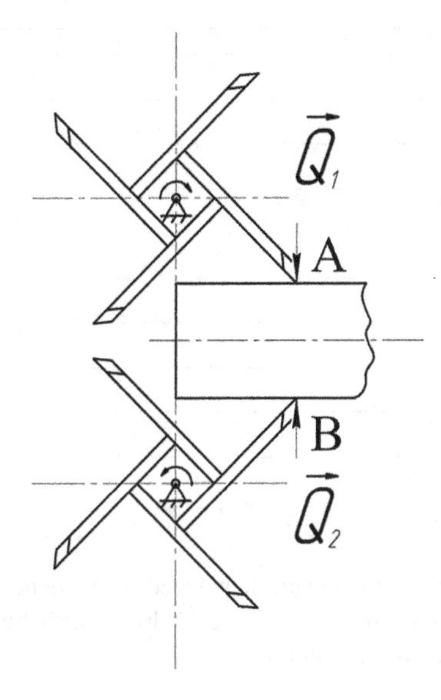

Fig. 5. The scheme of two-rotor timber grinder

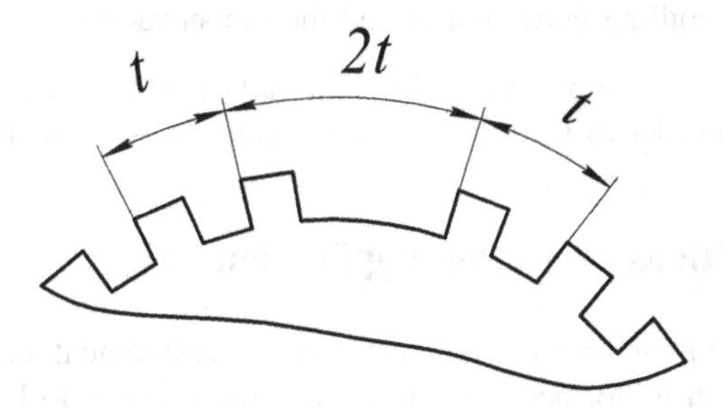

Fig. 6. The timing drive marker area

The rotor shaft torque mathematic model is designed to be used to grind any cross section timber:

$$M_{R(i)} = Q_{R(i)} R_{fp} \sin\left(\alpha \frac{n-i}{n}\right) \tag{3}$$

where: $Q_{Ri} = \left(S_{seg(i)} - S_{seg(i-1)}\right) \tau_{zp}$.

In particular, for the cylindric timber:

$$S_{seg(i)} = R_{bp}\left(\beta_i - \frac{1}{2}\sin 2\beta_i\right) \tag{4}$$

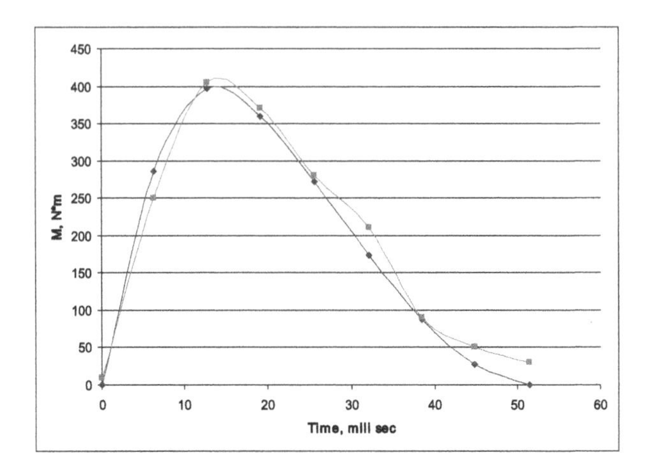

Fig. 7. The theoretical (blue) and the experimental (red) dependencies of the torque on the shaft of rotor timber grinder. The pine timber of the cylindrical form by the diameter 100 mm and voltage cutoff 9.2 MPa (Color figure online)

n – is the number of measurements per one operation cycle (in the given research $n = 9$, $Z = 36$),
R_{fp}, R_{bp} – is the milling cutter radius and the timber radius.

Two timing drives were placed on the shaft butts Fig. 1. One notch was distracted from every drive in order to identify the signals (Fig. 6) is the marker area (Fig. 7).

6 The Propositions as to the Application

The proposed system of transmission parameter measurement has some important advantages. Firstly, it is not necessary to change the scheme, to break the kinematic scheme and to find some extra space to assemble the additional elements of dynamometers. Moreover, every, among the mentioned machines, requires the essential re-equipping in order to attach the torque dynamometers. Such measurements are easy to conduct on the special stands. But, it is not always possible to reproduce the real operating values of the loading modes with such stands.

Some variants of technical solutions are offered applying our methodology:

Variant A. The long shaft (L/D > 50), low revolutions (n < 300 RPM) are the transmissions of stone processing and agricultural machines, screw transporters, mixers, etc. The measurement system (Ω M N) consists of two spot magnets (for example, of the cylindrical form and the diameter 1–2 mm and 1–3 mm high), two Hall sensors, two screen lines with USB ports to transmit the impulse signals to the computer to be registered, processed and stored. If it is needed, the spot magnets can be placed not on the surface of the shaft, but on the extra weight or special disk. It allows increasing the radius of the trajectory of the magnet movement and, correspondingly, providing the precision of the torque measurement. It has been shown, that the mentioned above types of machines allow finding some space 1–2 cm^3 and the space to assembly the screen lines by the diameter 4 mm and 5–10 m long. It is possible at the whole layout density and without any extra changes of the design.

Variant B. A shaft of any length, the low revolutions (n < 100 RPM) are the transmissions for either of variant A machines, or the machines used to mix building mixtures, grinding and distribution of feed, fertilization and others. The measurement system (Ω M N) consists of (as in previous variant) two spot magnets, two Hall sensors, two screen lines with USB ports to transmit impulse signals to the computer to be registered, processed and stored. It is also used to provide the assembly of the element of the low torsional rigity (special elastic coupling – Fig. 4). It localizes and significantly (in 10–100 times) rises the increased torque value between the left and right butt of the shift.

Variant C. If the variant B does not provide the possibility to assembly the low torsion rigity element, but, there is the access to the shaft butts, then the set drives are attached to the shaft butts (Fig. 1) made of ferromagnetic material and the number of notches Z, selected from the condition $Z = \frac{2\pi}{4\Delta\phi}$.

This variant practically eliminates the limits for the revolution velocity dependence of torque (t) and is universal one to be applied.

7 Conclusions

1. The methodology for the measurement of the angular velocity, torque and capacity (Ω M N) on rotating transmission is implemented for the modern systems of technological processes. It can be assembled directly in the technological machine without crucial changes. The set precision enables to measure the mentioned transmission parameters.
2. It has been shown, that to implement the process of the parameter (Ω M N) registration it is recommended to measure only three parameters: τ, $\Delta\tau$ and C which is the period between impulses, shaft deformation impulse shift and the active transmission part torsion rigity.

References

1. Kuo, B.C., Golnaraghi, F.: Automatic Control Systems, 609 p.Wiley, New York (2002)
2. Grabar, I.G.: Patents of Ukraine, № 83475, 85252, 73063, 91664
3. Yahno, O.M., Grabar, I.G., Taurit, T.G.: Wind Power: the Design and Calculation, 255 p. (2003). Zhitomir
4. Grabar, I.G., Bezvesilna, O.M., Kubrak, Y.O.: Precolation tensoconvertor of the linear deformation and the theoretical backgrounds of its operation. ZSTU Messenger. Tech. Sci. **1** (46) (2008). №3
5. Avrutov, V.V.: The angular velocity determination by means of circuit. NUNU KPI Messenger. Issue Instr. 28–32 (2014). №. 47

6. Korobiichuk, I.: Mathematical model of precision sensor for an automatic weapons stabilizer system. Measurement **89**, 151–158 (2016). doi:10.1016/j.measurement.2016.04.017147
7. Kravchenko, N.A., Yashkov, V.A., Soldatkin, V.M., Porunov, A.A.: Two-component Flapper Nozzle Transducer of the Angular Velocity, patent RU № 2462723
8. http://www.elec.ru/news/2008/12/25/sovremennye-sredstva-izmerenij-akip-novye-cifrovye.html. (Tachometers AKIP-9201, AKIP-9202)
9. Yakovlev, D.P.: The device for the torque measurement on the rotating shaft. The papers of Odessa Polytechnic University **1**(19) (2003)
10. Krimmel, V.: http://www.lorenz-m.ru/index.php. The Development and the Future of Torque
11. Gurinov, A.C., Dudnik, V.V., Gaponov, V.L., Kalashnikov, V.V.: http://ivdon.ru/magazine/archive/n2y2012/798. Don Engineering Messenger. №2 (2012). The torque Measurement on Rotating Shafts
12. Popov, A.P., Vinokurov, M.R., Moiseenko, A.A.: Microprocessor System of Non-Contact Control and Torque Measurement. DGU Messenger T.10. №2(45) 243–248 (2010)
13. Grabar, I.G., Grabar, O.I., Gutnichenko, O.A., Kubrak, Y.O.: Percolation-Fractal Materials, Monograph, 370 p. (2007). Zhitomir
14. Petrosyan, D.P., Grigoryan, S.M., Mikalean, S.E.: The comparative estimation of the known methods and the development of the optical method of rotating shaft dynamic parameter measurement. Agriculture Sci. News **9**(4), 1–5 (2011)
15. Korobiichuk, I., Bezvesilna, O., Ilchenko, A., Shadura, V., Nowicki, M., Szewczyk, R.: A mathematical model of the thermo-anemometric flowmeter. Sensors **15**, 22899–22913 (2015). doi:10.3390/s150922899
16. http://m.ustroistvo-avtomobilya.ru/
17. http://www.prom-tex.org/products/krmom.html

Analysis of the Phenomena Occurring During Initial Phase of Resistors Thermal Characteristics Measurement

Andrzej Juś[1(✉)], Paweł Nowak[1], Roman Szewczyk[2], and Weronika Radzikowska-Juś[3]

[1] Industrial Research Institute for Automation and Measurements PIAP, Al. Jerozolimskie 202, 02-486 Warsaw, Poland
ajus@piap.pl

[2] Faculty of Mechatronics, Warsaw University of Technology, sw. A. Boboli 8, 02-525 Warsaw, Poland

[3] Faculty of Civil Engineering and Geodesy, Military University of Technology, gen. Sylwestra Kaliskiego 2, 00-908 Warsaw, Poland

Abstract. Paper presents some of the research problems encountered during the development of the test stand for research of temperature characteristics of the ultra-precise reference resistors. Such problems includes phenomena which occur immediately after power supply of the test stand. Among identified phenomena are: influence of the data acquisition card, influence of the self-heating of tested and reference resistors, influence of the other electrical elements heating and self-heating. These effects causing significant changes of an indication during initial measurements phase. The paper not only identifies the sources of the errors, but also provides methods of their elimination or compensation, which have been used during the development of the test stand. Thanks to the presented solutions very high accuracy of the measurements was achieved and several measurement methods were proposed.

Keywords: Resistance measurement · Stable resistors · Temperature coefficients · TCR

1 Introduction

Increasing requirements for measurement accuracy, forced growing of the requirements to precision of electronical components. Often this involves the high cost of such precision components. So producers take the opportunity to costs reduction while maintaining their required precision.

In response to the current realities, the test stand for temperature properties of high stability reference resistors were developed [7, 9–12]. That tests stand allows selection of the resistors for the current-voltage processing e.g. for laboratory scales. Elements from which the reference resistors are selected are about 10–100 times cheaper than the equivalent elements from renowned manufacturers, which justifies carrying out such selection.

R. Szewczyk and M. Kaliczyńska (eds.), *Recent Advances in Systems, Control and Information Technology*, Advances in Intelligent Systems and Computing 543, DOI 10.1007/978-3-319-48923-0_50

However, to carry out described selection in industrial solutions, high accuracy and short time of measurements is required. To enable this, in addition to appropriate construction of the test stand, a number of experiments in the field of measurement methodology were conducted. Particularly high influence on the obtained results have the phenomena occurring in the initial phase of the tests. These phenomena, the results of their research and developed guidelines in the field of tests methodology are presented in the paper.

2 Test Stand and Measurement Method

Phenomena occurring in the initial phase of the measurements are related with the elements of the developed test stand [7]. Figure 1 presents the block diagram of the test stand.

Resistors: tested and reference are in the center of the tests stand. After power up of the measurement loop (containing the resistors) resistors heat up due to the current flow, but after a few minutes their temperature stabilizes [3–5]. Such effect can disturb the results of measurements (during which resistors are heated externally). It is also possible to use this effect for evaluation of resistors properties [11]. However irrespective of the measurements methodology, resistors self-heating effect has to be taken into account or eliminated during the evaluation of the tests results.

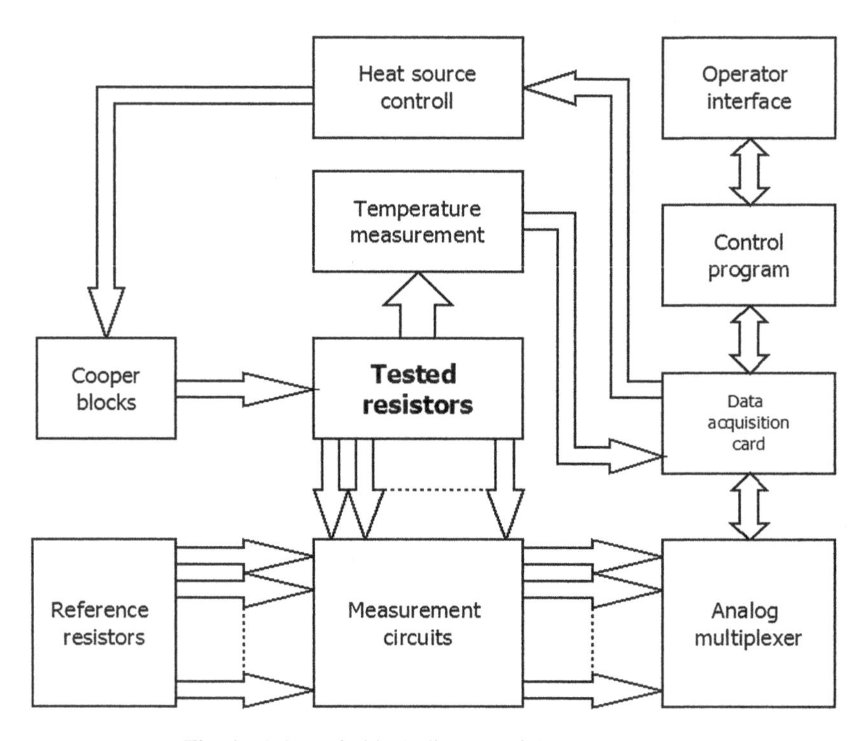

Fig. 1. Schematic block diagram of the test stand [7]

Another important element of the test stand is the measurement loop, based on the differential amplifier circuit. The measurement is based on a differential method, which ensures higher accuracy than absolute measurement methods [1, 13, 14]. Electronic components used in measurements loop, same as reference and tested resistors, are heated up after powering up system [1, 2, 6, 8]. Influence of this effect has to be compensated or eliminated. Analog signal from measurement loop goes to the data acquisition card and is converted to a digital signal. Next data are analyzed in the control software and logged. Data acquisition card can also disturb the measurement process. The device impact on the results and methods of its elimination have been identified in Sect. 3.1.

Another elements of the test stand are associated with external heating of resistors and control of that process and have no direct influence on the measurement signal value, especially in the initial phase of the measurements cycle, which is a matter of the paper.

3 Results of Investigation

The chapter presents study of the causes of the test stand signal's variation, directly after its supplying. Identified the reasons of such variation, its value and its changing in time. For the study 200 Ω resistors were used and their TCR values are summarized in the Table 1.

Table 1. Indicative values of TCR of resistors used for the tests

Resistor no.	25	31	36
Indicative value of TCR (ppm/°C)	–0.45	+0.7	–0.45

3.1 Influence of Data Acquisition Card

Initially influence of data acquisition card on the results was examined. The study was conducted by running tests after different time of device activity:

- directly after other test (typical operating conditions)
- after supplying data acquisition card previously not supplied for 30 min,
- directly after start measuring program (data acquisition card was launched for long time before it),
- after supplying data acquisition card previously not supplied for long time.

The study was conducted in both configurations of resistors (when tested and reference resistors were replaced with each other) with similar temperature coefficients of resistance (TCRs) (25 and 36) and with different values of this parameter (25 and 31). During the tests the ambient temperature was controlled and its influence was included in results analysis. Example waveforms are presented in Fig. 2.

Analyzing obtained results the following conclusions and guidelines were formulated:

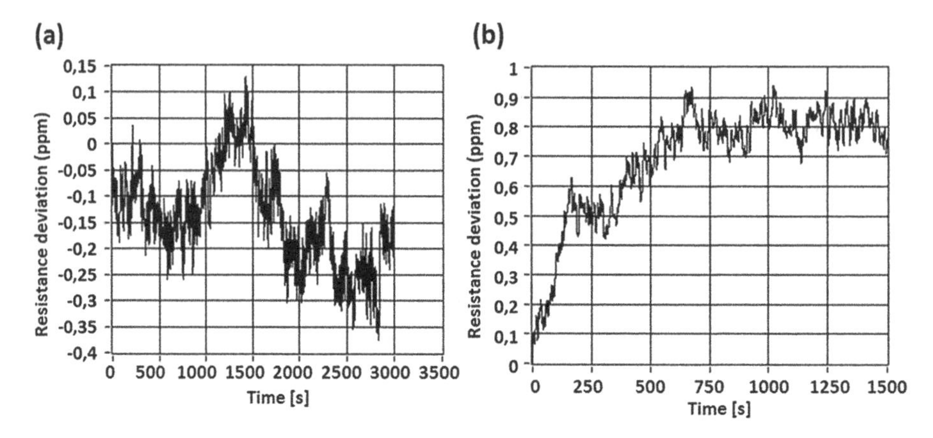

Fig. 2. Example waveforms – tested resistors 36, reference 25: (a) data acquisition card active for a long time before test, (b) data acquisition card off for 30 min before test

- during tests unused channels of data acquisition card should be connected to the ground,
- during all tests, after a long activity of device, high stability results were obtained (despite the small temperature fluctuations) – dispersion of results was at the level of several tenths of ppm (example Fig. 2a),
- during tests directly after launching data acquisition card after (both: long term and 30-minutes) its inaction significant growth of a measured value (at the level of 0.7–0.8 ppm) in time of 500–750 s was observed (example Fig. 2b),
- during tests directly after running measurement software (off for 30 min before it, but data acquisition card was on all this time) results analogical to results of tests conducted directly after previous tests was observed.

The study highlighted possible significant impact of data acquisition card on the measurement results. Its value didn't depend on tested resistors TCR's. To eliminate that influence the device have to be on for 10–15 min before the measurement starts. Running or not running measurement software before tests have no influence on the results.

3.2 Self-heating of Test Stand Elements

After supplying the test stand, both resistors (tested and reference) as well as elements on a circuit boards are heated as a result of current flow. The results of this phenomena observation were presented in Sect. 4.1 of the paper [9].

This phenomena may cause a shift of the results and consequently the TCR determination error. The easiest way to eliminate this impact is wait for indications to stabilize and conducting measurements after self-heating of circuit components.

Research were conducted in cases of resistors in the air and between the copper blocks. In second case the change of value was smaller what is a result of improved heat dissipation between copper blocks. The term of circuit board self-heating was similar in both cases and was about 500–1000 s. So to study other parameters of tested resistors it is necessary to wait this time (about 15 min) after supplying the circuit board.

3.3 Cooling Process of the Test Stand

This section contains observations of a circuit board cooling (with tested and reference resistors) after its self-heating. As in Sect. 3.1 the study was conducted in two configurations of resistors (when tested and reference resistors were replaced with each other) with different values of their TCRs (resistors no. 25 and 31). Temperature influence was included in results analysis.

Characteristic of the circuit board (with resistors) cooling could be particularly important when the operation of the test stand will be based on self-heating phenomena (changing of the indication immediately after measurement circuit power-up). Principle of such system operation is presented in [11]. This characteristic can be used to reduce time between the same resistors measurements thanks to the compensation of the varying level of the resistor and circuit board cooling process.

Moreover characteristic of the offset caused by measuring circuit in the function of power off time (before measurements) was determined. The characteristics allow to compensate influence of the level of circuit board cooling process on the results of measurements which allows significant reduction of a measurements time.

To determine these characteristics, tests after different terms of circuit with resistors power-off were conducted. Obtained waveforms are analogical to Fig. 2. While a summaries of the results are included in tables:

- for resistors 25 as tested and 31 as reference in Table 2,
- for resistors 31 as tested and 25 as reference in Table 3,
- the average of the values obtained in both resistors settings – offset generated by measurement system in Table 4.

Table 2. Changes of indication according to power-off time before test. Resistors: tested 25, reference 31

Unplugging time before test [s]	Indication changes after stabilisation [ppm]	The level of the heat at the start of the test [%]
0	0	100
60	–2.0	74.7
150	–4.0	49.4
300	–5.4	31.6
600	–6.9	12.6
900	–7.6	3.8
1500	–7.8	1.3
few days	–7.9	0

Table 3. Maximum change of indication according to the time of measuring system unplugging before test. Resistor no. 31 as a tested and no. 25 as a reference

Unplugging time before test [s]	Indication changes after stabilisation [ppm]	The level of the heat at the start of the test [%]
0	0	100
60	0.55	70.3
150	1	45.9
300	1.4	24.3
600	1.65	10.8
900	1.8	2.7
1500	1.85	0
few days	1.85	0

Table 4. Maximum change of indication according to the time of measuring system unplugging before test. In both setting of resistors 31 and 25 and average value which is offset of the measuring system itself

Unplugging time before test [s]	Indication changes after stabilisation 25(31) [ppm]	Indication changes after stabilisation 31(25) [ppm]	Average value [ppm]	The level of the heat at the start of the test [%]
0	0	0	0	100
60	–2	0.55	–0.725	76.0
150	–4	1	–1.5	50.4
300	–5.4	1.4	–2	33.8
600	–6.9	1.65	–2.625	13.2
900	–7.6	1.8	–2.9	4.1
1500	–7.9	1.85	–3.025	0

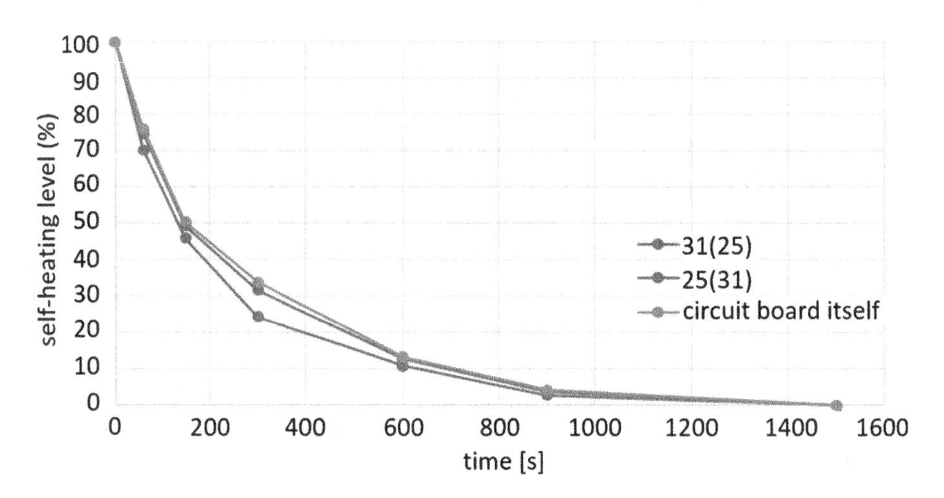

Fig. 3. Summary of cooling processes characteristics of measuring circuit with resistors in both settings and without them

Tables contains percentage values of the measurement circuit heat level at the start of test (in relation to maximum level of its heating – after stabilization of indication when test was conducted after system long term power-off).

Based on the data summaries in the Tables (2, 3 and 4) characteristics of measurement circuit cooling processes (in both settings of resistors (25 and 31) and cooling process of measuring circuit itself) was determined (Fig. 3).

Summary in Fig. 3 shows that the cooling processes in all cases has similar course. This is advantageous given the offset generating by measuring circuit itself is proportional to the changes of measurement circuit indication with resistors in any position regardless from the stage of cooling process.

4 Conclusion

Presented results allowed to formulate the following guidelines for the research methodology:

- data acquisition card have to be on for 10–15 min before starting measurements. During tests device's unused (but declared in software) channels should be connected to the ground to avoid its influence on the results,
- to avoid the influence of a circuit board self-heating it is necessary to wait about 15 min after supplying circuit board, in order to stabilize its indication,
- the measuring system needs about 15–25 min to reach a state of complete cooling after its unplugging. To carry out further tests based on a self-heating phenomena (after preceding test) it is necessary to: wait for the test stand to cool down or compensate the level of the test stand cooling level using the cooling characteristics (Fig. 3).

Compliance with the above guidelines allows to high accuracy measurements. Also the pace of measurements will be increased with the usage of cooling process characteristics determined in Sect. 3.3.

References

1. Horowitz, P., Hill, P.: The art of electronics, 2nd edn. Cambridge University Press, Cambridge (1989)
2. Buckingham, H., Price, E.N.: Principles of Electrical Measurements. English Universities, London (1966)
3. Golding, E.W.: Electrical Measurement and Measuring Instruments, 3rd edn. Sir Issac Pitman and Sons, London (1960)
4. Chwaleba A., Poniński M., Siedlecki A.: Metrologia elektryczna. Wyd. 8 zmienione. Wyd. Naukowo-Techniczne, Warszawa 1979 (2003)
5. Szałatkiewicz, J., Szewczyk, R., Budny, E., Missala, T., Winiarski, W.: Identification of thermal response, of plasmatron plasma reactor. Adv. Intell. Syst. Comput. **267**, 265–274 (2014). doi:10.1007/978-3-319-05353-0_26

6. Stepowicz, W.J., Górecki, K.: Materiały i elementy elektroniczne. Wyd. 2. Wydawnictwo Akademii Morskiej w Gdyni, Gdynia (2008)
7. Nowak, P., Juś, A., Szewczyk, R., Pijarski, R., Nowicki, M., Winiarski, W.: Test stand for temperature characteristics of ultra-precise resistors. In: Awrejcewicz, J., Szewczyk, R., Trojnacki, M., Kaliczyńska, M. (eds.) Mechatronics: Ideas for Industrial Applications, vol. 317, pp. 345–352. Springer, Heidelberg (2015). doi:10.1007/978-3-319-10990-9_32
8. Szałatkiewicz, J.: Metals content in printed circuit board waste. Pol. J. Environ. Stud. **23**(6), 2365–2369 (2014)
9. Juś, A., Nowak, P., Szewczyk, R., Nowicki, M., Winiarski, W., Radzikowska, W.: Assesment of temperature coefficient of extremely stable resistors for industrial applications. In: Awrejcewicz, J., Szewczyk, R., Trojnacki, M., Kaliczyńska, M. (eds.) Mechatronics: Ideas for Industrial Applications, vol. 317, pp. 297–306. Springer, Heidelberg (2015). doi:10.1007/978-3-319-10990-9_27
10. Bartnicki, A., Łopatka, J., Muszyński, T., Wrona, J.: Concept of IED/EOD operations (CONOPs) for engineer mission support robot team. J. KONES **22**(3), 269–274 (2015). doi:10.5604/12314005.1181703
11. Juś, A., Nowak, P., Szewczyk, R.: Automatic system for identification of temperature parameters of resistors based on self-heating phenomena. In: Szewczyk, R., Zielinski, C., Kaliczyńska, M. (eds.) Progress in Automation, Robotics and Measuring Techniques, vol. 352, pp. 91–100. Springer, Heidelberg (2015). doi:10.1007/978-3-319-15835-8_11
12. Juś, A., Nowak, P., Szewczyk, R., Radzikowska-Juś, W.: Research of metal film resistor's temperature stability according to their nominal wattage. In: Szewczyk, R., Kaliczyńska, M., Zieliński, C. (eds.) Challenges in Automation, Robotics and Measurement Techniques, vol. 440, pp. 807–815. Springer, Heidelberg (2016). doi:10.1007/978-3-319-29357-8_71
13. Korytkowski, J.: Digital controlled resistance synthesis electronic circuit for precise simulation of thermo-resistance sensors, Pomiary Automatyka Robotyka, Nr 5, pp. 86–92 (2013). (in Polish)
14. Warsza, Z.L.: New approach to the accuracy description of unbalanced bridge circuits with the example of Pt sensor resistance bridges. J. Autom. Mob. Robot. Intell. Syst. **4**(2), 8–15 (2010)

Two-Channel MEMS Gravimeter of the Automated Aircraft Gravimetric System

Igor Korobiichuk[1(✉)], Olena Bezvesilna[2], Maciej Kachniarz[3], Andrii Tkachuk[4], and Tetyana Chilchenko[4]

[1] Institute of Automatic Control and Robotics, Warsaw University of Technology, Warsaw, Poland
kiv_igor@list.ru

[2] National Technical University of Ukraine "Kyiv Polytechnic Institute", Kiev, Ukraine
bezvesilna@mail.ru

[3] Industrial Research Institute for Automation and Measurements PIAP, Warsaw, Poland
mkachniarz@piap.pl

[4] Zhytomyr State Technological University, Zhytomyr, Ukraine
andrew_tkachuk@i.ua, xvulunka@mail.ua

Abstract. A new two-channel gravimeter of the automated aircraft gravimetric systems (AGS), the design of which is based on MEMS technology, was considered. It was established that the accuracy and speed of the new gravimeter is higher as compared to the currently known counterparts. A new two-channel gravimeter operating principle was described and its main advantages were shown, i.e. the lack of the error signals in the two-channel gravimeter output signal from the impact of vertical acceleration and residual nonidentity of the structures of two capacitive elements. The issue of filtering the output gravimeter signal from high-frequency noise was solved. It was established that the AGS accuracy with a two-channel MEMS gravimeter to measure the anomalies of gravity acceleration is 1 mGal.

Keywords: Two-channel capacitive gravimeter · Aircraft gravimetric system · Gravitational acceleration · Sensor · MEMS

1 Introduction

Today it is extremely important to use the aircraft gravimetric system (AGS) for mineral exploration (geology, geophysics), adjustment of inertial navigation systems (aerospace complex), for location of moving objects in seas and oceans, to perform the tasks of archeology, earthquake prediction etc. The conventional air sea gravimetry possesses a somewhat outdated technology and insufficient accuracy, so it is characterized by low productivity, efficiency and detail and high costs, particularly due to the operating of the middle-class ships and aircraft with crew. At the same time, the problem can be solved by a tiny onboard navigation system [1] based on microelectromechanical systems and technologies (MEMS), combining the microelectronic and micromechanical components.

R. Szewczyk and M. Kaliczyńska (eds.), *Recent Advances in Systems, Control and Information Technology*, Advances in Intelligent Systems and Computing 543, DOI 10.1007/978-3-319-48923-0_51

The AGS effectiveness is largely determined by selection of the systems sensor, i.e. gravimeter. The AGS construction and study were made possible due to development and widespread use of the modern advances in inertial navigation, gyroscopes applied theory and gravimeters.

2 Setting the Task

The accuracy of the current gravimeters is insufficient, only 2–10 mGal [2, 3]. Most of them are non-automated. The measurement results are processed after the aircraft flight on the ground for months. Moreover, these gravimeters measure the gravity acceleration along with aircraft vertical acceleration $\ddot{h}$, which is a complex scientific and technical problem and requires the use of additional filters.

Today, there is no scientific-theoretical and practical work on the possibility and feasibility of using the two-channel capacitive MEMS of the gravimeter as an AGS gravimeter. Therefore, it is advisable to explore the gravimeters of this type.

The scope of study is the process of measuring the gravity acceleration (GA) using a capacitive MEMS gravimeter of an automated aircraft gravimetric systems.

The subject of the study is a new two-channel capacitive MEMS gravimeter of the automated aircraft gravimetric systems.

The article **objective** is to provide a rationale for feasibility of the study and description of design of a new two-channel capacitive MEMS gravimeter of the automated aircraft gravimetric systems.

Main tasks:

- To conduct an analytical review of papers on aircraft gravimetry and capacitive accelerometers and to indicate their strengths and weaknesses;
- To provide a description of the design and lay out the operating principle of a new two-channel capacitive MEMS gravimeter of the automated AGS.

3 Published Data Analysis

The studies have shown that a large contribution to the theory and practice of capacitive transducers was associated with the names of L. Bergman, H. Tiersten, A.A. Andreev, V.V. Malov, N.A. Shulga, V.V. Lavrinenko, S.I. Pugachov, A.P. Kramarov, A.Y. Kolesnikov, P.O. Hrybovskyi and others [4].

The well-known micromechanical capacitive accelerometer, consisting of a glass substrate with deposited aluminum film as a fixed plate of the capacitor and silicon frame with inertial mass in the form of a board as a movable plate of the capacitor [5]. The disadvantages of this accelerometer are low metrological characteristics due to the use of silicon and glass, materials having different temperature coefficients of linear expansion (TCLE).

Also, there is a known accelerometer based on the use of both silicon capacitor plates providing for plate connection at their full peripheral areas through glass insertions [6]. The disadvantages are, first, lower metrological characteristics due to the presence of glass inserts in the areas of plate and frame connections having TCLE different from silicon; and, secondly, the complexity of the technologies of making glass inserts on the board, which provides for etching the grooves for glass, local glass insertion into the grooves, and glass machining to the board plane [7].

Also, the material for sensor manufacture is the SOI structure, a substrate made of SiO_2 coated with a layer of glass subsequently etched to ensure a gap between the inertial mass and the stator, with a layer of low-resistance silicon on top of the glass, on which the rotor is eroded [8].

Today, one of the most promising gravimeters is a capacitive two-channel gravimeter (CG). It is the primary sensing element of the developed automated AGS [9, 10]. CG sensor settings are selected so that the frequency of its natural oscillations is equal to the greatest frequency of gravitational acceleration, which can be measured against the background noise. In other words, the gravimeter sensor also performs the function of a low-pass filter. This eliminates the impact of errors, the frequency of which is greater than the frequency of CG natural oscillations, on the CG output indications, and increases the measurement accuracy of the gravity acceleration. However, a single-channel CG does not provide for elimination of errors caused by the influence of vertical acceleration and instrumental errors.

4 Key Provisions

Since there is no analogue among the known gravimeter, to develop the structure design, we rely on analogues among capacitive MEMS accelerometers. The most urgent and most practical design is presented in (Fig. 1). It consists of a fixed base (1), often connected to the body and the frame (2) of the sensor (3) attached to the upper movable plates (4). Both plates have a dielectric layer deposited (5). The plates are electrodes conducting a certain form, located in the operating environment.

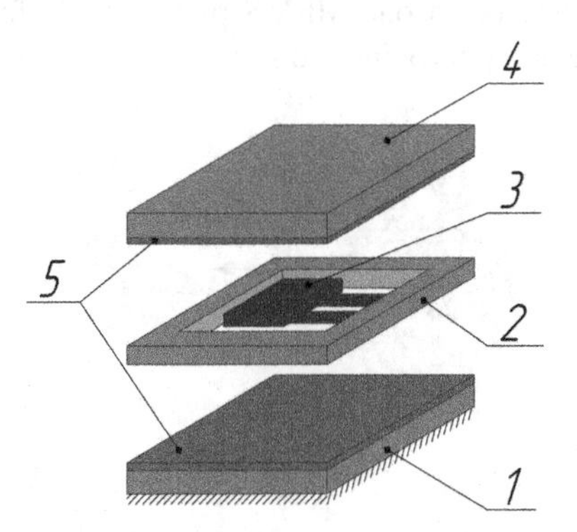

Fig. 1. Typical design of a capacitive MEMS accelerometer

The plates have a small thickness, and the surface effect phenomena can be manifested there at very high frequencies only, about 100 MHz.

The material used to manufacture the plates must meet the following requirements:

- a low electrical resistance of the plates, especially for high-frequency capacitors;
- TCLE close to TCLE of substrate and insulator;
- good adhesion to both substrate and previously formed films;
- have a low migration mobility of atoms and high corrosion resistance [11].

For plates, do not use materials with high mobility of atoms, such as copper or gold. The atoms of these metals, penetrating into the insulator, may form the bridges between plates. The plate material should have good adhesion to the substrate material and dielectric.

One of such materials is sapphire monocrystal. Sapphire monocrystal has the following properties: high mechanical hardness (9 according to Moss scale), high temperature according to Debye (On = 1040 K), and high Peierls barrier. It was proven experimentally that the dislocation velocity is much smaller in sapphire monocrystal than in quartz and metal, and the velocity of dislocations at low temperatures is decreased especially dramatically due to the high physical and chemical resistance of sapphire. Since sapphire has a high Debye temperature, the linear expansion coefficient and Young's modulus coefficient is reduced with a decreasing temperature much faster than those in elastic materials.

5 Problem Solution for a Specific Technological Object

Considering the issue of optimizing the design and reduction of the weight and dimensional characteristics, it is appropriate to seek to reduce the inertial mass magnitude, while reaching an increase in frequencies of the sensor natural oscillations. An expedient solution is to exclude the inertial mass and use only a distributed console mass. It is better to use the console with a distributed mass and variable thickness and width. All these conditions are easily achieved using silicon. For more efficient operation, it is proposed to use several capacitive sensors on one MEMS plate. The differential structure allows increasing the amplitude and improving the signal linearity (Fig. 2).

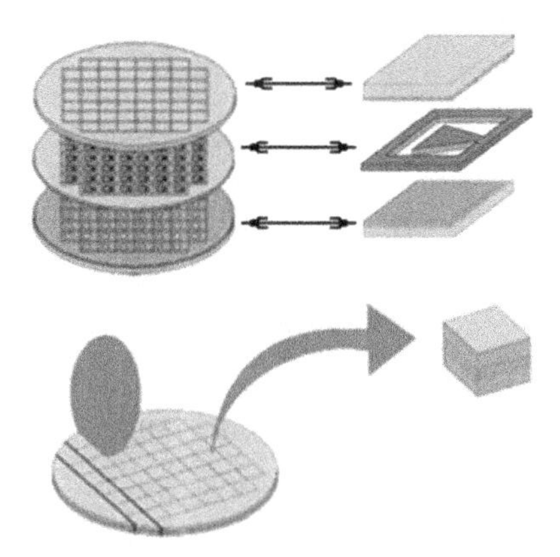

Fig. 2. Placing the capacitive MEMS transformers with an optimized sensor on one plate

To ensure the two-channel principle, two identical MEMS plates are installed symmetrically; their signals are summed in the adder and sent for further processing and amplification (Fig. 3). Sensor symmetry reduces dependence on temperature and sensitivity on the axis and improves linearity. The sealed transducer is provided by anodic connection of plates to each other. This facilitates the element enclosing and allows using the gas damping in a sensor element.

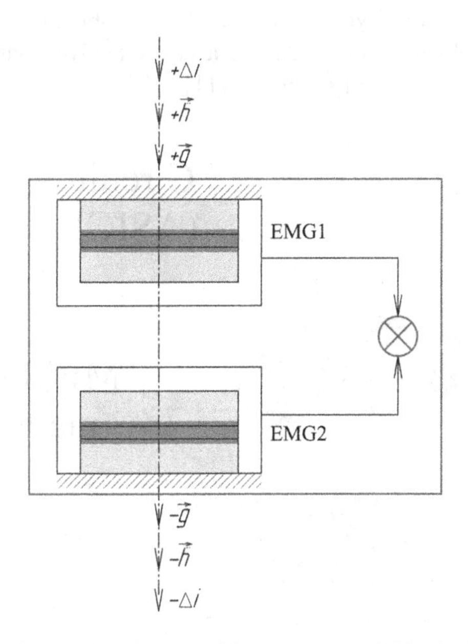

Fig. 3. Placing the capacitive sensors on MEMS plate

The capacitive elements of both channels are affected by gravity acceleration g, aircraft vertical acceleration $\ddot{h}$ and instrumental errors Δi from the influence of residual nonidentity of identical plates' structures and masses, from the influence of changes in temperature, humidity and ambient pressure, as well as boundary effects. If these impacts are projected on the measuring axis of the gravimeter, considering that the EMG1 capacity of one channel increases when EMG2 capacity is reduced by the same amount, we get:

$$\begin{aligned} u_1 &= k(mg_z + m_i\ddot{h} + \Delta i); \\ u_2 &= k(mg_z - m_i\ddot{h} - \Delta i), \end{aligned} \tag{1}$$

where u_1 is an output electrical signal of capacitive MEMS of EMG1 gravimeter of one channel; u_2 is an output electrical signal of capacitive MEMS of EMG2 gravimeter of the second channel; m_i is an inertial mass in each channel; k is an electrical constant.

Output electrical signals u_1 and u_2 of capacitive elements of both channels are added in the adder:

$$u_{\Sigma} = u_1 + u_2 = 2kmg_z, \tag{2}$$

where u_{Σ} is the output signal.

During production, a new concept of heterogeneous integration should be used to combine MEMS sensor and circuit (ASIC): "Chip-on-MEMS" or CoM. This concept is based on a combination of encapsulated wafer-level 3D-MEMS structures, as well as enclosing technology at the wafer and chip technology on the plate. All these processes have been existing for several years already. Their combination can address the most difficult enclosing problem, how to ensure a cost-effective combination of MEMS-components and integrated circuits (Fig. 4) [11].

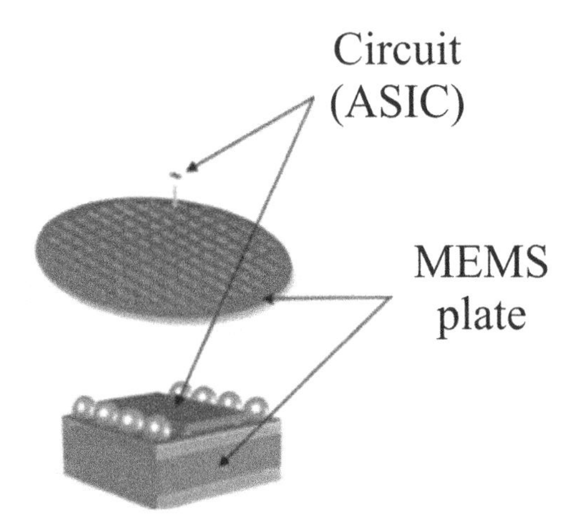

Fig. 4. Integration of MEMS elements on the plate

6 Conclusions

Therefore, to reduce the dependence on the effects of residual nonidentity in design, changes in temperature, humidity, pressure, boundary effects and to improve the sensitivity and linearity, it is advisable to install two identical MEMS plates symmetrically with a triangular sensor, the signals of which will be added and transferred for further processing and amplification.

References

1. Korobiichuk, I.: Mathematical model of precision sensor for an automatic weapons stabilizer system. Measurement **89**, 151–158 (2016). doi:10.1016/j.measurement.2016.04.017147
2. Korobiichuk, I., Bezvesilna, O., Tkachuk, A., Chilchenko, T., Nowicki, M., Szewczyk, R.: Design of piezoelectric gravimeter for automated aviation gravimetric system. J. Autom. Mob. Robot. Intell. Syst. (JAMRIS) **10**(1), 43–47 (2016). doi:10.14313/JAMRIS_1-2016/6

3. Korobiichuk, I., Bezvesilna, O., Tkachuk, A., Nowicki, M., Szewczyk, R., Shadura, V.: Aviation gravimetric system. Int. J. Sci. Eng. Res. **6**(7), 1122–1127 (2015)
4. Karachun, V., Mel'nick, V., Korobiichuk, I., Nowicki, M., Szewczyk, R., Kobzar, S.: The additional error of inertial sensor induced by hypersonic flight condition. Sensors **16**(3), 299 (2016). doi:10.3390/s1603029
5. A micromechanical capacitive accelerometerurith a turo - pointinertial – masssuspension. Sens. Actuators **83**, 4, N 2, 190–198
6. Pat. France 2564593, MKI 4; G 01 P 15/125, H 01 C 13/00, filing date 03.19.85; publ. in the Official Journal of Industrial Property, N 47, 22 November 1985
7. Kozin, S.A., Kolganov, V.N., Malkin, Y.M., Papko, A.A: Capacitive accelerometer and method of manufacture, Pat. 2114489 Russia, IPC G 01 V 7/02
8. Development of MEMS – capacitive sensor element of a comb-type accelerometer [electronic resource], The International Center for Science and Education. http://nauchforum.ru/node/8314. Accessed 21 Jan 2016. City - Caps. screen
9. Korobiichuk, I., Bezvesilna, O., Tkachuk, A., Nowicki, M., Szewczyk, R.: Piezoelectric gravimeter of the aviation gravimetric system. In: Proceeding of Automation-2016, Challenges in Automation, Robotics and Measurement Techniques, Warsaw, Poland, 2–4 March 2016, pp. 753–763 (2016). doi:10.1007/978-3-319-29357-8_65
10. Korobiichuk, I., Bezvesilna, O., Tkachuk, A., Chilchenko, T., Nowicki, M., Szewczyk, R.: Design of piezoelectric gravimeter for automated aviation gravimetric system. J. Autom. Mob. Robot. Intell. Syst. (JAMRIS) **10**(1), 8–16 (2016). doi:10.14313/JAMRIS_1-2016/6
11. Modern MEMS gyroscopes and accelerometers [electronic resource], "Sovtest ATE", LLC. http://www.sovtest.ru/news/publications/sovremennye-mems_giroskopy-i-akselerometry/. Accessed 21 Jan 2016. City - Caps. Screen

Measurement Setup for the Thermal and Line Regulation Characteristics of Reference Voltage Sources

Paweł Nowak[1(✉)], Andrzej Juś[1], and Roman Szewczyk[2]

[1] Industrial Research Institute for Automation and Measurements, Al. Jerozolimskie 202, 02-486 Warsaw, Poland
pnowak@piap.pl

[2] Faculty of Mechatronics, Warsaw University of Technology, sw. A. Boboli 8, 02-525 Warsaw, Poland

Abstract. Paper presents universal test stand utilized for testing functional parameters of a reference voltage source based on Zener diode. The stability of the standard is crucial during the operation of precise voltage A/D converters. Value of the output voltage from the reference source may change in the function of the operating temperature or due to the variation of power supply. Presented test stand allows to determine the thermal and line regulation characteristics of the tested element. Test stand is based on differential measurement, where the output voltage from the tested element is compared with the voltage from the reference element kept in a stable conditions. Tested element can be exposed to the temperature variation or to the changes of power supply voltage. Output voltage from the differential circuit as well as a value of an influential factor (operating temperature or supply voltage) is continuously measured and logged. Based on obtained characteristics optimal working conditions for tested elements can be identified. It also allows proper selection of the most stable specimens for the crucial applications.

Keywords: Voltage reference · Line regulation · Temperature coefficient

1 Introduction

The common method of precision industrial processes monitoring is based on electronical measurements. Biggest advantages of that method are reliability of signal, ease of its use and ease of digitalization, which allows further usage of the data in complex systems. In order to obtain the most reliable conversion, precise A/D converters are commonly used [1]. Highest conversion accuracy is obtained for Sigma-Delta converters. Most of the presented methods are based on differential measurement, where tested voltage is compared with standard voltage value. Stability of the reference is crucial for the proper work of the setup.

The most stable voltage references are based on buried Zener type, which consists Zener diode connected in series with precise resistor, which limits the diode current and thus protects the semiconductor structure from overheating. The biggest influence on voltage standards stability has their operating temperature which influences both

R. Szewczyk and M. Kaliczyńska (eds.), *Recent Advances in Systems, Control and Information Technology*, Advances in Intelligent Systems and Computing 543, DOI 10.1007/978-3-319-48923-0_52

diode's breakdown voltage and value of integrated resistor. Temperature influence on the standards voltage output (TCV_O – thermal coefficient of output voltage) is typically represented in ppm/°C and is calculated as (1) presents [2–4]:

$$TCV_O = \left(\frac{V_{max} - V_{min}}{V_{nominal} \cdot (T_{max} - T_{min})}\right) \cdot 10^6 \tag{1}$$

where:

- V_{max} is analyzed standards voltage value for T_{max},
- V_{min} is standards voltage for T_{min},
- $V_{nominal}$ is an nominal value of standards voltage output
- T_{max} and T_{min} are respectively maximal and minimal temperature in which reference was tested.

Thus (1) is unsuitable for description of elements which $\Delta V(\Delta T)$ characteristic may be approximated by even degree polynomial. Thus it is crucial to determine whole characteristic of temperature influence on the voltage output. Additionally, advanced measurement setup may be utilized for determination of thermal hysteresis of tested element.

Besides temperature, highest influence on voltage reference stability has line regulation, which is a change in output voltage caused by a change in input voltage (LRV_O – line regulation of voltage output). It is usually specified in ppm/V and is calculated as (2) presents [2]:

$$LRV_O = \left(\frac{V_{max} - V_{min}}{V_{nominal} \cdot (V_{supplymax} - V_{supplymin})}\right) \cdot 10^6 \tag{2}$$

where:

- V_{max} is analyzed standards voltage value for $V_{supply\ max}$,
- V_{min} is standards voltage for $V_{supply\ min}$,
- $V_{nominal}$ is an nominal value of standards voltage output
- $V_{supply\ max}$ and $V_{supply\ min}$ are respectively maximal and minimal supply voltage in which reference was tested.

Increment of input voltage generally causes an increase of the current which slightly changes operating point of buried Zener structure as well as causes self-heating phenomena which further increases the output voltage variation.

2 Main Idea of the Test Stand

In order to obtain highest measurement accuracy the developed test stand was based on differential measurement, where output voltage from tested voltage standard was compared with voltage from reference voltage standard kept in stable conditions. Tested voltage standard may be exposed to controlled operating temperature variation or power supply variation in order to determine $\Delta V(\Delta T)$ and $\Delta V_{out}(\Delta V_{in})$ characteristics as presented in Fig. 1. Output voltage from the differential amplifier circuit (described in Sect. 3), proportional to tested voltage reference output variation, is measured by data acquisition card. The value of influential factor (operating temperature or supply voltage) is measured so requested characteristics can also be obtained.

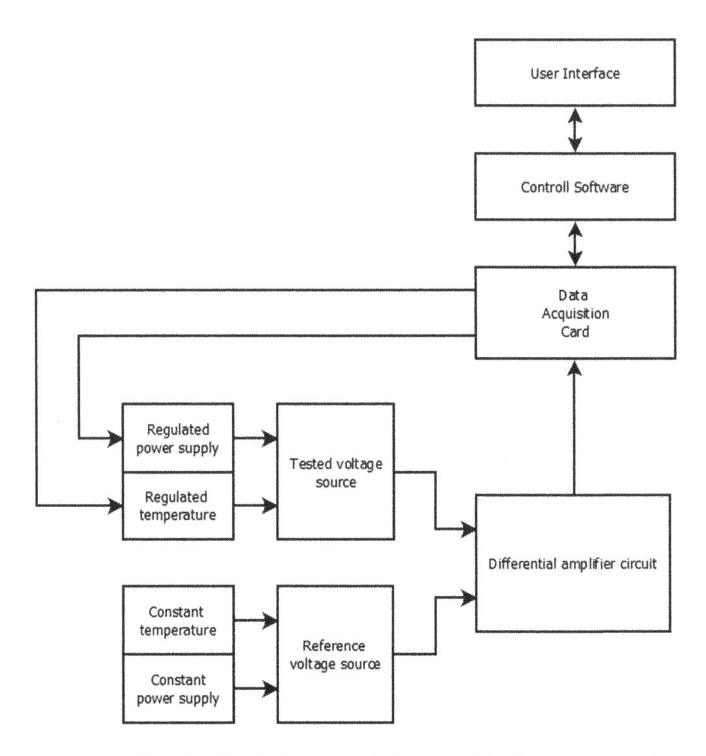

Fig. 1. Block diagram of developed test stand

3 Test Stand Implementation

Prototype of the test stand was implemented in Industrial Research Institute for Automation and Measurements PIAP. The whole setup was based on a data acquisition card – NIDAQ 6363 which ensures high measurement accuracy for voltage deviation signal as well as for influence factor signals (temperature and power supply voltage). Due to two analog outputs in the card, whole test procedure could be controlled automatically. Control software for the setup was developed in NI LabVIEW software and allows conducting single or multiply series of test for selected influence factor.

Temperature regulation is obtained by Joule heating of power resistor supplied by voltage-controlled current source. Operating temperature of the tested voltage reference is measured by a K-type thermocouple and signal conditioner with internal cold-junction compensation [5, 8] based on Pt 100 sensor.

Regulation of tested reference power supply is obtained by the usage of voltage controlled voltage source. This voltage is measured by DAQ card in order to obtain more reliable characteristic and to identify possible damage of the supplying circuit.

In order to keep reference voltage source in stable conditions it was placed in isolated container and supplied from stabilized power source.

Main circuit was based on differential amplifier circuit as presented in Fig. 2. Also two operational amplifiers in a voltage follower configuration were utilized in order to ensure constant and identical output impedance for both voltage references (tested – U4 and reference – U5). This limits influence of the load regulation on the measurement results.

Differential amplifier circuit, composed of single op-amp (U3) and four resistors (R1–R4) amplifies the difference of the input voltages, held by voltage followers at the output of op-amps U1 and U2. By using the same type reference voltage source like the tested one, very high measurement accuracy can be obtained (up to 0.1 ppm).

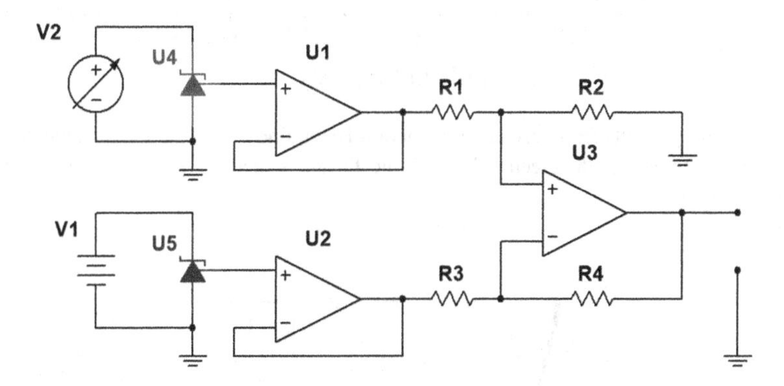

Fig. 2. Wiring diagram of measurement circuit

4 Exemplary Results

Test stand measurement accuracy and proper operation was validated on extremely stable reference voltage sources – Max6325 manufactured by Maxim Integrated [6]. Both types of test stand operation were validated – exemplary line regulation characteristic is presented in Fig. 3 whereas temperature characteristic is presented in Fig. 4 [7]. On both graphs black line represents maximal allowable deviation, determined by circuit manufacturer [6].

In Fig. 3 results of the hysteresis of the line regulation are presented as well. No significant signal hysteresis is visible. Those result confirm high repeatability of the test stand. Figure 3 presents additionally two possible approximations of obtained measurement results – violet curve is a result of second order polynomial fit, whereas green one presents logarithmic fitting. Both fits have extremely high R^2 coefficient (over 0.99), but logarithm fitting better describes physical phenomena occurring in the reference semiconductor structure [2].

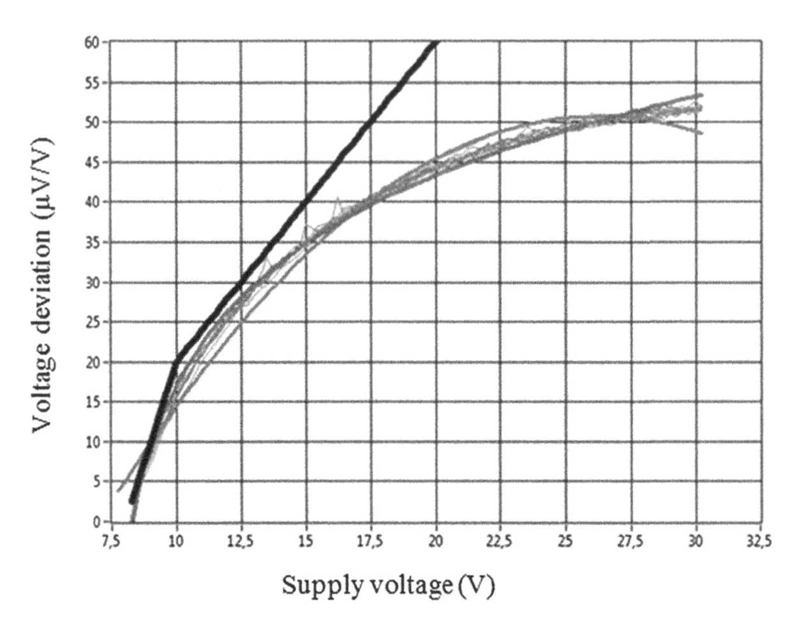

Fig. 3. Exemplary results from test stand validation Black line – maximum allowable deviation, orange – measurement result, green – logarithmic fit and violet – the second degree curve fit (Color figure online)

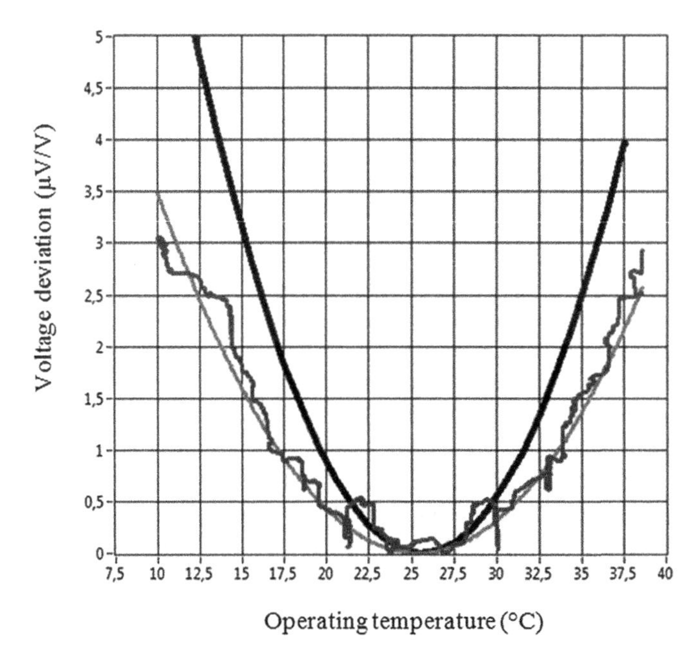

Fig. 4. Exemplary results from test stand validation Black line – maximum allowable deviation, blue – measurement result, red– the second degree curve fit (Color figure online)

Figure 4 presents exemplary results of $\Delta V(\Delta T)$ measurement. Test was also conducted on Max6325. As presented, tested reference was heated from 10°C to 37°C which is extremely expanded range, comparing to typical temperature range of precise electronic operation. On the other hand, due to wider range more data were obtained about the shape of the characteristic. Despite few points, which are probably caused by quantization error of DAQ card, obtained measurement results are bellow maximum allowable deviation declared by manufacturer. As before, measurement result were fitted with second degree polynomial with very good (over 0.99) R^2 coefficient.

This exemplary result legitimate the development of the test stand. During two point measurement (as described in the Sect. 1) and basing on the (1), obtained TCV_O would significantly vary, depending on the measurement point selection. For example, if measurement points would be placed symmetrically with respect the minimal value of the $\Delta V(\Delta T)$ characteristic (25°C), obtained TCV_O coefficient would be close to zero. On the other hand, basing on obtained characteristic, optimal (with the smallest value of local derivative) working temperature for the tested element can be determined.

5 Conclusions

As presented, universal, accurate and reliable test stand was designed and developed. Obtained measurement repeatability in single point is about 0.5 ppm, but during wider range measurements it does not influence results significantly. Further increase of measurement accuracy may be obtained by improvement of the thermal stabilization of the setup, especially for reference voltage source and differential amplifier circuit. Usage of high resolution voltmeter would also improve measurements accuracy, due to limitation of quantization error influence.

Based on the designed prototype, industrial partner implemented similar solution for testing new consignments of the utilized elements.

Acknowledgments. This work was partially supported by The National Center for Research and Development with PBS Program – Grant no. PBS 1/B3/8/2012 and partially from statutory funds.

References

1. Horowitz, P., Hill, P.: The Art of Electronics, 2nd edn, p. 3. Cambridge University Press, Cambridge (1989)
2. Harrison, L.T.: Current sources and voltage references. Elsevier Inc, Oxford (2005)
3. Bartnicki, A., Łopatka, J., Muszyński, T., Wrona, J.: Concept and development of engineer mission support robot. J. KONES **22**(1), 269–274 (2015)
4. Szałatkiewicz, J., Szewczyk, R., Budny, E., Missala, T., Winiarski, W.: Identification of thermal response, of plasmatron plasma reactor. Adv. Intell. Syst. Comput. **267**, 265–274 (2014). doi:10.1007/978-3-319-05353-0_26
5. Kerlin, T.W., Johnson, M.: Practical Thermocouple Thermometry, 2nd edn. Research Triangle Park: ISA, USA (2012)

6. MAX 6325 datasheet provided by Maxim Integrated
7. Nowak, P., Juś, A., Szewczyk, R.: Resistance of MAX 6325 reference voltage source on operating temperature variation. In: Szewczyk, R., Zieliński, C., Kaliczyńska, M. (eds.) Progress in Automation, Robotics and Measuring Techniques Volume 3 Measuring Techniques and Systems, vol. 352, pp. 189–196. Springer, Heidelberg (2015). doi:10.1007/978-3-319-15835-8_21
8. Szałatkiewicz, J., Szewczyk, R., Budny, E., Missala, T., Winiarski, W.: Measurement and control system of the plasmatron plasma reactor for recovery of metals from printed circuit board waste. Recent Adv. Autom. Robot. Measuring Tech. **267**, 687–695 (2014). doi:10.1007/978-3-319-05353-0_65

Development of Graphene Based Leak Detector

Marcin Safinowski[1(✉)], Krzysztof Trzcinka[1], Cezary Dziekoński[1], Andrzej Juś[2], Maciej Kachniarz[2], Roman Szewczyk[2], and Wojciech Winiarski[1]

[1] Industrial Research Institute for Automation and Measurements, Al. Jerozolimskie 202, 02-486 Warsaw, Poland
{msafinowski,ktrzcinka}@piap.pl
[2] Institute of Metrology and Biomedical Engineering, Warsaw University of Technology, sw. Andrzeja Boboli 8, 02-525 Warsaw, Poland

Abstract. The following paper presents one of the possible applications of the graphene based flow detector developed at Industrial Research Institute for Automation and Measurements in cooperation with Institute of Electronic Materials Technology, Faculty of Physics of Warsaw University and Apator Powogaz company. Application involves using flow detector to measure the tightness of the water installation fittings. The construction and principle of operation of the developed graphene based leak detector are described in the paper. Results of performed investigation on functional properties and electromagnetic compatibility (EMC) are also presented. The uncertainty of the tightness measurement was determined. Finally, guidelines that should be met while constructing a leak detector are set in the paper.

Keywords: Flow measurement · Tightness testing · Leak detection · Graphene

1 Introduction

During the implementation of research project "Graphene based, active flow sensors" within national GRAF-TECH program, prototype of graphene based sensor of volumetric flow rate (graphene flow sensor) was developed. Its principle of operation is generally based on generation of the voltage between polarized graphene electrodes under the influence of fluid flow. As a result, electric signal is obtained, which is proportional to the flow rate of the fluid when measurement is performed in short period of time [1–9].

Unique properties of the graphene flow sensor, like ability to detect very little changes in flow velocity of the fluid (from 0.6 mm/s), allowed its application as a leak detector utilized in testing of tightness of the closed valves at receiving stations of hydraulic fittings.

Exemplary system for testing tightness of closed hydraulic valves, which utilizes graphene leak detector, is presented in Fig. 1. After tested valve is mounted in the system for tightness testing between two sealing plates, hydraulic pump takes water from the reservoir and forces the flow in direction indicated in the Fig. 1 with an arrow.

R. Szewczyk and M. Kaliczyńska (eds.), *Recent Advances in Systems, Control and Information Technology*, Advances in Intelligent Systems and Computing 543, DOI 10.1007/978-3-319-48923-0_53

Behind the tested valve there is a separating valve and graphene leak detector. After the entire system is filled with water, the tested valve is closed. As a result, the hydraulic pressure acting on the closing element of the valve is created at the inlet of the valve, while at the outlet of the valve there is atmospheric pressure. If there is a leak in the tested hydraulic valve, there will be water flow at the outlet of the valve. This flow will be detected by graphene leak detector as a change of the electric potential difference between measurement electrodes connected to the graphene sensor from initial electric potential difference resulting from electric potential gradient within graphene layer introduced by polarizing electrode. Value of the change of electric potential difference is dependent on the fluid flow resulting from leak in the tested valve. Electric signal from the flow detector is processed by the signal processing system and the result is presented on the diode indicator. Number of glowing diodes is proportional to the fluid flow resulting from leak in the valve.

Even minimal flow detected with the leak detector indicates that tested valve is not tight. Detection threshold of the developed graphene leak detector is 60 ml/h (16.6 μl/s), which allows to detect very small leaks in the hydraulic valves.

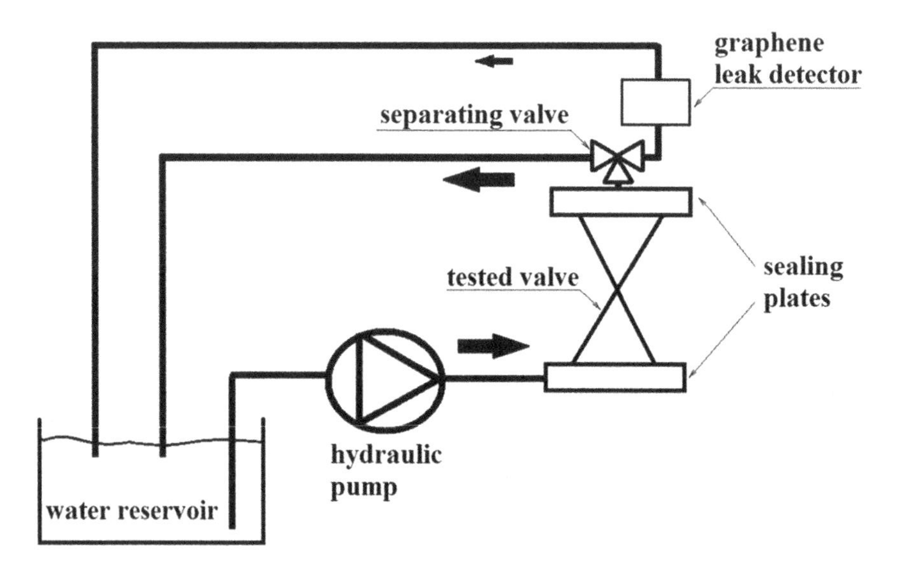

Fig. 1. Schematic diagram of the system for testing tightness of hydraulic valves

2 Construction of Graphene Leak Detector

Graphene leak detector is device containing two main parts:

- electronic signal processing system (Fig. 2a),
- small flow transducer with graphene sensor (Fig. 2b) through which the fluid flow.

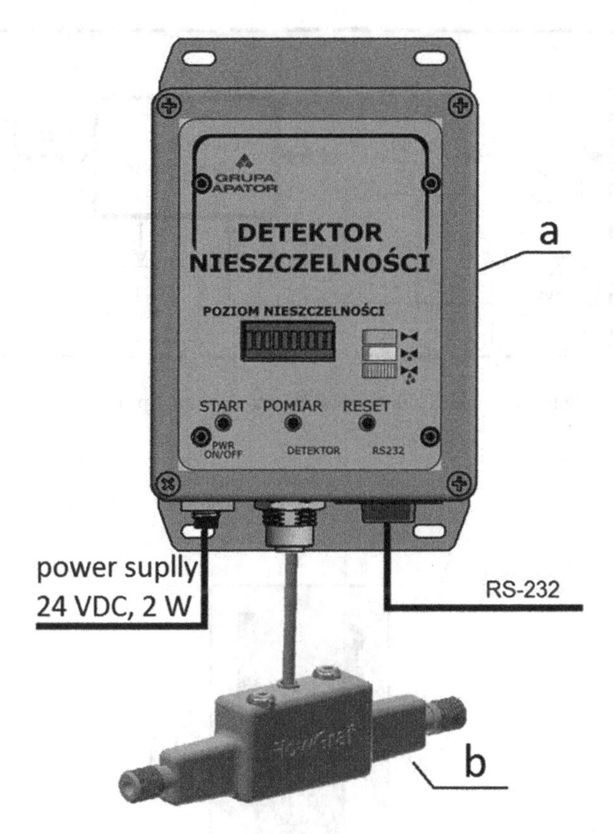

Fig. 2. Graphene leak detector: a – electronic signal processing system, b – small flow transducer with graphene sensor

2.1 Electronic Signal Processing System

Schematic block diagram of the electronic signal processing system, developed for graphene leak detector, is presented in Fig. 3. Voltage signal from graphene flow sensor is transmitted to the input of precise instrumental amplifier. After amplification signal is converted to the digital form with ADC. Result of the measurement is presented in the linear LED indicator of leakage level. If more than one diode of the indicator is glowing, leak in the tested hydraulic valve was detected. The developed signal processing system is equipped with RS-232 interface allowing communication with external devices.

2.2 Small Flow Transducer

The main component of the developed small flow transducer is graphene sensor presented in Fig. 4. It contains two graphene electrodes with dimensions 2 mm × 10 mm

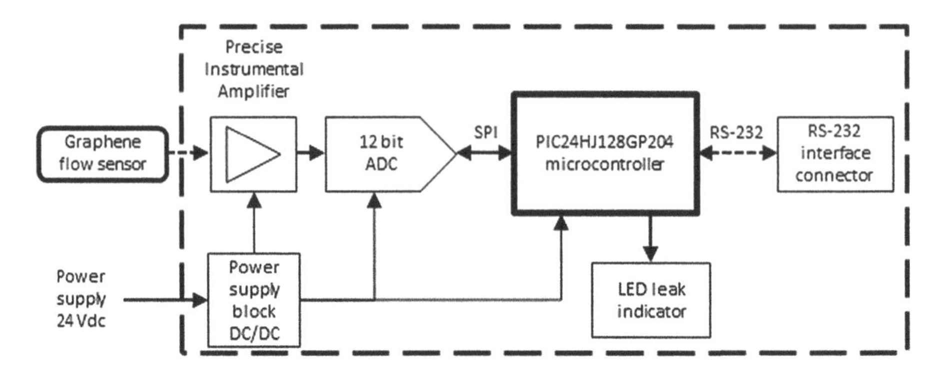

Fig. 3. Schematic block diagram of electronic signal processing system

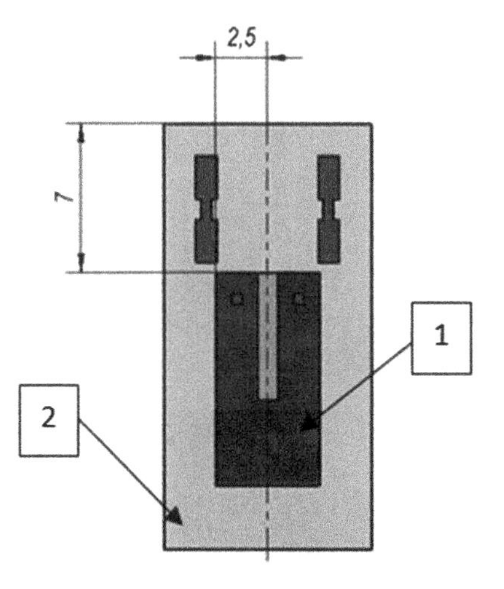

Fig. 4. Visualization of the graphene sensor (1) with substrate (2)

separated with area 1 mm × 10 mm created by oxygen plasma etching of graphene layer. In photography presented in Fig. 5 this area is highlighted with red dashed line. Metallic 20 nm_Ti/80 nm_Au contacts were formed with Chemical Vapour Deposition (CVD) method.

The cross-section view of the small flow transducer with graphene flow sensor is presented in Fig. 6.

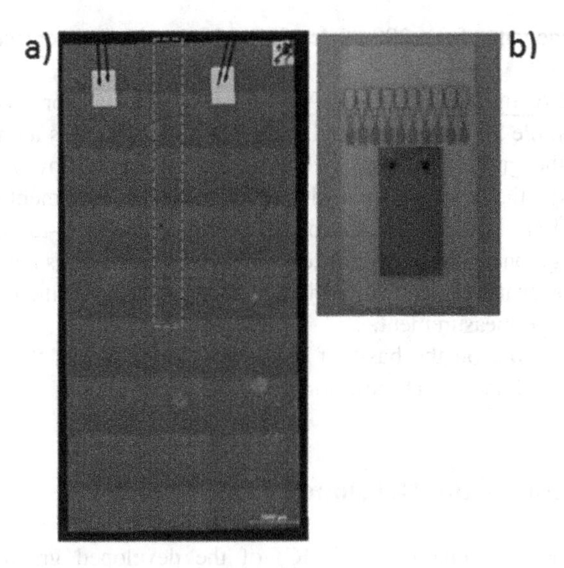

Fig. 5. Developed graphene flow sensor: (a) graphene structure, (b) complete sensor module

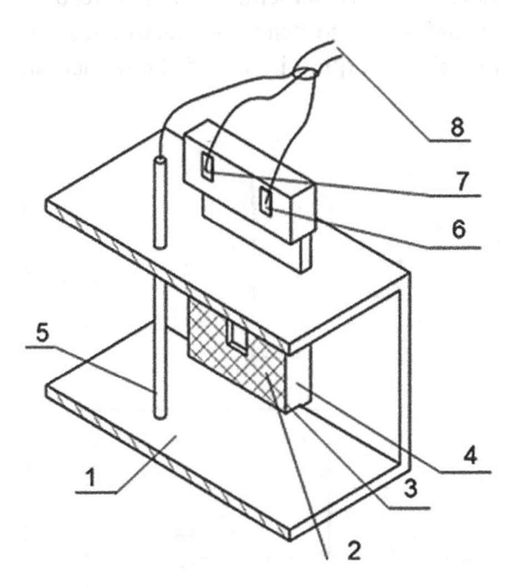

Fig. 6. Cross-section view of the small flow transducer: 1 – casing, 2 – graphene layer, 3 – graphene flow sensor, 4 – substrate of the sensor, 5 – polarizing electrode, 6, 7 – measurement electrodes, 8 – shielded cable

3 Algorithm of Tightness Testing

Algorithm utilized for testing of tightness of the hydraulic valve consists several stages:

– Calculation and storing the reference voltage: after the tested valve and small flow transducer are filled with water, signal processing system calculates and stores the average voltage value from 200 reads of signal from the graphene sensor connected to the ADC.

- Measurement of tightness of the valve: the valve is tested for a certain period of time (for example 30 s). If the tested valve has a leak, there is a change in voltage signal from the graphene sensor, which is dependent on flow ratio of the fluid flowing through the small flow transducer. Results of measurement of the change in voltage signal are filtered with the low-pass filter. Then peak-to-peak value of the voltage from graphene flow sensor is calculated and the result is compared with data base of sensor characteristics stored in EEPROM memory, which allows to obtain result of leakage measurement.
- Result presentation: on the basis of measured voltage value, the leakage level is presented in the linear LED indicator.

4 Investigation of the Developed Leak Detector

The electromagnetic compatibility (EMC) of the developed graphene based leak detector was investigated in order to verify its resistance to electromagnetic interferences found in the industrial environment. Obtained results are satisfying. Leak detector works correctly under the influence of electromagnetic interferences, which was achieved by filtration of all output circuits of the device and using only shielded cables.

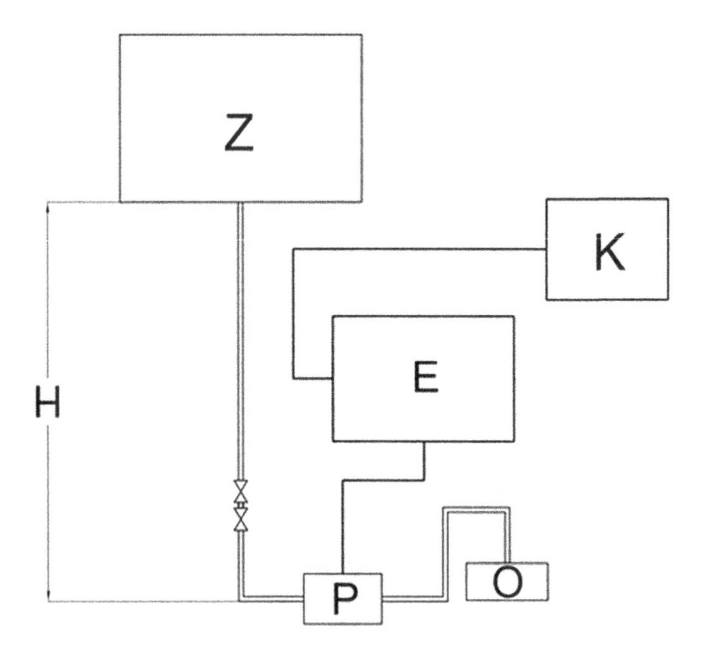

Fig. 7. Measurement setup block diagram: E – electronic signal processing system, P – small flow transducer with graphene sensor, Z – distilled water tank, O – water outflow, K – PC, H – height of the water column (about 1.7 m)

Verification of measurement characteristic of developed graphene leak detector required special measurement setup presented in Fig. 7. The liquid tank was located on the height allowing to obtain the liquid column height about 1.7 m in order to achieve stable flow through the small leak transducer with graphene sensor, which was important due to small cross-sectional area of both the hydraulic conduits and the graphene sensor itself. The fluid used during the investigation was distilled water produced in laboratory distiller exhibiting conductivity of (0.5–3.0) μS/cm.

Table 1 presents results of the investigation on metrological characteristic of the graphene based leak detector. For each applied value of volumetric flow rate at least 30 repeats of measurement were performed.

Table 1. Results of measurement of the developed leak detector characteristic

No.	Volumetric flow ratio (ml/h)	Standard deviation (ml/h)	Standard deviation (%)
1	54.2	12.8	24.4
2	329	24.9	7.6
3	458	23.8	5.2
4	898	44.1	4.9
5	1275	69.3	5.4
6	1426	57.6	4.0
7	1815	77.1	4.2
8	2310	104.3	4.5
9	2942	104.2	3.5
10	3342	1337.9	4.1
11	3731	97.0	2.6

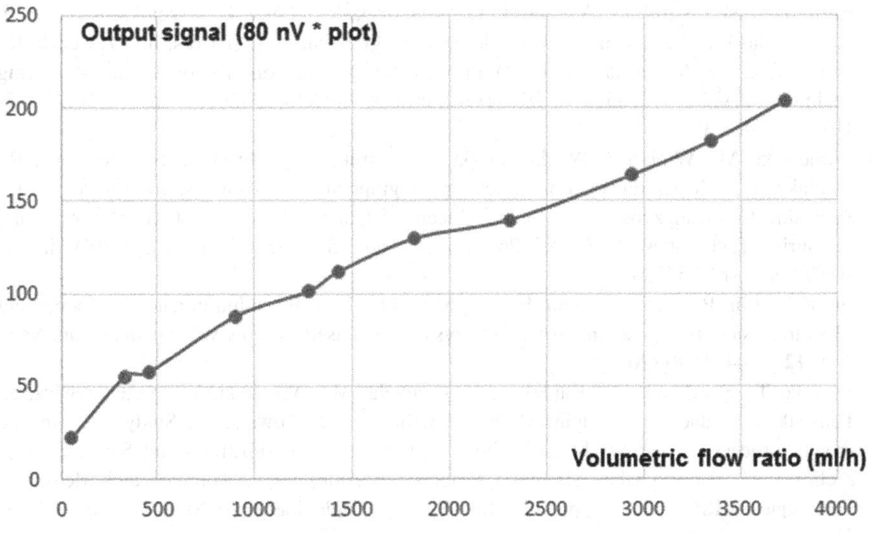

Fig. 8. Characteristic of volumetric ratio dependence of output voltage signal in developed graphene leak detector (1 elementary plot = 80 nV).

The highest value standard deviation of volumetric flow rate measurement expressed as a percentage is 24.4 % for volumetric flow rate value 54.2 ml/h. For higher values of the flow ratio obtained results are improving. For 329 ml/h of volumetric flow ratio, standard deviation was 7.5 % and for 3731 ml/h it was only 2.6 %. Obtained characteristic of volumetric ratio dependence of output voltage signal is presented in Fig. 8. One elementary plot on the vertical axis (output signal) corresponds to 80 nV.

5 Conclusion

Developed graphene leak detector presented in the following paper is able to detect leakage resulting in volumetric flow ratio 60 ml/h. Obtained measurement results are very satisfying and confirms that developed device can be used in testing of tightness at receiving stations of hydraulic fittings. Due to utilization of suitable algorithm for flow measurement in order to determine the leakage level it is enough that graphene sensor will react for sudden and temporary changes of flow ratio.

Acknowledgments. This work has been supported by the National Centre for Research and Development (NCBiR) within the GRAF-TECH program (no. GRAF-TECH/NCBR/02/19/2012). Project "Graphene based, active flow sensors" (acronym FlowGraf).

References

1. Newaz, A.K.M., Markov, D.A., Prasai, D., Bolotin, K.I.: Graphene transistor as a probe for streaming potential. Nano Lett. **2**, 2931–2935 (2012)
2. Safinowski, M., Winiarski, W., Petruk, O., Szewczyk, R., Gińko, O., Trzcinka, K., Maciąg, M., Łoboda, W.: Advancement in development of graphene flow sensors. In: Szewczyk, R., Zieliński, C., Kaliczyńska, M. (eds.) Progress in Automation, Robotics and Measuring Techniques. AISC, vol. 352, pp. 205–217. Springer, Heidelberg (2015). doi:10.1007/978-3-319-15835-8_23
3. Safinowski, M., Winiarski, W., Domański, K., Petruk, O., Dąbrowski, S., Szewczyk, R., Trzcinka, K.: Measuring station for testing of graphene flow sensors. In: Szewczyk, R., Zieliński, C., Kaliczyńska, M. (eds.) Recent Advances in Automation, Robotics and Measuring Techniques. AISC, vol. 267, pp. 649–663. Springer, Heidelberg (2014). doi:10.1007/978-3-319-05353-0_62
4. He, R.X., Lin, P., Liu, Z.K., Zhu, H.W., Chan, H.L., Yan, F.: Solution-gated graphene field effect transistors integrated in microfluidic systems and used for flow velocity detection. Nano Lett. **12**, 1404–1490 (2012)
5. Missala, T., Szewczyk, R., Kamiński, M., Hamela, M., Winiarski, W., Szałatkiewicz, J., Tomasik, J., Salach, J., Strupiński, W., Pasternak, I., Borkowski, Z.: Study on graphene growth process on various bronzes and copper-plated steel substrates. In: Szewczyk, R., Zieliński, C., Kaliczyńska, M. (eds.) Progress in Automation, Robotics and Measuring Techniques. AISC, vol. 352, pp. 171–180. Springer, Heidelberg (2015). doi:10.1007/978-3-319-15835-8_19

6. Gosh, S.H, Sood, A.K., Kumar, N.: Carbon Nanotube Flow Sensors. Science **299**, 1042–1044 (2003)
7. Bartnicki, A., Łopatka, J., Muszyński, T., Wrona, J.: Concept and development of engineer mission support robot. J. KONES **22**(1), 269–274 (2015)
8. Sklyar, R.: The Microfluidic Sensors of Liquids, Gases, and Tissues Based on the CNT or Organic FETs. J. Autom. Mob. Rob. Intell. Syst. **1**(2), 20–34 (2007)
9. Kowalski, A., Safinowski, M., Szewczyk, R., Winiarski, W.: Development of graphene based flow sensor. J. Autom. Mob. Robot. Intell. Syst. **9**(4), 55–57 (2015)

Earth Remote Sensing Satellite Navigation Based on Optical Trajectory Measurements

Ruslan Hryshchuk(✉) and Andriy Zavada

Cybersecurity Department of the Research Center,
Zhytomyr Military Institute Named After Sergey Korolyov, Zhytomyr, Ukraine
dr.hry@i.ua, androidmax@meta.ua

Abstract. The article determines motion parameters of earth remote sensing satellites of high spatial discrimination using target information. It offers a method enabling satellite autonomous navigation in case information from satellite navigation systems is unavailable. The feature of the proposed method is application of non-local approach to minimization of target function on the basis of Nelder-Mead method, allowing for expansion of convergence domain of boundary value problem related to determination of motion parameters of satellites based on trajectory measurement results and getting more useful information about the desired minimum of objective function at low volume of measurement information.

Keywords: Navigation · Satellite control · Measurement · Algorithm · Satellite motion parameters

1 Introduction

When determining motion parameters of satellites based on trajectory measurements, along with measurement errors, a key role in the mechanism of error creation is given to motion model errors that arise from inaccurate description of relationships between measured values and estimated parameters. These errors depend on the time interval between moments of measurement and estimate of satellite motion parameters. The time interval is determined by technological cycle of ballistic navigational support (BNS) of satellite control and relative to accumulation of the required amount of trajectory measurements for convergence of satellite motion parameter estimate algorithm.

For effective survey of target area of observation with high spatial discrimination, motion parameters (MP) of earth remote sensing satellites (RSS) should be determined as soon as possible before the moment of survey that allows motion parameter prediction error caused by satellite motion model error. It should be noted that satellite motion model error caused by perturbation error (influence of atmosphere, non-sphericity and geopotential anomalies of Earth, etc.) calculated as error of satellite position along the orbit, can exceed the frame size of target equipment of RSS that would lead to errors in determining survey moment and full divergence of target and shot area. One of the ways to improve the efficiency and accuracy of determining motion parameters of a satellite is to carry out autonomous navigation requiring additional measurement information.

R. Szewczyk and M. Kaliczyńska (eds.), *Recent Advances in Systems, Control and Information Technology*, Advances in Intelligent Systems and Computing 543, DOI 10.1007/978-3-319-48923-0_54

2 Analysis and Definition of a Problem

Analysis of literature sources has found that most of the proposed approaches to improve the accuracy of RSS navigation are focused on obtaining trajectory measurements by involving additional tools. Thus, well-known are approaches based on the use of satellite navigation systems and relay satellite for transmitting measurement data in the absence of a satellite in the area of radio vision of ground tools and providing pseudo-satellite navigation [1, 2]. Disadvantages of such approaches are difficulty of complex processing of measurement information obtained by various tools and considerable financial costs [3, 4]. Besides, it is not always possible to use information of satellite navigation systems due to technical or organizational reasons [1, 5].

The main contradiction in increasing accuracy of navigation RSS navigation in case of unavailability of the information of satellite navigation systems is the need to reduce time interval for satellite motion prediction for ensuring the required accuracy of ballistic calculations, on the one hand, and to increase time interval for accumulation of required volume of trajectory measurements (prediction time interval) for ensuring the convergence of boundary value problem, on the other hand.

One of the ways to improve the accuracy of satellite navigation in case of unavailability of the information of satellite navigation systems is implementation of autonomous navigation of electro-optical satellites using target information (using the so-called optical trajectory measurements). Coordinates for referencing the images to the optical field map serve as current navigation parameter measurements. However, this navigation method significantly limits the volume of trajectory measurements due to space-time transiency of optical fields and limited areas of earth's surface suitable for effective solution of correlation-extreme problems of referencing obtained images to the map.

Thus, algorithm for determining satellite motion parameters based on a reduced volume of trajectory measurements should be developed to implement RSS navigation using target information (optical trajectory measurements).

3 Presentation of Basic Material

Algorithm for determining satellite motion parameters using target information is shown structurally in Fig. 1.

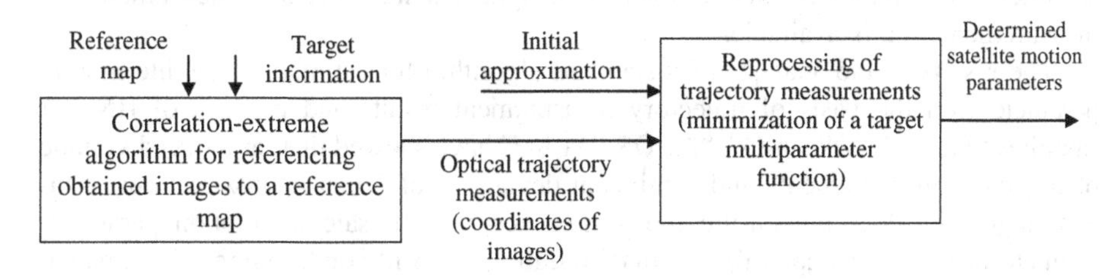

Fig. 1. Scheme of determining satellite motion parameters

To solve the problem of referencing images to reference maps, we should use correlation-extreme methods [6] based on spectral analysis and models of spatial distribution of navigation field intensity. However, in many cases, especially when processing fields that contain large objects, the second approach gives better results compared to the multidimensional spectral analysis [6, 7].

Let's consider referencing of the obtained image ($N \times N$) to the reference field in two-dimensional ($M \times M$) and three-dimensional ($M \times M \times M$) cases. Image size is much smaller than reference model size.

Navigation field measuring device is rigidly fixed to the satellite, which carries out two types of controlled motion: progressive (of the centre of mass) and rotational (relative to the centre of mass).

Besides, platform is affected by perturbation factors that lead to a shift of the platform both in corner $\Delta\alpha$, $\Delta\beta$ and linear Δx, Δy, Δz coordinates. As a consequence, there is a deviation of the current platform position from the expected one.

It is necessary to estimate values Δx, Δy (for surface field) and Δx, Δy, Δz (for spatial field) that characterize the shift of current image relative to the reference (Fig. 2), which makes it possible to estimate the position of satellite in orbit.

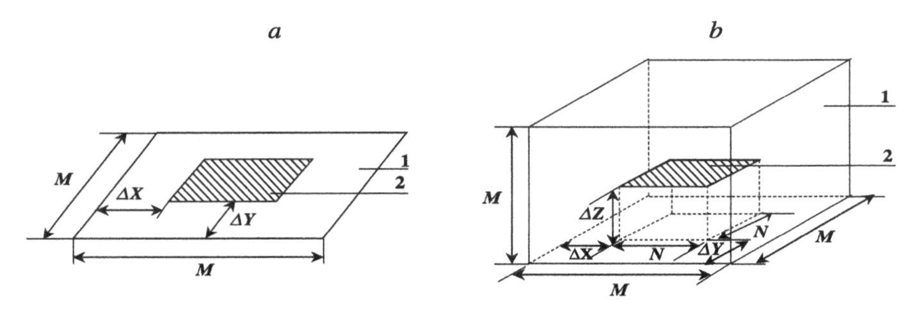

Fig. 2. Affixment of reference and measured navigation fields: a – two-dimensional shift, b – three-dimensional shift; 1 – reference field model, 2 – current measurement of navigation field

Solution of correlation-extreme problem on board of satellite is complicated both from algorithmic and technical points of view. It should also be noted that there are not many areas on Earth over which atmosphere transparency is better than 2 points, which meets the requirements of high-quality survey of earth's surface [7]. Thus, given the satellite speed and space-time transiency of optical field, quite a limited amount of measurement data is available.

The results of the analysis of a standard algorithm for estimating satellite motion parameters on the basis of trajectory measurement results and practice of BNS of satellites "Sich-1", "Okean-O", "AUOS-SM-KF" have showed that, at reduced volume of trajectory measurements and significant deviations of initial approximations, standard algorithm does not ensure the desired accuracy of satellite motion parameter estimate or makes it impossible to find solution due to ill-conditioning of estimation problem.

According to the principle of least squares, estimates of satellite motion parameters based on the trajectory measurement results are determined from the condition:

$$F(\vec{\tilde{q}}) = \sum_{i=1}^{n} \left(d_i(t) - \tilde{d}_i(\vec{\tilde{q}}, t) \right)^2 \Rightarrow \min \quad (1)$$

where $\vec{\tilde{q}} = \{\tilde{x}, \tilde{y}, \tilde{z}, \tilde{V}x, \tilde{V}y, \tilde{V}z\}$ is the vector of initial conditions of satellite motion in a Greenwich coordinate system;

$d_i(t)$ is the trajectory measurement obtained in time point t_i;

$\tilde{d}_i(\vec{\tilde{q}}, t)$ is the trajectory measurement calculated value obtained by predicting satellite motion using a satellite motion model.

The most common methods of expression minimization (1) are argument value search methods and gradient methods.

Newton's gradient method, based on local (around the current point) linear models of objective function, is used to minimize the target multiparameter function (1) in the standard algorithm. The advantage of Newton's method is relatively simple computational scheme and quadratic convergence in solution domain. Disadvantages of Newton's method are as follows: linearization of target function, which causes additional errors in satellite motion parameter estimation; locality of the approach conditions method criticality to the accuracy of initial approximation that does not allow efficient use of the standard algorithm when there are significant deviations of the calculated parameters of orbit from actual ones. Advantages of Newton's method are crucial for relatively unproductive computers, for which standard algorithms of BNS are calculated, and in case there are low requirements for accuracy of estimation results $\vec{q}$. Minimization methods deprived of the disadvantages should be used given the growing performance capacity of modern computers and growing demands on accuracy of solution problems related to BNS of RSS.

The paper contains a comparative analysis of the following optimization methods: Newton's method, simple iteration method, secant method, steepest descent method, Gauss-Seidel method (coordinate-wise descent), Hook-Jeeves method (direct search) and Nelder-Mead method (deformable polyhedron). The analysis was conducted by criteria: speed of convergence, computational efficiency, accuracy of initial approximations, dimension of initial approximation vector, calculation of derivatives, criticality to the volume of measurement information, potential accuracy of the result. Results of the analysis indicate that for minimization of the function (1) it is reasonably to use non-local approaches which enables getting greatest possible amount of information about the target function [8, 9] at small volume of trajectory measurements and significant deviations of initial approximation vector from actual motion parameters in terms of one-point control technology.

Given the advantages of the Neldera-Mead method (ease of implementation, lack of linearization of the target function, the feature of deformable polyhedron area to degenerate into a point only in achieving the minimum of a target function), its computational scheme is taken as a basis for the development of convergent algorithm for satellite motion parameter estimation based on reduced volume of trajectory measurements. Convergence is improved because at the beginning of minimization it is possible to obtain a significant amount relatively to a target function by examining a large number of tops of initial approximation domain. At each stage of optimization search for information required to implement the next iteration is obtained by examining seven tops of the domain.

The method for determination of satellite motion parameters using target information includes the following steps:

1. Obtaining optical trajectory measurements (referencing of the images to the map by solving a correlation-extreme problem).
2. Imaging an initial approximation domain.
   $w = 7$ initial approximation vector values (one unit more than the number of estimated parameters) are required to apply Neldera-Mead optimization method. There are two approaches proposed to form the initial approximation domain: pursuant to the first approach initial approximation domain is formed in accordance with the limit values of errors of orbit insertion (in the primary determining of motion parameters after launch); pursuant to the second approach domain is formed in accordance with the maximum deviation of calculated time of satellite at the traverse point from of the point in time determined by results of communication session with a satellite.

Thus, the domain of initial approximation is as follows:

$$\vec{\tilde{Q}}_{np}^{(k)} = \left\{ \vec{\tilde{q}}_{np1}^{(k)}, \ldots, \vec{\tilde{q}}_{np7}^{(k)} \right\},$$

where $\vec{\tilde{q}}_{npw}^{(k)} = \left\{ \tilde{x}_{npw}^{(k)}, \tilde{y}_{npw}^{(k)}, \tilde{z}_{npw}^{(k)}, \tilde{V}x_{npw}^{(k)}, \tilde{V}y_{npw}^{(k)}, \tilde{V}z_{npw}^{(k)} \right\}$, $w = \overline{1,7}$ is the index of a relevant top of initial approximation domain; k is the number of iteration.

3. Calculation of a minimized function according to the method of least squares for all the tops of initial approximation domain

$$F_w^{(k)}(\vec{\tilde{q}}_{npw}^{(k)}) = \sum_{i=1}^{n} \left(d_i - \tilde{d}_i(\vec{\tilde{q}}_{npw}^{(k)}) \right)^2, \; w = \overline{1,7}, \tag{2}$$

   where $\tilde{d}_i(\vec{\tilde{q}}_{npw}^{(k)})$ is the calculated value of the parameter measured.
4. Determination of $\vec{\tilde{q}}_{\min}^{(k)}$ and $\vec{\tilde{q}}_{\max}^{(k)}$ of initial approximation domain tops, in which the function (2) takes on the lowest and highest values respectively.
5. Determination of gravity center of all tops except $\vec{\tilde{q}}_{\max}$

$$\tilde{q}_{cj}^{(k)} = \frac{1}{6} \left[\left(\sum_{w=1, i \neq h}^{7} \tilde{q}_{npw,j}^{(k)} \right) \right], \; j = \overline{1,6},$$

   where index j is coordinate direction; h is the number of a top $\vec{\tilde{q}}_{\max}^{(k)}$.
6. Check of search end state

$$\left\{\frac{1}{7}\sum_{w=1}^{7}\left[F_w(\vec{q}_{npw}^{(k)}) - F(\vec{q}_c^{(k)})\right]^2\right\}^{\frac{1}{2}} \leq \varepsilon,$$

where ε is the threshold criterion of search end;

$F(\vec{q}_c^{(k)})$ is the target function value at the gravity center.

The number of iterations k in finding a minimum of the target function (1) and the accuracy of search for the minimum depend from the value ε. When choosing the value ε it should be taken into account that due to inevitable presence of systematic and random errors zero. Value of minimized function cannot be reached, so it is inappropriate to establish a very small value ε. Research results have showed that value $\varepsilon = 10^{-7}$ is the best in terms of computational cost and accuracy of the resulting solution.

7. If the end search term (3) is not met, according to the Neldera-Mead method initial approximation domain $\vec{Q}_{np}^{(k)}$ is converted using one of the operations: reflection, compression and reduction. Then, the algorithm repeats from p. 3, where $k = k + 1$.

As a result of the minimization algorithm initial approximation domain is compressed, approaching the point of minimum value of the function (2). That is in the result of minimization we get

$$\vec{q}_{npL} = \arg \min_{\vec{q}_{np} \in \vec{Q}_{np}} \sum_{i=1}^{n}\left[d_i - \tilde{d}_i(\vec{q}_{np})\right]^2,$$

where $\vec{q}_{npL} = \vec{q}$ is the result of determination of satellite motion parameters at a point in time t_0.

Unbiasedness of the developed method for determining satellite motion parameters is estimated using simulation modeling. Unbiasedness of the method means that the estimate $\vec{q}$ approaches the accurate value $\vec{q}^*$, when there are no systematic and random errors ($\eta = \xi = 0$). The estimation results are shown in Table 1.

Table 1. Estimate of unbiasedness of the developed method

Satellite coordinates	Actual parameters $\vec{q}^*$	Determined parameters $\vec{q}$	Deviation $\lvert\Delta\vec{q}\rvert$
x, km	6003.214325	6003.214272	0.000053
y, km	3623.164556	3623.164642	0.000086
z, km	0.000000	0.000030	0.000030

Deviation values $|\Delta\vec{q}|$ characterize potential accuracy of the developed method for determining satellite motion parameters. Analysis of the data shows that the spatial shift of satellite position in the absence of trajectory measurement errors and model position does not exceed the satellite size.

The paper presents experimental studies of the developed method. Figure 3 visualizes the process of determination of satellite motion parameters based on the results of trajectory measurements at different deviations of initial approximation from the desired result using the standard algorithm and the developed method.

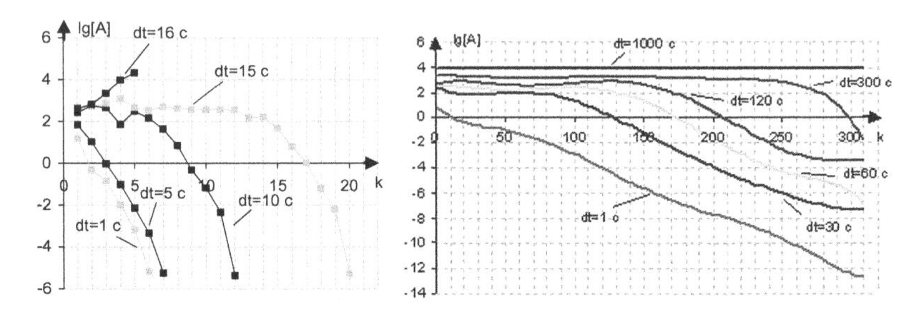

Fig. 3. Convergence of satellite motion parameter estimation process: (a) on the basis of Newton method (b) on the basis of Nelder-Mead method

Indications in Fig. 3 have the following meanings: k is the number of iterations of the target function minimization process; dt is temporal deviation of initial approximation vector from the result of determination describing the size of convergence domain; $A = |\vec{q}^* - \vec{q}|$.

Analysis of the results shows that the standard algorithm has a much greater rate of convergence, but the developed method has substantially broader convergence domain, which helps to solve the problem of determining the initial conditions of satellite motion at significant deviations of the calculated parameters of orbit from actual ones, confirming theoretical provisions.

The paper also studies the organization of RSS autonomous navigation using target information based on the developed algorithm. Affixment coordinates of images of high spatial discrimination can serve as trajectory measurements. Estimates of RSS motion parameters on the basis of target information are given in Table 2. Initial data were used in calculation: standard deviation of errors of determining geodetic latitude and longitude images $\sigma_B = \sigma_L = 4.3$ m; 3 images are obtained with coordinates (7.2240, 47.3710), (32.8070, 49.5070), (57.8050, 206.6080); image coordinates have north latitude.

Table 2. Estimates of satellite motion parameters

	Time, s	Along the orbit, km	On radius, km	On side, km
Odchylenie $\|\Delta\vec{q}\|$	0.021	0.157095	0.015110	0.032898

Analysis of the results shows that application of the developed method for determining RSS motion parameters provides a result based on a reduced volume of optical trajectory measurements with estimate accuracy meeting the accuracy requirements for the survey of target area observed by modern RSS.

4 Conclusions

The paper presents solution of an actual scientific task, which is to develop a method for determining RSS motion parameters on the basis of target information in terms of data unavailability of satellite navigation systems at reduced capacity optical trajectory measurements.

The method is based on a non-local optimization approach allowing for getting more information about the target function, which makes it possible to estimate satellite orbital parameters based on reduced volume of trajectory measurements. Lack of target function linearization in solving the boundary problem of determining satellite parameters can increase the guaranteed accuracy of RSS navigation.

Application of the method for determining RSS motion parameters based on the reduced volume of trajectory measurements can increase the guaranteed accuracy of satellite position estimation at the time of performance of target task due to reducing the influence of motion model errors traffic by reducing the time between cycles of determining satellite motion parameters.

The method allows several-fold expansion of convergence domain, expressed in a deviation of temporal parameters of initial approximations of satellite motion parameters from the desired estimates, as compared with the standard algorithm based on Newton's method, which allows its use in primary determination of satellite motion parameters after orbit insertion at significant deviations of actual orbital parameters from calculated ones in terms of standard algorithm impracticability.

References

1. Ivanov, N.M.: Ballistics and Navigation of Satellites: Textbook for Universities, 2 edn., 544 p. (2004). [in Russian,rev. and ext. - M.: Drofa]
2. Yastrebov, V.D.: Research of Methodological Issues and Development of Algorithms of Satellite Control Ballistic Support Using One-Point Measurement Scheme, 174 p. (1993). [in Russian, M.: TsNIIMash]
3. Korobiichuk, I., Bezvesilna, O., Ilchenko, A., Shadura, V., Nowicki, M., Szewczyk, R.: A mathematical model of the thermo-anemometric flowmeter. Sensors **15**, 22899–22913 (2015). doi:10.3390/s150922899
4. Korobiichuk, I.: Mathematical model of precision sensor for an automatic weapons stabilizer system. Measurement **89**, 151–158 (2016)
5. Chornyi, I.: USA Are Concerned About Plans of China. Space News, № 10, p. 23 (2003). [in Russian]
6. Androsov, V.A., Boyko, Y.V., Bochkarov, A.M., Odnorog, A.P.: Combining Images under Uncertainty. International Radio Electronics, № 4, pp. 54–70 (1985). [in Russian]
7. Lebedev, A.A., Nesterenko, O.P.: Space Surveillance Systems, Synthesis and Modeling, 346 p. (1991) [in Russian, M.: Mashynostroenie]
8. Himmelblau, D.: Applied Nonlinear Programming, 534 p. (1975). [in Russian, M.: Mir]
9. Nemirovskiy, A.S., Yudin, D.B.: Complexity of Tasks and Effectiveness of Optimization Methods, 429 p. (1980) [in Russian, M.: Nauka]

Time-Domain Reflectometry (TDR) Square and Pulse Test Signals Comparison

Sylwester Kostro[1], Michał Nowicki[2](✉), Roman Szewczyk[1], and Katarzyna Rzeplińska-Rykała[2]

[1] Institute of Metrology and Biomedical Engineering, Warsaw University of Technology, Św. Andrzeja Boboli 8, 02-525 Warsaw, Poland
szewczyk@mchtr.pw.edu.pl

[2] Industrial Research Institute for Automation and Measurements PIAP, Al. Jerozolimskie 202, 02-486 Warsaw, Poland
nowicki@mchtr.pw.edu.pl

Abstract. It is very important to ensure operational reliability of measurement systems. Every shutdown causes loses. The longer diagnostics and repair of the measurement system last, the higher are the loses and they often build up faster the longer the system is not working. The damage can be caused by many things and all of them have to be checked to be absolutely sure that the line is working correctly. This is where TDR reflectometry comes to the aid. It allows detecting, estimating and defining the type of the cable disruption. Disruptions can occur because of natural aging processes, environmental influence, but also because of careless excavation works taking place where the cables are laid underground or other situations like intentional vandalistic behaviors. In this work the theory of the coaxial cable construction and its connection with the used method was shown. Basics of TDR reflectometry and commonly used test signals were presented. Additionally, other methods of cable length measurements were mentioned. Measurements were conducted using suggested test stand consisting of oscilloscope and function generator. Screenshots from the oscilloscope were added to show typical view of the measurement signal. Discussion about existing errors was performed and comparison of two types of test signals, square and pulse, was conducted. Errors caused by each used signal and ease of their usage were taken under consideration too. It has also been checked if acquired data is statistically different depending on method. Lastly conclusions about uses of discussed method, signals and their limitations were expressed.

Keywords: Time-Domain Reflectometry · TDR · Square signal · Pulse signal · Coaxial cable

1 Important Properties of Coaxial Cable Using TDR Testing

Coaxial cable is very resistant to mechanical damage and noise immune, so it is one of the most commonly used type of cables in engineering. These properties result from its construction, which is shown in the Fig. 1 below.

R. Szewczyk and M. Kaliczyńska (eds.), *Recent Advances in Systems, Control and Information Technology*, Advances in Intelligent Systems and Computing 543, DOI 10.1007/978-3-319-48923-0_55

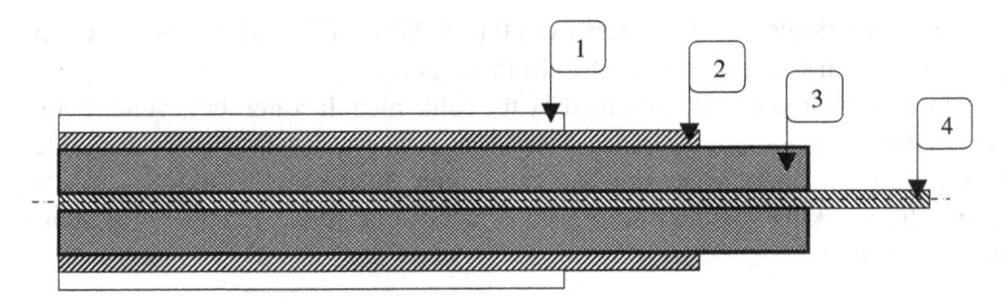

Fig. 1. Construction of unbalanced (coaxial) line

Typical coaxial cable, referring to Fig. 1, consists of outer sheath (1), shield (2), inner insulator (3) and core (4). The core and shield are usually made out of pure conducting metals or metals covered with layers. Used metals are usually copper or aluminum. Materials used for inner insulators are solid polyethylene, polyethylene foam and rarely PTFE (Teflon).

Because of specific construction, coaxial cable electrical characteristic is much more complicated comparing to typical copper cable. They are often referred to as substitute series resistance R, series inductance L, shunt conductance G and shunt capacitance C per length unit (represented by "1") [1]. These properties depend on the type of insulator and the diameters of core and shield. On schematic they are shown in Fig. 2.

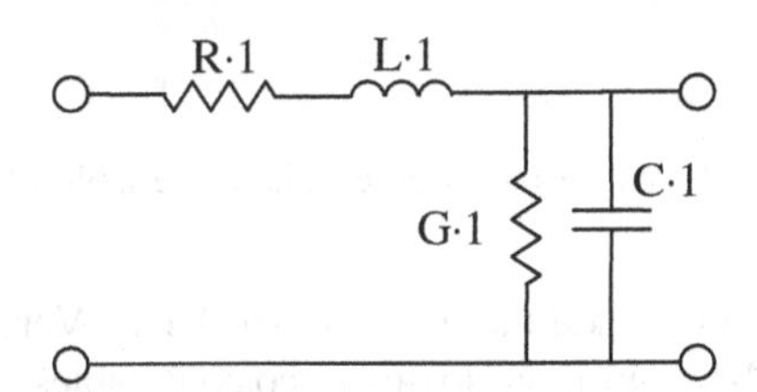

Fig. 2. Electrical schematic of coaxial cable

Values shown up are used to define certain parameters of coaxial cable, for example characteristic impedance Z, but they are not necessary to the disquisition, so they will not be described further.

Important parameter which is tied together directly in with structure of coaxial cable is the Velocity Factor (short: VF). It is described by the expression below:

$$VF = \frac{1}{\sqrt{\varepsilon_r \mu_r}} \approx \frac{1}{\sqrt{\varepsilon_r}}$$

ε_r – *dielectric constant of the insulator*
μ_r – *magnetic permeability of the insulator*

VF is dimensionless and it varies from 0 to 1. When $VF = 1$, the speed of wave in the cable is equal to c (the speed of light in vacuum).

Velocity Factor is usually provided by the cable manufacturer, but even if not it can be easily found by looking at the cable markings. These usually start with two letters RG followed by number meaning the cable type (for ex. RG6, RG58, etc.). When knowing the marking, VF and other parameters can be found in tables like shown in Fig. 3 (an extract).

Common Coaxial Cables					
		Loss in db per 100 feet			
Common Name	Impedance in Ohms	at 50MHz	at 1GHz	Velocity	Max voltage in RMS KV
RG6U	75	1.5	11 dB	.78 (foam)	0.6kv
RG8U	52	1.2	9	.66 (poly)	5kv
RG8U	50	1.1		.78 (foam)	0.6
RG8X	50		13.5	.84	2.5
RG9U	51	1.6		.66 "	5
RG11U	75	1.3	9	.66 (poly)	5
RG11U	75	1.0		.78 (foam)	0.6
22B/U	95		2.1	66 (poly)	?
RG55B/U	53.5		16.5		1.9kv
RG58U	53	3.1	20	.66 (poly)	1.9kv
RG58U	50	3.2		.78 (foam)	0.2kv
RG59U	73	2.4	11.5	.66 (poly)	2.3
RG59U	75	2.1		.78 (foam)	0.3kv

Fig. 3. An extract from coaxial cable table [2]

It is also very important to add that in VHF and UHF (Very High and Ultra High Frequencies) like 100 MHz and more losses in coaxial cables are getting way bigger than in low and medium frequencies. It happens because of the skin effect, dielectric polarization and sometimes even change of the propagation mode of the signal in the cable [3]. However frequencies used in standard TDR measurements are far lower than VHF and UHF, so mentioned phenomena will not take place.

2 TDR Theory

Time-Domain Reflectometry is a method of measuring the length of the cable from the signal source to a certain "disruption". Sending a signal through the cable when a disruption happens in a certain point makes it possible to analyze what type and where the disruption takes place, because some of the send energy is reflected back to the source in a specific way. When a cable has loose end the resistance is approaching infinity. Reflection from this type of disruption can be used to estimate the length of a cable [4–6].

TDR can be used for diagnostic testing of measurement systems cable connections, because it makes finding broken or damaged cables very easy and fast. For example, these damages can appear when careless excavation works are being conducted. Reflectometry can also detect illegal connections to the line.

The equation which allows calculating the length of the coaxial cable is:

$$l = 0.5 \cdot \Delta t \cdot c \cdot VF$$

l – cable length
t – time of propagation
c – speed of light in vacuum (299 792 458 $\frac{\text{m}}{\text{s}}$)
VF – Velocity Factor

As it can be easily deduced, the length of the coaxial cable is dependent only on one value, the time of propagation Δt. Other parts of the equation are constants when researching a certain type of cable. Time of propagation definition will be explained in the next point of the discussion. Worth mentioning is also the 0.5 number in the equation. It is there because Δt is the time which passes from signal entering the cable to getting out after reflection, so total way of the signal is $2l$. Therefore to get the right length the whole equation has to be divided by 2.

3 Theory of Square and Pulse Signals in TDR Measurements

In this experiment, square and pulse signals were used. Ideal graphs of these signals are shown in Fig. 4 (square on the left, pulse on the right).

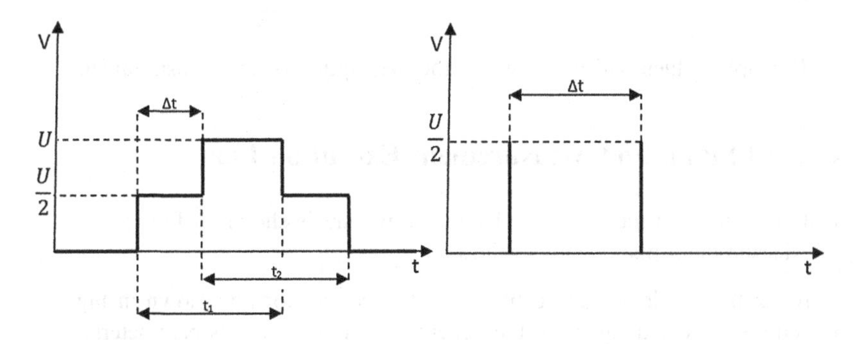

U – generated voltage
Δt – time of propagation
t_1 – time of the incident pulse
t_2 – time of reflected pulse

Fig. 4. Ideal graphs of used signals; V – voltage (V), t – time (s)

Incident pulse is the signal going straight from the generator to the receiver (ex. oscilloscope) and the reflected pulse is the signal which went doubled way from the beginning of the cable to the loose end. In pulse method these signals exist too, but in ideal example, their rise times approach 0 (incident on the left, reflected on the right).

Analyzing graphs above time of propagation can be defined for both of them (order being the same):

- Δt is the time from the rising edge of incident pulse to when incident and reflected pulses start to sum up,
- Δt is the time from when the first test pulse rise occurs to the time when reflected pulse returns to the receiver (shows up on screen).

There are some properties of used signals which have to be chosen carefully to make sure the measurement will be correct. Most important ones are in Table 1.

Table 1. Important properties of test signals

	Square signal	Pulse signal
Frequency	Used to set the "length" of the signal. Function period needs to be longer than summed incident and reflected pulses at least.	The frequency also has to ensure that the incident and reflected pulse will both be seen and will not interfere with the next period of the signal.
Filling	Does not play a big role in measurement because the length of states can be regulated by frequency. Usually set at 50%.	The smaller the better (typically set to a few percent); this helps to precisely position on pulses peaks and recognize them, especially with shorter cables.
Peak to peak voltage	Plays a minor role in measurement because the interesting thing is the time between voltage rises, not its value.	

Example values will be shown in the description of the measurement.

4 Test Stand and Measurement Execution Plan

Test stand used in the research ideological picture is shown in Fig. 5.

Measurement plan went as follows:

- realization of device and cable under test connections as shown in Fig. 5,
- setup of selected signal on the generator and choosing its parameters,
- using oscilloscope to observe the received graph and measuring the needed propagation time Δt,
- calculation of cable length,
- comparison of the calculated value with exact mechanical measured length,
- calculation of errors.

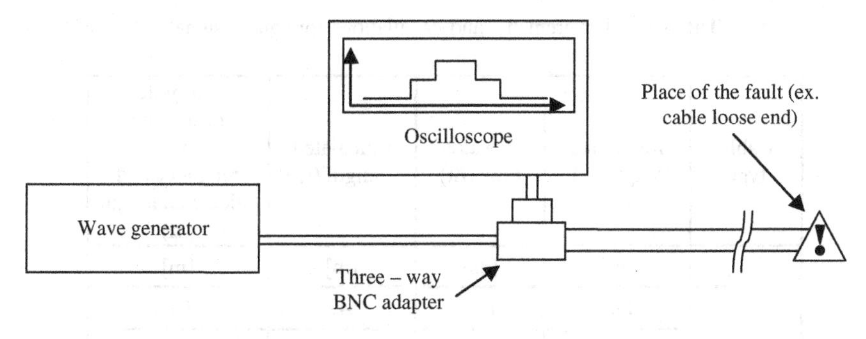

Fig. 5. Test stand

5 Measurement Results

Cables under test were RG6 and RG058 type coaxial cables. The RG6 insulator was polyethylene foam ($\varepsilon = 1.6$; $VF = 0.79$) and RG058 insulator was solid polyethylene ($\varepsilon = 2.3$; $VF = 0.66$) [3].

Settings of used signals are shown in the Table 2 below.

Table 2. Signal settings

	Square signal	Pulse signal
Frequency	250 kHz	1–2 MHz for the shortest cables (under 2 m)
		800 kHz for the rest
Amplitude	8 V_{pp}	8 V_{pp}
Filling	50%	1.6–3%

Higher frequency for shorter cables was used because the pulses could not be observed separately while using 800 kHz (signals were one on another). The filling needed to be set differently for shorter and longer cables too. All of these values were defined by research.

Measurements and calculations for both signals are shown in Tables 3 and 4.

Exemplary graphs from oscilloscope are shown in Figs. 6 and 7.

The rule of measurement for square signal was to set oscilloscope markers on two levels: 25% and 75% of max amplitude, which were 2 V and 6 V respectively. When doing that, the measurement is taking place exactly at half way between the voltage states for incident and summed waves, what is accordant to assumption that the signal has reached its high state when it gets over 50% of its max voltage value.

When considering pulse signal the measurement was second (reflected) pulse peaks.

As it can be easily seen. real graphs do not look simple positioning the markers on the highest voltage values on the first and the same as these in Sect. 3:

- The rises for square signal last some time (they are not instant like in ideal situation) and after rising signal voltage is not exactly divided into two equal parts. It also can be observed that there are some voltage fluctuations in both incident and incident

Table 3. Measurements and calculations for square signal

Cable type	Reference length (l_r)	Measured time (Δt)	Calculated length (l)	Absolute difference between reference and calculated length (Δl)
[–]	[m]	[ns]	[m]	[m]
RG6	1.44	12.6	1.493	0.0531
	6.43	53.6	6.352	0.0784
	10.83	91.2	10.807	0.0228
	20.17	170.0	20.145	0.0250
	30.48	256.0	30.336	0.1440
	50.32	422.0	50.007	0.3130
	80.81	678.0	80.343	0.4670
	100.97	848.0	100.488	0.4820
RG058	1.32	14.4	1.426	0.1106
	5.18	52.8	5.227	0.0522
	10.10	102.4	10.138	0.0376
	20.38	206.0	20.394	0.0140
	31.17	315.0	31.185	0.0150
	50.90	516.0	51.084	0.1840
	82.08	832.0	82.368	0.2880
	102.45	1050.0	103.950	1.5000

plus reflected pulse levels (ex. caused by marginal reflections, loses in cable). All of these made positioning the oscilloscope markers harder.

- The rise time for pulse signal is not instant and the pulses are not infinitely narrow. It can be seen that second pulse lost some of the voltage send thought the cable (in the shown case about 1.6 V). Some minor fluctuations can be observed on the “0” level of the signal.

6 Sources of Error and Signal Error Analysis

There were a few possible error sources:

- oscilloscope, t – connector and generator resistances.

Table 4. Measurements and calculations for pulse signal

Cable type	Reference length (l_r)	Measured time (Δt)	Calculated length (l)	Absolute difference between reference and calculated length (Δl)
[-]	[m]	[ns]	[m]	[m]
RG6	1.44	13.0	1.541	0.1005
	6.43	54.8	6.494	0.0638
	10.83	92.0	10.902	0.0720
	20.17	171.0	20.264	0.0935
	30.48	257.0	30.455	0.0255
	50.32	424.0	50.244	0.0760
	80.81	680.0	80.580	0.2300
	100.97	848.0	100.488	0.4820
RG058	1.32	14.2	1.406	0.0908
	5.18	52.6	5.207	0.0324
	10.10	102.4	10.138	0.0376
	20.38	205.0	20.295	0.0850
	31.17	313.0	30.987	0.1830
	50.90	512.0	50.688	0.2120
	82.08	826.0	81.774	0.3060
	102.45	1030.0	101.970	0.4800

- Resistances of used items could have affected the voltage numbers, but not the wideness of the time period. as the voltage was used only to position oscilloscope markers right to measure the propagation time.
- reference cable length measurement uncertainty.
- Mentioned in Tables 6 and 7. Not only summing up the minimal scale of the meter stick (max measure length was 10 m), but also included potential divergences from parallel positioning between stick and cable.
- experimenter error caused by positioning at wrong spots while using oscilloscope markers.
- It was decided that a good interpretation of experimenter error would be the minimal oscilloscope marker movement unit. It changed depending on the measured cable length. Values are shown in Table 5.

  The experimenter error was calculated from the equation:

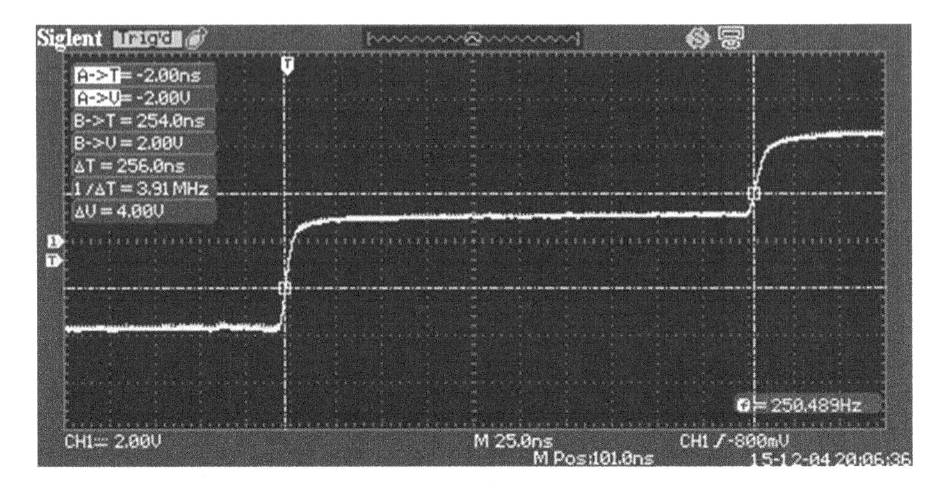

Fig. 6. Oscilloscope graph for square signal (RG6. 30.48 m)

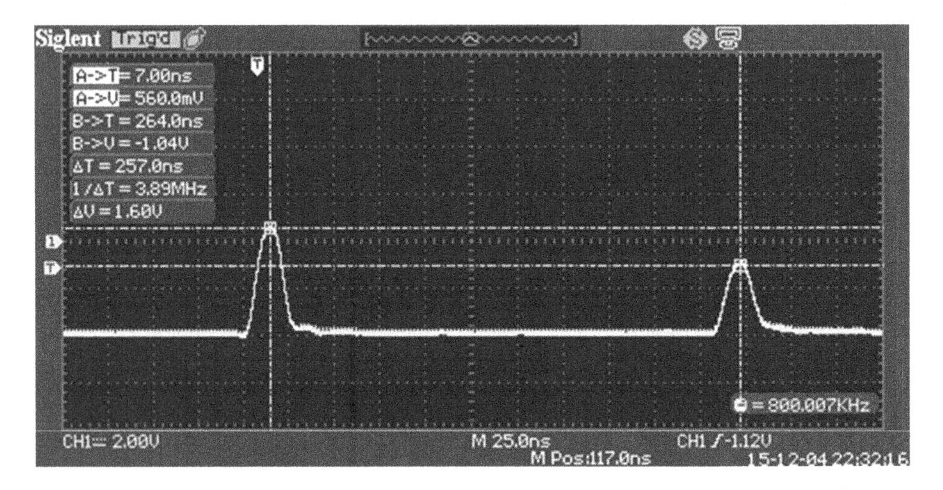

Fig. 7. Oscilloscope graph for pulse signal (RG6. 30.48 m)

$$e_e = 2 \cdot t_m \cdot VF \cdot c$$

It is doubled marker movement unit (the potential error can be made where the first and the second signal rise occurs). For very long cables (over 100 m in this experiment), the error was estimated to be only one minimal marker movement unit, because signal slope had already been so tight that it was far easier to position in the right spot. The calculation from nanoseconds corresponding to meters are shown in Tables 6 and 7.

- time measurement discretization oscilloscope error.
  Negligible because experimenter error was mentioned, so it would be far smaller than its value.

Table 5. Marker movement

Square signal and pulse signal		
Cable type	Reference length (l)	Minimal marker movement unit (t_m)
(-)	(m)	(ns)
RG6	1.44; 6.43	0.20
	10.83	0.40
	20.17; 30.48	1.00
	50.32; 80.81	2.00
	100.97	10.00
RG058	1.32; 5.18	0.20
	10.10	0.40
	20.38; 31.17	1.00
	50.90; 82.08	2.00
	102.45	10.00

Table 6. Summed up uncertainties for square signal

Cable type	Calculated length (l)	Reference cable length measurement uncertainty (e_{ref})	Experimenter error (e_e)	Absolute difference between reference and calculated length (Δl)	Total (summed) absolute error (Δl%)
(-)	(m)	(m)	(m)	(m)	(%)
RG6	1.493	0.0010	0.0948	0.0531	**9.97**
	6.352	0.0020	0.0948	0.0784	**2.76**
	10.807	0.0040	0.1896	0.0228	**2.00**
	20.145	0.0080	0.4740	0.0250	**2.52**
	30.336	0.0120	0.4740	0.1440	**2.08**
	50.007	0.0200	0.9480	0.3130	**2.56**
	80.343	0.0320	0.9480	0.4670	**1.80**
	100.488	0.0400	2.3700	0.4820	**2.88**
RG058	1.426	0.0010	0.0948	0.1106	**13.38**
	5.227	0.0020	0.0948	0.0522	**2.55**
	10.138	0.0040	0.1896	0.0376	**1.97**
	20.394	0.0080	0.4740	0.0140	**2.05**
	31.185	0.0120	0.4740	0.0150	**1.36**
	51.084	0.0200	0.9480	0.1840	**1.95**
	82.368	0.0320	0.9480	0.2880	**1.35**
	103.950	0.0400	2.3700	1.5000	**2.84**

Table 7. Summed up uncertainties for square signal

Cable type	Calculated length (l)	Reference cable length measurement uncertainty (e_{ref})	Experimenter error (e_e)	Absolute difference between reference and calculated length (Δl)	Total (summed) absolute error (Δl%)
(-)	(m)	(m)	(m)	(m)	(%)
RG6	1.493	0.0010	0.0948	0.1005	**12.74**
	6.352	0.0020	0.0948	0.0638	**2.47**
	10.807	0.0040	0.1896	0.0720	**2.44**
	20.145	0.0080	0.4740	0.0935	**2.84**
	30.336	0.0120	0.4740	0.0255	**1.68**
	50.007	0.0200	0.9480	0.0760	**2.08**
	80.343	0.0320	0.9480	0.2300	**1.50**
	100.488	0.0400	2.3700	0.4820	**2.88**
RG058	1.426	0.0010	0.0948	0.0908	**12.16**
	5.227	0.0020	0.0948	0.0324	**2.18**
	10.138	0.0040	0.1896	0.0376	**1.97**
	20.394	0.0080	0.4740	0.0850	**2.41**
	31.185	0.0120	0.4740	0.1830	**1.91**
	51.084	0.0200	0.9480	0.2120	**2.02**
	82.368	0.0320	0.9480	0.3060	**1.38**
	103.950	0.0400	2.3700	0.4800	**2.83**

- potential changes in generated signal frequency.
  Generated signal frequency affected the period of the signal. but because the frequency was set to fit the measuring range, potential small changes in the period did not matter.

Whole discussion about uncertainty is shown in Table 6 for square signal and Table 7 for pulse signal.

7 Statistical Analysis

To check if there are significant differences between test signals used in the experiment, probability Student's *t*-Test was conducted. This test is used to check if outcome of two paired samples have the same mean value.

The calculated lengths, depending on the used test signal. were put into statistical software and the test was performed. The null hypothesis was that the difference between mean values of one test and another was equal to zero.

The program clearly shown that **there is no significant differences (at 95% confidence level) between mean values of test signals.**

8 Conclusion

Simple test stand for TDR length measurements of coaxial cables has been introduced and measurements were conducted.

It has also been clearly shown (Tables 6 and 7) that this method should not be used to measure very short cables (to keep safety margin. about under 5 m) because errors are bigger the shorter the cable is (even more than 13%). These happen because it is very hard to operate in single nanoseconds range and position in the right spots.

While taking the statistical analysis under consideration, both tests seem to be the same when it comes to estimating the cable lengths, so used test will probably be chosen depending on the user preference. However, what is worth mentioning, while using pulse signal it is far easier to observe voltage losses in the cable (check Fig. 7). That was not important when measuring propagation time, as the voltage value was only informational, but it is an interesting deduction if there is a need to measure voltages.

References

1. http://www.repeater-builder.com/antenna/wa2ise-coaxial-cable.html. Accessed 01 Jan 2016
2. Strickland, J.A.: Time-Domain Reflectometry Measurements. TEKTRONIX
3. http://www.radio-electronics.com/info/antennas/coax/coax_velocity_factor.php. Accessed 01 Jan 2016
4. Gale, P.F., Yaozhong, G., Crossley, P.A., Cory, B.J., Bingyin, X., Barker, J.R.G.: Fault Location Based on Travelling Waves
5. Thompson, M.: Transmission Lines Physics 623 (1999)
6. Carr, J.J.: Practical Antenna Handbook. McGrav-Hill, New York (2001)

Transforming the Conversion Characteristic of a Measuring System Used for Technical Control

Eugenij Volodarsky[1], Zygmunt Warsza[2(✉)], Larysa A. Kosheva[3], and Adam Idźkowski[4]

[1] Department of Automation of Experimental Studies, National Technical University of Ukraine "KPI", Kiev, Ukraine
vet-l@ukr.net

[2] Industrial Research Institute of Automation and Measurement (PIAP), Warsaw, Poland
zlw@op.pl

[3] Department of Biocybernetics and Aerospace Medicine, National Aviation University of Ukraine, Kiev, Ukraine
l.kosh@ukr.net

[4] Faculty of Electrical Engineering, Bialystok University of Technology, Bialystok, Poland
a.idzkowski@pb.edu.pl

Abstract. As it is known, the real conversion characteristic of a measuring system is different than the nominal characteristic, e.g. due to lower sensitivity and (or) the presence of nonlinearity. In measurement control, particularly in the areas of the limiting values of tolerance interval, this can increase the effect of random variables, and consequently, this can increase a probability of erroneous decisions. A difference between the real and nominal conversion characteristic of a measuring system is corrected using the calibration procedure. But the real conversion characteristic remains unchanged. Therefore, during calibration it is necessary to transform a conversion characteristic and move it towards the nominal one considering the limiting values. The transformation of characteristic may proceed in several cycles, each one can consist of two phases – multiplicative and additive. In case of the multiplicative transformation, the cycle starts with the multiplicative phase (a change in sensitivity of the conversion characteristic). In this case, the probability of erroneous decisions at both stages, the multiplicative and additive, remains unchanged. In case of the additive transformation in the first (additive) phase of the cycle, there is a parallel displacement of the real conversion characteristic. In this case, in the second (multiplicative) stage, the probability of decision will be β times less than in additive stage cycle. For selecting the type of transformation and the number of cycles it is necessary, first of all, to inspect the relationship between components of errors of conversion measuring system and the length of the tolerance interval which is proportional to β. The relationships, which allow estimating the probability of erroneous decisions (both within a cycle and in the transition to next cycle of transformation) are obtained.

R. Szewczyk and M. Kaliczyńska (eds.), *Recent Advances in Systems, Control and Information Technology*, Advances in Intelligent Systems and Computing 543, DOI 10.1007/978-3-319-48923-0_56

Keywords: Calibration · Conversion characteristic · Additive transformation · Multiplicative transformation · Measurement inspection

1 Introduction

During production control and in diagnostics of equipment in industry, including chemical industry, decisions about the utility of a test object are made based on measurements of relevant parameters. A correctness of decision is affected by inaccuracy of a measurement system. The real characteristic of a measuring system $\varphi(x)$ differs from the nominal characteristic $\varphi_0(x)$. Its deviation from the nominal characteristic is determined by a calibration [1]. This calibration is carried out for the range of expected changes in the controlled parameter x, which covers the range of limiting values x_l, x_h but it do not need contain the value $x = 0$. For example, for this purpose a set of 5 to 7 values of reference quantity is used to determine a calibration curve. Then, with the method of least squares a calibration curve is determined [2], which reflects the characteristics of a measurement system. For example, the linear characteristic is obtained as

$$\varphi(x) = a + bx \text{ or } \varphi(x) = (x + \Delta)(1 + \gamma) \quad (1)$$

where: Δ – additive component and γ – multiplicative component of a system error.

In many cases, the ultimate goal of object inspection is to make a decision classifying an object to the set of objects that fit or unfit [7]. Usually, the values of tested quantity x are controlled in the limiting values of its tolerance interval (x_l, x_h) and a positive decision is taken on the basis of condition $x_l < x < x_h$. If the real characteristic $\varphi(x)$ of a measuring system deviates from the nominal one around the limits x_l, x_h (e.g. shifting, poor sensitivity, linearity) [4, 5], incorrect results of that inspection can be received. At control systems, as opposed to measurement systems, non-ideality of conversion characteristic influences correctness of accepted decisions (their reliability) only in some areas adjoining to the limiting values x_l, x_h of tolerance interval. In [3] are presented the boundaries of such areas, which sizes depend on a ratio of system error components. There are four different character combinations of error components Δ and γ and two special cases. In one of them, the effect of non-ideal conversion characteristic on the accuracy of control is equal to zero.

Elementary, from mathematical point of view, basing on the real characteristic, the settings $\varphi(x_l)$ and $\varphi(x_h)$ can be formed. The result of transformation $\varphi(x)$ is compared with them. At the same time the condition to recognize the validity of controlled object is $\varphi(x_l) < \varphi(x) < \varphi(x_h)$, which corresponds to the condition of serviceability $x_l < x < x_h$.

However, from a physical point of view the probability of correct decisions in most cases is reduced. It is explained that with a negative value of multiplicative component of error a sensitivity of system decreases. Thus, its resolution decreases and an influence of random variables increase. A limitation of dynamic range of input values occurs with the positive values of multiplicative component. It is due to the elements of a measurement system enter the saturation mode.

To reduce the incorrect decisions the actual processing characteristic have to be calibrated in two limit points, i.e. for the standard $x_{01} = x_h$ and $x_{02} = x_l$. However, the impact of randomness (x size distribution, random errors) causes that there is not enough to introduce only bias corrections. The real characteristic $\varphi(x)$ should be transformed in such a way that at the limiting values x_l, x_h it is moved towards the nominal characteristic $\varphi_0(x)$.

In further considerations, a quite common case is assumed when the random errors of a measuring system are much smaller than systematic errors (biases) determined during the study.

2 Principle of the Method

To implement this method the auxiliary values $x_h = x_{01}$ and $x_{02} = x_l$ are given alternately to the input of measurement system (Fig. 1). The results of their processing in the main system channel are used to determine a value proportional to the processing error and then they are applied for an appropriate transformation of conversion characteristic. The required change in the real characteristic is achieved by its linear shift with respect to the coordinate axes – additive transformation, or by changing its sensitivity (slope of an angle), i.e. multiplicative transformation. An additive shift and a change in angle of inclination can be made both at the input and at the output of measurement system. A block diagram of the control-measurement system according to this method is presented in Fig. 1.

To implement the first of these operations an additional sum block circuit in processing channel of the x is used. One of the inputs of sum block is connected to the

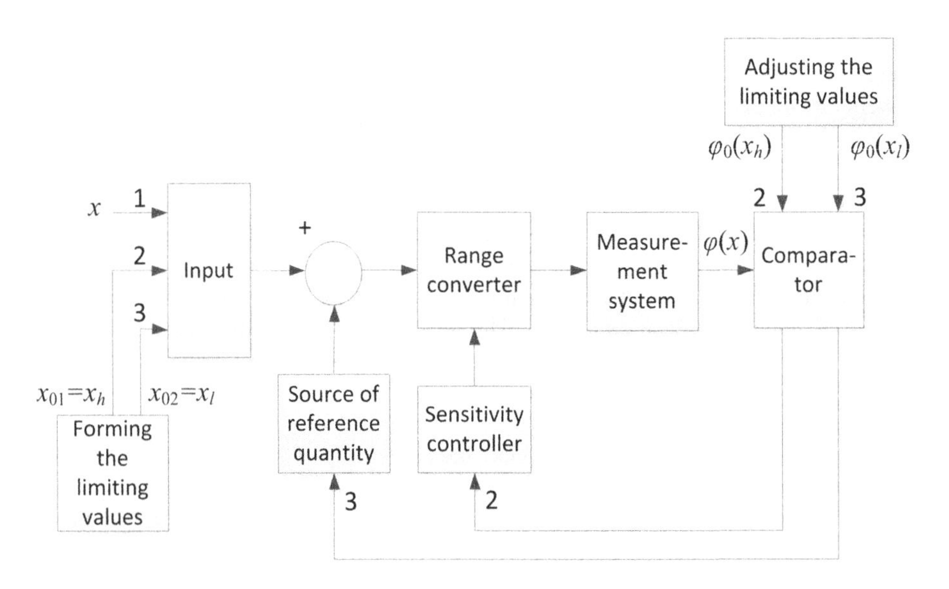

Fig. 1. The inspection system for determining suitability of an object, equipped with the transformation of conversion characteristic

input measured quantity x and another input is connected to a controlled source of reference quantity. At the second operation is carried by range converter, a block with controllable sensitivity, which is also introduced into the measuring channel (Fig. 1). A transformation algorithm of conversion characteristic comprises two successive cycles, each of two stages. For example, when the process of transformation begins with giving the upper limit $x_{01} = x_h$ of tolerance interval at input. Then, the result of processing $\varphi(x_h)$, which is obtained at the output of the measuring system, is compared with the nominal value $\varphi_0(x_h)$, which is set on the adjuster of the limit settings. This stage in Fig. 1 corresponds to the functional connections, indicated by number 2. The result of this comparison is used to set a new sensitivity range of the conversion characteristic with the use of range converter, i.e. to change its angle of inclination. Then, the input is linked to the lower limit $x_{02} = x_l$ and the result of processing $\varphi(x_l)$ is compared with the nominal value $\varphi_0(x_l)$. The result of this second comparison is used to control by a reference value, which is a shift, i.e. the additive transformation of characteristic. This step in Fig. 1 corresponds to the functional connections indicated by number 3. Each transformation cycle consists of two stages, called as multiplicative and additive. A processing cycle of characteristic, which begins from the multiplicative stage, will be also called as multiplicative cycle and vice versa.

3 Multiplicative Transformation

The processing result $\varphi(x_h) = (x_h + \Delta)(1 + \gamma)$ of the auxiliary quantity $x_{01} = x_h$ is used to carry out the activities that appropriately change the range ratio, i.e. the slope of conversion characteristic, to fulfil the condition:

$$\varphi_{11m}(x_h) = (x_h + \Delta)(1 + \gamma) = \varphi_0(x_h), \tag{2}$$

where: the first digit in φ_{11m} indicates a cycle number, the second digit – stage number, a letter – cycle type: m – multiplicative, a – additive.

The subsequent changes in the characteristic in multiplicative cycle 1 are presented in Fig. 2.

After the first stage of correction the equating of real conversion characteristic φ_{11} and nominal characteristic φ_{01} in point $[x_h, \varphi(x_h)]$ is obtained. The remainder, i.e. residual value of multiplicative error, is determined from the expression (2) as

$$\gamma_1 = \frac{\varphi_0(x_h)}{x_h + \Delta} - 1 \text{ or } \gamma_1 = \frac{\Delta}{x_h + \Delta}. \tag{3}$$

Here and hereinafter, for simplicity, it is assumed that the nominal transformation ratio is 1, this means that nominal characteristic is expressed as the parts of range and it is linear with slope equal to 1. Then $\varphi_0(x_h) = x_h$ and $\varphi_0(x_l) = x_l$.

In the first cycle and in the first stage of characteristic correction it is obtained

$$\varphi_{11m}(x) = (x + \Delta)\frac{x_h}{x_h + \Delta}. \tag{4}$$

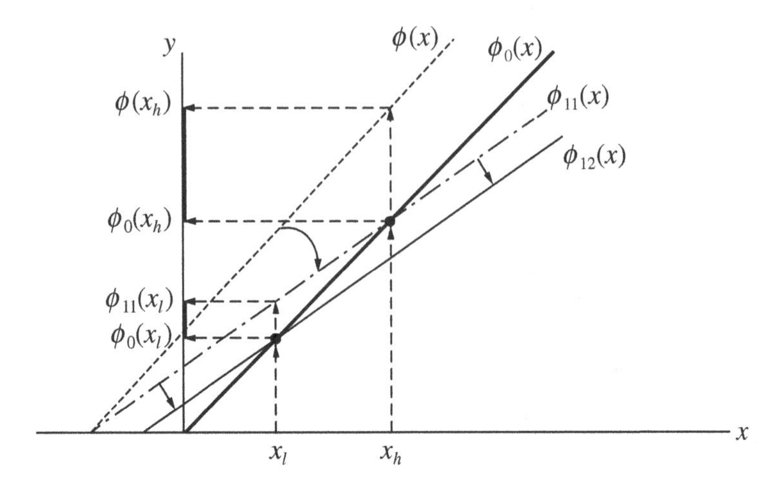

Fig. 2. Transformations of conversion characteristic in multiplicative cycle (φ_{11}, φ_{12})

Taking into account the impact of this correction a rule in decision-making on compliance of an object with the requirements is [6]:

$$x_l + \frac{\Delta}{x_h}(x_h - x_l) < x < x_h \text{ or } x_l - \theta_{l11} < x < x_h, \tag{5}$$

where: $\theta_{l11} = \Delta(\beta - 1)/\beta$, $\theta_{h11} = 0$ and $\frac{x_h}{x_l} = \beta$.

From the expression (5) it follows that after the first stage of transformation its influence is described as equivalent interval, located on the left of the lower limit x_l. Therefore, the objects whose parameter values are in the interval $[(x_l - \theta_{l11}) - x_l]$ will be regarded as usable, although in reality they are not suitable. The probability of incorrect decisions is proportional to the area under a curve of possible values of the analyzed quantity in this interval. The incorrect decisions do not occur at the upper limit of the tolerance range x_h.

In the second, i.e. additive stage of the first cycle, the result of reference value processing $x_{02} = x_l$ is used to make an additive shift of conversion characteristic that it should pass through the point $[x_l, \varphi_0(x_l)]$ belonging to the nominal characteristic. The additive stage ends with the fulfilment of condition

$$\varphi_{11m}(x_l) = (x_l + \Delta)\frac{x_h}{x_h + \Delta} = \varphi_0(x_l). \tag{6}$$

Hence, it can be found a final additive component Δ_1 of processing error:

$$\Delta_1 = \frac{x_l}{x_h}\Delta.$$

The conversion characteristic after the correction in the second phase of cycle 1 is described by the relationship

$$\varphi_{12m}(x) = \left(x + \Delta\frac{x_l}{x_h}\right)\frac{x_h}{x_h + \Delta} \tag{7}$$

a decision rule (algorithm) can be written as

$$x_l < x < x_h + \frac{\Delta}{x_h}(x_h - x_l) \text{ or } x_l < x < x_h + \theta_{h12}. \tag{8}$$

Ultimately, equivalent shift (offset) intervals are:

$$\theta_{l12} = 0, \quad \theta_{h12} = \Delta\frac{\beta - 1}{\beta}. \tag{9}$$

Thus, the same as in the first stage the objects, whose values of tested parameter are in the interval $[x_h - (x_h + \theta_{h12})]$, will be classified as fit even though in reality they do not fulfil the requirements. It results from Eqs. (5) and (9) that the equivalent length of the intervals, which reflect in the final rule the influence of the residual non-idealities of conversion characteristic, remains unchanged in both steps. Also, a type of bad decisions does not change.

If the probability of incorrect decisions in the first stage of cycle exceeds the limit then skipping the second stage of cycle is necessary to go on to the second cycle.

In general, the lengths of the end shift intervals in i-th stage (i = 1, 2) and n-th cycle are equal to:

$$\theta_{l\,ni} = \Delta\frac{\beta - 1}{\beta^n}(2 - i) \quad \text{and} \quad \theta_{h\,ni} = \Delta\frac{\beta - 1}{\beta^n}(i - 1), \tag{10}$$

and the corresponding probability of incorrect decisions is

$$\begin{aligned} P_{err_{nim}} &= F(x_l) - F\left[x_l - \Delta\frac{\beta - 1}{\beta^n}(2 - i)\right] \\ &+ F\left[x_h + \Delta\frac{\beta - 1}{\beta^n}(i - 1)\right] - F(x_h), \end{aligned} \tag{11}$$

where $F(x) = \int_{-\infty}^{x} f(x)dx$ – the cumulative distribution function of possible values of studied variable x.

In the second component of Eq. (11) for i = 1 the equivalent shift occurs and in the third component for i = 2 the shift does not occur. For i = 1 the third component the equivalent shift does not appear but for i = 2 it reappears.

It results from the expressions (10) that the multiplicative component of error does not affect the correctness of decision. However, changing a sign of the additive component of error will change the type of incorrect decisions and for distributions other than uniform – it also will change their probability.

4 Additive Transformation

It was mentioned above that in the multiplicative cycle of transforming the conversion characteristic the main cause of erroneous decisions was an additive error (shift). If the preliminary analysis of the characteristic shows that this shift is significant, it is advisable to begin its transformation from additive cycle. The subsequent stages in this cycle are shown in Fig. 3.

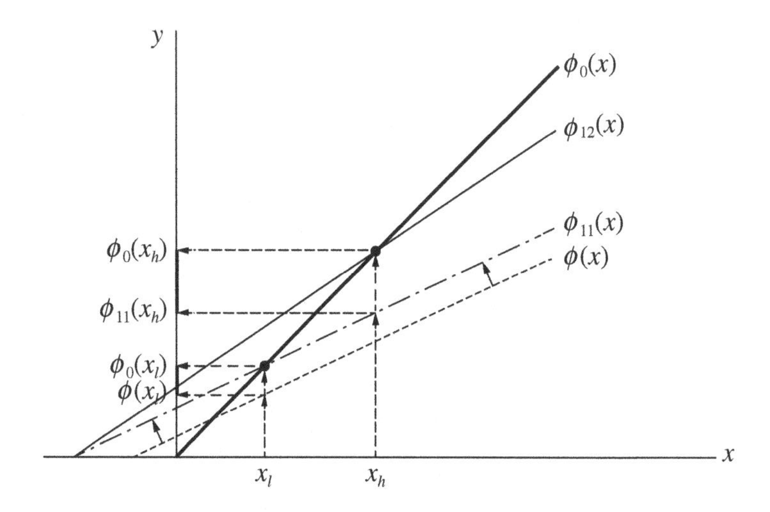

Fig. 3. Transformations of conversion characteristic in additive cycle

In the first stage the condition must be satisfied

$$\varphi_{11a}(x_l) = (x_l + \Delta_1)(1 + \gamma) = \varphi_0(x_l). \tag{12}$$

The additive error will be equal $\Delta_1 = -\alpha x_l$ and decision rule on compliance with the standard will be

$$x_l < x < x_h - \underbrace{\alpha(x_h - x_l)}_{\theta_{h11}}, \tag{13}$$

where $\alpha = \gamma/(1 + \gamma)$.

The upper interval of shift will be situated to the left of x_h, i.e. within the tolerance interval. This brings up a false result of inspection because the objects whose values are in the range $[(x_h - \theta_{h11}) - x_h]$ will be classified as unfit, although in reality they are fit [5].

In the second (multiplicative) stage of cycle for $x_{02} = x_h$ a change in the angle of characteristic occurs such as to fulfil the condition

$$\varphi_{12a}(x_h) = \varphi_0(x_h) \text{ or } (x_h - \alpha \cdot x_l)(1 + \gamma_1) = x_h. \tag{14}$$

Using this relationship, a new value of multiplicative error can be defined:

$$(1+\gamma_1) = \frac{x_h}{x_h - \alpha \cdot x_l} \tag{15}$$

Basing on the decision rule, which takes into account the last transformed characteristic, one can obtain:

$$\underbrace{x_l + \alpha \cdot \frac{x_l}{x_h}(x_h - x_l)}_{\theta_{l11}} < x < x_h \tag{16}$$

The lower interval of shift θ_{l11} in the second stage of the cycle will be located to the right of limit value x_l. As in the first step it means that the objects whose values are in the range $[(x_l, \theta_{l12})]$ is considered as unfit, although they are fit in fact.

Comparing the expressions (13) and (16) it can be concluded that the additive conversion in the second stage of cycle a length of equivalent shift range will be β times less than in the first stage of this cycle.

After a similar transformation, in the second and the next cycles, there is the general expression for the final shifts and the probabilities of erroneous decisions (false rejections):

$$\theta_{l\,ni} = \alpha \cdot x_l \frac{\beta - 1}{\beta}(i-1) \text{ or } \theta_{h\,ni} = \alpha \cdot x_l \frac{\beta - 1}{\beta^{n-1}}(2-i), \tag{17}$$

$$\begin{aligned} P_{\text{err}\,nia} = F\left[x_l + \alpha \cdot x_l \left(\tfrac{\beta-1}{\beta^n}\right)(i-1)\right] - F(x_l) + F(x_h) \\ \pm F\left[x_h - \alpha \cdot x_l \tfrac{\beta-1}{\beta^{n-1}}(2-i)\right]. \end{aligned} \tag{18}$$

Then, the additive error during processing does not influence on credibility of the decisions. However, changing a sign of multiplicative error does not lead only to change the type of incorrect decisions. But at the other data it also increases the length of equivalent intervals, as

$$\alpha_1 = \frac{\gamma}{1-\gamma} > \alpha = \frac{\gamma}{1+\gamma}$$

The purpose of use an additive or multiplicative transformation of conversion characteristic will be assessed. Assume that the distribution of possible values of the tested quantity is uniform with a density $f(x)$.

At first, it is considered the probability of incorrect decisions for the first stage $i = 1$ and in the first cycle $n = 1$. From expressions (11) and (18), it follows that equality of the probability of incorrect decisions is reached at the fulfillment of condition

$$\Delta(\beta - 1)/\beta = \alpha \cdot x_l(\beta - 1). \tag{19}$$

Assuming that $\gamma = \alpha/(1 + \alpha)$ it follows that

$$\gamma = \frac{\Delta}{x_l \cdot \beta - \Delta} \tag{20}$$

Expression (20) is a boundary condition of the application of transformation type. Thus, fulfilling one of the conditions $\gamma < \frac{\Delta}{x_l \cdot \beta - \Delta} = \mu$ or $\Delta < \alpha \cdot \beta \cdot x_l$ a transformation with additive cycle should be used because then the probability of incorrect decision will be lower. Otherwise, it is necessary to use a transformation with multiplicative cycle.

5 Numerical Example

Let us consider the general case when the values of tested quantity have a uniform distribution with a density $f(x)$. For unconverted characteristic of measuring system $\varphi(x) = (x + \Delta)(1 + \gamma)$ the probability of erroneous decisions is:

$$P_{err} = f(x) \cdot [\alpha \cdot x_l(\beta + 1) + 2\Delta].$$

Assume that $\varphi(x) = (x + 2)(1 + 0.1)$, i.e. $\Delta = 2$ and $\gamma = 0.1$ tested quantity is in the range (10–50) units and $f(x) = 0.025$. The suitability requirement is determined by the limiting values $x_l = 20$ and $x_h = 40$ units. On the basis of the above mentioned data it is obtained that $\alpha = 0.09$ and $\beta = 2$. To select the type of transformation characteristic it is necessary to calculate the μ value using the formula

$$\mu = \frac{\Delta}{x_l \cdot \beta - \Delta} = 0.053.$$

Since the value of μ is less than $\gamma = 0.1$, it is necessary to implement the multiplicative transformation algorithm of characteristic. In fact, in the first stage of transformation the probability of erroneous decisions P^m and the reliability D^m are respectively:

$$P^m = f(x) \cdot \Delta(\beta - 1)/\beta = 0.025$$

$$D^m = 1 - P^m = 0.975$$

With the additive transformation is obtained:

$$P^a = f(x) \cdot \alpha x_l \cdot (\beta - 1) = 0.045$$

$$D^a = 0.955$$

In the second stage of cycle the probability of erroneous decisions for the multiplicative transformation – does not change and for the additive transformation – is β-times less. And it is equal to 0.023, reliability is $D^a = 0.973$.

Table 1. Results

Cycles	Stages	Probability (multiplicative transformation) P^m	Probability (additive transformation) P^a	Reliability (multiplicative transformation) D^m	Reliability (additive transformation) D^a
Cycle 1	Stage 1	0.025	0.045	0.975	0.955
	Stage 2	0.025	0.023	0.975	0.973
Cycle 2	Stage 1	0.013	0.023	0.987	0.973
	Stage 2	0.013	0.012	0.987	0.988

If in the first cycle the obtained probability of erroneous decisions is greater than the limit then it is necessary to proceed to the second cycle. Obtained results are presented in Table 1.

It is noted that for large values of β (large length of tolerance interval) in the second step of the first cycle of additive transformation a less probability of erroneous decisions is achieved than it is for the multiplicative transformation. In practice to achieve the desired reliability of decisions only two cycles are usually sufficient.

However, if the initial characteristic has the form $\varphi(x) = (x + 5)(1 + 0.1)$, then $\mu = 0.143$ is greater than $\gamma = 0.1$ Therefore, the additive transformation should be used as the first.

The additive transformation excludes the influence of additive component in a measuring channel error. Its increase ($\Delta = 5$ instead of $\Delta = 2$) does not change the probability of erroneous decisions. In case of multiplicative transformation, the probability of erroneous decisions increases proportionally with an additive error. Also, the type of transformation depends on the absolute lower limit value.

Consider the real characteristic $\varphi(x) = (x + 2)(1 + 0.1)$. Assume that tested quantity is within the range (5–55) units and $f(x) = 0.02$. The limiting values are $x_l = 10$ and $x_h = 20$ units. Thus, $\beta = 2$ remains the same. For these values $\mu = \frac{2}{10 \cdot 2 - 2} = 0.11$ is greater than $\gamma = 0.1$. Therefore, in this case the additive transformation is preferable to use.

Considered example testifies that to select the type of transformation it is necessary, at all equal conditions, to take into account the lower limit value of tolerance interval, i.e. its location relative to the center of the distribution of possible values.

6 Conclusions

To ensure the reliability of decisions and the compatibility of tested parameter with requirements (standard) it is sufficient to calibrate a conversion characteristic of measuring system only in two points, which correspond to the limiting values of tolerance interval. Usually, a real conversion characteristic of measuring system has a sensitivity, which differs from a nominal characteristic. It may have a different initial value and other slope. These errors are systematic. They can also lead to increased influence both the random scatter of studied quantity and the random errors of measurement on the reliability of decisions.To eliminate this impact, it is not enough to make appropriate corrections. It is also necessary to transform the conversion characteristic of measuring system and to match it with a nominal characteristic, taking into account the limiting values of tolerance interval.

Such a transformation consists of several cycles, each of them comprises two steps: additive and multiplicative. In the cycle starting from the change in sensitivity (slope) of conversion characteristic (transformation multiplicative) the probability of incorrect decisions in both stages of a cycle remains unchanged. In the cycle of additive transformation, in its first stage, a parallel shift of characteristic is made and it results with a less probability in the second stage. The reduction depends on the length of tolerance interval of a controlled parameter.

Therefore, when choosing the type of transformation, it is necessary to take into account not only the size of systematic error components of conversion characteristic but also the length of tolerance interval of a controlled parameter.

References

1. International Vocabulary of Metrology – Basic and General Concepts and Associated Terms (VIM 3rd edn.) JCGM 200:2012
2. Freund, J.E., Walpole, R.E.: Mathematical Statistics, 4th edn. Prentice Hall, Englewood Cliffs (1987)
3. Volodarsky, E.T., Koshevaya, L.A., Warsza, Z.L.: Niepewność decyzji o zgodności mierzonego obiektu z wymaganiami (Uncertainty of the decision about the compliance of the measured object to given requirements). Pomiary Automatyka Kontrola (Meas. Autom. Monit.) **58**(4), 391–395 (2012). in Polish
4. Grous, A.: Analysis of Reliability and Quality Control. Fracture Mechanics 1. ISTE Ltd. and Wiley, New York (2013)
5. Nakamoto, H., Kojima, F., Kato, S.: Reliability assessment for thickness inspection of pipe wall using probability of detection. E-J. Adv. Maint. Jpn. Soc. Maintenol. **5**, 228–237 (2014)
6. Volodarsky, E., Warsza, Z., Kosheva, L., Idzkowski, A.: Method of upgrading the reliability of measurement inspection. In: Jabłoński, R., Brezina, T. (eds.) Advanced Mechatronics Solutions. AISC, vol. 393, pp. 431–438. Springer, Heidelberg (2016). doi:10.1007/978-3-319-23923-1_63
7. ISO 10576-1:2003. Statistical methods – Guidelines for the evaluation of conformity with specified requirements - Part 1: General principles

Factors of AC Field Inhomogeneity in Impedance Measurement of Cylindrical Conductors

Aleksandr A. Mikhal[1(✉)], Aleksandr I. Glukhenkyi[1], and Zygmunt L. Warsza[2]

[1] Institute of Electrodynamics, National Academy of Science of Ukraine, Kiev, Ukraine
a_mikhal@ukr.net
[2] Industrial Research Institute of Automation and Measurements (PIAP), Warsaw, Poland
zlw@op.pl

Abstract. The AC and quasi-DC models of an impedance of cylindrical conductor have been developed. The influence of inhomogeneity of field inside this conductor on measurement errors is considered. Heterogeneity current densities in metals are conditioned by the skin effect and proximity effect of the return conductor. The voltage drop on thin current electrodes conditions the heterogeneity current densities in liquids. Character of change of error is identical regardless of the type of field (electromagnetic or quasi-static). The electrical model of the bulk impedance should be the parallel capacitive equivalent scheme. The ratio of the imaginary and real components of the bulk impedance determines the frequency dependence of this impedance at higher frequencies. The characteristic frequency at which impedance falls twice is determined by the electrical conductivity and dielectric permeability of conducted liquids.

Keywords: Electrolytic conductivity · Electrical field · Measurement · Error

1 Introduction

The basis of the design of many primary measuring transducers (sensors) is the plot of cylindrical conductor. For example the metal conductor of such shape can be the shunt as current sensor in electrical measurements or temperature sensor in thermometry. Liquid conductors of such shape in the conductometric cells with the calculated value of constant are sensors of electrolytic conductivity (EC). In recent years the leading national metrology institutes (NMI) have established the standards for electrolytic conductivity (EC) k [S/m], which is based on usage of direct or "absolute" method for reproducing this physical value [1, 2]. Its essence is reduced to measuring of resistance R of a liquid column with length L and cross-section area S. There are several ways to implement such method in the global practice. However, in all variants of the implementation, the standard cell can be represented as one physical model in the form of conductor section resembling the right circular cylinder. In our earlier works [3, 4] we discussed the three major factors that lead to the violation of the electrical field

R. Szewczyk and M. Kaliczyńska (eds.), *Recent Advances in Systems, Control and Information Technology*, Advances in Intelligent Systems and Computing 543, DOI 10.1007/978-3-319-48923-0_57

homogeneity. These are primarily effects due to alternating character of the conductor's operating current. This effects lead to occurrence of a methodical error in resistance measurement. Such error is of the systematic type. It is convenient to consider it as a relative value

$$\delta = R_C/R_H - 1 \tag{1}$$

Where: R_H is resistance of cylindrical conductor for an ideal model with homogeneous field; R_C is the resistance for the real field.

2 Calculation of Cylindrical Conductor Impedance Components

2.1 AC Model of Impedance

Idealized physical model for calculating the cylindrical conductor resistance can be represented as a system of direct conductor 1 (of circular cross-section with diameter *D*) and linear (infinitely thin) return conductor 2 through which sinusoidal current flows in opposite directions, Fig. 1. Return conductor is located in parallel with direct conductor at a distance *h* from isolation 3. To obtain difference of potentials two points *a* and *b* are selected. These points are located at a distance *l* along the direction of current. One of the potential wires runs in parallel to the axis of the conductor with current to the connection with the second potential wire, and then plaits with the latter. Plaited wires are connected to measuring terminals *c*, *d*.

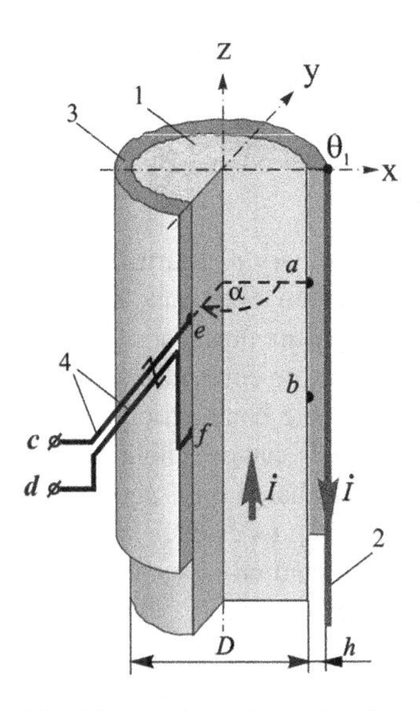

Fig. 1. The physical model of the conductor's section for measuring impedance [3]

Such configuration of conductors in the potential circuit will largely allow minimizing area, i.e. to ensure bifilarity and thereby to create conditions to form coaxial measuring circuits [5, 6]. In our works [3, 4], on the basis of Maxwell equations the expressions for the impedance components are obtain, which are given below:

$$R_C = \mathrm{Re}(Z) = R_H + \mathrm{Re}\,(\frac{j\omega l}{IS}\int\limits_{(S)} \mathbf{A}ds - j\omega \mathbf{A}l) \tag{2a}$$

$$X = \mathrm{Im}(Z) = \mathrm{Im}\,(\frac{j\omega l}{IS}\int\limits_{(S)} \mathbf{A}ds - j\omega \mathbf{A}l) \tag{2b}$$

$$tg\varphi = \mathrm{Im}(Z)/\mathrm{Re}(Z) \tag{2c}$$

Where $S = \pi D^2/4$ *is* cross-sectional area of conductor and $R_H = l/kS$ is resistance of the cylindrical conductor with homogeneous field which covers (1).

To assess an impact of the magnetic vector potential **A**, we have considered two types of conductors. These are metals of specific conductivity $k = (10^7, 10^8)$ S/m and electrolyte solutions with electrolytic conductivity $k = (10^{-4}, 10^2)$ S/m. Figures 2 and 3 show results of numerical calculations.

In the cylindrical metal conductors, the skin effect will lead to a noticeable change of the current density along the radius of the conductor. If the reverse wire is lying at a small distance h, then the axial symmetry of the electrical field is changed. As a result, the error (1) will be a function of the angle α on Fig. 1. As can be seen from Fig. 2, the error at a frequency 1 kHz is 100 times higher than at 50 Hz. The angle, at which the error is zero, can be used in calculations of non-inductive shunts for current and power measurements.

The national standard of Ukraine [2, 7] has the value of conductivity $k = 60$ S/m as the upper limit of the measuring range. The error (1) for electrolytic solutions is a linear

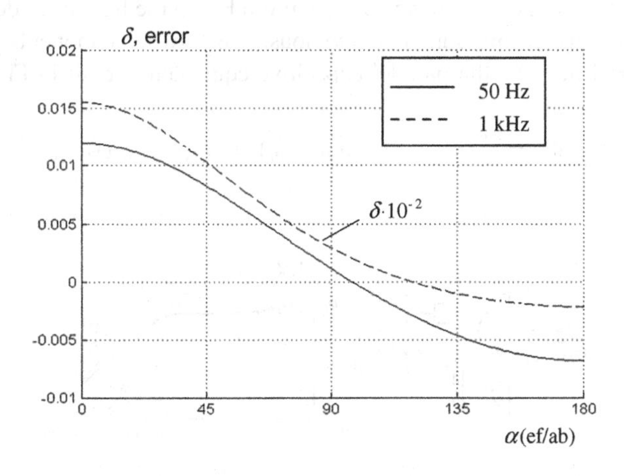

Fig. 2. Influence of the connection point on the error resistance of conductor, D = 10 mm, $h = 2.5$ mm, for copper $k = 6 \times 10^7$ S/m

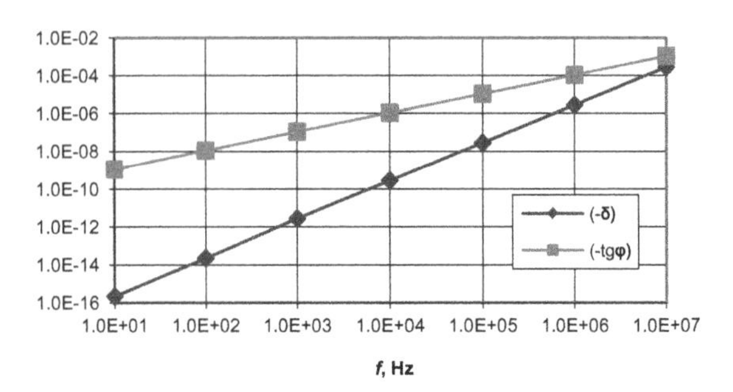

Fig. 3. Dependence of an error and tangent of phase angle of the cylindrical conductor resistance from frequency f. Geometrical dimensions: D = 10 mm, h = 2.5 mm, electrolyte solution k = 60 S/m

function of frequency and conductivity of electrolyte. Figure 3 shows that for concentrated aqueous solutions of electrolytes (of conductivity by 14 times higher than for seawater [8]) we can ignore the occurrence of electromagnetic induction. Relative error δ (Eq. 1) in such case shall not exceed 0.1 ppm for frequencies less than 100 kHz. At lower frequencies and in more diluted solutions an error will be even smaller. Such a small value can be disregarded in the budget of errors when assessing uncertainty even in the international comparisons.

The physical model given in Fig. 1 is a two-electrode measurement object. Connection non-dimensional points a and b are the mathematical abstractions. It is allowed for the metallic conductor. For the liquid conductor is necessary to introduce the concept of the electrode and the electrochemical interface. After including the structure of electrode-solution interface [9] the electric model given in Fig. 4a is obtain. It comprises two impedances connected in series, electrochemical impedance Z_E, which is described by the Randles equivalent circuit [10] and bulk impedance Z_B. For the physical model shown in Fig. 1 we will examine only the bulk impedance Z_B. Now let us consider the frequency dependence of tgφ. From Fig. 3 the frequency dependence of tgφ is a linearly increasing function. Obviously, the two components of electrical impedance model must be the parallel capacitive equivalent circuit in Fig. 4.

$$Y_B = \frac{1}{Z_B} = \frac{1}{R_B}(1 + j\omega C_B R_B) = G_B(1 + jtg\varphi) \tag{3}$$

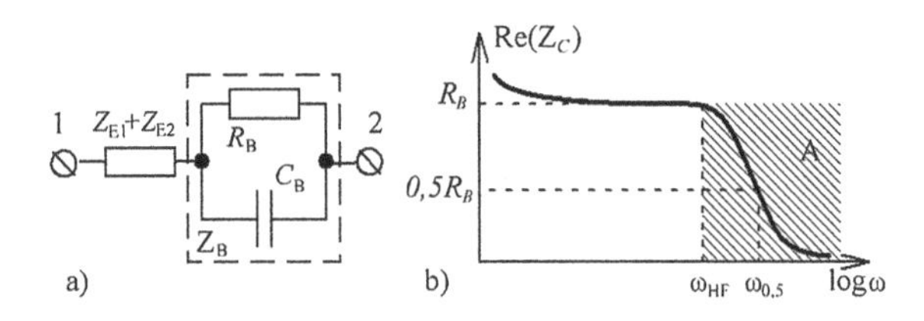

Fig. 4. (a) Electrical model of two-electrode conductometric object; (b) typical experimental dependence of the conductometric cell resistance from frequency

2.2 Quasi-DC Model of Impedance

Let us consider the physical model of a cylindrical liquid conductor with an electrode system for the application in measurements. The electrodes have a particular structure and dimensions. It is known [5, 6] that the best elimination of resistance in supply leads is achieved upon four-terminal connections to an object. These requirements have been implemented when creating four-electrode AC conductivity cells. Their specific feature is in the fact that current and potential electrodes are spaced over a distance (Fig. 5) [2].

Liquid is located inside the tube with inner diameter D and total length $L + 2\,l$. The tube serves for the fixation of the liquid conductor geometry. It consists three sections: central section 1 and two side sections 2. Metal circular potential electrodes 3 are located at the ends of the central section 1, these electrodes 3 have thickness h(P) and width which corresponds with the thickness of the tube wall. Two discs 4 are fixed at the edges of the tube. Inner surface of the discs is coated with metal films 5 with thickness $h(T)$ performing the function of current electrodes. The material for tubes and discs 4 is quartz glass, which has good insulating properties, temporal stability and minimal coefficient of thermal expansion. Platinum is used as a metal for electrodes since such material has minimal polarizing effect for most electrolytes.

The electrical model of four-electrode cell is shown in Fig. 6. Specialized devices (AC bridges) are used for measurement the parameters of bulk impedance Z_B, regardless of size and structure of lead impedances $Z_1 - Z_4$. Therefore, for the model in Fig. 5 the measured cell impedance is the bulk impedance as $Z_C = Z_B$.

In the audio frequency range upon isotropic nature of solution one should consider electrolytic conductivity as a complex value σ [3, 4]. This follows from the expression for the total current: conduction and displacement currents.

$$\mathbf{J}_{total} \stackrel{\sin}{=} (k + j\omega\varepsilon\varepsilon_0)\,\mathbf{E} = \sigma\,\mathbf{E} \tag{4}$$

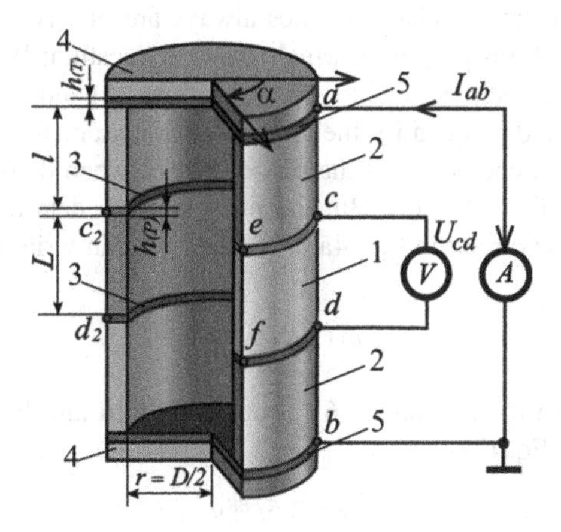

Fig. 5. Physical model of the four-electrode cell [3]

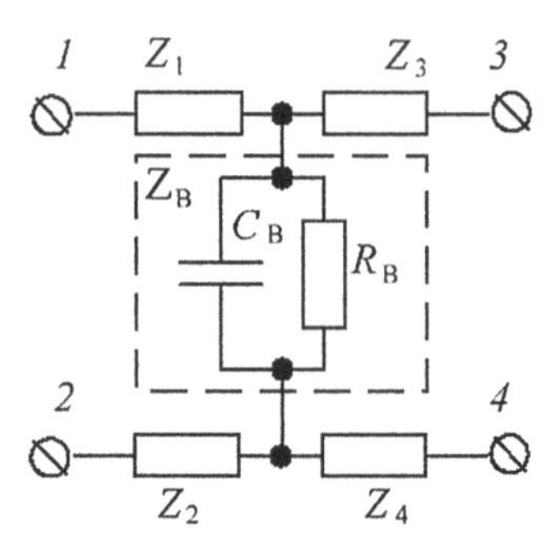

Fig. 6. The electric model of four-electrode conductometric object: Z_B – volumetric impedance of tube 1, Z_1 and Z_2 – total (including electrochemical impedance) impedances of tubes 2, Z_3 and Z_4 – electrochemical impedances at the potential electrodes 3

Impedance and admittance circuit models are used in practice of measurements.

$$Z_C = U_{cd}/I_{ab}, \quad Y_C = I_{ab}/U_{cd} \tag{5a, 5b}$$

Then for the impedance model we receive:

$$R_C = \mathrm{Re}\left(\frac{U_{cd}}{I_{ab}}\right) = \frac{K}{k}\mathrm{Re}\left(\frac{1}{1+j\omega\varepsilon\varepsilon_0/k}\right), \tag{6}$$

where: $K = 4L/\pi D^2$ - is the cell's constant, $K/k = R_H$ is the resistance of some liquid volume with uniform field.

The mathematical model used for the calculations is the Laplace equation.

In AC conductivity measurements, the interpolation method is used. Its essence is that the actual value of the measured resistance corresponds to the infinite value of frequency [11]. But this conflicts with the experimental data. Figure 4b shows a typical frequency dependence of the cell resistance in a wide range of frequencies. At frequencies over ω_{HF}, decline of characteristics always are observed. In [12] an electric model (similar to Fig. 4a) is proposed, which answers question: Why any resistance of fluid at infinite frequency should tend to zero. Equations (5) and (6) make it possible to explain this problem and to quantify the characteristic parameters.

Really, ratio $j\omega\varepsilon\varepsilon_0/k$ characterizes the phase shift between density **J** of current and strength **E** of electric field, Eq. (4). But the phase shift is determined by bulk impedance parameters of the circuit in Fig. 4a or 6. Then we can write the following useful expression.

$$tg\varphi = \omega C_B R_B \equiv \omega\varepsilon\varepsilon_0/k \tag{7}$$

When $tg\varphi = 1$ then the resistance (6) 2-fold decreased and frequency of that $\omega_{0,5}$ can be obtained from Eq. (7).

$$\omega_{0,5} = k/\varepsilon\varepsilon_0 \tag{8}$$

Frequency $\omega_{0,5}$ (Fig. 4b) is a calculated value. It is determined only by liquid parameters. Area A for very dilute solutions falls within the range of sound frequencies. The validity of the formula (8) is supported by studies of [13]. For distilled water with k = 0.14 mS/m calculated frequency is $\omega_{0.5}$ = 31 kHz.

The construction of the cell and especially current electrodes 5 and potential electrodes 3 (Fig. 5) must meet the following requirements: minimum polarization of electrodes, the minimum error in the calculation of constant, minimal stress of deformation for temperature changes. For the thinner electrodes, the above conditions are met more fully.

The error due to the complex nature of the electrolytic conductivity (4) was considered in detail in [4]. Below we describe the systematic component (1) of error, which occurs when electrodes are made in the form of thin films.

The full 3D task of calculating electromagnetic field can be reduced to 3D task of so-called approximation of small currents, which significantly simplifies the calculation. In this case the task is reduced to solving calculation subareas (electrolyte column and quartz tube) of Laplace equation [3, 4].

Upon conducting calculations the following requirements were set: the outer and inner D diameters of the quartz tube (15 mm and 10 mm correspondingly); the length L of the central tube section. (30 mm); electric conductivity of platinum (10^7 S/m) and of quartz (10^{-14} S/m); relative dielectric permittivity of electrolyte (81) and of quartz (3); operating frequency (1000 Hz). Variations undertook lengths of the side tube sections, thicknesses of current and polarizing electrodes, values of liquid's EC.

According to the results of calculation, the values were determined for electric potentials in point's c and d. Further, according to (6) and (1) we determined the relative error. The result of numerical analysis is shown in Fig. 7.

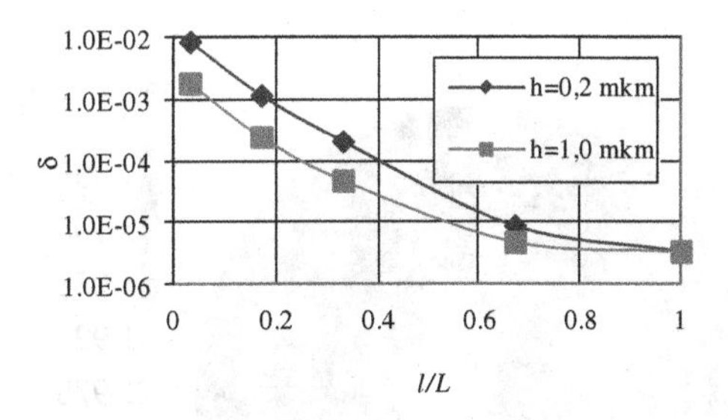

Fig. 7. Dependence of the error δ from ratio l/L

The cause of existing error is that the thin current electrodes have a finite resistance. The potential from the operating current is not the same everywhere on the electrode surface. With increasing thickness of electrodes, this effect is reduced. The ratio of the lengths of pipes l/L (Fig. 1) is also a damping factor. To reduce the error by several orders is necessary that tubes 1 and 2 (Fig. 5) have been the same length. In this case, the voltmeter V (Fig. 5) will operate with in-phase signal 200%. Then instruments measuring U_{cd} voltage should be special.

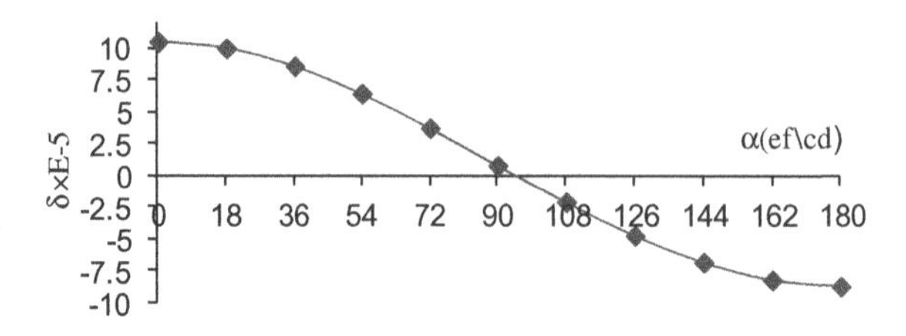

Fig. 8. Dependence of the error δ from angle α for thickness of electrodes $h_T = h_P = 0.3$ μm and electrolyte solution $k = 60$ S/m

Figure 8 shows a plot of the error (1) as function of the angle α. This angle is formed by moving the potential points from c d positions to e f positions. The function is changing the sign on $\pi/2$. This property is used in the cell construction.

3 Analysis of Non-uniformity of Current Density

One of manufacturing technologies use for making the potential and current electrodes (3 and 5 in Fig. 5) is the evaporation of chloroplatinic acid H_2 [$PtCl_6$] and organic oils suspension in temperature 660 °C. After 20–25 such cycles the platinum layer of thickness 0.25–0.35 μm remains on the surface. This method very securely holds the electrodes at both ends of the tube 1 and at the inner surface of the disk 4. However, to get the electrode thickness of more than 0.4 μm is almost impossible. As the result is a thin electrode with a finite resistance. The source of measuring current or voltage is connected to these electrodes located at points a and b. In the process of impedance measurement, the operating current creates the voltage drop on current electrode. This voltage

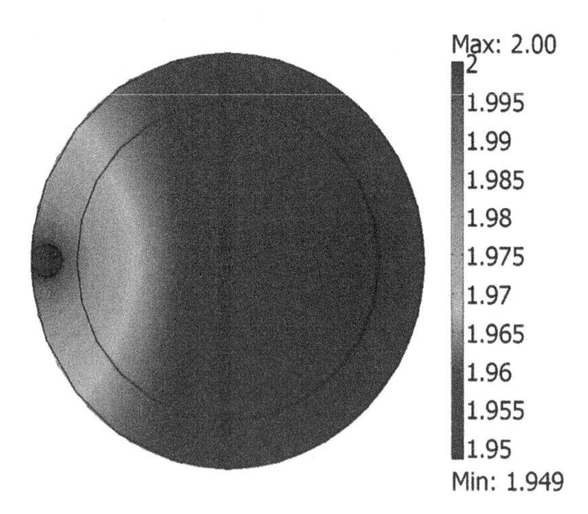

Fig. 9. The potential distribution on the surface of the current electrode 5 (Fig. 5) from the maximum voltage applied to the electrode $U_{Max} = 2$ V, frequency $f = 1$ kHz with dimensions: $D = 10$ mm, thickness of electrodes $h_T = h_P = 0.3$ μm and electrolyte solution $k = 60$ S/m,

distribution is shown in Fig. 9. From this figure is concluded that under given conditions the unevenness of the potential drop $U_{Max}/U_{Min} - 1$ on the electrode surface is about 2.5%. Obviously, the initial potential distribution causes also the unevenness of current density within the tube 1. This non-uniformity will be maximal near the current electrode and minimal near the potential electrode. This is confirmed by calculations given in Figs. 10 and 11. The maximal unevenness of current density $J_{Max} / J_{Min} - 1$ at the distance of 1 mm from the current electrode (Fig. 10) is more than 28%. While at 5 mm distance this unevenness is only slightly above 4.2%.

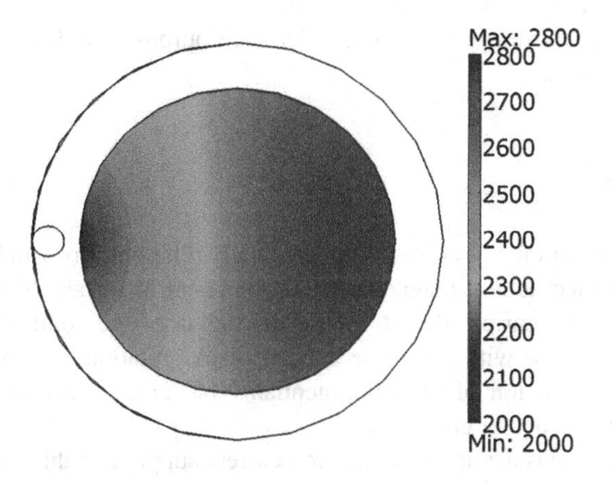

Fig. 10. The current density distribution in the cross section of the conductor at a distance of 1 mm from the current electrode 5 under the same (Fig. 9a) conditions. Maximum value of current density is $J_{Max} = 2800$ A/m^2

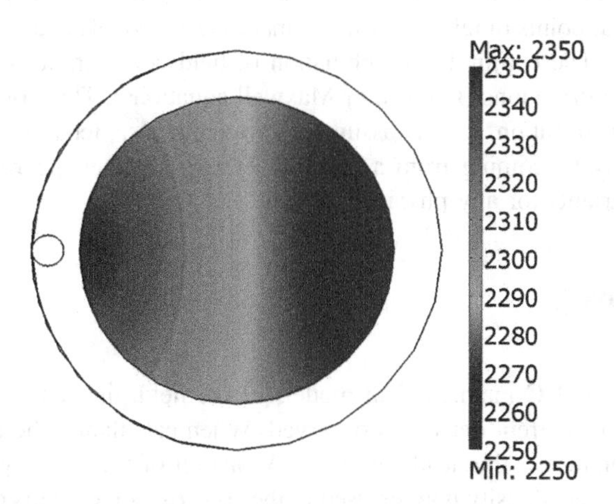

Fig. 11. The current density distribution in the cross section of the conductor at a distance of 5 mm from the current electrode 5 under the same (Fig. 9a) conditions. Maximum value of current density is $J_{Max} = 2350$ A/m^2

It shows that for such EC range the resistance of the liquid column between the current and potential electrode is low and becomes comparable with the resistance of the current electrode 5 (Fig. 5). As a result, the share of the drop in potential (voltage) on the current electrode increases in the overall distribution of the potential of the cell.

The surface of thin films 5 (Fig. 5) is not the equal potential surface. To reduce this component of the error δ_h we can suggest the following:

- Calculate the value of the error and consider it as a correction to the measurement results. However, in this case it is necessary to measure the thickness of electrodes.
- Increase the thickness of electrodes. But in such case electrode can become loose and its operational reliability can get worse.
- Change the manufacturing technology. For such purpose the discs 4 (Fig. 5) shall be made from metal.

4 Discussion

The nature of errors in Figs. 2 and 8 is fundamentally different. Error in Fig. 2 is due to the interaction of electric and magnetic field component in metals. Skin effect causes the unevenness of the current density along the radius of the conductor. The nearer presence of a conductor with a reverse current breaks rotational symmetry. Error in Fig. 8 is due to a violation of the equipotentiality of current electrodes. This occurs because of two given below conditions.

Firstly, when there is a zone of connection current supply to a thin current electrode as a point.

Secondly, when the resistance of the metal electrode 5 and the resistance of the liquid column 1 (Fig. 5) are very near each other.

Regardless of the causes, the errors in Figs. 2 and 8 are very similar by their appearance. Variations in the physical system in the form of a plane-parallel shear removing potential points (angle α) for mathematical AC models and quasi-DC models behave the same. The continuing combination of field configurations corresponds to each type of transformations (Laplace or Maxwell equations). This combination is the current density distribution. Let us assume as a hypothesis. Behavior of error (1) on Figs. 2 and 8 may be coming from a more general E. Noether theorem [14] - about existence of invariants for any physical process.

5 Conclusions

1. In AC and quasi-DC mathematical models of the field, the same character of the heterogeneity of current density is observed. When calculating the impedance of a cylindrical conductor, this leads to errors. Violation of the axial symmetry in distribution of current density may be used in the construction of shuts or four terminal conductivity cells. For this purpose, the current supply terminals and voltage terminals should be deployed in space at an angle close to $\pi/2$.

2. The electrolytic conductivity of a conductometric cell will be a complex value. The electrical model of volume bulk impedance should be represented in the form of two-element parallel capacitive equivalent circuit. The frequency characteristics of the impedance of the liquid conductor on the HF high frequencies are determined by the bulk impedance parameters: resistance and capacitance.
3. The characteristic frequency (the frequency at which the impedance is reduced by half) is a calculated value, which is determined by liquid parameters, i.e. electrolytic conductivity and permittivity.

References

1. Shreiner, R.H., Pratt, K.W.: Standard reference materials: primary standards and standard reference materials for electrolytic conductivity. NIST Special Publication 260-142, Washington (2004)
2. Brinkmann, F., et al.: Primary methods for the measurement of electrolytic conductivity. Accred. Qual. Assur. **8**, 346–353 (2003). doi:10.1007/s00769-003-0645-5
3. Mikhal, A.A., Warsza, Z.L.: Impact of AC electric field non-uniformity on impedance of the conductivity cell. In: XXI IMEKO World Congress Measurement in Research and Industry, Prague, Czech Republic, pp. 2266–2270 (2015)
4. Mikhal, A.A., Warsza, Z.L.: Influence of AC field distribution on impedance of the conductivity cell. Meas. Autom. Monit. MAM **11**, 521–525 (2015)
5. Kibble, B.R., Rainer, G.H.: Coaxial Alternative Current Bridges. Adam Hilder, Bristol (1984)
6. Awan, S., Kibble, B.R., Schurr, J.: Coaxial Electrical Circuits for Interference-Free Measurements. Institution of Engineering and Technology, London (2011)
7. Jensen, H.D.: Final Report of Key Comparison CCQM-K36 (2006). http://kcdb.bipm.org/AppendixB/appbresults/ccqm-k36/ccqm-k36_final_report.pdf
8. Seitz, S., Spitzer, P., Brown, R.: Consistency of practical salinity measurements traceable to primary conductivity standards: Euromet Project 918. Accred. Qual. Assur. (2008). doi:10.1007/s00769-008-0444-0
9. Brett, C., Brett, A.: Electrochemistry. Principles, Methods and Applications. Oxford University Press, Oxford (1994)
10. Bard, A.J., Faulkner, L.R.: Electrochemical Methods. Fundamentals and Applications, 2nd edn. Wiley, New York (2001)
11. Máriássy, M., Pratt, K.W., Spitzer, P.: Major applications of electrochemical techniques at national metrology institutes. Metrologia **46**, 199–213 (2009)
12. Xiaoping, S., Spitzer, P., Sudmeier, U.: Novel method for bulk resistance evaluation in conductivity measurement for high-purity water. Accred. Qual. Assur. **12**, 351–355 (2007)
13. Bottauscio, O., Capra, P.P., Durbiano, F., Manzin, A.: Modeling of cell for electrolytic conductivity measurements. IEEE Trans. Magn. **42**(4), 1423–1426 (2006)
14. Noether, E.: Invariante Variation probleme. Nachrichten von der Kön. Ges. Der Wissenschaften zu Göttingen. Math. Phys. K1 **2**, 235–258 (1918)

Radio Electronic System Elements Diagnostics by Means of Lissajous Curves with the Extended Database

Serhii Yehorov(✉)

National Aviation University, Kiev, Ukraine
sehorov@gmail.com

Abstract. The method of radio electronic component diagnostics by means of Lissajous curves is considered. The random samples modeling by the method of Monte-Carlo is conducted. It is performed to determine the probability of the sample infallible functioning. The statistic quality and the statistic series are analyzed. The series consist of random numbers by means of the criteria for approval by Anderson-Darlington (Ω^2 Mises) χ^2 Pearson. The method of modeling of the Gaussian distribution law is presented.

Keywords: Radio electronics · Diagnostics · The theory of probability · Mathematical statistics · Hypothesis believability testing · The probability of infallible functioning

1 Introduction

The problem of the existing diagnostics methods will always be of current interest, in spite of the deep researches in the field of the diagnostics of the radio electronic system components and devices. First of all, it is connected to development of science and, as a consequence, technology [1, 2]. New materials to manufacture the elements of radio technic, as well as new technologies appear [3]. As a result, radio electronic components are becoming more reliable, cheaper, and technological and they are reduced in sizes. All these consequences facilitate the development and the improvement of the existing diagnostics technologies and make new ones appear.

The rapid pace of science and technology development of the last century has penetrated into this century. This development has become possible thanks to the development of computers [4].

The society development causes the development in science. The improvement of the existing diagnostics methods for radio electronic apparatuses and creation new ones are the components of this development. Nowadays, the existing diagnostics methods for radio electronic apparatuses do not pay much attention to the effect of Lissajous curves. They are applied (mostly because of their properties) to tune one source of a signal up to another one [5]. Unfortunately, [6] has named the Lissajous curves practically useless.

Despite all these facts, the useful properties of the Lissajous curves can be applied to diagnosing of the electronic components such as capacitors, resistors, diodes and

R. Szewczyk and M. Kaliczyńska (eds.), *Recent Advances in Systems, Control and Information Technology*, Advances in Intelligent Systems and Computing 543, DOI 10.1007/978-3-319-48923-0_58

others. That is why, it is recommended to develop the diagnostics method for electronic components by means of the Lissajous curves using the oscilloscope as a sound card of the computer.

2 Materials and Methods

The Lissajous curves [7] (named after French physicist J. Lissajous) – are the closed points of the trajectory which makes two harmonic oscillatory movements simultaneously to two mutually perpendicular directions. The description of the Lissajous curves depends on the correlation among the periods (frequencies), phases and amplitudes of both oscillations and allows determining these correlations as well as the form of oscillations. The Lissajous curves can be observed on the screen of electron beam or digital oscillator if alternating current with the equal or aliquot periods flows through two pairs of the deflection plates. The trajectory of the movement will have the certain forms at different frequency correlations.

Let us consider the movement of a beam at the alternating current simultaneous flow through both pairs of plates. The trajectory of a beam movement will have the certain forms at different frequency correlations. The curves, made by a beam on the screen are called the Lissajous curves. Let us consider how the Lissajous curves are formed on the oscilloscope screen on the example of the sinusoid signal series which come to the inputs X and Y.

1. Both signals have the same frequency and phase ($f_1 = f_2$, $j_1 = j_2 = 0$). Let us consider the coherent state of a beam at different moments of time by splitting one period of voltage into the temporary intervals (Fig. 1). At the moments of time t_0, t_4 and t_8 ($U_x = 0$ i $U_y = 0$) the beam is situated in the middle of the screen. At the

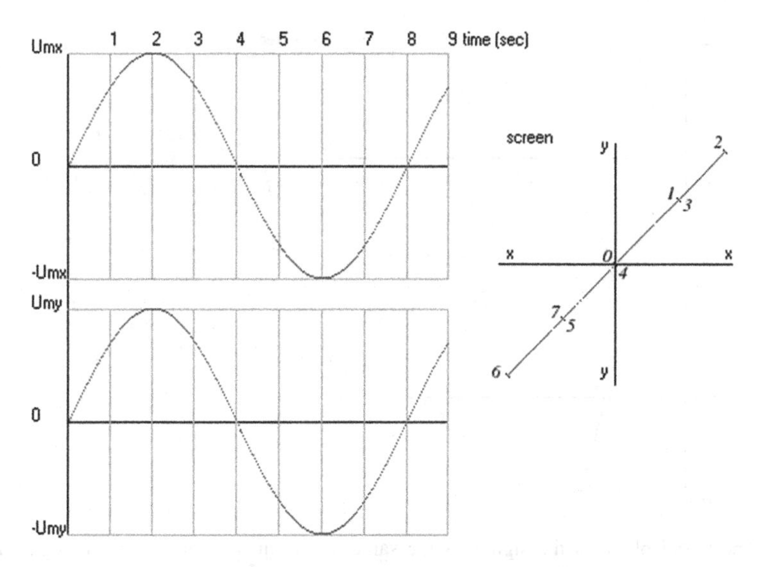

Fig. 1. Both signals have the same frequency and phase ($f_1 = f_2$, $j_1 = j_2 = 0$)

moments of time t_2 and t_6 the voltage of both signals has the maximum value and the beam deviates in a maximum way from the axes x and y: ($U_x = \pm U_{mx}$ and $U_y = \pm U_{my}$). Therefore, the beam moves at the same signal frequency and at the same phases on the inclined straight between points 2 and 6, passing the intermediate points 1, 3, 5 and 7. The tilt angle of the straight depends on the amplitude values U_{mx} and U_{my}.

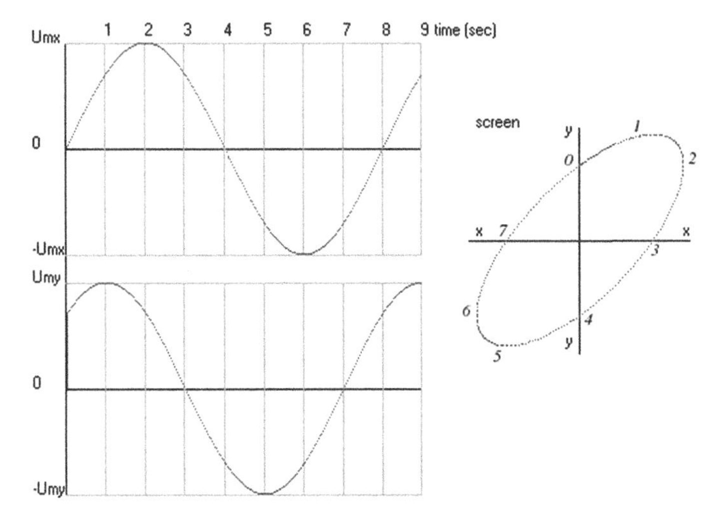

Fig. 2. Both signals have the same frequency, but they differ in phase ($f_1 = f_2$, $v_1 = 0$, $v_2 = $ p/4)

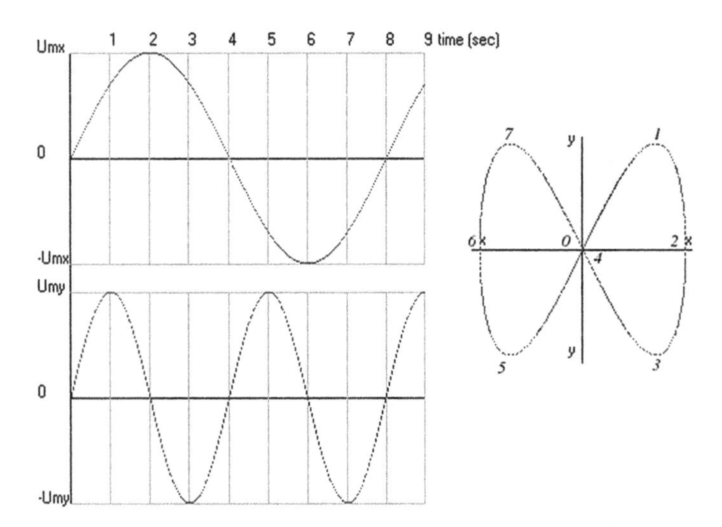

Fig. 3. The initial phase of the signals is the same, the frequency of the second signal is two times more ($f_2 = 2f_1$, $j_1 = j_2 = 0$)

2. Both signals have the same frequency, but they differ in phase ($f_1 = f_2$, $v_1 = 0$, $v_2 = p/4$). The Lissajous curves are formed the same way at the intervals of time and it is presented in Fig. 2. The Lissajous curves have the form of a sloping ellipse
3. The initial phase of the signals is the same, the frequency of the second signal is two times more ($f_2 = 2f_1$, $j_1 = j_2 = 0$). The Lissajous curves are formed at the intervals of time (Fig. 3) and possess the stable form of number "eight"

Therefore, the Lissajous curves can be presented in graphic form and reproduced on the oscilloscope screen. The curves, mentioned above, produce the stable image. The criteria for the image stability are the multiplicity of two signals (the frequency value ratio equals an integer) the phase permanence. This can be analytically proved.

3 Result and Discussion

The simplest case is when (provided that two period are equal) the curves are the ellipses which at the phase difference 0 or π obtain the form of line segments, but at the phase difference π/2 transform into circle. If there is no strict coincidence of two period oscillations, the phase difference constantly changes. As a result, the deformation of the ellipse takes place all the time. If the periods are extremely different, the Lissajous curves are not observed. It is due to the fact that the ellipse deformation is fast and the image becomes blurred. However, if the periods are the integers, the Lissajous curves of more complex form are observed. It occurs at the interval of time which is equal to the least aliquot of the both periods and the point that is moving turns to the same position. The Lissajous curves fit into the rectangle. The centers of this rectangle match the initial state of the coordinates and its sides are parallel to the coordinate axes and are placed on both sides from them at the distances equal to the oscillation amplitudes. Here, the number of contacts of the image with the rectangle sides makes the equation of two oscillation periods.

The mathematical expression for the Lissajous curve [8] is

$$\begin{cases} x(t) = U_x sin(f_x t + \varphi) \\ y(t) = U_y sin(f_y t + \varphi) \end{cases} \tag{1}$$

where U_x, U_y – are the oscillation amplitudes, f_x, f_y – are the frequencies, φ – is the phase shift.

The form of the curve significantly depends on the correlation f_x/f_y. When the correlation is equal to 1, the Lissajous curve looks like an ellipse. At certain conditions it obtains the form of a circle ($U_x = U_y$, $\varphi = \pi/2$ rad) and the line ($\varphi = 0$). Another example of the Lissajous curve is the parabola ($f_x/f_y = 2$, $\varphi = \pi/2$). Other correlations produce more complex figures which are closed, if f_x/f_y – is rational number. It can be assumed, that the visual form of these curves is often the three-dimensional node and it is true, that the projection on the surface of many nodes, including the Lissajous nodes, are the Lissajous curves.

The Lissajous figures, where $U_x = 1$, $U_y = N$ (N – the integer) and

$$\varphi = \frac{(N-1)\pi}{2N} \tag{2}$$

are Chebyshev polynomials of the first type of the degree N.

If signals come to the inputs «X» and «Y» of the oscilloscope and they are of multiple or equal (close) frequencies, it is possible to observe the Lissajous curves on the screen. This method is widely applied to compare the frequencies and phases of two sources of signals and tuning one source up to the frequency of another. If the oscillations produced by the point occur not in accordance with the harmonic, but more complex law and with the same period, then the closed trajectories look similar to the Lissajous curves but they are subjected to deformation. When the frequencies are similar, but not the same as each other, the image on the screen rotates and the period of the rotation cycle is the value conversed to the frequency difference. For example, the rotation period is equal to. If the frequencies are the same, the image freezes motionless at any phase. However, in practice, the image on the oscilloscope screen can tremble because of short non-stable signals. The frequencies to be compared do not have to be the same, but those which are multiple. For example, the ideal source can produce the frequency of 5 MHz only and the source which is being tuned – 2.5 MHz.

Visual Analiser 2014 was used as software. Before the sound card as the oscilloscope input is used (Line in), it is recommended to increase the input resistance of input Line in (the right and the left channels). This task is easy because it is needed to perform simple calculations of voltage divider. At the same time, the resistance of Line in (the right and the left channel) has to be taken into account. The standard oscilloscope probes can be used as probes. In order to obtain the Lissajous curves, the oscilloscope has to set up to XY mode (Fig. 4). The computer must be reliably grounded. The external clock

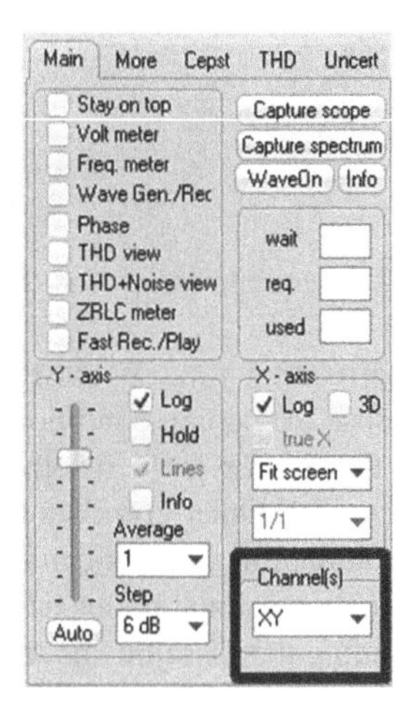

Fig. 4. The oscilloscope tuning to produce the Lissajous curves

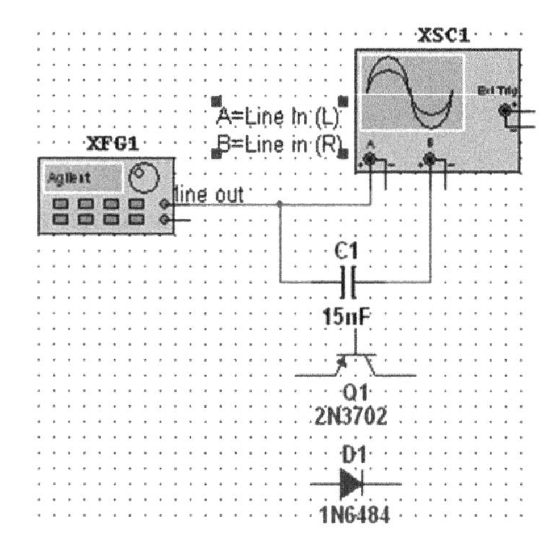

Fig. 5. The device connection circuit (for a computer connection)

can be used as a generator, otherwise the clock built-in Visual Analiser 2014 can be used as well.

Figure 5 presents the circuit of device connection. The inputs A and B of the oscilloscope is the linear input of the sound card. A signal comes to this input through the voltage dividers. They increase the income resistance of the sound card. The external clock (of low frequencies) can be taken or it is also possible to use that one, which comes with Visual Analiser. In this case the signal from the external clock comes from line out of the sound card.

The condenser, diode, Zener diode, transistor (Table 1) were tested by applying the mentioned above method.

When diagnosing the radio components it is necessary to accumulate the statistics of failures in order to determine the possibility of their further proper operation.

Every research of random phenomena performed by methods of reliability theory, which is based on the mathematical tools of the theory of probability and mathematical statistics, is either directly or indirectly grounded on the experimental data. The theory of probability allows determining the probability of some events by the probability of other ones, the distribution laws and numerical characteristics of some random values by the distribution laws and the numerical characteristics of others. It is performed by

Table 1. The waveforms of radio components at their proper operation

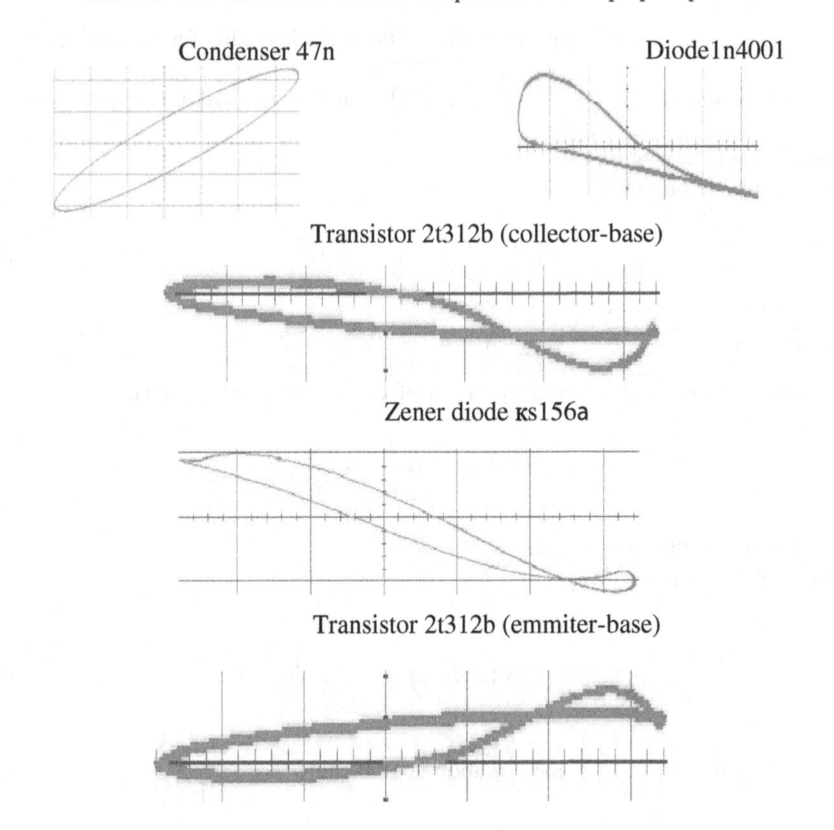

using such notions as the events and their probabilities. Such indirect methods significantly help to save time and money, spent on the experiment, but they do not exclude the experiment by itself. Every research in the sphere of random phenomena, no matter how abstract it is, is based on the experiment, on the experimental data, on the system of observations. Therefore, the following can be suggested.

Let us consider the task connected with the hypotheses believability testing, namely, the task of sequence of the theoretical and statistical distribution by means of the criteria for approval by Pearson χ^2 iω^2. Here the following distribution laws and mathematical expressions are used.

The investigated statistics has the form

$$\vartheta = \frac{r}{N}, \tag{3}$$

where r – is the number of samples which failed at the operating time t among N samples, being tested.

The normal distribution (N):

$$\mathrm{F}(\vartheta) = \mathrm{N}\left(\vartheta, \widehat{\mu}, \widehat{\sigma}\right) = \Phi\left(\frac{\vartheta - \widehat{\mu}}{\widehat{\sigma}}\right), \tag{4}$$

where $\widehat{\mu} = \frac{\sum_{i=1}^{n} \vartheta_i}{n}$ – is the sample mean; ϑ_i – is the statistics value for i – sample; n – is the number of samples; $\widehat{\sigma} = \sqrt{\frac{1}{1-n}\sum_{i=1}^{n}\left(\vartheta_i - \widehat{\mu}\right)^2}$ – is the standard deviation; Φ – is the rationed normal distribution.

Reley distribution (R):

$$F(\vartheta) = R(\vartheta, \widehat{\sigma}) = \frac{\vartheta}{\widehat{\sigma}^2} e^{\frac{-\vartheta^2}{2\widehat{\sigma}^2}} \tag{5}$$

where $\widehat{\sigma}^2 = \left(2 - \frac{\pi}{2}\right)D$, $D = \frac{1}{n-1}\sum_{i=1}^{n}(\vartheta_i - \hat{\mu})$

Kosha distribution (C) (the distribution of Lorents or Breit-Vigner):

$$F(\vartheta) = C(\vartheta, \hat{\mu}, \gamma) = \frac{1}{\pi} \frac{\gamma}{(\vartheta - \widehat{\mu})^2 + \gamma^2} \tag{6}$$

where γ is the coefficient of scale.

Veybull distribution (W):

$$\mathrm{F}(\vartheta) = \mathrm{W}\left(\vartheta; \widehat{b}, \widehat{a}\right) = 1 - e^{-\left(\frac{\vartheta}{\widehat{a}}\right)^{\widehat{b}}}, \tag{7}$$

where $\widehat{b} \cong \frac{1}{\widetilde{v}}$; where $\widetilde{v} = \frac{\widehat{\sigma}}{\widehat{\mu}}$ is the variation coefficient, $\widehat{a} = \left(\frac{1}{n}\sum_{i=1}^{n} \widehat{\vartheta_i^b}\right)^{\frac{1}{b}}$.

Log-normal distribution (LN):

$$F(\vartheta) = LN\left(\vartheta; \breve{\mu}, \breve{\sigma}\right) = \Phi\left(\frac{\ln(\vartheta) - \breve{\mu}}{\breve{\sigma}}\right), \quad (8)$$

where $\breve{\mu} = \ln \widehat{\mu} - \frac{1}{2}\ln\left(\frac{D}{\widehat{\mu}^2} + 1\right)$; $\breve{\sigma} = \left(\ln\left(\frac{D}{\widehat{\mu}^2} + 1\right)\right)^{\frac{1}{2}}$.

The criteria for approval were also used:

- χ^2 by Pearson:

$$\chi^2 = \sum_{j=1}^{r'} \frac{\left(m_j - n_{p_j}\right)^2}{n_{p_j}}, \quad (9)$$

where r' – is the number of intervals after their combining; m_j – is the number of statistics ϑ elements which occurred in j – interval; $n_{p_j} = p_j n$; p_j – the probability by the theoretical distribution; n- is the number of samples.

The application of the criteria of type χ^2 assumes the split of area for the determination of a random value into k intervals with the calculation the number of observations n_i, which occurred within these intervals the probability of being included in the intervals $P_i(\theta)$, where θ – is the known value of the parameter (scalar or vector) corresponding to the theoretical law.

- Anderson-Darling (Ω^2 Mises):
- $\Omega_N^2 = -n - 2\sum_{j=1}^{n}\left\{\frac{2j-1}{2n} \ln F(\vartheta_i) + \left(1 - \frac{2j-1}{2n}\right) \ln(1 - F(\vartheta_i))\right\}$.

In order to obtain the assessment of adequacy of the mathematical tools there is not always the possibility to conduct the real experiment. The various methodologies of modeling can be used to solve this problem. The modeling by Monte-Carlo method can be suggested for the given case. It was performed as following. One hundred samples were made in order to test the believability of hypotheses. Every sample contained one hundred elements. The samples were modeled by Monte-Carlo method.

The idea of the method is as following. Let us select the universal set with the number (the number of elements) $N = 463$. Let us perform the modeling. It is needed to limit the time of the experiment or the operating time of every sample, that has not failed. In order to do this, it is recommended to accept, for example, the empiric probability of failure $F = 0.1$. And let us solve the following equation:

$$\frac{x}{N} = F \Rightarrow \frac{x}{463} = 0,1 \Rightarrow x \approx 46, \quad (10)$$

where x is the sequence number of a sample element which has to be used to limit the sample for the given value F.

Further, we create the values of sequence numbers in the range from 1 to N (we have to generate one hundred elements) by means of the random number generator.

After that, we trace the numbers which are $\leq x$. The number of generated numbers is equal to the number of failures.

The modeling of the universal set to test the statistics under the investigation was performed by the mentioned above order for the accepted level F. The size of every sample consisted of 100 values. The modeling was performed in Microsoft Excel. If to accept $F > 0.1$, then the random number generator uniformity of Microsoft Excel will become obvious. The modeling in this case is not recommended. Therefore, it makes no sense to accept $F > 0.1$ (if to conduct the modeling in Microsoft Excel).

The results of the researches, which were obtained after the hypothesis believability testing, are presented in Tables 2 and 3.

Table 2. The observed values of χ^2 and the corresponding to them levels of significance ρ for $F = 0.1$

N	R	C	LN	W
ρ = 0.01	The hypothesis is refuted	The hypothesis is refuted	ρ = 0.05	The hypothesis is refuted

Table 3. The observed values of Ω^2 and corresponding to them levels of significance ρ for $F = 0.1$

N	R	C	LN	W
ρ = 0.615	The hypothesis is refuted	The hypothesis is refuted	ρ = 0.7	The hypothesis is refuted

The proposed method of diagnostics is quite simple and comfortable to use. Using computer significantly saves money for purchasing the diagnostics equipment and service of the apparatuses of sound frequency.

It is possible to state, that normal and log-normal distribution do not contradict the experimental data by analyzing the results of modeling for various types of empiric distribution of operating time (Tables 2 and 3). The normal distribution is the best to be used for the processing of the statistical data which has been modeled.

By means of the criterion χ^2, ω^2 (any other can be applied as well) it is possible to refute the selected hypothesis only in some cases. It will not conform to the experimental data.

4 Conclusions

In order to select the most adequate theoretical function of distribution, it is recommended to hypothize about several competing hypotheses and to perform the hypothesis believability testing. This allows correct selecting of the theoretical distribution to align the statistical series. It can also provide the error assessment, which is very important. That error can be made when taking one or another theoretical distribution.

The advantage of Mises criterion is the fast convergence to the margin law. 40 observations as minimum are needed (but not some hundreds as for the criterion χ^2) [9].

References

1. Korobiichuk, I.: Mathematical model of precision sensor for an automatic weapons stabilizer system. Measurement **89**, 151–158 (2016)
2. Korobiichuk, I., Bezvesilna, O., Tkachuk, A., Nowicki, M., Szewczyk, R.: Piezoelectric gravimeter of the aviation gravimetric system. In: Szewczyk, R., Zieliński, C., Kaliczyńska, M. (eds.) Challenges in Automation, Robotics and Measurement Techniques. AISC, vol. 440, pp. 753–761. Springer, Heidelberg (2016). doi:10.1007/978-3-319-29357-8_65
3. Salach, J.L., Szewczyk, R.J., Nowicki, M.S., Korobiichuk, I.V.: Metallic glass core utilization as the magnetoelastic torque sensor. East. Eur. J. Enterp. Technol. **5/11**(77), 4–7 (2015). doi:10.15587/1729-4061.2015.50153
4. Korobiichuk, I., Podchashinskiy, Y., Shapovalova, O., Shadura, V., Nowicki, M., Szewczyk, R.: Precision increase in automated digital image measurement systems of geometric values. In: Jabłoński, R., Brezina, T. (eds.) Advanced Mechatronics Solutions. AISC, vol. 393, pp. 335–340. Springer, Heidelberg (2016). doi:10.1007/978-3-319-23923-1_51
5. Access mode: https://uk.wikipedia.org/wiki/Фігури_Ліссажу – The name from the screen
6. Horovits, P., Hill, U.: The Art of Circuitry, vol. 2, p. 598. M.: Publishing House «Mir», Moscow (1986)
7. New polytechnic dictionary. In: Ishlinskiy, A.U. (ed.) M.: The Great Russian Encyclopedia, p. 671 (2000)
8. Yavorskiy, B.M., Detlaf, A.A.: Reference Book on Physics for Engineers and Students of High Schools, p. 940. M.: Publishing House «Nauka», Moscow (1968). The main edition of literature on physics and mathematics
9. Saint Petersburg State University of Telecommunication named after Prof. M.A. Bonch-Bruevich. Access mode Режим. http://dvo.sut.ru/libr/opds/i130hodo_part1/3.htm – The name from the screen

Investigation of Temperature Measurement Uncertainty Components for Infrared Radiation Thermometry

Nataliya Hots(✉)

Department of Metrology, Standardization and Certification,
Lviv Polytechnic National University, Lviv, Ukraine
natalia.gots@lp.edu.ua

Abstract. Analysis of uncertainty components of temperature measurement results by infrared radiation is given. The influence of such parameters of radiation thermometer is examined: the noise of detector; the spectral sensitivity changes in optical receiving system and non-linearity of conversion; the influence of external temperature; changes in temperature sensitivity; changes in spatial sensitivity. Analysis of such components of uncertainty caused by the measurement method of infrared radiation thermometry is made. Factors that lead to the emergence of methodological uncertainty component of radiation temperature measurement are explained: mutual dependence of radiation and emissivity of the surface; the impact of background radiation reflected by other objects, such as the sun, surrounding objects; impact of transmission intermediate environment. Coefficients of influence of each factor are identified. Temperature measurement uncertainty budget is formed. As a result the quantitative information of influence parameters increases the reliability of measurement procedures. Evaluation of measurement uncertainty optimizes measuring procedures through a better understanding of physical processes in infrared radiation thermometry.

Keywords: Radiation thermometry · Measurement uncertainty components · Influence factor · Emissivity of object surface · Background radiation · Transmission of intermediate environment

1 Introduction

Temperature measurement in industry and science constitute about 40% of all measurements [7]. Modern technologies require high speed from measurement, no distortion of the thermal object field, possibility of temperature measurement of moving objects in an aggressive and radiation environment. And the only method that meets these requirements is a radiation thermometry.

Infrared (IR) radiation thermometers are used often for temperature control of technical devices in industry and scientific researches. They are used for obtaining information about the temperature of different technical objects and technological process. The temperature of object surface is the informative parameter of technical

R. Szewczyk and M. Kaliczyńska (eds.), *Recent Advances in Systems, Control and Information Technology*, Advances in Intelligent Systems and Computing 543, DOI 10.1007/978-3-319-48923-0_59

state of different devices. The results of the temperature measurements are used for preventive maintenance and repair of technical equipment.

Therefore, the ensuring of accuracy and reproducibility of results of radiation temperature measurements of the industrial objects is actual problem.

It should be noted that the two types of infrared radiation devises exist: radiation thermometer and infrared cameras. The differences between them are as follows:

- infrared radiation thermometer measures the temperature T_{ob} at a point on the object surface;
- infrared camera measures the temperature T_{iob} and the temperature gradient ∇T_{ob} between the two points on the object surface and shows the temperature field of surface area,

The ideal input signal of infrared radiation thermometer detector $L_{LC}(\lambda, T)$ is presented by equation:

$$L_{LC}(\lambda, T_{bb}) = \int_{\lambda_1}^{\lambda_2} \left(k \cdot C_1 \lambda^{-5} (e^{\frac{C_2}{\lambda T_{bb}}} - 1)^{-1}\right) d\lambda \qquad (1)$$

where: λ_1–λ_2 – working spectral band of optical system of infrared thermometer; $R(\lambda, T)$ – spectral sensitivity; T_{bb} – temperature of reference black body source; C_1 and C_2 – constants, k – coefficient of the distance to the object, angle, area of the surface.

Measurement equation for the output signal of the infrared radiation thermometer in real conditions of measurement $S_{IRT}(\lambda, T_{ob})$ is presented by equation:

$$S_{IRT}(\lambda, T_{ob}) = K \int_{\lambda_1}^{\lambda_2} R(\lambda) \cdot \tau_{atm}(\lambda, T) \begin{bmatrix} \varepsilon_{ob}(\lambda, T) \cdot C_1 \lambda^{-5} (e^{\frac{C_2}{\lambda T_{ob}}} - 1)^{-1} \\ + (1 - \varepsilon_{ob}(\lambda, T)) L(\lambda, T_f) x \end{bmatrix} d\lambda + S_N, \qquad (2)$$

where: $\tau_{atm}(\lambda, T)$ – transmission coefficient of intermediate environment; $\varepsilon_{ob}(\lambda, T)$ – coefficient of emissivity of the object's surface; $S(\lambda, T_f)$ – the flow of background radiation of other objects in real conditions; K – conversion coefficient; S_N – signal of noise.

Measurement equation for output signal of the infrared cameras (signal of the temperature gradient ∇T_{ob} between the two points on the object surface $T1_{ob}$ and $T2_{ob}$) in real conditions of measurement $S_{IRT}(\lambda, \nabla T_{ob})$ is presented by equation:

$$\begin{aligned} S_{IRC}(\lambda, \nabla T_{ob}) &= S_{IRT}(\lambda, T1_{ob}) - S_{IRT}(\lambda, T2_{ob}) \\ &= K \int_{\lambda_1}^{\lambda_2} R(\lambda) \cdot \tau_{atm}(\lambda, T) \begin{bmatrix} \varepsilon_{ob}(\lambda, T) \cdot C_1 \lambda^{-5} (e^{\frac{C_2}{\lambda T1_{ob}}} - 1)^{-1} \\ -\varepsilon_{ob}(\lambda, T) \cdot C_1 \lambda^{-5} (e^{\frac{C_2}{\lambda T2_{ob}}} - 1)^{-1} \end{bmatrix} d\lambda + S_N \end{aligned} \qquad (3)$$

The combined standard uncertainty of temperature measurement $u(S_{IRT}(T_{ob}))$ by infrared radiation thermometer is the positive square root of the combined variance, depends on the following components, which is given by

$$u(S_{IRT}(T_{ob})) = \sqrt{\begin{array}{l} c_{T_{ob}}^2 u^2(T_{ob}) + c_K^2 u^2(K) + c_{\Delta\lambda}^2 u^2(\Delta\lambda) + c_{R(\lambda)}^2 u^2(R(\lambda)) \\ + c_{\tau_{atm}(\lambda,T)}^2 u^2(\tau_{atm}(\lambda,T)) + c_{\varepsilon_{ob}(\lambda,T)}^2 u^2(\varepsilon_{ob}(\lambda,T)) \\ + c_{T_{ob}}^2 u^2(T_f) + c_{S_N}^2 u^2(S_N) \end{array}} \quad (4)$$

where: $u(T_{ob}), u(K), u(\Delta\lambda), u(R(\lambda), , u(\tau_{atm}(\lambda,T), u(\varepsilon_{ob}(\lambda,T), u(S_N)u(T_f)$, uncertainty components in the corresponding values in mathematical model, $c_{T_{ob}}, c_K, c_{\Delta\lambda}, c_{R(\lambda)}, c_{S_N}, c_{\varepsilon_{ob}(\lambda,T)}, c_{T_{ob}}, c_{S_N}$ – sensitivity coefficients in the corresponding values in mathematical model.

The combined standard uncertainty of gradient temperature measurement $u(S_{IRT}(\nabla T_{ob}))$ by infrared camera is the positive square root of the combined variance, depends on the following components, which is given by

$$u(S_{IRT}(\nabla T_{ob})) = \sqrt{\begin{array}{l} c_{\nabla T_{ob}}^2 u^2(\nabla T_{ob}) + c_K^2 u^2(K) + c_{\Delta\lambda}^2 u^2(\Delta\lambda) + c_{R(\lambda)}^2 u^2(R(\lambda)) \\ + c_{\tau_{atm}(\lambda,T)}^2 u^2(\tau_{atm}(\lambda,T)) + c_{\varepsilon_{ob}(\lambda,T)}^2 u^2(\varepsilon_{ob}(\lambda,T)) \end{array}} \quad (5)$$

The differences between Eqs. (4) and (5) case the difference in accuracy of temperature measurement and temperature gradient measurement. In practice the uncertainty of radiation temperature measurement method can be the same order with the measured value. The uncertainty of radiation temperature gradient measurement can be less to 1 °C [6]. In the article the features of formation of uncertainty radiation thermometry are investigated.

2 Principle of the Method

"Guide to the expression of uncertainty in measurement" recommends the uncertainty as measure of measurement accuracy and presents methods of its calculations [3]. Also, according to the international standard ISO 17025 (General requirements for the competence of testing and calibration laboratories) it is required to use reliable measuring devices and to evaluate an uncertainty of measurements and present it in the measurement results and certificates.

The combined standard uncertainty of radiation temperature measurement can be calculated using the square root of the sum of squares partial uncertainties due to possible random variations of the different components of radiation thermometry method using the following formula:

$$u_c(S(T)) = \sqrt{u_A^2(S(T)) + u_B^2(S(T))} \quad (6)$$

where: $u_A(S(T))$ – type A of uncertainty of infrared radiation temperature measurement; $u_B(S(T))$ – type B of uncertainty of infrared radiation temperature measurement [4].

3 Type a Radiation Thermometry Measurement Uncertainty

Type A of uncertainty of infrared radiation temperature measurement is calculated according to next equations.

The best available estimate of expected value of temperature signal $S(T)$ that varies randomly, and for which n independent observations under the same conditions of measurement, is the arithmetic mean of the n observations:

$$\overline{S(T)} = \frac{1}{n}\sum_{i=1}^{n} S(T)_i. \tag{7}$$

The experimental variance of the observations, which estimates the variance σ^2 of the probability distribution of $S(T)$, is given by equation:

$$S_n^2 = \frac{1}{n-1}\sum_{i=1}^{n} (S(T)_i - \overline{S(T)})^2. \tag{8}$$

The best estimate is given by:

$$u_A(T) = \sqrt{\frac{S_n^2}{n}}. \tag{9}$$

That is why the particular properties of formation of type B uncertainty discussed in the article.

4 Type B Radiation Thermometry Measurement Uncertainty

But for infrared radiation temperature measurement the particular properties of formation of type B of uncertainty are existed. It is advisable to classify the uncertainty components of type B in such way using the Eq. (10):

- instrumental measurement uncertainty $u_{Bj(IRK)}(S(T))$;
- uncertainty of the method of temperature measurement $u_{Bl(MM)}(S(T))$;
- uncertainty of reference devices and calibration methods $u_{Bj(RD)}(S(T)$.

$$u_B(S(T)) = \sqrt{\sum_{j=1}^{J} u_{Bj(IRK)}^2(S(T)) + \sum_{l=1}^{L} u_{Bl(MM)}^2(S(T)) + \sum_{m=1}^{M} u_{Bm(RD)}^2(S(T))}. \tag{10}$$

But it should be noted, the accuracy of radiation temperature measurements is low. This problem determines the high value of uncertainty of the method of measurement of radiation temperature in real condition in industry. Uncertainty of the method of measurement in radiation thermometry is the most difficult uncertainty component to evaluate that is associated with the method of temperature radiation measurement.

5 Instrumental Component of Measurement Uncertainty (IMU)

Instrumental measurement uncertainty is the component of measurement uncertainty arising from a measuring instrument in use. Instrumental measurement uncertainty of infrared radiation temperature measurements usually is obtained through calibration or testing of a measuring instrument in calibration laboratory, except for a primary measurement with reference black body sources or reference radiation thermometers.

Components of uncertainty type B are caused by the influence of parameters of the infrared radiation device. To assess the type B uncertainty it is advisable to identify the main functional and technical parameters of the infrared radiation device, changes of which affect the accuracy of measurement results. Based on the results of research to the basic parameters we can attribute:

- radiation receiver noise $u(U_{noise})$;
- changes in the spectral sensitivity of the optical-receiver system, nonlinearity of conversion $u(R(\lambda))$;
- temperature impact $u(T_{amb})$;
- changes in the temperature sensitivity $u(\Delta T)$;
- changes in the spatial sensitivity $u(\Delta S)$.

Therefore, the uncertainty of type B $u_{B(IRK)}(T)$ is described by considering the above factors:

$$u_{B(IRK)}(T) = \sqrt{\sum_{j=1}^{J} u_{j(IRK)}^{2}(T)} = \sqrt{u^2(U_{noise}) + u^2(R(\lambda)) + u^2(T_{amb}) + u^2(\Delta T) + u^2(\Delta T)}. \tag{11}$$

Let us consider the impact of these factors on the measurement uncertainty.

During the measurement of temperature, particularly below 500 °C, the accuracy of temperature measurement is significantly influenced by the noises caused by physical processes in radiation detectors because their values may be comparable with the value of the useful signal. The presence of noises makes difficult to detect the weak signals of low temperature radiation sources and also causes the measurement errors of signal parameters: amplitude, frequency, phase, time of occurrence of radiation pulses.

The main components of the noise of radiation detectors of infrared radiation device are current noise (flicker noise) S_{Nf}, thermal noise (noise of Nyquist) S_{NT} and generation-recombination noise (shot noise) S_{Nd}, and to characterize them it is advisable to use the mean-square value of the fluctuation of the total signal at the receiver output radiation:

$$S_N = \sqrt{S_{Nf}^2 + S_{NT}^2 + S_{Nd}^2} = \sqrt{\frac{A \cdot \overline{I^2 R_d^2}}{f} \cdot \Delta f + 4 \cdot k \cdot T_d \cdot R_d \cdot \Delta f_{ef} + 2 \cdot e \cdot \bar{I} \cdot R_d^2 \cdot \Delta f_{ef}}, \quad (12)$$

where: A – constant depending on the material of the radiation receiver ($A = 3 \cdot 10^{-12} \ldots 3 \cdot 10^{-7}$; for sulfur-lead photoresistors $A \approx 10^{-11}$); I – the average value of radiation receiver current; f – average frequency in the frequency band; R_d – radiation receiver resistance; T_d – temperature of radiation receiver; e – charge of the electron ($e = 1{,}6 \cdot 10^{-19}$ Kl). Each component of the noise signal is characteristic of a certain type of radiation receiver.

The spectral and the temperature dependences of the radiation power of the object, the optical elements of the system, and radiation receiver make influence on the change of the spectral sensitivity.

The perception of receiver's radiation occurs selectively in a certain spectral range. Within this range, the receiver has a different sensitivity to radiation of different wavelengths, which is characterized as a spectral sensitivity $R_{\Pi Bi}(\lambda)$ – the function that associates the spectral energy density of radiation at that wavelength and spectral review radiation receiver (changing physical quantities such as voltage, current, resistance) on the sensed radiation. Furthermore, the sensitivity of radiation receiver is determined by the spectral transmittance of the optical system of the radiation device $\varphi(\lambda)$ – lenses, mirrors and so on. Also different optical filters are used. They are characterized by the transmission functions $o(\lambda)$ that produce radiation in different spectral ranges. As a result, we obtain the spectral transfer function, which will be determined by the following factors $R(\lambda) = R_{\Pi B}(\lambda) \cdot o(\lambda) \cdot \varphi(\lambda)$.

In the operating conditions the measurement results may be determined also by the temperature influence $u(T_{amb})$ of the ambient temperature on the infrared radiation device, in particular in a high temperature production or influence of solar radiation exposure outdoors. Temperature sensitivity is equal to the difference between the lowest temperature levels that can be assigned due to the limited resolution of the digital channel of the specific infrared radiation device. Its change causes measurement uncertainty. According to studies carried out in (8) we can assume that the parameters discussed above are characterized by equiprobable distribution law, so the uncertainty for each of them will be determined by the formula:

$$u_{Bj(IRK)}(T) = \frac{u_j}{\sqrt{12}}, \quad (13)$$

where: $u_{Bj(IRK)}(T)$ – the uncertainty of type B for a separate parameter; u_j – the uncertainty of the effect of the radiation device's individual characteristics.

In the literature, in general, this component of the uncertainty is estimated by the expression (14):

$$u_{B(IRK)}(S(T)) = \frac{\Delta_{IRK}}{\sqrt{3}}, \quad (14)$$

where: Δ_{IRK} – the main uncertainty of measurement of the radiation device temperature that is indicated in the technical documentation for measuring instruments.

Today, the possibility of microprocessor technology, the development of radiation detectors and research accumulated in the world open up new means for the development of radiation thermometry and improve the accuracy of radiation temperature measurement. Instrumental measurement uncertainty of infrared radiation devices is nearly 0.1–1% [2].

6 Uncertainty of the Method of Radiation Temperature Measurement (UMM)

The most difficult uncertainty component to evaluate is that associated with the method of radiation temperature measurement [5]. Thus, even though the uncertainty of the radiation temperature method may be the dominant one, the only information often available for evaluating its standard uncertainty is one's existing knowledge of the physical world, measurement conditions and influence factors.

It should be noted that the uncertainty of the radiation temperature measurement method in industry is high. Its value can be up to tens of percent. The uncertainty of the method of measurement in radiation thermometry is caused by following factors:

- the unknown of the value of emissivity of technical object surface;
- the influence of intermediate environment transmission;
- the influence of the background radiation of other objects.

Thus, such components of uncertainty are described in equation:

$$u_{BJ(MM)}(S(T)) = \sqrt{C'^2_{\varepsilon_{ob}(\lambda,T)} u^2(\varepsilon_{ob}(\lambda,T)) + C'^2_{\tau_{atm}(\lambda)} u^2(\tau_{atm}(\lambda)) + C'^2_{T_f} u^2(T_f)}, \quad (15)$$

The partial derivatives, often called sensitivity coefficients, describe how the output signal estimates with changes in the values of the input estimates:

$$C'_{\varepsilon_{ob}(\lambda,T)} = \frac{\partial(S(T))}{\partial \varepsilon_{ob}(\lambda,T)}, \quad C'_{\varepsilon_{ob}(\lambda,T)} = C_1 K \left(\frac{1}{\lambda^5 \left(e^{\frac{C_2}{\lambda T}} - 1 \right)} - \frac{1}{\lambda^5 \left(e^{\frac{C_2}{\lambda T_{BG}}} - 1 \right)} \right), \quad (16)$$

$$C'_{\tau_{atm}(\lambda)} = \frac{\partial(S(T))}{\partial \tau_{atm}(\lambda)}, \quad C'_{\tau_{atm}(\lambda)} = -C_1 C_2 K \tau(\lambda) \frac{e^{\frac{C_2}{\lambda T_{BG}}} \cdot (\varepsilon(\lambda) - 1)}{\lambda^6 T_{BG} \left(e^{\frac{C_2}{\lambda T_{BG}}} - 1 \right)^2}, \quad (17)$$

$$C'_{T_f} = \frac{\partial(S(T))}{\partial T_f}, C'_{T_f} = -C_1 K \left(\frac{(\varepsilon - 1)}{\lambda^5 \left(e^{\frac{C_2}{\lambda T_{BG}}} - 1 \right)} - \frac{\varepsilon}{\lambda^5 \left(e^{\frac{C_2}{\lambda T}} - 1 \right)} \right). \quad (18)$$

The relative uncertainty is estimated by equations granted of normal probability:

$$u(\varepsilon_{ob}(\lambda)) = \frac{\delta(\varepsilon_{ob}(\lambda)) \cdot S(T_{ob})}{\sqrt{3} \cdot 100\%}, \tag{19}$$

$$u(\tau_{atm}(\lambda)) = \frac{\delta(\tau_{atm}(\lambda)) \cdot S(T_{ob})}{\sqrt{3} \cdot 100\%}, \tag{20}$$

$$u(T_{BG}) = \frac{\delta(T_f) \cdot S(T_{ob})}{\sqrt{3} \cdot 100\%}. \tag{21}$$

where: $\delta(\varepsilon_{ob}(\lambda))$, $\delta(\tau_{atm}(\lambda))$, $\delta(T_f)$ – maximum relative values of the multiplicative deviations.

7 Measurement Uncertainty of Reference Device (URD)

The calibration methods of infrared radiation thermometers are more complicated than for contact thermometers or high-temperature radiation thermometers. In this method the reference devices are used – reference infrared radiation thermometer and reference black body source [1, 2].

The relationship between detector signal and temperature is given by a thermometer response interpolation function, which is well approximated by the equations:

$$S_{\text{int}}(T) = \frac{C}{\exp\left(\frac{C_2}{AT+B}\right) - 1}, T = \frac{C_2}{A \ln(C/S(T_{ob})) + 1} - \frac{B}{A} \tag{22}$$

where: A, B and C – constants related to the properties of infrared radiation thermometer.

The values of these uncertainties depend of the properties of reference infrared radiation thermometer, reference black body source and calibrated infrared radiation thermometer. Thus, components of such uncertainties are described in equation:

$$u_{Bm(RD)}(S(T)) = \sqrt{C'^2_{Rbb} u^2_{Rbb}(T) + C'^2_{RRT} u^2_{RRT}(T) + C'^2_{\text{int}} u^2_{\text{int}}(T)}, \tag{23}$$

where: $u^2_{Rbb}(T)$ – uncertainty component of reference black body source; $u^2_{RRT}(T)$ – uncertainty component of reference infrared radiation thermometer; $u^2_{\text{int}}(T)$ – uncertainty component of interpolation function. The sensitivity coefficient of interpolation function is:

$$C'_{int} = \frac{\partial(S_{int}(T))}{\partial T},$$
$$C_{int} = \frac{A \cdot C \cdot C_2 \cdot \exp\left(\frac{C_2}{B+AT}\right)}{(B+AT)^2 \cdot \left(\exp\left(\frac{C_2}{B+AT} - 1\right)\right)^2} \tag{24}$$

8 Graphic Simulation and Uncertainty Budget of Radiation Temperature Measurement

The simulation examples of influence of different factors on the radiation temperature measurement accuracy are presented in Fig. 1.

Uncertainty budget of infrared radiation temperature measurement is presented in the Table 1.

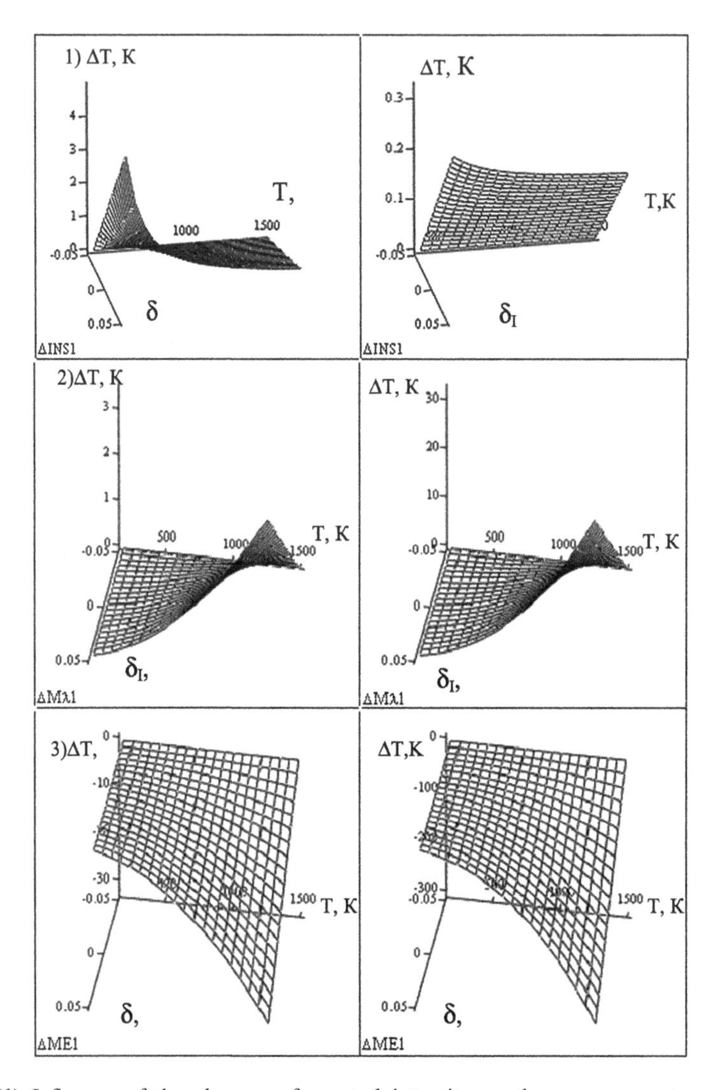

Fig. 1. (**1**) Influence of the changes of spectral intensity on the measurement result in the spectral range 1–2 μm and 8–12 μm; (**2**) Influence of the wavelength values changes on the measurement result in the spectral ranges 1–2 μm and 8–12 μm; (**3**) Influence of the changes of emissivity values in the spectral ranges 1–2 μm and 8–12 μm

Expanded uncertainty of output estimate:

$$U_p(S(T)) = k_p \cdot u_c(S(T)), \tag{25}$$

where: k_p – expansion coefficient.

Table 1. Uncertainty budget of radiation temperature measurement

Source of uncertainty	Estimate of expected value	Standard uncertainty component	Type of uncertainty	Probability distribution	Sensitivity coefficient	Standard uncertainty
Value of temperature signal	t	$u_A(T)$	A	Normal	–	$\sqrt{\frac{S_n^2}{n}}$
Noise of detector	U_{noise}	$u(U_{noise})$	B (IMU)	Uniform	–	$\frac{u_B(U_{noise})}{\sqrt{12}}$
Nonlinearity of optical system	$R(\lambda)$	$u(R(\lambda))$	B (IMU)	Uniform	–	$\frac{u(R(\lambda))}{\sqrt{12}}$
Temperature influence	T_{amb}	$u(T_{amb})$	B (IMU)	Uniform	–	$\frac{u(T_{amb})}{\sqrt{12}}$
Temperature sensitivity	ΔT	$u(\Delta T)$	B (IMU)	Uniform	–	$\frac{u(\Delta T)}{\sqrt{12}}$
Variations of spatial sensitivity	ΔS	$u(\Delta S)$	B (IMU)	Uniform	–	$\frac{u(\Delta S)}{\sqrt{12}}$
Emissivity of surface	$\varepsilon_{ob}(\lambda)$	$u(\varepsilon_{ob}(\lambda))$	B (UMM)	Normal	$\frac{\partial(S(T))}{\partial \varepsilon_{ob}(\lambda)}$	$C_{\varepsilon_{ob}(\lambda)} u(\varepsilon_{ob}(\lambda))$
Influence of intermediate environment transmission	$\tau(\lambda)$	$u(\tau(\lambda))$	B (UMM)	Normal	$\frac{\partial(S(T))}{\partial \tau}$	$C_{\tau(\lambda)} u(\tau(\lambda))$
Influence of background radiation	T_f	$u(T_f)$	B (UMM)	Normal	$\frac{\partial(S(T))}{\partial T_f}$	$C_{T_f} u(T_f)$
Reference black body source	ΔT_{bb}	$u(\Delta T_{bb})$	B (URD)	Uniform	–	$\frac{u(\Delta T_{bb})}{\sqrt{12}}$
Reference radiation thermometer	ΔT_{RP}	$u(\Delta T_R)$	B (URD)	Uniform	–	$\frac{u(\Delta T_{RP})}{\sqrt{12}}$
Interpolation function	$S_{int}(T)$	$u\ (S_{int}(T))$	B (URD)	Uniform	$\frac{\partial(S_{int}(T))}{\partial T}$	$C_{S_{int}(T)} u(S_{int}(T))$

9 Conclusions

International standards and metrology document recommends the characterizing of the infrared radiation measurement systems and temperature measurement results using their uncertainties. The mathematical models of radiation thermometry uncertainty of temperature measurement results are presented in this paper.

The article examined all influence factors if infrared radiation temperature measurements accuracy. In practice the uncertainty of radiation temperature measurement method can be the same order with the measured value. The uncertainty of radiation temperature gradient measurement can be less to 1 °C.

The analysis of all components of uncertainty helps identify the dominant factors. This information helps to choose methods to improve the accuracy of temperature measurement and apply or introduce amendments.

References

1. Expression of the uncertainty of measurement in calibration, EAL-R2 (1997)
2. Expression of uncertainty of measurement in calibration, Supplement to EAL-R2
3. Guide to the expression of uncertainty in measurement, International Organization for Standardization-International Electrotechnical Commission-International Organization of Legal Metrology-International Bureau of Weights and Measures, TAG 4/WG 3 (1993)
4. International Vocabulary of Metrology – Basic and General Concepts and Associated Terms (VIM 3rd edition) JCGM 200:2012
5. Chrzanowski, K., Fischer, J., Matyszkiel, M.: Testing and evaluation of thermal cameras for absolute temperature measurement. Opt. Eng. **39**, 9 (2000)
6. Minkina, W., Dudzik, S.: Infrared Thermography – Errors and Uncertainties. Wiley, Chichester (2009). ISBN 978-0-470-74718-6, Online Books TM ISBN 978-0- 470-68223-4
7. Więcek B., De May G.: Termowizja w podczerwieni. Podstawy i zastosowania, Wydawnictwo PAK, Warszawa, (2011)

Accuracy of Reconstruction of the Spatial Temperature Distribution Based on Surface Temperature Measurements by Resistance Sensors

Mykhaylo Dorozhovets[1], Mariana Burdega[2], and Zygmunt L. Warsza[3](✉)

[1] Rzeszow University of Technology, Rzeszow, Poland
michdor@prz.edu.pl

[2] National University – Lviv Politechnic, Lviv, Ukraine
burdegamariana@gmail.com

[3] Industrial Research Institute of Automation and Measurement (PIAP), Warsaw, Poland
zlw@op.pl

Abstract. In this paper, the method and instrumental aspects of the reconstruction of 2D temperature distribution are investigated. It is based on measurements of the resistance of planar temperature sensors. The first component depends on number of sensors and of the used 2D approximation of temperature distribution. The second component depends on the parameters (additive and multiplicative) of accuracy of used resistance measurements. The mean, root mean squared and maximum, minimum error parameters are determined and analyzed using Monte Carlo method.

Keywords: Temperature distribution · Resistive sensors · Approximation and instrumental error

1 Introduction

Temperature measurements are the most often used in industry for the controlling parameters of various technology processes [1]. Measurement of spatial distribution of bulk and surface temperature at various industrial facilities is the important issue. In particular, determination of the spatial surface temperature of silicon wafers, during the various deep ultraviolet lithographic steps is needed, because temperature no-uniformity has direct impact on the critical dimensions and other characteristics of electronic elements [2]. Study of the thermal conductivity of materials and their construction (walls) needs measurement of the heat flow by measurement the surface temperature distribution. For this purpose on the investigated surface (for example plate or wall), the set (array) of planar resistance sensors can be used [3, 4]. During measurements, the wires connecting sensors should not affect the thermal field. Therefore, preference should be provided to planar sensors with measuring electrodes which are arranged at the edges

R. Szewczyk and M. Kaliczyńska (eds.), *Recent Advances in Systems, Control and Information Technology*, Advances in Intelligent Systems and Computing 543, DOI 10.1007/978-3-319-48923-0_60

but not inside of such object. Due to this the measuring wires do not influenced on the surface temperature field.

There are two main components of the uncertainty of results of temperature distribution measurements: the methodical component connected with the limited numbers of sensing elements (spatial sampling) and instrumental ones (of random and systematic nature with additive and multiplicative components) which is dependent from influences on the resistance of sensing elements used in measurements. This article investigates and analyzes above aspects.

2 Resistance of Planar Sensing Element Located in Uniform Temperature Field

If sensing element is placed in surface with the spatial temperature distribution $\theta(x, y)$, then in surface points (x, y) is the resistivity $\rho(x, y)$ and along the line l_j on the surface is the resistivity $\rho(l)$, i.e.

$$\rho(x,y) = \rho[\theta(x,y)], \quad \rho(l) = \rho[\theta(l)]. \tag{1}$$

Dependency of the resistance of linear sensors on the temperature is determined as the integral (Fig. 1):

$$R_j(l) = \frac{1}{S}\int_{l_j} \rho(l)dl, \tag{2}$$

where: S – is the cross-section area of the sensors. The number of line (linear sensor) is m $(j = 1 \ldots m)$.

For the line l_j:

$$y(l_j) = a_{0,j} + a_{1,j}x, \tag{3}$$

where: $a_{0j} = y_{1j} - a_{1j} \cdot x_{1j}$ – is the y – intercept of the line l_j, $a_{1j} = \frac{y_{2j} - y_{1j}}{x_{2j} - x_{1j}}$ – is the slop of the line l_j and

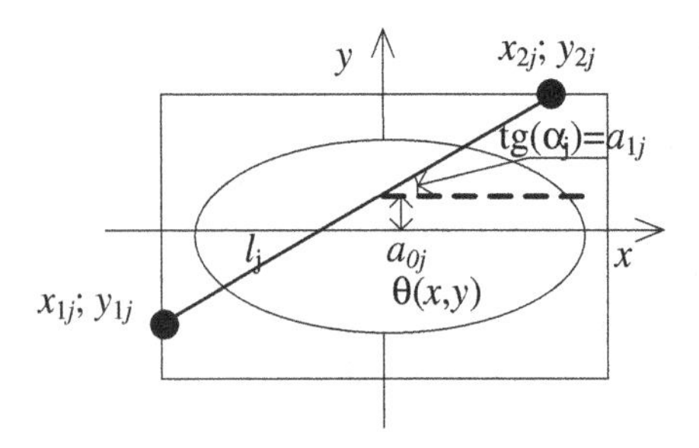

Fig. 1. The line of resistance (sensitive element) along which temperature is changed

$$dl = \sqrt{dx^2 + dy^2} = \sqrt{1 + a_{1,j}^2}dx = b_j dx, \tag{4}$$

where: $b_j = \sqrt{1 + a_{1,j}^2}$.

Then from (2) and (3) is:

$$R_j = \frac{|l_j|}{S} \cdot \frac{1}{|x_{2j} - x_{1j}|} \int_{x_{1j}}^{x_{2j}} \rho(x,\ a_{0j} + a_{1j} \cdot x)dx, \tag{5}$$

where $|l_j| = |b_j| \cdot |x_{2j} - x_{1j}|$ is a length of j sensor.

3 Reconstruction of 2D Temperature Distribution

Temperature distribution $\theta(x,\ y)$ can be reconstructed after set of m measured resistances using electrical resistance tomography techniques [5, 6]. For this purpose the resistivity distribution $\rho(x,\ y)$, which is depended from temperature distribution $\theta(x,y)$, should be approximated by known two-dimensional basic functions $\varphi_i(x, y)$:

$$\rho(x, y) = \sum_{i=1}^{n} \mathbf{C}_{\varphi_i} \varphi_i(x, y), \tag{6}$$

where $\mathbf{C}_{\varphi_i}$ – is a vector of the n unknown coefficients of basic functions $\varphi_i(x, y)$.

Then resistance $R_j(l)$ along the line l_j can be calculated from formula:

$$R_j^{calc} = \frac{|l_j|}{S} \cdot \frac{1}{|x_{2j} - x_{1j}|} \int_{x_{1j}}^{x_{2j}} \sum_{i=1}^{n} \mathbf{C}_{\varphi_i} \varphi_i(x,\ a_{0j} + a_{1j} \cdot x)\ dx. \tag{7}$$

In the simple case, the basic function defined as two-dimensional algebraic polynomial:

$$\varphi_i(x, y) = \sum_{i+k=0}^{i+k \le n} x^i \cdot y^k, \tag{8}$$

Using binomial theorem [7]:

$$y^k = (a_0 + a_1 x)^k = \sum_{r=0}^{k} C_k^r \cdot a_0^r \cdot a_1^{k-r} \cdot x^{k-r}, \tag{9}$$

we get

$$R_j^{calc} = \frac{|l_j|}{S} \sum_{i=1}^{n} C_{\varphi,i} \psi_{i,j}, \tag{10}$$

$$\text{where: } \psi_{i,j} = \sum_{i+k=0}^{i+k \le n} \sum_{r=0}^{k} \frac{a_{0,j}^{r} a_{1,j}^{k-r}}{i+k-r+1} \cdot \frac{x_{2,j}^{i+k-r+1} - x_{1,j}^{i+k-r+1}}{x_{2,j} - x_{1,j}} \tag{11}$$

Parameters $\psi_{i,j}$ are the elements of matrix $\boldsymbol{\psi}$ of size $n \times m$, which depended only the spatial location of linear resistivity sensors:

$$\boldsymbol{\psi} = \begin{vmatrix} \psi_{0,0} & \psi_{0,1} & \cdots & \psi_{0,m} \\ \psi_{1,0} & \psi_{1,1} & \cdots & \psi_{1,m} \\ \vdots & \vdots & \ddots & \vdots \\ \psi_{n,0} & \psi_{n,1} & \cdots & \psi_{n,m} \end{vmatrix}. \tag{12}$$

After normalization of measurement results to S/b_j we obtain

$$R_{meas,j}^{*} = R_{meas,j} \frac{S}{|l_j|} = \rho_{meas,j}. \tag{13}$$

Then using approximation (6) and previously calculated values ψ_{ij} (11) all measurement can be presented in the matrix form as:

$$\boldsymbol{\rho}_{meas} = \boldsymbol{\psi} \cdot \mathbf{C}_{\varphi}. \tag{14}$$

The reconstructed vector $\mathbf{C}_{rec_{\varphi}}$ of coefficients in (6) can be obtained after solving the matrix Eq. (14) using the least squares method with Levenberg-Marquardt regularization [8]:

$$\mathbf{C}_{rec_{\varphi}} = \left(\boldsymbol{\psi}^{T} \boldsymbol{\psi} + \mu \cdot \mathbf{I}\right)^{-1} \cdot \boldsymbol{\psi}^{T} \cdot \boldsymbol{\rho}_{meas}, \tag{15}$$

where: μ - is a parameter of regularization in Levenberg-Marquardt method, **I** – is the identity matrix.

For the temperature measurement very often the copper (Cu) and platinum (Pt) sensors are used. For these sensors, the transfer function $\rho(\theta) = F(\theta)$ is [1]:

$$\rho_{Cu}(\theta) = \rho_{0,Cu}(1 + \alpha_{Cu}\theta); \tag{16}$$

$$\rho_{Pt}(\theta) = \rho_{0,Pt}\left(1 + \alpha_{Pt}\theta + \beta_{Pt}\theta^{2}\right), \tag{17}$$

where: $\rho_{0,Cu}, \rho_{0,Pt}$ – are copper and platinum sensor resistances at reference temperature 0°C and α_{Cu}, α_{Pt}, β_{Pt} – the temperature coefficients for copper and platinum respectively.

Then, using coefficients $C_{rec_{\varphi,i}}$ of (15), the spatial temperature distribution can be determined from (6) as:

$$\theta_{rec}(x,y) = F^{-1}[\rho_{rec}(x,y)] = F^{-1}\left[\sum_{i=1}^{n} C_{rec_{\varphi,i}} \cdot \varphi_i(x,y)\right], \tag{18}$$

For example, the spatial temperature distribution of copper transducer (16) can be calculated as:

$$\theta_{rec}(x,y)_{Cu} = \frac{\frac{1}{\rho_{0,Cu}}\sum_{i=1}^{n} C_{rec_{\varphi,i}} \cdot \varphi_i(x,y) - 1}{\alpha_{Cu}}. \tag{19}$$

In the same way, the spatial temperature distribution of platinum transducer (17) is:

$$\theta_{rec}(x,y)_{Pt} = \frac{\frac{2}{\alpha_{Pt}} \cdot \left[\frac{1}{\rho_{0,Pt}}\sum_{i=1}^{n} C_{rec_{\varphi,i}} \cdot \varphi_i(x,y) - 1\right]}{1 + \sqrt{1 + \frac{4\beta_{Pt}}{\alpha_{Pt}^2}\left[\frac{1}{\rho_{0,Pt}}\sum_{i=1}^{n} C_{rec_{\varphi,i}} \cdot \varphi_i(x,y) - 1\right]}}. \tag{20}$$

4 Simulations

The Monte-Carlo method was used for simulations. The number of simulations is $M = 10^4$. The following scheme of the sensitive elements placing on the surface of the rectangular object (Fig. 2) was selected

The sensitive elements used in simulations have following parameters:

- Material – copper, resistivity $\rho_{0,Cu}$ = 0.01724 μΩ•m, temperature coefficient of resistance $\alpha_{Cu} = 4.3 \cdot 10^{-3}$ 1/ °C, and
- Material – platinum, resistivity $\rho_{0,Pt}$ = 0.105 $\Omega \cdot mm^2/m$, temperature coefficient of resistance $\alpha_{Pt} = 3.9702 \cdot 10^{-3}$ 1/ °C and second order coefficient $\beta_{Pt} = -5.8893 \cdot 10^{-7}$ 1/ °C^2.

Sensitive elements can be in the form of wire of diameter d = 0.2 mm (Fig. 3a) or in the rectangular form of size $a \times b$ (Fig. 3b). Then in first case cross-section is $S = \pi d^2/4$, and in second case – $S = a \times b$.

The investigated object has the size 6 × 6 m × m. The approximation model used for the temperature distribution has the mathematical form of two-dimension cosine function with initial temperature θ_0 = 100 °C and maximal interval of temperature changes θ_m = 75 °C. It is represented by the formula:

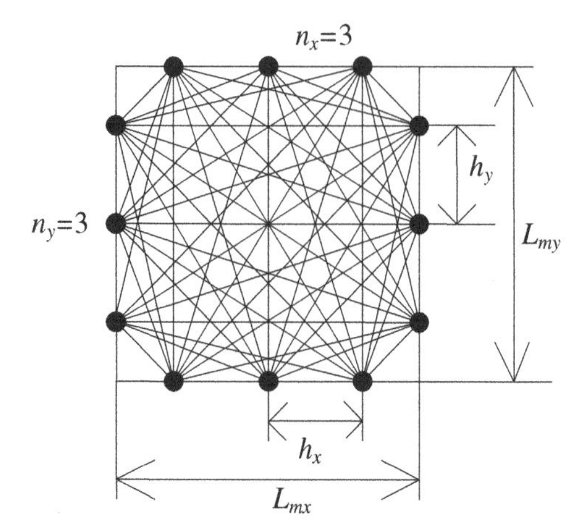

Fig. 2. Possible accommodation of linear resistive sensors on the surface of rectangular object for the number $n_x = 3$, $n_y = 3$ connections at the edges of wall

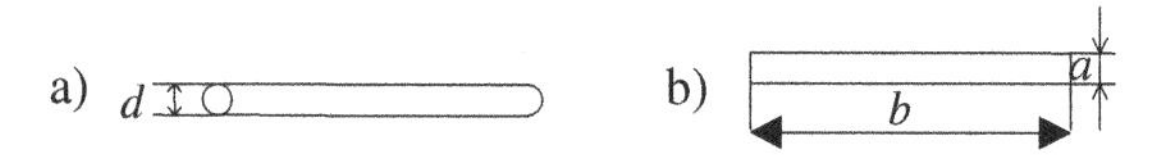

Fig. 3. Resistive sensor in the form of wire (a) and rectangular form (b)

$$\theta(x,y) = \theta_0 + \theta_m \cdot \cos\left(\pi \frac{x}{L_{mx}}\right) \cdot \cos\left(\pi \frac{y}{L_{my}}\right), \tag{21}$$

where: x, y are changed independently in ranges: $0 \leq |x| \leq \frac{L_{mx}}{2}$, $0 \leq |y| \leq \frac{L_{my}}{2}$.

Simulation was performed for $m = 54$ ($n_x = n_y = 3$ connection points at one side) and $m = 96$ sensitive elements ($n_x = n_y = 4$ connections at one side) and algebraic polynomial of order $k = 5$ and order $k = 6$.

In the first stage of simulation, the temperature distribution $\theta(x,y)$ from (21) at first is used in (16) and (17) and then corresponding resistivity distributions are used in (5) and m results of resistances ($R_{j,\mathrm{meas}}$, $j = 1.. m$) of all sensing elements are calculated.

In second stage, these results (vector $\mathbf{R}_{\mathrm{meas}}$) after normalization are used in (13) and vector $\mathbf{C}_{rec_\varphi}$ of reconstructed coefficients is calculated.

In the third stage, with using these reconstructed coefficients the spatial temperature distribution $\theta_{rec}(x, y)$ is determined by formula (19) for the copper sensors and by formula (20) for the platinum sensors.

In fourth stage, the reconstructed errors are determined and analyzed.

The normalized to the maximum temperature ($\theta_0 + \theta_{\mathrm{m}}$) error of the reconstructed spatial temperature distribution are calculated by formula:

$$\gamma_{\theta\, rec}(x,y) = \frac{\theta_{rec}(x,y) - \theta(x,y)}{\theta_0 + \theta_m}, \tag{22}$$

where $\theta_0(x,y)$ – is reconstructed the spatial temperature distribution without the influence of the instrumental error.

In each simulation the surface average value $\overline{\gamma}$ (23), its variation s_γ^2 (24), minimum $\gamma_\theta(x,y)_{\min}$ and maximum $\gamma_\theta(x,y)_{\max}$ errors (22) and $\max = \max(\gamma_\theta(x,y)_{\max}, |\gamma_\theta(x,y)_{\min}|)$ are determined from formulas:

$$\overline{\gamma} = \frac{1}{L_{mx}L_{my}} \int_{-Lmx/2}^{Lmx/2} \int_{-Lmy/2}^{Lmy/2} \gamma_{\theta rec}(x,y)\, dy\, dx; \tag{23}$$

$$s_\gamma^2 = \frac{1}{L_{mx}L_{my}} \int_{-Lmx/2}^{Lmx/2} \int_{-Lmy/2}^{Lmy/2} [\gamma_{\theta rec}(x,y) - \overline{\gamma}]^2 dy\, dx. \tag{24}$$

The surface distributions of reconstructed errors: for k = 5 and k = 6 and $n_x = n_y = 4$ are presented in Fig. 4. The standard deviation $\sqrt{s_\gamma^2}$ (24) and maximal value $\max(\gamma)$ of normalized approximation error $\gamma_{\theta rec}(x,y)$ of the reconstructed spatial temperature distribution (22) for the k = 5 and k = 6 orders of two-dimensional algebraic polynomial (8), number $n_x = n_y = 3$ and $n_x = n_y = 4$ of connected points the edges of wall for the cooper sensors are presented in Table 1.

The approximation errors of the platinum sensors are obtained by the same way. The surface average error $\overline{\gamma}$ is less than a few hundredths of a percent; therefore the value of average error is not shown in Table 1. Maximal values of the error are located in the corner of object, because in these parts there are no measuring paths (Fig. 2). From Table 1 we can see that two-dimensional algebraic polynomial (8) of k = 6 order is sufficient from the point of view of error of approximation of the temperature distribution (21).

As the next problem, how the influences of such types: additive random Δ_{rnd}, systematic additive $\Delta_{0,\text{syst}}$ and multiplicative δ_m, changed the measurement results is analyzed. The standard deviations of additive random errors normalized to the measurement range are: $\gamma_{\sigma,rnd}$ = 1%, 0.316%, 0.1%, normalized to the range limited value

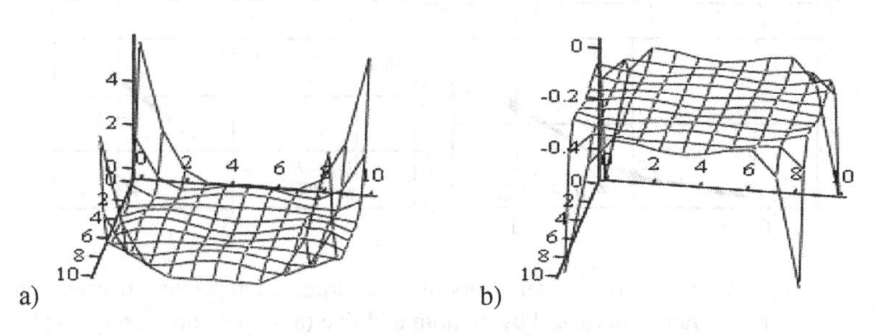

Fig. 4. The surface distributions of reconstructed errors: for k = 5 (a) and k = 6 (b) $n_x = n_y = 4$

Table 1. Characteristics of approximation error of the temperature distribution (for cooper sensors) versus of order (k) of two-dimensional polynomials

	$k = 5$		$k = 6$	
	$\sqrt{\overline{s^2_\gamma}}, \%$	max.(γ), %	$\sqrt{\overline{s^2_\gamma}}, \%$	max.(γ), %
$n_x = n_y = 3$	1.2	6.0	0.11	0.58
$n_x = n_y = 4$	1.0	5.3	0.093	0.51

of additive systematic errors are: $\gamma_{0,lim} = \pm 1\%, \pm 0.316\%, \pm 0.1\%$, and limited value of relative multiplicative errors are: $\delta_{m,lim} = \pm 1\%, \pm 0.3\%, \pm 0.1\%$ are obtained.

Using M realizations of these errors in measured results (5), the reconstructed temperature distribution is calculated by (15), (19) and (20). After this, the M realizations of surface errors of reconstruction are determined by formula (22). Then the mean $\overline{\overline{\gamma}}$ and variation $s^2_{\overline{\gamma}}$ of average error $\overline{\gamma}$ (23) and mean of variation $\overline{s^2_\gamma}$ of errors (24) and maximal (max) values of errors were also determined.

The value of standard deviation $\sqrt{\overline{s^2_\gamma}}$ and maximum (max) of errors of local temperature distribution determined for the number of connectors on one side $n_x = n_y = 3$ for 54 copper sensors and for the number $n_x = n_y = 4$ and for 96 copper sensors and order $k = 5$ two-dimensional polynomials are presented in Fig. 5.

Standard deviation and maximal errors of reconstructed temperature distributions using platinum sensors are nearly similar as for the cooper sensors.

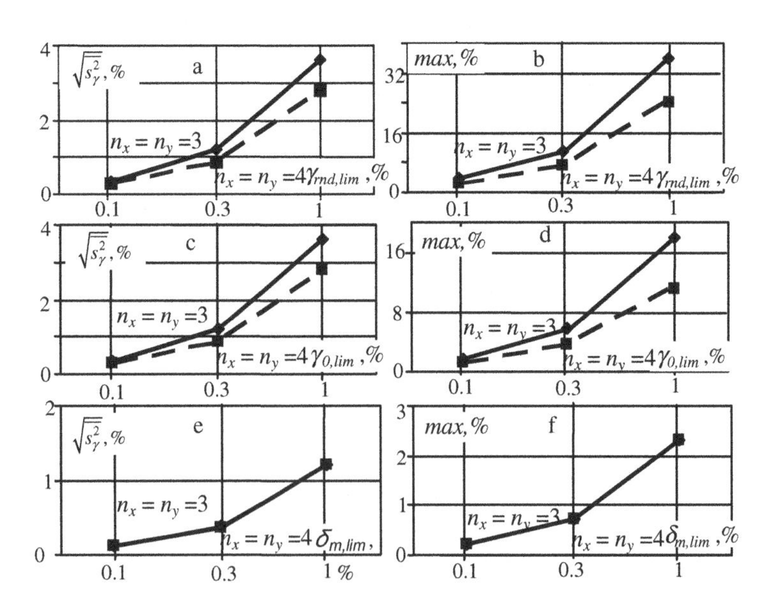

Fig. 5. Standard deviation and maximal errors of reconstructed temperature distributions using copper sensors: error parameters caused by random additive (a, b), systematic additive (c, d) and systematic multiplicative (e, f) influences

5 Conclusions

1. The proposed method of the measurement of spatial temperature distribution can be used for various applications.
2. Reconstruction of surface temperature (field) is based on:

   - approximation of searched temperature distribution by set of two-dimensional known base functions;
   - determination (for the used number and spatial location of sensing elements) of reconstruction matrix;
   - measurements of all *m* resistance sensors;
   - determination (using reconstructed matrix for the known measured sensor's resistances) the values of coefficients of base function and
   - calculation of approximated temperature distribution.

3. The results of simulations of the reconstruction of temperature distribution from measurements of the resistance of copper and platinum sensors (for number 54 and 96) and using polynomial of 5 and 6 order provide to the following conclusions:

   - the $k = 6$ order of two-dimensional algebraic polynomial and it is sufficient as the methodical error of approximation of the temperature distribution;
   - the mean and standard deviation of average error is less than a few hundredths of a percent, therefore the mean temperature is reconstructed very precisely;
   - on the accuracy of temperature field reconstruction the most negatively affects the additive random component in measurement results of sensor's resistances. To obtain a standard deviation of the normalized to maximum temperature error less than 1% the standard deviation of additive random component in the sensor resistance measurements should not exceed 0.1–0.2% of the range;
   - the influence of additive systematic component of approximately twice smaller from the influence of random;
   - maximum errors are in the object corners, in which sensing elements are absent;
   - on the accuracy of the temperature distribution lowest affects the multiplicative component in the measurement of sensor's resistances;
   - accuracy increases with increasing of a number of sensing elements and order of approximated functions.

References

1. Bernhard, F. (ed.): Technische Temperaturmessung. Springer, Berlin (2004)
2. Parker, J., Renken, W.: Temperature metrology for CD control in DUV lithography. Semiconduct. Int. **20**(10), 111–117 (1997)
3. Lichtenwalner, D.J., Hydrick, A.E., Kingon, A.I.: Flexible thin film temperature and strain sensor array utilizing a novel sensing concept. Sens. Actuat. A **135**, 593–597 (2007)
4. Mocikat, H., Herwig, H.: Heat transfer measurements with surface mounted foil-sensors in an active mode: a comprehensive review and a new design. Sensors **9** (2009). (www.mdpi.com/journal/sensors)

5. Dickin, F.J., Zhao, X.J, Abdulach, M.Z., et al.: Tomographic imaging of industrial process equipment using electrical impedance sensors, Sensors VI: Technology, Systems and Applications, pp. 215–220. In: Grattan, K.T.V. (ed.) 1991 Proceedings of the 5th Conference on Sensors and Their Application. Adam Higler (1991)
6. Dorozhovets, M., Kowalczyk, A., Stadnyk, B.: Measurement of a temperature non-uniformity using resistance tomography method. In: Proceedings of 8th International Symposium on Temperature and Thermal Measurements in Industry and Science, Tempmeko (2001)
7. Korn, A., Korn, T.: Mathematical handbook for scientists and engineers: definitions, theorems and formulas for reference and review (1968)
8. Dennis, J.E., Shnabel, R.B.: Numerical Method for Unconstrained Optimization in Nonlinear Equations. Prentice Hall Inc., Englewood Clifs (1983)

Possibility of Sensors Application of 2714A Type Amorphous Alloys

Jacek Salach[1(✉)], Dorota Jackiewicz[2], and Magdalena Krześniak[1]

[1] Institute of Metrology and Biomedical Engineering,
Warsaw University of Technology,
Warsaw, Poland
j.salach@mchtr.pw.edu.pl
[2] Industrial Research Institute for Automation and Measurements PIAP,
Warsaw, Poland

Abstract. The paper presents the measurement results of the influence of the thermal annealing carried out in order to obtain better parameters of the amorphous cores. Amorphous alloys were examined in as quenched and annealed state. The results of investigation on influence of external stresses on magnetic characteristics of those cores have been done. First core was in as-quenched state, whereas others were annealed in 350 °C for one hour, annealed in 355 °C for one hour, and annealed in 360 °C for one hour. Presented results confirm the high magnetoelastic sensitivity of 2714A type amorphous alloy in as-quenched and annealed states in case of influence of compressive stresses from 0 MPa to 10 MPa. While in the case of tensile stress effect was negligible for stresses from 0 MPa to 3 MPa.

Keywords: Magnetoelastic · Amorphous materials · Force sensors

1 Introduction

Knowledge about the magnetoelastic properties of soft amorphous alloys is very important from both practical and theoretical points of view. Thermodynamically reverse effect connected with the influence of the external tensile stresses on the magnetic properties of this alloys, so called magnetoelastic Villari effect [2, 4, 5], has also significant, technical consequences.

From the other hand knowledge about the influence of stresses on the magnetic properties of magnetic materials, despite the many years of development, is still low. There are many studies on the effects of compressive stresses on both the ceramic magnetic materials [7], as well as amorphous magnetic materials [2, 6, 9]. Effect of the shear stresses which is a composite effect of the compressive and tensile stresses acting at 45 degrees to the direction of the magnetic circuit is also known from the literature [8]. However, there is a lack of knowledge about the effects of tensile stresses on the magnetic properties of magnetic materials, both crystalline and amorphous materials. The paper is designed to fill this gap, presenting the method to obtain a uniform state of stresses in the material as well as obtaining a closed magnetic circuit. The tested

R. Szewczyk and M. Kaliczyńska (eds.), *Recent Advances in Systems, Control and Information Technology*, Advances in Intelligent Systems and Computing 543, DOI 10.1007/978-3-319-48923-0_61

material was also subjected to thermal relaxation aimed at demonstrating the applicability of amorphous materials as the core of tension sensors.

2 Methodology of Investigation

For proper testing of the magnetoelastic properties of magnetic materials two conditions should be obtained: samples with closed magnetic circuit should be used and uniform distribution of stresses in the tested sample should be obtained. As a result in case of strip-shaped samples, due to the appearance of demagnetization energy in total free energy of the sample, results of investigation were connected with dimensions of the sample. On the other hand, when ring-shaped samples are loaded in the direction of diameter [12], the distribution of stresses in the sample is highly non-uniform. Investigation on the magnetoelastic characteristics requires a methodology enabling achievement of the uniform compressive and tensile stresses σ in the ring shaped core. Moreover, the investigated core has to be wounded by a magnetizing and sensing winding, to measure the magnetic hysteresis loop. A method enabling magnetoelastic tests of the ring-shaped amorphous alloy core was developed previously [10, 11]. The device for practical realization of magnetoelastic investigation is presented in Fig. 1. This method allows measurement of magnetoelastic characteristics of both bulk material rings and ribbon ring cores. It should be stressed, that device based on the idea presented in Fig. 1 enables also measurements of the influence of torque on magnetic characteristics of the ring shaped cores [13]. The principle of the novel method of applying the tensile stresses to the ring-shaped sample is presented in Fig. 2a. The tensile force F is applied to the ring-shaped sample perpendicularly to its bases. In such case uniform, tensile stress σ can be obtained on the whole length of the magnetic

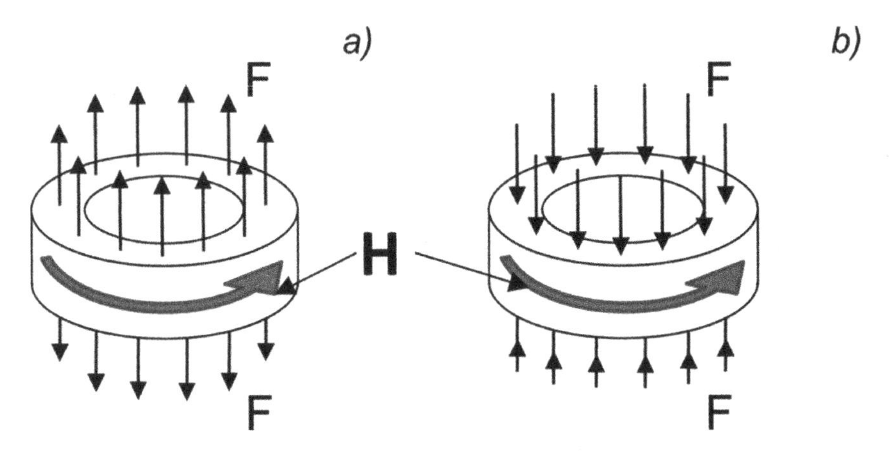

Fig. 1. Idea of applying stresses to amorphous ring core

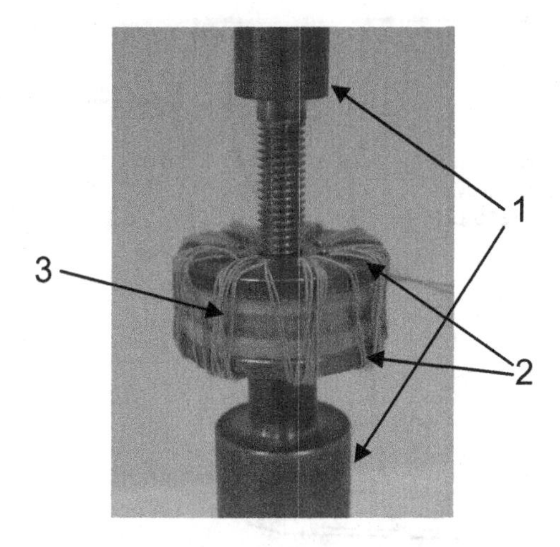

Fig. 2. Device for applying stresses to amorphous ring core

circuit of the sample. It should be indicated, that this methodology allows for testing of all commercially available, ring shaped cores made of soft ferrites, amorphous and nanocrystalline alloys. Device for practical realization of developed methodology is presented in Fig. 2. Tested core (3) is fixed to core backings (2). In backing (2) special holes were drilled, to enable core to be wound by magnetizing and sensing windings. Compressive force F generated by oil press is applied to the shaft (1). The magnetic properties of the samples, as well as the changes of these properties as a function of compressive and tensile stresses, were measured by digitally controlled measuring system HBPL-3.0. HBPL-3.0 system consist of digitally controlled current source and precise flux meter. Moreover this system works together with hydraulic press which enable application of axial stresses to the magnetic sample.

3 Results

Results of research show the effect of compressive and tensile stresses on the magnetic properties of amorphous cores. This is presented in Figs. 3, 4, 5, 6 and 7. The tested cores were heat treated in order to get rid of stresses generated in the amorphous ribbon during manufacture. One core to which no heat treatment was applied was taken for comparison. The cores were heated for one hour in argon protective atmosphere at 350 °C, 355 °C, 360 °C and one core was left in as-quenched state.

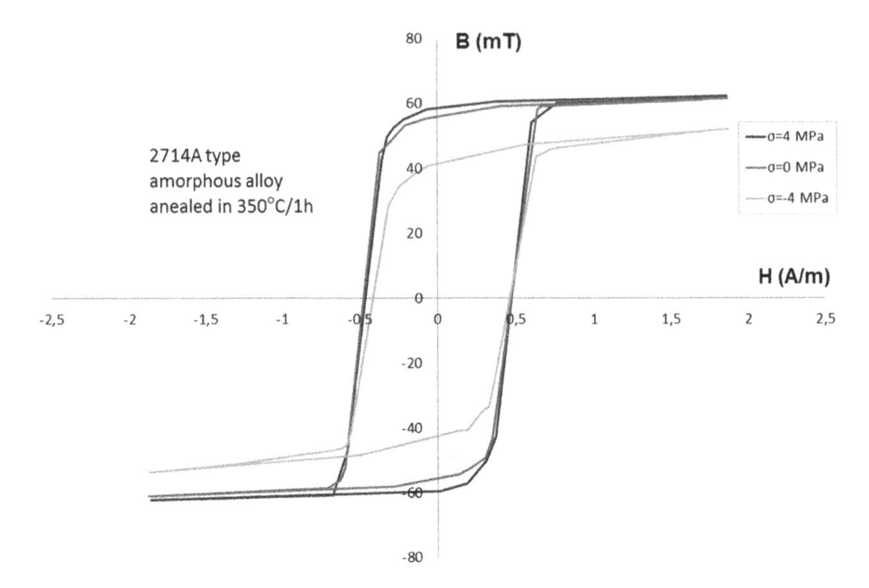

Fig. 3. Influence of compressive and tensile stresses σ on hysteresis loop $B(H)_\sigma$ ring-shaped core made of 2714a amorphous alloy after annealing in 350 °C. Values of amplitude of magnetizing field H_m is equal to coercive field H_c = 1,8 A/m

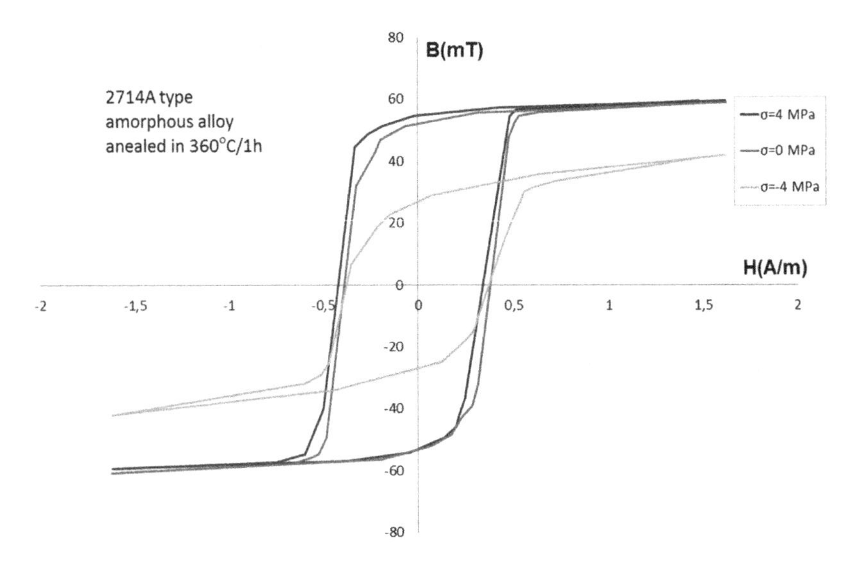

Fig. 4. Influence of compressive and tensile stresses σ on hysteresis loop $B(H)_\sigma$ ring-shaped core made of 2714a amorphous alloy after annealing in 360 °C. Values of amplitude of magnetizing field H_m is equal to coercive field H_c = 1,6 A/m

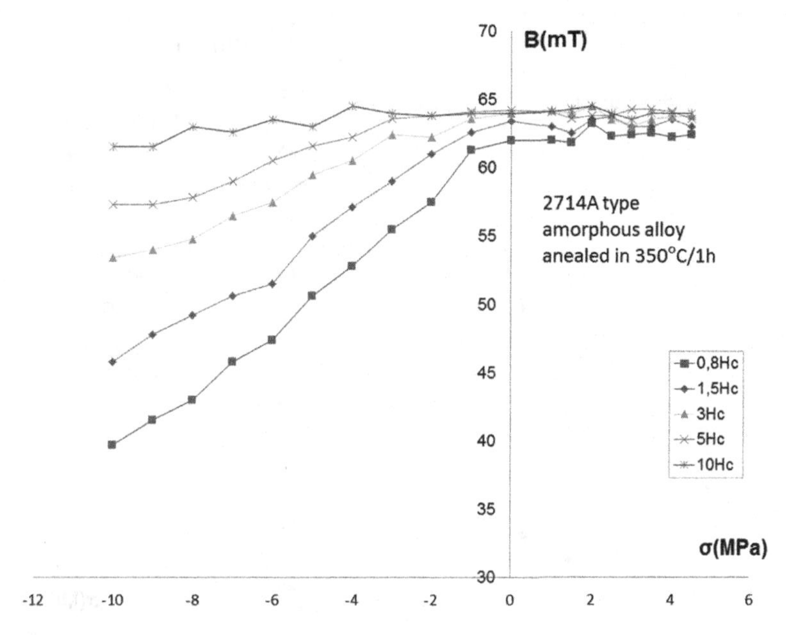

Fig. 5. Influence of both compressive and tensile stresses σ on maximal value of flux density B achieved for different values of amplitude of magnetizing field H_m in ring-shaped core made of 2714A amorphous alloy after thermal treatment in 350 °C. Values of amplitude of magnetizing field H_m is equal to multiplicities coercive field H_c = 1.8 A/m

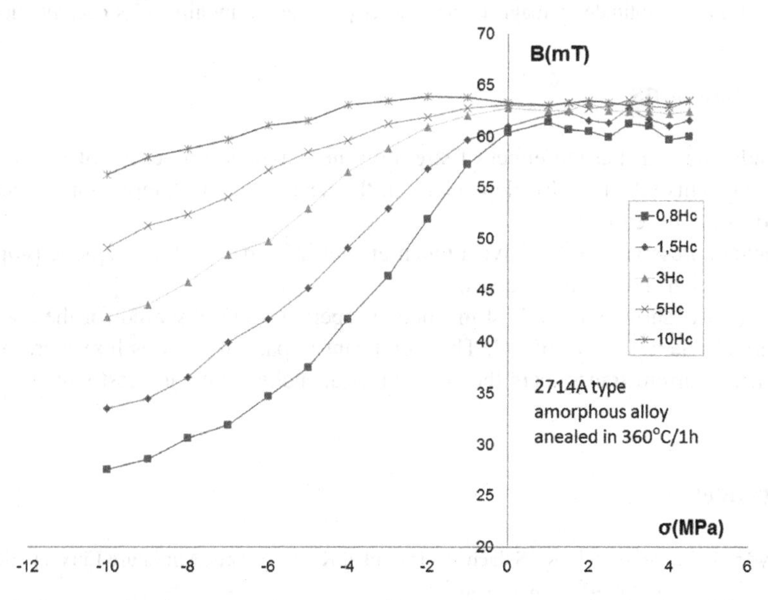

Fig. 6. Influence of both compressive and tensile stresses σ on maximal value of flux density B achieved for different values of amplitude of magnetizing field H_m in ring-shaped core made of 2714A amorphous alloy after thermal treatment in 360 °C. Values of amplitude of magnetizing field H_m is equal to multiplicities coercive field H_c = 1.6 A/m

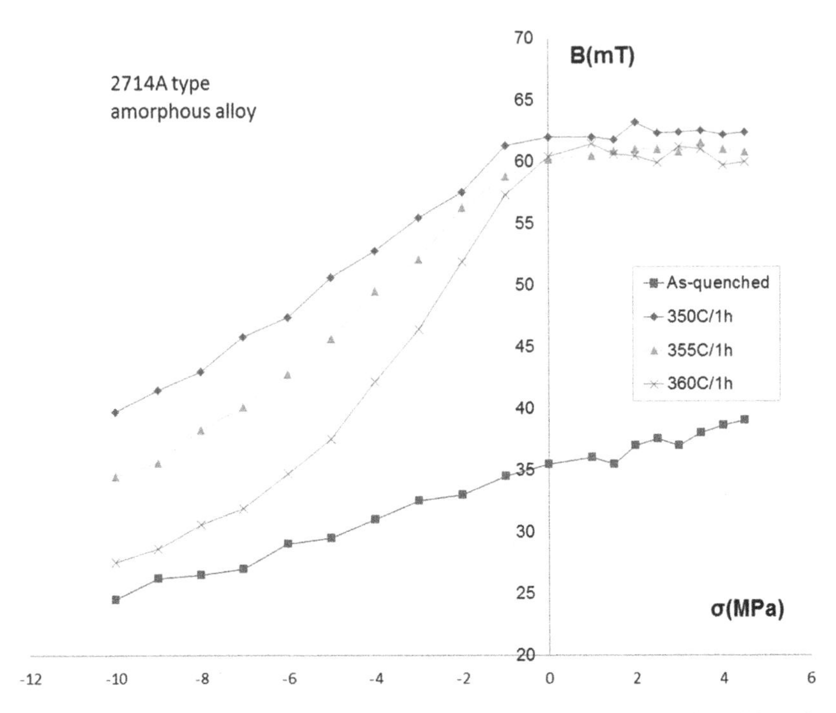

Fig. 7. Influence of both compressive and tensile stresses σ on maximal value of flux density B achieved for different temperature of relaxation ring-shaped core made of 2714A amorphous alloy. Values of amplitude of magnetizing field H_m is equal to multiplicities coercive field H_c.

4 Conclusions

The study showed that the effect of stress on the magnetic properties of ring cores is significant. This effect is also dependent on the heat treatment temperature which was applied to the tested core.

The compressive stresses have a much greater influence on the magnetic properties of the core than the tensile stresses.

The results show that the best magnetic properties were observed for the core after the thermal relaxation in 365 °C. The significant impact of stresses however, indicate that for the current transducers the cores in as-quenched state are best suited.

References

1. Frydrych, P., Szewczyk, R., Salach, J., Trzcinka, K.: Two-Axis, Miniature Fluxgate Sensors. IEEE Trans. Magn. **48**, 1485–1488 (2012)
2. Salach, J., Szewczyk, R., Bieńkowski, A., Frydrych, P.: Methodology of testing the magnetoelastic characteristics of ring-shaped cores under uniform compressive and tensile stresses. J. Electr. Eng. **61**, 93–95 (2010)

3. Szewczyk, R., Svec Sr., P., Svec, P., Salach, J., Jackiewicz, D., Bienkowski, A., Hosko, J., Kaminski, M., Winiarski, W.: Thermal annealing of soft magnetic materials and measurements of its magnetoelastic properties. Meas. Autom. Robot. **2**, 513–518 (2013)
4. Kolano-Burian, A., Varga, L.K., Kolano, R., Kulik, T., Szynowski, J.: High-frequency soft magnetic properties of Finemet modified with Co. J. Magn. Magn. Mater. **316**, e820–e822 (2007)
5. O'Handley, R.: Modern Magnetic Materials – Principles and Applications. John Wiley & Sons, New York (2000)
6. Shi, Y., Zang, Y.: Application of the nanocrystalline alloy in lubrication oil pressure-measuring for auto engines. Adv. Mat. Res. **661**, 7–10 (2013)
7. Bieńkowski, A., Rozniatowski, K., Szewczyk, R.: Effects of stress and its dependence on microstructure in Mn-Zn ferrite for power applications. J. Magn. Magn. Mater. **254–255**, 547–549 (2003)
8. Salach, J., Bieńkowski, A., Szewczyk, R.: The ring-shaped magnetoelastic torque sensors utilizing soft amorphous magnetic materials. **316**, e607–e609 (2007)
9. Liu, K.-H., Lu, Z.-C., Liu, T.-C., Li, D.-R.: Magnetoelastic anisotropy of FeSiB glass-coated amorphous microwires. Chin. Phys. Lett. **30**, 017501 (2013)
10. Salach, J., Bienkowski, A., Szewczyk, R.: Magnetoelastic, ring-shaped torque sensors with the uniform stress distribution. J. Autom. Mob. Robot. Intell. Syst. **1**(1), 66–68 (2007)
11. Szałatkiewicz, J., Szewczyk, R., Budny, E., Missala, T., Winiarski, W.: Identification of thermal response, of plasmatron plasma reactor. In: Szewczyk, R., Zieliński, C., Kaliczyńska, M. (eds.) Recent Advances in Automation, Robotics and Measuring Techniques. AISC, vol. 267, pp. 265–274. Springer, Heidelberg (2014). doi:10.1007/978-3-319-05353-0_26
12. Pressman, A.: Switching Mode Power Supply Design. McGraw-Hill, New York (1998)
13. Szewczyk, R.: Pramana. J. Phys. **67**, 1165 (2006)

New Type of the Test Stand for Surfaces and Lubricant Tribological Properties Test

Marcin Kamiński[1(✉)], Dawid Pogorzelski[2], Andrzej Juś[1], Tadeusz Missala[1], Roman Szewczyk[2], Wojciech Winiarski[1], Marcin Safinowski[1], and Marek Hamela[1]

[1] Industrial Research Institute for Automation and Measurements, Al. Jerozolimskie 202, 02-486 Warsaw, Poland
{mkaminski,ajus}@piap.pl

[2] Faculty of Mechatronics, Warsaw University of Technology, sw. A. Boboli 8, 02-525 Warsaw, Poland

Abstract. The paper presents a new type of test stand developed for surfaces and lubricants tribological properties testing. The utilized method, the test stand construction, principles of its operation and its capabilities are described in comparison to the standardized testing methods features and commercially available tribometers. In particular, the range of the input parameters (such as: relative speed between surfaces of the samples, surface pressure, test duration, types of material for tribological pairs, types of friction) and the results to be obtained during tests on the described stand (such as friction coefficient, temperature) are presented. Furthermore, metrological parameters of the test stand as well as its software responsible for the control of the measurement process and recording of its result are described. Based on tests carried out on the stand exemplary results and their interpretation are shown.

Keywords: Tribological properties · Tribology · Tribometer · Graphene lubricant · Graphene coatings

1 Introduction

The issue of friction has great economic importance. It causes energy loss and wearing of the surfaces of kinematic pairs (there are estimations that over 1/3 of all generated energy is wasted because of friction) so in most applications there is a need to decrease it (there are also products where high friction is essential, e.g. breaking pads). The most popular solutions to reduce friction and wear are additional coatings or modifications of the friction pairs surfaces as well as introducing between such surfaces substance with the desired properties – lubricant [1–3]. To determine changes of tribological properties (as a result of such actions) the tests should be carried out. There are a few typical methods of such tests and a lot of devices in the stock. Despite this, a new own-constructed test stand was developed to increase functionality in relation to commercially available devices.

R. Szewczyk and M. Kaliczyńska (eds.), *Recent Advances in Systems, Control and Information Technology*, Advances in Intelligent Systems and Computing 543, DOI 10.1007/978-3-319-48923-0_62

2 Base Method

Devices for measuring tribological properties are tribometers. The design of such instruments is directly related to the used test method. Typical constructions of tribometers (therefore typical methods) are e.g.:

- four ball,
- pin on disc,
- block on ring,
- fretting test machine,
- twin disc [4].

However, individually constructed test stands are often used, especially in the wearing studies [1, 4, 5] to reproduce real operating conditions of determined application.

Design of the developed test stand is based on pin on disc tribometers. In such tribometers stationary pin presses the sample (stationary, too) to the rotating track with a known force P. The track rotates with rotational speed n, and the pin contacts with the track in the radius R. The sample is spherical or cylindrical [4].

The basic similarities of the construction of the designed test stand and pin-on-disc tribometers (Fig. 1) is stationary placing of samples pressed to the rotating track. Accurate description of the test stand is presented in Sect. 3.

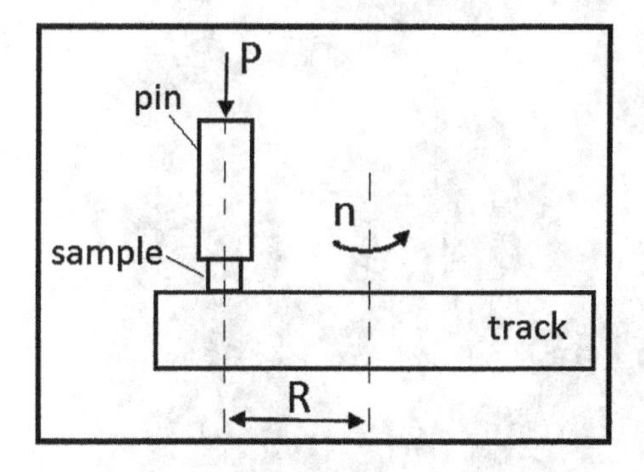

Fig. 1. Schematic of the principle of operation pin-on-disc tribometers [1, 4]

3 Test Stand Realization

Design of the mechanical part of the test stand is presented in Fig. 2. Friction pairs in this stand consist of a ring-shaped replaceable track (1) and three samples (2) to be tested. The shape of the samples (2) is determined by the shape of the holders (3) which retain samples in place. In presented version of the test stand samples (2) have to be rectangular with dimensions of 20 mm × 20 mm and thickness about 5 mm (other dimensions and shapes are possible after changing the holders (2)). Samples placed in the holders (2) before tests are presented in Fig. 3. During tests the track (1) rotates

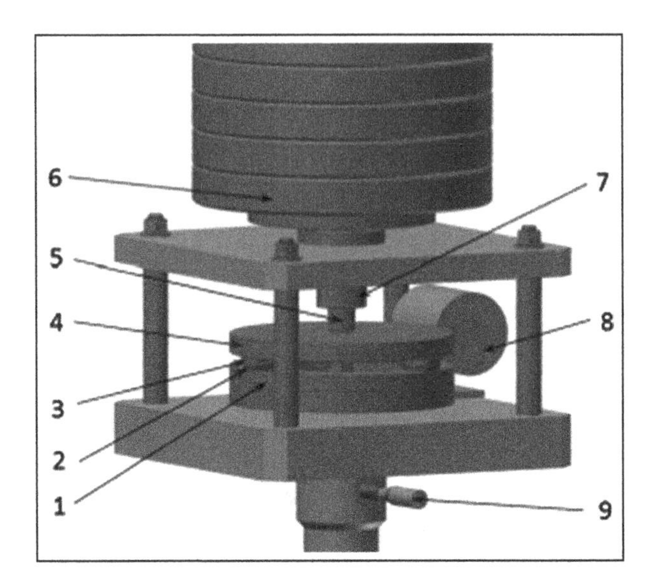

Fig. 2. Schematic of the test stand for tribological properties measurement

Fig. 3. Samples placed in holders before covering them by a disc

driven by the motor located below it with the rotational speed measured by the rotation sensor (9). Specimens (2) are gravitationally pressed to the track (1) by weight of the disc (4), the pin (5) and the weights (6). Rotation of such elements (2–6) is blocked by the force sensor (8). Force measured by this sensor (8) is reaction to friction forces between the track (1) and samples (2).

In Fig. 4 block diagram including all the subsystems of the test stand is presented. It consists of two measuring circuits. The first one is responsible for digital readout of the value from the force sensor. It includes strain gauge force sensor CL-14 with range

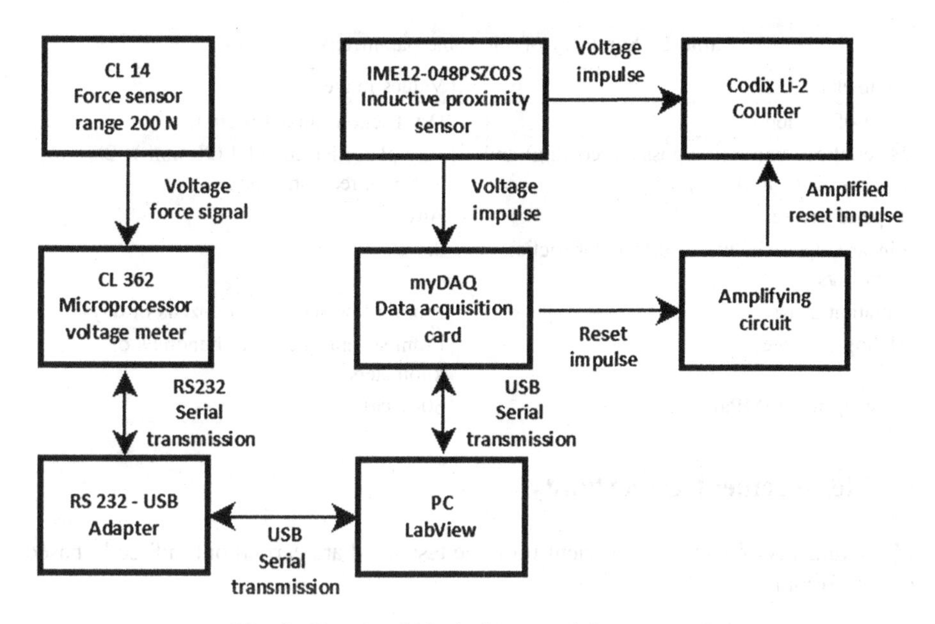

Fig. 4. Functional block diagram of the test stand

of 0–200 N, which changes of resistance are converted by microprocessor-strength meter CL362 and uploaded to a computer via serial port USB. Second measurement circuit is responsible for recording voltage impulses, representing the full rotations completed by the track. An inductive proximity sensor IME12-04BPSZC0S from SICK detects presence of the measuring point in the form of the tongue on the shaft. Voltage pulse is recorded via data acquisition card MyDaq from National Instruments, which allows to connect indicated values of force in time with rotations of the track in time. At the same time pulses are counted by electronical counter Codix Li-2 in order to verify correctness of their recording. Every time the new test starts a voltage impulse resetting electronical counter is generated.

The software of the test stand was developed in LabVIEW environment from National Instruments. Typical program in this software includes two panels: front panel and block diagram. Front panel is particularly important for users. In the developed software it is divided into the setting section and the display section. In the first one parameters of the tests are set, while in the second one actual results of the tests are presented.

The test stand allows conducting tests with a wide range of parameters. The basic parameters are: type of friction, type of grease, pressing force, linear velocity between surfaces of friction pairs, sliding distance, materials of friction pairs (their coatings) and process of their machining. Such parameters along with limits of their values (or their exemplary values) are summarized in Table 1.

Table 1. Summary of allowable parameters of the tests

Parameter	Values range
Type of friction	Dry friction, mixed friction
Material of sample (it's possible coating) and process of its machining	Any plate shaped solid (20 mm × 20 mm × 5 mm recommended)
Type of grease	Any
Linear velocity between surfaces of friction pairs [m/s]	0.2–2
Duration of test	From a few seconds to dozens of days
Sliding distance	From several meters to hundreds of kilometers
Pressing force [MPa]	50–1050

4 Measurement Uncertainty

Measurements of friction coefficient μ on the test stand are carried out indirectly based on the formula (1) [4].

$$\mu = \frac{FR_1 \sin \alpha}{R_2 N} \tag{1}$$

where:

F = (20–200) N ± 2 N – the force measured by the force sensor (ZEPWN CL-14),

R_1 = (123 ± 0, 2) mm – distance between the force sensor and the rotation axis of the test stand's track,

α = (90 ± 1) – angle between the direction of force measurement and the plane in which the measurement is carried out,

R_2 = (75 ± 0, 2) mm – distance between the samples and the rotation axis of the test stand's track,

N = (550 ± 5) N – average pressing force that can be applied

All components of the formula (1) are graphically presented in Fig. 5.

To determine uncertainty of measurements on the test stand, propagation of uncertainty law were used. As a result equation of combined standard uncertainty (2) of such measurements was obtained [6, 7].

$$u(\mu) = \sqrt{c_F^2 u^2(F) + c_{R1}^2 u^2(R_1) + c_{R2}^2 u^2(R_2) + c_N^2 u^2(N) + c_\alpha^2 u^2(\alpha)} \tag{2}$$

where [6, 7]:

$u(X) = \frac{x}{k_X}$ – standard uncertainty of particular elements of the formula (2)

$k_X = \sqrt{3}$ – coverage factor of particular elements of the formula (2) (value adopted assuming rectangular distribution of all components)

x – half-width of the variability of the particular components of the Eq. (2)

$c_X = \frac{\partial \mu}{\partial X}$ – sensitivity coefficient of particular elements of the formula (2):

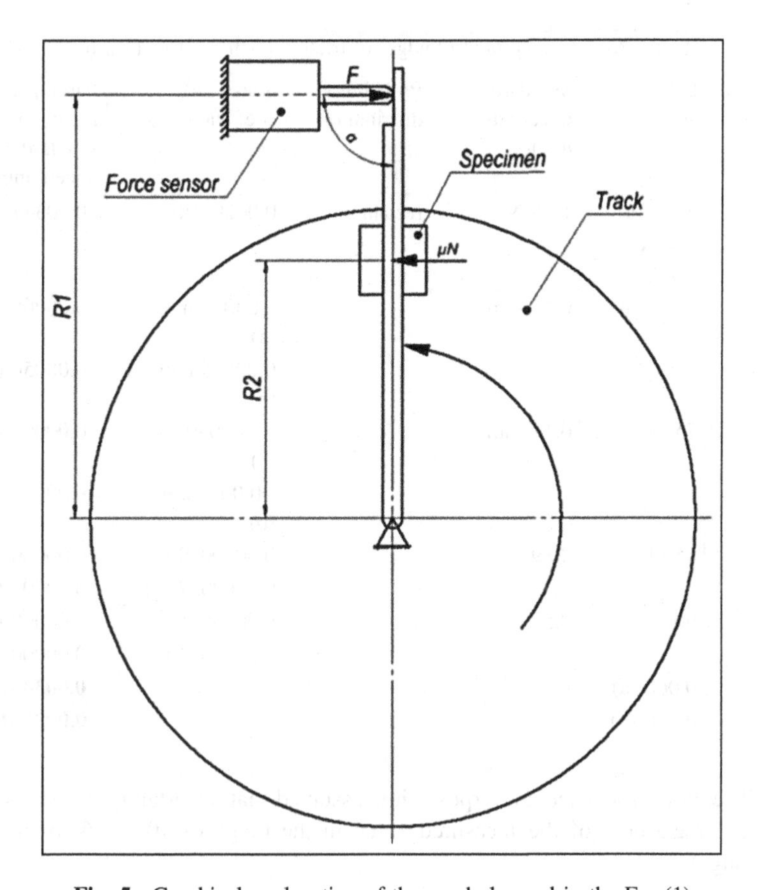

Fig. 5. Graphical explanation of the symbols used in the Eq. (1)

$$c_F = \frac{R_1 \sin\alpha}{R_2 N},\ c_{R1} = \frac{F \sin\alpha}{R_2 N},\ c_{R2} = -\frac{FR_1 \sin\alpha}{R_2^2 N},\ c_N = -\frac{FR_1 \sin\alpha}{R_2 N^2},\ c_\alpha = \frac{FR_1 \cos\alpha}{R_2 N}$$

The calculated values are summarized in Table 2 in the form of uncertainty budget of friction coefficient measuring using the tests stand. The values in the table are given for the two extreme values of the force sensor range (a – 20 N, b – 200 N) and for the average pressing force that can be applied.

Assuming a confidence level of 0.95 (coverage factor k = 2) the following values of the expanded uncertainty were obtained [6, 7]:

$$U(\mu) = k \cdot u(\mu) = 2 \cdot 0.0034 = 0.0068,\ \text{if}\ F = 20\,N$$
$$U(\mu) = k \cdot u(\mu) = 2 \cdot 0.0036 = 0.0072,\ \text{if}\ F = 200\,N$$

Since the main component of the combined uncertainty (derived from the force sensor) is constant over the measuring range, the variability of combined uncertainty

Table 2. The uncertainty budget of friction coefficient measuring

Quantity (X)	Estimate (x)	Standard uncertainty $u(X)$	Probability distribution	Sensitivity coefficient C_X	Participation in the combined standard uncertainty
F	20 N (a)	1.15 N	rectangular	0.00298/N	0.00344
	200 N (b)				
R_1	123 mm	0.12 mm		0.00048/mm (a)	0.00006 (a)
				0.00485/mm (b)	0.00056 (b)
R_2	75 mm	0.12 mm		−0.00080/mm (a)	−0.00009 (a)
				−0.00795/mm (b)	−0.00092 (b)
N	650 N	2.89 N		0.00000/N (a)	0.00000 (a)
				0.00000/N (b)	0.00000 (b)
α	90°	0.58°		0.00000 (a)	0.00000 (a)
				0.00000 (b)	0.00000 (b)
μ	0.060 (a)	–	–	–	**0.0034 (a)**
	0.596 (b)				**0.0036 (b)**

is low. Therefore, for practical purposes it is assumed that uncertainty of the test stand is ±0.007 (regardless of the measured value in the range of 10–100% of the force sensor range).

5 Exemplary Results of Measurements

Figures 6 and 7 present exemplary results of measurements conducted on the developed measuring system. These are partial results of the study on the graphene coatings influence on the tribological properties of surfaces and they are described in [8]. The mentioned figures present in particular characteristics of friction coefficient and temperature changes versus distance (such tests lasted for 24 h).

The exemplary results confirm clarity and valuability of results for further tribological analysis, especially in connection with results of samples wear research using other devices (for example profilometers or microscopes to determine surface roughness changes, scales for mass loss determination). Moreover there is possibility to analyze temperature distribution in friction pairs using thermograms of contact areas.

Results of some other tests conducted on the test stand are presented in the papers [8–10].

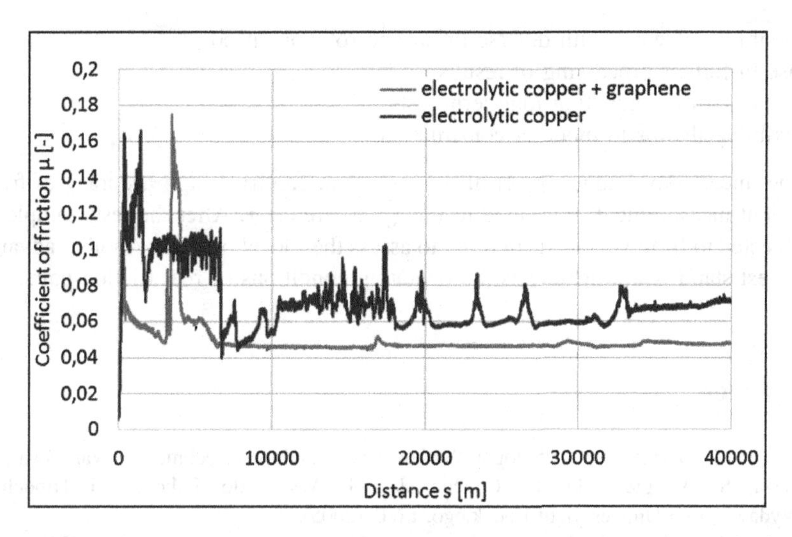

Fig. 6. Characteristics of friction coefficient in function of distance changes for graphenized and not graphenized samples [8]

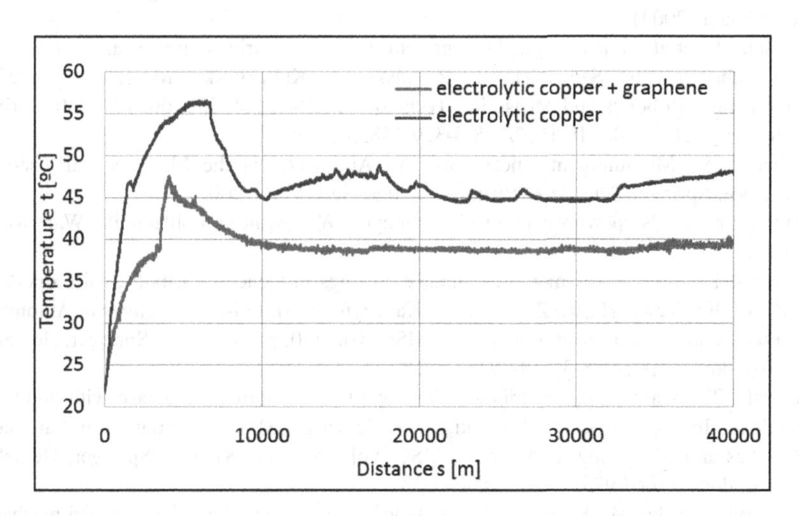

Fig. 7. Characteristics of temperature in function of distance changes for graphenized and not graphenized samples [8]

6 Summary

The test stand for tribological properties of solid materials tests was developed. Its main features are:

- possibility of testing dry and mixed friction,
- possibility of determining the coefficient of friction and temperature in the area of friction pairs in function of time or distance,

- possibility of work with diverse linear and rotation speeds,
- ease in further processing of results,
- possibility of conducting long-term tests,
- flexibility thanks to modular construction.

The uncertainty budget (p. 4) of the test stand confirms high accuracy of friction coefficient measurement. Presented results (p. 5) are clear. After the tests samples can be subjected to further analysis in order to assess their level of wear. Another advantage of the test stand is that it reflects the operating conditions of real friction pairs.

References

1. Hebda, M., Wachal, A.: Trybologia. Wydawnictwa Naukowo-Techniczne, Warszawa (1980)
2. Płaza, S., Margielewski, L., Celichowski, G.: Wstęp do Tribologii i Tribochemia. Wydawnictwo Uniwersytetu Łódzkiego, Łódź (2005)
3. Booser, E.R.: CRC Handbook of Lubrication. Theory and Design, vol. II. CRC Press LLC, Boca Raton (1983)
4. Stachowiak, G., Batchelor, A.W.: Experimental Methods in Tribology, vol. 44. Elsevier, Amsterdam (2004)
5. Missala, T., et al.: Study on graphene growth process on various bronzes and copper-plated steel substrates. In: Szewczyk, R., Zieliński, C., Kaliczyńska, M. (eds.) Progress in Automation, Robotics and Measuring Techniques. AISC, vol. 352, pp. 171–180. Springer, Heidelberg (2015). doi:10.1007/978-3-319-15835-8_19
6. Salicone, S.: Measurement Uncertainty: An Approach via the Mathematical Theory of Evidence. Springer Science & Business Media, New York (2007)
7. Arendarski, J.: Niepewność pomiarów, Oficyna Wydawnicza Politechniki Warszawskiej (2013)
8. Missala, T., et al.: Assessment of graphene coatings influence on tribological properties of surfaces. In: Szewczyk, R., Zieliński, C., Kaliczyńska, M. (eds.) Challenges in Automation, Robotics and Measurement Techniques. AISC, vol. 440, pp. 781–788. Springer, Heidelberg (2016). doi:10.1007/978-3-319-29357-8_68
9. Missala, T., et al.: Study on tribological properties of lubricating grease with additive of graphene. In: Szewczyk, R., Zieliński, C., Kaliczyńska, M. (eds.) Progress in Automation, Robotics and Measuring Techniques. AISC, vol. 352, pp. 181–187. Springer, Heidelberg (2015). doi:10.1007/978-3-319-15835-8_20
10. T, J., Wiśniewska, M., Kamiński, M.: Tribological behavior of graphene-coated mechanical elements. In: Jabłoński, R., Brezina, T. (eds.) Advanced Mechatronics Solutions. AISC, vol. 393, pp. 521–526. Springer, Heidelberg (2016). doi:10.1007/978-3-319-23923-1_76

Analysis of Automated Ferromagnetic Measurement System

Tomasz Charubin[1(✉)], Michał Urbański[2], and Michał Nowicki[2]

[1] Institute of Metrology and Biomedical Engineering, Warsaw University of Technology, Warsaw, Poland
t.charubin@mchtr.pw.edu.pl

[2] Industrial Research Institute for Automation and Measurement, Warsaw, Poland
{murbanski,mnowicki}@piap.pl

Abstract. Paper presents an automated measuring station for testing of ferromagnetic materials. The system allows for conducting a series of preconfigured measurements in various settings by producing a graphical output of hysteresis loops. Presented system utilises a high resolution U/I converter and a fluxmeter controlled by a PC class computer with National Instruments Data Acquisition Device, running under LabVIEW software. Measuring system is able to obtain full characteristics for magnetoelastic crystal magnetics, ceramic magnetics and amorphous magnetics. The aim of this paper is to present the setup of measuring station along with the measurement process itself. To ensure the quality of measurement conducted on developed station, the results for given reference material were compared to results obtained from standard hysteresis measurement system, as well as with material manufacturer datasheet. This paper aims to present simple setup for ferromagnetic measurement system for obtaining reliable and accurate results, in automated process.

Keywords: Magnetic measurement · Automated testing · Non-destructive testing · Hysteresis graph · Measurement setup

1 Introduction

In recent years, magnetic measurements became one of the most popular method for non-destructive testing [1, 2]. This leads to development of new and improved methods of utilizing magnetic parameters measurements in various cases. For example, ferromagnetic materials, which are present in almost all electronical devices as instances of magnetic cores in filters, transformers, chokes and other inductive elements [3], the investigation of magnetic parameters can be supported by hysteresis graph. Measurements provided by such device in form of hysteresis loops acquired during measurements can provides almost full characteristic for the investigated material. Based on basic concept of measuring hysteresis loops, a range of various test cases can be branched-out. Measurements can be acquired for example for stress- and temperature-driven tests, etc. [4]. Full specification of tested material can only be derived from full spectrum measurements, and in case of ferromagnetic samples, those consist of analysis

R. Szewczyk and M. Kaliczyńska (eds.), *Recent Advances in Systems, Control and Information Technology*, Advances in Intelligent Systems and Computing 543, DOI 10.1007/978-3-319-48923-0_63

in function of frequency and amplitude for magnetizing field, as well as different initial state of the material.

For this purpose, an automated testing stand for measuring magnetic materials was build. The purpose of this system is to allow setting up a series of autonomous tests to be conducted over a set period of time with various input parameters [5]. The aim of designed system was to allow for a user friendly usage and data representation, as well as basic analysis and recording.

The paper presents resulting configuration of such testing station, presents complete process of measurement and data acquired from a reference sample.

2 System Setup

The core system element is a PC class computer with National Instruments LabVIEW Software. LabVIEW (Laboratory Virtual Instrument Engineering Workbench) is a system-design platform and development environment for a visual programming language from National Instruments. The software present on the PC is designed to control National Instruments Data Acquisition Device, which is responsible for generating signal. Generated voltage signal from DAQ device is then processed by a high resolution voltage-to-current converter and will be thereafter referred to as magnetizing current. The magnetization current is then driven through the magnetization winding on sample, causing anomalies in voltage induced in the sample's sensing winding (Fig. 1).

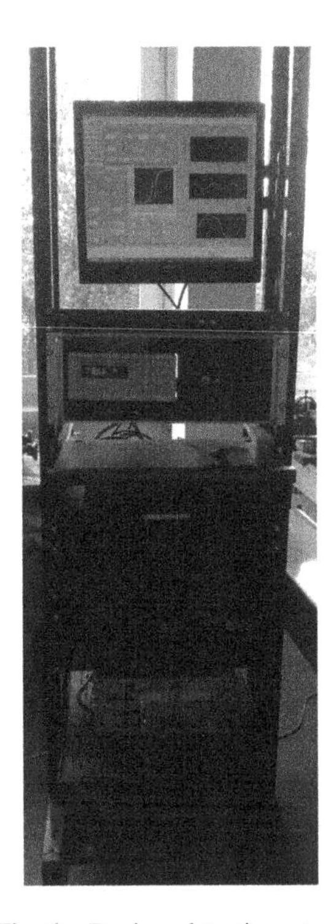

Fig. 1. Designed testing stand

The voltage signal present on the sensing winding is recorded by a Lakeshore 480 Fluxmeter. The voltage measurement from sensing winding is integrated into flux density value. The signal is integrated to a flux density value and back to NI Data Acquisition Device.

The software design in LabVIEW environment is responsible for all core calculations, acquisition of all generated and sensed data, and control over all sensing devices integrated into testing stand.

Another feature of the designed testing stand is temperature monitoring, developed with a type K thermocouple, supported natively by a National Instruments Data Acquisition Device (Fig. 2).

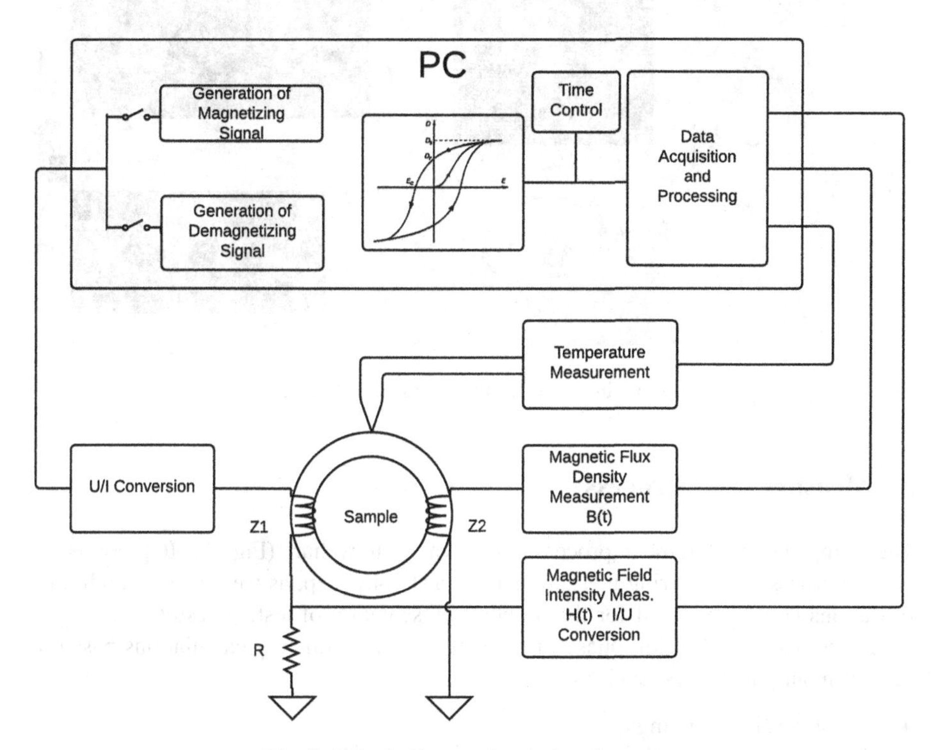

Fig. 2. Block diagram for designed system

The modules of the testing stand are summarized in Table 1.

Table 1. List of devices used in the testing stand

Module	Device name	Additional info
U/I converter	Kepco BOP 36-6M	Max. current: ± 6 A
Data acquisition device	NI PCI 6221	Max. voltage: ± 10 V
Fluxmeter	Lakeshore model 480	N/A

The interface for the designed software is presented in Fig. 3.

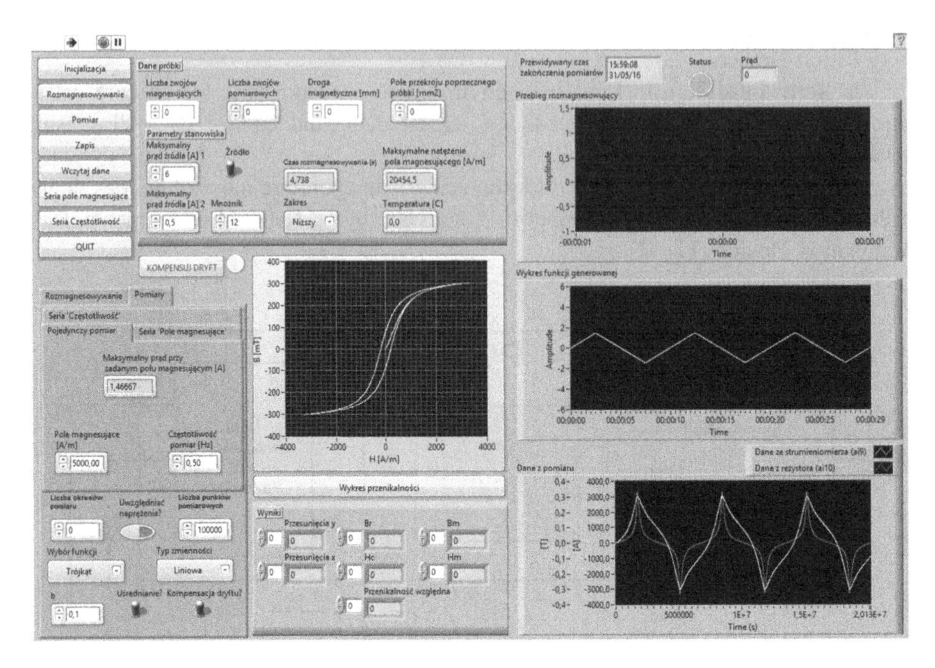

Fig. 3. Interface for the designed testing stand

3 Measurement Process

The complete measurement process is shown as flowchart (Fig. 4). It presents the general process of acquiring measurement data step-by-step, as well as most additional operations being processed for the proper representation of testing results.

In the first Initialization phase of the software is conducting calculations based on several input parameters, such as:

- No. of sensing windings
- No. of magnetizing windings
- Sample's magnetic length
- Sample's cross-sectional area

With accordance to user inputs and the setting of U/I conversion ratio, the maximal available value of magnetic field intensity is being calculated. In this step system initializes the communication over RS-232 interface with Lakeshore 480 fluxmeter, all input data is fed to the device to ensure proper voltage signal output for magnetic flux density values.

During the measurement there is an additional, optional step in which the sample can be demagnetized. This step is run in order to remove any residual magnetism,

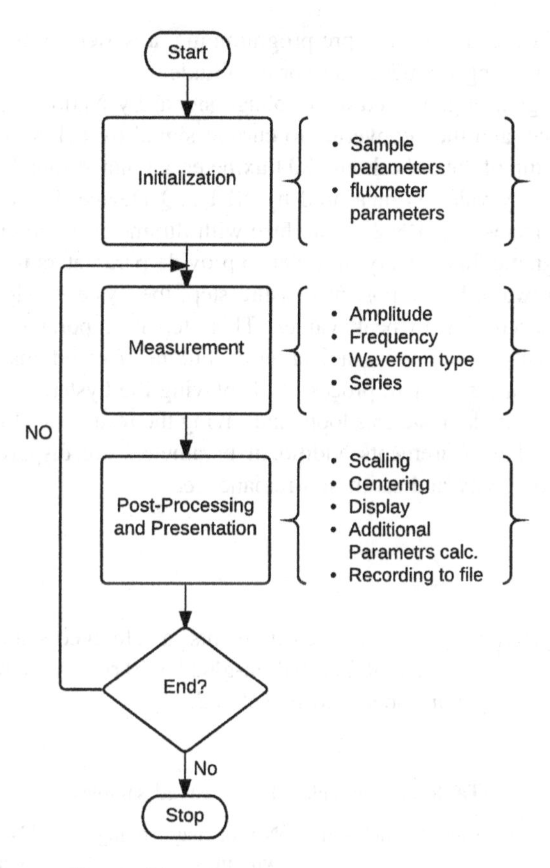

Fig. 4. Flowchart for measurement process

which is the magnetization left behind in a ferromagnetic material after an external magnetic field is removed. Demagnetization is done by generating proper current signal that follows the formula 1.

$$I = I_0 \cdot \sin(2\pi ft) \cdot e^{\left(-\frac{t}{\tau}\right)} \tag{1}$$

where: I – Current Value, I_0 – max. Current Value, f – frequency, t – time, τ – time constant.

The values presented in formula above can be changed between measurements. The measurement starts with initial computation of magnetizing waveform which is a representation of generated signal, with accordance to the inputs of desired magnetic field intensity, and frequency. The options available for magnetizing waveform are as follows:

- Sinusoidal waveform,
- Sawtooth waveform
- Custom waveform (generated by user with LabVIEW's waveform tools)

Designed software allows for pre-programming a series of fully customizable measurements of varying frequency and/or amplitude.

The driving signal is generated as a voltage signal by National Instruments Data Acquisition Device, and then converted to current signal by U/I converter.

After initial setup of the Lakeshore 480 fluxmeter, a voltage signal corresponding to magnetic flux density value is measured by NI DAQ Device. For the next step, the software communicates over RS-232 interface with fluxmeter, in order to capture peak values of the magnetic flux density, in order to provide parameters for post-processing module of the software. In the post-processing step, the hysteresis loop is being centered and scaled according to peak values. This step is important due to measured magnetic flux density value, as its indication on fluxmeter could change over time.

The last step of measurement process is displaying the hysteresis loop, calculating basic parameters of single hysteresis loop, and saving the results to a formatted text file. In case of series of measurements additional parameters are displayed, e.g. relative permeability or coercivity and saturation remanence.

4 Results

In order to verify the correctness of measurements, a reference sample made of carbonyl iron, which core consisted of 2 collodion-glued K40 rings was measured [6]. The parameters of this sample are presented in Table 2.

Table 2. Parameters for measured samples

Magnetic Length (cm)	Cross-sectional Area (cm^2)	No. of magnetizing windings	No. of sensing windings
10.21	0.974	178	80

The measurement results of reference sample are presented in Figs. 5 and 6. The x axis presents the value of magnetizing field (Amperes per Meter), while the y axis presents the magnetic flux density value (10^{-3} Teslas). The charts shown are exported directly from LabVIEW application.

Figure 5 represents low magnetizing field values 1000–5000 A/m. The loops have specific lens-like shape, which indicates the sample operates in Rayleigh area [7, 8]. Figure 6 represents high magnetizing field values 8000–10000 A/m. These hysteretic loops are deformed from the lens-like shape, which indicates the sample is no longer in Rayleigh area.

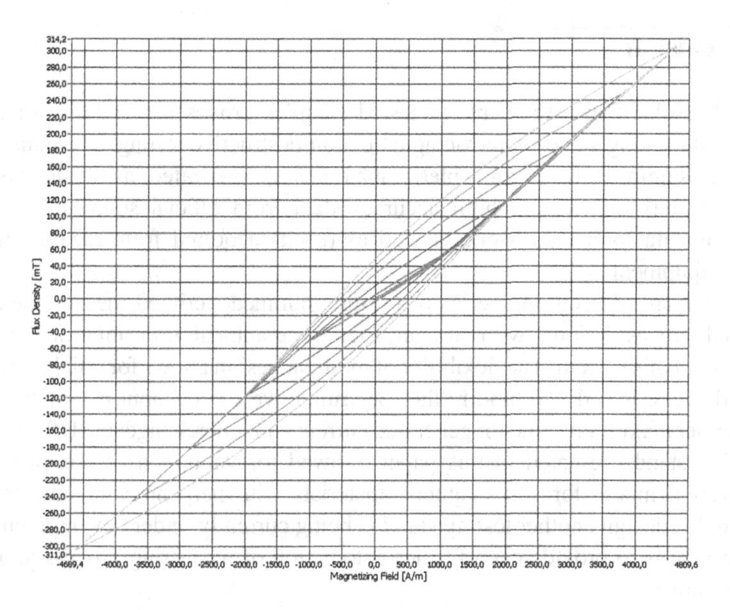

Fig. 5. Magnetizing fields H: (1000, 2000, 3000, 4000, 5000) A/m

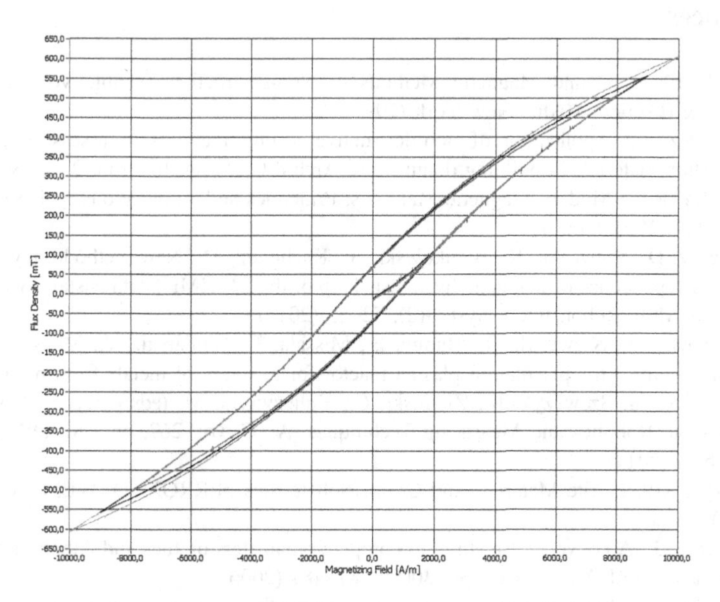

Fig. 6. Magnetizing fields H: (8000, 9000, 10000) A/m

5 Conclusion

Presented result of measurement for tested samples proves the viability of utilizing custom build testing station if ferromagnetic, non-destructive testing. The data acquired during measurement of ferromagnetic materials is presented as hysteresis loops. Obtained results match with high accuracy data from official sources (e.g. sample material manufacturer datasheet), as well as results acquired from measurements on different equipment.

Main feature of acquiring series of pre-programmed, scripted measurements in an automated process, along with human friendly graphical user interface, are main advantages of this system. The flexibility of modular design allows for quick exchange of utilized devices, in order to modify the spectrum coverage of input parameters (e.g. to ensure higher maximum value for generated current with interchangeable U/I converters).

Testing stand presented in this paper allowed for validation of the an hysteretic magnetization model for soft magnetic materials with perpendicular anisotropy [9]. Based on this design another testing stand is being currently under development, where the main goal is to provide even further automation of measuring process, along with higher accuracy.

References

1. Blitz, J.: Electrical and Magnetic Methods of Non-destructive Testing, vol. 3. Springer Science & Business Media, New York (1997)
2. Runkiewicz, L.: Application of non-destructive testing methods to assess properties of construction materials in building diagnostics. Archit. Civ. Eng. Environ. **2**, 79–86 (2009)
3. O'handley, R.C.: Modern Magnetic Materials: Principles and Applications, vol. 830622677. Wiley, New York (2000)
4. Jackiewicz, D., Szewczyk, R., Bieńkowski, A., Kachniarz, M.: New methodology of testing the stress dependence of magnetic hysteresis loop of the L17HMF heat resistant steel casting. J. Autom. Mob. Robot. Intell. Syst. **9**(2), 52–55 (2015)
5. Szałatkiewicz, J., Szewczyk, R., Budny, E., Missala, T., Winiarski, W.: Measurement and control system of the plasmatron plasma reactor for recovery of metals from printed circuit board waste. In: Szewczyk, R., Zieliński, C., Kaliczyńska, M. (eds.) Recent Advances in Automation, Robotics and Measuring Techniques. AISC, vol. 267, pp. 687–696. Springer, Heidelberg (2014)
6. Catalogue R-9, Ferrite Materials and Cores, Soft-ferrites FERROXYD, WEMA (1971). (in Polish)
7. Izydorczyk, J.: A convenient method to compute parameters of Jiles and Atherton model for ferrite materials. J. Magn. Soc. Jpn. **30**(5), 481–487 (2006)
8. Kachniarz, M.: Measurement system for investigating magnetic characteristics of soft magnetic materials in Rayleigh region. In: Jabłoński, R., Brezina, T. (eds.) Advanced Mechatronics Solutions. AISC, vol. 393, pp. 473–480. Springer, Heidelberg (2016). doi:10.1007/978-3-319-23923-1_69
9. Szewczyk, R.: Validation of the anhysteretic magnetization model for soft magnetic materials with perpendicular anisotropy. Materials **7**(7), 5109–5116 (2014)

Experimental Research of Improved Sensor of Atomic Force Microscope

Vytautas Bučinskas(✉), Andrius Dzedzickis, Ernestas Šutinys, Nikolaj Šešok, and Igor Iljin

Vilnius Gediminas Technical University, Vilnius, Lithuania
{vytautas.bucinskas, andrius.dzedzickis, ernestas.sutinys, nikolaj.sesok, igor.iljin}@vgtu.lt

Abstract. Atomic force microscope (AFM) – is device widely used in many scientific fields for nano-scale surface scanning. AFM also can be used to probe mechanical stiffness, electrical conductance, resistivity, magnetism and other properties. The main limitation of AFM implementation is relatively low scanning speed. This speed depends from dynamical characteristics of AFM sensor and from surface roughness of scanned sample. Our research is focused on increasing scanning speed of AFM microscope assuming AFM mechanical sensor as sensitive dynamic system. Our proposed method enables increase of scanning speed by modifying some features of mechanical sensor by adding non-linear force to the surface of cantilever of AFM sensor. Proposed method is modelled theoretically using Simulink features. This paper presents research of mechanical sensor of AFM. After performed research, obtained results are presented on graphical form. At the end of paper discussion presented and conclusions are drawn.

Keywords: Atomic force microscope · Cantilever · Modelling · Nonlinear stiffness

1 Introduction

Atomic force microscopy (AFM) invented by Binning et al. in 1986 as a type of scanning probe microscopy. Since that time, AFM used in various research disciplines as versatile imaging technique for visualizing sample surfaces at very high resolution. Ability of atomic force microscope to measure force in nanometric scale range has made this microscope attractive tool suitable for measurement of various properties like friction, adhesion, density, viscoelasticity and even intermolecular forces [1, 2].

AFM working principle based on measurement of interaction forces between scanning probe tip and sample surface. Mechanical part of AFM sensor consist of two main elements – probe and micro cantilever. Sensor probe is attached to the free end of micro cantilever; other end of cantilever is attached to the AFM base. During scanning process probe interacts with imperfections of sample surface and creates deflections of cantilever. The surface properties obtained by observing the cantilever deflections.

Initially, AFM designed to operate in contact mode. Contact mode is scanning mode, in which the probe's tip scans the sample surface by contacting it. From

R. Szewczyk and M. Kaliczyńska (eds.), *Recent Advances in Systems, Control and Information Technology*, Advances in Intelligent Systems and Computing 543, DOI 10.1007/978-3-319-48923-0_64

deflection of the micro cantilever due to the interaction force between probe's tip and sample's surface, topography of the sample being obtained [3]. Applications of this scanning method limited by hardness of sample surface. In order to avoid this limitation and increase application possibilities in 1987 was developed non-contact mode AFM. Non-contact AFM scanning mode is suitable to obtain image from the soft biological tissues where the contact mode may not be applicable due to the damage to the surface. Nonetheless, in non-contact AFM mode, the probe needs to be excited at near its resonant frequency while the distance between the tip and sample's surface must be kept constant [4]. In this mode, the surface properties are revealed by observing dynamic changes of the vibration parameters (amplitude, resonance frequency and phase angle) due to tip–sample interaction [3].

Tapping mode is another AFM mode, which has characteristics typical to contact, and non-contact modes. Tapping mode was developed in 1993 [5]. In this technique, the probe's tip can hover over the sample's surface while the micro cantilever is oscillating at amplitudes mainly higher than the amplitudes in the non-contact mode. However, the nanoscale interaction forces may cause the amplitude of the oscillation decreases when the probe's tip approaches the surface [3, 4, 6].

In addition to conventional AFM techniques, which are capable of achieving images of the sample's surface, most recent innovations have been carried out in revealing the structure formation under the surface [3].

Despite all improvements and numerous studies carried out after the AFM invention exists many challenges related with such AFM characteristics like resolution or scanning speed. Usually when AFM works in contact mode, its resolution is limited by finite size of probe tip, scanning speed is limited by dynamic characteristics of cantilever [7–9]. Oscillation frequency of mechanical sensor cantilever during scanning process depends from scanning speed and from roughness of sample surface. If cantilever oscillation frequency becomes too high and goes close to the resonant one, contact between probe and sample surface became unstable and scanning results are distorted. Resonant frequency of cantilever of the mechanical sensor is determined by its material properties and geometric parameters [10]. Despite numerous work done on analyzing relationships between geometry, stiffness and resonant frequencies of cantilever versatile solution of increasing atomic force microscope scanning speed is still requested.

In this paper will be presented important intermediate results of research aimed to deliver solution of increase AFM scanning speed in contact mode.

2 Object of Research

From engineering point of view, atomic force microscope is typical mechatronic system with feedback loop. During scanning process probe with initial force is pressed to the sample surface. Then piezoelectric actuators of the AFM platform are activated and scanning starts; deflection of cantilever follows topography of sample surface [9], thus allowing to record surface parameters. Deflection of cantilever is measured using non-contact optical interferometry method [1–3] or using piezo resistive or piezoelectric sensors installed on cantilever surface [6, 11–14]. Moreover, from cantilever deflection normal interaction force between probe and sample is determined. This force is kept in specified range using actuator, operating in the direction normal to surface of sample. This AFM platform actuator is controlled by feedback loop [15].

For easier control of complex AFM structure, al AFM components are classified into functional electrical and mechanical systems such as scanned sample positioning system, mechanical sensor with scanning tip and cantilever displacement measurement system. All systems within AFM are interdependent on each other [7, 8].

Our research is focused on increasing of speed of sample scanning process by affecting to mechanical part of AFM sensor. Increase of scanning speed could be achieved by altering dynamic characteristics of AFM cantilever. Idea of improvement of AFM sensor lays in method of increase of resonance frequency of mechanical sensor cantilever. In this case, cantilever displacement response for cinematic excitation generated by roughness of sample surface will not create distortions, this will allow scan samples at higher speed not losing contact between probe and sample. Cantilever resonance frequency can be increased by changing geometric parameters of cantilever like described in [10, 16] but in this case cantilever stiffness is changed irreversibly. Cantilever spring constant is very important parameter if cantilever is to stiff probe will damage sample surface and scanning results will became inaccurate.

In this paper proposed method of increasing AFM scanning speed is based on idea to add to AFM sensor system additional controllable stiffness element [7–9]. Implementation of this element not require essential changes in AFM construction, moreover this method is suitable for use with different types of cantilevers. Additional stiffness to the cantilever added as external nonlinear force. This force created using stream of compressed air [7–9]. Directed stream of gases creates nonlinear aerodynamics force which depend from size of cantilever surface area, gases pressure, diameter of air duct, gap size between cantilever surface and air duct. Main advantage of aerodynamic force that it is distance dependent [7] and on cantilever surface acts as complex spring, which prevents probe bounce from sample surface in critical moments, but not pressing probe to much then contact between probe and sample surface is stable. Size of additional force is controlled by gas stream. Whereas size of additional stiffness depends on parameters related with probe, cantilever, sample surface and other properties, for each configuration individual gas pressure identified and set during the initial scanning.

For practical implementation of this idea, standard AFM construction requires design change. Only specialized cantilever holder with installed micro air duct should replace the standard holder. Function of gas pressure control then installed to the AFM control system, or pressure control performed by additional independent system. Schematic illustration of proposed AFM sensor improvement presented in Fig. 1.

In order to realize proposed improvement was carried out modelling of dynamic system of mechanical sensor. There were analysed most common AFM cantilever modelling methods [17]. There was build mathematical model of mechanical part of AFM sensor, which proves effect of aerodynamic force to cantilever frequency response [8]. Characteristics of aerodynamic force acting on cantilever surface were investigated [7]. In addition, initial experimental research, which showed that cantilever displacement could be controlled using aerodynamic force, was performed [9].

Purpose of our research presented in this paper is to modify initial mathematical model for better detection of probe and surface contact loss.

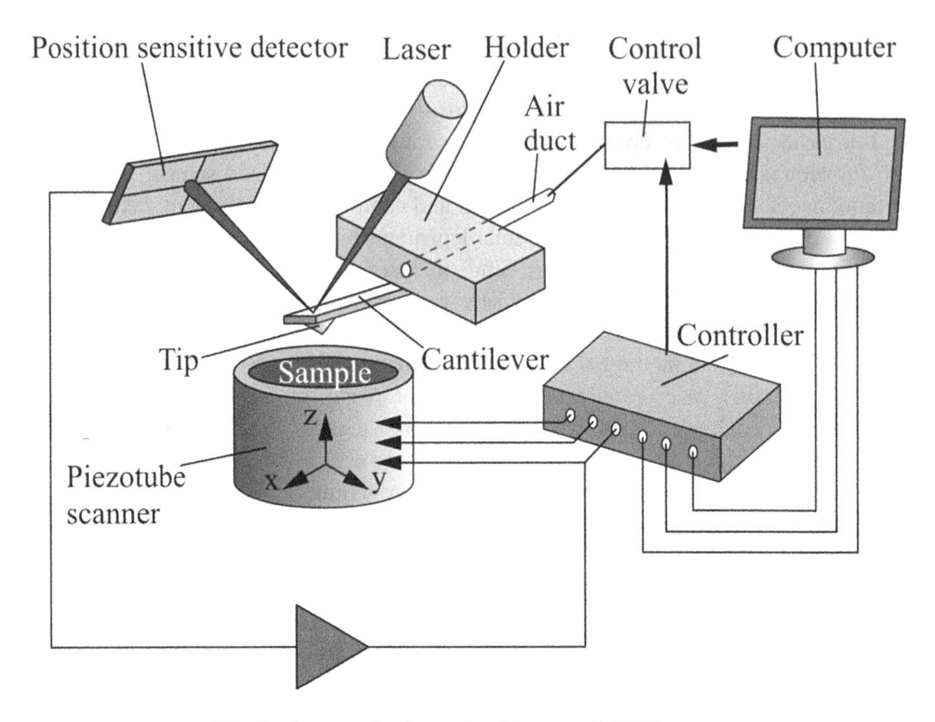

Fig. 1. Structural schematic of improved AFM sensor

3 Mathematical Modelling of Mechanical System of AFM Sensor

There are big variety of mathematical models of AFM cantilevers and probe – surface interactions, different methods of analysis are applied as presented in [17] also in [1–5, 10]. The focus in modelling of the AFM sensor – correct representation of probe – sample surface interaction. In the majority of papers, this interaction is approximated as one degree of freedom system with stiffness and damping elements of various linearity.

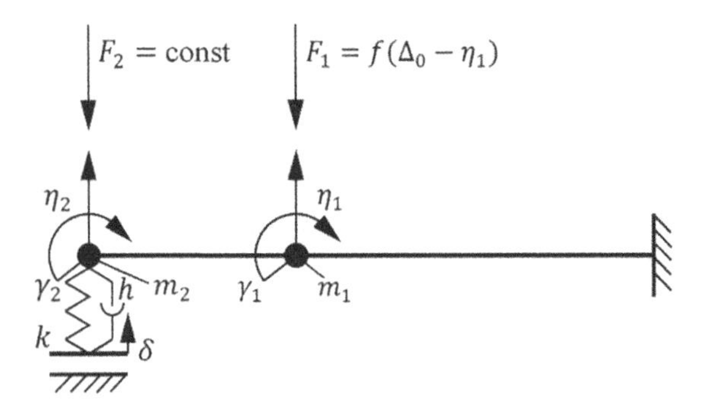

Fig. 2. Mathematical model of improved AFM sensor [7–9]

Proposed dynamic model of modified AFM sensor presented in Fig. 2. Mathematical model and corresponding equation are presented in details in [8, 9].

Analyzing results of model, presented in Fig. 2 we obtain that moment contact loss between probe and sample surface is not clearly detected. Problem is in approximation of probe – surface interaction forces. When these forces are modelled as spring in corresponding equations are created virtual relationship between probe and surface, or point on which cinematic excitation is applied. This relationship disturb simulation of moment then probe – surface contact is lost because probe and surface all the time are connected by spring. We improved our model by adding additional function for control stiffness k in moments then conditions of probe – sample contact loss met. Calculation scheme for determination probe – surface contact loss conditions presented in Fig. 3.

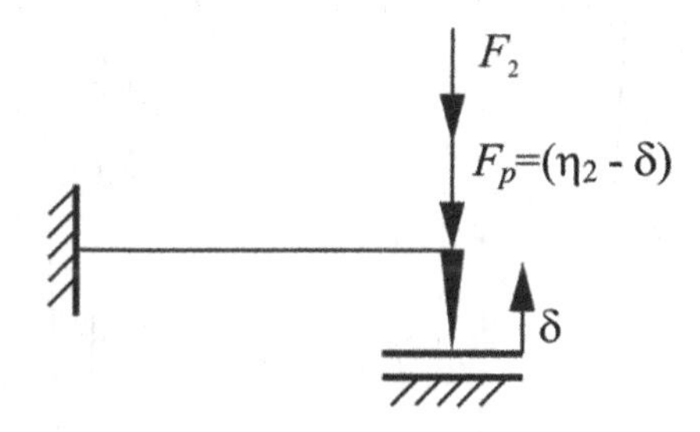

Fig. 3. Scheme for determination probe – surface contact loss conditions

In scheme, F_2 represents initial probe clamping force, F_p represents contact resistance force. Kinematic excitation, which simulates surface roughness represented by δ. From scheme presented in Fig. 2 we obtain main condition which corresponds to situation when probe detach sample surface:

$$m\ddot{\eta}_2 > F_2 + F_p + mg \tag{1}$$

First equation transformed to corresponding form.

$$\ddot{\eta}_2 > \frac{F_2}{m} + \frac{k(\eta_2 - \delta)}{m} + g \tag{2}$$

Second condition is determined from geometric properties it describes case then probe is in contact with surface.

$$\eta_2 > \delta \tag{3}$$

Second and third conditions transformed to corresponding Simulink diagram are added to stiffness control block. This block generate control signal, which allows to eliminate effect of contact stiffness k then cantilever moves up to peak of surface roughness. Then cantilever moves down our contact spring is pressed and have no effect to results. Improved mathematical model was tested using various excitation parameters.

4 Results of Research of AFM Sensor

Results of the Simulink model simulation with different excitation frequency were obtained. Also we calculate cantilever response for stochastic excitation. Obtained results presented in Figs. 4, 5 and 6 respectively.

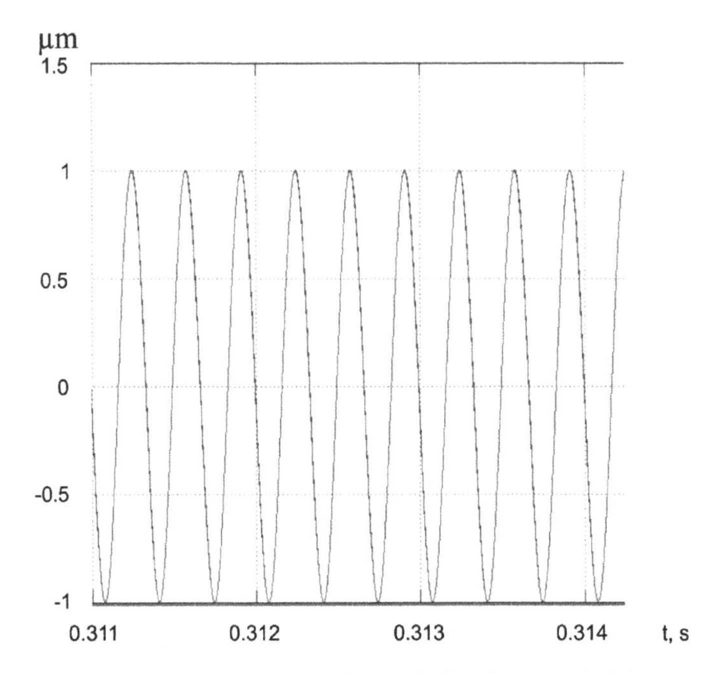

Fig. 4. Cantilever response when excitation frequency is 3 kHz

From graph presented in Fig. 4 it is seen that AFM probe trajectory ideally corresponds to the excitation signal which frequency is 3 kHz. This case simulates situation then AFM sensor works in normal conditions and contact between probe and sample surface is stable.

In case then excitation frequency is increased to 15 kHz probe starts bouncing and shape of excitation signal is not repeated properly. This case corresponds to situation then in real AFM scanning speed is increased and probe sample contact is interrupted. This case presented in Fig. 5.

Graph presented in Fig. 6 shows cantilever response for stochastic signal. Line represents 2 chaotic excitation signal, line 2 represents behavior of AFM probe. Stochastic signal is more suitable for simulation of sample surface topography. From graph it seen that situation is similar to presented in Fig. 5. In case then excitation frequency becoming to higher probe stops repeating topography of sample surface. Improved mathematical model is suitable to detect moment of tip – surface loss and this is desired result of our research.

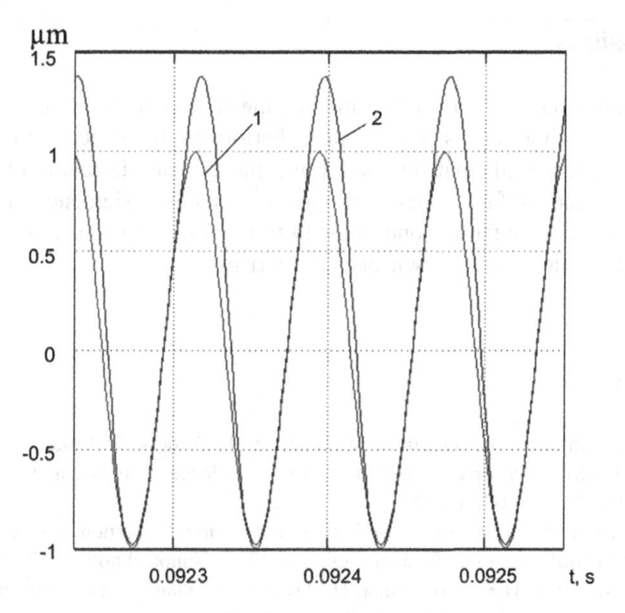

Fig. 5. Cantilever response when excitation frequency is 15 kHz: 1 – excitation signal; 2 – cantilever response

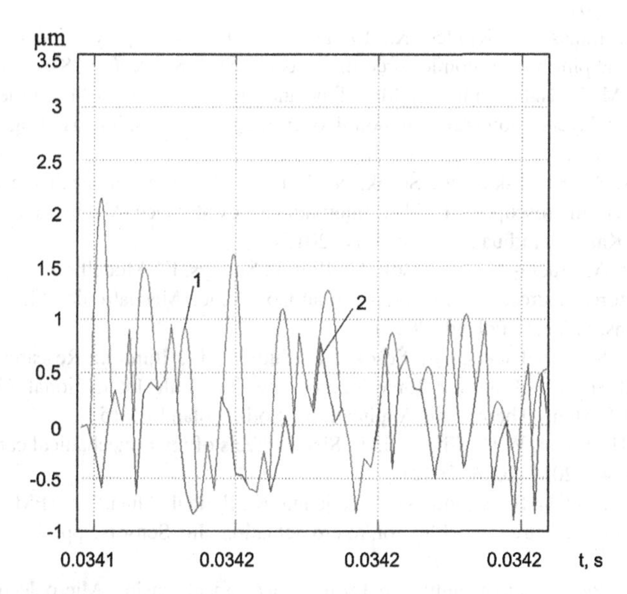

Fig. 6. Cantilever response for stochastic excitation frequency: 1 – cantilever response; 2 – excitation signal

5 Conclusions

After performing the theoretical research of the mechanical sensor of AFM, some useful and interesting results were found. Performed theoretical research gave us possibility improve mathematical model for more clear detection of contact loss between probe and surface. Improved model is suitable simulate AFM cantilever behavior at different excitation conditions. In future researches this model will be used for simulation of various AFM sensor characteristics.

References

1. Eichel, A., Schlecker, B., Ortamns, M., Fantner, J., Anders, J.: Modeling and design of high-speed FM-AFM driver electronics using cadence virtuoso and simulink. IFAC Pap. OnLine **48**, 671–672 (2015)
2. Joseph, A., Wiehn, S.: Sensitivity of flexural and torsional vibration modes of atomic force microscope cantilevers to surface stiffness variations. Nanotechnology **12**, 322–330 (2001)
3. Martin, M.J., Fathy, H.K., Houston, B.H.: Dynamic simulation of atomic force microscope cantilevers oscillating in liquid. J. Appl. Phys. (2008)
4. Sohrab, E., Nader, J.: A comprehensive modeling and vibration analysis of AFM microcantilevers subjected to nonlinear tip-sample interaction forces. Ultramicroscopy **117**, 31–45 (2012)
5. Zhong, Q., Inniss, D., Kjoller, K., Elings, V.B.: Fractured polymer/silica fiber surface studied by tapping mode atomic force microscopy. Surf. Sci. **290**, 688–692 (1993)
6. Korayem, M.H., Nahavandi, A.: Modeling and simulation of AFM cantilever with two piezoelectric layers submerged in liquid over rough surfaces. Precis. Eng. **42**, 261–275 (2015)
7. Dzedzickis, A., Bučinskas, V., Šešok, N., Iljin, I.: Modelling of mechanical structure, of atomic force microscope. In: 11th International Conference Mechatronic Systems and Materials, Kaunas, Lithuania, pp. 63–64 (2015)
8. Dzedzickis, A., Bučinskas, V., Šešok, N., Iljin, I., Šutinys, E.: Modelling of dynamic system of atomic force microscope. In: International Conference Mechatronics Ideas for Industrial Applications, Gdansk, Poland (2015)
9. Bučinskas, V., Dzedzickis, A., Šešok, N., Šutinys, E., Iljin, I.: Research of modified mechanical sensor of atomic force microscope. In: 13th International Conference on Dynamical Systems Theory and Applications, Lodz, Poland (2015)
10. Hocheng, H., Weng, W.H., Chang, J.H.: Shape effects of micromechanical cantilever sensor Measurem. **45**, 2081–2088 (2012)
11. Viannie, L.R., Joshi, S., Jayanth, G.R., Rajanna, K., Radhakrishna, V.: AFM cantilever with integrated piezoelectric thin film for micro-actuation, In: Sensors, pp. 1–4. IEEE, Taipei (2012)
12. Bausells, J.: Piezoresistive cantilevers for nanomechanical sensing. Microelectron. Eng. **145**, 9–20 (2015)
13. Su, Y., Brunnschweiler, A., Evans, A.G.R., Ensell, G.: Piezoresistive silicon V-AFM cantilevers for high-speed imaging. Sens. Actuators **76**, 139–144 (1999)
14. Rogers, B., Manning, L., Sulchek, T., Adams, J.D.: Improving tapping mode atomic force microscopy with piezoelectric cantilevers. Ultramicroscopy **100**, 267–276 (2004)

15. Humphris, A.D.L., Miles, M.J., Hobbsb, J.K.: A mechanical microscope: high-speed atomic force microscopy. Appl. Phys. Lett. **86** (2005)
16. Hosaka, S., Etoh, K., Kikukawa, A., Koyanagi, H.: Megahertz silicon atomic force microscopy AFM cantilever and high-speed readout in AFM-based re-cording. J. Vac. Sci. Technol. Microelectron. Nanometer Struct. **18**, 94–99 (2000)
17. Dzedzickis, A., Bučinskas, V.: Analysis of mechanical structure of atomic force microscope. Sci. Future Lith. **6**, 589–594 (2014)

Survey on River Water Level Measuring Technologies: Case Study for Flood Management Purposes of the C2-SENSE Project

Anna Bączyk[1(✉)], Jan Piwiński[2], Rafał Kłoda[2], and Mateusz Grygoruk[1]

[1] Department of Hydraulic Engineering, Warsaw University of Life Sciences – SGGW, ul. Nowoursynowska 159, 02-776 Warsaw, Poland
{a.baczyk,m.grygoruk}@levis.sggw.pl
[2] Industrial Research Institute for Automation and Measurements, Al. Jerozolimskie 202, 02-486 Warsaw, Poland
{jpiwinski,rkloda}@piap.pl

Abstract. In our survey we point out the most relevant features of water level monitoring methods applicable in Flood Warning Systems with respect to the framework of C2-SENSE project. We discuss the most common approaches to water level measurements regarding their spatial and temporal distribution and continuity of data recording and transfer. In detail we refer to three the widest applied methods of water level measurements, namely standard water gauge-based water level assessment, automatic pressure transducers and radar measurements of water table elevation. We refer to the most common problems of water level measurements related to the uncertainty of measurements and flaws of technical solutions applied. We revealed that the most critical features of water level monitoring frameworks are (i) assuring high quality of measurements by permanent quality check of automatic measurements, (ii) avoiding human-based measurements of water levels as being inappropriate in critical and dangerous flood events and (iii) assuring – if possible – continuity of data transfer from monitoring stations to final user along with avoiding public GSM networks being overloaded by private users during floods. In this survey the contribution of the C2-SENSE project to improvement of water level measurements for flood warning systems.

Keywords: Hydrometry · Water level · Hydrology · Flood · Measurements · River · Warning system

1 Introduction

Measurements of water levels constitute one of the simplest, yet still demanding test allowing quantitative analysis of hydrological dynamics of the water body (directly at the place of the measurement) and of the whole system of the catchment (where hydrological processes determine the availability and dynamics of water in the place of the measurement. The accuracy of water level measurements is essential for

R. Szewczyk and M. Kaliczyńska (eds.), *Recent Advances in Systems, Control and Information Technology*, Advances in Intelligent Systems and Computing 543, DOI 10.1007/978-3-319-48923-0_65

hydrological forecasts regardless their spatial and temporal resolutions and time horizons of the forecasts. Directly, accurate water level measurements provide useful information applicable in many areas of science, economy, technology and planning.

Reliable information about the spatial distribution of open surface water is critically important in various scientific disciplines, such as the assessment of present and future water resources, climate models, agriculture suitability, river dynamics, wetland inventory, watershed analysis, surface water survey and management, flood mapping and environment monitoring [1]. Particularly important in this regard are water level measurements done with the use of various devices and measurement procedures for the purposes of flood warning systems (FWS). Diagnosed and proved, ongoing climatic changes that in Central Europe are likely to be expressed by increasing frequencies of extreme meteorological and hydrological events [2, 3]. Along with exponentially increasing pressures on intensification of use of river valleys on one hand, and on the other – constraints for river and floodplain management originating from the legal requirements and environmental policy (e.g. EU's Water Framework Directive), the search for appropriately working flood warning systems (FWS) using accurate water level monitoring systems remains a crucial challenge for hydrologists, technicians and river managers worldwide [2, 4–6].

Traditional techniques and procedures of water level measurements are based on standard, optical and manual measurements done on a daily (in critical situations of floods – hourly) basis by trained staff recruited from the group of inhabitants of areas located adjacent to water bodies being monitored [7]. One of the main features of water level monitoring is to assure continuity of measurements over time and to keep appropriate accuracy of each measurement. Bearing in mind that monitoring systems capable to be applicable in FWS, which main goal is to prevent deadly injuries and mortality among the local communities, traditional water level monitoring methods fail to assure these features (e.g. evacuation of people from areas exposed to flooding – absence of the staff responsible for measurements). In order to fulfil the requirements of applicability of FWS namely: (i) detecting and forecasting hazards and developing flood hazard warning messages; (ii) assessing potential risks and integrating risk information into warning messages; (iii) disseminating timely, reliable and understandable warning messages to authorities and local communities and (iv) community-based emergency planning eliciting an effective response to warnings and reducing the impact of flooding on lives and livelihoods, water level monitoring networks have to be fully automated and interconnected [8].

The C2-SENSE project aims to develop a profile-based Interoperability Framework by integrating existing standards and semantically enriched web services to expose the functionalities of Command & Control (C2) Systems and Sensing Systems involved in the prevention and management of disasters and emergency situations. Main aim of our chapter is to present and discuss exemplary methodologies of water level measurements applicable in FWS to be applicable in the framework of C2-SENSE project.

2 Local Flood Warning Systems – Data Collection Assumptions and Procedures

2.1 General Information

FWS are implemented worldwide in order to prevent loss of life and reduce the economic and material impact of flood disasters. To be effective, FWS need to actively involve the communities at risk, facilitate public education and awareness of risks, effectively disseminate alerts, and warnings and ensure there is constant state of preparedness [9]. FWS are based on the real-time collected data on meteorological and hydrological features of the area.

After reaching defined threshold values, these information trigger the reaction of the system by transferring the output message (warning) to the target group of users (local inhabitants, public officials, emergency units) eventually (Fig. 1). Once the triggering thresholds decline below the specified level (most frequently the value that corresponds to the flood level of particular recurrence time, e.g. 10%), the system terminates sending the warning. Among FWS, similarly to the other data-collecting and measurement-related algorithms, manual (Fig. 1), semi-automated and automated systems regarding triggering threshold data collection can be distinguished. In each of the cases, water level monitoring points (water gauges, automatic sensors) are the first element of the reaction chain (Fig. 2).

Depending on the scale and spatial extent of the FWS that corresponds to hydrological dynamics of catchments, its sizes and numbers of tributaries which outflow induces flood wave propagation, FWS consist of up to some 20–50 points of water

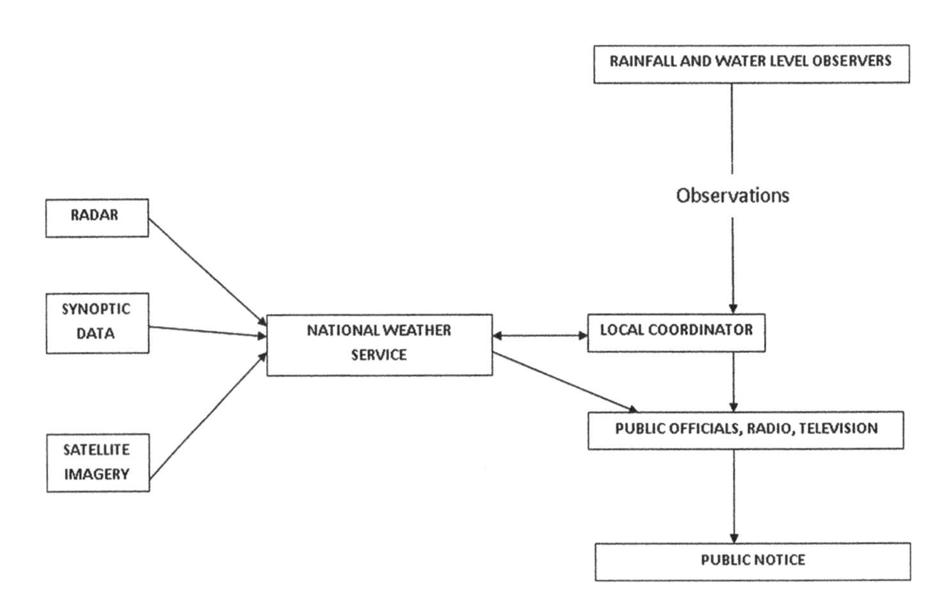

Fig. 1. Exemplary structure of a typical manual (human-made-observation-based) FWS. Source: US Department of Commerce (1985), modified and adjusted

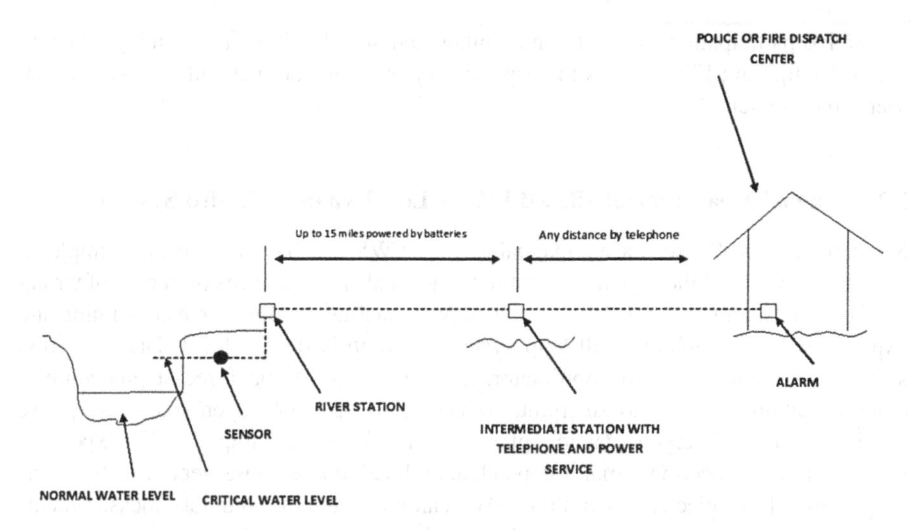

Fig. 2. Exemplary structure of a typical automated FWS. Source: US Department of Commerce (1985), modified and adjusted

level control randomly distributed within the river basin [13]. Development of FWS in areas of flash flood occurrence (floods resulting from storm rainfalls, in regions of high climatic variability and presence of episodic rivers; e.g. Mediterranean zone, semi-arid regions and big cities regardless their location) require automatic and quick transfer of triggering information as the time from flood initiation to the occurrence of final hazard is counted in minutes and hours. FWS developed in temperate regions where flood phenomena are not that dynamic does not require that immediate data transfer. However, in these areas it is the spatial distribution of the tributaries and flood wave propagation changes in space of the river basin. There are a few types of FWS being used around the world e.g. Local Flood Warning Subsystems (LFWS), The Flash Flood Guidance Subsystem (FFGS), The Flash Flood Monitoring Program or Global Flash Flood Guidance (NOAA, 2010). LFWS and FFGS remain the most commonly and well-tested subsystems applied in the most hazardous catchments. Early warning systems have four interlocking parts: risk knowledge, monitoring, response capability and warning communication. Each part must function efficiently for the system to be successful:

- Risk knowledge builds the baseline understanding about risks (hazards and vulnerabilities) and priorities at a given level.
- Monitoring is the logical follow-on activity to keep up-to-date on how those risks and vulnerabilities change through time.
- Response capability insists on each level being able to reduce risk once trends are spotted and announced – this may be through pre-season mitigation activities, evacuation or duck-and-cover reflexes, depending on the lead-time of a warning.
- Warning communication packages the monitoring information into actionable messages understood by those that need, and are prepared, to hear them [9, 11].

Active participation of local communities and stakeholders in sustaining, shaping and re-configuring FWS allows to keep these systems operational and tailor-fit to final user's requirements [12].

2.2 Manual Measurements-Based FWS – Low Dynamics Hydro-Systems

Standard, human-Senses-based manual (optic) FWS are inexpensive and simple to operate. However, if the structure is not fit to the local circumstances or in case of wrong level of staff training, such FWS can remain personnel-demanding, time consuming and expensive. Typical manual self-help system is comprised of a local data collection system, a community flood coordinator, a simple to use flood forecast procedure, a communication network to distribute warnings to appropriate emergency/response officials, and an emergency/action plan (Fig. 1). Standard, simple and inexpensive method of data collection remains dependent on local, trained observers of water levels on gauges [7]. In selected cases, these FWS can be related on voluntary measurements done by random people equipped with special (survey-supplied) measurement instruments such as standard water gauges [10]. Although manual gauge reports are less prone to errors, they are also less able to provide high temporal resolution. It is typically much easier to obtain rainfall rate information or short-duration accumulation from automated gauges. In such systems not only the measurement is done one a manual basis but also the other remaining elements of these systems are based on human actions (e.g. passing the information from the water (rain) gauge observer to final user of FWS). Hence, in case of human mistakes, the system can generate inappropriate information. Though such systems are nowadays implemented in big catchments of low dynamics of river discharge, where the reaction time plays a secondary role.

2.3 Automated FWS – High-Dynamics Hydro-Systems

Facing the substantial progress in technology and the decrease in the cost of computer and remote-controlled systems, the ongoing development of automated highly user-friendly and nearly faultless procedures FWS can be observed. Automated systems (water level measurements, information transfer, and generation of the final warning) require less attention during the most critical periods of time when ongoing and increasing hazard does not allow the extension of reaction time. However, similar to manual systems, automated and semi-automated FWS require complex maintenance and regular calibration, including geodetic measurements of the reference point's elevation. Flash flood alarm gauges consist of water level sensor (one or a few) connected to an alarm or light located at a community agency with 24-hour operation.

2.4 Data Transfer and Systematic Response of FWS

Water levels exceeding one or more preset levels trigger the alarm. The alarm is located upstream of the community. The lead-time information (warning signal) is provided by the system when water level exceed certain threshold value and sent along to the intermediate station collecting data from multiple monitoring locations (see Fig. 2). Communication between the monitoring location at the river and the intermediate station is assured by the wire connection or wireless technologies (most frequently via radio waves or GSM-GPRS network) [8, 10]. Certain flood alarm gauges used in FWS can also be used as part of the manual and automated river monitoring systems belonging to the regional or national water level control networks. From the intermediate station, the information is sent through to appropriate units that – after analysing the signal and regional situation – sends warning messages. Warnings are normally sent through the local radio broadcasts, local TV stations and GSM networks. The latter is disseminated to pre-registered users only.

3 Overview of Measurement Technologies and Procedures

3.1 Overview

An essential component of early warning systems are river level observations. Monitoring networks are normally composed of standard gauges where additional automatic devices are used such as pressure transducers and radar sensors. These three monitoring methodologies are the most frequently applied and though – as the most applicable in FWS – will be discussed. Water level changes within a certain period can also be analysed through long-time records of the software, in order to make the timely preparedness [14]. Monitoring, which includes instrument installation and data communication and analysis, is a crucial activity that must be performed throughout the life of the early warning systems [15].

3.2 Standard Water Level Observations

Standard water-gauge based observation provide an accuracy dependent on the scale of the measurement device (the gauge) and sensitivity of Senses of the observer. Normally, according to the standard procedures, water level observations are done with the 0.01 m accuracy, which is a reasonable minimum bearing in mind turbulences of the river flow, wave effects and resources required to do a measurement. Given accuracy allows reasonable uncertainty of water level assessment, including interpolation between particular water gauges. Measurements of water levels, depending on actual hydrological situation in the catchment, are done in 24-, but seldom also in 12- and 6-hour intervals. The latter is normally implemented, when hydrological situation is dynamic (floods). Due to the fact that these measurements can be done only by trained staff, these measurements can hardly be done in cases of floods assuring appropriate level of accuracy. Moreover, the limitations coming along with the human-made measurements related to relativity of one's Senses (e.g. during flood the floodplain is filled with water; though – accessibility to the installed water gauges is very limited) may affect the final result of the measurement or even make the results untrustworthy. Hence, gauge-based water level observations have evolved through the years and modern types of measurement devices were developed. Among them there are so-called precise water gauges, where water level measurements – although still related to certain metric scale – are done (and sometimes recorded). Examples of those devices, such as the “needle gauge”, “plate gauge” and limnigraphs (recording devices) proved their higher applicability in FWS than the standard optical-measurement-based water gauges [7]. Standard water gauges and their precise-measuring variations are being produced by multiple manufacturers and – in principle – used as the elements of national water level monitoring networks. Due to the high level of complexity of these semi-automatic monitoring devices and due to relatively low applicability of standard, human-Senses-based water level measurements on traditional water gauges, these devices are seldom used in modern FWS.

3.3 Automatic Pressure Transducers

The most commonly-used, automatic water level recorders are based on hydrostatic pressure transduction [16]. Automatic pressure transducers convert pressure-induced mechanical changes into an electric signal. Their construction allows full submersion in water, typically as the elements attached to the firmly established element (e.g. steel stick installed in the river bottom). The most important element in the design and construction of automatic pressure transducers is the membrane capable to react on changing hydrostatic pressure of water. Multiple producers of these devices keep similar outline of the products: relatively short (up to 20 cm) and thin (diameter up to 3 cm) metal sticks of some 200 g of weight. Such dimensions of the device allows installing them in piezometric pipes and tubes. Among the contemporary produced and used automatic pressure transducers there are the ones capable to record data and data download while being fully submersed under water. In such cases, the logger is hung at the wire becoming at the same time the connection cable. The socket of the connection cable is firmly attached to the top of the piezometric pipe, allowing connection of the device to data recording and data transfer station. In most cases, such design of the water level monitoring system requires the top of the piezometric pipe (including the socket) not to be submersed at any time. However, waterproof connections of the logger to the recording station allows high applicability of automatic pressure transducers in modern and comprehensive FWS systems.

Measuring water levels with automatic pressure transducers requires continuous compensation of recorded data. This is due to the fact that the measurement of pressure done with the use of pressure transducers submersed in water consists of the sum of hydrostatic pressure of water and actual atmospheric pressure of the air. Hence, to compensate the measurements, additional measurements of air pressure at the location of monitoring station is required, so the final result of water level measurements could be reduced by the actual pressure of the air.

Measurements done by automatic pressure transducers can be accurate as high as up to some 0.001 m which allows using these devices in a fine-scale hydrological measurements [17]. However, in purposes of FWS, accuracies reaching 1 cm are acceptable, which – in case of pressure transducers – allows programming them for the high range of water level changes. Record interval can be set very accurate, but normally 1-minute temporal resolution is fine enough to meet the requirements of modern FWS. Automatic pressure transducers used for water level monitoring purposes

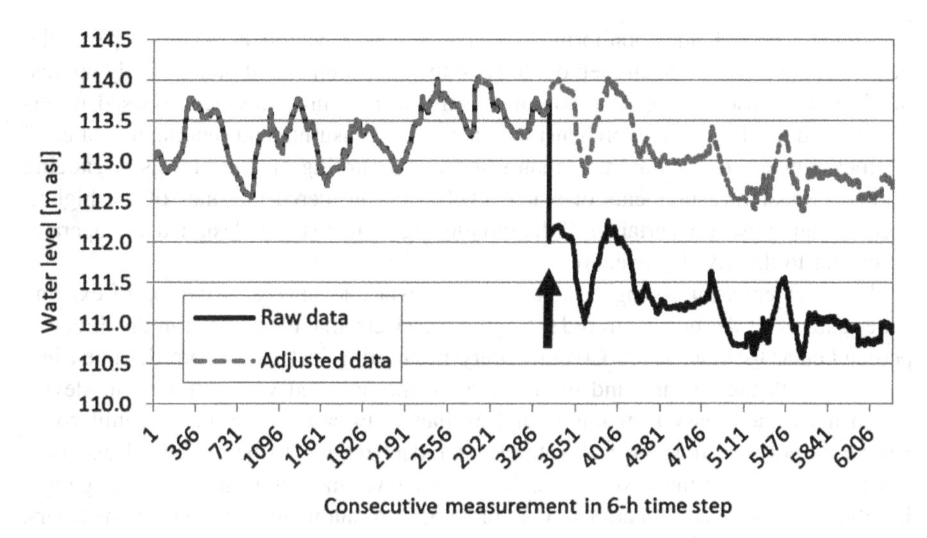

Fig. 3. Example of water level data series recorded with automatic pressure transducer that was moved towards the river surface in 3300 time step of the observation (arrow marker). Raw data has had to be adjusted with the linear function applied to represent actual water levels (Water levels of river Jegrznia, own study)

require calibration based on the standard measurements of water level on water gauge compared to the records of the device. Calibration allows accurate transfer of recorded values of compensated pressure into water level, expressed as the total hydraulic head given in meters. Once installed, transducers require permanent control of the level of their installation. If the level of installation was changed due to some environmental features (monitoring point was degraded by flood, e.g. solids flowing downstream the river can destroy monitoring location) or in result of human intervention (e.g. vandalism), water levels recorded with the device do not reflect the actual situation (Fig. 3). This is due to the fact that e.g. moving the device closer to the surface of water, hydrostatic pressure is being reduced. Hence – water levels measured in such situation by automatic pressure transducer can remain too low (as shown at the Fig. 3). FWS-related water level measurements with automatic pressure transducers should be equipped with the system preventing such situations or – at least – capable to provide information of such malfunctions to the operator of the FWS. Wrong interpretation of water levels may result in the lack of signal provided when water levels exceed the threshold value.

3.4 Radar Measurements

Radar measurements of water levels are based on the signal-reflect concept, where the signal (electromagnetic wave) after being emitted from certain point is being reflected by certain body and then received by the sensor. On the basis of difference of time between the emission and reception of the signal, as well as analysing the strength of the signal received, the monitoring device assesses elevation of water in particular measurement location. Such methodology of the measurements if being widely applied in FWS worldwide, being less complex than the pressure-transduction-based observations and much more comprehensive in a way of measuring and sending the value of the measurement to the final user than standard monitoring of water levels in gauging stations. Radar measurements of water levels are considered the ones of the highest accuracy and lowest uncertainty, if the monitoring system is well designed and operates according to the given procedure.

Radar antennas are being installed over the water level (e.g. at the top of existing water gauges, at the bridge) in order to send the electromagnetic signal in the direction perpendicular to water level. Contemporary monitoring instruments are designed in a way that both the antenna and the receiver of the reflected wave remain one device which make them easy to maintain and exchange, in case of any flaws. Similarly to automatic pressure transducers, water level radars require to be calibrated and controlled over time on the basis of standard water level measurements on water gauges installed at the same locations. Contradictory to automatic pressure transducers, installation of radar sensors requires general knowledge on water level magnitudes in particular monitoring point: radar antenna and signal receiver have to be installed in a way to be kept always above water level, even during the highest floods. It is because the electromagnetic wave of the radar has to be transferred in the air. Submersion of the antenna/receiver ends up with irrational measurements (dispersion of radar wave in liquids in different than in the air). It was reported that due to the wrong installation of radar antennas, water level measurements in critical moments of the flood can be wrong and – eventually – the function of the whole FWS can be negatively affected [18].

4 Discussion and Recommendations

Presented methodologies and assumptions of water level measurements to be applied in FWS are likely to be useful in different setups of data transfer, relatively to dynamics of floods and density of monitoring networks. Summarized features of the measurements methodologies discussed in the Sect. 3 can successfully be applied in FWS after considering their major advantages and flaws (Table 1). Standard manual measurements still remain the methodology which, although being not automated and uncertain as to the final results, remains the cheapest and applicable. Moreover, standard measurements are always needed even if the automated monitoring is implemented in order to verify the monitoring results. As a system itself, standard water level measurements in gauging stations are the most adaptive to changing hydrological conditions (immediate response to some failures and malfunctions is possible). Despite this facts, automated and calibrated monitoring devices are preferred over human-Senses-based measurements. In case of a flood hazard, human resources are limited and their capacity of implementing actions related to monitoring during floods can remain very low (e.g. due to evacuation). Both automated monitoring methodologies discussed require continuous quality check. In case of automatic pressure transducers, assuring constant level of their installation is a key to provide reliable and comprehensive datasets and trigger the FWS at the appropriate moment, when threshold water levels are exceeded.

In case of any flaws and malfunctions of devices, they should be easy to be exchanged. This also corresponds to the fact that in case of any damages in monitoring devices, the replacement of damaged devices should quick and not demanding for the long public procurement procedures. It is foreseen that long and complex public procurement procedures for purchase additional measurement devices by bodies responsible for supervising FWS (mostly regional authorities) can block the functionality of FWS over time. This aspect of FWS function, however, is to be addressed by appropriate management plans. As the transfer of data should be assured especially during critical periods of floods, public GSM channels should be avoided in transfer of data between the monitoring point and the station receiving and processing the threshold signal.

Public channels of GSM are overloaded during floods, which poses significant threat to the final use of the FWS triggering information coming from water level monitoring network.

5 C2-SENSE Project – Facing Challenges in Water Level Monitoring and FWS

C2-SENSE is a tool for Disaster and Emergency Management. All types of disasters and emergencies are meant for managing. The key words describing the tool are interoperability, extensibility, data analysis and presentation, and decision making support. C2-SENSE is primarily to be used by all actors that participate in an organized emergency management. The main goal with the tool is to gather all relevant information, analyze it, present it to the actor in form of maps and features and provide help for decision making.

When/before a disaster strikes (flood, fire, landslide, etc.), the C2-SENSE system is activated by appropriate users. The sensor networks are initialized, and the data are streamed.

The C2-SENSE users are government authorities, police department, hospitals, ambulances, fire department, emergency teams, NGOs, etc. Each is provided the tailor-made information particular for the user by C2-SENSE system.

The C2-SENSE system will provide only relevant information for the particular user. For example:

- Fire department will be provided with information showing where the fires are, or with the location of somebody who is in need of evacuating
- Hospitals will be given information on how many injured persons are heading their way
- Ambulances will provide information on injured people, and will be given information on which hospitals with available resources are closest

One of the goal of this project is to design the pilot application, to verify the actual usefulness in the field of pluviometric emergency management. For this reason it has been defined a pilot scenario that will be implemented in the territory of the Puglia region in Italy.

Table 1. Collection of positive and negative features of river level monitoring methods in the aspect of their applicability in FWS

Monitoring method	Positive aspects and benefits	Negative aspects and risk of flaws and failures
Standard water level observations	- high adaptation capacity, - immediate response to malfunctions of measurement devices, - continuous control of measurements, - low costs of the network establishment and maintenance,	- measurement device (water gauge) is easy to be destroyed by extreme phenomena (e.g. ice sheet in winter, solids transported, - measurement uncertainty (e.g. wrong read of the data by an observer) is considerably higher than in the case of automatic devices, - in critical periods of flood peaks local inhabitants are asked/forced to leave the area of flood, giving up at the same time their observations,
Automatic pressure transducers	- high accuracy of the measurements, - measurements done in full submersion – water levels do not affect function of the device, - ease of collecting long sets of data, - ease of connecting sets of data, - possibility of recording data with a high temporal resolutions (lower than 1 s),	- installation of devices recording air-pressure compensation data is required, - high costs of the monitoring devices, - high costs and effort-consuming maintenance of monitoring network, - standard measurements required, - additional measurements of water level (water gauges) required to calibrate the results, - additional infrastructure required (pipes, piezometers, reference points), - in case of some malfunctions in the system, wrong data can be recorded and sent to FWS,
Radar measurements	- moderate costs of devices and network, - possibility of using existing infrastructure to install monitoring devices	- too low installation of antenna and its submersion does not allow monitoring with sufficiently high metric accuracy, - additional measurements of water level (water gauges) required to calibrate the results, - continuous supervision of monitoring network required, - in case of some malfunctions in the system, wrong data are being recorded and sent to FWS,

The C2-SENSE will be built to enable the exchange of data using prominent wired and wireless communication mediums in the emergency domain as well as will provide interoperability with the several different wireless transmission technologies and enhance the connectivity between various networks and devices. The basic scenario during the pilot will be based on two main actions:

(1) Integration of different radio networks (GSM, CB-radio, ZigBee) and acquisition the data from these networks to one local server to send these data directly to C2-Sense system. In this case, if SQL server in the central site in Bari, will have the connection with C2-Sense system, it will be aware about our aforesaid data.
(2) Establish connection between local server with one monitoring station (over data logger) – for example 'FIUME FORTORE A PONTE FORTORE (CASALNUOVO MONTEROTARO)'. In this case SQL server in the central site in Bari will acquire the data from our above channels over licensed radio network of Puglia. In case of radio network collapse, this is an improvement of critical infrastructure, where there are alternative channels for communication and sensor data acquisition.

Sensors system in physical layer will be controlled with a microcontroller and will be based on prevalent communication standards, namely ZigBee protocol or IEEE 802.11 protocol on 2.4 GHz band. Aforesaid system can perform many environmental variables acquisition, like temperature, humidity, air pressure, water height and velocity, etc. from the integrated and calibrated sensors. With additional GPS receiver system can get the data with 2.5 m accuracy, elevation (Meters above sea level), velocity as well as equips such system with radio communication, enable to create a point to point sensor network on any area.

Dependently on density of sensors deployment, developer can use compact radio transmitters with 1 km coverage. Furthermore the usage of micro power technology, enables the power consumption and enable to build autonomic, battery-powered measurements system. Dependently on repetitions of performed measurements and the quantity of data transmission, it is possible to sustain the power for aforesaid measurement system even for 2 years' time. In the case of equipping such system with solar batteries, it is possible to build power self-sufficient system.

All gathered data will be processed by a central decision support system, based on web-based and GIS technology, and will be made available into the data warehouse.

Information collected will be integrated with geo-data and forecast data in order to identify and assess degrees of risk and critical events. The system will also allow the management and storage of reports and events related to all environmental risks (fire, hydrogeological, floods, weather, etc.) that could be related to the census data and stored in the data warehouse.

Overall, this project has the primary goal to design a cooperation platform that promotes spatial data interchange involving the end user, and provides support for the assessment and decision on the territory to determine how to deal with in operating mode and systematic different types of events from the knowledge of the location, by the compliance Geographic, from environmental data and historical information.

This goal represents an improvement from a basic approach focused on damage management, to a culture of prevention and prediction, spread out at various levels, based on the identification of risk conditions and the adoption of measures aimed at minimizing the impact of events.

References

1. Rokni, K., Musa, T.A., Hazini, S., Ahmad, A., Solaimani, K.: Investigating the application of pixel-level and product-level image fusion approaches for monitoring surface water changes. Nat. Hazards **78**, 219–230 (2015)
2. Becker, A., Grünewald, U.: Flood risk in Central Europe. Science **300**, 1099 (2003)
3. Grygoruk, M., Biereżnoj-Bazille, U., Mazgajski, M., Sienkiewicz, J.: Climate-induced challenges for wetlands: revealing the background for the adaptive ecosystem management in the Biebrza Valley, Poland. In: Rannow, S., Neubert, M. (eds.) Managing Protected Areas in Central and Eastern Europe Under Climate Change. AGCR, vol. 58, pp. 209–232. Springer, Heidelberg (2014). doi:10.1007/978-94-007-7960-0_14
4. Neal, J.C., Bates, P.D., Fewtrell, T.J., Hunter, N.M., Wilson, M.D., Horrit, M.S.: Distributed whole city water level measurements from the Carlisle 2005 urban flood event and comparison with hydraulic model simulations. J. Hydrol. **368**, 42–55 (2009)
5. Gaume, E., Livet, M., Desbordes, M., Villeneuvem, J.-P.: Hydrological analysis of the river Aude, France, flash flood on 12 and 13 November 1999. J. Hydrol. **286**, 135–154 (2004)
6. Jasper, K., Gurtz, J., Lang, H.: Advanced flood forecasting in Alpine watersheds by coupling meteorological observations and forecasts with a distributed hydrological model. J. Hydrol. **267**, 40–52 (2002)
7. Byczkowski, A.: Hydrology. SGGW Press, Warsaw (1999)
8. NOAA: Flash flood early warning system reference guide 2010. University corporation for Atmospheric Research, USA (2010). http://www.meted.ucar.edu/hazwarnsys/haz_fflood.php
9. Wiltshire, A.: Developing early warning systems: a checklist. In: Proceedings of the 3rd International Conference on Early Warning EWC III, Bonn (2006)
10. US Department of Commerce: local flood warning systems. Operations manual. National Weather Service (1985). http://www.nws.noaa.gov/wsom/manual/archives/NE408506.HTML. Accessed 25 May 2016
11. Community early warning systems: guiding principles. International Federation of Red Cross and Red Crescent. http://www.ifrc.org/PageFiles/103323/1227800-IFRC-CEWS-Guiding-Principles-EN.pdf. Accessed 16 May 2016
12. Abon, C., David, C.: Community-based monitoring for flood early warning system. An example in central Bicol River basin, Philippines. Disaster Prev. Manag. **21**, 85–96 (2012)
13. Local Flood Warning System for Klodzko Region. http://www.lsop.powiat.klodzko.pl. Accessed 14 May 2016
14. Li, L., Li, Z.: Water Level Intelligent Monitoring System Based on Mobile Terminal and PC Terminal. Appl. Mech. Mater. **475–476**, 198–203 (2014)
15. Intrieri, E., Gigli, G., Casagli, N., Nadim, F.: Landslide early warning system: toolbox and general concepts. Nat. Hazards Earth Syst. Sci. **13**, 85–90 (2013)
16. Freeman, L.A., Carpenter, M.C., Rosenberry, D.O., Rousseau, J.P., Unger, R., McLean, J.S.: Use of submersible pressure transducers in water-resources investigations. Techniques of Water-Resources Investigations 8-A3, U.S. Department of the Interior, Reston, VA (2004)

17. Grygoruk, M., Batelaan, O., Mirosław-Świątek, D., Szatyłowicz, J., Okruszko, T.: Evapotranspiration of bush encroachments on a temperate mire meadow - a nonlinear function of landscape composition and groundwater flow. Ecol. Eng. **73**, 598–609 (2014)
18. http://thelensnola.org/2012/10/10/outfall-canal-gauge-failures/. Accessed 12 May 2016

Error Ratio of a Measuring Instrument Under Calibration and the Reference Standard: Conditions and Possibilities of Decrease

Valerii A. Granovskii(✉) and Mikhail D. Kudryavtsev

Concern CSRI Elektropribor, JSC, 30 Malaya Posadskaya Street, Saint Petersburg 197046, Russia
vgranovsky@eprib.ru, q-007@yandex.ru

Abstract. The paper is aimed at the problem of the accuracy relation choice for a measuring instrument under calibration and the reference standard used. The study is fulfilled in the framework of the new approach based on analyze of the measuring instrument and reference standard actual errors instead of their norms. We show early that the problem can be solved if the error of the reference standard is known with sufficient accuracy. This study is focused on the initial data features which are necessary to get proper estimates of the measuring instrument and reference standard errors.

Keywords: Calibration · Measuring instrument · Reference standard · Accuracy relation · Error · Statistical estimate · Data stationarity · Data ergoditicy

1 Introduction

This work is the second publication consecrated to the problem of accuracy relation of a measuring instrument (MI) under calibration and the reference standard (RS). Efficient calibration of a MI requires selecting a RS with the appropriate accuracy. It is usually assumed that nominal error δ_{rs} of the RS should be much less than nominal error δ of calibrated MI. It is important to underline that the point is just nominal errors, i.e. error norms are under comparison. In conformity with this approach, a RS is interpreted as a device much more accurate than the sensor under test. The more accurate – so much the better. The general error model used for each compared devices is a random value δ equal the sum of unbiased random error ξ and systematic error a. Usually, the ratio $k = \xi / \xi_{rs} = 3$ is considered to be sufficient while a is supposed to be near zero. This ratio can be obtained using the specific negligible random error variance of the RS (about 5%) in comparison with the same characteristic of MI [1]. Other particular conditions lead to other ratios. The problem solution is pointed to expanding a range of acceptable accuracy relations towards their decreasing.

The new approach was proposed in [2] for the problem solution of a MI under calibration and the RS accuracy relation. Its idea is as follows. We consider a relation of actual accuracy characteristics of the sensor and a RS, for instance, sample estimates of error standard deviations. We have been able to show that the relation can be

R. Szewczyk and M. Kaliczyńska (eds.), *Recent Advances in Systems, Control and Information Technology*, Advances in Intelligent Systems and Computing 543, DOI 10.1007/978-3-319-48923-0_66

reduced right up to slightly more than unity. In accordance with this conception, a RS has to be interpreted as a device which has been studied in depth to get very good estimates of its accuracy characteristics. In other words, a RS is a device with perfectly known accuracy characteristics (not necessarily perfect characteristics).

2 Problem Formulation

Above-mentioned new approach is not free of charge. We need a big size of data to study a RS and to get good estimates of the sensor under test accuracy characteristics. And, surely, the conditions for statistical stability have to be ensured.

In any case, it is clear that no universal ratio k can exist – it depends on the structure of errors of both calibrated MI and the RS, and the ratio uncertainty is caused by the quality (including the amount) of data used for error estimation.

The paper deals with calibration by comparing the MI and the RS, i.e. multiple paired synchronous measurements of the same parameter (maybe a variable) by the MI and by the RS. According to the mentioned error model as a sum of unbiased random error and systematic one, the formal calibration (comparison) problem falls into several variant depending on a priori information on parameters of compared devices errors, as shown in Table 1 [2]. Notation $a \in [a_{low}, a_{up}]$ means that sample average changes from sample to sample but remains unchanged within a sample. Notation $a = \text{const}$ means that the sample average remains also unchanged from sample to sample.

Table 1. Variants of problem conditions

Variant	RS sample parameter $\{\xi_{rs,i}\}$	MI sample parameter $\{\xi_i\}$	Analyzed parameters		Note
			RS	MI	
A	$a_{rs} = 0$	$a = 0$	σ_{rs}	σ	–
B	$a_{rs} = 0$	$a = \text{const}$	σ_{rs}	a, σ	–
C	$a_{rs} = 0$	$a \in [a_{low}, a_{up}]$	σ_{rs}	$a(j)$, σ	$j = 1, \ldots, m$ – sample no.

Article [2] proposes the simple criterion to determine the MI fitness (serviceability). The MI is known *bad* if its error sample variance is:

$$\tilde{D}_x > \sigma^2_{max} + \sigma^2_{rs,\,max}, \tag{1}$$

where $\tilde{D}_x$ is sample variance of pairwise differences $x_i = \varphi_i - \varphi_{rs,\,i}$ between readings φ_i (MI) and $\varphi_{rs,\,i}$ (RS); σ_{max}, $\sigma_{rs,\,max}$ are the norms of standard deviation of MI and RS.

To confirm the fitness of the MI, this limit should be toughened:

$$\tilde{D}_x \leq \sigma^2_{max} + M^2 \cdot \sigma^2_{rs,\,max}, \tag{2}$$

where coefficient $M = \inf\{\sigma_{rs}/\sigma_{rs,\max}\} \leq 1$ determines the "dead zone" criterion:

$$M < \frac{\sqrt{\tilde{D}_x - \sigma_{\max}^2}}{\sigma_{rs,\max}} \leq 1. \tag{3}$$

This zone has a relative width 1–M and characterizes the acceptable risks of the manufacturer and the customer. Equation (3) can be transformed into

$$M < \sqrt{(\alpha^2 - 1)k^2 + \beta^2}, \tag{4}$$

where parameters $\alpha = \sigma/\sigma_{\max}$, $\beta = \sigma_{rs}/\sigma_{rs,\max}$ characterize the degree of closeness of MI and RS error to their normal values.

So, as it follows from (4), the "dead zone" defines the domain of the criterion applicability or the range of possible MI and RS accuracy ratios (the least accessible permissible value k):

$$k > \sqrt{\frac{\beta^2 - M^2}{1 - \alpha^2}}. \tag{5}$$

For random errors of the MI and the RS, which are characterized by estimates $\tilde{\sigma}$, $\tilde{\sigma}_{rs}$ of standard deviations σ, σ_{rs}, the following expression is derived:

$$\varepsilon = \varepsilon(k, \varepsilon_{rs}) = \frac{1}{k^2}\sqrt{\frac{1}{2n}(1 + k^2)^2 + \varepsilon_{rs}^2} + O(n^{-3/2}), \tag{6}$$

where $\varepsilon = \sqrt{D(\tilde{\sigma})}/\sigma$, $\varepsilon_{rs} = \sqrt{D(\tilde{\sigma}_{rs})}/\sigma_{rs}$ are the relative errors of estimates of analyzed parameters σ, σ_{rs}; n is the number of repeated pairs of MI and RS readings.

Formula (5) determines the minimum permissible ratio $k = \sigma/\sigma_{rs}$, which allows obtaining the desired standard deviation σ of calibrated MI with the required ε over sample of amount n, if RMSE of the known estimate of standard deviation σ_{rs} (in relative form) is ε_{rs}. For example, by sampling $n = 600$ and standard deviation σ_{rs}, which is known accurate to $\varepsilon_{rs} = 0.01$, we may estimate standard deviation σ with the required error $\varepsilon = 0.05$ with minimum ratio $k = 1.2$.

The paper is aimed at expanding the possibilities for MI calibration. For this purpose we have to consider some factors which determine the required ratio k and hence the minimum values that can be achieved. In other words, we focus on conditions for decreasing the error ratio of calibrated the MI and the RS.

In what follows the above mentioned factors are considered, which provide for possible decrease of k.

3 Problem Solution

3.1 Structure of MI and RS Errors

The paper classifies and analyzes the ratios between the levels of random and systematic errors, separately for the MI under calibration and the RS. Let a, a_{rs} designate systematic errors, and $\xi(t)$, $\xi_{rs}(t)$ – random errors with variances σ^2, σ_{rs}^2 accordingly. At that the error parameters are independent of t in view of the process $\{x_i\}$ stationarity being discussed below.

It is shown that an important (but not necessary) condition for possible k decrease is that the RS does not have any significant systematic error, against its random errors: $|a_{rs}| \ll \sigma_{rs}$. In this case the calibration problem is successfully solved for the following typical models of MI error:

(a) with zero systematic error: $a = 0$ (more exactly, $|a| \ll \sigma$),
(b) with constant unknown systematic error: $a = \text{const}$,
(c) with unknown systematic error within the known boundaries, but constant for every time sample: $a \in [a_{low}, a_{up}]$.

The problem solutions obtained for the typical models (a)–(c) of MI systematic error remain valid in conditions when random components of MI errors and the RS belong to a rather wide class of distributions. It is shown that the mentioned class contains the normal (N) and equally probable (U) distributions, and their various compositions $N*N$, $N*U$, $U*U$.

It is shown also that the confidence limits $a \pm K(P) \cdot \sigma$ of the MI under calibration total error, in the case $|a_{rs}| \ll \sigma_{rs}$, are

$$\begin{cases} \Delta_{conf,\,low} = \bar{x} - K\sqrt{\tilde{D}_x - \sigma_{rs}^2} \\ \Delta_{conf,\,up} = \bar{x} + K\sqrt{\tilde{D}_x - \sigma_{rs}^2} \end{cases} \tag{7}$$

where $K = K(P)$ is determined by fixed probability distributions of random errors $\xi(t)$ and $\xi_{rs}(t)$ (for example $K = 3$ in normal distribution case and for $P = 0.997$).

Taking into account the well-known extremum behavior of two above-mentioned typical distributions [1, 2], we can make a conclusion about feasibility of almost any distribution with finite variance, with model approximation of random components of MI and RS errors.

3.2 Data Quality

Since setting of rational k is based on statistical knowledge of MI and RS errors, the quality of initial data (entire assembly) and representativity of sampling are the determining factors for solving the problem. To be able to use the data, the data stream should be stationary. Criteria for testing the stationarity hypothesis are known but they require the samples for fixed time points. However in real experiments we usually have only one time sample of the process. It can be used as an adequate carrier of the process statistical properties only if the process is ergodic. To check this condition, first it should be established that the process is stationary. Thus we come to the endless circle. The only way out is to check if the time sample data are homogeneous [3]. For checking homogeneity, many proper tests can be used [4].

Here we refer to ergodicity with respect to the average and the variance. Slutskiy condition can serve as a good ergodicity criterion for the average [5]:

$$\lim_{T\to\infty}\frac{1}{T}\int_0^T \psi(\tau)d\tau = 0, \tag{8}$$

where $\psi(\tau)$ is the correlation function of stationary random process $\{x_i\}$.

The ergodicity criterion for the variance is a generalized expression (8), which takes the following form [6]:

$$\lim_{T\to\infty}\frac{1}{T}\int_0^T \psi^2(\tau)d\tau = 0, \tag{9}$$

Since correlation function $\psi(\tau)$ is estimated by discrete time sequence $\{x_i\}$, it is advisable to use the analogs of formulas (8), (9) for discrete stationary random processes:

$$\lim_{N\to\infty}\frac{1}{N}\sum_0^N \psi[m] = 0, \tag{8a}$$

$$\lim_{N\to\infty}\frac{1}{N}\sum_0^N \psi^2[m] = 0. \tag{9a}$$

If we mean both stationarity and ergodicity, then we may use the sequence of processing the time series for the data of compared devices, as it is recommended in [7]. Check of stationarity can be performed by calculation of the average and the mean-square value for each of the sections into which the data series is divided. Comparing the corresponding averages, we determine the absence of the trend line and the presence of deviations within the reasonable limits. With the positive result of comparison, the data may be considered stationary. Certainly, the check of stationarity can be enhanced by using different statistical criteria, primarily distribution-free (for unknown probability distributions). As it is correctly emphasized in [7], the absence of harmonic components in the analyzed data is worthy of special check which is performed by means of data filtering or correlation analysis.

4 Conclusions

Analysis and formal characterization of the problem of MI calibration by comparing it with the RS show that the problem can be solved, i.e. the estimates of characteristics of the MI under calibration may be obtained with the required quality. The traditional calibration condition, i.e. a certain ratio k of accuracies of MI and the RS, is define more exactly and expanded. It is shown that the important (but not necessary) condition for possible k decrease is that the RS does not have any significant systematic error, against its random errors. In this case we note that the necessary and sufficient condition for the problem solution is the presence of estimates of the RS error characteristics, the accuracy of which is not less than the predetermined level depending on the accuracy ratio of the compared devices. This makes it possible to extend the range of acceptable accuracy ratios towards its decrease. The obtained theoretical ratios and dependencies may be the basis for development of calibration procedures for certain MI which impose significantly less strict requirements to the quality of the available RS, under the determined conditions.

Above-mentioned solution results of the accuracy relation problem of MI under calibration and RS are reasonable only if initial data meet the requirements of stationarity and ergodicity. The approach stated above can be used for checking the requirements compliance.

References

1. Malikov, M.F.: Basics of Metrology. Kommerpribor, Moscow (1949)
2. Granovskii, V.A., Kudryavtsev, M.D.: Comparisons of inertial heading sensors: reference sensor with zero systematic error. In: The First International Conference on Advances of Sensors, Actuators, Metering and Sensing (ALLSENSORS 2016), IARIA, Venice, pp. 46–50 (2016)
3. Cramer, H.: Mathematical Methods of Statistics. Almqvist and Wiksells, Uppsala (1945)
4. Lemeshko, B.Y., Sataeva, T.S.: On the properties and application of tests for homogeneity of variance in the problems of metrology and control. In: International Conference on Systems, Control and Information Technology (SCIT 2016), Warsaw (2016)
5. Yaglom, A.M.: Introduction to the theory of stationary random functions. UMN, vol. 7, pp. 3–168. USSR Academy of Sciences, Moscow (1952)
6. Kosarev, E.L.: Methods of Experimental Data Processing. Fizmatlit, Moscow (2008)
7. Bendat, J.S., Piersol, A.G.: Measurement and Analysis of Random Data. John Wiley & Sons, Inc., Hoboken (1967)

Generalized Description of the Frequency Characteristics of Resistors

Stefan Kubisa and Zygmunt L. Warsza(✉)

Industrial Research Institute of Automation
and Measurement (PIAP), Warsaw, Poland
zlw@op.pl

Abstract. The real resistor in AC current can be described as an impedance Z of components changed with frequency. The real part R_s of this impedance is not equal to the resistance R in the DC current, and there is the imaginary component $X_{s,}$ both frequency dependent. A three-element equivalent circuit including the resistance R as the main parameter and the residual parameters: serial inductance L and a parallel capacitance C is the basic AC model of a real resistor. Frequency characteristics of impedance Z components of this model in relative to R values r_s, x_s are given. Patterns of these components in generalized form, i.e. as functions of the relative values of characteristic resistance $\rho = R\sqrt{C/L}$ and characteristic frequency $\eta = \omega\sqrt{CL}$ of this equivalent circuit are determined and their frequency curves for a few values of ρ are shown. Variants of this model with four models of connections are analyzed. Relative errors δR, δX and $\delta Z \stackrel{\text{def}}{=} \sqrt{\delta R^2 + \delta X^2}$ are also find. Few numerical examples are given and considerations are of changes of errors in the frequency band. Some conclusions are included.

Keywords: Resistor · AC models · Equivalent circuit · Impedance · Characteristic parameters · Frequency errors

1 Introduction

The flow of electrical current in the real objects is described by the electromagnetic field equations as functions of space and time. If current is direct in time (DC) the ratio of voltage and current of any two-terminal object is determinated as its resistance R. If geometric sizes of real physical object are much smaller than the shortest wavelength of AC current waveform, then the simpler description by theory of the electrical circuits containing ideal dimensionless elements can be used. In models of AC circuits of the such low frequency f beside the ideal resistance R, the ideal reactive elements: capacitance C, inductance L, mutual inductance M and also an ideal stationary or controlled sources of voltage E and current J are applied. If for the permissible ranges of voltages and currents the values of these parameters are constant, then modeled circuit is linear. The method of analysis frequency characteristics of the practical two-terminal element with constant parameters, described in relative values is presented and analyzed below on the example of bulk and metal foil resistors used in measurement and electronics [1, 2].

R. Szewczyk and M. Kaliczyńska (eds.), *Recent Advances in Systems, Control and Information Technology*,
Advances in Intelligent Systems and Computing 543, DOI 10.1007/978-3-319-48923-0_67

For the alternating current (AC) of frequency f the complex impedance $\underline{Z}$ describes any passive two-terminal device

$$\underline{Z}(\mathrm{j}\,\omega) = \mathrm{Re}\;\underline{Z} + \mathrm{jIm}\;\underline{Z} = Z\,e^{j\varphi}$$

Either rectangular or polar components are functions of pulsation $\omega = 2\pi f$. Impedance $\underline{Z}$[1] in the circuit equivalent scheme can be presented as ideal resistance $R_s(\omega) \equiv \mathrm{Re}\;\underline{Z}$ and reactance $\mathrm{j}\,X_s(\omega) = \mathrm{j}\,\mathrm{Im}\;\underline{Z}$ connected in series. For very high frequencies, in which wavelengths of AC current and geometrical sizes of the real object are comparable, description with distributed parameters and the wave impedance Z as for transmission lines, is used.

As the simple example of the two-terminal object working in AC circuits is the real resistor. In addition to the its basic parameter R the impact of the residual parameters, the inductance L and capacitance C, called also as parasitic parameters, must be taken into account. These parameters are distributed in a volume of analyzed device and depend on its structure, shape and sizes. An impedance $\underline{Z}$ measured on terminals depends on combination of all $R\;L\;C$ values of its equivalent scheme.

Properties of real resistors used in AC circuits usually are described by equivalent schemes with values of $R\;L\;C$ parameters independent on frequency ω in some range of it. The basic 3-element model has the structure given on Fig. 1a. In the following text it is referred as model type Γ (gamma). The similar model, but with a different values of parameters can be used also for coils without a ferromagnetic core. The equivalent to the three-element model type Γ are two-element models (Fig. 1(b)) as serial R_s, L_s and parallel R_r, C_r circuits. In the general case their parameters depend on the frequency ω. The serial model is preferred for $\mathrm{Im}(\underline{Z}) > 0$ as then $L_s > 0$. If $\mathrm{Im}(\underline{Z}) < 0$ then $C_r > 0$ and resistor is modeled more simply by the parallel circuit with admittance $\underline{Y}(\mathrm{j}\omega) = 1/R_r + \mathrm{j}\omega C_r$.

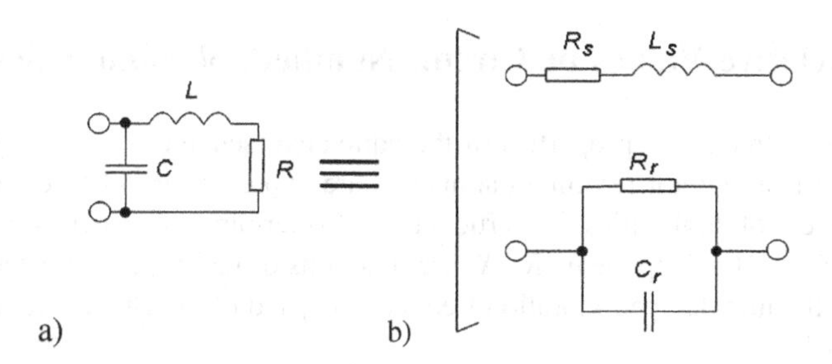

Fig. 1. The basic models of a real two terminal resistors: (a) type **Γ** of three constant elements $R\;L\;C$ and (b) its two-parameter equivalents: serial $R_s\;L_s$ or parallel $R_r\;C_r$ with both parameters dependent on frequency ω

[1] For simplicity in the following text as frequency are called both f and pulsation $\omega = 2\pi f$ also known as the circular frequency. Often in text a parameter dependence on the frequency is not marked as it is written shorter, e.g. Z instead of $Z(\mathrm{j}\omega)$, R_s instead of $R_s(\omega)$, etc.

2 Assumptions

The real resistor can fulfill properly the function of an ideal single resistance R in AC circuits only when $R_s(\omega) = R$ and $X_s(\omega) = 0$. If not, then the difference $\Delta R = R_s - R$ is the resistive component of its frequency error, and X_s is the reactive component of this error. A resistor, for which $\Delta R \approx 0$ and $X_s \approx 0$ can be called as *good* resistor.

In considerations below is assumed that:

- the fixed values of $R\ L\ C$ and $\underline{Z}$ parameters can be used throughout the full range of permissible values of currents and voltages and also in considered frequency. This applies to a such electrical devices which structures do not include the non-linear ferromagnetic and dielectrics materials and semiconductor junctions.
- resistor is treated as a single element, i.e. not directly or electromagnetically coupled, to other components of the circuit,
- $L\ C$ parameters of the resistor depend only on the resistor geometry including connections, but do not depend directly on the value of its resistance R,
- $R\ L\ C$ parameters also do not depend on the frequency, i.e. the skin effect is negligible. As example are the thin film resistors made from materials with high resistivity.

Constructions and values of parameters of three types actually produced resistors are described in [2]. The method for measurement values of parasitic parameters $L\ C$ is given in [3].

Numerical example and resulting conclusions can be given in this work for a resistors of the flat thin film plotted on the insulating substrate. At a given geometrical shape and sizes, their resistance R depends only on the thickness of the conductive layer. Parameters L, C of such resistors with identical shape and size, including the leads, are the same for the various R. This condition does not meet the resistors with a complex form of resistance track as meander or helix and for wire wounded resistors.

3 The Relative Values or Circuit Numbers of Similarities

The analysis of frequency properties of the equivalent schemes of real objects can be more general if all circuit terminal parameters are expressed in relative values named also as numbers of similarities. In particular of two-terminal object from Fig. 1a, their impedance $\underline{Z}$ and its components R_s, X_s are functions of resistance R for DC current or functions of the non damped vibration frequency ω_0 and of the characteristic resistance R_0 of circuit [4]:

$$\omega_0 = \frac{1}{\sqrt{L\,C}} \tag{1}$$

$$R_0 = \sqrt{L/C} \tag{2}$$

From (1) and (2) it is seen that ω_0 and R_0 are linked themselves by L and C. The frequency ω_0 is when $R \to 0$. The value of resistance R_0 results from the equality of capacitive and inductive time constants $\tau_C = R_0C$ and $\tau_L = L/R_0$ respectively – Eq. (14).

The frequency $f_0 = \omega_0/2\pi$ inversely depends to a geometric sizes of resistor, e.g. for flat resistors of the centimeter sizes f_0 is of the GHz order, and for micrometer sizes – of the THz order. If the resistor is made of materials with the magnetic and dielectric relative permeability equal to 1 and is located in the environment of such both permeabilities, then the characteristic resistance R_0 is very close to the wave resistance of vacuum, approx. 377 Ω. In the presence of materials with greater than vacuum permeabilities the value of R_0 is reduces, usually to the range (50–300) Ω – see the numerical example in Sect. 5.1.

Generalization of the impedance model description may be achieved by the use of so-called similarity numbers. These dimensionless parameters include, for example ratios of two quantities of the same dimensions. They are referred often as the relative values. In describing the frequency characteristics of resistors, the similarity numbers given below are used:

- relative frequency η as ratio of the considered frequency f and the characteristic ones f_0 of the model:

$$\eta \stackrel{\text{def}}{=} \omega/\omega_0 = f/f_0 \tag{3}$$

where: $\omega = \eta\,\omega_0$ and $f = \eta\, f_0$.
- relative resistance ρ as the ratio of resistance R for DC current and characteristic resistance R_0:

$$\rho \stackrel{\text{def}}{=} R/R_0 \quad \text{and hence} \quad R = \rho\, R_0 \tag{4}$$

Similarity numbers of the impedance components related to R of the serial circuit model on Fig. 1b, i.e.:
- serial resistance:

$$r_s \stackrel{\text{def}}{=} \frac{R_s}{R} \quad \text{and from it} \quad R_s = r_s\, R \tag{5}$$

- relative serial reactance:

$$x_s \stackrel{\text{def}}{=} X_s/R \quad \text{and from it} \quad X_s = x_s R \tag{6}$$

- Figure 2 shows on the Gaussian plane the vector diagram of impedance Z components of the serial model in Fig. 1b and components ΔR and jX_s of a vector error $\Delta \underline{Z}$ of the resistor. Directions of arrowheads indicate the signs of components.
  From this graph is also possible to determine the relationship between the relative errors frequency δR δX δZ of resistor and components R_s X_s of its impedance $\underline{Z}$, i.e.:
- relative resistance error

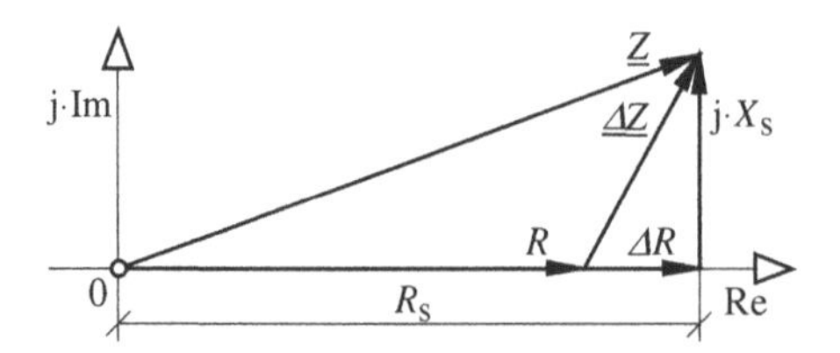

Fig. 2. Graph vector of serial impedance components of the resistor

$$\delta R \stackrel{\mathrm{def}}{=} \frac{\mathrm{Re}\underline{Z} - R}{R} = \frac{R_S - R}{R} = r_S - 1 \tag{7}$$

– relative reactance error

$$\delta X \stackrel{\mathrm{def}}{=} \frac{\mathrm{Im}\underline{Z}}{R} = \frac{X_S}{R} = x_S \tag{8}$$

– relative impedance error

$$\delta Z \stackrel{\mathrm{def}}{=} |\Delta \underline{Z}|/R = \sqrt{\delta R^2 + \delta X^2} \tag{9}$$

4 Frequency Properties of the Resistor Model Γ

The Laplace operator impedance Z(s) of the model in Fig. 1a can be described by following patterns

$$Z(\mathrm{s}) = \frac{(R + \mathrm{s}L)/\mathrm{s}\,C}{R + \mathrm{s}L + 1/\mathrm{s}\,C} = R\,\frac{1 + \mathrm{s}\,L/R}{\mathrm{s}^2 LC + \mathrm{s}\,R\,C + 1} = R\,\frac{1 + s\tau_L}{\frac{\mathrm{s}^2}{\omega_0^2} + \frac{R}{R_0}\frac{\mathrm{s}}{\omega_0} + 1} \tag{10}$$

The impedance $Z(s)$ formulas (10) are products of two transfer functions: the non-ideal first order differentiator (in nominator) with time constant $\tau_L = L/R$ and the second order inertia element with transfer function reciprocal to the denominator of (10). For $R < 2R_0$ (i.e. $\rho < 2$) second one is a attenuated oscillating element of the damping coefficient $\zeta = R/2R_0$. Then for $R \geq 2R_0$ its characteristic equation has two real roots:

$$s_{1,2} = \omega_0\,[\frac{-R}{2R_0} \pm \sqrt{(\frac{R}{R_0})^2 - 4}\,] = \omega_0\,[-\frac{\rho}{2} \pm \sqrt{\rho^2 - 4}\,] \tag{10a}$$

and transfer function is the product of two first degree members.

From (10) if $s \to \mathrm{j}\omega$ the following complex impedance $Z(\mathrm{j}\omega)$ is obtained

$$\underline{Z}(j\omega) = R\frac{1 + j\omega L/R}{(1 - \omega^2 LC) + j\omega RC} = R_s(\omega) + jX_s(\omega) = Z(\omega)\cdot e^{j\varphi(\omega)} \tag{11}$$

where: Z – magnitude of impedance $\underline{Z}$, φ – the phase angle of $\underline{Z}$.

$$Z \equiv |\underline{Z}(j\omega)| = R\frac{\sqrt{1 + (\omega\tau_L)^2}}{\sqrt{(1 - \omega^2 LC)^2 + (\omega RC)^2}} \tag{12}$$

The phase angle φ of $\underline{Z}$ depends on equivalent time constant τ, i.e.:

$$\varphi = \text{arctg}\frac{\omega L_s}{R_s} = \text{arctg}\ \tau \tag{13}$$

$$\tau = L_s/R_s = \tau_L(1 - \omega^2/\omega_0^2) - \tau_C \tag{14}$$

The constant τ can be positive or negative, depending on whether the predominant is the first component of (14) with τ_L or τ_C.

For $\omega^2/\omega_0^2 << 1$

$$\tau \approx \tau_L - \tau_C = (L/R) - RC \tag{14a}$$

For the relative frequency $\eta \equiv \omega/\omega_0 = \omega\sqrt{CL}$ and relative resistance $\rho \equiv R/R_0 = R\sqrt{C/L}$ from (10) follows

$$\underline{Z}(j\eta) = R\frac{1 + j\eta/\rho}{1 - \eta^2 + j\eta\rho} \tag{15}$$

After elementary transformations, the relative impedance $\underline{z}(j\eta) \equiv \underline{Z}(j\eta)/R$ and its rectangular components r_s, x_s are [4]:

$$\underline{z}(j\eta) = \frac{1 + j\eta/\rho}{1 - \eta^2 + j\eta\rho} \equiv r_s + jx_s \tag{16}$$

$$r_s = \frac{1}{(1 - \eta^2)^2 + \eta^2\rho^2}; \qquad x_s = \frac{\eta\left(\frac{1-\eta^2}{\rho} - \rho\right)}{(1 - \eta^2)^2 + \eta^2\rho^2} \tag{17a, b}$$

Functions of $r_s(\eta)$ and $x_s(\eta)$ for three values of the relative resistance ρ are given in Fig. 3.

The resistor is considered to be *good* if $r_s \approx 1$ and $x_s \approx 0$. Figure 3 shows that this can occur only for low relative frequencies, i.e. $\eta < 0.2$ and when the relative resistance $\rho \approx 1$. For example, if $\eta = 0.1$ and $\rho = 1$ from (17a, b) is $r_s = 1$ and $x_s = 0.0001$.

As it is proposed before, the resistor of $\rho = 1$, i.e. of the resistance $R = R_0$, is called below as *compensated*.

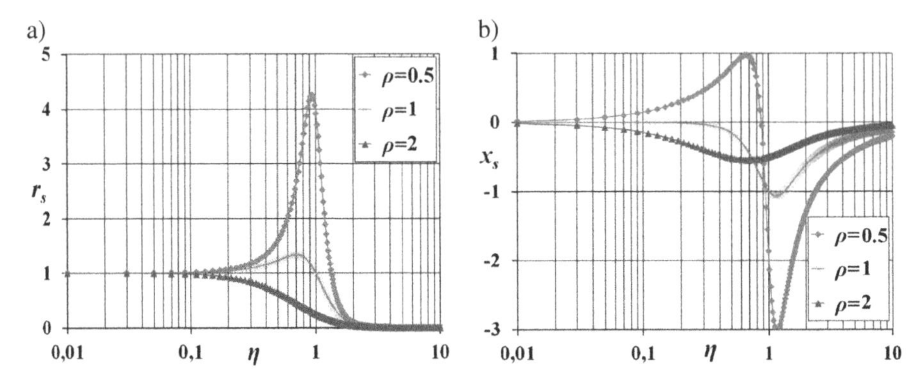

Fig. 3. Serial relative components (17a, b) as function of relative frequency η for three values of $\rho = R/R_0$: (a) relative resistance r_s, (b) relative reactance x_s.

The relative frequency errors of the circuit model in Fig. 1a are resulting from definitions (7), (8) and described by patterns

$$\delta R = r_s - 1 = -\frac{\eta^2(\rho^2 - 2) + \eta^4}{1 + \eta^2(\rho^2 - 2) + \eta^4}; \qquad \delta X = x_s \tag{18a, b}$$

For $\eta^{4} << 1$ i $\eta^{4}\rho^{4} << 1$ simplified formulas can be used, i.e.:

$$\delta R \approx -\eta^2(2 - \rho^2), \qquad x_s \approx \omega(\tau_L - \tau_C) = \eta\,(\rho^{-1} - \rho) \approx \delta Z \tag{19}$$

where: $\tau_L = L/R$, $\tau_C = RC$, $\eta^2 = \omega^2 LC = \omega^2\ \tau_L\ \tau_C$, $\rho^2 = R^2\ \tau_C/\tau_L = C/L$.

Further considerations concern on the scope of relatively low frequency, i.e. $\eta < 0.1$. Formulas (18a, b) of frequency error plots to η in the range 0.1 = 0.001 … 0.1 and several values ρ are given in Fig. 4 [4]. For the frequency range η = 0.001 … 0.01 and relative resistance ρ = 0.1 … 1 the resistance error δR (Fig. 4a) is very near to zero. When $\eta = 0.1$ the error δR is ≈ 2% for $\rho = 0.1$, $\delta R \approx +1\%$, for $\rho = 1$, $\delta R \approx -2\%$ for $\rho = 2$, $\delta R \approx -20\%$ for $\rho = 5$ and for $\rho = 10$ $\delta R \approx -50\%$.

From the Fig. 4a, for a given value of ρ the upper limit of frequency η_{gr} can be estimated. It is the relative frequency, at which the relative error δR does not exceed the permissible value. For example, if the resistor $\rho = 10$ (i.e. the resistance R is 10 R_0) and the limited error is 5%, then the upper relative frequency η_{gr} is approx. 0.023 (see in Fig. 4a point with the blue circular outline). Graphs in Fig. 4b also allows you by the same way to assess the upper frequency η_{gx} for which the reactance error δX does not exceed its limited value. So for the permissible value 5% of the δX error, the upper relative frequency amounts to approx. 0.005 for $\rho = 0.1$ (point with a blue border), and for $\rho = 10$ (point with a red border). For intermediate values of ρ between 0.1 and 10 a relative frequency limits are higher. These graphs also show that for $x_s = 0$, a good approximation can be taken by (14a) rule that resistors with a resistance R smaller than the characteristic resistance R_0 (i.e. $\rho < 1$) are inductive type ($\delta X > 0$), and at the resistance $R > R_0$ ($\rho > 1$) – capacitive type ($\delta X < 0$). This can be exactly recognized from pattern (14) for $\tau = 0$ and was specified in [4].

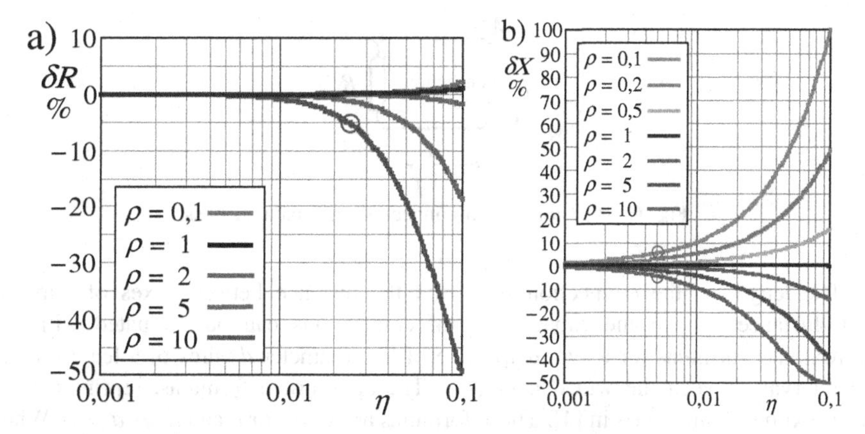

Fig. 4. Frequency errors of the resistor model Γ as function of relative frequency η for several values of relative resistance ρ: (a) relative resistive error δR, (b) relative reactance error δX (Color figure online)

Figures 4a and b also show that the requirement of upper frequency due to the permissible reactance error is higher than for the maximum value of resistive error. Therefore, to apply in practice as a more convenient the relative impedance δZ is proposed – formula (9). This error is only slightly larger than the larger of errors δR or δX. Functions of error δZ in double logarithmic scale are given in Fig. 5 [4].

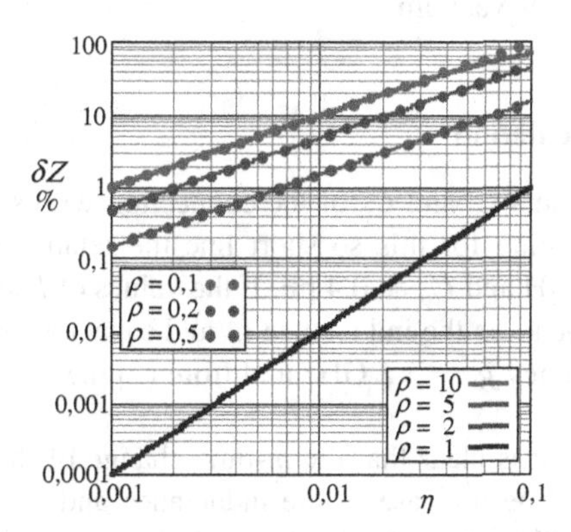

Fig. 5. Dependence of the relative impedance error δZ of resistor on the relative frequency $\eta \leq 0.1$ for several relative resistance values ρ

5 Assessment of Residual Parameters

A significant influence on the frequency characteristic of the resistor has its own geometry and geometry of its connection with other elements of the electric circuit. Let us assumed the simplest situation that these connection are two conductors of the diameter d, with negligible resistance arranged parallely, as shown in Fig. 6.

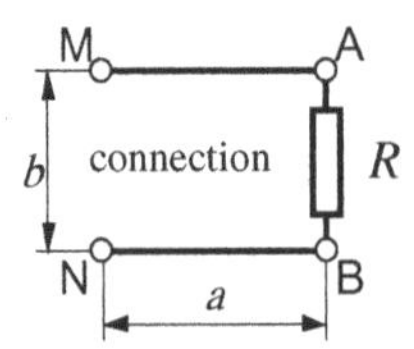

Fig. 6. Resistor R with connection of 2 parallel lines

The length of the connection is a, and the distance between axes of wires is b. Inductance L_p and the capacity C_p of connections can be evaluated with the approximate formulas for a symmetric line of two diameter d wires of a length a and the interval between the wires' centers b. The appropriate formulas, reprinted from older text books are given in [4]. These formulas are valid for b and $d \gg a \gg b$. When the second condition is not fulfilled the obtained values L_p and C_p are undercounting, because they do not take into account the increase in inductance and capacitance dependent on the shape of the magnetic and electric field on the ends of the connection line. This increase is taken into account by entering the additional coefficient $\psi > 1$ and then is obtained:

$$L_p = \psi \cdot \frac{\mu_0}{\pi} \cdot a \cdot \ln\frac{2 \cdot b}{d} \; ; \qquad C_p = \psi \cdot \pi \cdot \varepsilon_0 \cdot a \cdot \left(\ln\frac{2 \cdot b}{d}\right)^{-1} \tag{20}$$

We assume here that the connection is in the environment of magnetic and dielectric permeabilities of vacuum.

5.1 Example of Calculation

Let us calculate parameters L_p and C_p of the connection with sizes of $a = b = 10$ mm and $d = 1.0$ mm. Assuming for this so short line the relatively large $\psi \approx 1.5$ in [2] obtained is: $L_p \approx 0.18$ μH and $C_p \approx 0.4$ pF. If the values of L and C of the model Γ of real resistor mainly depend on the inductance L_p and capacitance C_p of this connection, then from (1) the frequency $f_0 \approx 3.2$ GHz and from (2) the resistance $R_0 \approx 360\ \Omega$ are obtained.

From formulas (12) it is clear that the m-time change of the geometric size of the connection give the m-time increase of the inductance and capacitance values. Thus, from the Eq. (1), that will be the m-fold decrease in frequency of ω_0. E.g. if leads sizes are $a = b = 100$ mm and $d = 10$ mm then the two terminal circuit would have a natural frequency $f_0 \approx 320$ MHz. For connections with sizes of $a = b = 1.0$ mm and $d = 0.10$ mm if $f_0 \approx 32$ GHz. This qualitatively explains why electronic systems due to the miniaturization of sizes achieved the higher operating frequencies. Additionally from formula (2) is seen that m-fold change of the resistor sizes does not change the characteristic resistance R_0.

6 Other Models of Resistor with Connections

The two wire connection of resistor (Fig. 7) has an inductance L_p and the capacitance C_p and can be considered as a lossless line ($R_p << R$). If the frequency is so small that the electromagnetic wave length is far greater than the length of a connection, this can be modeled by two-port type **Π** or type **T** [4] as shown in Fig. 7.

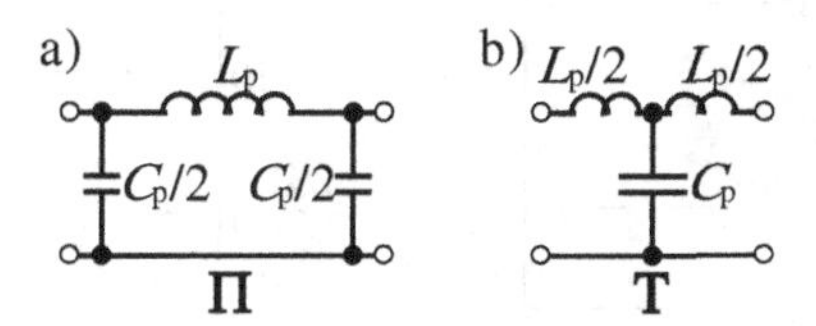

Fig. 7. Models of connections: (a) type **Π**, (b) type **T**

Connection of leads models type **Π** or **T** given in Fig. 7 and model Γ of resistor from Fig. 1a with residual parameters L_r, C_r gives models with 5-parameters L_p, C_p, L_R, C_R, R. Separate determination the values of L_p, L_R and C_p, C_R is very difficult in practice, and such models would not convenient to use. Four simpler models given in Fig. 8 with the resultant parameters L, C as sums $L = L_p + L_R$; $C = C_p + C_R$ are therefore applied. We shall show that for the relative frequency $\eta < 0.1$, these models are roughly equivalent each other and have the same frequency ω_0.

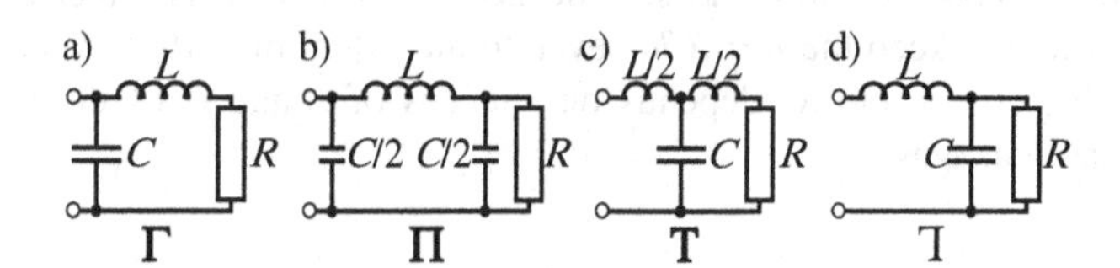

Fig. 8. The equivalent 3-parameters circuits as models of the resistor with connection leads: (a) the type **Γ** (gamma as on Fig. 1(a)); (b) **Π** type, (c) **T** type, (d) **Ꞁ** type (reversed gamma)

Frequency characteristics of the relative resistance r_s (η) and the relative reactance x_s (η) of two terminal models in Fig. 8 differ from each other, but significant differences exist only for higher values of the relative frequency approaching $\eta = 1$.

The relative values of resistance r_s and reactance x_s of type Γ model are represented by the formulas (17a, b). For a model Ꞁ these formulas are different [42]:

$$r_s = \frac{1}{1 + \eta^2 \cdot \rho^2}; \qquad x_s = \frac{\eta}{\rho} \cdot \frac{1 - \rho^2 \cdot (1 - \eta^2)}{1 + \eta^2 \cdot \rho^2} \tag{21a, b}$$

To obtain the similar formulas for **Π** and **T** models is quite cumbersome. More easy is find the pattern of relative complex impedance $\underline{z} = \underline{Z}/R$ and by the application of appropriate mathematical formulas from $\underline{z}$ numerically calculate $r_s = \text{Re } \underline{z}$ and $x_s = \text{Im } \underline{z}$. For relative frequencies $\eta < 0.1$, and the relative resistances of $0.1 < \rho < 10$ models Γ and ˥ have very similar values of impedance error δZ. They are slightly larger than for the models **Π** and **T** – see Figs. 9 and 10 [4].

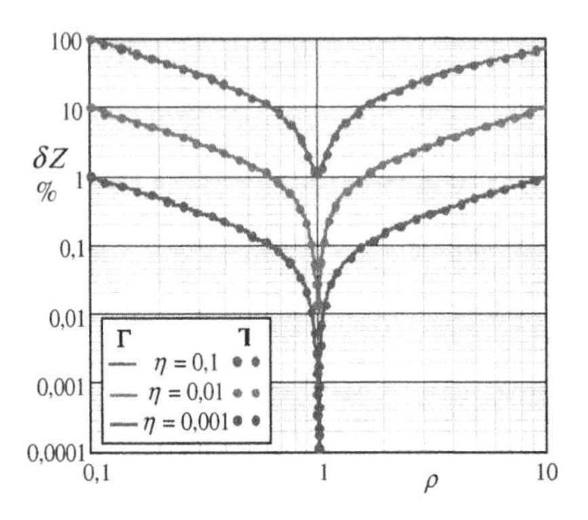

Fig. 9. The impedance error δZ of the resistor models **Γ** and dependence on η and ρ

Fig. 10. The impedance error δZ of the resistor models **Γ** and **Π** dependence on η and ρ

The small differences in the impedance error δZ between different models for the relative resistance value ρ around 1 is demonstrated by drawings in Fig. 11. Such small differences of δZ however, have no significance in practice. It is rather a theoretical curiosity. Possible to keep the $\rho = R/R_0$, even to the extent of $\pm 1\%$ is very difficult or impossible, since the value R_0 depends on changes of L and C of the resistor with connection and leakages.

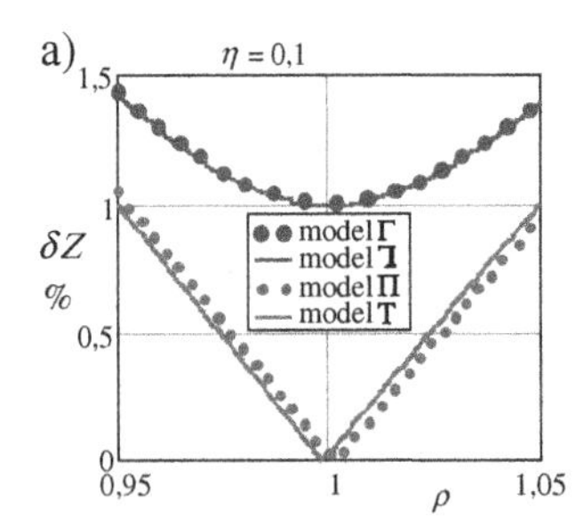

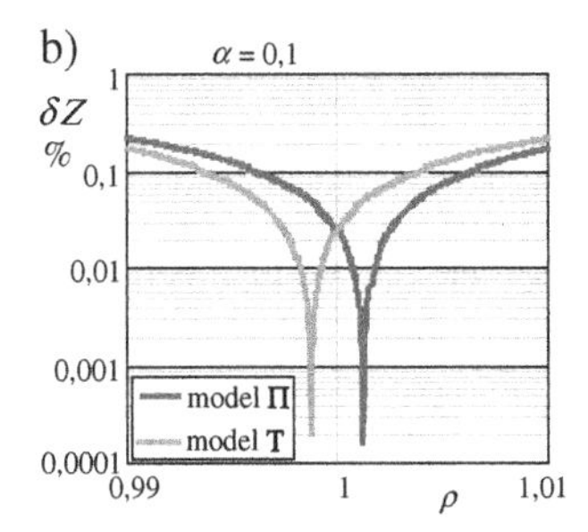

Fig. 11. Details of the impedance error δZ of resistor dependence on the relative resistance ρ (around $\rho = 1$) for models shown in Fig. 8: (a) $\rho = 0{,}95 \ldots 1{,}05$, (b) $\rho = 0{,}99 \ldots 1{,}01$

There is no possibility of deciding in practice which model in Fig. 8 is the most appropriate in particular situation. This leads to choice the model of the simplest formulas to calculations. For the estimation of the limited values of the frequency error δZ of the resistor impedance, the model ٦ seems to be the most convenient. The formula for this model has fairly simple form:

$$\delta Z(\eta, \rho) = \eta \, \frac{\sqrt{\eta^2 \rho^4 + [\rho^{-1} - \rho\,(1 - \eta^2)\,]^2}}{1 + \eta^2 \rho^2} \tag{22}$$

The possibility of correction the frequency characteristics of resistors used in practice are discussed in the part 2 of work [4]. The correction of the residual reactance $X_r > 0$ by small C only can be successfully implemented, e.g. for standard resistors or resistance sensors used in AC circuits.

7 Accuracy of Resistor in Frequency Band

The foregoing considerations, including the drawings on Figs. 3, 4 and 5 show that for the relative frequency $\eta < 0.1$ (i.e. at a frequency f up to approx. 0.1 of f_0), the equivalent resistance of the resistor R_s Fig. 1b is equal to the resistance value R for direct current. If the relative resistance $\rho \approx 1$, i.e. when the resistor is compensated, in a wide frequency band is $R_s \approx R = R_0$.

Dependence of resistance R_0 characteristic of the resultant residual parameters – the capacitor C and the inductance L of the resistor with the connection is defined by the formula (1). Even if the values of L and C resulting from the construction of resistor are constant, other parameters may be connection in each of the systems. Furthermore, the L, C and R_0 values are known with the low accuracy, but the value R can be only measured accurately.

Specific values of L and C can be taken as the nominal inductance and the nominal capacity of the resistor. Calculated according to the formula (1) of the value of the frequency ω_0 – is the nominal own frequency of the resistor. Resistor with resistance $R = R_0$ defined by the formula (2) is the nominally compensated resistor. To describe changes of parameters of the resistor with a connection the concept of multiplier $\boldsymbol{\kappa}$ as a number greater than 1 ($\boldsymbol{\kappa} > 1$) is now introduced. The C and L will be in the ranges:

- Capacity: between the bottom value of C_d κ-fold less than the nominal value C, up to a upper value C_g κ-fold greater than C; then
- Inductance: between the lower value L_d κ times less than the nominal value L up to a upper value L_g which is κ L.

It is described as follows:

$$C_d = \frac{C}{\kappa}\ ; \quad C_g = C \cdot \kappa\ ; \qquad L_d = \frac{L}{\kappa}\ ; \qquad L_g = L \cdot \kappa\ ; \ \text{ if } \kappa > 1 \tag{23}$$

Table 1 summarizes the four cases when the capacitance and inductance take the extreme (lower or upper) values.

Table 1. Relative frequency and relative resistance in the four situations of extreme values of inductance and capacitance

Case	Relative frequency η	Relative resistance ρ
I. C_d, L_d	$\eta_I = \omega \cdot \sqrt{C_d \cdot L_d} = \frac{\eta}{\kappa}$	$\rho_I = R \cdot \sqrt{\frac{C_d}{L_d}} = \rho$
II. C_d, L_g	$\eta_{II} = \omega \cdot \sqrt{C_d \cdot L_g} = \eta$	$\rho_{II} = R \cdot \sqrt{\frac{C_d}{L_g}} = \frac{\rho}{\kappa}$
III. C_g, L_d	$\eta_{III} = \omega \cdot \sqrt{C_g \cdot L_d} = \eta$	$\rho_{III} = R \cdot \sqrt{\frac{C_g}{L_d}} = \rho \cdot \kappa$
IV. C_g, L_g	$\eta_{IV} = \omega \cdot \sqrt{C_g \cdot L_g} = \eta \cdot \kappa$	$\rho_{IV} = R \cdot \sqrt{\frac{C_g}{L_g}} = \rho$

In the cases given Table 1, values of relative frequency and relative resistance should be rescaled respectively as given below.

- Case I: Relative frequency η_I decreases κ-fold, and the relative resistance ρ_I remains unchanged. Frequency characteristics shown in Figs. 3, 4 and 5 require only sharing relative frequency by κ η. In a logarithmic scale on the abscissa, it means the shift of graphs by logκ to the left.
- Case II: frequency relative η_{II} does not change, and the resistance decreases relative ρ_{II} κ times. That mean that characteristics run for the κ times smaller values of resistance ρ.
- Case III: Relative frequency η_{III} remains unchanged, and the relative resistance ρ_{III} growing κ times. This means that the frequency response must be calculated for κ times larger values of resistance ρ.
- Case IV: The relative frequency η_{IV} increases κ-fold and the relative resistance ρ remains unchanged. This means that the frequency characteristics shown in Figs. 3, 4 and 5 require multiplying the relative frequency η by κ. In a logarithmic scale on the abscissa this corresponds to a shift of plots to right on logκ.

Measures of the resistance R inaccuracy for alternating current are its relative frequency errors, i.e. an resistance error δR – formulas (5) and (22), the reactance error δX – formulas (6) and (22) and the impedance error δZ – patterns (7) and (22).

In cases II and III the frequency axis rescaling is not require. As the result of derogations of capacitance C and inductance L from their nominal values is the inaccuracy of resistance R, which depends on changes of the relative resistance ρ value. According to Table 1, the resistance ρ of the nominally compensated resistor can take extreme values of $1/\kappa$ and κ. Figure 12 shows the changes of the impedance error δZ of resistor as function of relative frequency η for four values of multiplierκ.

When changing η from 0.001 to 0.1 then δZ varies in the range:

- from approx. 1×10^{-4}% to approx. 1% at $\kappa = 1$. That is the perfectly compensated resistor,
- from approx. 2×10^{-3}% to approx. 1% at $\kappa = 1.01$. That is when deviations of L and C from their nominal values are within the range from –1% to 1%,
- from approx. 0.01% to approx. 1.4% at $\kappa = 1.05$. That is if deviations of L and C are in the range of approx. 4.8% to 5%,

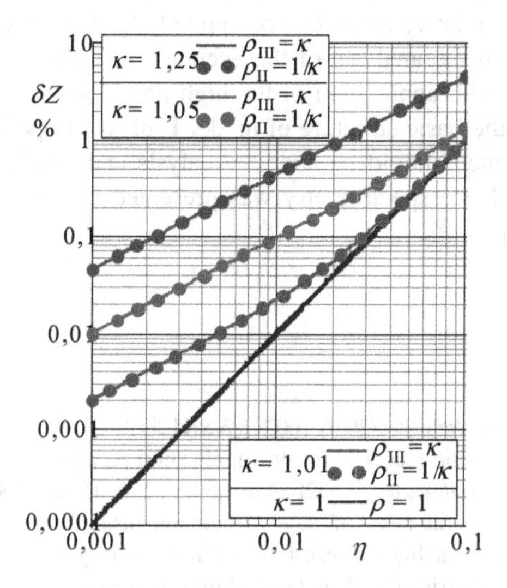

Fig. 12. The impedance error δZ of nominally compensated resistor ($\rho = 1$) as function of the relative frequency η for several values of multiplier κ.

- from approx. 0.045% to approx. 4.5% at $\kappa = 1.25$. That is if the deviations of L and C are in the range of -20% to 25%.

From these considerations is resulted that the best accuracy of the resistance R at higher frequencies has the compensated resistor, i.e. satisfying the relationship $X_s \approx 0$. However, even a small deviation determined by multiplier κ, causes a considerable error. For example, for $\kappa = 1.05$, i.e. for approx. 5% deviations of the capacity C and the inductance L from their nominal values, the impedance error δZ increases from approx. 0.0001% to approx. 0.01% with relative frequency $\eta = 0.001$ and from approx. 1% to approx. 1.4% at $\eta = 0.1$. Maintaining the C and L values in the range of 5% is not easy in practice. More realistic is the range of -20% to $+25\%$ ($\kappa = 1.25$). Then the impedance error δZ may reach values in the range of 0.05% for $\eta = 0.001$ up to $\sim$ 5% if $\eta = 0.1$.

Anticipating the distribution $p(\delta Z)$ in a given range of the relative frequency η, the uncertainty B-type of the resistor with connection can be also estimated. Randomness is coming from random changes of parasitic C and L and frequency f.

8 Summary

The generalized description of frequency characteristics of resistors is presented in this work. These characteristics are expressed in numbers of similarity that is in relative terms. Such description is not meet in literature and can be used for of the frequency analysis of the equivalent schemes of different electrical devices. As the examples are resistors, including wounded and performed by other the newest technologies, models of other passive objects with the dominant parameter L or C etc.

Generalized description in relative terms can be applied also for standard platinum temperature sensors (SPRT) in the analysis of the accuracy of temperature measurement between control points with application of the high precision AC bridges [5].

At higher frequencies to the basic structure of model **Γ** of real impedance element an additional parallel RC branch should be added. Analysis and application of such extended model in the impedance spectrometry was presented at the recent IMEKO 2015 Congress in Prague [6].

References

1. www.vishay.com/company/press/releases/2016/160216WSK1206/
2. Pieńkowski, D.: Impact of parasitic reactance on the application parameters of resistors and capacitors (Edition on line). http://elektronikab2b.pl/technika/1236. 15 Jan 2007
3. https://www.ieee.li/pdf/essay/s_parameters_of_passive_components.pdf
4. Kubisa, S., Warsza, Z.L.: Analiza błędów częstotliwościowych rezystorów. Część 1 i 2. (Analysis of frequency errors of resistors). Przegląd Elektrotechniczny – Electr. Rev. **92,** Part 1 no 6, 211–216, Part 2 no 7, 211–216 (2016). ISSN 0033-2097, in Polish
5. Mikhal, A.A., Warsza, Z.L.: Electromagnetic protection in high precision tri-axial thermometric AC bridge. In: Szewczyk, R., Zieliński, C., Kaliczyńska, M. (eds.) Progress in Automation, Robotics and Measuring Techniques. AISC, vol. 352, pp. 147–156. Springer, Heidelberg (2015). doi:10.1007/978-3-319-15835-8_17
6. Baltianski, S.: Impedance spectroscopy: separation and asymptotic model interpretation. In: Proceedings of XXI IMEKO World Congress "Measurement in Research and Industry" Prague, Czech Republic, 30 August–4 September, pp. 493–497 (2015)

Engineering Materials

A New Magnetic Method for Stress Monitoring in Steels

Evangelos V. Hristoforou[1(✉)], Aphrodite Ktena[2], Polyxeni Vourna[1(✉)], Eleni Mangiorou[1], and Stelios Mores[3]

[1] National TU of Athens, Athens, Greece
eh@metal.ntua.gr, xenia.vourna@gmail.com, mangiorou.eleni@gmail.com
[2] Technological Education Institution of Sterea Ellada, Evia, Greece
aktena@teiste.gr
[3] NOMASICO Ltd., Nicosia, Cyprus
steliosmrs@gmail.com

Abstract. Technology developments in analytic techniques as well as in data acquisition and management are already revolutionizing the ways that steel industry assets are managed. Their adoption in several fields of steel operations promises to a competitive advantage and sustainability to the steel industry. Real-time monitoring of steel production lines as well as steel structures using non destructive inspection is such a field. In this paper we present a sensor developed in our laboratory which when used in the non-destructive evaluation procedure also described in this work, it offers real-time monitoring of the structural integrity of magnetic steels, going beyond the state of the art by allowing for the correlation of macroscopic magnetic and magnetoelastic parameters to the total of the hydraulic and residual stresses as a function of position.

Keywords: Magnetoelasticity · Magnetostriction · Ferromagnetic steels · Stresses tensor

1 Introduction

Ferromagnetic steels have a wide range of applications, such as in nuclear reactors, transformers and automotive gas industries. The efficiency of these structures is strongly related to the distribution of the internal stresses, which exist within a material. The residual stresses are responsible for lattice distortions, which in turn lead to a rearrangement of the material's magnetic domains [1–3].

The total stress tensor (I, II and III type) is increased when a mechanical field is applied in a ferromagnetic material. This increase in internal energy affects the microstructural configuration, as well as the magnetic, mechanical and electrical properties of the material. Hence, the residual stress state in a material should be evaluated, in order to effectively monitor the health of a ferromagnetic steel structure.

R. Szewczyk and M. Kaliczyńska (eds.), *Recent Advances in Systems, Control and Information Technology*, Advances in Intelligent Systems and Computing 543, DOI 10.1007/978-3-319-48923-0_68

Magnetic non-destructive methods use the combined knowledge of the structural and the magnetic configuration for testing and evaluating metallurgical, microstructural and, mechanical properties of various ferromagnetic steel grades [6–13].

A large number of studies has been devoted to the exploitation of semi- and non-destructive techniques in measuring residual stresses, such as X-ray and neutron diffraction and quasi-dc magnetic permeability measurement [4, 5, 14, 15].

The above-mentioned measurements require the development of proper magnetic sensors to capture the desired data and process them in order to retrieve the desired information.

This paper aims at presenting our new technology concerning the development and application of a holistic methodology for the online monitoring of the structural integrity and annihilation/control of stresses in critical steel structures in various applications. In this paper, a well-developed new method is illustrated in order to monitor the stress tensor distribution on the surface and in the bulk of magnetic steels, as well as to provide initial experimental evidence on our relatively new electromagnetic forming method.

2 Materials and Methodology

The sensor presented here is related to a method and apparatus for monitoring the distribution of hydraulic (stresses type I) and residual stresses (stresses types II and III) in ferromagnetic steels within the elastic region as well as the plastic deformation region.

The sensor set up (Fig. 1) has the ability to measure both the magnetic permeability, μ(H), and magnetostriction, λ(H), properties of the material as a function of the applied field at the measuring point. These magnetic parameters are directly related to the algebraic sum of the residual (type II and III) and hydraulic stresses (type stresses I) at the point under test, when the shape of the under test steel allows for an acoustic waveguide – type response, i.e. small square cross-section rods of steel, I- and/or H-shape, where the length of the sample is significantly longer than the cross section area.

As illustrated in Fig. 1, the excitation/receiving coil (3a) and (3b) and the bias coils (2a) and (2b) are placed around at the large steel axis (5). In this arrangement, it is possible to measure the magnetic permeability, μ(H), and magnetostriction, λ(H), at the measuring point along the axis of the excitation/receiving coils.

One of the two excitation/receiving coils (3a) is acting as the excitation coil, generating a low frequency magnetic field which penetrates through the under measurement

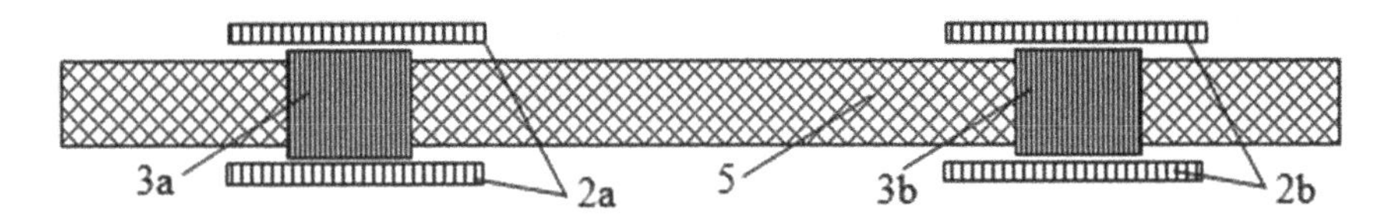

Fig. 1. Sensor arrangement in under test steel.

steel sample (5), which in turn creates elastic waves due to the movement of the 90° magnetic domain walls. These elastic waves are basically transmitted longitudinally in the direction of the long axis of the examined steel (5), in several modes. The longitudinal mode of the elastic wave, being the fastest to propagate, reaches first and quite isolated from the rest of the elastic waves in the steel region (5) beneath the other excitation/receiving coil (3b) which senses the mentioned elastic wave as an output voltage, due to the inverse magnetostrictive effect and thus functions as the receiving coil.

Thereafter, the supply of adjustable intensity bias current in the biasing coil (2a) located around the excitation/receiving coil (3a) generates a bias field at the one end of the examined steel (5), resulting in the modulation of the output voltage (Vo) of the excitation/receiving coil (3b), the maximum value of which is proportional to the mean value of the magnetostriction (λ) of the volume of the under test steel sample (5) enclosed by the excitation/receiving coil (3a), according to the formula:

$$V_o = C_1 \frac{d\lambda}{dH} \tag{1}$$

where C_1 is a constant, determined from the experimental data.

Similarly, the generation of a biasing field from the biasing coil (2b) located around the excitation/receiving coil (3b) leads to the modulation of the peak of the output voltage (V_o) at the excitation/receiving coil (3b), the peak of which is proportional to the differential magnetic permeability (μ) of the volume of the steel, enclosed by the excitation/receiving coil (3b):

$$V_o = C_2 \frac{dM}{dH} = C_2 \mu \tag{2}$$

where C_2 is a constant, determined from experimental data.

In this way, the magnetostrictive function $\lambda(H)$ of the steel volume, inside the excitation/receiving coil (3a) and the magnetic permeability function $\mu(H)$ of the steel inside the excitation/receiving coil (3b), are determined.

Reversing the role and the function of the two coils (3a) and (3b), using (3b) as the excitation coil and (3a) as the receiving coil, the measurement of the magnetostrictive function $\lambda(H)$ in the steel volume enclosed by the coil (3b) and the magnetic permeability function $\mu(H)$ of the steel volume enclosed by coil (3a) is achieved. Thus, the measurement of the $\lambda(H)$ and $\mu(H)$ functions of the steel volume located beneath the two excitation/receiving coils (3a) and (3b) is obtained.

In this way and by moving the coils (2a)–(3a) and (2b)–(3b) along the sample length (5), the measurement of the distribution of the magnetic permeability $\mu(H)$, and the magnetostriction $\lambda(H)$ along the steel and hence the distribution of the mechanical stress along the under measurement steel (5), is obtained.

3 Discussion

The measurement of the tensor of the algebraic sum of the residual (type II and III stresses) and the hydraulic stresses (type I stresses) through its correlation with the tensor of magnetic permeability μ(H) and magnetostriction λ(H) is as follows:

The μ(H) function, follows a monotonic response up to the Villari point of the under-measurement steel, wherein up to that point, the corresponding λ(H) function also follows a monotonic change; after Villari point, the slope of μ(H) function changes and the corresponding λ(H) function decreases to negative values. In steels with low energy of stacking errors and a positive magnetostriction coefficient in the elastic region, the μ(H) function is continuously increasingwhile the λ(H) function is continuously decreasing up to the Villari point for positive (tensile) values of applied stress; after the Villari point, the μ(H) function is continuously decreasing and the λ(H) function is getting negative values up to the Ultimate Tensile Stress (UTS) point. The reverse is valid for negative (compressive) stress values. In steels with low energy of stacking errors and negative magnetostriction coefficient in the elastic region, the μ(H) function is continuously decreasing for positive (tensile) stress values and the λ(H) function has negative values up to the Villari point, while after the Villari point, the μ(H) function is continuously increasing and the λ(H) function is getting positive values.

In Fig. 2, a typical response of the μ(σ) function with the applied positive mechanical stress and the waveform of the λ(H) curve as a function of the stress of the examined steel with positive magnetostriction coefficient in the elastic region, is shown. The reverse response is observed for negative values (compressive) of stress, where the μ(H) function will be decreasing until the negative Villari point and subsequently it will be increasing until the UTS point. The exactly opposite behavior will be observed on a steel with a negative magnetostrictive function in the elastic region.

Consequently, the monotonic correlation of the local value of the magnetic permeability μ with the local value of the algebraic sum of the residual stresses (type II and III stresses) and the hydraulic stress (type I stresses) along the measuring axis, in the region of the excitation and reception of the elastic waves, becomes possible, by

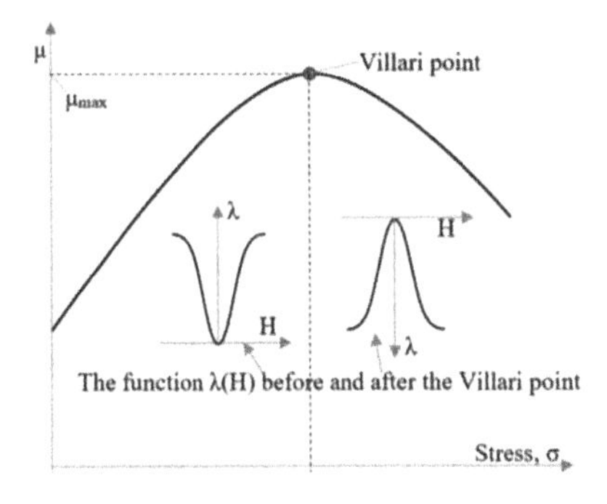

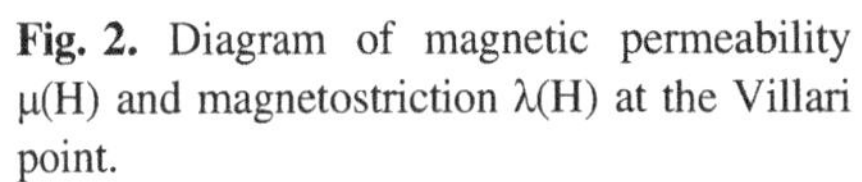
Fig. 2. Diagram of magnetic permeability μ(H) and magnetostriction λ(H) at the Villari point.

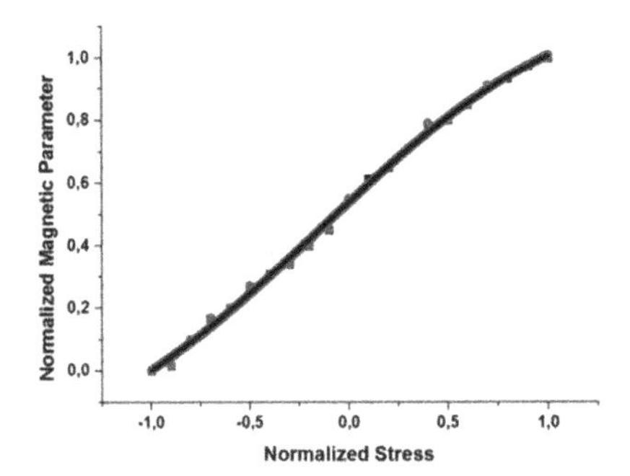

Fig. 3. Normalized stresses calibration curve.

selecting one of the two μ(H) functions, based on the measured value of the λ(H) function. Note that the mechanical residual stresses (type II and III stresses) are measured by standard techniques such as X-ray diffraction with the Bragg - Brentano technique (XRDBB) and neutron diffraction (ND) for surface and bulk measurements, respectively, while the hydraulic stress (type I stresses) measurement corresponds to the macroscopic stress, applied throughout the length of the steel. Thus, the Magnetic Stress Calibration Curve (MASC) up to the UTS point of the under-test steel is obtained.

Having obtained the appropriate MASC, through correlating the magnetic permeability with the mechanical stress along the axis of the examined steel, it becomes possible to measure the algebraic sum of the residual stresses and the hydraulic stress in the steel volume (5) enclosed by excitation and receiving coils (Fig. 1).

The normalization of the calibration curve with respect to the permeability value, μ, at the Villari point and the stress value at the yield point and UTS point, leads to a global MASC curve, the same for any type of steel, such as the curve shown in Fig. 3. Based on this curve, it is not necessary to know the type of the examined steel.

In this way, there is no need of knowledge of the stoichiometry and the phase of each steel: the measurement of the permeability function μ(H) and the hysteretic function λ(H), allows the finding of the normalized value of the magnetic parameters and hence, the value of the component of the stress in the measuring axis.

4 Conclusions

The proposed system allows the measurement of the local magnetic differential permeability μ(H) and magnetostriction λ(H) as a function of an externally applied magnetic field.

The sensor presented has been developed to solve the problem of providing a measuring system which can determine the distribution of the overall stresses, both on the steel surface and through the cross-section, by means of correlating the magnetic and magnetoelastic properties to the algebraic sum of the hydraulic and residual stresses in a steel sample.

References

1. Iordache, V.E., Hug, E., Buiron, N.: Magnetic behaviour versus tensile deformation mechanism in a non-oriented Fe-(3 wt.%) Si steel. Mater. Sci. Eng. A **359**, 62–74 (2003)
2. Perevertov, O.: Influence of the residual stress on the magnetization process in mild steel. J. Phys. D Appl. Phys. **40**, 949–954 (2007)
3. Cullity, B.D., Graham, C.D.: Introduction to Magnetic Material, 2nd edn. Addison-Wesley, London (1972)
4. Vourna, P., Hervoches, C., Vrána, M., Ktena, A., Hristoforou, E.: Correlation of magnetic properties and residual stress distribution monitored by X-ray & neutron diffraction in welded AISI 1008 steel sheets. IEEE Trans. Magn. **51**(1), 1–4 (2015)

5. Vourna, P., Ktena, A., Tsakiridis, P.E., Hristoforou, E.: A novel approach of accurately evaluating residual stress and microstructure of welded electrical steels. NDT&E Int. **71**, 33–42 (2015)
6. Martinez-de-Guerenu, A., Gurruchaga, K., Arizti, F.: Nondestructive characterization of recovery and recrystallization in cold rolled low carbon steel by magnetic hysteresis loops. J. Magn. Magn. Mater. **316**, e842–e845 (2007)
7. Piotrowski, L., Augustyniak, B., Chmielewski, M., Labanowski, J., Lech-Grega, M.: Study on the applicability of the measurements of magnetoelastic properties for a nondestructive evaluation of thermally induced microstructure changes in the P91 grade steel. NDT&E Int. **47**, 157–162 (2012)
8. Kobayashi, S., Kikuchi, N., Takahashi, S., Kamada, Y., Kikuchi, H.: Magnetic properties of α' martensite in austenitic stainless steel studied by a minor-loop scaling law. J. Appl. Phys. **108**, 043904-1–043904-8 (2010)
9. Vértesy, G., Mészáros, I., Tomáš, I.: Nondestructive magnetic characterization of TRIP steels. NDT&E Int. **54**, 107–114 (2013)
10. Stupakov, O.: Controllable magnetic hysteresis measurement of electrical steels in a single-yoke open configuration. IEEE Trans. Magn. **48**, 4718–4726 (2012)
11. Stupakov, O., Perevertov, O., Stoyka, V., Wood, R.: Correlation between hysteresis and barkhausen noise parameters of electrical steels. IEEE Trans. Magn. **46**, 517–520 (2010)
12. Kikuchi, H., Ara, K., Kamada, Y., Kobayashi, S.: Effect of microstructure changes on barkhausen noise properties and hysteresis loop in cold rolled low carbon steel. IEEE Trans. Magn. **45**, 2744–2747 (2009)
13. Tomáš, I., Kadlecová, J., Vértesy, G.: Measurement of flat samples with rough surfaces by magnetic adaptive testing. IEEE Trans. Magn. **48**, 1441–1444 (2012)
14. Rossini, N.S., Dassisti, M., Benyounis, K.Y., Olabi, A.G.: Methods of measuring residual stresses in components. Mater. Des. **35**, 572–588 (2012)
15. Vourna, P., Ktena, A., Svec, P., Hristoforou, E.: Universality law on the dependence of magnetic parameters on residual stresses in steels. IEEE Trans. Magn. **52**(5) (2016) doi:10.1109/TMAG.2015.2509642

Study of Ultrasonic Characteristics of Ukraine Red Granites at Low Temperatures

Valentyn Korobiichuk(✉)

National Technical University of Ukraine "Kiev Polytechnic Institute", Kiev, Ukraine
korobiichykv@gmail.com

Abstract. An experimental study of frozen fine-grained red granite was performed ultrasonically, and the ultrasonic wave propagation dependences on the natural stone temperature were established. The study showed that the natural stone porosity affects the acoustic wave propagation in the permafrost significantly. The ranges of the ultrasonic superficial wave propagation in monolithic granite samples are established.

Keywords: Ultrasound · Natural stone · Negative temperatures · Natural stone porosity

1 Introduction

The mining is associated with excavations in the rocks, breaking the existing balance in the array and forming some stress field, together with the tectonic and gravitational stresses. This field is additionally imposed by an ever-changing field of thermal stress and strain caused by mechanical impact on the rock array [1, 2], the intensity of which depends on the environment temperature [3, 4]. To study the complete picture of processes occurring in the rock array, it is required to choose a non-destructive testing method [5]. The authors chose an ultrasonic method, since the growth of stresses in the array increases the speed of elastic waves passage through the same, so, measuring the latter, the stressed condition and the nature of its temporal and spatial dynamics can be determined.

This issue was studied by a number of scientists, i.e. Stavros K. Kourkoulis [6], E. Molina, G. Cultrone, E. Sebastián and F.J. Alonso [7], Mauro Bramanti, Edoardo Bozzi [8], L.M. del Río, F. López, F.J. Esteban, J.J. Tejado, M. Mota, I. González, J.L. San Emeterio, A. Ramos [9], V.S. Yamshchikov, M.P. Nisnevich [10], M.P. Mokhnachov [11], A.P. Dmitriev [12], M.T. Bakka, I.S. Redchits, and V.S. Redchits [13]. These scientists investigated d the behavior of ultrasonic waves in different rocks. Besides, the review of their papers showed that each of the acoustic characteristics is uniquely associated with a certain kind of rock deformation; the velocity of ultrasound propagation is related to elastic deformations, while the damping factor – with the plastic ones. The variation factor of the estimated compressive strength stress by means of ultrasound is ±10–25%, which is much less than the adopted value of the variation factor for the rock strength index determined by mechanical tests on the samples.

R. Szewczyk and M. Kaliczyńska (eds.), *Recent Advances in Systems, Control and Information Technology*, Advances in Intelligent Systems and Computing 543, DOI 10.1007/978-3-319-48923-0_69

The possibility of determining the elastic and strength properties of rocks by ultrasonic method with higher reliability compared to the static ones open up broad prospects for the field research methods. The analysis of scientific publications shows that heating of most rocks leads to changes in the acoustic characteristics [14], while the temperature changes in the different rocks vary.

2 Materials and Methods

The effect of temperature on the change of velocity in the rock is the most obvious at 150–200 °C and higher, i.e. in the range where the presence of mineral and external moisture in the rock has no effect. Still, different rocks have three most typical relationship:

- granite type rocks feature a noticeable, almost monotonous speed reduction with the temperature;
- gabbro and peridotite type rocks feature a speed reduction up to 535–550 °C, after which it shows a smooth increase;
- quartzite type rocks show an even transformation up to 573 °C, and thereafter it becomes sharp (polymorphic).

The main factors affecting the change in velocity of elastic waves is change in the crystal elasticity, forming the body, and relaxation processes at the boundaries between grains. It is considered that the deviation from the monotony upon temperature-related speed reduction is mainly due to the relaxation processes [15–17]. Identification of speed dependence on the temperature and its use as an informative sign is associated with a number of special problems of the mining practice:

- taking into account the influence of temperature when monitoring the properties and status of rocks at great depths;
- with direct monitoring of the process and quality of rock hardening when heated or frozen (since this is the only method of studying the dynamics of change, it can be used to monitor the mining process);
- using the dependence of the rock properties in the simulation of stressed array.

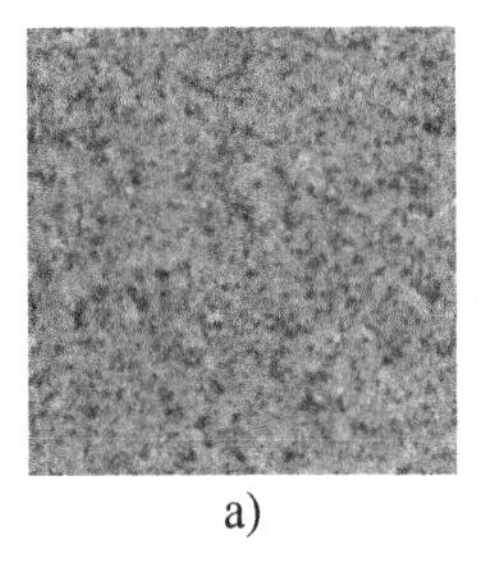
a)

b)

c)

Fig. 1. (a) Mezhdurechensk granite deposit (Flower of Ukraine), (b) Tokovskiy granite deposit (Carpazi), (c) Leznikovskiy granite deposit (Maple Red) (Color figure online)

Table 1. Properties and mineral composition of granites

Indicators	Mezhdurechenskiy	Tokovskiy	Leznikovskiy
Bulk density, kg/m^3	2600–2650	2670	2650
Compressive strength, MPa	138.3–203	225	135–260
Abrasion, g/cm^2	0.42–0.53	0.26	0.24
Water absorption,%	0.18–0.51	0.55	0.18
Porosity, %	1.88	0.56	0.37
Granite mineral composition			
Plagioclase, %	20	35	
Quartz, %	15	25	15
Microcline, %	60	35	80
Biotite, %	5	4	5
Other minerals, %	<1	1	

Since the Ukrainian granites are unexplored, the authors decided to test three rocks of fine-grained red granites (Fig. 1) – Mezhdurechenskiy (Flower of Ukraine), Tokovskiy (Carpazi), and Leznikovskiy (Maple Red).

The properties and mineral composition of the studied granites are shown in Table 1.

3 Results and Discussion

In order to study these rocks, the cubes were produced. The cubes were laid in the freezer for a day, where the temperature dropped from 1 °C to –40 °C for 10 min. The cubes contain process holes for temperature sensors.

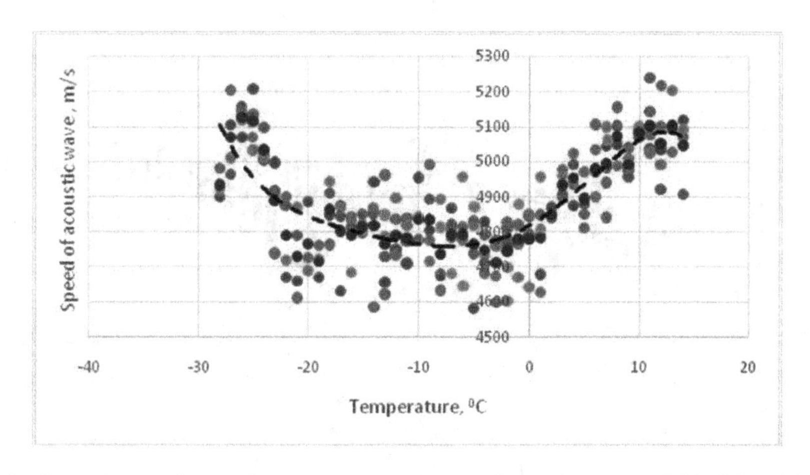

Fig. 2. Dependence of acoustic wave propagation on the temperature of Mezhdurechenskiy granite deposit (acoustic wave velocity)

Next, the cubes were removed from the chamber and sonicated by Pulsar 2.2 [18] with the handle containing a 120 mm sensor base. After each degree, the readings of surface sounding were recorded. The results are shown in Figs. 2, 3 and 4.

The dependence of the acoustic wave propagation on the temperature of Mezhdurechenskiy granite deposit can be described as follows:

$V = -0.0001t^5 - 0.004t^4 - 0.0195t^3 + 1.3445t^2 + 17.788t + 4818.7$, m/s

$R^2 = 0.7643$

The dependence of the acoustic wave propagation on the temperature of Leznikovskogo granite deposit can be described as follows:

$V = -0.00007t^5 - 0.0027t^4 + 0.0192t^3 + 0.8916t^{2-3}.982t + 5370.4$, m/s

$R^2 = 0.6383$

The dependence of the acoustic wave propagation on the temperature of Tokovskiy granite deposit can be described as follows:

$V = -0.0002t^5 - 0.0103t^4 - 0.0721t^3 + 1.3915t^2 - 10.772t + 5505.5$, m/s

$R^2 = 0.8228$

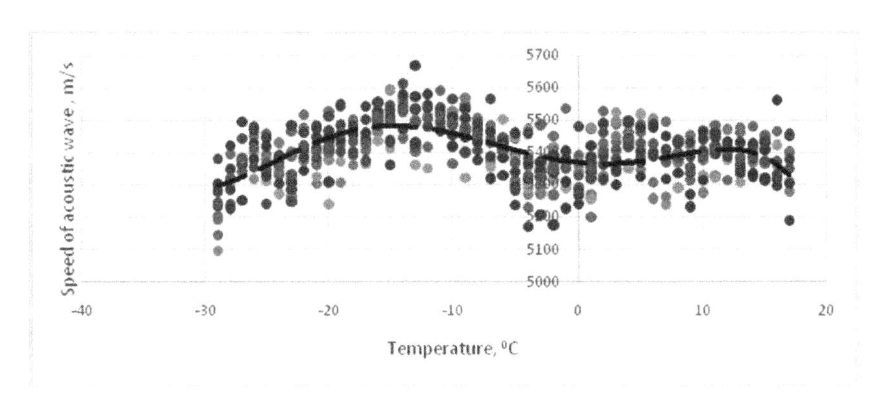

Fig. 3. Dependence of acoustic wave propagation on the temperature of Leznikovskyi granite deposit

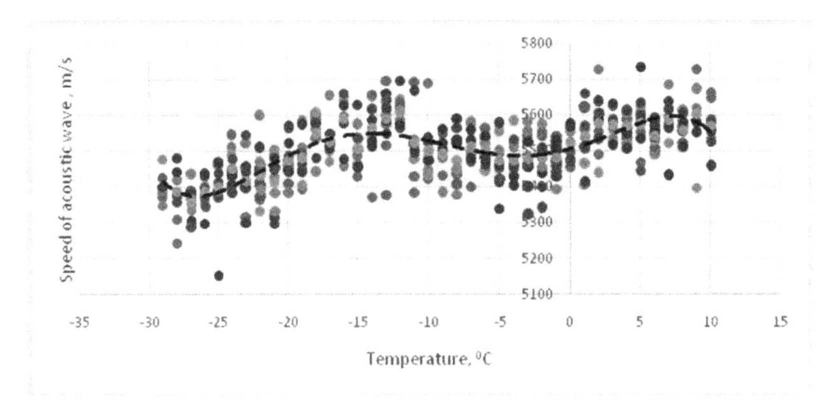

Fig. 4. Dependence of acoustic wave propagation on the temperature of Tokovskiy granite deposit

4 Conclusions

Experimental study of frozen rock showed that a decrease in temperature even within a relatively small range (0 to –20 °C) leads to a significant change in the velocity of elastic waves. As the temperature decreases from positive to negative subject to zero crossing, the speed is changed abruptly. This is due to the internal structure of permafrost, which can be represented as two-phase systems consisting of a solid skeleton and substance filling the pores.

The chart analysis shows that the behavior of rocks from Tokovskiy and Leznikovskyi deposits during freezing and heating is the same, while the chart of changes in acoustic wave velocity at Mezhdurechenskiy granite deposit is different. This can be explained by a twofold increase in the number of pores compared to the above rocks.

References

1. Kosolapov, A.I.: Research and substantiation of the mining technology for upland fields of inlaid marble. Thesis of the Doctor of Engineering, Krasnoyarsk, p. 284 (1993)
2. Korobiichuk, I., Korobiichuk, V., Nowicki, M., Shamrai, V., Skyba, G., Szewczyk, R.: The study of corrosion resistance of Pokostivskiy granodiorites after processing by various chemical and mechanical methods. Construct. Build. Mater. **114**, 241–247 (2016)
3. Kaderov, M.Y.: Substantiation of the technology of combined marble deposit development in the harsh climatic conditions: thesis of the Ph.D. in Engineering, Krasnoyarsk, p. 163 (2009)
4. Wen, L., Li, X., Su, W.: Study of physico-mechanical characteristics of slope hard rocks of metal mine influenced by freeze-thaw cycles. J. Min. Safety Eng. **32**(4), 689–696 (2015). doi:10.13545/j.cnki.jmse.2015.04.027. Caikuangyu Anquan Gongcheng Xuebao
5. Korobiichuk, I., Podchashinskiy, Y., Shapovalova, O., Shadura, V., Nowicki, M., Szewczyk, R.: Precision increase in automated digital image measurement systems of geometric values. In: Jabłoński, R., Brezina, T. (eds.) Advanced Mechatronics Solutions. AISC, vol. 393, pp. 335–340. Springer, Heidelberg (2016). doi:10.1007/978-3-319-23923-1_51
6. Kourkoulis, S.K.: Fracture and Failure of Natural Building Stones – Applications in the Restoration of Ancient Monuments, p. 592. Springer, Netherlands (2007)
7. Molina, E., Cultrone, G., Sebastián, E.: Alonso, F.J: Evaluation of stone durability using a combination of ultrasound, mechanical and accelerated aging tests. J. Geophys. Eng. **10**(3), 035003 (2013)
8. Bramanti, M., Bozzi, E.: A procedure to detect flaws inside large sized marble blocks by ultrasound. Subsurf. Sens. Technol. Appl. **2**(1), 1–13 (2001)
9. del Río, L.M., López, F., Esteban, F.J., Tejado, J.J., Mota, M., González, I., Emeterio, J.L.S., Ramos, A.: Ultrasonic characterization of granites obtained from industrial quarries of Extremadura (Spain). Ultrasonics **44**, 1057–1061 (2006)
10. Yamshchikov, V.S., Nisnevich, M.P.: Quality Control at the Enterprises of Non-metallic Building Materials, p. 264. Stroyizdat, Leningrad department, Moscow (1981)
11. Mokhnachev, M.P.: Rock Fatigue, p. 152. M.: Nauka, Moscow (1979)
12. Dmitriev, A.P.: Thermodynamic Processes in Rocks: Textbook for Universities, 2nd edn, p. 360. M.: Nedra, Moscow (1990)

13. Bakka, M.T., Redchits, I.S., Redchits, V.S.: Thermodynamics Fundamentals in Mining: Manual, p. 210. ZHITI, Zhytomyr (2000)
14. Karachun, V., Mel'nick, V., Korobiichuk, I., Nowicki, M., Szewczyk, R., Kobzar, S.: The additional error of inertial sensor induced by hypersonic flight condition. Sensors **16**(3), 299 (2016). doi:10.3390/s1603029. (ISSN 1424-8220)
15. Rzhevskiy, V.V., Yamshchikov, V.S.: Acoustic Methods of Investigation and Monitoring of Rocks in the Array, p. 223. M.: Nauka, Moscow (1973)
16. Ye, X.-F.: Mineralization and metallogenic model of fluorite deposits in the Zhejiang area. Northwest. Geol. **47**(1), 208–220 (2014)
17. Zhu, Y., Liu, Q., Kang, Y., Liu, K.: Study of creep damage constitutive relation of granite considering thermal effect. Chin. J. Rock Mech. Eng. **30**(9), 1882–1888 (2011). Yanshilixue Yu GongchengXuebao
18. http://interpribor.com/index.php/products-m/ultrasonic-devices/pulsar22

Study of the Durability of Reinforced Concrete Structures of Engineering Buildings

Lyudmyla Kuzmych[1(✉)] and Volodymyr Kvasnikov[2]

[1] National University of Water and Environmental Engineering, Rivne, Ukraine
l.v.kuzmych@nuwm.edu.ua
[2] National Aviation University, Kiev, Ukraine
kvp@nau.edu.ua

Abstract. The degradation process of the long-operated facilities, which include the reinforced concrete structures of engineering buildings, in particular the waterworks based on concrete structures, are characterized by physical and moral degradation processes. The physical degradation is caused by material wear and destruction as well as aging of concrete structures. As for the moral degradation, it is only 2 to 3% for waterworks. The durability of technical structures was assessed by testing the strength of concrete elements of hydraulic structures. The strength tests were carried out on the elements located below the ground surface level, which are in direct contact with moist environment. The tests were carried out using a standard Kashkarov hammer. To ensure and obtain the reliable findings, the trials were conducted in the upper and lower zones of elements with 10 prints in each. Each test series was verified for its affiliation to a sample by removal of abnormal research findings and determination of the required number of prints in the research. Based on research findings, the graph of the elements strength dependence on time was plotted. It was found that the strength of concrete hydraulic structures varies between 35 MPa and 44 MPa. The strength of concrete structures, the actual life of which is 31 years, while the standard life is 20 years, fell by only 10%, while the designed strength margin is 40%.

Keywords: Reliability · Durability · Strength · Engineering structures · Concrete structures

1 Introduction

The current level of technical facilities brings the issue of ensuring the technical level, quality, reliability and efficiency of engineering objects to the fore. Improvement of the calculation methods during design, aimed at reduction of material consumption and energy intensity, continues to be one of the most pressing issues [1–6].

The need to resolve these issues is also due to the fact that assessment of the engineering facilities in terms of reliability instead of an insufficiently reasoned intuitive-empirical approach allows endowing the elements and materials with properties required in specific conditions and time intervals [7].

During the operation of technical facilities, the influence of various factors cause wear of material resource, aging and damage of elements, resulting in deterioration of

R. Szewczyk and M. Kaliczyńska (eds.), *Recent Advances in Systems, Control and Information Technology*, Advances in Intelligent Systems and Computing 543, DOI 10.1007/978-3-319-48923-0_70

the building functionality [8]. To consider these factors, it is required to analyze the objects from the point of view of reliability. Such research allow specifying the lifetime of engineering structures from the standpoint of their durability, specifying the terms of preventive maintenance, technical maintenance and supervision of such facilities.

2 Methods and Techniques

One of the main properties of the technical system reliability is its durability.

A durable technical facility is an object able to perform the specified functions, preserving the value of the set performance indicators over time within the established limit to set limits corresponding to specific modes and conditions of use, maintenance, repairs, storage and transportation [9]. Object means both separate elements and system as a whole.

Reliability, durability and sparing requirements are, therefore, in conflict [4, 5]. An increased reliability requires less maintenance costs with an increase in the system cost during construction, and vice versa, a cheaper system leads to reduced reliability and requires higher operating costs. The optimal reliability level H_{opt} is determined by comparing the costs to improve its reliability and the amount of losses from its reduction (Fig. 1).

The allowable minimum reliability level $H_{\min}$ is set by comparing the cost $B(t)$ of maintenance and repair with damage $Z(t)$ caused by reduced reliability [10]. If $Z(t) > B(t)$, it is required to carry out the repair and restoration works, and if $Z(t) < B(t)$, then the repairs are unnecessary (Fig. 1).

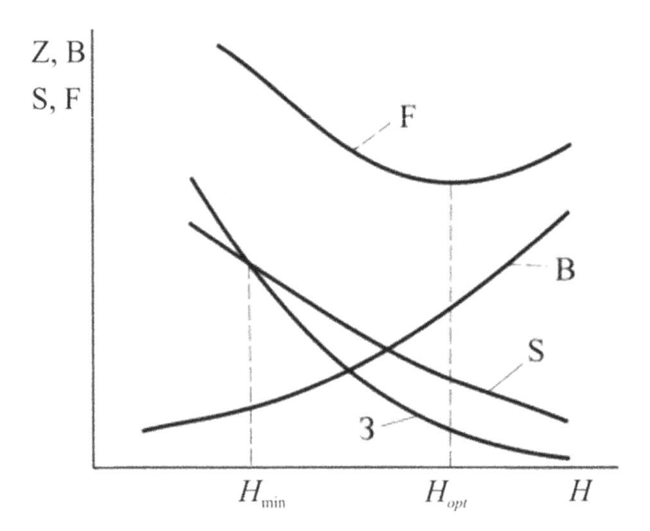

Fig. 1. Dependence of the technical facility reliability on the losses, maintenance costs, and reliability improvement costs: S – maintenance costs; B – reliability improvement cost; F – total expenditures; C – losses from reduced reliability

3 Results and Discussion

The analysis of publications and standards has shown that processes of degradation of the long-operated objects, which include the reinforced concrete structures of engineering buildings, in particular the waterworks based on concrete structures, feature the processes of physical and moral degradation. The physical degradation is formed as a result of material wear and destruction, as well as aging of concrete structures. As for the moral degradation, for waterworks, it is only 2–3% [6].

The macroscopic representation of waterworks operation in drainage system shows that the draining well is a part of the hydro-reclamation system dividing the drainage manifold into two parts connected with it in series, i.e. inlet and outlet manifolds. The drainage route reliability depends on each of these elements. Therefore, the requirements for operation of the drainage-regulating network, according to the system analysis, become the supersystem requirements for the drainage well operation [11].

The durability of drainage wells is assessed based on strength tests of the basic elements, i.e. concrete well rings. The strength tests were carried out on the elements located below the ground level in direct contact with moist environment.

The tests were carried out using a standard Kashkarov hammer [12].

To ensure reliable results, the tests were conducted in the upper and lower zones of the rings with 10 prints on each [13].

Each series of tests was verified for its belonging to the same sample by removal of abnormal test results and determination of the required number of prints in trials $[n]$ as follows [14, 15]:

$$[n] = 400K^2(R^*_{i,\max} - R^*_{i,\min})^2/\bar{R}^{*2}, \quad (1)$$

where K is a coefficient chosen depending on the number of prints. In our case, for 10 prints $K = 0.325$;

$\bar{R}^*$ is an arithmetic average strength of concrete based on individual print results (* – R concrete strength according to [9], however, according to [4, 5], R means the probability of failure-free operation);

$R^*_{i,\max}, R^*_{i,\min}$ are the maximum and minimum values of concrete strength among individual prints, respectively.

The objects of study were chosen so that to take into account the modern requirements for facilities, quality of construction and operation of facilities, different years of construction, typical technical solutions, and typical soil, hydrogeological and economic conditions.

The soil-forming rocks are water-glacial and mergelized deposits. The soil is represented by sod-podzolic and sod-gley sandy and loamy soils with humus horizon 18 cm to 20 cm thick. The sod gley soils occur in places of closed depressions with humus horizon of 20 cm to 30 cm. Soil waterlogging is mainly due to excessive rainfalls and close occurrence of the groundwater, which plays an important role, as in soils with high permeability at the groundwater depth of 1.2 m to 1.5 m, no waterlogging is observed even in extremely wet periods.

These soils are the most common in the Western Polissya of Ukraine [16]. Besides, the soil and hydrogeological conditions of the objects are typical for the region. [17]

The concrete strength tests were carried out for drainage wells with the lifetime of 18, 21, 24 and 31 years.

200 prints were taken in different areas of facilities subject to verification of their belonging to one sample.

The calculations determined that prints in two areas are abnormal, so they were excluded from further calculations.

The corresponding concrete strength R_m^* was found for each area.

Based on research, a graph of the concrete strength of drainage well rings over time was build (see Fig. 2).

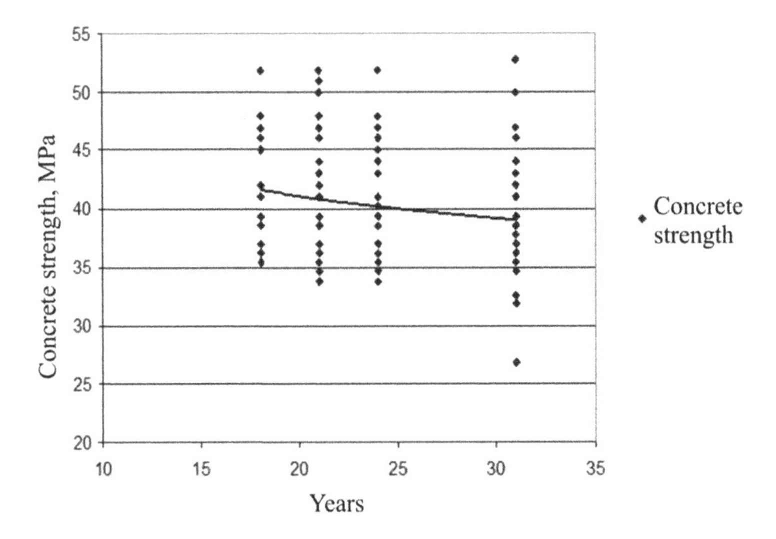

Fig. 2. Graph of concrete strength over time

4 Conclusions

It was found that the concrete strength of the drainage well rings ranges from 35 MPa to 44 MPa. The strength of concrete structures, the actual life of which is 31 years, while the standard lifetime is 20 years, fell by 10% only, while the design margin is 40% [12].

Thus, we can conclude that the durability of concrete rings material is sufficient to extend the service life of the drainage wells to the life of a closed drainage. The strength of concrete structures R_m^* meets the operation requirements.

Acknowledgement. I hereby express my gratitude to Rivne regional management of water resources, represented by its manager, Sergiy Vasilev, for the opportunity to conduct the test of concrete structure strength in the hydraulic structures on water management systems, as well as to my peers from the department of architectural design, construction and graphics fundamentals, in particular prof. Vasyl Romashko.

References

1. Kuzmych, L.: Prognostication of the hydrotechnical structures reliability. Water conservation and hydraulic engineering. Sci. Res. J. (27) (2002). Rivne
2. Chybowski, L., Żółkiewski, S.: Basic Reliability Structures of Complex Technical Systems. Springer International Publishing Switzerland (2015)
3. Korobiichuk, I., Bezvesilna, O., Ilchenko, A., Shadura, V., Nowicki, M., Szewczyk, R.: A mathematical model of the thermo-anemometric flowmeter. Sensors **15**, 22899–22913 (2015). doi:10.3390/s150922899
4. Halyamyn, E.P.: Principles of managing the process of hydro-reclamation system maintenance in the good working order: improvement of the reclaimed land effectiveness and use of water resources in land reclamation/MGMI.-M., pp. 178–187 (1988)
5. Kasharin, D.V.: Reliability assessment of lightweight hydrotechnical structures formed from composite materials. Power Technol. Eng. **43**(4), 230–236 (2009)
6. Krzysztof, K., Joanna, S.-B.: Reliability and Safety of Complex Technical Systems and Processes. Springer Science & Business Media, 405 p. (2011)
7. Korobiichuk, I., Shostachuk, A., Shostachuk, D., Shadura, V., Nowicki, M., Szewczyk, R.: Development of the operation algorithm for a automated system assessing the high-rise building. Solid State Phenom. **251**, 230–236. doi:10.4028/www.scientific.net/SSP.251.230
8. Korobiichuk, I., Korobiichuk, V., Nowicki, M., Shamrai, V., Skyba, G., Szewczyk, R.: The study of corrosion resistance of Pokostivskiy granodiorites after processing by various chemical and mechanical methods. Constr. Build. Mater. **114**, 241–247 (2016). doi:10.1016/j.conbuildmat.2016.03.147
9. GOST 2470-94. Machinery reliability. Engineering systems. Terms and definitions. State Standard of Ukraine. K. (1994)
10. Wu, S., Clements-Croome, D., Fairey, V., et al.: Reliability in the whole life cycle of building systems. Eng. Constr. Architectural Manage. 13(2), 136–153 (2006)
11. Kuzmych, L.: Development of organizational and technical measures for keeping items of hydrotechnical structures in the good condition. Materials of scientific conference of young scientists. The role of reclamation sustainable development of agriculture. IHM, Kyiv (2007)
12. Romashko, V.M.: Inspection and Testing of Facilities: Praktykum, 117 p. RDTU, Rivne (1999)
13. Avyrom, L.S.: Design Reliability of Prefabricated Buildings and Structures, 215 p. Publishing house on construction, L. (2001)
14. DSTU 2861-94. Machinery reliability. Reliability analysis. Main provisions. State Standard of Ukraine. K. (1994)
15. DSTU 2864-94. Machinery reliability. Experimental evaluation and reliability control. Main provisions. State Standard of Ukraine. K. (1994)
16. Krupskii, N.K., Polunin, N.I.: Atlas of Soils of the Ukrainian SSR, 160 p. Urozhai, K. (1979)
17. Sozinov, A.A., Kovalenko, P.I. (eds.): Fundamentals of Ecological and Land Reclamation Monitoring of the Ukrainian Polissya. Lutsk, Kyiv (1992)

Comparison and Analysis of Steel and Tungsten Carbide Rockwell B Hardness Ball Indenters Utilizing a General Purpose Finite Element Approach

Vladimir Skliarov[1(✉)] and Maxim Zalohin[2]

[1] National Scientific Centre "Institute of Metrology", 42, Myronosytska Str., Kharkiv 61002, Ukraine
vladimir.skliarov@metrology.kharkov.ua
[2] Kharkiv National Automobile and Highway University, 25, Petrovskogo Str., Kharkiv 61002, Ukraine
zalogin_maxim@mail.ru

Abstract. This paper considers the comparison and evaluation of Rockwell B (HRB) hardness tests using the steel and tungsten carbide ball indenters. There are differences observed in Rockwell B hardness scale tests when using 1.588 mm diameter steel and tungsten carbide indenters. The modeling was made with HRB hardness reference blocks and widespread soft materials. During the simulation it was determined that the tungsten carbide indenter balls had an advantage of being less likely to flatten due to the repeated use, and the use of hard metal indenters might have different measurement results than tests using steel indenters. For further analysis of the results of comparisons it is interesting to conduct researches for Superficial Rockwell scales with the study of very soft materials and the use of spherical indenters with the larger sizes. The obtained results are important for evaluation of the results of international comparisons of the national standards.

Keywords: International comparison · National standard · Steel ball indenter · Tungsten carbide ball indenter · Rockwell B hardness scale · Finite element method · Experimental researches

1 Introduction

The international recognition of the state primary standard of physical units and the confirmation of its metrological characteristics can only take place on the basis of international comparisons with similar national standards (standards) of other countries. The algorithm for international comparisons is determined by the technical protocol of comparisons and it must comply with the requirements [1–3]. When analyzing the results of international comparisons is it often necessary to explain the reasons for the significant differences between the hardness measurement results of the participants of comparison.

R. Szewczyk and M. Kaliczyńska (eds.), *Recent Advances in Systems, Control and Information Technology*, Advances in Intelligent Systems and Computing 543, DOI 10.1007/978-3-319-48923-0_71

The specific feature of the hardness measurement using Rockwell scale is the availability of two types of indenters–spheroconical one (HRA, HRC) and spherical one (HRB scale). Determination of hardness depends on the result of measurement of the greatest number of different parameters, such as load value (for preliminary and total load), the diameter of the spherical indenter, the radius and angle spheroconical indenter, the penetration depth of the indenter, load application time (for preliminary and total load). Moreover, the spherical indenter can be made of steel or tungsten carbide. The study and comparison of the results of measurements using steel and tungsten carbide indenters are presented in the works [4–6]. The present work confirms the importance of compliance with the technical protocol of measurements when conducting the international comparisons and many years of experience in operating the National standard of Rockwell hardness in the NSC "Institute of Metrology".

2 Modeling of Indentation Process

The modeling of indentation process is performed by finite element method in the academic version of ANSYS software product, which is available for free use on the website [7]. Table 1 shows the modeling of indentation for materials by different types of indenters [8].

Table 1. Materials for modeling

Specimens	Hardness reference block	Copper M1	Aluminum D16	Aluminum B95	Steel 20
European analog	AISI 1000 Series Steel	UNS C11000	Aluminum 2024-T 361	Aluminum 7075-T73	AISI 1020 Steel
Brinell hardness, HB	86.0–500	–	130	135	143
Rockwell hardness, HRB	49.0–100	35	80	82	78

The indenter ball with diameter of 1.588 mm was used for determining the hardness by Rockwell scale (HRB). In order to determine the influence of the indenter material on the uncertainty of measurement of the specimen hardness, modeling was performed with three different indenters, mechanical properties of which are shown in Table 2 [4].

Table 2. Mechanical properties of indenters

Indenters	Young's Modulus, E, GPa	Poisson's ratio, υ
Steel ball indenter	203.4	0.3
Tungsten carbide ball indenter	633	0.22
Rigid ball indenter	∞	N/A

It should be noted that the numerical values of the parameters determining the mechanical properties of the materials, were found from [8] and are given in Table 3.

Figure 1 shows the representation of a non-linear stress-strain material behavior and the graphic determination of the parameters of materials [9, 10]. From Fig. 1 it is seen that the strain diagram is nonlinear all over the range of deformation. To simulate the behavior of the material in the framework of ANSYS Academic, it is reasonable to approximate the strain diagram by several straight-line segments.

Table 3. Mechanical properties of materials

Mechanical properties	Materials				
	AISI 1000 Series Steel	UNS C11000	Aluminum 2024-T 361	Aluminum 7075-T73	AISI 1020 Steel
Modulus of Elasticity, E_0, GPa	200	120	73.1	72	186
Poissons Ratio, ν	0.29	0.33	0.33	0.33	0.29
Tensile Strength, Yield, $\sigma_{0,2}$, MPa	305	220	393	386	330
Tangential Elastic Module, E_t, MPa	1650	1200	2400	2000	1800

Considering that there are small plastic deformations when carrying out indentation of specimen by ball, in ANSYS Academic framework it was accepted to use a physical model of the material, which takes into account the bilinear approximation of diagram with two straight sections (model BISO) [4] (Fig. 1b).

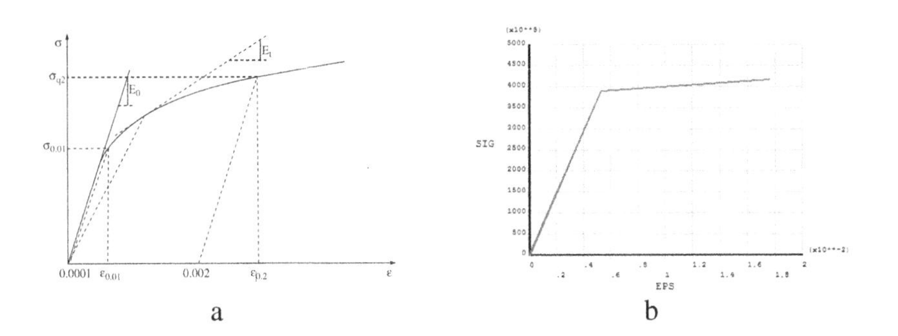

Fig. 1. Typical non-linear stress-strain curve

The slope of the first curve section is determined from the elastic properties of the material, such as elastic modulus E_0 and Poisson's ratio ν, and reflects the elastic deformations area. The slope of the second section is defined by the tangent modulus of elasticity E_t and represents the area of plastic deformations. In this connection, the tangent modulus of elasticity E_t is determined according to the tensile yield strength $\sigma_{0.2}$ of the material by a known method presented in [9, 10].

3 Statement of the Contact Problem

In order to reduce the amount of computing resources and increase the rate of solving the problem, the geometry of spherical indenter, which is in contact with the investigated specimen, is reasonable to represent as a two-dimensional axially symmetrical model projected on the X-Y plane, which passes through the axis of the ball. Thus, Y axis is a symmetry axis and the indenter axis. To create a finite element model of objects, 4-node type of 2-D finite element PLANE 182 was used (Fig. 2a). To create a contact interaction between the surfaces of two bodies, the contact 2-D elements such as Surface-to-Surface CONTA TARGE 171 and 169 were used (Fig. 2b).

The contact modified Lagrangian method was used as an algorithm, since it is less sensitive to the magnitude of the contact stiffness. Thus, the contact stiffness was determined by setting the depth of indenter penetration rate FKN = 0.1 into a specimen and contact stiffness factor FTOLN = 0.1. Given that in the statement of the problem the two contact surfaces comprise the shearing stresses τ determining slipping in the contact area, also the maximum shearing stress limit $\tau_{max} = \sigma_{0.2}/3^{1/2}$ [12, 13] was preset.

Requirements for the measurement of hardness by HRB scale, described in [1–3], stipulate the method of fixing the specimen and the nature of the application of the load on a spherical indenter. According to the above, in the physical model ANSYS the force F(t) (98.07 N then 980.7 N) acts on the indenter, and the specimen is fixed on the bottom surface in all directions. The coefficient of friction between the specimen and a spherical indenter is $\mu = 0.1$. Taking into account the geometric nonlinearities, in order to avoid the weak convergence of the simulation result, the full Newton-Raphson method was used for the solution of the contact problem [11, 12].

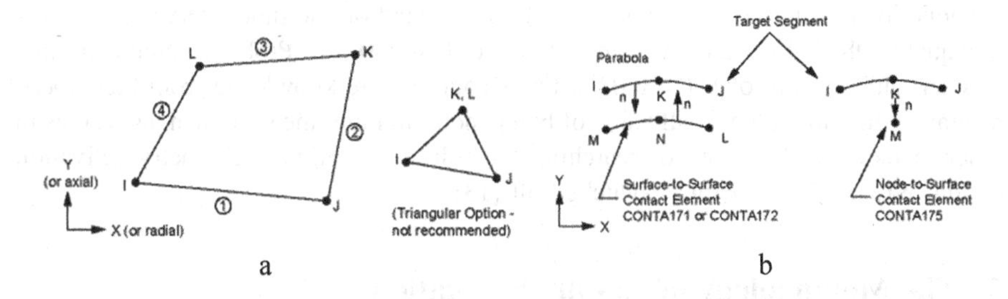

Fig. 2. PLANE 182 (a), CONTA 171 and TARGE 169 geometry (b)

4 Building and Solving 2-D Finite Element Model

In the presented form, the solution of the contact problem is a classical solution of Heinrich Hertz problem for flat surface and sphere. The classical problem of Heinrich Hertz had found a practical solution for determining the differences in the measurement of hardness. In this regard, using the capabilities of ANSYS Academic, an input-file was created in APDL programming language, the content of which reflects: building of geometry, setting the type of finite elements and material properties, net partition into finite elements, the application of the boundary conditions and solver setup options.

Using the above finite elements, automatic meshing of two bodies with a given finite element size of 0.05 mm was made (see Fig. 3a). For better convergence of the problem solution, finite element mesh refinement was performed in the contact area. As a result, 2002 finite elements and 1968 nodes were created in the mesh. Considering axially symmetric of geometry, for clarity, it is possible to represent 1/2 of the model by turning the plane around the axis of symmetry (Fig. 3b).

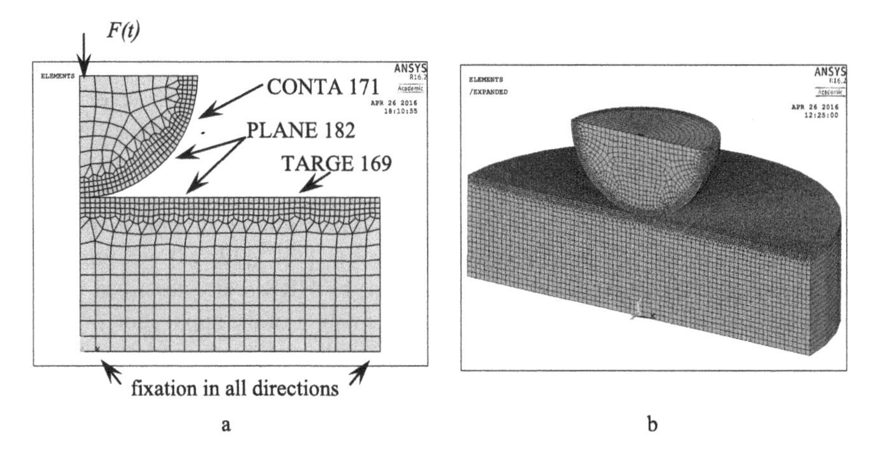

Fig. 3. FE model for material and indenter

The solution of contact problem was obtained with the response of the system to a dynamic load change F(t). In this connection, the load on the time interval of 3 s was set equal to 98.07 H, and on the time interval of 6 s–980.7 H. Problem simulation time was set with spacing of DT = 0.02 s. On account of pre-knowledge from the experimental studies about the small value of ball penetration into the specimen, as well as in order to increase the speed of searching the solution, nonlinear geometry activation option (NLGEOM, OFF) was switched off [13].

5 The Methodology of Results Evaluation

According to the classic definition and [4], the hardness measurement is carried under the influence of preliminary and total force (load). It is measured the depth of the indentation of the preliminary force and the depth indentation after removing the total force. The difference between the depth of the indentation $\Delta h = h(unloading) - h(loading)$ (see Fig. 4) is substituted into the formula (1).

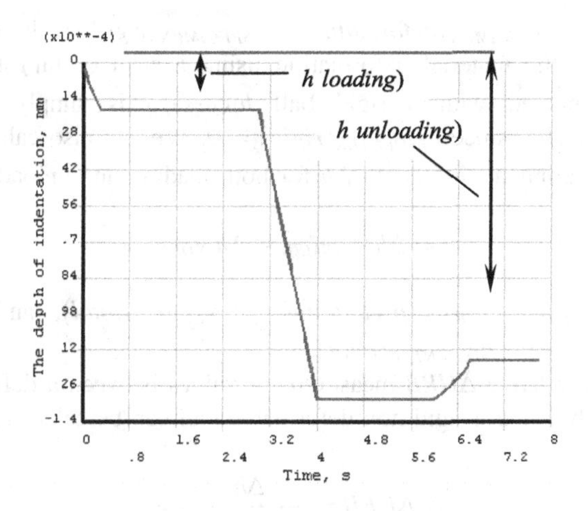

Fig. 4. The levels of loading and unloading

$$HRB = 130 - \frac{\Delta h(mm)}{0.002(\mathrm{mm})}, \tag{1}$$

Since all indenter material deforms under loading, the measurement of indentation depth h (for both $h(loading)$ and $h(unloading)$) includes both the depth, $h_{SPECIMEN}$, due to the deformation of the material being indented, as well as the depth, $h_{INDENTER}$, due to the deformation of the ball in the direction of the loading, as

$$h = h_{SPECIMEN} + h_{INDENTER}. \tag{2}$$

For the non-deformable rigid indenter $h_{INDENTER}(rigid) = 0$. To study the effect of deformable indenters on the indentation depth, we calculate the difference between the indentation depth of a steel or WC ball as compared with the indentation depth that would occur with a rigid indenter, and refer to this difference as a relative indentation depth $\tilde{h}$ as,

$$\tilde{h} = h(deformable) - h(rigid) \tag{3}$$

where $h(rigid) = h_{SPECIMEN}(rigid) + h_{INDENTER}(rigid)$ is the indentation depth modeled for a rigid ball, and:

$$h(deformable) = h_{SPECIMEN}(deformable) + h_{INDENTER}(deformable). \tag{4}$$

Thus $h(deformable)$ it is the depth of the indentation for steel or tungsten carbide ball indenters. Using (2) and (3) we obtain:

$$\tilde{h} = \tilde{h}_{SPECIMEN} + \tilde{h}_{INDENTER}, \tag{5}$$

where $\tilde{h}_{SPECIMEN} = h_{SPECIMEN}(deformable) - h_{SPECIMEN}(rigid)$ is the relative indentation depth due to the material deformation using a steel or tungsten carbide ball indenters as compared with a rigid ball $\tilde{h}_{INDENTER}$ is simply represented by $h_{INDENTER}(deformable)$, since $h_{INDENTER}(rigid) = 0$. We can also calculate the difference in the relative indentation depth $\Delta\tilde{h}$ for both loading and unloading periods:

$$\Delta\tilde{h} = \Delta\tilde{h}_{SPECIMEN} + \Delta\tilde{h}_{INDENTER} \tag{6}$$

where $\Delta\tilde{h}_{SPECIMEN} = \tilde{h}_{SPECIMEN}(preloading) - \tilde{h}_{SPECIMEN}(total)$, and $\Delta\tilde{h}_{INDENTER} = \tilde{h}_{INDENTER}(preloading) - \tilde{h}_{INDENTER}(total)$.

Therefore, the relative ΔHRB measurement values between a deformable steel or tungsten carbide ball and a rigid non-deformable ball can be calculated from:

$$\Delta HRB = \frac{\Delta\tilde{h}}{0,002\,\text{mm}}, \tag{7}$$

where $\Delta\tilde{h}$ – is in mm.

The simulation of the loading process with different characteristics of indenters and materials for indentation will allowed to determine the $\Delta\tilde{h}$. The calculated by (7) value of ΔHRB allows to compare the simulated hardness materials.

6 The Results and Analysis of the Modeling

The result of the contact problem solutions for the different types of indenters (steel ball indenter, tungsten carbide ball indenter and rigid ball indenter) for Aluminum specimen B95 (European analog – Aluminum 2024-T 361) is shown in Fig. 5. The hardness values are determined in the modeling in accordance with formulas (1)–(7). For a better overview and analysis, the obtained values of hardness for different simulation variations are summarized in Table 4.

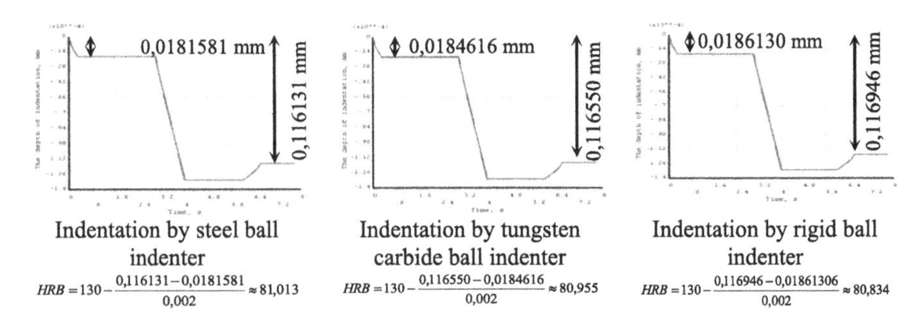

Fig. 5. Final graphs of simulation for Aluminum 2024-T 361

From Table 4 it is seen that the maximum and minimum hardness values are calculated for the steel indenter and absolutely solid one, respectively. This effect is explained by the greater deformation of both the indenter and the specimen, which is confirmed by calculations for absolutely rigid plate.

Figure 6 shows a curve for the deformation of steel and tungsten carbide indenter according to the applied force.

Table 4. The results of finite element modeling in ANSYS Academic

Specimens	Hardness for Steel ball indenter, HRB	Hardness for Tungsten carbide ball indenter, HRB	Hardness for Rigid ball indenter, HRB	Differences in hardness for steel and Rigid ball indenters, HRB	Expanded uncertantiy, HRBW
AISI 1000 Series Steel	51.965	51.908	51.778	0.187	0.323
UNS C11000	35.804	35.652	35.551	0.253	
Aluminum 2024-T 361	81.013	80.955	80.834	0.179	
Aluminum 7075-T73	82.851	82.793	82.682	0.169	
AISI 1020 Steel	78.371	78.342	78.215	0.156	

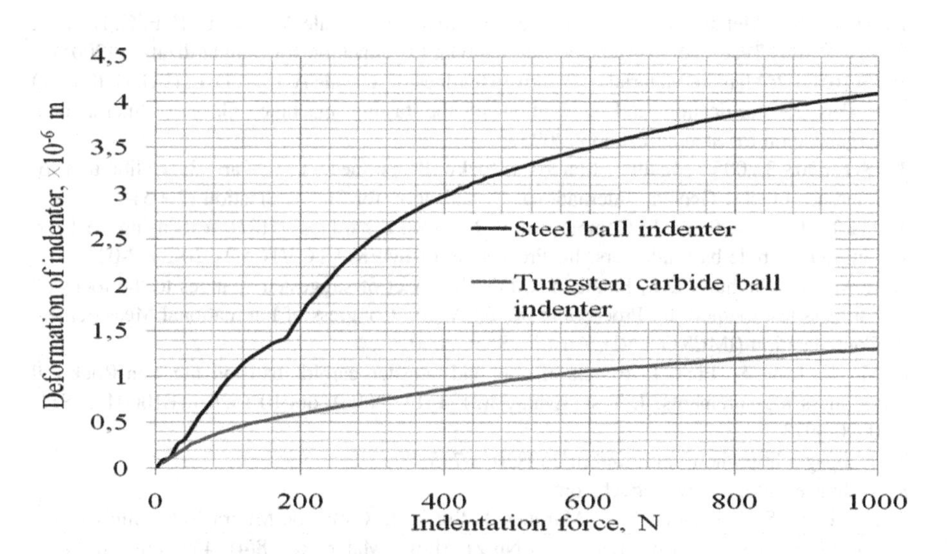

Fig. 6. The curve of deformation of indenter

7 Summary

Based on the performed modeling, it was established that the depth of penetration of a steel spherical indenter is not more than 2.8 μm less than the depth of the indentation of a tungsten carbide indenter under the same conditions of the measurement simulation (for different metals). These hardness values are different by no more than 0.3 HRB, which is within the expanded uncertainty for HRB scale. This agrees with experimental results.

As perspective directions for further research, the indentation process of modeling, analysis and influence of deformation of the indenter and the material on the measurement result are determined. For further analysis of the results of comparisons it is interesting to conduct researches for Superficial Rockwell scales with the study of very soft materials and the use of spherical ball indenters with the larger sizes. The obtained results are important for evaluation of the results of international comparisons of the national standards.

As a practical application, the methodology can be applied in calculation of the hardness modeling of materials with known mechanical characteristics.

Acknowledgement. The authors are grateful to Dr. Eduard Aslanyan for his valuable discussion and suggestions concerning the differences between the values in the measurement of steel and tungsten carbide Rockwell B hardness ball indenters. Special thanks to Dipl.-Eng. Konstantin Bragin for his assistance in the installation of an academic version of the FEM software.

References

1. ISO 6508-1: Metallic Materials – Rockwell hardness test (scale A, B, C, D, E, F, G, H, K, N, T) – Part 1: Test method, Geneva, International Organization for Standardization (2005)
2. ISO 6508-2: Metallic Materials – Rockwell hardness test (scale A, B, C, D, E, F, G, H, K, N, T) – Part 2: Verification and calibration of the tasting machine, Geneva, International Organization for Standardization (2005)
3. ISO 6508-3:2005: Metallic Materials-Rockwell hardness test- Part 3: Calibration of reference blocks, Geneva, International Organization for Standardization (2005)
4. Ma, L., Low, S., Song. J.: Comparison of Rockwell B hardness (HRB) tests using steel and tungsten carbide ball indenters. In: Proceeding of IMEKO TC3/TC5/TC20-2002-022
5. Low, S.R., Pitchure, D.J., Flanigan, C.D.: The effect of suggested changes to the rockwell hardness test method. In: Proceeding of 16th World Congress of International Measurement Confederation (IMEKO-XVI), Austria, Sept. 2000
6. Ma, L., Low, S., Fink, J.: Effects of steel and tungsten carbide ball indenters on Rockwell hardness measurements. J. Test. Eval. **34**(3), 1–13 (2006). doi:10.1520/JTE100048. ISSN 0090-3973
7. Webpage: http://news.plm-ural.ru/20082015-715
8. Webpage: http://www.matweb.com
9. Schneider, S., Schneider, S.G.: Marques da Silva, H., Carlos de Muora Neto: Study of the non-linear stress-strain behavior in Ti-Nb-Zr Alloys. Mater. Res. **8**(4), 435–438 (2005)
10. Rasmussen, K.J.R.: Full-range stress-strain curves for stainless steels alloys. www.sydney.edu.au/engineering/civil/publications/2001/r811.pdf

11. Kabanov, Y.: The contact technology in the action. ANSYS Solution, pp. 5–10. Autumn (2007). Russian version
12. Lukyanova, A.N.: Modeling of the axisymmetric contact problem using the ANSYS, [electronic edition] of Lab. work/A.N. Lukyanov, 52 p. Samara. State. Tehn. University, Samara (2014)
13. Webpage: http://www.cadfem.ru

Modeling of Freeform Reflecting Surface for LED Device

Natasha Kulik(✉)

National Aviation University, Kiev, Ukraine
natasha_artyuh@mail.ru

Abstract. Today the Illumination most promising is the development of LED lighting devices. Therefore, urgent issues is an effective method of describing reflecting surface. In computer graphics recently started using free-form surface based on function disturbances. In this paper, the possibility of modeling free-form reflecting surface using functions disturbances. Freeform surfaces from scalar function disturbances do not require time-consuming for their modeling, regardless of the complexity of the form. The feature of this method is that the surface of that model, the surface should be zero of a function. So having trouble simulations surfaces of revolution. However, this problem can be solved by breaking surface modeling into several parts (patches) so that each point on the base surface meets only one point on the surface of the free-form, and then combining these parts. Methods describe the surface reflectance functions disturbances prefers correction form reflector only required site without changing the whole mathematical model in general. The work performed simulations reflecting surface parabolic form. For the performance of the functions of the disturbances, the surface was divided into two equal parts, which were described by the scalar function disturbances. The next step was the consolidation of reflecting surfaces of two parts one by logical operations.

Keywords: Surface free form · Function disturbance LED device

1 Introduction

At the present stage of lighting is increasingly a question of energy efficiency lighting devices. Considering electric characteristics, the most promising area of research is the LED device.

In turn, for the best performance of intensity redistribution lighting emitted diodes requires appropriate secondary optics. The most modern types of secondary optics is to use reflective surfaces of arbitrary shape. The use of such surfaces requires new methods for calculating optical systems.

Secondary optics for developers facing two major problems:

1. Minimizing the loss of light in the optical system.
2. Obtaining the required lighting characteristics (distribution and intensity of illumination).

R. Szewczyk and M. Kaliczyńska (eds.), *Recent Advances in Systems, Control and Information Technology*, Advances in Intelligent Systems and Computing 543, DOI 10.1007/978-3-319-48923-0_72

One of the modern methods of calculation is SMS – a method using Bezier curves, spline representation of surfaces (including B-splines), and triangulation [1–6]. SMS – method to date is not entered in modern automated calculation program never received widespread use, unlike the spline representations and triangulation, in which the authors have made automated calculation program. The main requirement to the reflecting surface is to provide the necessary level of smoothness or continuity function of the second derivative. This requirement is necessary for a finding normal to the surface, which is a prerequisite for the calculation of the rays through the optical system. In this paper the possibility of modeling reflecting elements of LED devices using function disturbances. the proposed method meets the high degree of smoothness of functions on par with spline representation, I also arbitrary shape with a small number of perturbation functions. The mathematical model describes function of the second order function and has a smooth C2 within one node elevation map. Free form surface can have the procedure, significantly higher than the third, but wonder one card heights.

2 Materials and Methods

To describe the shape of the surface in general use surface theory [7]. May D – restricted area and $\overline{D}$ – its closure, and (u, v) – coordinate system in the plane. Let x, y, z – Rectangular Cartesian coordinates in Euclidean space E3.

Define on the set of three continuous functions

$$x = \varphi(u, v),\ y = \psi(u, v),\ z = \chi(u, v),\ (u, v) \in \overline{D}. \tag{1}$$

Let (u', v') and (u'', v'') – different points of the set $\overline{D}$, then point M'(x', y', z') and M"(x'', y'', z'') E^3 space whose coordinates solves by (1)

$$\begin{aligned} x' &= \varphi(u', v'),\ y' = \psi(u', v'),\ z' = \chi(u', v'), \\ x'' &= \varphi(u'', v''),\ y' = \psi(u'', v''),\ z'' = \chi(u'', v''), \end{aligned} \tag{2}$$

also different value (2) is called parametric equations, and u, v.

Parameters S of the set of points M(x, y, z) coordinates x, y, z are determined by (1), where the functions φ, ψ, χ in the closure of D correspond to the characteristics described above is called a simple surface. There is also a vector equation surface S. Let i, j, k – orts of axes. Then the setting surface S of a predetermined one vector function

$$r = r(u, v) = \varphi(u, v)\vec{i} + \psi(u, v)\vec{j} + \chi(u, v)\vec{k}, \tag{3}$$

defined in D. Vector function $r = r(u, v)$ is called the radius vector or vector surface S. Freeform surface F is called simple surface, which is a graph functions defined in three dimensions $z = f(x, y)$. Freeform surface based on the scalar field is a collection of some of the base surface P, which is in the same coordinate system as the F, and associated elevation map [8, 9]. As the base surface can be used any surface, such as a plane, ellipsoid, cylinder.

Map heights is a two-dimensional rectangle, called perturbation region P DP base surface and inside which a given function disturbance $h(u, v)$. Map heights defines the disturbance. The scope of the function $h(u, v)$ is $Dh(u, v) = \{U, V\}$, where U and V – the size of the rectangle. Map heights associated with the base surface as follows: There are conversion $G = (\mathfrak{R}^3 \Rightarrow \mathfrak{R}^2)$ the system of coordinates, which are F and P in elevation map coordinates. Typically, such a transformation is carried out using a parallel projection.

Perturbation function value set in the cross section of the box two-dimensional table heights. The direction of the normal base plane must coincide with the longitudinal direction of the box – the area of the function disturbance (Fig. 1).

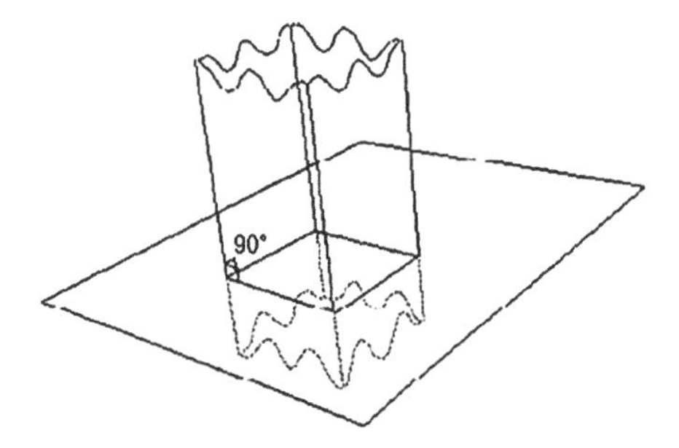

Fig. 1. Scope of the function disturbance

In Cartesian any point of space provided vector-matrix form (x y z) [10]. We will be using the homogeneous coordinates of the point in space (x y z 1).

As the picture plane plane choose XY, which is described by the equation Z = 0. The projection of a point on this plane obtained by multiplying (x y z 1) * A where

$$A = \begin{bmatrix} 1 & 0 & 0 & 0 \\ 0 & 1 & 0 & 0 \\ 0 & 0 & 0 & 0 \\ 0 & 0 & 0 & 1 \end{bmatrix} \tag{4}$$

asks conversion projecting on the plane XY. The value of the function $h(G(d_F))$ characterized deviation point d_F, which on the surface F from the point d_P, which is the projection of this point on the surface P. That function $h(G(d_F))$ equal to the magnitude of the vector

$$v = (d_F - d_P). \tag{5}$$

Region freeform surfaces can be defined as a set of points, determined by the vector equation

$$F = G(v) + n \cdot h(G(v)) \ \forall v \in \Re^3$$

where n – the normal to the surface.

If the vector v is outside the field disturbances, the vector = 0 and F is the vector of the base plane.

To set form surface disturbance can use table numbers, as well as the function h - interpolation function on the key points to be taken from the table (Fig. 2)

In this case, we can assume that in a given disturbance D_P scalar field.

The function h is calculated as follows

$$h(u, v) = f_0 + (f_1 - f_0)(v - m\_v), \quad (6)$$

where

$$f_0 = (1 - (u - m\_u))table[m\_u, m\_v] + (u - m\_u)table[m\_u + 1][m\_v],$$
$$f_1 = (1 - (u - m\_u))table[m\_u, m\_v + 1] + (u - m\_u)table[m\_u + 1][m\_v + 1],$$

m_u – the integral part of u, m_v – integer part of v, and is m_u- and m_v-th element of the table (Fig. 2).

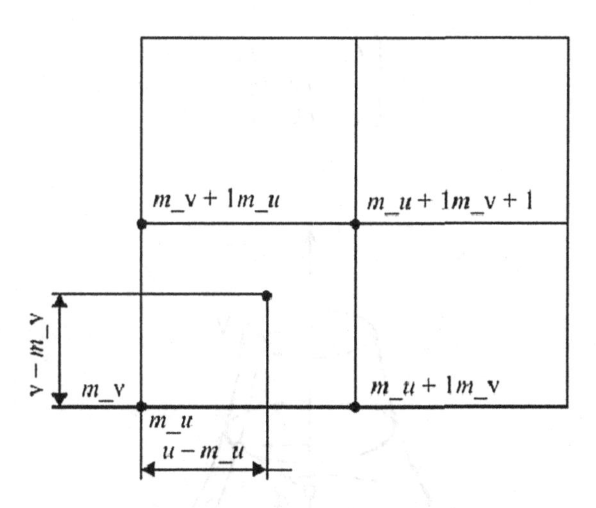

Fig. 2. Number elevation map

In the construction of freeform surfaces from scalar functions disturbance should define criteria belongings space and point threshold, which is located in the node table.

$$\begin{cases} F = G(v) + n \cdot h(G(v)), \\ h(G(v)) \geq threshold, \quad \forall v \in \Re^3, \\ G(v) \in D_p, \end{cases} \quad (7)$$

The introduction of requirements for supplies image vector $v(G(v))$ to the area of disturbance allows D_P cut off the portion of space that does not belong to the disturbance area. For test items clipped spaces non object modeling.

For surfaces from scalar functions disturbance restricts the class of objects for modeling. The surface should be the surface F zero level of a function, it means that each point of the base surface can be put in line just one point on the surface F. That rotation model so not because one point the plane will fit two points on the surface. In this case, the surface is divided into patches. So is building elevation maps for each of the planes and integration using logic operations.

3 Results

As for surface modeling was chosen paraboloid. This surface is a surface profile curve X0Z rotation axis Z.

In general, the surface of revolution described by [11] (Fig. 3).

$$r = r(u, v) = \varphi(u)\cos(v)\vec{i} + \varphi(u)\sin(v)\vec{j} + \psi(u)\vec{k}, \tag{8}$$

where $x = \varphi(u)$, $z = \psi(u)$.

The coordinates are as follows parabola

$$\begin{cases} r = \frac{2f}{1+\cos(\varphi)}, \\ x = r\sin(\varphi), \\ z = r\cos(\varphi), \end{cases} \tag{9}$$

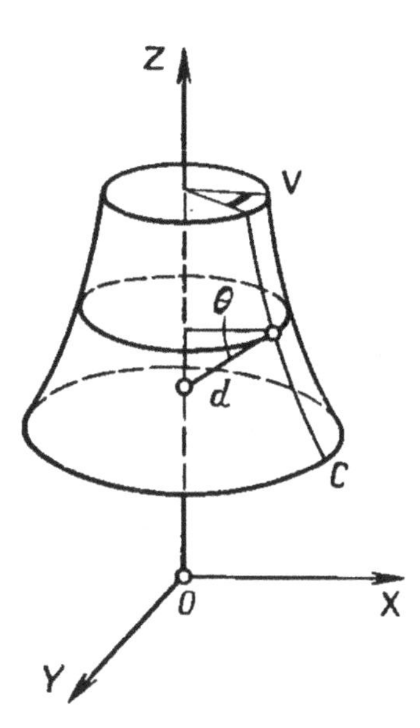

Fig. 3. Lines of the curves by surface of revolution

where r – radius vector, f – focal length, φ – angle between r and the axis 0Z.

Due to (9), Eq. (8) can be rewritten

$$r = r\sin(\varphi)\cos(\theta)\vec{i} + r\sin(\varphi)\sin(\theta)\vec{j} + r\cos(\varphi)\vec{k} \tag{10}$$

where θ – angle of rotation of the profile curve around an axis 0Z.

Profile paraboloid curve parameters: the focal length $f = 0.00383$ m full circumference angle φ max – 90°.

Based on the initial data in Mathematica software environment was created a mathematical model of the surface in parametric form (Fig. 4).

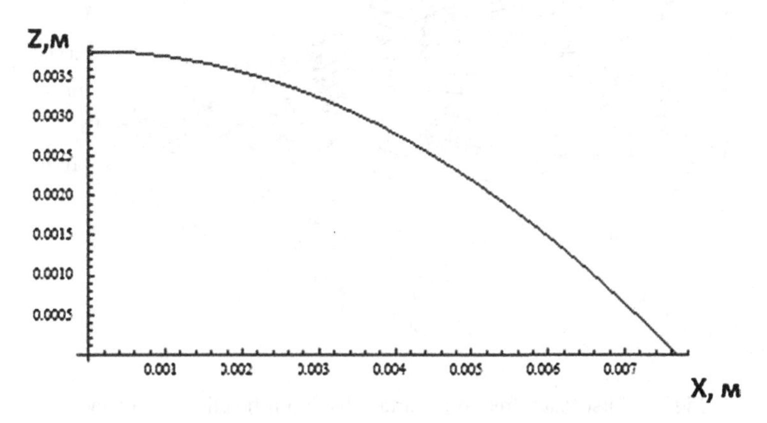

Fig. 4. Construction surface parabolic rotation in parametric form, the profile curve surface rotation

Further, according to the condition setting function disturbance should be implemented parallel projection coordinate set of points $M_i(XI\ y_i\ z_i)$ to the base plane. Provided that the basic plane is a plane X0Y, using the transformation matrix A and $i = 10$, we get a point on the base plane

$$M \times A = \begin{bmatrix} x1 & y1 & z1 & 1 \\ x2 & y2 & z2 & 1 \\ x3 & y3 & z3 & 1 \\ x4 & y4 & z4 & 1 \\ x5 & y5 & z5 & 1 \\ x6 & y6 & z6 & 1 \\ x7 & y7 & z7 & 1 \\ x8 & y8 & z8 & 1 \\ x9 & y9 & z9 & 1 \\ x10 & y10 & z10 & 1 \end{bmatrix} \times \begin{bmatrix} 1 & 0 & 0 & 0 \\ 0 & 1 & 0 & 0 \\ 0 & 0 & 0 & 0 \\ 0 & 0 & 0 & 1 \end{bmatrix} \tag{11}$$

Thus, we see that the coordinate XI corresponds u_i, y_i and responsible v_i. Get the coordinates of points on the base plane $[m\_u_i, m\_v_i]$, $i = 1..10$.

Since the surface is a paraboloid of revolution, and these types of surfaces should be split into patches to each point of the base surface to meet one point freeform surface. In this paper paraboloid surface is divided into four patches on the corner θ = 0–90, 90–180, 180–270, 270–360. Figure 5 is represented by the construction of the surface using disturbance at θ = 0–90.

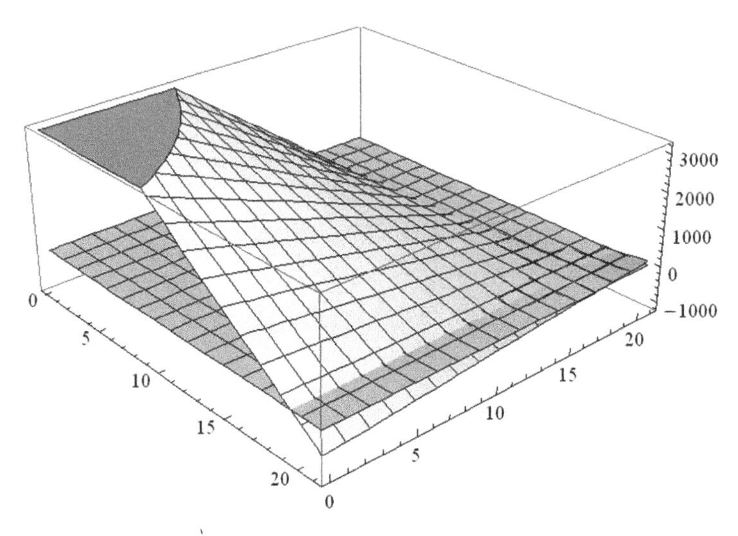

Fig. 5. First Patch free-form surface based on function disturbance

To complete the construction of the reflector surface were obtained 4 and patches sewn into one function with logical operation association. The result is a three-dimensional model presented in Fig. 6.

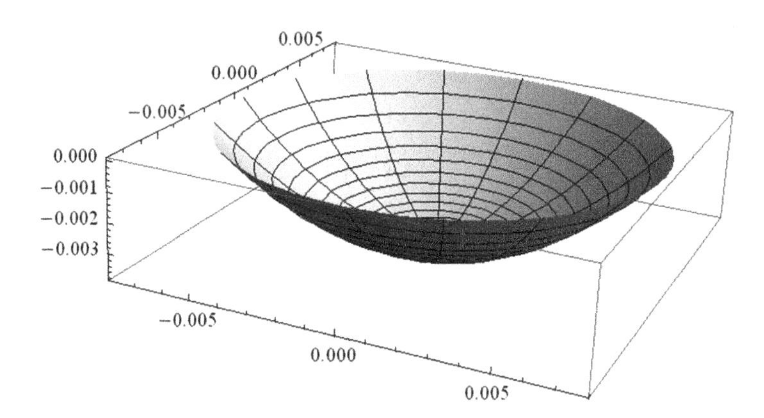

Fig. 6. Three-dimensional surface model paraboloid rotation

4 Conclusions

During the research it was constructed a mathematical model for LED reflector device based parabolic surface rotation. To construct the surface was chosen method of setting free forms using scalar functions disturbance. Scalar functions disturbances have advantages over other methods of setting surfaces: reduce the number of functions to describe complex surfaces, fewer operations for geometric transformations, high smoothness of spline functions at the level of representation. All these benefits are a prerequisite for function selection disturbances to model reflecting surface for LED devices.

References

1. Benítez, P., Miñano, J.C., Bien, J., Mohedano, R., Chaves, J., Dross, O., Hernández, M., Alvarez, J.L., Falicoff, W.: SMS design method in 3D geometry: examples and applications. In: Proceedings of SPIE, vol. 5185 (2003)
2. Trafimuk, A.: How to Create Surfaces of Revolution via User Defined Objects. Zemax knowledge base
3. Budak, V.P., Mukhanov, P.V.: Modelirovaniye svetovykh priborov na zadannoye svetoraspredeleniye s pomoshch'yu V-splayn poverkhnostey. Rossiyskaya svetotekhnicheskaya internet-konferentsiya "Svet bez granits", Moskva (2009)
4. Ivliyev, S.N., Mikayeva, S.A., Shibaykin, S.D.: Razrabotka trekhmernoy triangulyatsionnoy modeli osvetitel'nykh priborov. Vestnik Moskovskogo gosudarstvennogo universiteta priborostroyeniya i informatiki. M. Izd. MGUPI (29), 16–22 (2010)
5. Korobiichuk, I., Bezvesilna, O., Ilchenko, A., Shadura, V., Nowicki, M., Szewczyk, R.: A mathematical model of the thermo-anemometric flowmeter. Sensors **15**, 22899–22913 (2015). doi:10.3390/s150922899
6. Korobiichuk, I.: Mathematical model of precision sensor for an automatic weapons stabilizer system. Measurement **89**, 151–158 (2016)
7. Kreyszig, E.: Differential Geometry (1991). ISBN: 0-486-66721-9
8. Vyatkin, S.: Complex surface modeling using perturbation functions. Optoelectron. Instrum. Data Process. **43**(3), 226–231 (2007)
9. Vyatkin, S., Dolgovesov, B., Guimaoutdinov, O.: Synthesis of virtual environment using perturbation functions. In: Proceedings of the World Multiconference on Systemics, Cybernetics and Informatics, Orlando, FL, USA, vol. III, p. 350 (2001)
10. Craig, T.: A Treatise on Projections. University of Michigan, Historical Math Collection (1882)
11. Kushch, O.K.: Opticheskiy raschet svetovykh i obluchatel'nykh priborov na EVM. Energoatomizdat, Moscow (1991)

Investigation on Functional Properties of Hall-Effect Sensor Made of Graphene

Oleg Petruk[1], Maciej Kachniarz[1(✉)], Roman Szewczyk[2], and Adam Bieńkowski[2]

[1] Industrial Research Institute for Automation and Measurements PIAP, al. Jerozolimskie 202, 02-486 Warsaw, Poland
{opetruk, mkachniarz}@piap.pl

[2] Institute of Metrology and Biomedical Engineering, Warsaw University of Technology, sw. Andrzeja Boboli 8, 02-525 Warsaw, Poland

Abstract. Hall-effect sensors are commonly used in many industrial applications as position detectors, incremental counters and sensing elements of DC current transformers. Previously developed Hall-effect sensors utilize InAs or InSb as a sensing material. Recently, graphene seems to be the most promising material. Monolayer graphene was grown using the Chemical Vapor Deposition (CVD) method on the substrates. Prior to the growth, in situ etching of the SiC surface was carried out in hydrogen atmosphere. Functional properties of developed Hall-effect sensors based on graphene were tested on special experimental setup. Results presented in the paper indicate, that graphene is very promising material for development of Hall-effect sensors. Such sensors exhibit high magnetic field sensitivity and linear characteristic. It was also observed, that functional parameters of graphene Hall-effect sensors are diversified from the point of view of its transport properties. This phenomenon creates important possibilities of further optimization of Hall-effect sensors during their production.

Keywords: Graphene · Hall-effect sensor · Magnetic field measurement

1 Introduction

Hall-effect sensor are one of the most commonly utilized magnetic field sensors in the industrial measurements of magnetic field. Main areas of their application are position detectors, incremental counters and current transformers [1]. For all this applications, high sensitivity and stability of Hall-effect sensor are required. So far, InAS and InSb are considered as the best sensing materials for Hall-effect sensors [2]. However, recent studies indicate that graphene can exhibit even better properties when utilized as a sensing material in Hall-effect sensors.

Graphene is a recently discovered 2-dimensional material. It is crystalline allotrope of carbon, where atoms are organized in sp^2-bonded hexagonal pattern [3]. One of the most important parameters of material for Hall-effect sensor is electron mobility, which determines final sensitivity of sensor. Due to its high electron mobility, reaching values over 15 000 cm^2/Vs, graphene is considered to be good material for Hall-effect sensors

R. Szewczyk and M. Kaliczyńska (eds.), *Recent Advances in Systems, Control and Information Technology*, Advances in Intelligent Systems and Computing 543, DOI 10.1007/978-3-319-48923-0_73

[4, 5]. Especially, monolayer graphene structure, with theoretical electron mobility 200 000 cm^2/Vs at a carrier density of 10^{12} cm^{-2}, creates a possibility to obtain sensitivity of the Hall-effect sensor much higher, that any other previously used material.

2 Sample Preparation

Investigated graphene monolayer structures were grown using the Chemical Vapor Deposition (CVD) method on the Si face of semi-insulating on-axis 4H-SiC(0001) substrates in a standard hot-wall CVD Aixtron VP508 reactor. Before the growth process was initiated, in situ etching of the SiC surface was carried out in hydrogen atmosphere. The epitaxial CVD growth of graphene was realized under dynamic flow conditions that simultaneously inhibit Si sublimation and promote the mass transport of propane molecules to SiC substrate [6]. After completion of the growth process, as grown samples were characterized by Hall-effect measurements in Van der Pauw geometry (0.55T Ecopia HMS-3000 setup) with the four golden probes placed in the corners of the 10 mm × 10 mm substrates. The substrates were photolitographically patterned to form five graphene Hall-effect structures in the shape of symmetrical, equal-arm crosses. 20 nm_Ti/80 nm_Au ohmic contacts (200 μm × 200 μm) were e-beam deposited. Developed graphene Hall-effect structures are presented in Fig. 1.

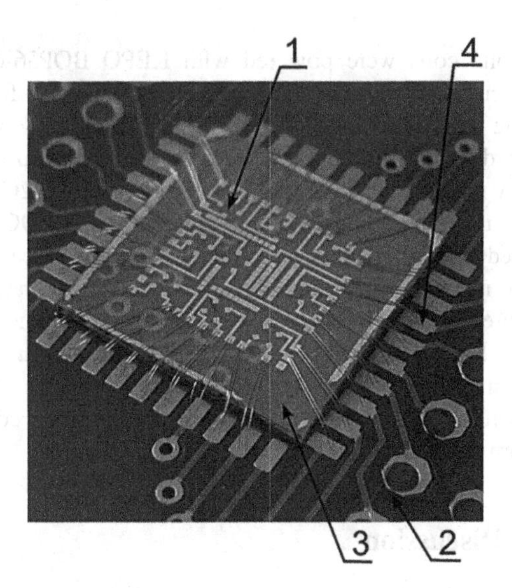

Fig. 1. Developed graphene based Hall-effect sensor: 1 – graphene cross (transparent), 2 – epoxy PCB board, 3 – SiC substrate, 4 – bonding

Paper presents characteristics of Hall-effect sensor made of graphene with electron concentration equal $6.24 \cdot 10^{11}$ cm^{-2}, electron mobility 1330 cm^2/Vs and sheet resistance 77.1 Ω/sq. The length of the Hall-effect cross beam was 800 μm. The sample consists of one layer of graphene.

3 Experimental Setup

Functional properties of developed Hall-effect sensors based on graphene were tested with specially developed measurement system. High precision Helmholtz coils were used as a source of reference magnetic field. The schematic diagram of the measurement system is presented in the Fig. 2.

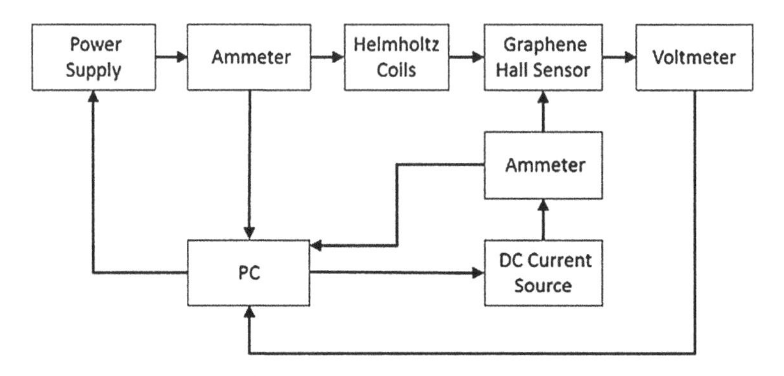

Fig. 2. Developed experimental setup

Utilized Helmholtz coils were powered with KEPO BOP36-6M bipolar power supply. Supplying current range allowed to obtain magnetic field acting on the investigated structure within the range of ±6.5 mT. Power supply was controlled with Data Acquisition Card installed in the PC. Current in the Helmholtz coils was measured with FLUKE 8808A multimeter connected to the PC via RS-232C interface.

Hall-effect structure was powered with precise INMEL 60 DC current calibrator. Structure was placed between Helmholtz coils with the surface of graphene layer perpendicular to the magnetic field direction. Hall voltage generated in the graphene structure was measured with high precision FLUKE 8846A voltmeter, which was also connected to the PC. Special application developed in NI LabVIEW environment was controlling the system and collecting measurement data.

Hall effect structures were tested under normal laboratory conditions, in room temperature and normal atmospheric pressure.

4 Results and Discussion

The characteristics of developed Hall-effect sensor made of graphene are presented in Fig. 3. Hall effect structures were tested under three different bias currents: 50 μA, 200 μA, 600 μA, and under perpendicular highly uniform DC magnetic field in the range of ±6.5 mT.

The obtained characteristics exhibit high linearity across the range tested. Linearity improves with increasing bias current. Exceptionally high sensitivity of 1200 V/AT

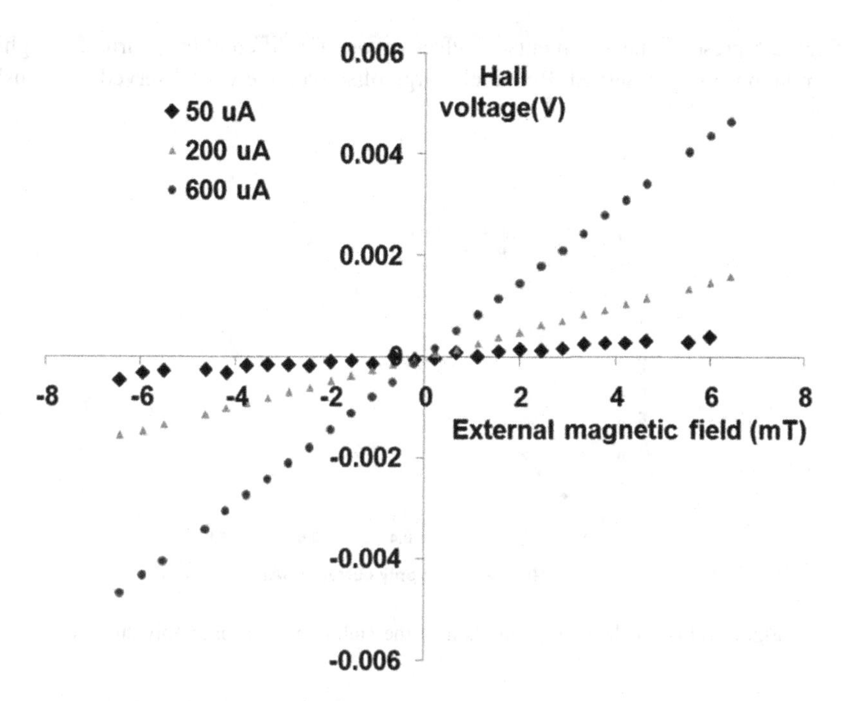

Fig. 3. Characteristics of developed graphene Hall-effect sensor at supply current 50 μA, 200 μA and 600 μA

was obtained. There was a linear relationship between supply current and current-related sensitivity and voltage offset observed.

Figure 4 presents current-related sensitivity for various bias currents used during testing phase. Rather constant sensitivity, higher than 1200 V/AT is observed, which is comparable with results in [7, 8].

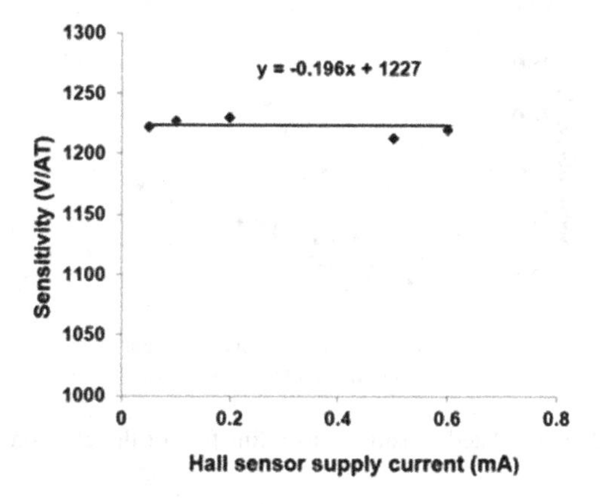

Fig. 4. Current-related sensitivity for various supply currents

Figure 5 presents measurements of offset voltage for different bias currents. Highly linear relationship is observed. Relatively large offset voltage was observed previously in other research [9].

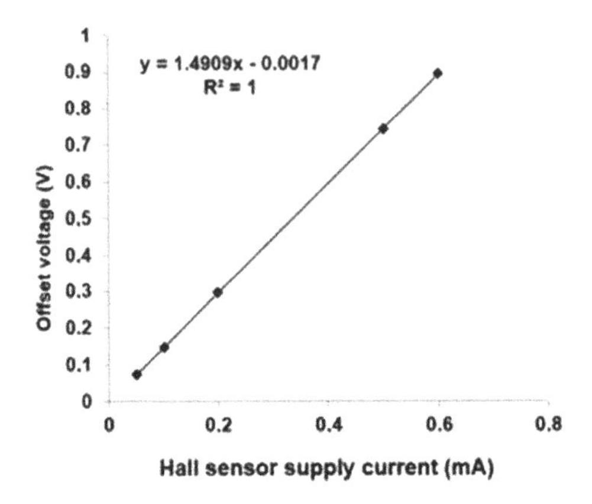

Fig. 5. Offset voltage as a function of the Hall-effect sensor supply current

This is due to inaccuracies in geometry reproduction of the Hall Effect structure [10], and due to substrate surface inhomogeneities resulting from the existence of terraces, which are specific for SiC. There is also the possibility of photoresist influence on offset voltage. Offset voltage was measured and compensated during the measurements of the Hall voltage.

Figures 6 and 7 present the dependence of sensitivity for the series of structures with varying electron mobility and surface resistance respectively.

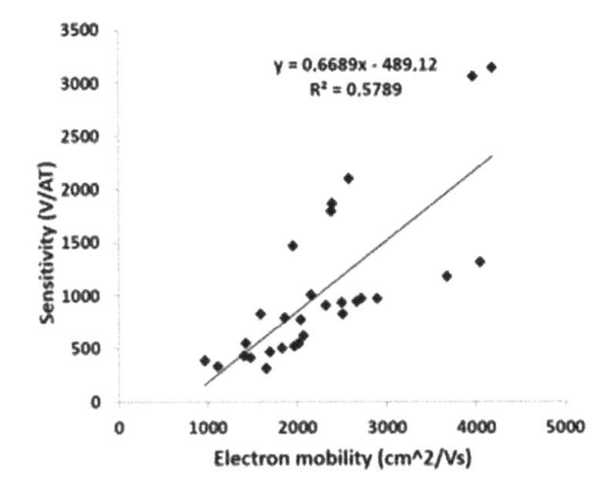

Fig. 6. Current-related sensitivity as a function of the electron mobility

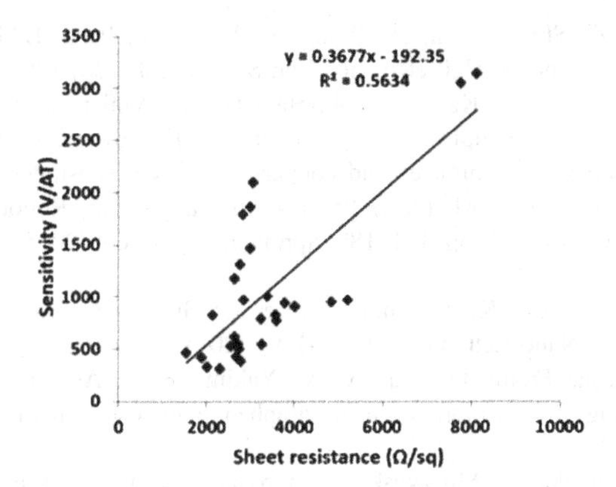

Fig. 7. Current-related sensitivity as a function of the sheet resistance

In both cases it may be observed, that R^2 determination coefficient is about 0.5. This mean, that other influence factors should be determined. Probably, changes of sensitivity are connected with internal discontinuities in graphene layer as well as connection between graphene and electrodes.

However, sensitivity for individual Hall-effect sensor is constant, which is most important from technical point of view. Due to the random distribution of discontinuities in graphene layer, each graphene Hall sensor has to be calibrated separately.

5 Conclusion

Results presented in the paper indicate, that graphene is very promising material for development of Hall-effect sensors. Such sensors exhibit linear characteristics and high magnetic field sensitivity. It was also observed, that functional parameters of graphene Hall-effect sensors are diversified from the point of view of its transport properties. This phenomenon creates important possibilities of further optimization of Hall-effect sensors during their development and production.

Acknowledgement. This work was partially supported by The National Center of Research and Development (Poland) within GRAF-TECH program.

References

1. Ai, X., Bao, H., Song, Y.H.: Novel method of error current compensation for Hall-effect-based high-accuracy current transformer. IEEE Trans. Power Deliv. **20**(1), 11–14 (2005)
2. Ramsden, E.: Hall-effect Sensors: Theory and Application. Newness (2006)
3. Krupka, J., Strupiński, W.: Measurements of the sheet resistance and conductivity of thin epitaxial graphene and SiC films. Appl. Phys. Letts. **96**, 082101 (2010)

4. Xu, H., Zhang, Z., Shi, R., Liu, H., Wang, Z., Wang, S., Peng, L.M.: Batch-fabricated high-performance graphene Hall elements. Nat. Sci. Rep. **3**, 1207 (2013)
5. Missala, T., Szewczyk, R., Kamiński, M., Hamela, M., Winiarski, W., Szałatkiewicz, J., Tomasik, J., Salach, J., Strupiński, W., Pasternak, I., Borkowski, Z.: Study on graphene growth process on various bronzes and copper-plated steel substrates. In: Szewczyk, R., Zieliński, C., Kaliczyńska, M. (eds.) Progress in Automation, Robotics and Measuring Techniques. AISC, vol. 352, pp. 171–180. Springer, Heidelberg (2015). doi:10.1007/978-3-319-15835-8_19
6. Strupiński, W., Grodecki, K., Wysmołek, A., et al.: Graphene epitaxy by chemical vapor deposition on SiC. Nano Lett. **11**(4), 1786–1791 (2011)
7. Panchal, V., Iglesias-Freire, O., Lartsev, A., Yakimova, R., Asensjo, A., Kazakova, O.: Magnetic scanning probe calibration using graphene hall sensor. IEEE Trans. Mag. **49**(7), 3520–3523 (2013)
8. Bartnicki, A., Łopatka, J., Muszyński, T., Wrona, J.: Concept of IED/EOD Operations (CONOPs) for engineer mission support robot team. J. Kones **22**(1), 269–274 (2015)
9. Panchal, V., Cedergren, K., Yakimova, R., Tzalenchuk, A., Kubatkin, S., Kazakova, O.: Small epitaxial graphene devices for magnetosensing applications. J. Appl. Phys. **111**, 07E509 (2012)
10. Paun, M.A., Sallese, J.M., Kayal, M.: Offset, Drift Dependence of Hall Cells with their Designed Geometry. Int. J. Electron. Telecommun. **59**, 169–175 (2013)

Possibilities of Application of the Magnetoelastic Effect for Stress Assessment in Construction Elements Made of Steel Considering Rayleigh Region

Dorota Jackiewicz[1(✉)], Maciej Kachniarz[1], Adam Bieńkowski[2], and Roman Szewczyk[2]

[1] Industrial Research Institute for Automation and Measurements, al. Jerozolimskie 202, 02-486 Warsaw, Poland
d.jackiewicz@mchtr.pw.edu.pl, mkachniarz@piap.pl
[2] Institute of Metrology and Biomedical Engineering, Warsaw University of Technology, sw. Andrzeja Boboli 8, 02-525 Warsaw, Poland

Abstract. This paper presents the results of investigation of X30Cr13 construction steel in a wide range of magnetizing fields from the Rayleigh region to the near-saturation region. The measurement system with hydraulic press as a source of external tensile stress is described as well as investigated samples of the X30Cr13 steel. Obtained results are presented in the paper separately for Rayleigh region and for higher magnetizing fields. Occurrence of the Villari effect is confirmed for entire range of magnetizing filed applied during the investigation. On the basis of obtained results, conclusion about possible application of the results in the stress assessment in constructional steel was formulated, which is included in the last section of the paper.

Keywords: Magnetism · Magnetoelastic effect · Constructional steel · Stress assessment · Rayleigh region

1 Introduction

Steel is a fundamental material used in modern construction industry. Great variety of types exhibiting different physical and chemical properties makes steel one of the most widespread materials in the world. Many types of steel are utilized for architecture purposes, where steel elements are working under large mechanical stress. Steel is also present in energy industry, where it also must meet high requirements regarding mechanical strength. For public safety reasons, there is a great need for development of methods allowing to constantly monitor the steel elements working in such hard conditions. Development of the Non-Destructive Testing (NDT) methodology seems to be the best possible solution [1].

Steel, as a ferromagnetic material, exhibits strong magnetic properties [2]. As it was previously proven, magnetic properties of steel are strongly influenced by the mechanical stress [3, 4, 6, 13]. This phenomenon was discovered in 1865 and is known

R. Szewczyk and M. Kaliczyńska (eds.), *Recent Advances in Systems, Control and Information Technology*, Advances in Intelligent Systems and Computing 543, DOI 10.1007/978-3-319-48923-0_74

as magnetoelastic Villari effect [7, 8]. This creates the possibility to assess mechanical stress level within the material by measuring its magnetic parameters. The most important problem connected with application of this method is lack of measurement results presenting magnetoelastic characteristics of different types of steels as well as lack of knowledge about magnetizing field level, which allows to obtain the highest sensitivity of the discussed method. Optimal value of magnetizing field should allow to obtain the highest possible change of magnetic parameters of the investigated material under the influence of mechanical stress. This will ensure high sensitivity of the stress measurement.

The following paper presents results of investigation of magnetoelastic properties of X30Cr13 constructional steel in wide range of magnetizing fields from very low values in so-called Rayleigh region [9] to the highest saturation fields. On the basis of the presented results the conclusion was formulated about the optimal value of magnetizing field, for which change of magnetic properties of the material under the influence of mechanical stress is the highest.

2 Investigated Material

For the performed investigation, X30Cr13 martensitic corrosion resistant steel with addition of chromium was used [5, 10, 14, 15]. It is widely utilized in energetic industry as a material for elements of turbines. It exhibits yield strength over 350 MPa and tensile strength within the range 650–880 MPa. Maximum magnetic flux density of the material is about 1.05 T, magnetic remanence is 0.72 and saturation coercive field is about 800 A/m. Material exhibits low relative magnetic permeability about 200.

To allow measurement of magnetoelastic characteristics of the steel, it was formed into frame-shaped sample presented in Fig. 1. Main reason for choosing this shape was to provide uniform distribution of the applied stress in the columns of the sample as well as to obtain closed magnetic circuit within the sample.

Before material was investigated, three sets of windings were made on the columns of the sample. First 200 turns of sensing winding were made. On top of them there were two sets of magnetizing windings: one counting 300 turns and second counting 40 turns, which allowed to cover the entire range of magnetizing fields from low fields in Rayleigh region (winding with 40 turns) to high fields in saturation region (300 turns).

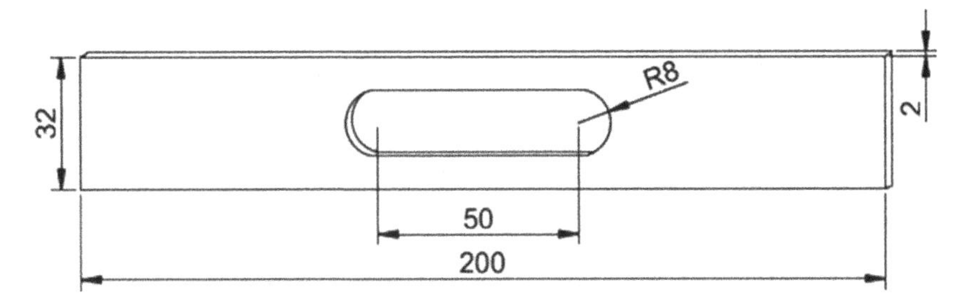

Fig. 1. Outline of the frame-shaped sample of the investigated X30Cr13 steel

3 Measurement System

Special measurement system, with hydraulic press as a source of mechanical stress, was developed for performed investigation. The schematic block diagram of the system is presented in Fig. 2.

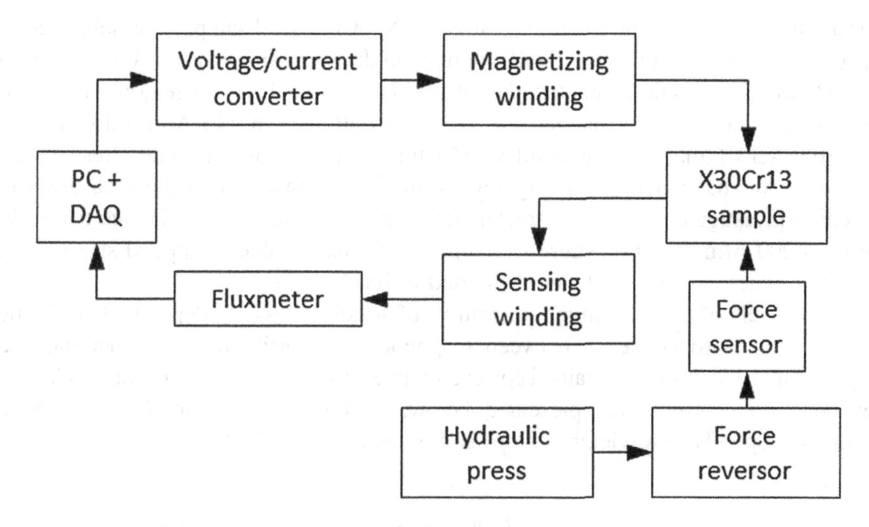

Fig. 2. Schematic block diagram of developed measurement system

System was digitally controlled by PC with Data Acquisition Card (DAQ) installed. Sinusoidal waveform from DAQ was converted into current waveform driving magnetizing winding by KEPCO BOP 36-6 M bipolar power supply working as a voltage/current converter. Current flowing through the magnetizing winding resulted in magnetizing filed acting on the sample, whose value can be calculated as [10, 11]:

$$H = \frac{n_m I}{l}, \tag{1}$$

where n_m is number of turns of magnetizing winding, I is magnetizing current and l is average flow path of magnetic flux in the sample. Current range of the utilized voltage/current converter allowed to investigate material in the wide range of magnetizing fields, but due to obtain correct results for low magnetizing fields, individual, less numerous set of magnetizing winding had to be made for this investigation. For low currents amplitude waveforms obtained from the converter were distorted, so it was better to made second, less numerous set of magnetizing winding than performing investigation in Rayleigh region with 300 turns of magnetizing winding using very low currents from the converter. Voltage induced in the secondary winding was measured with Lakeshore 480 fluxmeter and results were transmitted to the PC.

For mechanical tensile stress generation hydraulic press was utilized equipped with special force reversor converting compressive force from the press into tensile force acting on the sample. Generated force was measured with strain gauge force sensor.

4 Experimental Results

During the performed experiment, investigated X30Cr13 steel sample was subjected to the tensile stress σ_T from 0 to 575 MPa applied in the increasing manner. First, for each applied stress, magnetic characteristics of the material in Rayliegh region were measured within the range of magnetizing field from 50 A/m to 450 A/m, utilizing less numerous set of magnetizing winding (40 turns). Then, for the same stress value, second set of magnetizing winding (300 turns) was connected to the measurement system to investigate magnetic characteristics in high magnetizing filed range from 640 A/m to 4000 A/m. The procedure was repeated for each value of applied stress up to 575 MPa, where investigated sample started to break.

As a result of the investigation, family of tensile stress σ_T dependent magnetic hysteresis loops (dependence between magnetic flux density B in the material and magnetizing filed H) was obtained for each value of magnetizing field amplitude H_m. Several chosen families are presented below. Family obtained for H_m = 300 A/m, which corresponds to Rayleigh region, is presented in Fig. 3.

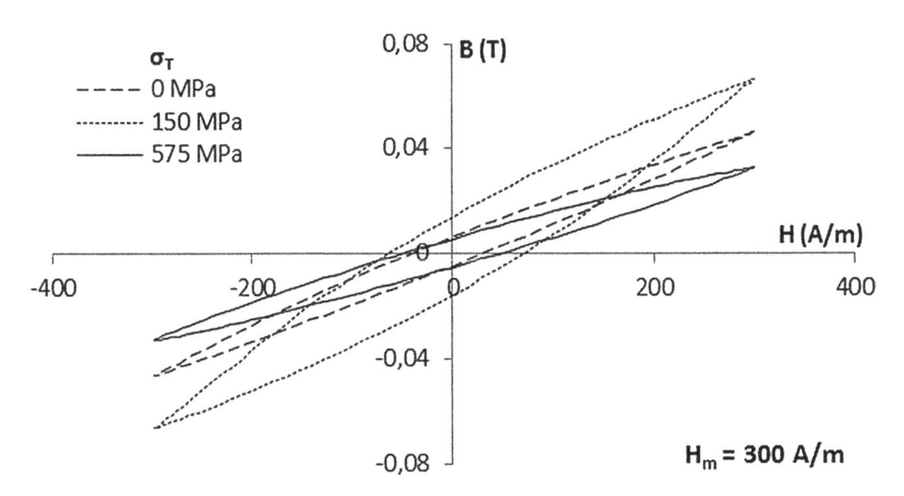

Fig. 3. Family of selected magnetic hysteresis loops of X30Cr13 steel under the influence of tensile stress σ_T for H_m = 300 A/m

Characteristics obtained for Rayleigh region are consistent with expectations. As it was investigated before, in Rayleigh region hysteresis loop exhibits lenticular shape being a combination of two symmetrical intersecting parabolic curves, which can be described with system of two second order polynomial equations [12, 14]:

$$B(H) = \mu_0[(\mu_i + \alpha_R H_m)H \pm \frac{\alpha_R}{2}(H_m^2 - H^2)], \tag{2}$$

where μ_0 is magnetic permeability of free space (constant), μ_i is initial relative magnetic permeability and α_R is so-called Rayleigh coefficient (dependent on the material). All magnetic characteristics of the investigated material obtained for low magnetizing fields (up to 450 A/m) are consistent with this description.

Figure 4 presents hysteresis loops for H_m = 640 A/m, which is amplitude form transition region between Rayleigh region and near-saturation region. As it can be seen, for higher magnetizing field hysteresis loop starts to lose its lenticular shape characteristic for Rayleigh region, however it is still possible to approximate it with second order polynomial equations.

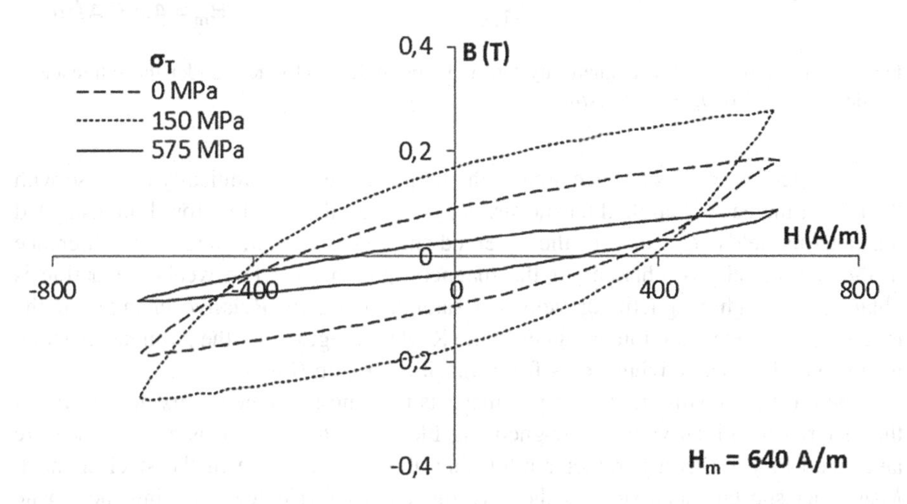

Fig. 4. Family of selected magnetic hysteresis loops of X30Cr13 steel under the influence of tensile stress σ_T for H_m = 640 A/m

In Fig. 5 hysteresis loops from saturation region (H_m = 4000 A/m) are shown. They exhibit well-known shape characteristic for hysteresis loop in near saturation region which cannot be described with simple mathematical equations.

In each presented figures three hysteresis loops obtained for constant magnetizing field amplitude H_m and three different values of applied tensile stress σ_T are visible. The σ_T values are 0 MPa (initial value), 150 MPa and 575 MPa (rupture point). Value σ_T = 150 MPa corresponds to the so-called Villari reversal point in the investigated material, where magnetic flux density B of the material reaches its maximum value [6]. For all investigated values of magnetizing field H_m influence of the tensile stress σ_T on the magnetic characteristics is similar. For initial values of stress within the range from 0 MPa to 150 MPa, all basic magnetic parameters: maximum magnetic flux density, magnetic remanence and coercive filed are increasing their values, which results in the increase of surface area of the hysteresis loop. After reaching the Villari point,

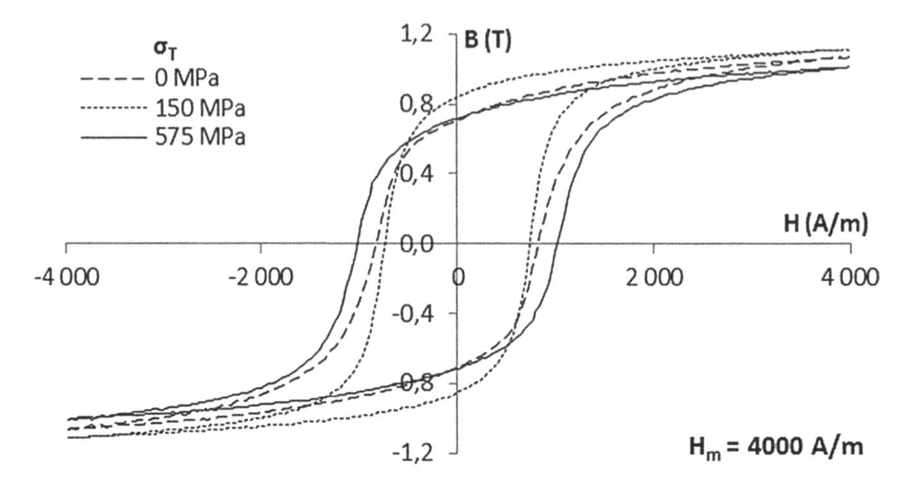

Fig. 5. Family of selected magnetic hysteresis loops of X30Cr13 steel under the influence of tensile stress σ_T for H_m = 4000 A/m

all parameters as well as surface area of the loop starts to monotonically decrease with the further increase of applied tensile stress σ_T. Despite the fact, that for all investigated magnetizing fields H_m nature of the observed changes is similar, there is the difference in the size of relative changes of the magnetic parameters. The weakest reaction is obtained for high magnetizing fields, in saturation region. Relative changes of the magnetic parameters are much higher in the Rayleigh region, but the strongest reaction is observed for magnetizing fields from transition region (Fig. 4).

The most interesting magnetic parameter is maximum magnetic flux density B_m of the material for given value of magnetizing filed amplitude H_m. It is easy to measure and could be used as a parameter informing about stress value in the steel element. Magnetoelastic characteristics of the investigated X30Cr13 steel showing the tensile stress σ_T dependence of maximum magnetic flux density B_m are presented in Fig. 6 (Rayleigh region) and Fig. 7 (higher magnetizing fields).

For all investigated magnetizing fields H_m, dependence of maximum magnetic flux density B_m on the tensile stress σ_T is similar. For initial values of stress there is an increase of the maximum magnetic flux density up to tensile stress value σ_T = 150 MPa (Villari point). Then further increase of the tensile stress σ_T results in monotonic decrease of the B_m parameter. However, as it was stated before, relative changes of the maximum magnetic flux density B_m under applied stress are different for each value of the magnetizing field H_m.

On the basis of the obtained $B_m(\sigma_T)$ characteristics, the magnetoelastic sensitivity coefficient was determined according to the equation:

$$S = \frac{\Delta B_m}{\Delta \sigma_T} = \frac{B_V - B_0}{\sigma_V - \sigma_0}, \tag{3}$$

where B_V is maximum magnetic flux density in the Villari point, B_0 is maximum magnetic flux density for initial stress value (σ_0 = 0 MPa) and σ_V is value of tensile

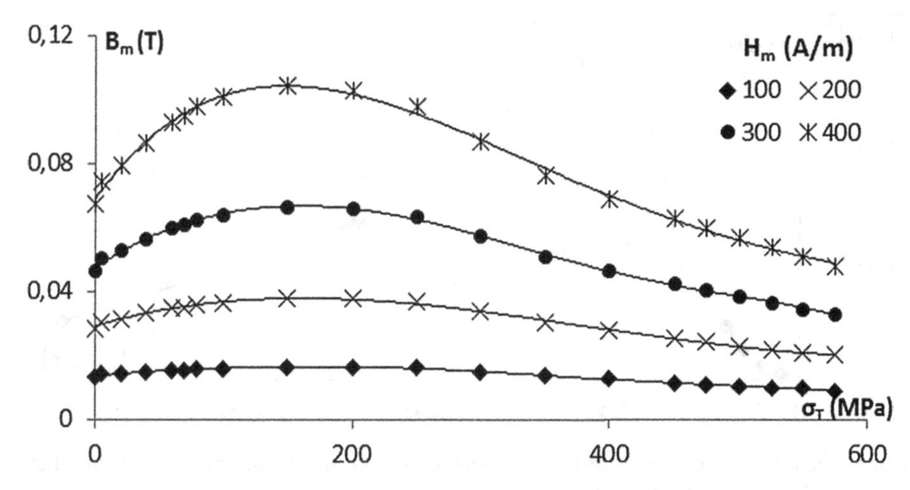

Fig. 6. Selected magnetoelastic characteristics $B_m(\sigma_T)$ of investigated X30Cr13 steel for Rayleigh region (H_m = (50–450) A/m)

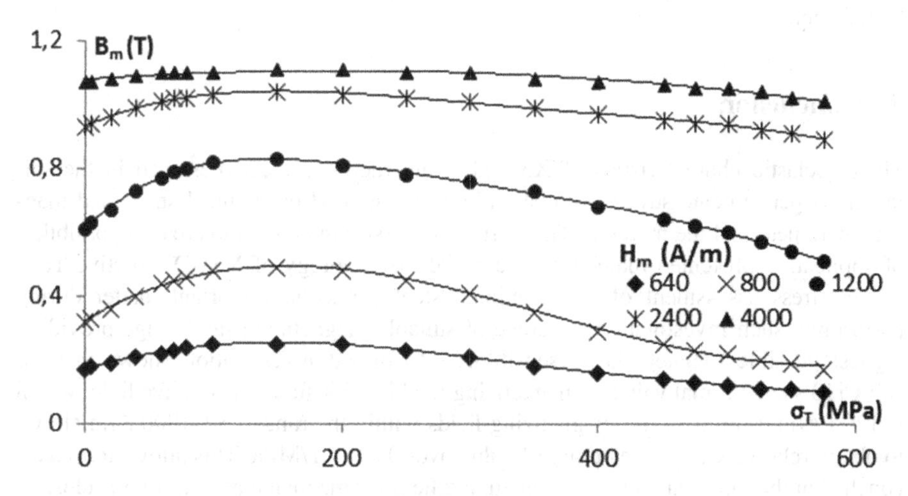

Fig. 7. Selected magnetoelastic characteristics $B_m(\sigma_T)$ of investigated X30Cr13 steel for high magnetizing field region (H_m = (640–4000) A/m)

stress in Villari reversal point. The parameter expresses, how much the value of maximum magnetic flux density is changing for given increase of the stress value. Value of the sensitivity coefficient was calculated for each investigated magnetizing field H_m. The obtained results are presented in Fig. 8. It is clearly visible that the highest magnetoelastic sensitivity S = 0.0011 T/MPa is obtained for magnetizing field H_m value 1200 A/m, which is about 1.5 of saturation coercive filed H_c for investigated material. Also values obtained for H_m = 800 A/m (1.0 H_c) and H_m = 2400 A/m (3.0 H_c) are relatively high. The magnetoelastic sensitivity coefficient value over

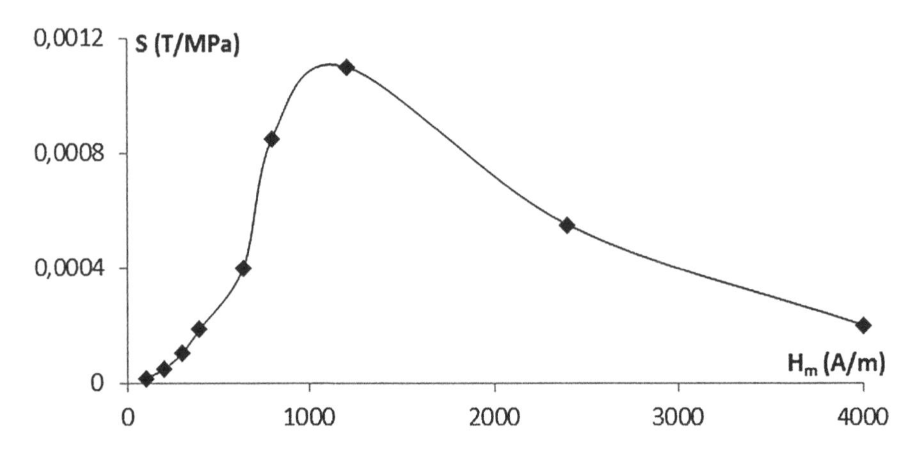

Fig. 8. Dependence of magnetoelastic sensitivity coefficient S on the magnetizing field amplitude H_m for investigated X30Cr13 steel

0.0008 T/MPa is obtained for magnetizing fields from about 750 A/m to about 1800 A/m, which is optimal range for magnetoelastic testing of X30Cr13 constructional steel.

5 Conclusion

Magnetoelastic characteristics of X30Cr13 constructional steel presented in the following paper indicate strong correlation between applied mechanical stress and magnetic parameters of the material. This correlation is strong enough to create a possibility of application presented magnetoelastic based methodology of Non-Destructive Testing in stress assessment of constructional steel elements. Important matter during performing such investigation is choice of suitable magnetizing field range providing highest possible magnetoelastic sensitivity. Performed investigation shows that for X30Cr13 steel optimal value of magnetizing field is 1.5 saturation coercive field, which is 1200 A/m. Moreover, all magnetizing fields within the range 750–1800 A/m allows to obtain relatively high sensitivity of value over 0.0008 T/MPa. This allows to make a conclusion that optimal value of magnetizing field for magnetoelastic testing is close to the value of saturation coercive filed H_c (from about 0.8 H_c to 2.0 H_c).

It also should be stated that Rayleigh region seems to be interesting from the point of view of magnetoelastic based NDT methodology. Magnetoelastic sensitivity obtained for magnetizing fields in Rayleigh region is lower, but results are still giving clear information about stress in the material. Moreover, Rayleigh region is a region of low magnetizing fields, so it does not demand high magnetizing currents or very numerous sets of magnetizing winding, which reduces costs of application of this method.

Acknowledgements. This work was partially supported by The National Centre of Research and Development (Poland) within grant no. PBS/B4/6/2012 and partially supported by the statutory founds of Institute of Metrology and Biomedical Engineering, Warsaw University of Technology (Poland).

References

1. Runkiewicz, L.: Application of non-destructive testing methods to assess properties of construction materials in building diagnostics. Archit. Civ. Eng. Environ. **2**(2), 79–86 (2009)
2. O'Handley, R.: Modern Magnetic Materials – Principles and Applications. John Wiley & Sons, New York (2000)
3. Jackiewicz, D., Kachniarz, M., Rożniatowski, K., et al.: Temperature Resistance of Magnetoelastic Characteristics of 13CrMo4-5 Constructional Steel. Acta Phys. Pol. A **127** (2), 614–616 (2015)
4. Shi, Y., Fan, S.: Application of Magnetoelastic Effect of Ferromagnetic Material in Stress Measurement. Adv. Mater. Res. **496**, 306–309 (2012)
5. Szałatkiewicz, J., Szewczyk, R., Budny, E., Missala, T., Winiarski, W.: Identification of thermal response, of plasmatron plasma reactor. In: Szewczyk, R., Zieliński, C., Kaliczyńska, M. (eds.) Recent Advances in Automation, Robotics and Measuring Techniques. AISC, vol. 267, pp. 265–274. Springer, Heidelberg (2014). doi:10.1007/978-3-319-05353-0_26
6. Ktena, A., Hristoforou, E.: Stress Dependent Magnetization and Vector Preisach Modeling in Low Carbon Steels. IEEE Trans. Mag. **48**(4), 1433–1436 (2012)
7. Bieńkowski, A.: Magnetoelastic Villari effect in Mn-Zn ferrites. J. Magn. Magn. Matter. **215-216**, 231–233 (2000)
8. Bieńkowski, A., Szewczyk, R., Kulik, T., Ferenc, J., Salach, J.: Magnetoelastic properties of HITPERM-type $Fe_{41,5}Co_{41,5}Cu_1Nb_3B_{13}$ nanocrystalline alloy. J. Magn. Magn. Matter. **304** (2), E624–E626 (2006)
9. Ponomarev, Y.F.: One the Rayleigh law of magnetization: a new mathematical model of hysteresis loops. Phys. Met. Metallogr. **104**(5), 469–477 (2007)
10. EN 10088-1:2005 Stainless steels. List of stainless steels
11. Szałatkiewicz, J., Szewczyk, R., Budny, E., Missala, T., Winiarski, W.: Measurement and control system of the plasmatron plasma reactor for recovery of metals from printed circuit board waste. In: Szewczyk, R., Zieliński, C., Kaliczyńska, M. (eds.) Recent Advances in Automation, Robotics and Measuring Techniques. AISC, vol. 267, pp. 687–695. Springer, Heidelberg (2014). doi:10.1007/978-3-319-05353-0_65
12. Kachniarz, M., Bieńkowski, A., Szewczyk, R.: Modelling magnetoelastic characteristics of manganese-zinc ferrite material in rayleigh region. In: Proceedings of the 21st International Conference on Applied Physics of Condensed Matter (APCOM 2015), pp. 84–88 (2015)
13. Bartnicki, A., Łopatka, J., Muszyński, T., Wrona, J.: Concept of IED/EOD operations (CONOPs) for engineer mission support robot team. J. KONES **22**(3), 269–274 (2015). doi:10.5604/12314005.1181703
14. Missala, T., et al.: Study on graphene growth process on various bronzes and copper-plated steel substrates. In: Szewczyk, R., Zieliński, C., Kaliczyńska, M. (eds.) Progress in Automation, Robotics and Measuring Techniques. AISC, vol. 352, pp. 171–180. Springer, Heidelberg (2015). doi:10.1007/978-3-319-15835-8_19
15. Bartnicki, A., Łopatka, M.J., Śnieżek, L., Wrona, J., Nawrat, A.M.: Concept of implementation of remote control systems into manned armoured ground tracked vehicles. In: Nawrat. M, A.M. (ed.) Innovative Control Systems for Tracked Vehicle Platforms. SSDC, vol. 2, pp. 19–37. Springer, Heidelberg (2014). doi:10.1007/978-3-319-04624-2_2

Influence of Temperature and Magnetizing Field on the Magnetic Permeability of Soft Ferrite Materials

Maciej Kachniarz[1(✉)] and Jacek Salach[2]

[1] Industrial Research Institute for Automation and Measurements PIAP, Al. Jerozolimskie 202, 02-486 Warsaw, Poland
mkachniarz@piap.pl

[2] Institute of Metrology and Biomedical Engineering, Warsaw University of Technology, sw. Andrzeja Boboli 8, 02-525 Warsaw, Poland

Abstract. Soft ferrite materials are commonly used as magnetic cores of inductive components utilized in electronic industry. Magnetic B-H characteristics of ferrites exhibit strong temperature dependence. It is connected with changes of magnetic permeability of ferrite material subjected to different temperatures. In ferromagnetic materials, also ferrites, magnetic permeability depends on the value of magnetizing field. In this paper, results of investigation on the influence of temperature and magnetizing field on the magnetic permeability of soft ferrite materials are presented. Several ring shaped cores made of different manganese-zinc ferrites were investigated on developed test stand utilizing cryostat to set the temperature in the range form −20 °C to 60 °C. Obtained results indicate that there is a strong correlation between temperature of ferrite material and its magnetic permeability, which cannot be neglected in technical applications of ferrite magnetic cores.

Keywords: Magnetism · Soft magnetic material · Ferrite · Magnetic permeability · Magnetic characteristics · Temperature

1 Introduction

Magnetic materials are very important for modern electronic industry. They are widely utilized as magnetic cores in inductive components like coils, filters, chokes, transformers, etc., which are present in all modern electronic devices. Magnetic properties of the material used for manufacturing the core are strongly influencing the characteristics of inductive component and, therefore, it is very important to select proper material for a particular application. Among many existing ferromagnetic materials, ferrites are very interesting group. They are often utilized in inductive components for the purposes of electronic industry, especially in EMC (Electro-Magnetic Compatibility) applications [1].

Ferrites are ceramic magnetic materials. They are composed of iron oxide Fe_2O_3 and one or more metallic elements [2, 4]. This results in double crystal lattice in the structure of material created of atoms of opposite magnetic moments. This opposite moments are not equal, so they do not compensate each other, which is the source of spontaneous magnetization of the material. Such materials are known as ferrimagnetics [3].

R. Szewczyk and M. Kaliczyńska (eds.), *Recent Advances in Systems, Control and Information Technology*, Advances in Intelligent Systems and Computing 543, DOI 10.1007/978-3-319-48923-0_75

Among many parameters used to characterize magnetic properties of the materials, magnetic permeability is one of the most significant. Magnetic permeability is a physical quantity determining the ability of material to produce magnetic field within itself as a response to the external magnetizing field. Magnetic properties of the material are usually characterized by relative magnetic permeability, which is magnetic permeability of the material in relation to magnetic permeability of free space μ_O [5]. In ferromagnetic materials, also ferrites, magnetic permeability depends on the value of magnetizing field. Magnetic properties of the material, also its magnetic permeability, are dependent on environmental conditions, especially temperature [6–10]. From technical point of view it is very important to determine the influence of magnetizing field and temperature on the magnetic permeability of ferrite materials due to their possible applications. The following paper presents results of investigation on the influence of temperature and magnetizing field on the magnetic permeability of soft ferrite materials utilized in technical applications.

2 Investigated Materials

Among many ferrite materials developed so far, one of the most important group are Mn-Zn ferrites, containing magnesium and zinc as metallic elements. Their chemical composition is characterized by general formula $Mn_{x-1}Zn_xFe_2O_4$. For performed investigation, four samples of Mn-Zn ferrite materials were prepared. Each of them had different content ratio of magnesium and zinc (*x* parameter in the general formula). Materials F-3001 and F-807 were fabricated by POLFER (Poland), while T38 and N41 by TDK-EPC Epcos (Germany). Investigated materials were formed into ring-shaped magnetic cores, which is presented in Fig. 1.

The flow path of magnetic flux density in the magnetic circuit and cross-sectional area of the core were calculated for each sample, which was necessary to determine

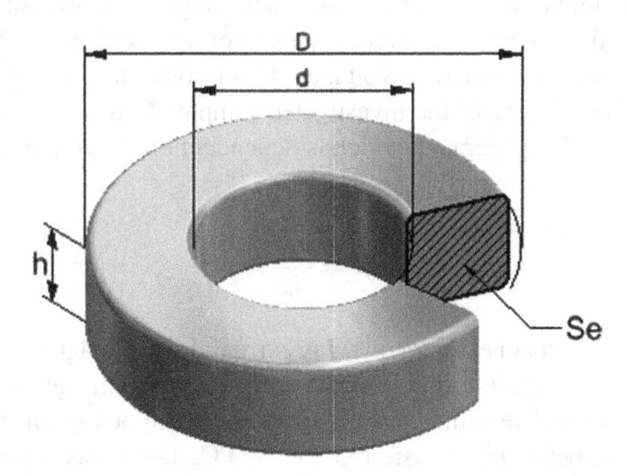

Fig. 1. Investigated Mn-Zn ferrite core, *D* – outer diameter, *d* – inner diameter, *h* – thickness, S_e – cross-sectional area

values of magnetizing filed and magnetic flux density during measurements. Magnetizing and sensing windings were made on each investigated core.

3 Measurement Methodology

For performed experiment, special measurement system was developed with cryostat for stabilization of the temperature. Schematic block diagram of the system is presented in Fig. 2. System was digitally controlled by the Personal Computer with Data Acquisition Card (DAQ) installed.

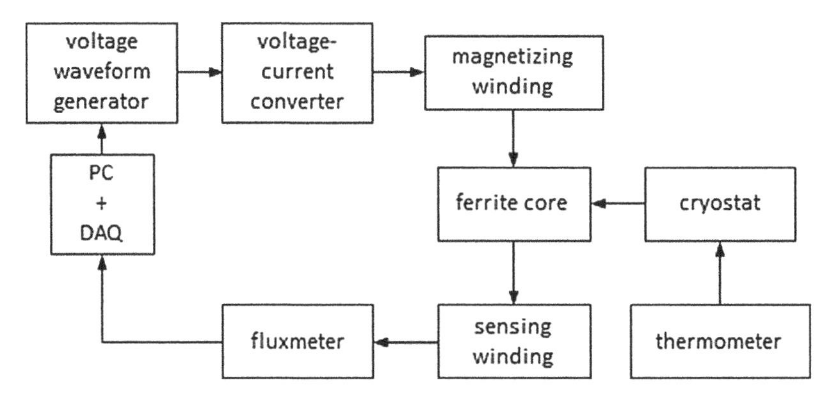

Fig. 2. Schematic block diagram of the developed measurement system

Main part of the developed system were devices allowing investigation of magnetic *B-H* characteristics of the material (*B* – magnetic flux density, *H* – magnetizing field, *B-H* characteristic is initial magnetization curve and magnetic hysteresis loop). To produce magnetizing field, voltage waveform generator was utilized. Voltage waveform from generator was transformed by voltage-current converter into current waveform driving magnetizing winding of the investigated sample. As a result, magnetizing filed acting on the sample was generated, which is dependent on the current according to the Eq. (1):

$$H = \frac{n_m I}{l_e} \tag{1}$$

where n_m is number of magnetizing coils, I is current driving magnetizing winding and l_e is the flow path of magnetic flux density in the sample. Changes of magnetic flux density B in the material resulting from magnetizing field acting on the sample were measured with fluxmeter and transferred to the PC. Developed system allowed to investigate *B-H* characteristics for given value of magnetizing field amplitude and frequency.

Investigated samples were placed in cryostat chamber filled with thermally conductive fluid, which allowed to obtain stable values of temperature within the range from −20 °C to 60 °C. APPA 207 multimeter with K-type thermocouple was used to measure the temperature inside main chamber of the cryostat.

For all investigated samples family of 16 *B-H* characteristics was measured in given temperature. For each obtained *B-H* characteristic relative magnetic permeability was calculated according to the Eq. (2):

$$\mu = \frac{1}{\mu_0}\frac{dB}{dH} \tag{2}$$

where $\mu_0 = 4\pi \cdot 10^{-7}$ H/m is vacuum magnetic permeability (physical constant). As a result, 16 points on each μ-H characteristic (μ – relative magnetic permeability, H – magnetizing field) were obtained for given temperature. The μ-H characteristics were investigated within temperature range from −20 °C to 60 °C with step of 10 °C.

4 Experimental Results

As a result of performed investigation, family of 9 μ-H characteristics was obtained for each sample, each characteristic composed of 16 points. Results are presented in Figs. 3, 4, 5 and 6. For clear presentation only three μ-H characteristics obtained for temperatures −20 °C, 20 °C and 60 °C are shown in each figure.

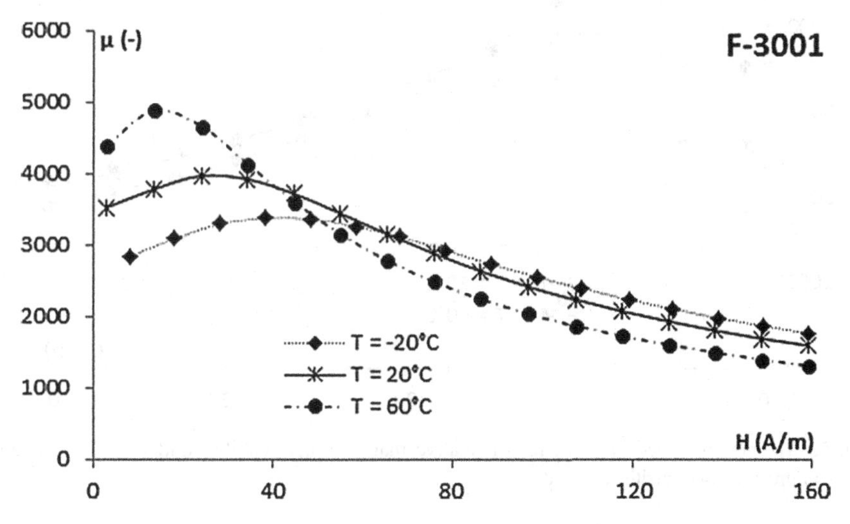

Fig. 3. Magnetizing filed dependence of relative magnetic permeability under the influence of temperature for F-3001 ferrite material

As it can be seen in the figures, all obtained μ-H characteristics have similar shape. Each one starts with initial value for low magnetizing field and then relative permeability is increasing with the magnetizing field to reach its maximum value. Then permeability value starts to decrease with further increase of magnetizing field and is

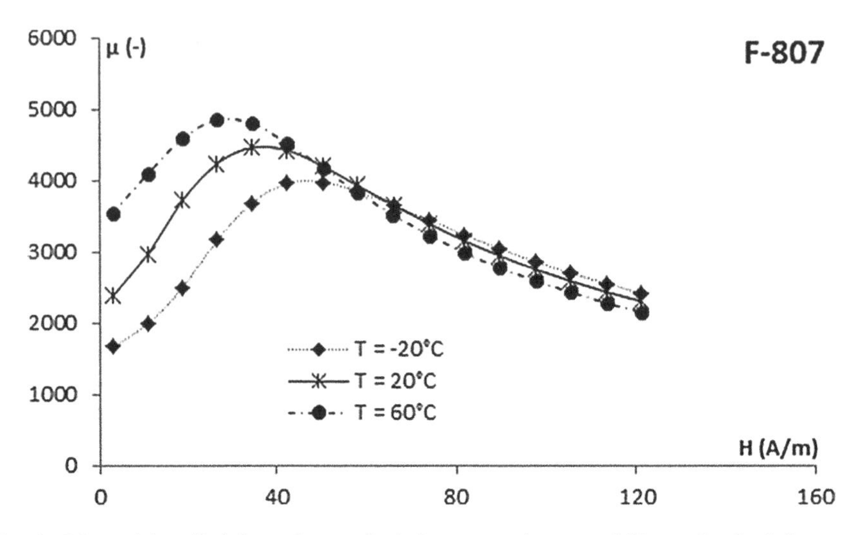

Fig. 4. Magnetizing filed dependence of relative magnetic permeability under the influence of temperature for F-807 ferrite material

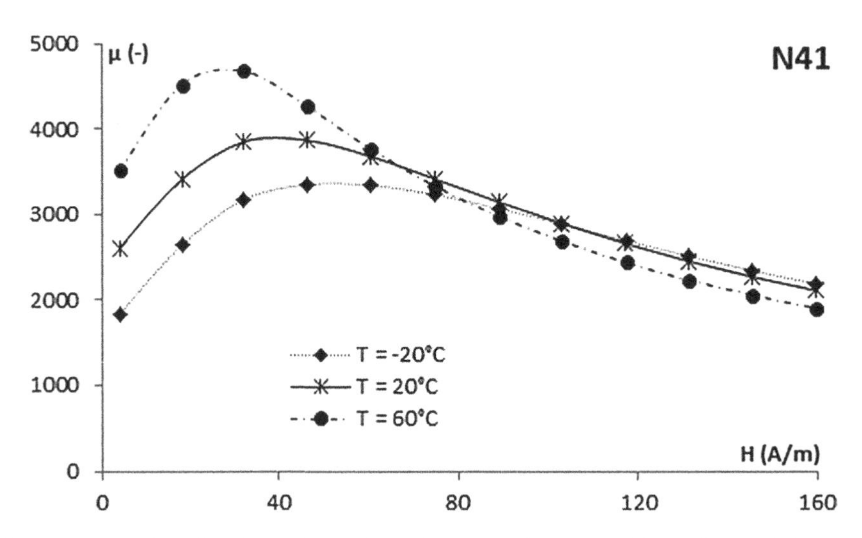

Fig. 5. Magnetizing filed dependence of relative magnetic permeability under the influence of temperature for N41 ferrite material

asymptotically tending to 1. In room temperature, about 20 °C, three of investigated materials have maximum relative permeability value within the range 4000–5000. Only T38 ferrite exhibit significantly higher maximum permeability with the value almost 16 000. This difference is the result of different chemical composition of T38 material (different content ratio of magnesium and zinc).

Temperature influence on the values of relative magnetic permeability is similar for all investigated cores. With the temperature growth value of maximum permeability is

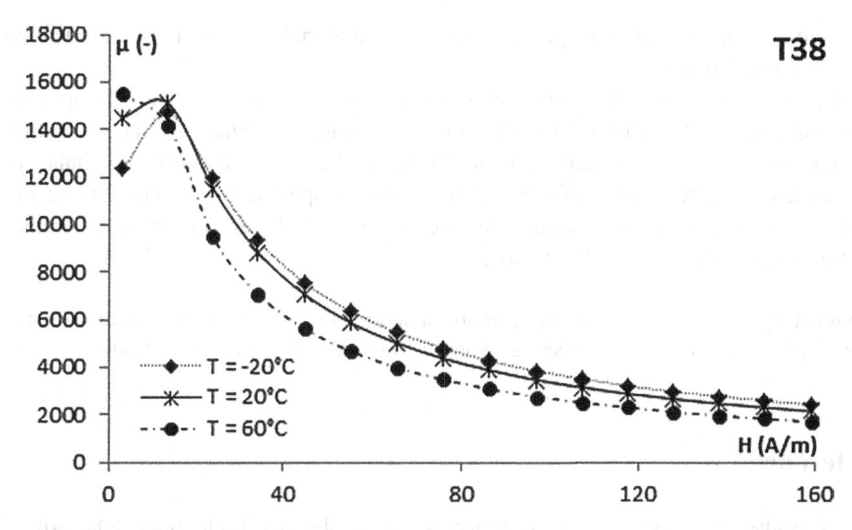

Fig. 6. Magnetizing filed dependence of relative magnetic permeability under the influence of temperature for T38 ferrite material

increasing. For 60 °C maximum permeability is substantially higher than for −20 °C for all investigated materials. Moreover, with temperature increase, maximum value of permeability is reached for lower magnetizing field. For T38 material at 60 °C maximum permeability is reached for magnetizing field so low, that it was not registered during the experiment, which explains, why the initial permeability is invisible in the chart. The decreasing part of the μ-H characteristic is also affected by the temperature. The decrease is greater for higher temperature for all investigated ferrite materials, so the values of relative permeability in the saturation region (high magnetizing filed) are lower for high temperatures. N41 material seems to be the most susceptible for the temperature influence. For this one the differences between the values of maximum permeability are the greatest. F-3001 and F-807 materials also exhibit significant temperature susceptibility. T38 ferrite is definitely the most temperature resistive among investigated materials.

5 Conclusion

For all investigated materials temperature dependence of magnetic permeability is similar. Higher temperature value results in higher maximum relative permeability and lower magnetizing filed necessary to obtain this maximum. It is caused by the increase of thermal energy of atoms in the crystal structure of material resulting from higher temperature. Thermal vibrations of atoms in the crystal lattice are becoming stronger and it is easier for their magnetic moments to redirect their vector to the direction of magnetizing field, which can be observed as increase of the magnetic permeability of the material. At the same time in saturation region, where all atomic magnetic moments should be directed according to the magnetizing field vector, thermal vibrations interfere their arrangement, so permeability of the material in saturation region is lower in higher temperatures.

Results presented in the paper show significant correlation between temperature and magnetic permeability of Mn-Zn soft ferrite materials. This correlation is strong enough to be taken into account in technical applications of ferrite materials. The operating temperature range of ferrite core should be specified and taken into account during designing electronics with inductive components based on ferrite materials to ensure proper operation of the device.

Acknowledgements. This work was partially supported by the statutory founds of Institute of Metrology and Biomedical Engineering, Warsaw University of Technology (Poland).

References

1. O'Handley, R.: Modern Magnetic Materials – Principles and Applications. John Wiley & Sons, New York (2000)
2. Carter, C.B., Norton, M.G.: Ceramic Materials: Science and Engineering. Springer, New York (2013)
3. Cullity, B.D., Graham, C.D.: Introduction to Magnetic Materials. John Wiley & Sons, New York (2009)
4. Szałatkiewicz, J.: Metals content in printed circuit board waste. Pol. J. Environ. Stud. **23**(6), 2365–2369 (2014)
5. Tumański, S.: Handbook of Magnetic Measurements. CRC Press, New York (2011)
6. Szewczyk, R., Švec Sr., P., Švec, P., et al.: Thermal annealing of soft magnetic materials and measurements of its magnetoelastic properties. Pomiary Automatyka Robotyka **2**, 513–518 (2013)
7. Szałatkiewicz, J., Szewczyk, R., Budny, E., Missala, T., Winiarski, W.: Measurement and control system of the plasmatron plasma reactor for recovery of metals from printed circuit board waste. In: Szewczyk, R., Zieliński, C., Kaliczyńska, M. (eds.) Recent Advances in Automation, Robotics and Measuring Techniques. AISC, vol. 267, pp. 687–695. Springer, Heidelberg (2014). doi:10.1007/978-3-319-05353-0_65
8. Szewczyk, R.: Modeling the influence of temperature on the magnetic characteristics of $Fe_{40}Ni_{38}Mo_4B_{18}$ amorphous alloy for magnetoelastic sensors. In: Jabłoński, R., Turkowski, M., Szewczyk, R. (eds.) Recent Advances in Mechatronics, pp. 586–590. Springer, Heidelberg (2007). doi:10.1007/978-3-540-73956-2_115
9. Tsepelev, V., Starodubtsev, Y., Zelenin, V., et al.: Temperature affecting the magnetic properties of the $Co_{79-x}Fe_3Cr_3Si_{15}B_x$ amorphous alloy. J. Alloys Compd. **643**, 280–282 (2015)
10. Kulikowski, J., Bieńkowski, A.: Field, temperature and stress dependence of magnetostriction in Ni-Zn ferrites containing cobalt. Phys. Scripta. **44**, 382–383 (1991)

Selected Trends in New Rapidly Quenched Soft Magnetic Materials

Peter Švec[1(✉)], Irena Janotová[1], Juraj Zigo[1], Igor Matko[1], Dušan Janičkovič[1], Jozef Marcin[2], Ivan Škorvánek[2], and Peter Švec Sr.[1]

[1] Institute of Physics, Slovak Academy of Sciences, Bratislava, Slovakia
peter.svec@savba.sk

[2] Institute of Experimental Physics, Slovak Academy of Sciences, Kosice, Slovakia

Abstract. Classical rapidly quenched nanocrystalline soft magnetic materials such as FINEMET, NANOPERM and HITPERM are seconded by new intensely investigated systems where special attention is put on materials with high saturation magnetization while preserving low coercivity. Diverse systems based on Fe-B with additions of Co, Cu, C, P and other elements were developed and tested. Successful compositional and processing design have lead to a new class of soft magnetic materials fulfilling these requirements based on Fe-Co-B, where the role of nanocrystal-forming element, namely nanograin size control, is taken up by Sn. Selected results will be presented on the case of rapidly quenched amorphous Fe-Sn-B alloy and its nanocrystallization by thermal treatment.

Keywords: Amorphous alloys · Nanocrystallization · Soft magnetic materials · High saturation magnetization · Fe-Sn-B alloys

1 Introduction

Nowadays the well-known classical rapidly quenched nanocrystalline soft magnetic materials such as FINEMET [1], NANOPERM [2] and HITPERM [3] are seconded by new intensely investigated systems. Among these a special attention is put on materials with increased saturation magnetic flux density exceeding the values of 1.8 T while preserving low coercivity [4, 5]. Technically it is also important that the materials can be exposed to long-term operating temperatures ranging well above ambient, exceeding 500 K or even more. Diverse new systems based on Fe-B with additions of Co, Cu, C, P and other elements were developed and tested. Successful compositional and processing design has lead to a new class of so-called NANOMET alloys which fulfill requirements posed on modern soft magnetic materials [6–9].

The potential of replacement of rare-earth elements in FINEMET, NANOPERM and HITPERM alloy systems by more abundant elements has been successfully shown on NANOPERM-like Fe-B system with small additions of Sn [10, 11]. The drive for high saturation magnetic flux density is inevitably related to increase of content of ferromagnetic elements in the alloy, in this case Fe, to the limit of glass formability given by the minimal content of glass-forming element, i.e. B. Thus our work is

R. Szewczyk and M. Kaliczyńska (eds.), *Recent Advances in Systems, Control and Information Technology*, Advances in Intelligent Systems and Computing 543, DOI 10.1007/978-3-319-48923-0_76

focused on rapidly quenched Fe-Sn-B systems with decreased B content and addition of Sn at the expense of Fe in order to form nanocrystalline grains from as-cast amorphous structure by suitable annealing.

The reason for selection of Fe-B system is the fact that it represents one of the most studied amorphous systems. In addition, atomic radii of Fe and Sn atoms are favorable for incorporation of Sn in the Fe-rich amorphous phase in spite of a rather low melting point of Sn as compared to Fe. Another reason for interest in Fe-Sn-B or Fe-Co-Sn-B based systems is the elimination of conventional nanocrystal forming elements belonging to the rare-earth group (Nb, Nd, etc.), which is in accord with recent EU policy related to the use of critical raw elements; furthermore, Sn is an abundant and inexpensive alloying element.

Hereby we present selected results related to the preparation, structure and some properties of rapidly quenched $Fe_{78}Sn_7B_{15}$ and $Fe_{81}Sn_7B_{12}$ with emphasis on $Fe_{81}Sn_7B_{12}$.

2 Experimental

Amorphous ribbons were prepared by the planar flow casting from previously prepared master alloys with selected chemical composition. The samples 6 mm wide and ~20 μm thick were linearly heated with 10 K/min heating rate and isothermally annealed at several selected temperatures. The transformation from as-cast amorphous state was monitored using the measurements of temperature dependencies of magnetic weight in Perkin Elmer TGA-7 thermogravimeter with small (~20 mT) applied magnetic field. Contact furnace with vacuum was used for thermal treatment of samples for transmission electron microscopy (JEOL 2000FX at 200 kV) and X-ray diffraction using Bruker D8 Advance diffractometer with parallel incident beam (Cu K_alpha radiation 40 kV/35 mA) and LiF monochromator in the diffracted beam. Magnetic hysteresis loops were acquired using a Forster type B-H loop tracer based on flux-gate magnetometer.

3 Results

Transformation of as-cast Fe-Sn-B with different boron content is shown in Fig. 1 using temperature dependence of magnetic weight during linear heating. The decrease of magnetic weight from room temperature up to about 650 K indicates the approach to the Curie temperature of the amorphous sample. The increase of magnetic weight above 650 K reflects the formation of a magnetic phase from amorphous matrix, namely the formation of bcc-Fe(Sn). This behaviour is superseded by a decrease again due to the approach to the Curie temperature of the crystallized phase. Crystallization of the remaining amorphous matrix takes place above 750 K and is reflected by a sharp rise of magnetic weight. It is to be noted that the temperature interval between the first and the second crystallization reactions is higher than 100 K, which provides a conveniently broad interval for thermal processing and controlled formation of ferromagnetic bcc-Fe(Sn) phase.

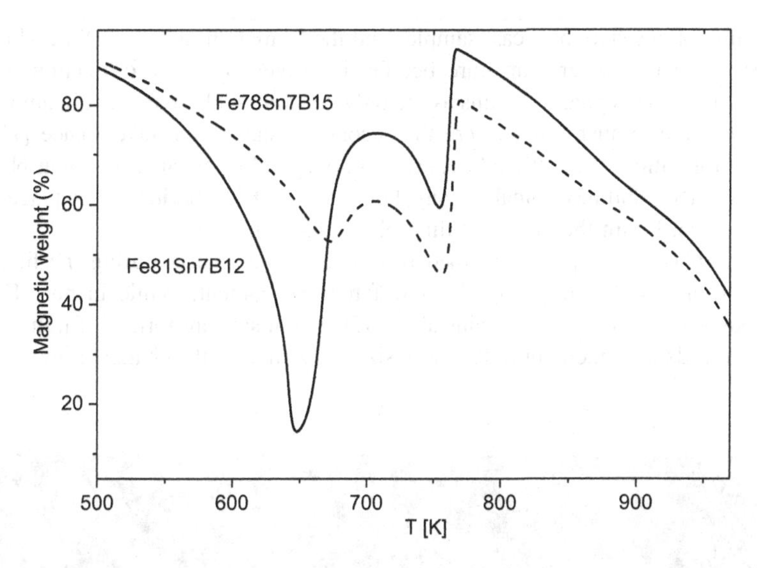

Fig. 1. Temperature dependence of evolution of magnetic weight (M-TGA) of rapidly quenched $Fe_{78}Sn_7B_{15}$ and $Fe_{81}Sn_7B_{12}$ during linear heating with rate 10 K/min

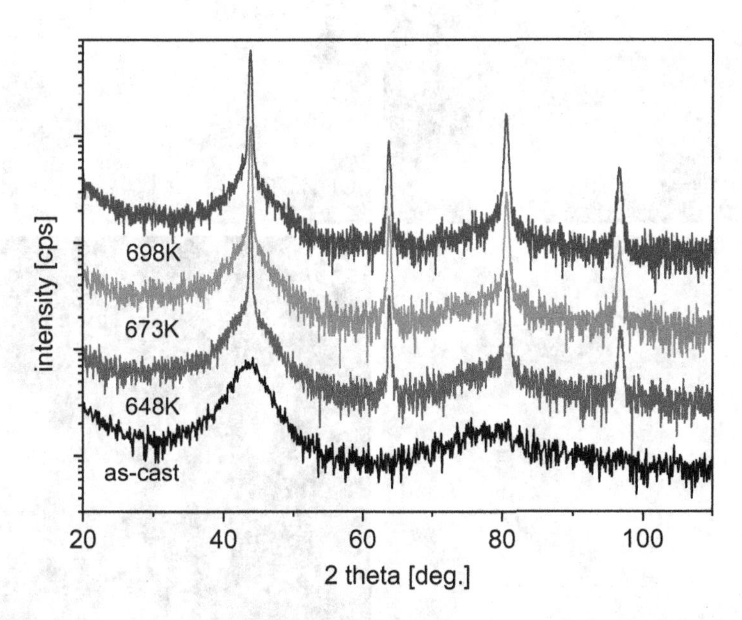

Fig. 2. X-ray diffraction patterns of $Fe_{81}Sn_7B_{12}$ in as-cast state and after isothermal annealing for 30 min at indicated temperatures. Maxima formed at the indicated annealing temperatures correspond to bcc-Fe(Sn) with the lattice parameter $\sim$0.2921 nm

The amorphous state of a-cast samples and the formation of bcc-Fe(Sn), which has a higher lattice parameter than pure bcc-Fe, is proven by x-ray diffraction patterns shown in Fig. 2. This phase remains as the only crystalline phase present in amorphous matrix up to the temperatures where the second crystallization takes place [11, 12], leading to formation of additional peaks, all corresponding to Sn-containing phases – hexagonal Fe(Sn) and hexagonal Fe_3Sn_2. Formation of Fe-B borides takes place above 950 K, in accord with the corresponding phase diagram [13].

The morphology of phases formed from amorphous matrix during the first crystallization stage is shown in Fig. 3 for different Sn content. While in pure $Fe_{85}B_{15}$ amorphous alloy dendritic-like grains about 150 nm in size are formed [5], increase of Sn content leads to refinement of the grain sizes as well as to the change of morphology

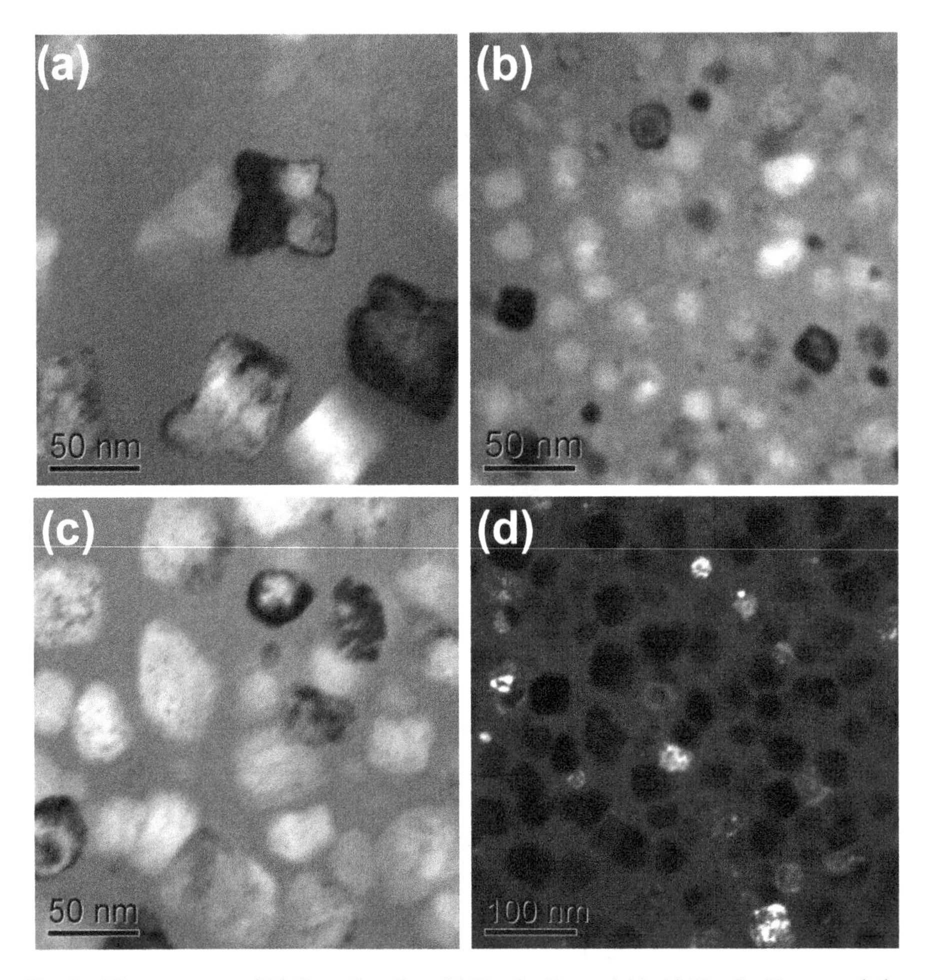

Fig. 3. Microstructure of (a) $Fe_{81.5}Sn_{3.5}B_{15}$, (b) $Fe_{78}Sn_7B_{15}$ and (c), (d) $Fe_{81}Sn_7B_{12}$ annealed at 678 K for 30 min. Note the change of scale bar in (d) showing larger area of the sample and the distribution of the grains in amorphous matrix in a dark-field image

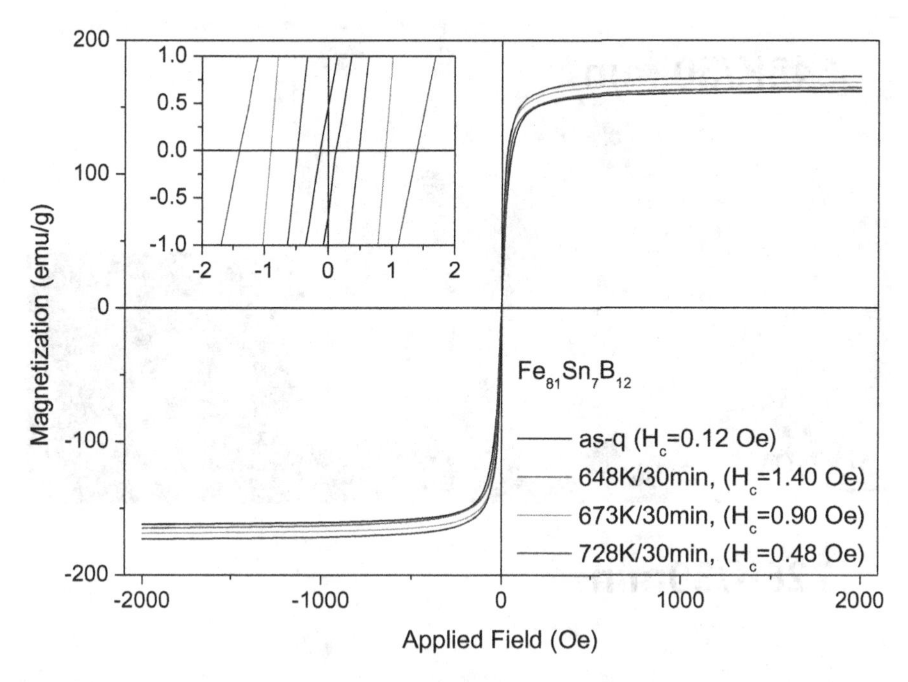

Fig. 4. Quasistatic hysteresis loops of as-cast and annealed $Fe_{81}Sn_7B_{12}$; inset shows the behaviour of magnetization at low applied fields for determination of coercivity H_c

of the grains. Tiny dendritic-like ~50 nm large crystals observed for 3.5 at. % Sn (Fig. 3a) change into significantly smaller grains of rhomboedric shape for 5 at. % Sn [12]. Nearly spherical 25–50 nm sized structures are formed for 7 at. % Sn (Fig. 3b, c). The morphology is similar to that observed in classical nanocrystalline alloys mentioned in the Introduction, however, without rare-earth grain refiners [14]. Very even distribution of the nanograins in the remaining amorphous matrix can be observed, especially in dark-field TEM (Fig. 3d).

Magnetic hysteresis loops shown for $Fe_{81}Sn_7B_{12}$ in Fig. 4 exhibit at the same time high values of magnetization for as-cast and nanocrystalline samples as well as low magnetic coercivity values. It is interesting to note that the coercivity value decreases only slightly with proceeding nanocrystallization. However, samples annealed to nanocrystalline state are structurally stable and, with respect to high Curie temperature of the annealed alloy, which lies above the temperature of the second crystallization stage, allow applications at temperatures exceeding 600 K. Higher annealing temperature (TEM images in Fig. 5) leads to further refinement of the nano-sized grains without their tendency to coarsening even after prolonged annealing. This effect suggests that the formation of nanocrystalline grains is thermodynamic and not kinetic effect; the grain growth stops after attaining a definite size and is not simply kinetically arrested due to annealing for a short time only, which is the case of Fe-B-Cu [4, 5].

A detailed analysis of the evolution of size of nanocrystalline bcc-Fe(Sn) is shown in Fig. 6. It can be seen that the width of the [110] peak slightly decreased with

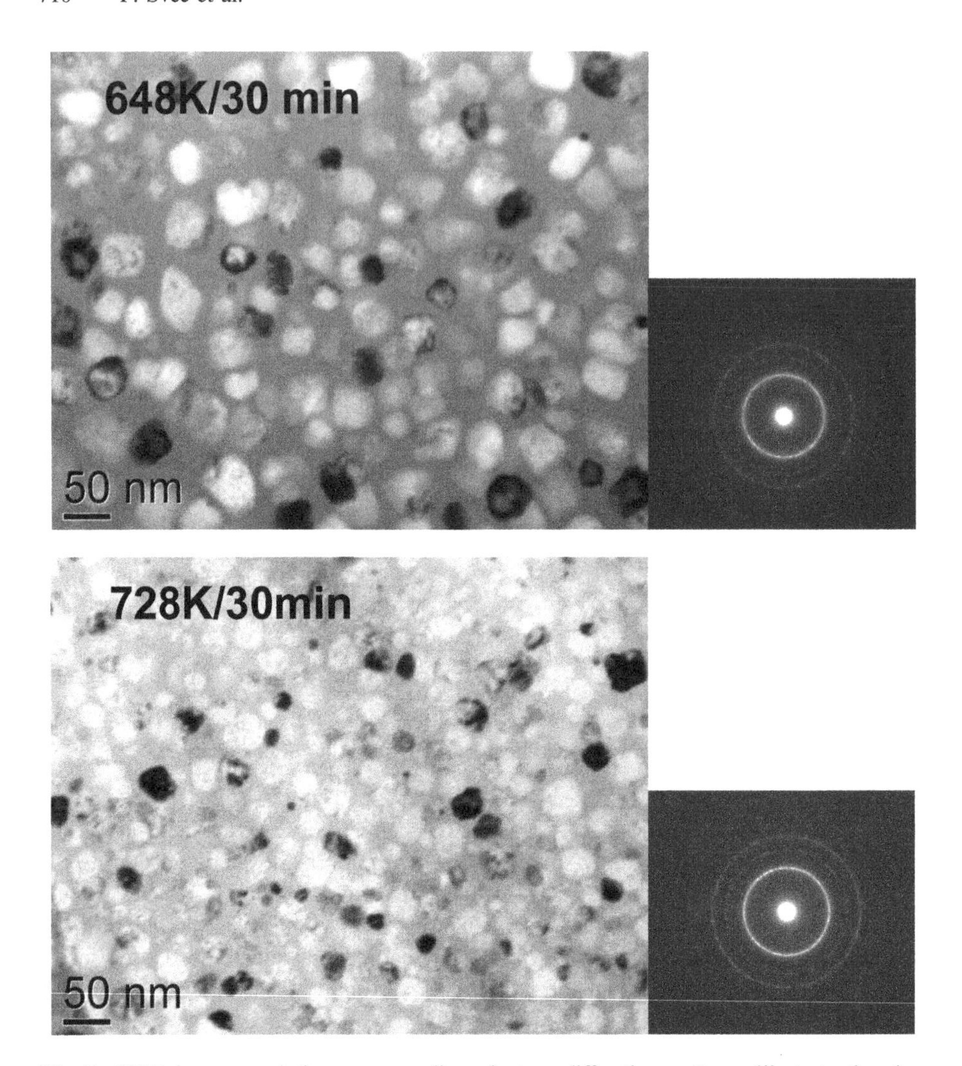

Fig. 5. TEM images and the corresponding electron diffraction patterns illustrate the size, content and distribution of the nanocrystalline bcc-Fe(Sn) grains in amorphous matrix formed after isothermal annealing in argon at indicated temperatures

annealing time, as can be expected due to slight grain growth of the nanocrystals in time. However, in the temperature range investigated the width increases with increasing temperature, suggesting formation of smaller nanograins at higher temperatures. This phenomenon allows stable tuning of the proper grain size in a wide temperature interval, which can conveniently be used to tuning magnetic properties strongly dependent on this value [15]. The values of FWHM suggest the sizes of the grain decrease from ~50 nm to ~25 nm with increasing annealing temperature. This trend follows well the values of Hc in Fig. 4 and the grain sizes observed by TEM in Fig. 5. It is also worthwhile to mention that a specific evolution of lattice parameter of

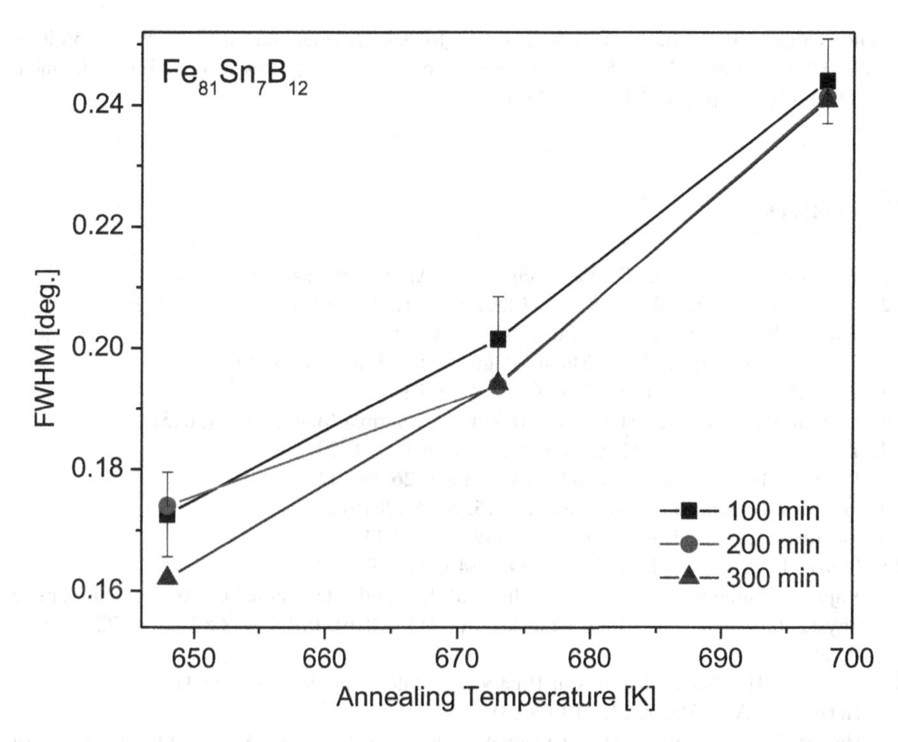

Fig. 6. The full width at half maximum (FWHM) of the first diffraction peak of the bcc-Fe(Sn) phase in the course of isothermal annealing. The lines connecting isochronal FWHM values are to guide the eye only

the bcc-Fe(Sn) phase with Sn and B content and with annealing as well as complex microstructural morphology of the nanograins have been observed [11] – decrease of the bcc-Fe(Sn) lattice parameter from 0.2903 nm down to 0.2886 nm during annealing between 723 and 873 K has been observed, accompanied by formation of small amount of hexagonal FeSn. Detailed investigations of these phenomena are in progress.

4 Conclusions

Substitution of Fe by small amounts of Sn in amorphous Fe-B system leads to formation of amorphous structure which transforms partly into nanocrystalline grains stable until 750 K. Increased Sn content up to 7 at. % in $(Fe\text{-}Sn)_{88}B_{12}$ system enhances nanocrystallization and leads to a large number of spherical grains distributed evenly in the remaining amorphous matrix. The nanocrystalline phase formed in the first crystallization stage was identified as bcc-Fe(Sn). In combination with amorphous remains the alloy forms a composite structure with high saturation magnetization and low values of magnetic coercivity which can be further tuned by proper thermal treatment. The results suggest potential application of the Fe-Sn-B system as nanocrystalline soft magnetic material eliminating the use of strategic elements as grain refiners and nanocrystal formers.

Acknowledgement. The authors would like to acknowledge the support of the projects APVV-0460-12 and APVV-15-0621 supported by the Slovak Research and Development Agency and by the project VEGA 2/0189/14.

References

1. Yoshizawa, Y., Oguma, S., Yamauchi, K.: J. Appl. Phys. **64**, 6044 (1988)
2. Suzuki, K., Makino, A., Inoue, A., Masumoto, T.: J. Appl. Phys. **70**, 6232 (1991)
3. Willard, M.A., et al.: J. Appl. Phys. **84**, 6773 (1998)
4. Ohta, M., Yoshizawa, Y.: J. Magn. Magn. Mater. **320**, 750 (2008)
5. Svec, P., et al.: IEEE Trans. Magn. **46**, 408 (2010)
6. Sharma, P., Zhang, X., Zhang, Y., Makino, A.: Scripta Mater. **95**, 3 (2015)
7. Takenaka, K., et al.: J. Magn. Magn. Mater. **401**, 479 (2016)
8. Janotova, I., et al.: J. Supercond. Novel Magn. **26**, 793 (2013)
9. Janotova, I., et al.: J. Alloys Compd. **615**, S198 (2015)
10. Illekova, E., et al.: J. Alloys Compd. **509**, S46 (2011)
11. Matko, I., et al.: J. Alloys Compd. **615**, S462 (2015)
12. Vajda, J., Jamnicky, I. (eds.) Proceedings of the 22nd International Conference on Applied Physics of Condensed Matter, Bratislava, p. 311 (2016). ISBN: 978-80-227-4572-7. Svec, P., et al.
13. Okamoto, H.: Phase Diagrams of Binary Iron Alloys, pp. 93–101 (1993)
14. Herzer, G.: Acta Mater. **61**, 718 (2013)
15. Herzer, G.: Nanocrystalline soft magnetic alloys. In: Buschow, K.H.J. (ed.) Handbook of Magnetic Materials, vol. 10, pp. 415–462. Elsevier (1997)

The Comparison of Rapidly Quenched Co-Sn-B and Fe-Sn-B Alloys

Irena Janotová[1(✉)], Peter Švec Sr.[1], Peter Švec[1], Igor Mako[1], Dušan Janičkovič[1], Juraj Zigo[1], Jozef Marcin[2], and Ivan Škorvánek[2]

[1] Institute of Physics, Slovak Academy of Sciences, Bratislava, Slovakia
irena.janotova@savba.sk

[2] Institute of Experimental Physics, Slovak Academy of Sciences, Košice, Slovakia

Abstract. The ferromagnetic systems based on Fe-Sn-B, Co-Sn-B and Fe-Co-Sn-B were studied in nanocrystalline state due to their interesting magnetic properties that are caused by the homogeneous and ultrafine structure. The alloying of Co-B and/or Fe-B by Sn 3.5 and/or 5 at. % improves the properties of resulting structure composed of the ferromagnetic grains in the amorphous matrix as proved by the XRD and TEM methods. The structure transformation from amorphous to (nano)crystalline state was investigated by DSC and TGA methods and the resulting phase and morphology of crystalline products were analyzed.

Keywords: Soft magnetic materials · Nanocrystalline structure · Iron-boron alloys

1 Introduction

The metastable Fe-B, Co-B and Fe-Co-B alloys are frequently studied and commonly used as a base for different soft magnetic metallic glasses [1–8]. To improve the nanocrystallization of these systems the rare-earth elements are often used. To eliminate the need of these expensive and strategic elements the alloying by the post-transition metals could offer "the fail-back". So that the addition of small amounts of Sn into the Fe-B, Co-B and Fe-Co-B based metastable systems [5–11] was studied. The difference of the atomic radii of Fe and/or Co and Sn atoms are favorable for the incorporation of Sn into the ferromagnetic-rich amorphous phase in spite of a rather low melting point of Sn as compared to Fe or Co ones. Furthermore, Sn is an abundant and inexpensive alloying element [4, 5, 12–15]. In this study the results of the structural and thermodynamic investigation of rapidly quenched $(Fe/Co)_{(85-x)}Sn_xB_{15}$ systems, for the ratio Fe/Co = 1/0, 0/1 and 1/1 and for Sn x = 3.5, 5 at. %.are presented. The structure evolution from amorphous state into the crystalline one was observed by the calorimetric (DSC) and thermogravimetric (TGA) measurements. The formation of ferromagnetic grains from the amorphous matrix is shown by the direct structure observation using transmission electron microscopy as well as by the X-ray diffraction methods. The results of in-situ phase analysis at a pre-defined temperature regime are reviewed, too.

R. Szewczyk and M. Kaliczyńska (eds.), *Recent Advances in Systems, Control and Information Technology*, Advances in Intelligent Systems and Computing 543, DOI 10.1007/978-3-319-48923-0_77

2 Experimental Procedure

The amorphous alloys under investigation were prepared in the form of ribbons 6 mm wide and ~20 μm thick by the planar flow casting method. The samples were linearly heated with 10 K/min heating rate and also isothermally annealed at several selected temperatures. The sequence and products of crystallization stages of the amorphous structure at given time and temperature were thus observed. The measurements of temperature dependencies of normalized heat flow and magnetic weight was used to obtain the basic information about the transformation behavior of the studied metallic systems. These methods enable us to define the beginning of the crystallization (temperature of onset of transformation) T_x and to observe the character of the transformations, too. The kinetic parameters (T_x(10 K/min), T_c(10 K/min)) were investigated by differential scanning calorimetry (DSC7 Perkin Elmer) and by thermogravimetry with small applied magnetic field (TGA7 Perkin Elmer), both in the protective argon atmosphere. X-ray diffraction (XRD) using Bruker D8 diffractometer and transmission electron microscopy (TEM) using JEOL 2000FX were used for microstructural characterization of as-cast and isothermally annealed samples. The parameters for the heat-treatment (T_x − 20 K for 30 min) were selected according to the resistivity measurements. Magnetic hysteresis loop was acquired using a Forster type B-H loop tracer based on flux-gate magnetometer.

3 Results and Discussion

The rapidly quenched systems usually exhibit typical two-stage transformation from amorphous to nanocrystalline state. The additions of small amounts of alloying element can change this process dramatically. For the systems under investigation it can be seen as the exothermic reactions on the normalized heat flow (Fig. 1a): the two major falls of the heat flow values indicate the two different system structure changes separated in the temperature. Figures 1a and 1b show samples with different Fe/Sn, Fe-Co/Sn and Co/Sn ratio for the constant B. The onset of transformation in the samples containing Fe is shifted towards lower temperature in comparison with the Co based samples. The reduction of the temperature interval between both transformations onsets $-\Delta T$ (Fig. 1a) for the Fe-based samples decrease/disappear with the increasing Co content in the Fe-Co-Sn-B samples. The onsets of the crystallizations stages for the Fe-based samples depend only weakly on the Sn content. For the Co-based samples Sn content is more significant, the transformation pattern is obviously changed into the one step – "polymorphous" crystallization, and for higher −5 at. % Sn content to tree-step transformation; also the shift of T_x to the higher temperatures is seen. For the Fe-Co base sample (in the Fe/Co = 1/1 ratio) the heat flow curve shows two steps of crystallization desirably separated in the time and temperature, while the crystallization temperature T_x is higher than 690 K. The TGA measurements (Fig. 1b) show the Curie temperature of amorphous samples, for Fe based samples in the range of 630 K and 668 K, whereas higher content of Sn affects also higher T_c. Co based samples indicated T_c above the first crystallization stage, but the magnetic weight curves show the decreasing of weight caused by changes in magnetic properties at higher temperatures

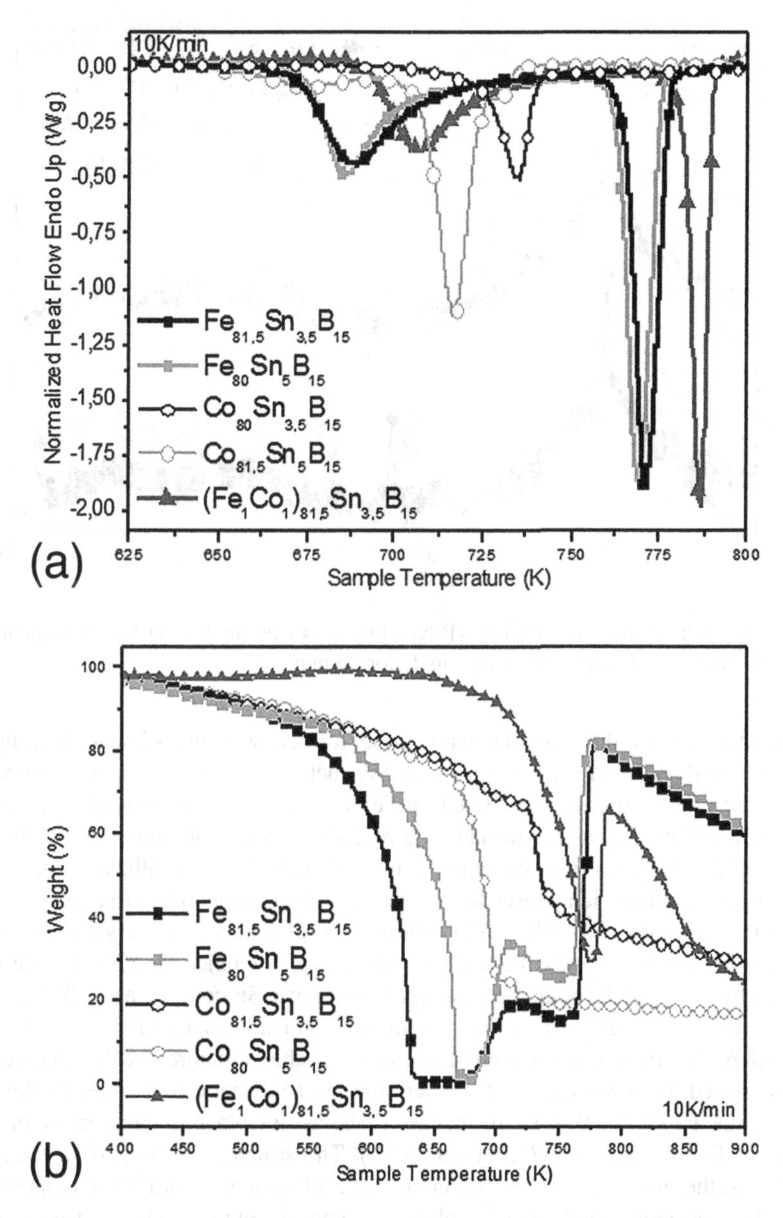

Fig. 1. Temperature dependence of: (a) the normalized heat flow from DSC measurement for the systems with different Fe/Co ratio; (b) the magnetic weight for systems with different Sn content and with different Fe/Co ratio

above 670 K; however, formation of ferromagnetic phase in this temperature region prevents accurate determination of T_c of the amorphous phase. The temperature for isothermal annealing was chosen according to the results of the thermal analysis, with respect to the expected final structure after regulated heat treatment.

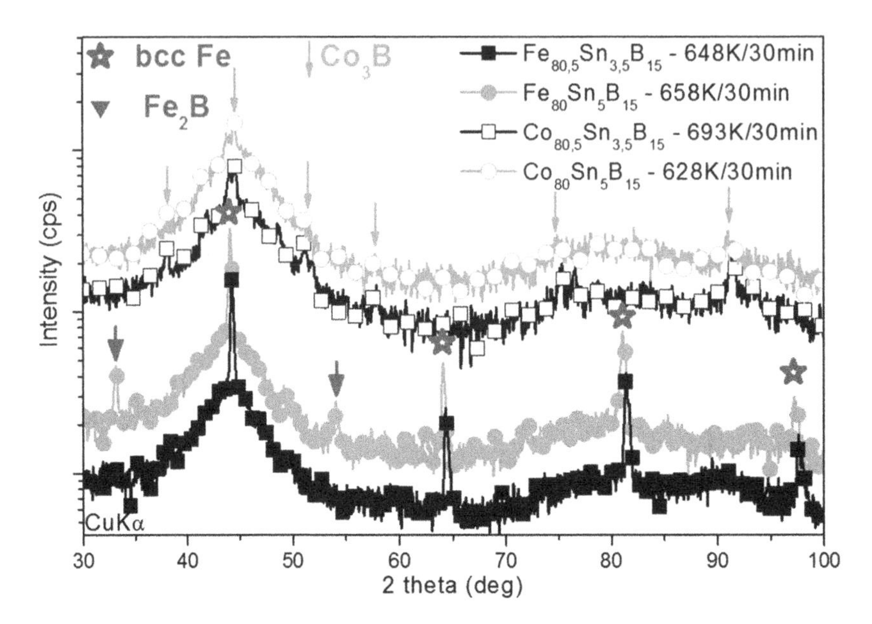

Fig. 2. XRD patterns from Fe-B and Co-B based samples with Sn 3.5 and 5 at. % content after isothermal annealing at temperatures near to T_x for 30 min

The structure of the ferromagnetic phase formed was investigated thoroughly. The XRD analysis of the early stages of the metal rich phase formation from the amorphous state for the samples based on Fe or Co isothermally annealed at temperatures 20 K below T_x was performed. Figure 2 shows the evolution of the metal-rich bcc-Fe and Co_3B phases from the amorphous state in the first crystallization stage. The formation of metalloid-rich phases from the remaining amorphous matrix takes place in the second stage (Fig. 1a). The XRD pattern (Fig. 2) indicates the crystalline bcc-Fe formation for the heat treated Fe-Sn-B samples. The difference in ΔT is visible here, followed by the Fe_2B formation for sample with higher Sn content, as well. Co based samples exhibit one for 3.5 at. % Sn, and three crystallization stages for 5 at. % Sn. The failure of ΔT for these compositions causes an unstable structure and high content of borides created, both responsible for deterioration of magnetic properties. In this perspective, the Fe-Co-Sn-B system appears to be a suitable combination of thermal parameters like T_x, T_c and ΔT (Fig. 1a and b). The structure of $(Fe_1Co_1)_{81.5}Sn_{3.5}B_{15}$ after the isothermal annealing in different stage of structure formation is shown in Fig. 3. The formation of the bcc-Fe phase is shown in Fig. 3a. The measurement of XRD at in situ annealing reveals the ferromagnetic phase evolution from amorphous phase during linear annealing at 2 K/min (Fig. 3b).

Structure after annealing at the temperature of the end of the first transformation (723 K/30 min) exhibits the presence phase in matrix (Fig. 3a), similarly to the Fe-based samples (Fig. 2). From Fig. 4 it is of the bcc-Fe grains in the amorphous matrix (Fig. 3b). Sample annealed at the temperature in the first half of the second transformation (783 K/30 min) exhibit $Fe(Co)_2B$ obvious that the structure after

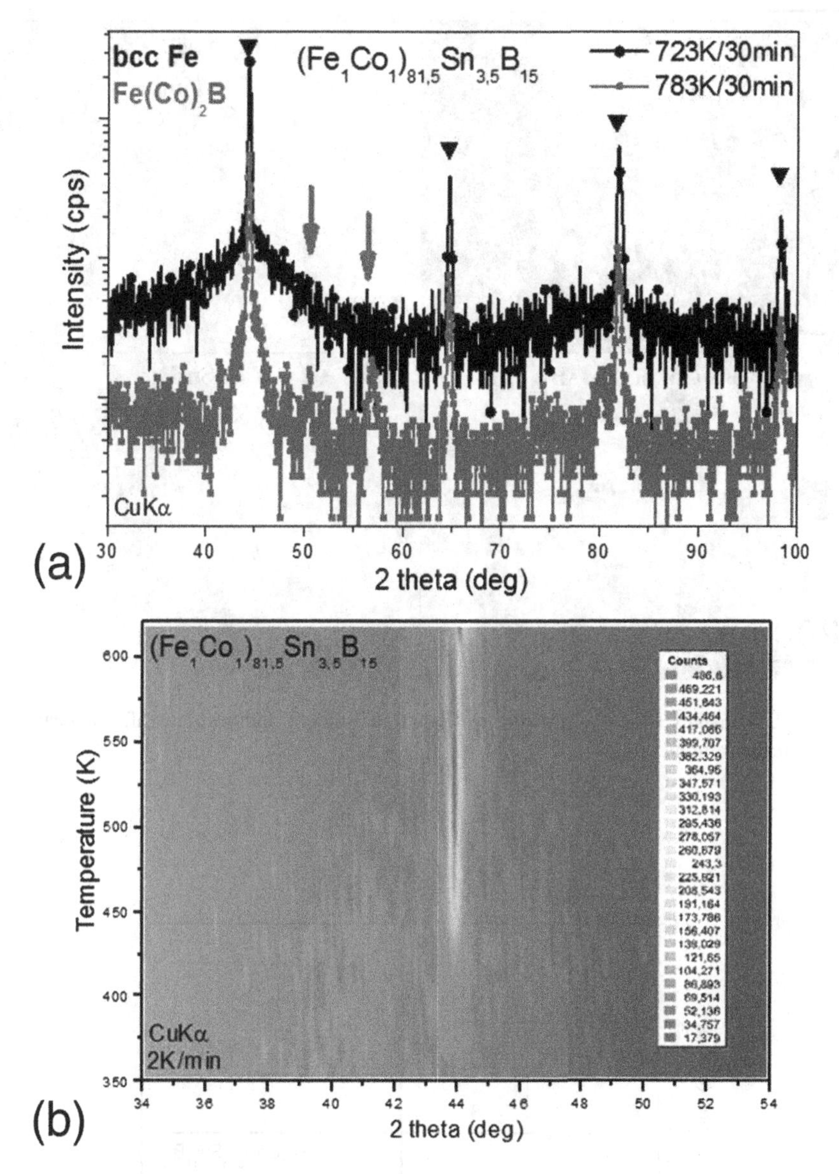

Fig. 3. XRD patterns from Fe-Co-B based samples with Sn 3.5 at. % content: (a) XRD after isothermal annealing at temperatures near to T_x for 30 min; (b) XRD in-situ linear annealing from AQ state

annealing at the temperatures chosen for phase analysis consists of standard small (up to 80 nm) polyhedral bcc-Fe grains for Fe-based samples while more regular Co_2B grains for Co-based samples, surrounded by the amorphous matrix in both cases. TEM images suggest that the difference in ferromagnetic element (Fe or Co) content leads only to a small change in the morphology and crystallinity. For the Fe-Co-based sample

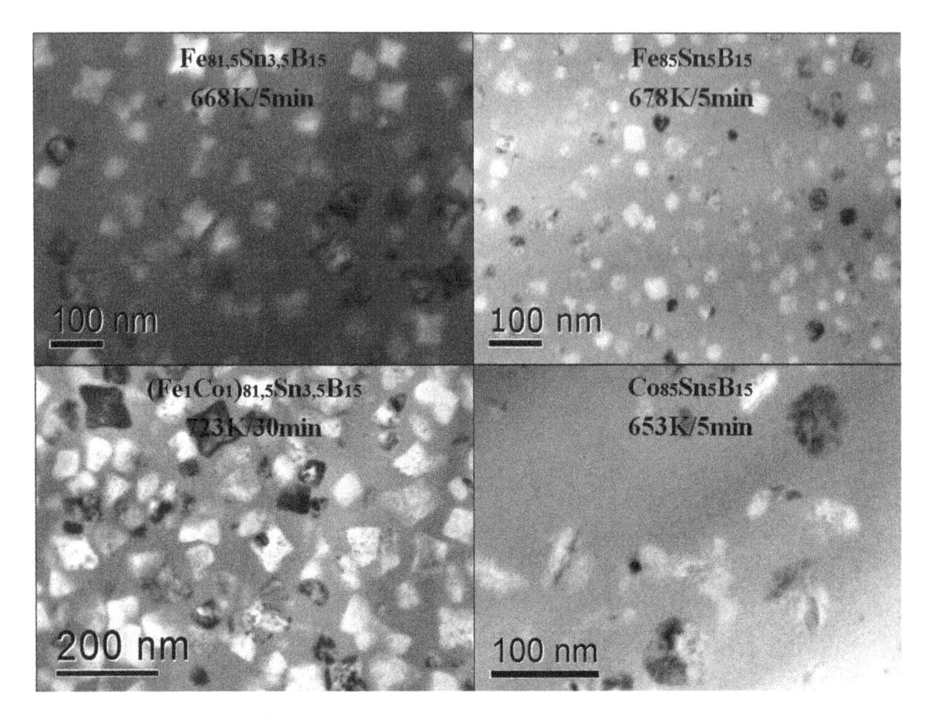

Fig. 4. TEM images showing structure evolution for student compositions after annealing at different temperatures

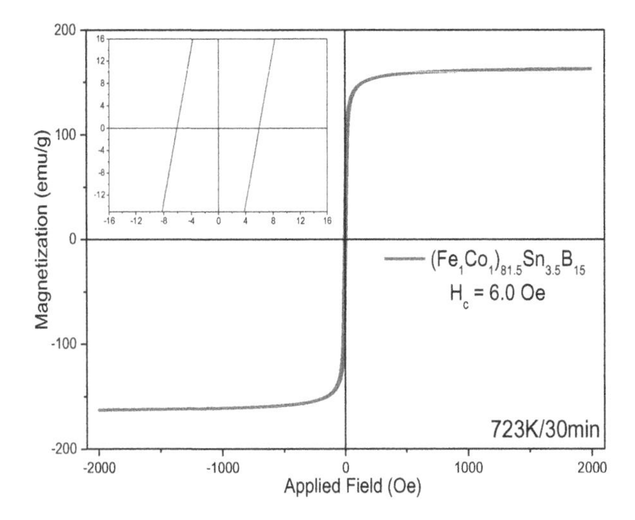

Fig. 5. Quasistatic B-H loop of Fe-Co-Sn-B ribbon annealed at 723 K for 30 min showing the enhanced value of saturation magnetization and coercivity H_c

the grain size and morphology of bcc-Fe(Co) remains unaffected in comparison with the Fe-based samples, but the influence on crystallinity is more obvious. For this sample the observed structure of the ferromagnetic phase could be that one of the expected magnetic properties: the measurement of B-H loop (Fig. 5) exhibits the coercive force $H_c = 6$ Oe after annealing on 723 K/30 min.

4 Conclusion

Microstructure and compositional dependence of the first transformation stage of Fe-Sn-B, Co-Sn-B and (Fe/Co)-Sn-B based systems were studied. The dependencies of the temperatures of crystallization onsets and of the temperature intervals between the first and next crystallization as well as the position of the Curie temperature relative to the first transformation were investigated. For the Fe-based alloys the crystallization temperature increases with increased Sn content, but this effect is unfavorable for the stability of the resulting ferromagnetic phase. Thus the main interest was focused on the first transformation and the ferromagnetic products n amorphous matrix of (Fe/Co)-Sn-B systems. Diffraction patterns annealed in the vicinity of the first-stage transformation exhibit bcc-Fe (bcc-Fe(Co)) peaks for compositions containing Fe. Materials based on Co only transform into Co_3B containing structures from the beginning – this and the stability of remaining amorphous matrix are not suitable to achieve required magnetic structure. The investigation of combined Fe-Co-Sn-B system with 3.5 at. % of Sn content is more promising. The Fe-Co base has the most suitable influence on the stability of the remaining amorphous phase and its thermal properties, necessary to achieve desirable magnetic properties. The measured B-H loop of (Fe/Co)-Sn-B based system shows high enough saturation magnetization and low values of magnetic coercivity which can be further tuned by proper thermal treatment.

Acknowledgement. This work was supported by the projects VEGA 2/0189/14, APVV-0460-12 and APVV-15-0621 and by the CEX FUN-MAT.

References

1. Herzer, G.: Acta Mater. **61**, 718 (2013)
2. Willard, M.A., Daniil, M.: Handbook of Magnetic Materials, vol. 21, p. 173 (2013)
3. Kemeny, T., et al.: Phys. Rev. B **20**, 476 (1979)
4. Makino, A., et al.: J. Appl. Phys. **91**(10), 8420 (2002)
5. Yoshizawa, Y., et al.: Mater. Sci. Eng. A **375–7**, 207 (2007)
6. Švec, P., et al.: Preparation, processing and selected properties of modern melt-quenched alloys. In: Awrejcewicz, J., Szewczyk, R., Trojnacki, M., Kaliczyńska, M. (eds.) Mechatronics - Ideas for Industrial Application. AISC, vol. 317, pp. 381–396. Springer, Heidelberg (2015). doi:10.1007/978-3-319-10990-9_36
7. Janotova, I., et al.: J. Electr. Eng. **66**, 297–300 (2015)
8. Janotova, I., et al.: J. Alloy. Compd. **615**, 198 (2014)
9. Suzuki, K., et al.: J. Appl. Phys. **70**, 6232 (1991)

10. Ohta, M., Yoshizawa, Y.: J. Magn. Magn. Mater. **320**, 750 (2008)
11. Svec, P., et al.: IEEE Trans. Magn. **46**, 408 (2010)
12. Sharma, P., Zhang, X., Zhang, Y., Makino, A.: Scripta Mater. **95**, 3 (2015)
13. Janotova, I., et al.: J. Supercond. Novel Magn. **26**, 793 (2013)
14. Illekova, E., et al.: J. Alloy. Compd. **509**, S46 (2011)
15. Matko, I., et al.: J. Alloy. Compd. **615**, S462 (2015)

General Science and Technology

Determining Prospects of European Countries' Positive Experience Implementation into the Water Consumption Process in Ukraine

Valerij Shygonskyy(✉)

Zhytomyr National Agroecological University, 7, Stary Blvd, Zhytomyr, Ukraine
shigonsky@gmail.com

Abstract. At the stage of Ukrainian transition to the European living standards, one of the main tasks is to prevent the further deterioration of water resources, along with their protection and enhancement. Since 2006 Water Framework Directive has become the benchmark for purposeful improvements towards solving the problems of water resources in Ukraine. Forming the mechanism of sustainable water consumption is one of the conditions of its development. Its main principles are organic and cohesive combination of social and economic problems with ecological ones, the confirmed solving of which will help the society to come at an increased level of living standard: quality of public services, people's health, life expectancy etc. The objective of this paper is the system analysis of current water consumption state in rural areas of Ukraine, as well as new directions formulation of improving the social and ecological situation. Securing Ukrainian population with drinking water, especially in rural areas, is among the worst in Europe. While conducting the research the economics and statistics methods were used (comparison, statistical monitoring) as well as sociological (interviewing, sociometric polling). Following the statistical data we can conclude that the situation connected with providing rural population with standard quality water by means of arranging the systems of centralized water supply, accompanied by essential engineering constructions necessary for guaranteeing treated, standard quality water does not have positive dynamics. Conducted analysis enables to identify main problems of rural areas water consumption in Ukraine, and determine directions of their solving based on centralized water supply systems. Applying effective mechanisms based on the complex approaches such as active participation of Ukrainian society subjects are necessary to overcome it. It is essential to develop the practice of involving the Ukrainian citizens into solving the socially significant problems while using the European experience, along with making provisions in the budget for citizens initiative support that is aimed at improving water supply on the basis of co-financing.

Keywords: Drinking water · Rural areas · Water consumption · Water supply

R. Szewczyk and M. Kaliczyńska (eds.), *Recent Advances in Systems, Control and Information Technology*,
Advances in Intelligent Systems and Computing 543, DOI 10.1007/978-3-319-48923-0_78

1 Introduction

Ecological factor within the managerial process of Command and Administration System as a rule was left out of account while calculating the industrial efficiency. Unfortunately, the same can be referred to the current market system in Ukraine. During the long period of existing marked-based economy its main scientific task was considered to be looking for new ways of using limited nature resources to satisfy peoples growing needs. The necessity of introducing generally accepted and world-wide used concept of sustainable development in Ukraine demands elaborating the economic system that will be in stark contrast to Command and Administration and market-based economies. Nowadays there came the era of severe economy: needs and resource demands are to be restricted while environmental resources – restored and saved [1–3].

At the stage of Ukrainian transition to the European living standards one of the main tasks is to prevent further deterioration of water resources, along with their protection and enhancement. Since 2006 Water Framework Directive has become the benchmark for purposeful improvement towards solving the problems of water resources utilization in Ukraine. The given document has defined the main principles of administrating water resources, and the ways of improving water and land ecosystems state. Rural areas development as well as health and welfare of people living there is closely interconnected with water resources state. Forming mechanisms of sustainable water consumption is one of the conditions for its upgrading. Its main principles are organic and cohesive combination of social and economic problems with ecological ones, solving of which can help the society to come at a totally new level of living: quality of public services, people's health, life expectancy etc.

Rural economy development problems in Ukraine are deeply investigated by many scientists. In particular, H. Borodina [4, 5], T. Zinchuk [6], V. Jurchishyn [5] substantiated new approaches as for understanding the essence of rural development; I. Buzdalov [7, 8], A. Petrikov [9] presented the methodology of using system approach for elaborating rural territories sustained development concept; G. Green [10, 11], J. Flora [11], S. Deller [10] depicted present-day vision of village development within the globalization risks.

Water consumption lean optimization topicality connected with the necessity of water resources economy and water usage processes ecologization on different stages of water production has been grounded in the works of N. Hvesyk [12, 18–21], O. Jarotzka [13], A. Jatzyk [14, 15]. The problems of water usage has been actively investigated in the scientific papers of T. Galushkina [16], V. Golian [17], B. Danylyshyn [18], N. Zinovchuk [22], V. Stashuk [23], N. Hagemann [34] and others. However, rural areas water consumption problems in Ukraine are insufficiently explored. In the majority of cases, water objects are integrally exploited by different users who in their turn push different claims to water quality. Drinking water quality must meet the highest standards, while for irrigation or technical usage water can be diluted, partly treated and even unsafe.

To our opinion, solving the given problem lays in local community initiative. Former Eastern Bloc countries and post-Soviet republics – members of European

Union nowadays – Poland, Czech Republic, Slovakia, Latvia, Lithuania and Estonia, experience can be used here [27–39]. Further development of rural areas is impossible without population's active participation, raising its positive internal motives turned to change for the better social life-level. In order to get that we must, from one hand – essentially change legislative basics of municipal government – from the other – sufficiently raise population's civil responsibility for solving local problems.

Swiss-Ukrainian project "Decentralization support in Ukraine" DESPRO is of special interest and practice. Due to its support 78 projects with the general fund 37.3 million UA were realized, as well as 40 thousand of rural inhabitants, who live in 16 thousands of households were provided with sustainable water supply. Swiss Confederation contribution was almost 13 million UA, local budget – 6.3 million UA, community investment – 18 million UA [40].

2 Objective

The research goal lies in the system analysis of Ukrainian rural areas water consumption state in the current context, as well as guaranteeing economical demands of rural population with available water resources and identifying the attitude of Zhytomyr district population towards water consumption problems. In the given context we also consider the dynamics of state programs realization, particularly concerning the water use in rural settlements. On the basis of results analysis we formulate prospects of improving social and ecological water consumption situation in Ukraine with regard to current European experience.

3 Methods

While conducting the research the economics and statistics methods were used (comparison, statistical monitoring) as well as sociological (interviewing, sociometric polling). Systematization and results' presentation were carried out by applying graphical approach and table procedure that helped to rational presenting numeric data.

When calculating the percentage of rural settlements provided with centralized water supply and sewage systems we used the formula of finding relative statistical value [41, p. 59]:

$$\% RTS = (NSCS/TNRS) * 100 \quad (1)$$

% RTS – percentage of rural territory settlements provided with the centralized water supply and sewage systems in certain region;
NSCS – number of settlements with the centralized water supply and sewage systems in certain region;
TNRS – total number of rural territory settlements in the certain region.

The calculations were made in view of supplying Zhytomyr region and Ukraine, in general, with the water supply and sewage systems. All the data for the numerical calculations was taken from the government authorities' official reports: Ministry of

Environment and Mineral Resources of Ukraine [1], Ministry of Regional Development, Construction, Housing and Utility of Ukraine [24], State Statistics Service of Ukraine [26]. Calculations and analyses resulted time series criterions that allow tracing the dynamics of changes within the given problem.

Thus, we've conducted the sociological research aimed at revealing rural population's attitude towards certain problems of water consumption, and the necessity of sustainable water usage approaches implementation. The research has been conducted during 2011–2013 in the villages of Zhytomyr region. The amount of respondents was more that 800 (223238 – the total number of households in the region). Opinion poll was carried out by interviewing one representative from each household.

4 Results and Discussion

Ukrainian legal system, the Constitution in particular, guarantees every citizen the right for health-safe environment (art. 50), ecological safety (art. 16) and sanitary-epidemiological welfare (art. 49). According to the Act of Ukraine "Ensuring the people's sanitary and epidemiological welfare" the citizens also have the right for health-safe drinking water (art. 4). Therewith regional executive and self-governing authorities are responsible for guaranteeing the quality of drinking water for Ukrainian citizens that meets the State sanitary rules and norms (art. 18).

Ukrainian water resources concentrated in surface water reservoirs are also supplied with underground pools that play an important role in forming river run-off as well as rural population's household activity. In total, Ukrainian predicted resources make 61689.2 th. m^3/day and are inhomogeneously spread over the whole territory – the majority of them is focused in Northern and Western regions [24, pp. 5–16]. We support the conclusion for the necessity of water resources economy mechanisms active implementation, their rational usage and preservation.

Providing Ukrainian population with drinking water, especially in rural areas, is among the worst in Europe. Just the quarter of rural population has access to centralized water delivery systems, the rest prefer decentralized sources – wells, self-made wells, makeshift coalmines, imported water. 1274 rural-type settlements counting more than 700 th-s of inhabitants (14 regions of Ukraine and ARC) partly or entirely use imported or low-grade water [24, pp. 25–28].

Sewage system is poorly developed for less than 10% of the rural population. In fact, more than 14 million people make use of pit privy and septic tanks, thus being responsible for spew and waste water disposal. Besides, only 40% of wastewater is previously treated [24, pp. 4–56].

Monitoring chemical and bacteriological surface and ground water pollution has shown that sanitary and ecological state of water-supply sources in rural settlements is coming close to critical. Only 6.4 th-d settlements from 28.6 th-d have correctly engineered water supply systems, the half of which can hardly secure population with standard quality water due to incorrect and long-termed exploitation [24, p. 49].

The part of samples for special drinking water tests that have been taken from centralized water supply sources during the last ten years makes - 12% of sanitary and chemical indexes and nearly 5% epidemiological safety indexes. Samples taken from the non-centralized water supply sources – 30% do not meet sanitary and chemical norms, 20% - do not meet the norms of epidemiological safety.

Partial solution of the defined problems is conditioned by implementation the State program "Drinking water of Ukraine" (further – Program) that is aimed at improving water supply of Ukrainian population with high quality drinking water within the scientifically grounded norms of drinking water supply; reforming and developing water grid and sewerage facilities, rising the efficiency and ecological reliability of its functioning, thus improving the population health and environmental sanitation in Ukraine, together with renewal, protection and rational use of drinking water sources.

According to the Program, its financing must be carried out at the expense of state funds, local government financing, enterprises budgets, grants and international programs funds etc. General money amount during 2011–2020 makes 9.5 billion UA, including – 3 billion from government budget (that is 31.7%), other sources – 6.5 billion (68.3%).

The first stage of the Program (2006-2010, from 03.03.2005) was not entirely completed. During that period only 16.9% of funds was spent. It should be also mentioned that in 2008–2009 money was spend just for scientific projects financing [25].

During 2010–2011 the second money transfer of 288 million UA from planned 423.3 million UA came and just 175.1 million UA or 41.4% were used. It's worth mentioning that in 2011 Ministry of Finance homogeneously scheduled monthly budgetary appropriation of 268 million (67%) to IV quarter that resulted unfavorable conditions for project developing. That meant that the majority of planned issues had to be done in very tight time schedule accompanied by cross weather [25, pp. 14–15].

Some regions delayed regional projects elaboration and approval. For example, in Zhytomyr region up to January 1, 2015 – 171 rural settlements, or 10.6% have water supply, while only 27 of them (1.7%) sewerage systems (Fig. 1). The percentage of rural settlements in Ukraine that have water supply system is 21.9%, and 2.5% have sewerage systems [26, p. 202].

Following the statistical data we can conclude that the situation connected with providing rural population with standard quality water by means of arranging centralized water supply systems accompanied by essential engineering constructions necessary for guaranteeing treated standard quality water does not have positive dynamics (Fig. 1). The state program designed to solve the given problems was not successful, and that initiated search for new ways of its solving and new sources of its financing.

Current regulations even now allow for lobbying water supply problems with the community participation. Similar initiatives can be oriented at creating consumers' unities (non-commercial activity) through functioning servicing cooperatives or remodeling of income-generating activities. Though, it's necessary to change administrators, water consumers and society general consciousness to implement such approaches into life. That hinted the importance of exploring Ukrainian citizens' attitude towards the problem of water consumption in rural settlements.

The results testified that 95% of respondents use drinking water from non-centralized sources, only 10% of them use underground springs, just 5% (Fig. 2) use centralized water supply systems. The considerable part of water sources are interfered with ground polluted waters.

It's worth saying that 58% of respondents are quite satisfied with the water quality (Table 1) though nearly 40% of them never checked the quality in the laboratories.

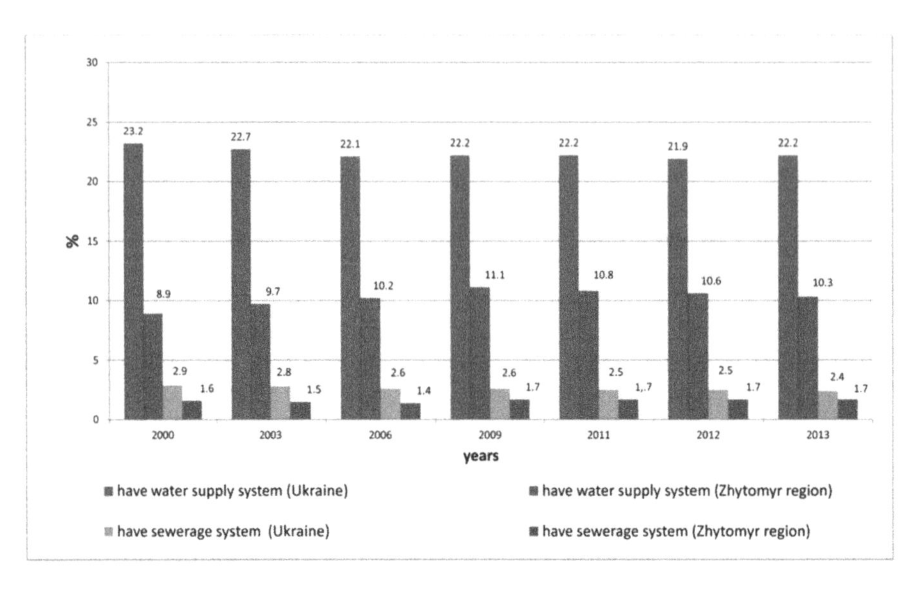

Fig. 1. Ratio of Ukrainian rural settlements that have water supply system or sewerage system, Ukraine and Zhytomyr region

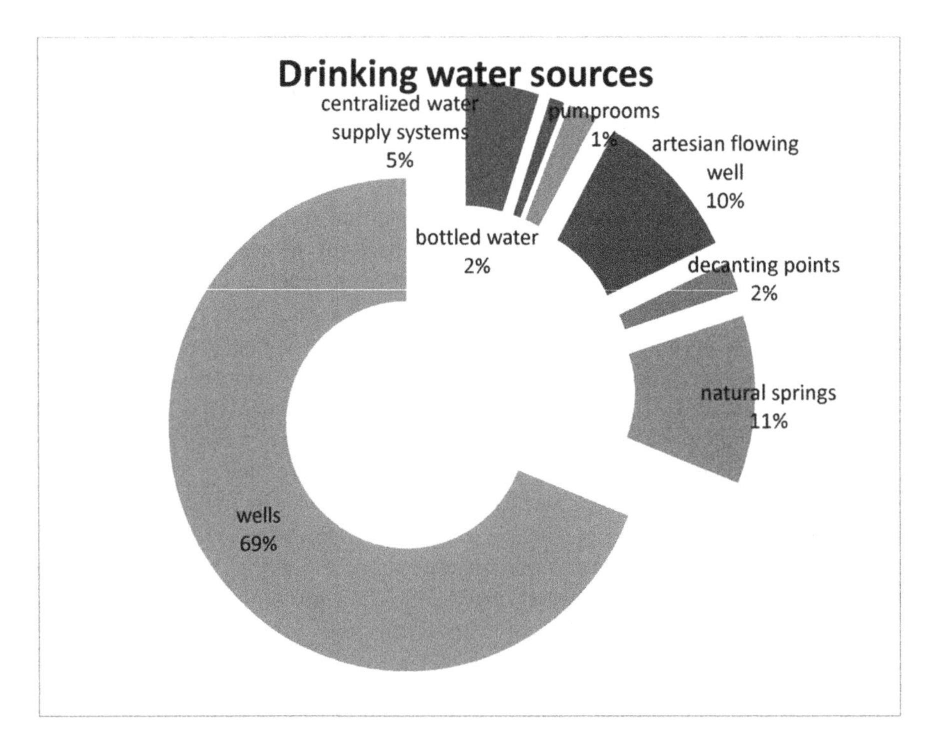

Fig. 2. Respondents' answers distribution to the question: "What water source do you use for drinking?"

At the same time 17% could hardly answer that question and 25% are completely dissatisfied with the quality of consumed water.

We also want to accent that village inhabitants not always can solve the problem of necessary quality household water supply on their own. It can have financial and technical ground. Consumer uniting could allow capital accumulation for solving the necessary tasks, and simultaneously cut burden of individual costs. 37% of the respondents were ready to enter such society (Table 1). This indicates to high motivation and good example for other community members.

Rural areas development is of crucial importance for further building national economy strategy. It guarantees not only state's food supply security, but helps to save

Table 1. Statistics of the respondents' answers

Inquiries	%, respondents' answers		
	Yes	No	Cannot say
Are you satisfied with the drinking water quality?	58	25	17
Do you check the quality of used drinking water?	59	41	– *
Do you treat water before using?	59	41	– *
Do you think that usage of low-quality drinking water negatively influences you and your family health?	77	8	15
Do you think that it is necessary to make centralized water supply system in your settlement?	45	30	25
Do you agree to become a shareholder (member of cooperative society) for building centralized water systems?	37	63	– *

Note:* = no response answer used.

Ukrainian national identity. It also is the first-order condition for balancing between economic, social and ecological constituents, element of forming favorable vital environment and leveling rural and urban population life activity.

However, rural areas as the system, in the context of all its elements and components, are still not the object of state interests. Every one of them pursues their own pragmatic objectives. The state is just interested in guaranteeing food supply security by means of functioning agriculture, putting aside social sphere and natural surroundings. Its significant feature is also poor level of financing of state special-purpose programs, or their full absence, which leads to low efficiency of their implementation. Water resources are of interest to business structures only as the means of profit taking. The considerable part of labor pool, especially young ones, doesn't see itself in villages in the future.

As a consequence, rural population is characterized by absence of motivation for developing water infrastructure. Rural civil-society organizations, due to objective reasons, have minor influence upon functioning processes. They become active only when it's necessary to solve regional or state problems and rarely local ones.

5 Conclusions

The conducted analysis testifies to absence of logics in the state policy towards water supply security in rural areas settlements, along with increasing housing and utilities sector commercialization. Implementing state programs doesn't ensure solving issues related to concentration of financial, material, technical and other resources that in its turn results in inability to resolve tasks aimed at securing the water supply for rural population.

In the meantime, rural areas further development is the warrant of country's food security and ought to be of paramount importance within national economic strategy. To large extent it's the guarantee of saving Ukrainian identity and sovereignty, necessary condition of maintaining the balance between economic, social and ecological constituents, the element of forming favorable vital environment and leveling-off rural and urban folk life-sustaining activity. However, rural areas as the system, in the context of all its components and elements, is still not the object of public and state interest, as well as agricultural subjects, rural people and public associations.

The state is just interested in providing food security owing exceptionally to agricultural sector, leaving out of view social and natural environment. The key factor is low level of financing of state purpose-oriented programs, and sometimes their absence at all, which results in low efficiency of their implementation. Water resources are of interest for business structures only as the means of profitability. The considerable amount of labor pool, especially the youngsters do not see themselves in rural areas. Consequently we have missing motivation in attracting proprietary funds for further water consumption infrastructure development. Due to objective reasons public organizations in villages have reduced influence. In most cases, with the support from national or regional organizations, they start to work for solving country-wide or regional problems and very seldom the local ones.

All presented above, proves that unsatisfactory condition of rural areas is the consequence of the continuous crisis. Applying effective mechanisms based on the complex approaches as surely as active participation of Ukrainian society subjects are necessary to overcome it. It is essential to develop the practice of involving the Ukrainian citizens into solving the socially significant problems while using the European experience along with making provisions in the budget for citizens' initiative support that is aimed at improving water supply on the basis of co-financing. The primary direction of water consumption development in Ukrainian rural areas in ought to become advancing the water supply and waste-water removal systems that include:

- complete certification of centralized and non-centralized sources of drinking water;
- graduate converting ground water runoff water supply into centralized deep subsurface water supply;
- total modernization of old water pipes as well as their upgrading with the systems of final water treatment at consumption areas;
- upgrading technical service for water supply systems;
- appropriate monitoring of drinking water quality;
- implementation of the final drinking water treatment system;

- implementation of the water economy stimulatory factors – the technological (individual water meters) and economical ones (introducing differentiated payment tariffs);
- using the proper quality water for different kinds of business activity;
- rainwater collection and its usage for technological purposes;
- investments into innovative water-efficient and dry technologies for water-using production activities based on the economical motivation system;
- water losses minimization during transport by means of networks upgrading;
- improving the state monitoring system of the drinking water market;
- reducing the volume of current water consumption and increasing the volume of water supply;
- introduction recycling and circulating water supply systems;
- applying drip irrigation that reduces the water requirements by 30–40% and increases crop-production in comparison to the traditional one;
- permanent monitoring of the waste-water treatment system;
- wide use of biological forms of water waste treatment including individual objects.

References

1. National report about environmental situation in Ukraine 2012. Ministry of Environment and Mineral Resources of Ukraine, Kyjiv (2013). (in Ukrainian)
2. Bigas, H. (ed.) The Global Water Crisis: Addressing an Urgent Security Issue. Papers for the InterAction Council, 2011-2012. Hamilton, Canada: UNU-INWEN 2012 (2012)
3. Water Ethics and Water Resource Management. Ethics and Climate Change in Asia and the Pacific (ECCAP) Project. Working Group 14 Report. UNESCO, Bangkok, Thailand (2011)
4. Borodina, O.M.: Agricultural policy in Ukraine: basics, current state and new possibilities in the context of institutionalism and global challenges. Ukrainian Econ. **10**, 94–111 (2008). (in Ukrainian)
5. Heyets', V., Yurchyshyn, V., Borodina, O.: Socieconomic modernization of Ukrainian agricultural sector. Ukrainian Econ. **12**, 4–14 (2011). (in Ukrainian)
6. Zinchuk, T.O.: Agrocentralism' transformation into the rural economy policy within global challenges. Bull. ZHNAEU **2**, 3–12 (2010). (in Ukrainian)
7. Buzdalov, Y.: Minor forms of economy management as the factor of sustainability the agricultural development. Int. Agric. J. **2**, 3–9 (2012). (in Russian)
8. Buzdalov, Y.N.: Main direction of guaranteeing sustainability of rural development. Econ. Agric. Process. Enterprises **7**, 1–8 (2013). (in Russian)
9. Petrykov, A.: Developmental tasks in agricultural sector. Economist **3**, 3–5 (2010). (in Russian)
10. Green, G.P., Deller, S.C., Marcouiller, D.W. (eds.) Amenities and Rural Development: Theory, Methods and Public Policy. Edward Elgar Publishing, Northampton (2005)
11. Flora, J.L., Green, G.P., Gale, E.A., Schmidt, F.E., Flora, C.B.: Self development: a viable rural development option? Policy Stud. J. **20**, 276–288 (1992)
12. Khvesyk, M.A.: State policy priorities in sustainable use, protection and renewing water resources in Ukraine. Reg. Policy **1**, 184–197 (2002). (in Ukrainian)
13. Yarotska, O.V.: Ecologinomical assessment of water management in multiunit water-basin-complexes. Kyjiv (2007). (in Ukrainian)

14. Yatsyk, A.V.: The horizons of Ukrainian water management. State Messenger **94**, 9 (2003). (in Ukrainian)
15. Yatsyk, A.V.: Water-supply ecology, Kyjiv (2004). (in Ukrainian)
16. Galushkina, T.: Drinking water supply: quality and improving ways assessment (2010). (in Ukrainian)
17. Holyan, V.A.: Institutional water consumption environment: the current state and the mechanisms of improvement. Odessa-Saki, Lutsk (2009). (in Ukrainian)
18. Bystryakov, I.K. Koval', Y.V., Khvesyk, M.A.: Natural-resources sphere in Ukraine: the problems of sustainable development and transformations, Kyjiv (2006). (in Ukrainian)
19. Dorohuntsov, S.I.: Water management optimization, Kyjiv (2004). (in Ukrainian)
20. Dorohuntsov, S.I., Khvesyk, M.A.: Ecoenvironment and modern era: current natural environment, Kyjiv (2006). (in Ukrainian)
21. Dorohuntsov, S.I., Khvesyk, M.A., Horbach, L.M., Pastushenko, P.P.: Ecoenvironment and modern era: Ukrainian ecological legislation and its harmonization with European legal frameworks. Kyjiv (2007). (in Ukrainian)
22. Zinovchuk, N.V.: Ecological policy in agro-industrial complex: economical aspect. LSAU, Lviv (2007). (in Ukrainian)
23. Stashuk, V.: Ukraine on the way to watershed management. Ukrainian Water Econ. **4**, 6–10 (2007). (in Ukrainian)
24. National report about drinking water quality and current water supply state in Ukraine 2012. Ministry of Regional Development, Construction, Housing and Utility of Ukraine, Kyjiv (2013). (in Ukrainian)
25. Zaremba, I.M. Shakh, H.A.: About the audit results of efficient using governmental money budgeted to State programme "Ukrainian Drinking Water" for 2006–2020. http://www.acrada.gov.ua/control/main/uk/publish/category/412;jsessionid=309C20757204B1DFEEEEFE625459AAF9. Accessed 8 October 2013. (in Ukrainian)
26. Osaulenk, O.H.: Ukraine in figures 2012: Statistical collected volume, Kyjiv (2013). (in Ukrainian)
27. Dzyarski, H., Halabuda, H., Zahaynyy, V.: Polish-Ukrainian academy of utility services and environmental protection, Regional Democracy Development Funding, Warsaw-Lviv (2011). (in Ukrainian)
28. Lofrano, G., Carotenuto, M., Maffettone, R., Todaro, P., Sammataro, S., Kalavrouziotis, I. K.: Water collection and distribution systems in the Palermo Plain during the Middle Ages. Water **5**, 1662–1676 (2013)
29. Mattas, C., Voudouris, K.S., Panagopoulos, A.: Integrated groundwater resources management using the DPSIR approach in a GIS environment context: a case study from the Gallikos River Basin. North Greece. Water **6**, 1043–1068 (2014)
30. Perin, R.C., Casalini, D.: Water property models as sovereignty prerogatives: European legal perspectives in comparison. Water **2**, 429–438 (2010)
31. Hoekstra, A.Y.: The global dimension of water governance: why the River Basin approach is no longer sufficient and why cooperative action at global level is needed. Water **3**, 21–46 (2011)
32. Kanakoudis, V., Tsitsifli, S., Papadopoulou, A.: Integrating the Carbon and Water footprints' costs in the water framework directive 2000/60/EC full water cost recovery concept: basic principles towards their reliable calculation and socially just allocation. Water **4**, 45–62 (2012)
33. Conte, G., Bolognesi, A., Bragalli, C., Branchini, S., Carli, A.D., Lenzi, C., Masi, F., Massarutto, A., Pollastri, M., Principi, I.: Innovative urban water management as a climate change adaptation strategy: results from the implementation of the project "water against climate change (WATACLIC)". Water **4**, 1025–1038 (2012)

34. Hagemann, N., Klauer, B., Moynihan, R.M., Leidel, M., Scheifhacken, N.: The role of institutional and legal constraints on river water quality monitoring in Ukraine. Environ. Earth Sci. **72**, 4745–4756 (2014)
35. Ready, R.C., Malzubris, J., Senkane, S.: The relationship between environmental values and income in a transition economy: surface water quality in Latvia. Environ. Dev. Econ. **7**, 147–156 (2002). doi:10.1017/S1355770X02000086
36. Bontemps, C., Couture, S.: Irrigation water demand for the decision maker. Environ. Dev. Econ. **7**, 643–657 (2002). doi:10.1017/S1355770X02000396
37. Howe, C.W.: Policy issues and institutional impediments in the management of groundwater: lessons from case studies. Environ. Dev. Econ. **7**, 625–641 (2002). doi:10.1017/S1355770X02000384
38. Sivakumar, B.: Water crisis: from conflict to cooperation—an overview. Hydrol. Sci. J. **56** (4), 531–552 (2011). doi:10.1080/02626667.2011.580747
39. Kundzewicz, Z.W.: Water problems of central and eastern Europe-a region in transition. Hydrol. Sci. J. **46**(6), 883–896 (2001). doi:10.1080/02626660109492883
40. Sorokovsky, V.Y.: Decentralization going: rising community capacity in rendering services, Kyjiv (2009). (in Ukrainian)
41. Tarasova, V.V.: Ecological statistics, Kyjiv (2008). (in Ukrainian)

Diesel Exhaust Gases Centrifugal-Jet Filter-Converter

Andrii Ilchenko(✉), Vladislav Balyuk, and Neonila Kosnitskaya

Zhytomyr State Technological University, Zhytomyr, Ukraine
avi_7@rambler.ru, mailbvu@gmail.com,
n.kosnitskaya@gmail.com

Abstract. Gas-dynamic analysis of diesel exhaust gases filter-converter performance has been conducted. The design of the centrifugal-jet swirler that provides an increase in the area of distribution of the exhaust gases to the filter element has been suggested; its optimal geometric parameters in terms of the distribution of the exhaust gases have been determined.

Keywords: Exhaust gases · Filter-converter · Swirler

1 Introduction

The studies are devoted to the problem of reducing emissions of solid particles along with exhaust gases (EG) of diesel engines, namely their cleaning in the filter-converter using the exhaust gases flow swirler that contributes to their better distribution in the filter element.

Analysis of this issue has shown [1–5] that various kinds of exhaust gases flow swirlers are widely used to improve the quality of cleaning and to increase the life cycle of the filter-converters of diesel engines. The main task of this element is the uniform distribution of the exhaust gases on the filter surfaces of a porous filter-converter material. At the same time, the presence of the swirler in the exhaust system of a diesel engine, in any case, will increase its overall hydraulic resistance. In [6, 7] the evaluation of the ways of improving the efficiency of the filter-converters using swirlers is given, the construction of which is described in [1, 8]. The use of the swirler is effective and allows to achieve EG distribution on 98% of the input end area of the filter element. At the same time the hydraulic resistance of the filter-converter is 1.361 kPa [8].

Formulation of the problem. To perform simulation of the well-known exhaust gases filter-converter of a diesel engine in COSMOSFloWorks CFD-complex, to define its effectiveness at different variants of centrifugal-jet swirler design, to determine the best geometrical parameters of the swirler from the point of view of cleaning exhaust gases from solid particles.

R. Szewczyk and M. Kaliczyńska (eds.), *Recent Advances in Systems, Control and Information Technology*, Advances in Intelligent Systems and Computing 543, DOI 10.1007/978-3-319-48923-0_79

2 Materials and Results of the Study

As a prototype, a widely-spread filter-converter used in many models of cars with diesel engines, for example, FIAT DOBLO, VOLKSWAGEN CADDY, and so on, has been chosen. To achieve the best possible exhaust gases distribution in the volume of the filter material with a relatively small increase in the hydraulic resistance of the filter-converter, it is proposed to install the swirler directly into the filter housing, where the cross-sectional area is larger and the exhaust gases pressure is correspondingly less. To create a turbulent flow of exhaust gases, it is proposed to use the swirler on the basis of a centrifugal-jet nozzle [9].

Figure 1 shows a perspective view and a longitudinal section of the swirler with the designation of dimensions that affect the efficiency of the filter-converter.

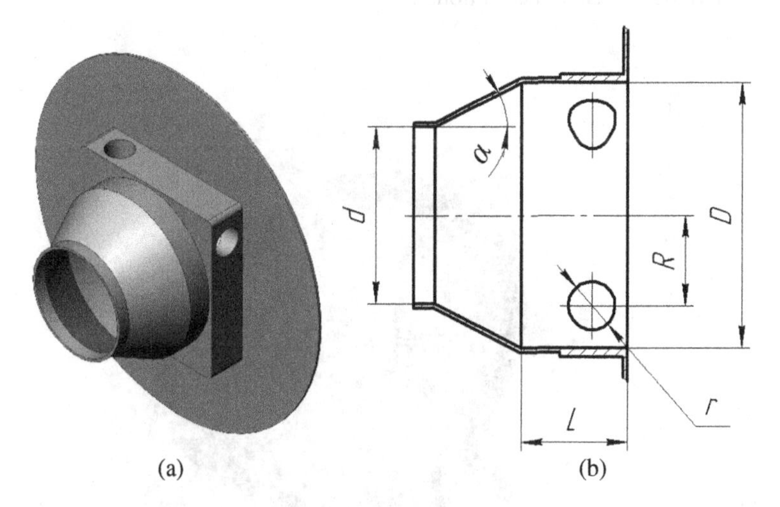

Fig. 1. Centrifugal-jet swirler: (a) perspective view; (b) longitudinal section

The swirler operates as follows. Exhaust gases are coming through the tangential and central apertures into the swirler chamber. Tangential openings direct the exhaust gases flows along the swirler chamber. Uniform annular flow interacts with the central flow, which creates a uniform filling of the exhaust gases on the cross section of the filter.

For comparison, operation of the filter with an elongated body without a swirler was also simulated.

The calculated model had the following conditions:

- body and swirler material was stainless steel;
- exhaust gases parameters corresponded to [10, 11];
- porosity of the filter element was determined by [12].

The main parameters under study are:

- the percentage of the input end area of the filter element, on which the projection of the velocity vector onto the longitudinal axis is positive – η, %;
- the total hydraulic resistance of the filter-converter – ΔP, Pascal.

The studies were conducted with the use of the following procedure:

1. The image of the fields of velocity projections on the longitudinal axis of the filter in the plane perpendicular to the filter axis at the distance of 5 mm from the filter element was obtained for different geometrical parameters of exhaust gases filter-converter. All velocity values were divided into 10 ranges. The first range corresponded to all the negative values of velocity. The other 9 ranges divided the whole velocity range into equal parts (Fig. 2 shows an example of one of the obtained velocity fields distributions).

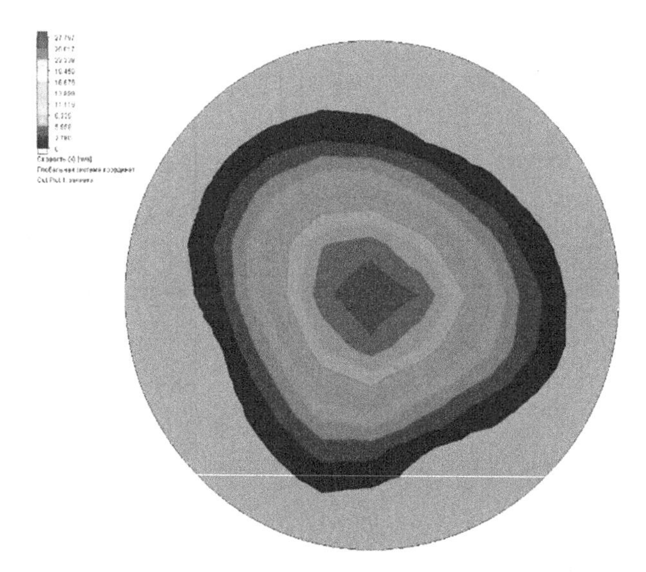

Fig. 2. Velocity projections on a longitudinal axis in the cross section at the distance of 5 mm from the filter element without a swirler. (Velocity. The global coordinate system. Color filling.)

2. The area of each field for all variants of the geometric parameters of the swirler was determined.
3. The study results were summarized in Table 1 and the conclusion was made about the best geometrical parameters of the swirler.

Based on these results the conclusion can be made that to improve the filter-converter used in many modern automotive diesels, the best in terms of exhaust gases flow distribution on the filter element is a filter-converter with a swirler having such geometrical parameters: $d = 20$ mm, $D = 35$ mm, $r = 11$ mm, $R = 29.5$ mm,

Table 1. Effect of the swirler parameters on the percentage of the area of the filter element input end and the total hydraulic resistance of the filter-converter

Swirler parameters						Simulation parameters	
d, mm	D, mm	r, mm	R, mm	α, degrees	L, mm	η, %	ΔP, Pa
10	30	11	20.5	45	25	73%	11164
15	30	11	20.5	45	25	80%	2971
20	30	11	20.5	45	25	55%	1059
25	30	11	20.5	45	25	49%	674
30	30	11	20.5	45	25	57%	608
20	20	11	14.5	45	25	46%	1490
20	25	11	19.5	45	25	58%	1050
20	30	11	20.5	45	25	55%	1059
20	35	11	29.5	45	25	96%	1076
20	40	11	34.5	45	25	94%	1084
20	30	9	20.5	45	25	63%	1115
20	30	10	20.5	45	25	56%	1089
20	30	11	20.5	45	25	55%	1059
20	30	12	20.5	45	25	56%	1028
20	30	13	20.5	45	25	57%	996
20	30	11	18.5	45	25	55%	1047
20	30	11	19.5	45	25	56%	1051
20	30	11	20.5	45	25	55%	1059
20	30	11	21.5	45	25	54%	1062
20	30	11	22.5	45	25	50%	1053
20	30	11	20.5	35	25	55%	1060
20	30	11	20.5	40	25	53%	1023
20	30	11	20.5	45	25	55%	1059
20	30	11	20.5	50	25	52%	1032
20	30	11	20.5	55	25	54%	1050
20	30	11	20.5	45	20	52%	1019
20	30	11	20.5	45	22.5	54%	1043
20	30	11	20.5	45	25	55%	1059
20	30	11	20.5	45	27.5	54%	1020
20	30	11	20.5	45	30	55%	1056
Without a swirler						53%	611

$\alpha = 45°$, $L = 25$ mm. Exhaust gas distribution area increases by 82%, while the hydraulic resistance increases by 465 Pa relatively to similar indicators of the filter-converter without a swirler.

Thus, the use of a swirler makes it possible to improve the exhaust gases cleaning from the solid particles in filters-converters used in many modern vehicles with diesel engines.

3 Conclusions

The new design of the centrifugal-jet swirler that provides an increase in exhaust gases distribution area on the filter element by 82% while the hydraulic resistance increases by 0.465 kPa is proposed.

With the help of simulation of the processes of exhaust gases distribution in filter-converter in COSMOSFloWorks CFD-complex, rationality of using the swirler in the filters that are used in many modern cars with diesel engines was proved, provided it has the following geometrical parameters: the height of the screw d = 20 mm, D = 35 mm, r = 11 mm, R = 29.5 mm, $\alpha = 45°$, L = 25 mm.

References

1. Polivaev, O.I., Baybarin, V.A., Bozhko, A.V., Mozheyko, A.V.: Neutralizer of the exhaust gas for diesel: Pat. 2280177 Russian Federation: IPC F01N3 / 035. The applicant and the patentee - Federal State Educational Institution of Higher Professional Education, Voronezh State Agriculture University named after K.D. Glinka - 2005105377/06; appl. 24.02.2005; 4 pp., publ. 20.07.2006
2. Korobiichuk, I., Bezvesilna, O., Ilchenko, A., Shadura, V., Nowicki, M., Szewczyk, R.: A mathematical model of the thermo-anemometric flowmeter. Sensors **15**, 22899–22913 (2015). doi:10.3390/s150922899
3. Korobiichuk, I., Shavursky, Y., Nowicki, M., Szewczyk, R.: Research of the thermal parameters and the accuracy of flow measurement of the biological fuel. J. Mech. Eng. Autom. **5**, 415–419 (2015). doi:10.17265/2159-5275/2015.07.006
4. Gordievskiy, V.N., Shestakov, S.V., Zalyubovskiy, A.F., Medvedev, Y.S.: Neutralizer of the exhaust gas: Pat. 2175391 Russian Federation: IPC F01N3 / 02, the applicant and the patentee - Military Automobile Institute - 2000100245/06; appl. 05.01.2000; 3 pp., publ. 27.10.2001
5. Nosyrev, D.Y., Pletnev, A.I.: Neutralizer of the exhaust gas: Pat. 2433285 Russian Federation: IPC F01N3 / 02, the applicant and the patentee State Educational Institution of Higher Professional Education, Samara State University of Railway Transport - 2010105462/06; appl. 15.02.2010; 3 pp., publ. 27.03.2011
6. Baluk, V.Y., Ilchenko, A.V., Trostenyuk, Y.V.: Evaluation of the method of improving the diesel exhaust gases filter-converter. Bulletin of National Technical University "KhPI". Collected Works. Series: Automobile and Tractor Industry, vol. 29(1002), pp. 79–85 (2013)
7. Baluk, V.Y., Ilchenko, A.V., Trostenyuk, Y.V.: Analysis of ways to improve the efficiency of the diesel exhaust gases filter-converter. Bull. SevNTU **143**, 62–65 (2013)
8. Markova, T.V., Tishyn, A.P.: Flow Swirler: Patent 2323386 Russian Federation: F23D14/24, the applicant and the patentee Tishyn A.P. 2006128198/06; appl. 03.08.2006; 2 pp., publ. 27.04.2008
9. Ibragimov, I.G., Tumanova, Y.Y. (RF).: AS 2271872 RF, IPC B05B1/34. Centrifugal-jet nozzle, № 2004119943/12; appl. 29.06.2004; publ. 20.03.2006
10. Nosyrev, D.Y., Prosvirov, Y.Y., Roslyakov, A.D., Frolov, S.G.: Guidelines for carrying out the student course work №1 on the discipline, Locomotives (general course). Samara: SamIIT, 24 pp. (2001)

11. Yevstigneev, V.V., Novoselov, A.L., Prolubnikov, V.I., Tubalov, N.P.: Modelling of processes of the exhaust gases cleaning of chemical plants and diesel units from solid particles by SHS filters. Bull. Tomsk Polytech. Univ. **308**(1), 138–143 (2005)
12. Ilchenko, A.V., Baluk, V.Y.: Method for determining the hydraulic resistance of porous material of exhaust gas filter element. Automobile Transp. **29**, 148–151 (2011)

Precautionary Statistical Criteria in the Monitoring Quality of Technological Process

Eugenij Volodarsky[1(✉)], Zygmunt Warsza[2(✉)], Larysa A. Kosheva[3], and Adam Idźkowski[4]

[1] Department of Automation of Experimental Studies, National Technical University of Ukraine "KPI", Kiev, Ukraine
vet-l@ukr.net

[2] Industrial Research Institute of Automation and Measurement (PIAP), Warsaw, Poland
zlw@op.pl

[3] Department of Biocybernetics and Aerospace Medicine, National Aviation University of Ukraine, Kiev, Ukraine
l.kosh@ukr.net

[4] Faculty of Electrical Engineering, Bialystok University of Technology, Bialystok, Poland
a.idzkowski@pb.edu.pl

Abstract. The quality of products depends on a stability of production process. In practice to identify reasons of the process degradation, the Shewhart control chart is typically used. At the construction of control X-charts, it is supposed that the dispersion of sample means in subgroups of data measured in process is caused by the influence of random factors and the limited sample size. In such cases it is an improbable event to obtain the output sample values, which are outside the interval $\pm 3\sigma$. Its appearance indicates the presence of systematic influence and it is the need to adjust the controlled parameters of technological process. Based on practical experience in the ISO 7870-2: 2013 standard it is recommended to pay attention to "… any unusual structure of data points, which may indicate about a manifestation of special (non-random) reasons". In the numerical example presented in this work the analysis of a structure of points on the control chart showed the presence of non-random values, although if the sample mean values were within interval $\pm 3\sigma$. The indicator of existence of non-randomness was the probability that the minimum number of consecutive selective averages, which got to a certain area did not exceed 0.003. As a result of executed analysis the criteria are established and the algorithm is developed. It helped to identify the dysfunction of technological process at an early stage.

Keywords: Quality control · Technological process · Shewhart control chart

R. Szewczyk and M. Kaliczyńska (eds.), *Recent Advances in Systems, Control and Information Technology*, Advances in Intelligent Systems and Computing 543, DOI 10.1007/978-3-319-48923-0_80

1 Introduction

A quality of production depends on the stability and absence of changes in a technological process. Therefore, it is important to identify timely the reasons of dysfunction of technological process and the signals corresponding to them. If the process is carried out in normal conditions then the scattering of parameters, which characterize the properties and quality of a product, depends only on the influence of random variables. The possible dispersion of parameters (their standard deviations) are usually standardized in the normalized conditions.

A stability control of results is based on a series of control procedures. In this case a number of observations and the intervals between them have to be established on the basis of relations between the rates of changes of measured statistical characteristics under the influence of different random variables. This should be realized in such a way that the influence of deviations in carrying out the technological process could be neglected.

As proven practice the control charts are widely used for the statistical control of stability and quality of processes. This method was developed by Shewhart [1]. The main idea of control charts is to divide observations in subgroups, in which variations due to random causes are only permitted. The differences between these subgroups may not only be caused by random specific causes and this must be identified by control charts [2]. For this purpose, a reference value is established. The deviations from this value are detected as observations.

Depending on a control algorithm, the warning and action signals are selected and the control limits are calculated. A warning signal is an event which testifies a confidence level greater than (0.95 ... 0.99) about the withdrawal of process from the statistically controlled conditions and about the requirement of technological process correction. An action signal is an event, which indicates to withdraw the process from the statistically controlled conditions with a confidence level of 0.997. The monitored results of a process are presented on the control charts after recording each current observation on them.

The practice of using charts for the control of process has shown that acceptable results are obtained when a number of elements in the subgroups is not more than 4 to 5. The number of elements has to be identical in the assumption of their normal distribution. Under this condition, the coefficients for calculating the control limits are derived. Since the control limits are used as empirical criteria for decision-making, it is allowed to ignore small deviations from normality.

If the volumes of experimental data are larger ($n > 10$) then the standard deviation (SD) adequately displays a scattering of results. It characterizes a stability of the controlled process. When the samples of small volume are considered then a sample range R_n (the absolute difference between the highest and lowest values of subgroup sample) gives better estimate of the scattering of results than the standard deviation and it is calculated more quickly [1]. In addition, for the evaluation of process stability it allows to have only two observations in each subgroup. That should often be enough due to a dynamics of process or an economic feasibility.

The sample range R_n and standard deviation σ is statistically connected. For normal distribution it is [3]

$$M\left(\frac{R_n}{\sigma}\right) = \alpha_n, \tag{1}$$

where: α_n – tabulated value depending on a number of elements n in sample, R_n - range of n-element sample.

Due to these statistics the mean value of R_n can be identified as

$$M(R_n) = \alpha_n \sigma. \tag{2}$$

In addition, as it can be seen from the expression (1), α_n is an unbiased estimate, and consequently, also, an unbiased estimate M (R_n) is important and it may be taken as the centre of possible scattering of inspection results.

The range of possible values of ratio R_n/σ at a fixed value of n due to the influence of random variables and the limited sample size is also tabulated. Thus, there is a relationship

$$var\left(\frac{R_n}{\sigma}\right) = \beta_n, \tag{3}$$

which allows, for a given σ, to set the possible values of scattering amplitude R_n with respect to M (R_n) in the form

$$var(R_n) = \beta_n \sigma. \tag{4}$$

The most common charts in the monitoring of process stability are:

- average value ($\overline{X}$-chart) and range ($\overline{R}$-chart) or sample standard deviation s,
- individual measured values (X-chart) and moving range (R-chart).

2 Average Value $\overline{X}$ Control Chart and Range $\overline{R}$ Control Chart

The control chart of average values $\overline{X}$ is used to demonstrate what the average value of process is and what its stability is. Moreover, it allows identifying variations between subgroups that cannot be explained only by the influence of random variables and their relation to the total variation of the mean. If this type of control has to be reliable, the samples should be stable for a time period between repeated measurements.

Range control chart $\overline{R}$ identifies any undesirable variation within a subgroup and it is an indicator of variability of a controlled process. If $\overline{R}$ – chart shows that the variation within a subgroup are not changed then it informs about the uniformity of process. It is necessary to analyze $\overline{R}$ – chart prior to the analysis of $\overline{X}$ – chart.

Since the average value $\overline{X}$ control chart and the range $\overline{R}$ control chart (sample or standard deviations) reflect the state of process through the spread (variability from unit to unit) and through the centre of location (the average value of process), they are always used inseparably. Thus, to ensure the stability of process is necessary, firstly, to monitor the changes of σ in time which can be caused by an influence of random variables and a dysfunctional process.

By using the expressions (2) and (3) one can determine the absolute value of the variation quantile span R_n as

$$M(R_n) \pm k(P)\mathrm{var}(R_n), \tag{5}$$

where: $k(P)$ – coefficient depending on a value of confidence interval.

If the range is positive number then for the left quantile the condition should be satisfied

$$[M(R_n) - k(P)\mathrm{var}(R_n)] > 0. \tag{6}$$

These relations form the basis for the construction of Shewhart control charts. To make a decision on the stability of technological process, there are introduced precautionary warning limits $k(P) = 2$ and action limits $k(P) = 3$. It corresponds to the probability of decision $P = 95\%$ and $R = 99.7\%$.

When a range control chart $\overline{R}$ is constructed a tabulated value α_n is used as a centre line which in the standard [4] is indicated as d_2. For example: when $n = 2$ in accordance with [3] $d_2 = 1.128$ and $M(R_n) = 1.128\sigma$ is taken as a centre line CL. The action limits in the $\overline{R}$ control chart must be separated from the centre line by $\pm 3\mathrm{var}(R_n)$. For example, the upper action limit is

$$UCL_a(n) = d_2\sigma + 3d_3\sigma = D_2\sigma, \tag{7}$$

where: d_3 corresponds to the value where β_n, taken from the same table, for $n = 3$, $d_3 = 0.853$.

Thus, calculated values $D_2 = (d_2 + 3d_3)$ are provided in the table [4]. Similarly, one can obtain for the lower action limit

$$LCL_a = d_2\sigma - 3d_3\sigma, \tag{8}$$

or

$$LCL_a = D_1\sigma, \tag{9}$$

where $D_1 = d_2 - 3d_3$.

After analyzing the relation (6) one can conclude that for $k(P) = 3$ it will be executed if $n \geq 7$ only. For $n < 7$ as the lower action limit of Shewhart chart is taken zero line.

To calculate the upper and lower warning limits, one can use the expressions (5) and (6). For example, if $k(P) = 2$

$$UCL_w = D_2(2)\sigma \tag{10}$$

where:$D_2(2) = d_2 + 2d_3$,

$$LCL_w = D_1(2)\sigma \tag{11}$$

where: $D_1(2) = d_2 - 2d_3$.

In this case Eq. (6) will only be carried out when $n \geq 4$. In other cases the numerical value of the lower warning limit is absent – it is replaced by 0.

The calculated values for the centre line $CL = d_2\sigma$, upper and lower action limits $LCL_a = D_1\sigma$, as well as upper and lower warning limits and $LCL_w = D_1(2)$, respectively, are used to build the Shewhart charts.

In Fig. 1 an example of construction $\overline{R}$ – charts is presented. The ordinate axis represents the value of magnitude and on the horizontal axis are a number of observations of subgroup.

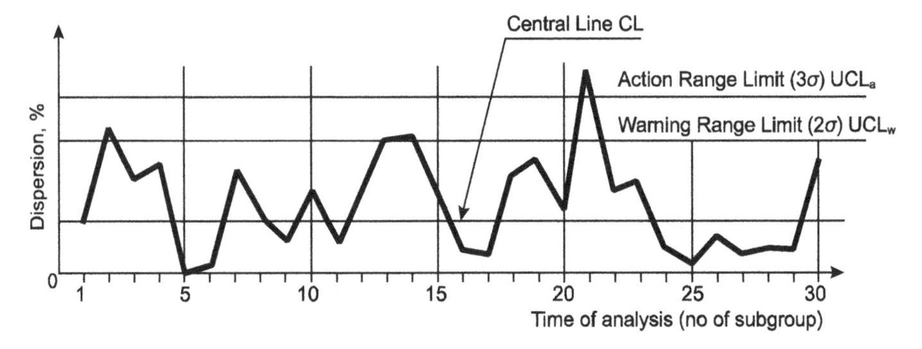

Fig. 1. Example of construction – Shewhart chart

The values of coefficients used to determine the control limits of Shewhart charts are presented in Table 1 [4]. The coefficients for calculating the warning limits are derived according to formulas

$$D_1(2) = d_2 - 2d_3, \tag{12}$$

$$D_2(2) = d_2 + 2d_3. \tag{13}$$

Assessing the bias stability of technological process should be carried out on $\overline{X}$ – chart using a standard sample with a SD value of μ. As mentioned before, a requirement is the invariance of its characteristics in time. A monitoring is carried out with a standard sample several times ($n \geq 2$) and average $\bar{x}_i$ is calculated for each subgroup.

In this case as the centre line in the construction of the $\overline{X}$ – chart is used $CL = \mu$, in relation to which changes $\bar{x}_i$ are considered in i-subgroups. The warning and action limits of are defined as

Table 1. The values of coefficients to determine the limit values of Shewhart charts

The coefficients to calculate the median line and the limits of efficiency			Coefficients for warning ranges		
Number of observations in subgroup	Coefficient for the centre line	Coefficient for the action range			
n	d_2	D_2	d_3	$D_1(2)$	$D_2(2)$
2	1.128	3.686	0.853	–	2.834
3	1.693	4.358	0.888	–	3.469
4	2.059	4.698	0.880	0.299	3.819
5	2.326	4.918	0.864	0.598	4.054

$$UCL = \mu + k(P)\frac{\sigma}{\sqrt{n}}, \tag{14}$$

$$LCL = \mu - k(P)\frac{\sigma}{\sqrt{n}}, \tag{15}$$

where: σ – standard deviation (normalized value) of estimated results in the control due to the influence of random variables.

3 Individuals *X* Control Chart and Moving Range *R* Control Chart

In some cases for practical or economic reasons there is no possibility to carry out the multiple observations and $n = 1$. In such situation, the moving range R_i may be used to monitor the stability of process. The μ value of standard sample does not need to be known accurately. However, the values of the samples must be stable. So-called initial "reference" measurement of performance of standard sample is realized with the result y_0. Then, there is the difference between the result of the first measurements y_1 and y_0, i.e. it is calculated the first implementation of bias (offset) results $\hat{\delta}_i = y_1 - y_0$. Subsequently, the offset is determined as the difference values obtained in the current and previous times. Thus, the moving range of i-th control chart estimated as

$$R_i = \left|\hat{\delta}_i - \hat{\delta}_{i-1}\right|. \tag{16}$$

A fragment of the moving range *R*-chart with the results of calculations [5] using the standard object with $\mu = 10.29$ is presented in Table 2.

Since the centre line CL at moving range *R* chart is the zero line then the previously received formula to calculate the action and warning limits should be modified. The starting point is known and the SD value of the process $\sigma_I = 0.06645$ is used. Then the action limits and warning limits are defined as

Table 2. Data to build the charts of moving range R

No of subgroup i	Result of analysis y_i	Estimate of bias $\hat{\delta}_i$	Moving range R_i
1	10.30	0.01	0.01
2	10.29	0.00	0.01
3	10.28	–0.01	0.02
4	10.30	0.01	0.01
5	10.29	0.00	0.00
6	10.29	0.00	0.09

$$UCL_a = +3\sigma_1 = 0.1994, \quad LCL_a = -3\sigma_I = -0.1994,$$
$$UCL_w = +2\sigma_I = 0.1329, \quad LCL_w = -2\sigma_I = -0.1329.$$

In this case unlike the $\overline{R}$ – charts, lower warning and action limits cannot be equal to zero because deviation from μ value can be positive or negative.

The centre line $M(\hat{\delta}) = 0$ is taken as the zero line to create the X-chart. Symmetrically in relation to it, the upper and lower warning limits are determined with factor $k(P) = 2$ and upper and lower action limits with factor $k(P) = 3$. Further actions and solutions taken for X-chart are similar as for R-chart. The Shewhart chart for the example from Table 2 is presented in Fig. 2 a,b.

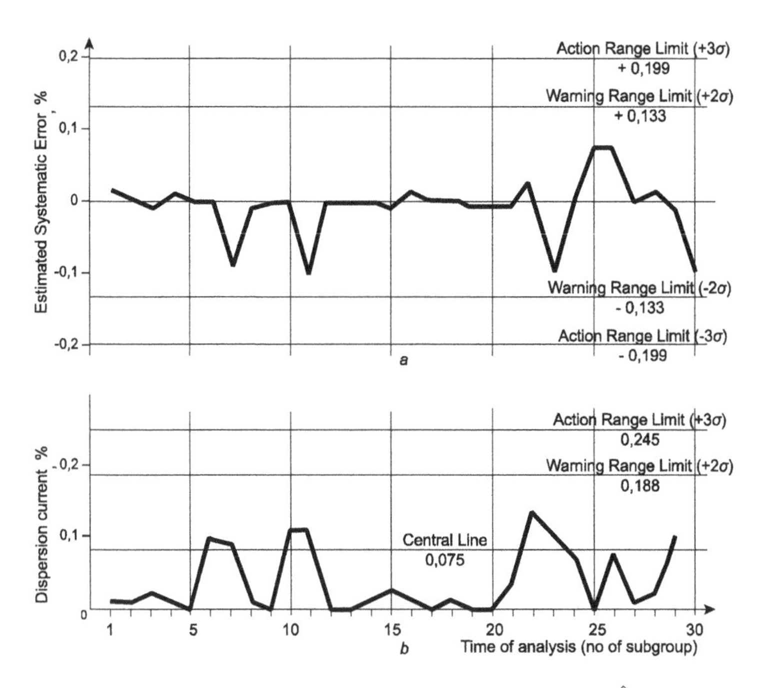

Fig. 2. Shewhart control charts for: (a) assessment of displacement $\hat{\delta}$ stability (b) current dispersion of results

4 Display of Process Instability

As it is presented in Fig. 2, there are some periods of time when the bias of process and the changes of range are low. There are other periods where results indicate an increased instability. This situation requires the identification of reasons that caused an increase of process instabilities over a certain period although the current values of the parameters are within the required limits.

When creating a control chart it is assumed that during monitoring the change in the X value (the dispersion of sample means as an estimate of bias) results from the influence of random factors and the limited sample size. In this case, the value outside $\pm 3\sigma$ is an low probable event. Its appearance indicates the presence of systematic influence, which leads to a process dysfunction and a change of its control. Based on practical experience in [4] is recommended to pay attention to "… any unusual structure of data points, which may indicate about a manifestation of the special (non-random) reasons". Such an event corresponds to the probability of 0.003 [4]. The observed situation can be attributed as "critical" which constitutes a violation of the conditions of process.

Figure 3 presents the probability of getting the results displayed in control zones A, B and C. It characterizes the relationship between σ and the number of observations n.

An indication of the influence of random variables is chaotic incidence of sub-groups results in all areas A, B and C of the Shewhart chart. Tracking the emergence of a systematic trend in a distribution of points corresponding to the mean values of samples (bias estimate) can be used as evidence of the trend of mean value of controlled process parameter.

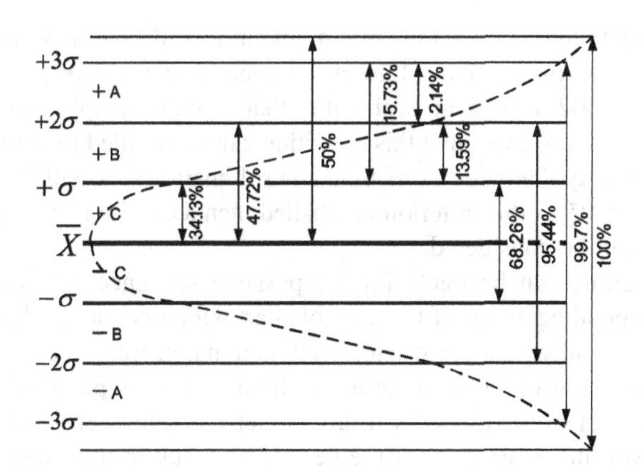

Fig. 3. The probability of getting the results into control zones A, B, C

The trends on the control chart, which emerged under the influence of special causes, can lead to a breakdown of process. They will be called as series criteria. In this case, the manifestation of a systematic influence on the background of random dispersion is to find a sequence of a certain number of points on the control chart in one of zones or in area covering several zones. This approach allows to justify theoretically that the incidence of trend reports at an early stage about the possibility of violation of a correct process. In this way a sequence of control points on the chart located within the warning or action limit, the probability of which is less than 0.003, can be regarded as a manifestation of joint influence of random and systematic reasons. This property can be used as the basis for the creation of preventive warning criteria. It would allow to adjust the progress of process without waiting for situation when it could be disordered that the results would be outside the control limits.

Checkpoints should be sufficiently distant from each other in time and space that the effect of autocorrelation cannot be included [7]. In accordance with the multiplication theorem for independent events the probability of getting a normally distributed random variable in area A, B, C is equal to the product of individual probabilities

$$P(A_1, A_2, \ldots, A_s) = P\left(\prod_{i=1}^{s} P(A_i)\right). \tag{17}$$

During analysis the criteria were established. According to them it can be found the probability of several independent, consecutive values in a particular field of control chart. It may indicate a trend of dysfunctional process. For example if we assume that the average value (bias estimate) of consecutive subgroups are independent random variables then the probability of getting the result for any subgroup above (or below) the centre line in each of zones A, B, C is 0.4986 approximately 0.5 (Fig. 3). Probability, that two consecutive sample values are e.g. above the centre line, is equal to 0.5 • 0.5 = 0.25.

It is necessary to find out what the minimum number of successive results, arranged in a row on one side of the center line of the control chart, corresponds to the probability of 0.003 (0.0027). It is the probability that a single sample value is not within the control limits $\pm 3\sigma$. It turns that this condition can be fulfilled by a sequence of nine points, i.e. probability that a series of nine control chart points will be on one side of central line is 0.00195. If this criterion is satisfied then a change in average value of the overall process can be considered.

Similar reasoning can be made for the possible sequences of points, which are located in corresponding zones of the control chart with a certain probability. One can define a set of preventive criteria when a dysfunctional process begins but results still do not go outside the warning limits or action limits. An example would be a situation when finding 15 consecutive results in a completely "safe" zones ±C cannot cause concern. However, the emergence of the next 16-th result in this zone corresponds to the probability 0.0023, which exceeds the value of 3σ. Therefore, this criterion is also "critical" as shown in Fig. 4.

With the use of similar analysis one can define a number of simple criteria shown in Table 3, which demonstrate that the mean value of controlled process is biased.

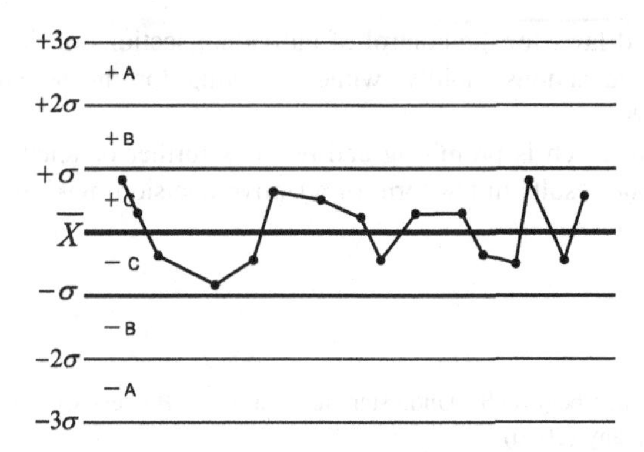

Fig. 4. The probability of warning signals and symptoms of the action

Table 3. Examples criteria that indicate the shift of mean value of controlled process variable

Situation	Number of consecutive points	Location of points	Probability of an elementary event in %
1	2	Only in zone plus A or minus A	2.14
2	3	Only in zone plus B or minus B	13.59
3	6	Only in zone plus C or minus C	34.13
	6	Permanent increase or decrease in points of any zone[a]	
4	7	Only in two zones[+A; +B] or [−A; −B]	15.73
5	8	Series arrangement only in zones [+A; +B] and [−A; −B], passing zones +C and −C	15.46
6	9	In any zones [+A; +B; +C] or [−A; −B; −C]	49.86
7	16	Only in zone + C or −C	68.26

[a] Sample values are independent variables, which have the same distribution. When comparing the current value with the previous ones we will have two outcomes: more than "+" or less "−". Under the influence of the random variable, signs will have the same frequency. Wald-Wolfowitz run test [6] establishes a relationship between the number of identical characters in a series and probability magnitude of the systematic influences.

5 Conclusions

In evaluating the stability of process it is possible an very early detection of trends leading to dysfunction of the process.

The considered approach and the resulting criteria for identifying trends can be treated as the basis of sequential analysis and the identification of the "critical" situation for the controlled process, depending on a location of area and a number of points in their sequence on the control chart.

The established laws for the control of individual sections of the X chart allow to introduce corrective actions rapidly, without waiting for the actual dysfunction of process parameters.

This area of research is promising and requires further detailed development and formalization of the results in the form of adaptive decision-making algorithms.

References

1. Wheeler, D.J., Chambers, D.S.: Understanding Statistical Process Control. Addison-Wesley Publishing Company (2010)
2. Nishina, K., Kuzuya, K., Ishi, N.: Reconsideration of Control Charts in Japan. Front. Stat. Qual. Control **8**, 136–150 (2005)
3. Gatti, P.L.: Probability Theory and Mathematical Statistics for Engineers. Taylor & Francis (2004)
4. ISO 7870-1, -2 … -6: 2014. Shewhart control charts – part 1-6
5. ISO 5725-6: 1994 (reviewed in 2012) Accuracy (trueness and precision) of measurement methods and results. Part 6: Use in practice of accuracy values
6. Alhakim, A., Hooper, W.: A non-parametric test for several independent samples. J. Nonparametric Stat. **20**(3), 253–261 (2008)
7. Warsza, Z.L.: Evaluation of the type A uncertainty in measurements with autocorrelated observations. J. Phys. Conf. Ser. **459**(1), Article no 012035 (2013)

Principles of Implementing an Electronic Progress Log at "Zhytomyr Nursing Institute" KVNZ

Svetlana Gordiichuk(✉)

Zhytomyr Nursing Institute, Zhytomyr, Ukraine
stepanovasv77@mail.ru

Abstract. The use of information and communication technologies in the management of higher education institutions is one of the main tasks of modern high school, since the timeliness and adequacy of management decisions determine the effectiveness of the education system as a whole. The article aims at analyzing the basic principles of implementing the electronic progress log at Zhytomyr Nursing Institute KVNZ. The study used the method of analyzing the results of implementation of an information system for attendance and progress registration based on web-technology, using the PHP 5.4 programming language, MySQL 5.1 database server, and HTML 4.01 web-page markup language; the interface was developed under Web 2.0. Electronic Progress Log standards, introduced in academic year 2014/2015. It allows recording the current progress and final knowledge testing with automatic conversion into ECTS scale, counting total points per module and calculating the final score for a training course. Thus, implementation of an automated system of progress and attendance registration ensures an efficient monitoring of all parties to the educational process.

Keywords: Electronic progress · Education system · Monitoring quality

1 Introduction

There are fundamental changes in the management system of higher medical educational institutions happening nowadays in Ukraine, that affect the reorganization of an educational process and the use of information and communication technologies, as the timeliness and adequacy of management decisions determine the effectiveness of the educational system as a whole. At the same time the productivity of management decisions depends on the monitoring quality and efficiency regarding the processes that take place in an educational institution. One of the priority processes requiring a continuous diagnostic supervision, analysis, synthesis and influence is an educational process [1].

Among the national priorities of the state educational informatization policy is the creation of infrastructure, information resources, new information technologies, information systems, automated databases in order to ensure free access to the computer network resources that determine the content and structure, the choice of forms, methods, means and management technologies in an educational institution.

R. Szewczyk and M. Kaliczyńska (eds.), *Recent Advances in Systems, Control and Information Technology*, Advances in Intelligent Systems and Computing 543, DOI 10.1007/978-3-319-48923-0_81

Incorporating software in a management process as a part of education informatization is not only the development factor of information society in Ukraine; it is also a prerequisite to ensure a higher level of effective management and the quality of education through the creation of information resources, introduction of information technologies, networks, etc. [2, 3].

Objective. To describe the content of an electronic progress log and its principles of implementation at "Zhytomyr Nursing Institute" KVNZ.

2 Materials and Methods

The study used the method of analyzing the results of implementing the information system for attendance and progress registration based on the web-technologies, using the PHP 5.4 programming language, MySQL 5.1 database server, and HTML 4.01 web-page markup language; the interface was developed under Web 2.0. standards.

3 Results and Discussion

The introduction of the credit accumulation and transferring system into an educational process has caused some difficulties for the teaching staff of higher educational institutions. First of all, these problems are associated with keeping track of a student's learning progress. It is necessary to assess student performance at every training session; to evaluate individual tasks; to count the total points per module considering the results of a current progress and final module tests; to calculate the final grade for a training course taking into account the points from all modules; to convert the grades from a traditional 4-point scale into the ECTS rating scale, etc. [4]. Thus, the detailed indicators of the student learning quality, which should be the ground for an objective evaluation of the educational performance, are derived from a progress log, and require rather significant efforts from the teacher's side, even if taking into consideration only one group of students.

Another essential problem is the fact that the assessment and attendance information is kept in a teacher's individual register (the register of students' practical classes attendance and learning progress), then transferred to the monthly reports (assessment sheets) and passed on to an administrative arm of the institute. Later the staff process the data and create the summary reports for academic groups, courses etc. Only after this work is done, the results become available to the institute authorities. Therefore, a classical model of monitoring a current progress and attendance is a laborious procedure, that requires a large quantity of routine transactions, which quite often may lead to errors [5, 6]. In addition, this process is time-consuming, which significantly affects the monitoring efficiency and, accordingly, the timeliness and quality of management decisions. The problems outlined above, the complexity of the procedure and the need for an objective evaluation determine the necessity to develop a new approach for automating the tracking of a student learning progress, and create a centralized repository for this kind of information – an electronic system for student attendance and learning progress [1].

One more prerequisite for creating such an information system is ensuring the transparency of an educational process by means of providing the students with the Internet access to the results of their progress, displaying their personal rating, generating a feedback between the students, the teaching staff and the administration of the institute.

To address this issue, we need to examine the software of different companies based on the criteria of "price-quality" and the benefits of using it in the educational practice [7].

While developing and implementing the electronic progress log at "Zhytomyr Nursing Institute" KVNZ, we applied a number of methods, including analysis, generalization, systematization and classification, that gave us the opportunity to properly examine the corresponding laws and regulations, scientific and educational materials, publications and electronic resources. On this basis we formulated the feature and parameter requirements to be met by an efficient electronic progress log [8].

Thus, it was determined that an electronic progress log should be [9]:

- objective and impartial: it should meet a basic structure and content of a standard printed version of a register;
- flexible: it should be possible to further improve and adapt the log to the curriculum peculiarities for each training course [10];
- accessible: it should be free of charge, with minimum skills required for use by teachers [11];
- convenient: it should be easy to use, demonstrate a clear logic of calculation formulas, have other useful functions (generate reports, predict learning results), calculate all possible operations automatically, be ready for use in any place with no tie to a user workplace [12];
- secure and possible to be saved: it should provide the possibility to create a progress log archive and access it when necessary [13].

Taking into consideration the above-mentioned information, "Zhytomyr Nursing Institute" KVNZ implemented the software module "PS – Progress Log – Web", which is a part of the software package "Dekanat (Dean's Office)" developed by PE "Politek – SOFT".

The module provides the possibility to [14]:

- register in real-time the students' current progress and attendance by the teaching staff of the institute;
- view the class schedule;
- give access to these data to all the students of the institute;
- generate a set of reports and summary figures in order to perform a comprehensive data analysis regarding a current student performance, and to make management decisions respectively;
- review the teaching load;
- convert the data on a current performance in order to automatically generate the semester progress final figures in the database and to print them in the form of credit and examination records (Figs. 1 and 2).

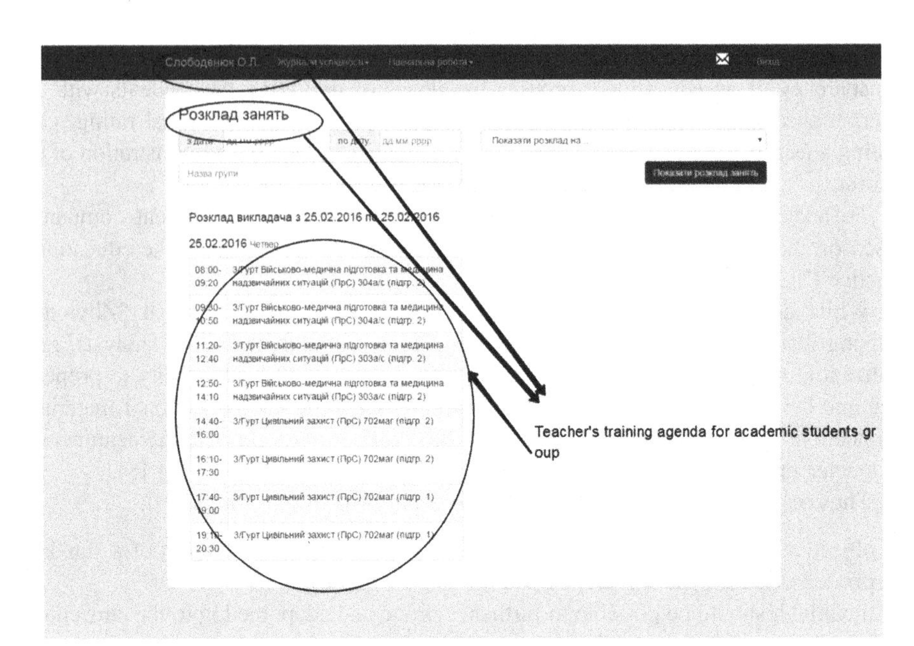

Fig. 1. The example of a class schedule of an academic group

Слободенюк О.Л.

Головна сторінка / Педагогічне навантаження

Педагогічне навантаження ← Yearly pedagogical load teacher of chair

Факультет Спеціальність	Форма навчання Група/підгрупа	Потік Збірні групи	Предмет	Всього годин
Акушерська справа Акушерська справа (2011)	Денна 301а/с	Потік 301а/с, 302а/с, 303а/с, 304а/с,	Неврологія та психіатрія з наркологією	51
Акушерська справа Акушерська справа (2011)	Денна 301а/с/2		Неврологія та психіатрія з наркологією	31
Акушерська справа Акушерська справа (2011)	Денна 301а/с/3		Неврологія та психіатрія з наркологією	31
Акушерська справа Акушерська справа (2011)	Денна 302а/с		Неврологія та психіатрія з наркологією	31
Акушерська справа Акушерська справа (2011)	Денна 302а/с/2		Неврологія та психіатрія з наркологією	31
Акушерська справа Акушерська справа (2011)	Денна 302а/с/3		Неврологія та психіатрія з наркологією	31
Акушерська справа Акушерська справа (2011)	Денна 303а/с		Неврологія та психіатрія з наркологією	31
Акушерська справа Акушерська справа (2011)	Денна 303а/с/2		Неврологія та психіатрія з наркологією	31
Акушерська справа Акушерська справа (2011)	Денна 303а/с/3		Неврологія та психіатрія з наркологією	31
Акушерська справа Акушерська справа (2011)	Денна 304а/с		Неврологія та психіатрія з наркологією	31
Акушерська справа Акушерська справа (2011)	Денна 304а/с/2		Неврологія та психіатрія з наркологією	31
Акушерська справа Акушерська справа (2011)	Денна 304а/с/3		Неврологія та психіатрія з наркологією	31
Стоматологія Стоматологія 3 роки (2011)	Денна 404стом		Основи неврології, психіатрії та наркології	12
Стоматологія	Денна	Потік 401стом	Основи неврології, психіатрії та	37

Fig. 2. The example of a teaching load "Zhytomyr Nursing Institute" KVNZ

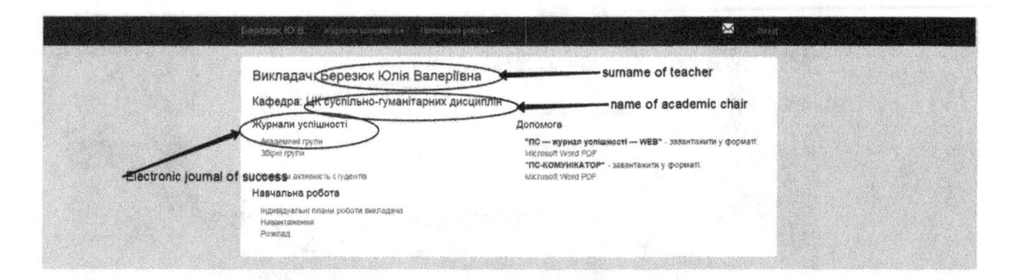

Fig. 3. The example of a teacher account in the electronic progress log

The key Module features and properties are the following [14]:

- creating accounts by the teachers using the search form and choosing oneself from the list of registered teachers of a corresponding department, setting a login and password. A registered teacher cannot change the data on a students' current progress without special confirmation (Fig. 3);
- saving in the database the encrypted usernames and passwords of the teachers in order to protect the information;
- a special mode of Module operation: the responsible department authorities have access to the teacher accounts, which enables them to provide or restrict the access of teachers to the data editing, and, if necessary, to change the corresponding logins and passwords;
- a special mode of operation of the "PS-Student-Web" program: the secretaries of the administrative arm can create (edit) the data on appointing the classes to the teachers;
- a teacher enters a login and password, then their credentials and the right to data editing of a students' current progress are verified; then the teacher sees the list of academic groups with the course, direction (speciality), faculty information, where they conduct the classes. After selecting a certain group, the teacher gets the access to data editing of a students' current progress on the subject they teach;
- in the mode of data editing of a students' current progress a teacher can:
  - register the fact of conducting a certain type of class in a certain subject in a certain day of a selected semester;
  - change the attributes of a class (date and type);
  - delete a registered class as one entered by mistake (in this case all the data on this class are stored in the database with a special status);
  - register (edit) the points a student received in class;
  - create (edit) the data on the absence of certain students in class and specify the reason;
  - review the students' semester final points that are formed by the Module automatically;
  - review a total number of missed class hours in a given subject and those missed without a valid excuse since the beginning of the semester;
  - create mixed groups in order to develop their own interface;

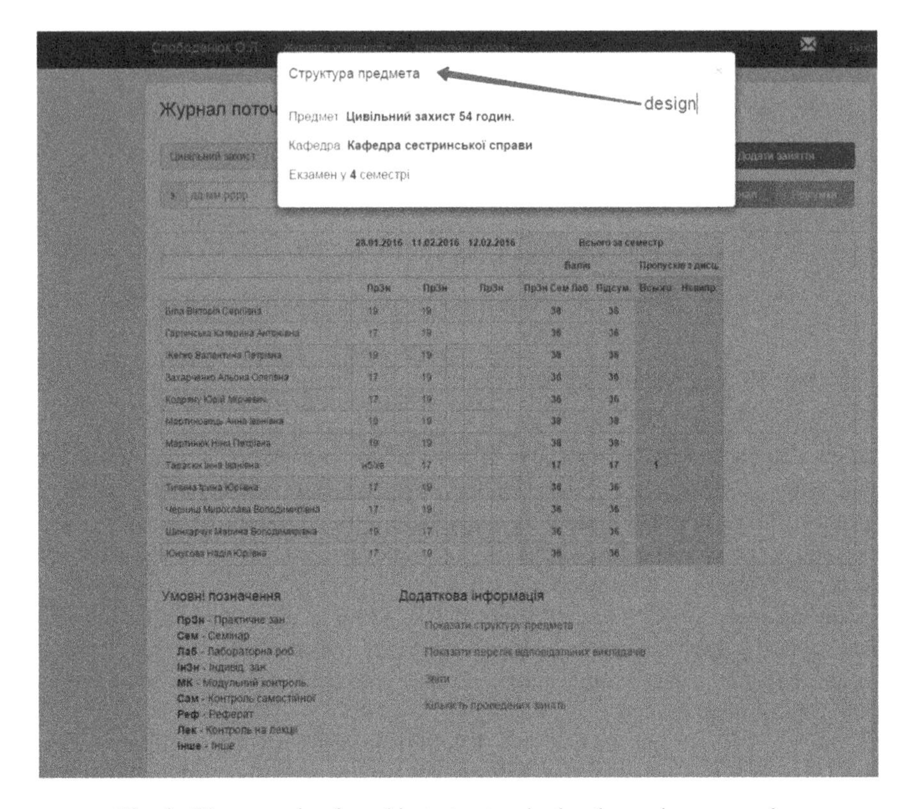

Fig. 4. The example of a subject structure in the electronic progress log

- all the changes made in the database by the teaching staff and secretaries of the administrative arm concerning the data of a students' current progress are automatically recorded in a special table. In this case a corresponding table title, field name, the value of data before editing, user ID, the date of changes are registered. If necessary, the analysis of the editing history can be performed (Fig. 4);
- a secretary of the administrative arm, after reviewing the data on a students' current progress recorded by the teachers, can activate an automatic conversion of this information into the final module points. After the procedure is executed, the teachers' access to the data change in a corresponding semester is blocked. But the secretary still has the right to manually edit the automatically formed final module points;
- the Module allows the students to see their current progress. To log in, they need to enter their last name and the number of a record book. A student can see only their own performance score;
- the registration of a students' current progress is performed in compliance with the regulations "On the knowledge assessment of the students of "Zhytomyr Nursing Institute" KVNZ";

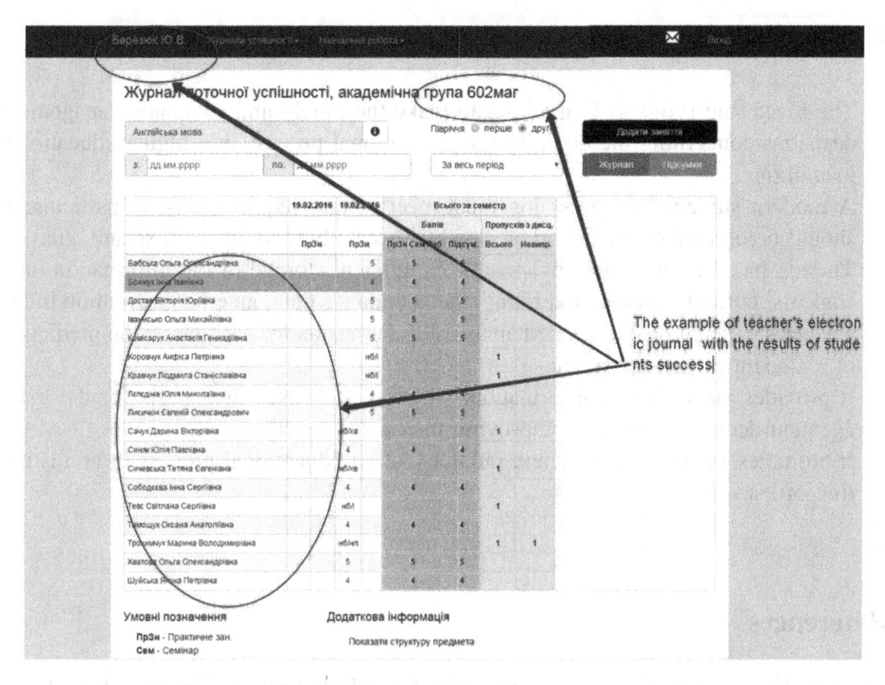

Fig. 5. The example of the electronic progress log with entered results

- the data is saved on a current progress of the students who are registered in the database throughout an academic year. In the end of the year after the conversion into the semester final grades, the data are deleted;
- the Module uses Web-interface and the Internet technologies (a CGI script). This eliminates the need to install any special software on the user computers [14] (Fig. 5).

The implementation of the Electronic Progress Log into the learning process has received favorable reviews among the teaching staff and students. The results of the survey among the institute teachers indicate, that the introduction of this information-communication technology positively affects a cognitive activity of the students (in this case, the activity aimed at achieving specific learning objectives) and motivates them to get higher grades. According to the survey among the students, about 75% feel positive about the system; 45% use the electronic progress log every day, 32% – 203 times per week, 15% – several times a week, and only 10.5% do not use the log. More than a half of the respondents (64%) believe that the log makes the educational process more transparent, 22% of respondents gave a negative answer to this question.

4 Conclusions

- The Electronic Progress Log allows to make the timely and adequate management decisions concerning the quality of an educational process in a higher educational institution.
- A modern students' progress log must meet certain requirements, in particular it should be objective and impartial, flexible, accessible, convenient, secure, etc.
- The log has the following advantages: an efficient storage of the information on a students' current progress and rating, the attendance data; an error prevention in the teacher's calculations; a detailed analysis of the results for each class; the prediction of a student performance, etc.
- It provides an analytical information, which can be used as a basis for the management decisions aimed at improving the quality of education.
- It promotes the transparency and publicity of an educational process in an institution of learning.

References

1. Babenko, V.: Electronic form account of student achievement. Taurian Med. Biol. J. Sci. Pract. Mag. **15**(14–60), 40–42 (2012). in Ukrainian
2. Boichuk, T., Herush, I., Khodorovsky, V., Barbie, A.: The first experience of implementing electronic journal of the success in Bucovina State University. Med. Educ. Sci. Pract. Mag. **2**(2), 67–72 (2015). (in Ukrainian)
3. Vasilev, V., Kostyukevich, R., Khomenko, I.: Scientific and practical approaches to project management account automation results of students work. New Pedagogical Thought Sci. Pract. Mag. **1**(1), 234–240 (2013). http://nbuv.gov.ua/UJRN/Npd_2013_1_62. (in Ukrainian)
4. Didenko, O., Kupriyenko, D.: Electronic magazine of the success of listeners (cadets, students) as a means of rationalizing the educational process and IT training. Inf. Technol. Educ. Means **47**(3) (2015). (in Ukrainian)
5. Didenko, O., et al.: Rationalization of educational process through the introduction electronic magazine of the success of listeners (cadets and students. Actual problems of increasing the quality of training in the field of economy in the sustainable economic development of Ukraine: II National Scientific and practical Conf. 25–26 Oct. 2012 ext.: thesis. Khmelnitsky: KNU, 23–26 (2012). (in Ukrainian)
6. The program module PS-Magazine achievement-Web PE, Politek SOFT. http://www.politeksoft.kiev.ua/index.php?do=newdevelopments&product=ps-gradebook-web. (in Ukrainian)
7. Kalinina, L.M.: The system of information implementation management of general educational institution State. HI. Teach. Institution, University of Education Management. C., 41 pp. (2008). (in Ukrainian)
8. Kalinina, L.M.: Information processes in management activity Head: nature, specificity and characteristic features of education and management. Educ. Manag. **8**(2), 35–44 (2005). (in Ukrainian)

9. Kalinina, L.M., et al.: Analysis and prospects of using cloud technologies in managerial process by general educational institutions Ukrainian pedagogical magazine. Ukrainian Pedagogical J. **2**(2), 44–64 (2015). (in Ukrainian)
10. Kernosova, A.: Development ontology of subsystem operational control of the current of student performance: the problems of information technologies. Sci. Pract. Mag. **8**(15), Article 177–185 (2014) (in Ukrainian)
11. Kupriyenko, D.A.: Electronic magazine of the success of listeners (cadets, students) of the National Academy of State Border Service of Ukraine – a tool of rationalization educational process. J. Nat. Acad. State Border Serv. Ukraine: the Electron. Science. Profession. Publishing, head. Ed. Gryaznov IO **3** (2012). http://www.nbuv.gov.ua/e-journals/Vnadps/2012_3/12kdanvp.pdf. (in Ukrainian)
12. Kupriyenko, D.A., Strelbitskyy, M.A.: Software account of the current and final success of and listeners cadets Of National Academy of State Border Service of Ukraine named after Bohdan Khmelnytsky certificate on the rationalization proposal №221; submitted 10.02.10; recognized 17.02.10. (in Ukrainian)
13. On enactment Provisions about the system of current and final assessment of students' knowledge KVNZ, Zhytomyr Nursing Institute: Prescript of rector KVNZ, Zhytomyr Nursing Institute, № 12. on 09.17.2015. (in Ukrainian)
14. On enactment Provisions about on electronic accounting journal of academic students work KVNZ, Zhytomyr Nursing Institute: Prescript of rector KVNZ, Zhytomyr Nursing Institute, number of 17. 22.09.2015. (in Ukrainian)

Chi-Squared Goodness-of-Fit Tests: The Optimal Choice of Grouping Intervals

Ekaterina V. Chimitova(✉) and Boris Yu. Lemeshko

Novosibirsk State Technical University, Novosibirsk, Russia
chimitova@corp.nstu.ru

Abstract. When using the chi-squared goodness-of-fit tests, the problem of choosing boundary points and the number of grouping intervals is always urgent, as the power of these tests considerably depends on the grouping method used. In this paper, the investigation of the power of the Pearson and Nikulin-Rao-Robson chi-squared tests has been carried out for various numbers of intervals and grouping methods. The partition of the real line into equiprobable intervals is not an optimal grouping method, as a rule. It has been shown that asymptotically optimal grouping, for which the loss of the Fisher information from grouping is minimized, enables to maximize the power of the Pearson test against close competing hypotheses. In order to find the asymptotically optimal boundary points, it is possible to maximize some functional (the determinant, the trace or the minimum eigenvalue) of the Fisher information matrix for grouped data. The versions of asymptotically optimal grouping method maximize the test power relative to a set of close competing hypotheses, but they do not insure the largest power against some given competing hypothesis. For the given competing hypothesis H_1, it is possible to construct the chi-squared test, which has the largest power for testing hypothesis H_0 against H_1. For example, in the case of the Pearson chi-squared test, it is possible to maximize the non-centrality parameter for the given number of intervals. So, the purpose of this paper is to give the methods for the choice of optimal grouping intervals for chi-squared goodness-of-fit tests.

Keywords: Chi-squared goodness-of-fit tests · Optimal grouping · Fisher information · Test power

1 Introduction

The χ^2 Pearson goodness-of-fit test is very popular in various applications, including the investigation of distributions of measurement error in problems of metrological support.

The correct usage of the Pearson χ^2 test for composite hypotheses (including testing normality) provides estimation of unknown parameters by the grouped data, as in the case of calculating parameter estimates by the original non-grouped sample the test statistic distribution differs from the χ^2 – distribution significantly [1, 2]. By this reason, a series of modified χ^2 tests has been offered, the most famous of which is the Nikulin-Rao-Robson test [3–5]. Moreover, it is necessary to take into account, that the power of Pearson test depends on the number of grouping intervals [6] and the grouping method used [7].

R. Szewczyk and M. Kaliczyńska (eds.), *Recent Advances in Systems, Control and Information Technology*,
Advances in Intelligent Systems and Computing 543, DOI 10.1007/978-3-319-48923-0_82

2 The Pearson χ^2 Test of Goodness-of-Fit

The procedure for hypothesis testing using χ^2 type tests assumes grouping an original sample $X_1, X_2, \ldots, X_n$ of size n. The domain of definition of the random variable is divided into k non-overlapping intervals bounded by the points:

$$x_0 < x_1 < \ldots < x_{k-1} < x_k,$$

where x_0, x_k are the lower and upper boundaries of the random variable domain. The number of observations n_i, in the i-th interval is counted in accordance with this partition, and the probability of falling into this interval,

$$P_i(\theta) = \int_{x_{i-1}}^{x_i} f(x, \theta)dx,$$

corresponds to the theoretical distribution law with the density function $f(x, \theta)$, where

$$n = \sum_{i=1}^{k} n_i, \sum_{i=1}^{k} P_i(\theta) = 1.$$

Measurements of the deviations n_i/n on $P_i(\theta)$ form the basis of the statistics used in χ^2 type goodness-of-fit tests.

The statistic of Pearson χ^2 test is calculated using the formula

$$X_n^2 = n \sum_{i=1}^{k} \frac{(n_i/n - P_i(\theta))^2}{P_i(\theta)}. \quad (1)$$

When a simple hypothesis H_0 is true (i.e., all the parameters of the theoretical law are known), this statistic obeys the χ_r^2 distribution with $r = k - 1$ degrees of freedom with $n \to \infty$. The χ_r^2 distribution has the density function

$$g(s) = \frac{1}{2^{r/2}\Gamma(r/2)} s^{r/2-1} e^{-S/2},$$

where $\Gamma(\cdot)$ is the Euler gamma function.

The test hypothesis H_0 is not rejected if the achieved significance level (p-value) exceeds a specified level of significance α, i.e., if the following inequality holds:

$$P\{X_n^2 > X_n^{2*}\} = \frac{1}{2^{r/2}\Gamma(r/2)} \int_{X_n^{2*}}^{\infty} s^{r/2-1} e^{-s/2} ds > \alpha,$$

where X_n^{2*} is the statistic calculated in (1).

When testing a composite hypothesis and estimating parameters by minimizing the statistic X_n^2 basing on the same sample, this statistic asymptotically obeys the χ_r^2 distribution with $r = k - m - 1$ degrees of freedom, where m is the number of parameters estimated.

The statistic X_n^2 has the same distribution if parameter estimate is obtained by the maximum likelihood method from grouped data by maximizing the likelihood function with respect to θ:

$$L(\theta) = \gamma \prod_{i=1}^{k} P_i^{n_i}(\theta), \tag{2}$$

where γ is a constant and

$$P_i(\theta) = \int_{x_{i-1}}^{x_i} f(x, \theta) dx$$

is the probability that an observation falls into i-th interval. This result remains for any estimation technique based on grouped data leading to asymptotically effective estimates.

If unknown parameters are estimated by the maximum likelihood method basing on non-grouped data, then the Pearson statistic is distributed as the sum of independent terms [1] $\chi_{k-m-1}^2 + \sum_{j=1}^{m} \lambda_j \xi_j^2$, where $\xi_1, \ldots, \xi_m$ are standard normal random quantities that are independent from each other and from χ_{k-m-1}^2; $\lambda_1, \ldots, \lambda_m$ are numbers between 0 and 1, representing the roots of the equation

$$\left|(1 - \lambda)\mathbf{J}(\theta) - \mathbf{J}_g(\theta)\right| = 0.$$

Here $\mathbf{J}(\theta)$ is the Fisher information matrix with respect to the non-grouped observations with elements

$$J(\theta_l, \theta_j) = \int \left(\frac{\partial f(x, \theta)}{\partial \theta_l} \frac{\partial f(x, \theta)}{\partial \theta_j} \right) f(x, \theta) dx;$$

$\mathbf{J}_g(\theta)$ is the Fisher information matrix with respect to the grouped observations with elements

$$\mathbf{J}_g(\theta) = \sum_{i=1}^{k} \frac{\nabla P_i(\theta) \nabla^\tau P_i(\theta)}{P_i(\theta)}.$$

In other words, the distribution of statistic (1), based on maximum likelihood estimates (MLE) calculated by non-grouped data, is unknown and depends, in particular, on the grouping method [2].

3 The Choice of Grouping Intervals

When using chi-squared goodness-of-fit tests, the problem of choosing boundary points and the number of grouping intervals is always important, as the power of these tests considerably depends on the grouping method used. In the case of complete samples (without censored observations), this problem was investigated in [7–10]. In particular, in [11], the investigation of the power of the Pearson and NRR tests for complete samples has been carried out for various numbers of intervals and grouping methods. The partition of the real line into equiprobable intervals (EPG) is not an optimal grouping method, as a rule. In [12], it was shown for the first time that asymptotically optimal grouping, for which the loss of the Fisher information from grouping is minimized, enables us to maximize the power of the Pearson test against close competing hypotheses. For example, it is possible to maximize the determinant of the Fisher information matrix for grouped data $\mathbf{J}_g(\theta)$, i.e. to solve the problem of D-optimal grouping

$$\max_{x_0 < x_1 < \ldots < x_{k-1} < x_k} \det\left(\mathbf{J}_g(\theta)\right). \tag{3}$$

In the case of the A-optimality criterion, the trace of the information matrix $\mathbf{J}_g(\theta)$ is maximized by the boundary points

$$\max_{x_0 < x_1 < \ldots < x_{k-1} < x_k} \mathrm{Tr}\left(\mathbf{J}_g(\theta)\right), \tag{4}$$

and the E-optimality criterion maximizes the minimum eigenvalue of the information matrix:

$$\max_{x_0 < x_1 < \ldots < x_{k-1} < x_k} \min_{i=1,2} \lambda_i\left(\mathbf{J}_g(\theta)\right). \tag{5}$$

The problem of asymptotically optimal grouping by the A- and E-optimality criteria has been solved for certain distribution families, and the tables of A-optimal grouping are given in [13]. The versions of asymptotically optimal grouping maximize the test power relative to a set of close competing hypotheses, but they do not ensure the highest power against some given competing hypothesis. For the given competing hypothesis H_1, it is possible to construct the χ^2 test, which has the highest power for testing hypothesis H_0 against H_1. For example, in the case of χ^2 Pearson test, it is possible to maximize the non-centrality parameter for the given number of intervals k:

$$\max_{x_0 < x_1 < \ldots < x_{k-1} < x_k} n \sum_{j=1}^{k} \frac{\left(p_j^1(\theta^1) - p_j^0(\theta^0)\right)^2}{p_j^0(\theta^0)}, \tag{6}$$

where $p_j^0(\theta^0) = \int\limits_{x_{j-1}}^{x_j} f_0(u, \theta^0)du$, $p_j^1(\theta^1) = \int\limits_{x_{j-1}}^{x_j} f_1(u, \theta^1)du$ are the probabilities to fall into j-th interval according to the hypotheses H_0 and H_1, respectively. Let us refer this grouping method to as optimal grouping.

Asymptotically optimal boundary points, corresponding to different optimality criteria, as well as the optimal points, corresponding to (6), are considerably different from each other.

For example, the boundary points maximizing criteria (3)–(6) for the following pair of competing hypotheses are given in Table 1. The null hypothesis H_0 is the normal distribution with density function

$$f(x) = \frac{1}{\sigma\sqrt{2\pi}} \exp\left\{ -\frac{(x-\mu)^2}{2\sigma^2} \right\}, \tag{7}$$

and parameters $\mu = 0$, $\sigma = 1$ and the competing hypothesis H_1 is the logistic distribution with density function

$$f(x) = \frac{\pi}{\theta_1\sqrt{3}} \exp\left\{ -\frac{\pi(x-\theta_0)}{\theta_1\sqrt{3}} \right\} \Big/ \left[1 + \exp\left\{ -\frac{\pi(x-\theta_0)}{\theta_1\sqrt{3}} \right\} \right]^2, \tag{8}$$

and parameters $\theta_0 = 0$, $\theta_1 = 1$.

Table 1. Optimal boundary points for $k = 9$

Optimality criterion	x_1	x_2	x_3	x_4	x_5	x_6	x_7	x_8
A-optimum	2.3758	1.6915	1.1047	0.4667	0.4667	1.1047	1.6915	2.3758
D-optimum	2.3188	1.6218	1.0223	0.3828	0.3828	1.0223	1.6218	2.3188
E-optimum	1.8638	1.1965	0.6805	0.2216	0.2216	0.6805	1.1965	1.8638
Optimal grouping	3.1616	2.0856	1.2676	0.4601	0.4601	1.2676	2.0856	3.1616

Moreover, in the case of the given competing hypothesis, we can use the so-called Neyman-Pearson classes [14], for which the random variable domain is partitioned into intervals of two types, according to the inequalities $f_0(t) < f_1(t)$ and $f_0(t) > f_1(t)$, where $f_0(t)$ and $f_1(t)$ are the density functions, corresponding to the competing hypotheses. For H_0 and H_1 from our example, we have the first-type intervals

$$(-\infty; 2.3747], (0.6828; 0.6828], (2.3747; \infty),$$

and the second-type intervals

$$(2.3747; 0.6828], (0.6828; 2.3747].$$

Figures 1 and 2 illustrate the power of the Pearson χ^2 test for the hypotheses H_0 and H_1 of our example in the case of different grouping methods, depending on the number of intervals ($\alpha = 0.1$, $n = 500$). The powers of the well-known nonparametric Kolmogorov, Cramer-von Mises-Smirnov and Anderson-Darling goodness-of-fit tests are given for the comparison.

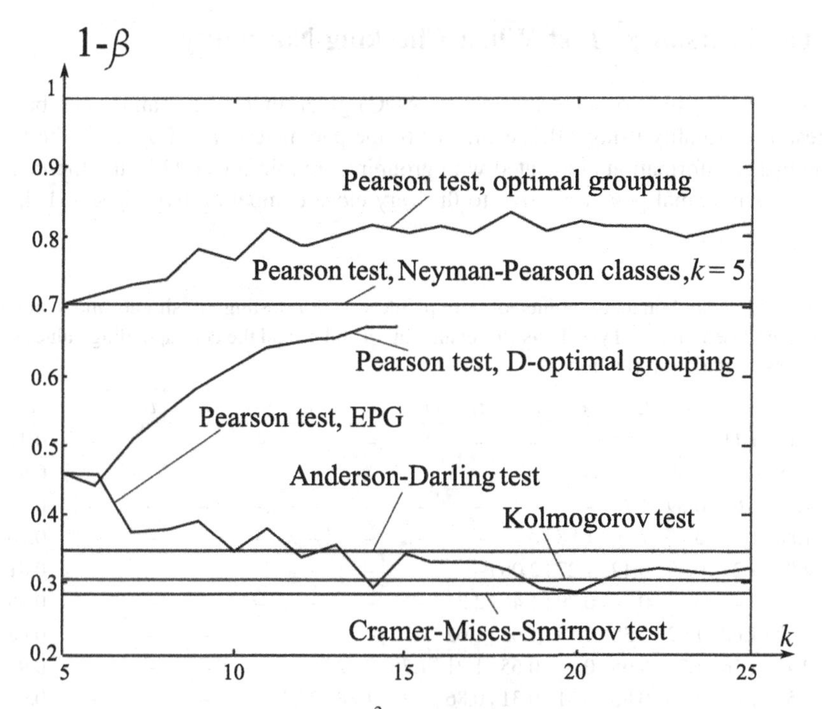

Fig. 1. The power of the χ^2 Pearson test for simple hypothesis

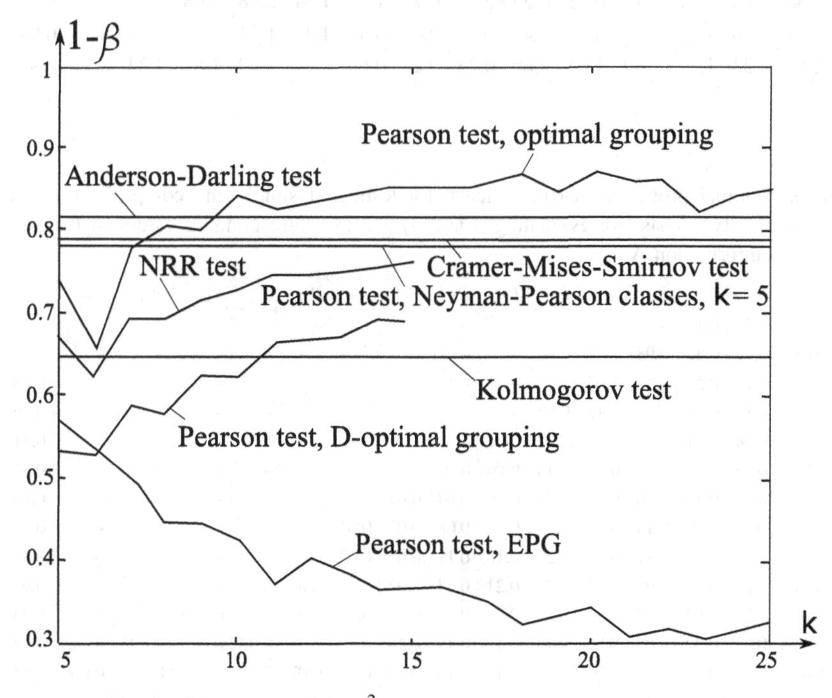

Fig. 2. The power of the χ^2 Pearson test for composite hypothesis

4 The Pearson χ^2 Test When Checking Normality

The asymptotically D-optimal groupings (AOG) given in Tables 2 and 3 can be used for testing normality using MLE estimates of the parameters μ and σ. Here, the losses in the Fisher information associated with grouping are minimized [13] and the Pearson χ^2 test has maximal power relative to the very close competing hypotheses [13].

Table 2. Optimal boundary points of group intervals for testing of simple and composite hypotheses based on χ^2 – Type Tests (for evaluating μ and σ) and the corresponding values of the relative asymptotic information A

k	t_1	t_2	t_3	t_4	t_5	t_6	t_7	t_8	t_9	t_{10}	t_{11}	t_{12}	t_{13}	t_{14}	A
3	1.11	1.11	–	–	–	–	–	–	–	–	–	–	–	–	0.41
4	1.38	0	1.38	–	–	–	–	–	–	–	–	–	–	–	0.55
5	1.7	0.69	0.69	1.7	–	–	–	–	–	–	–	–	–	–	0.68
6	1.88	1	0	1	1.88	–	–	–	–	–	–	–	–	–	0.76
7	2.06	1.27	0.49	0.49	1.27	2.06	–	–	–	–	–	–	–	–	0.81
8	2.2	1.46	0.79	0	0.79	1.46	2.2	–	–	–	–	–	–	–	0.85
9	2.32	1.62	1.02	0.38	0.38	1.02	1.62	2.32	–	–	–	–	–	–	0.88
10	2.42	1.76	1.21	0.65	0	0.65	1.21	1.76	2.42	–	–	–	–	–	0.90
11	2.52	1.88	1.36	0.86	0.31	0.31	0.86	1.36	1.88	2.52	–	–	–	–	0.91
12	2.6	1.9	1.49	1.03	0.53	0	0.53	1.03	1.49	1.9	2.6	–	–	–	0.93
13	2.68	2.08	1.61	1.18	0.75	0.27	0.27	0.75	1.18	1.61	2.08	2.68	–	–	0.94
14	2.74	2.16	1.71	1.3	0.91	0.48	0	0.48	0.91	1.3	1.71	2.16	2.74	–	0.943
15	2.81	2.24	1.8	1.42	1.04	0.66	0.23	0.23	0.66	1.04	1.42	1.8	2.24	2.81	0.95

Table 3. Optimal probabilities (frequencies) for testing of simple and composite hypotheses based on χ^2 – type tests (for evaluating μ and σ) and the corresponding values of the relative asymptotic information A

k	P_1	P_2	P_3	P_4	P_5	P_6	P_7	P_8	P_9	P_{10}	P_{11}	P_{12}	P_{13}	P_{14}	P_{15}	A
3	0.13	0.73	0.13	–	–	–	–	–	–	–	–	–	–	–	–	0.41
4	0.08	0.42	0.42	0.08	–	–	–	–	–	–	–	–	–	–	–	0.55
5	0.05	0.20	0.51	0.20	0.05	–	–	–	–	–	–	–	–	–	–	0.68
6	0.03	0.13	0.34	0.34	0.13	0.03	–	–	–	–	–	–	–	–	–	0.76
7	0.02	0.08	0.21	0.38	0.21	0.08	0.02	–	–	–	–	–	–	–	–	0.81
8	0.01	0.06	0.14	0.28	0.28	0.14	0.06	0.01	–	–	–	–	–	–	–	0.85
9	0.01	0.04	0.10	0.20	0.30	0.20	0.10	0.04	0.01	–	–	–	–	–	–	0.88
10	0.01	0.03	0.08	0.14	0.24	0.24	0.14	0.08	0.03	0.01	–	–	–	–	–	0.90
11	0.01	0.02	0.06	0.11	0.18	0.25	0.18	0.11	0.06	0.02	0.01	–	–	–	–	0.91
12	0.01	0.02	0.04	0.08	0.14	0.21	0.21	0.14	0.08	0.04	0.02	0.01	–	–	–	0.93
13	0.004	0.02	0.04	0.07	0.11	0.17	0.21	0.17	0.11	0.07	0.04	0.02	0.004	–	–	0.94
14	0.003	0.01	0.03	0.05	0.09	0.13	0.19	0.19	0.13	0.09	0.05	0.03	0.01	0.003	–	0.94
15	0.003	0.01	0.02	0.04	0.07	0.11	0.15	0.18	0.15	0.11	0.07	0.04	0.02	0.01	0.003	0.95

In Table 2, the boundary points t_i, $i = 1, \ldots, k-1$ are listed in a form that is invariant with respect to the parameters μ and σ for a normal distribution. For calculating the statistic (1), the boundaries x_i separating the intervals for specified k are found using the values of t_i taken from the corresponding row of the table: $x_i = \hat{\sigma}t_i + \hat{\mu}_i$, where $\hat{\mu}$ and $\hat{\sigma}$ are the MLE of the parameters derived from the given sample. Then, the number of observations n_i within each interval are used. The probabilities of falling into a given interval for evaluating the statistic (1) are taken from the corresponding row of Table 3.

Table 4. Percentage points $\tilde{\chi}^2_{k,\alpha}$ for the Pearson test statistic when evaluating the parameters μ and σ

k	$p = 1 - \alpha$					Limiting distribution model
	0.85	0.9	0.95	0.975	0.99	
4	2.74	3.37	4.48	5.66	7.26	BIII (1:2463; 3:8690; 4:6352; 19:20; 0:005)
5	4.18	5	6.39	7.77	9.59	BIII (1:7377; 3:8338; 5:5721; 26:00; 0:005)
6	5.61	6.54	8.09	9.61	11.62	BIII (2:1007; 4:1518; 4:1369; 26:00; 0:005)
7	6.95	7.98	9.67	11.31	13.43	BIII (2:5019; 4:6186; 3:4966; 28:00; 0:005)
8	8.28	9.4	11.21	12.95	15.22	BIII (2:9487; 5:8348; 3:1706; 34:50; 0:005)
9	9.56	10.76	12.69	14.53	16.87	BIII (3:5145; 6:3582; 3:2450; 39:00; 0:005)
10	10.84	12.11	14.16	16.12	18.58	BIII (3:9756; 6:7972; 3:0692; 41:50; 0:005)
11	12.08	13.42	15.55	17.59	20.19	BIII (4:4971; 6:9597; 3:0145; 43:00; 0:005)
12	13.34	14.74	16.98	19.1	21.77	BIII (5:1055; 7:0049; 3:1130; 45:00; 0:005)
13	14.56	16.01	18.34	20.53	23.3	BIII (5:7809; 7:0217; 3:2658; 47:00; 0:005)
14	15.78	17.29	19.68	21.96	24.81	BIII (6:6673; 6:9116; 3:5932; 49:00; 0:005)
15	16.98	18.54	21.04	23.4	26.37	BIII (7:0919; 7:2961; 3:4314:51:50; 0:005)

When AOG is used in the Pearson χ^2 test, the resulting percentage points $\tilde{\chi}^2_{k,\alpha}$ of the distributions of the statistic (1) and the models of limiting distributions constructed in this paper are shown in Table 4, where $\beta_{\mathrm{III}}(\theta_0, \theta_1, \theta_2, \theta_3, \theta_4)$ is the type III beta distribution with these parameters and the density

$$f(x) = \frac{\theta_2^{\theta_0}}{\theta_3 \beta(\theta_0, \theta_1)} \frac{[(x-\theta_4)/\theta_3]^{\theta_0 - 1}[1-(x-\theta_4)/\theta_3]^{\theta_1 - 1}}{[1+(\theta_2 - 1)(x-\theta_4)/\theta_3]^{\theta_0+\theta_1}}.$$

To make a decision regarding testing the hypothesis H_0, the value of the statistic X_n^{2*} is compared with the critical value $\tilde{\chi}^2_{k,\alpha}$ from the corresponding row of Table 4, or the attained level of significance $P\{X_n^2 > X_n^{2*}\}$, determined using the limiting distribution model in the same row of the table, is compared with a specified level of significance α.

The difference between the real distributions $G(X_n^2|H_0)$ of statistic (1) and the corresponding χ^2_{k-m-1} distributions, when hypothesis H_0 is true, is shown in Fig. 3.

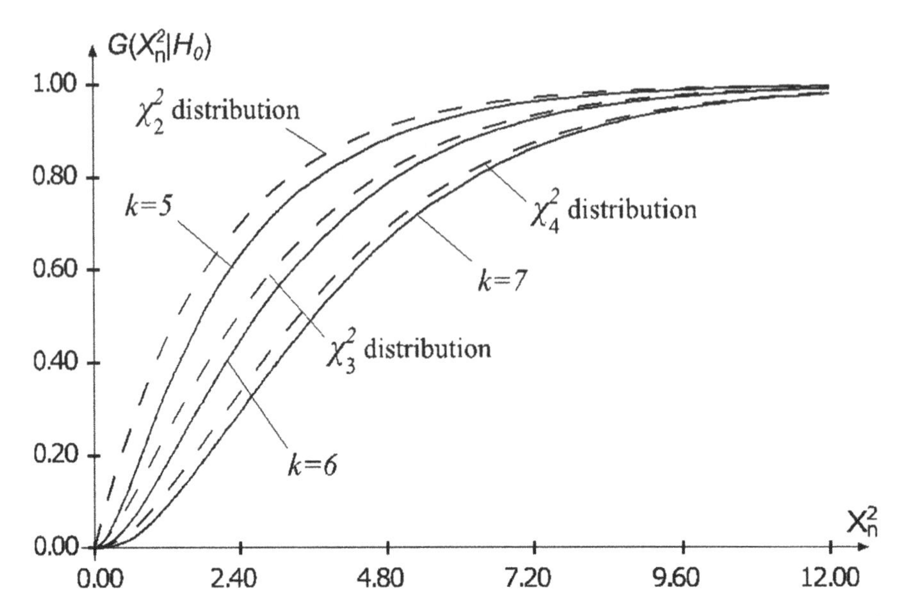

Fig. 3. Distributions of statistic (1) for maximum likelihood estimates of the parameters of a normal distribution based on non-grouped data together with the corresponding χ^2_{k-m-1} distributions

Tables 2 and 3 give Fisher asymptotic information:

$$A = \det \mathbf{J}_\Gamma \,/\, \det \mathbf{J}.$$

For tests of normality with calculations of an MLE based on the non-grouped sample, only the parameters μ or σ, the required AOG tables, percentage points, and the limiting distribution models can be found in [15].

For AOG relative to the parameter vector and $k = 15$ intervals in the grouped sample, about 95 % of the information is preserved. Further increases in the number k of intervals are insignificant; it should be chosen based on the following considerations. For an optimal grouping, the probabilities of falling into an interval are not generally equal (usually these probabilities are minimal for the outermost intervals), so that k should be chosen on the basis of the condition $nP_i(\theta) \geq 5 \ldots 10$ for any interval. At least, in choosing k the recommendation

$$\min_i \left\{ nP_i(\theta) \middle| i = \overline{1,k} \right\} > 1$$

should be followed. When this condition holds, in the case where the tested hypothesis H_0 is valid, a discrete distribution of the statistic in (1) differs insignificantly from the corresponding asymptotic limiting distribution. If this condition is violated, then the difference between the true distribution of the statistic and the limiting distribution will lead to an increase in the probability of a type I error relative to the specified significance level α. It should also be noted that for small sample sizes, n = 10–20, discrete distributions of the statistics differ substantially from the asymptotic distributions. This condition on the choice of k sets an upper bound estimate on the number of intervals ($k \leq k_{max}$). The number of grouping intervals affects the power of the Pearson χ^2 test [6]. It is absolutely unnecessary that its power against a competing distribution (hypothesis) should be maximal for $k = k_{max}$.

In order to compare the power of the Pearson χ^2 test for checking normality with the power of special normality tests and nonparametric goodness-of-fit tests, the power has been estimated relative to the same competing distributions (hypotheses) as in [15].

The test hypothesis H_0 is taken to be that the observed sample obeys the normal distribution (7).

As competing hypotheses for studying the power of the χ^2 test, we have considered adherence of the analyzed sample to the following distributions: competing hypothesis H_1 corresponds to a generalized normal distribution (family of distributions) with the density

$$f(x) = \frac{\theta_2}{2\theta_1\Gamma(1/\theta_2)}\exp\left\{-\left(\frac{|x-\theta_0|}{\theta_1}\right)^{\theta_2}\right\}$$

and a shape parameter $\theta_2 = 4$; hypothesis H_2 is the Laplace distribution with the density

$$f(x) = \frac{1}{2\theta_1}\exp\left\{-\frac{|x-\theta_0|}{\theta_1}\right\};$$

and hypothesis H_3 is the logistic distribution with the density (8), which is very close to a normal distribution. Figure 4 shows the densities of the distributions corresponding to hypotheses H_1, H_2 and H_3 with scale parameters such that they are the closest to the

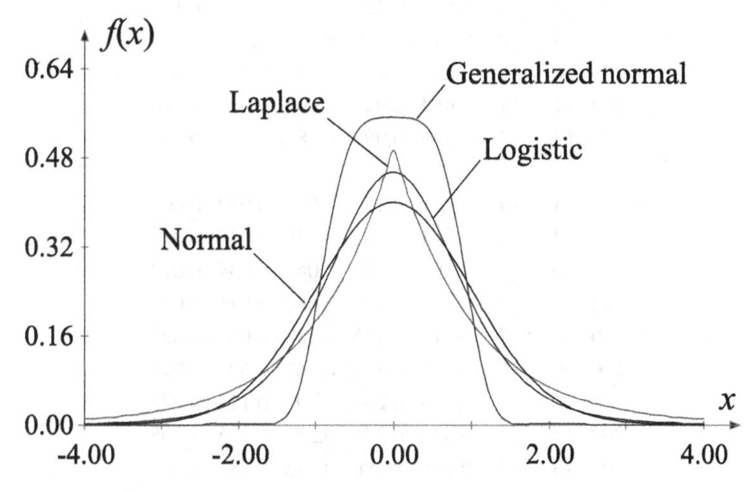

Fig. 4. Probability density functions corresponding to the considered hypotheses H_i

standard normal law. This choice of hypotheses has a certain justification. Hypothesis H_2, corresponding to a Laplace distribution, is the most distant from H_0. Distinguishing them usually presents no problem. The logistic distribution (hypothesis H_3) is very close to normal and it is generally difficult to distinguish them by goodness-of-fit tests.

The competing hypothesis H_1, which corresponds to a generalized normal distribution with a shape factor $\theta_2 = 4$, is a "litmus test" for detection of hidden deficiencies in some tests [15–17]. It turned out that for small sample sizes n and small specified

Table 5. Power of the Pearson χ^2 test with respect to hypotheses H_1, H_2 and H_3

n	k_{max}	k_{opt}	α				
			0.15	0.1	0.05	0.025	0.01
H_1							
10	4	4	0.235	0.146	0.043	0.032	0.002
20	4	5	0.262	0.177	0.1	0.058	0.021
30	5	5	0.312	0.216	0.136	0.079	0.043
40	6	5	0.336	0.267	0.168	0.111	0.061
50	6	5	0.401	0.311	0.204	0.129	0.068
100	9	5	0.558	0.479	0.352	0.254	0.158
150	10	7	0.722	0.634	0.486	0.353	0.217
200	11	9	0.783	0.695	0.548	0.417	0.279
300	13	11	0.907	0.858	0.756	0.646	0.492
H_2							
10	4	4	0,267	0,206	0,074	0,058	0,01
20	4	4	0,264	0,177	0,104	0,067	0,037
		5	0,247	0,189	0,116	0,061	0,024
30	5	5	0,312	0,261	0,153	0,103	0,044
40	6	7	0,443	0,358	0,25	0,167	0,101
50	6	7	0,5	0,423	0,312	0,225	0,138
100	9	9	0,77	0,708	0,596	0,494	0,379
150	10	9	0,899	0,86	0,785	0,705	0,596
200	11	11	0,964	0,946	0,908	0,88	0,786
300	13	13	0,996	0,993	0,985	0,974	0,95
H_3							
10	4	4	0.221	0.15	0.046	0.034	0.003
20	4	5	0.194	0.125	0.059	0.038	0.016
30	5	6	0.169	0.125	0.062	0.034	0.012
40	6	7	0.204	0.143	0.082	0.045	0.02
50	6	7	0.214	0.155	0.088	0.05	0.023
100	9	10	0.303	0.231	0.146	0.09	0.047
150	10	10	0.359	0.284	0.191	0.124	0.072
200	11	11	0.432	0.355	0.25	0.175	0.105
300	13	13	0.566	0.486	0.373	0.28	0.19

probabilities α of type I error, a number of tests employed for testing goodness-of-fit to normal are not able to distinguish close distributions from normal. In these cases, the power $1 - \beta$ with respect to hypothesis H_1, where β is the probability of a type II error, is smaller than α. This means that the distribution corresponding to H_1 is "more normal" than the normal law and indicates that the tests are biased.

The power of the Pearson χ^2 test was studied with different number of intervals $k \leq k_{\max}$ for specified sample sizes n. Table 5 lists the maximum powers of the χ^2 test relative to the competing hypotheses H_1, H_2 and H_3, and corresponding to the optimal number k_{opt} of grouping intervals. To a certain extent, it is possible to orient oneself in choosing k on the basis of the values of k_{opt} as a function of n listed in Table 5.

5 The Nikulin-Rao-Robson Goodness-of-Fit Test

A modification of the standard statistic X_n^2 was proposed [3–5] in which the limiting distribution of the modified statistic is a χ_{k-1}^2 distribution (the number of degrees of freedom is independent of the number of parameters to be estimated). The unknown parameters of the distribution $F(x, \theta)$ have, in this case, must be estimated on the basis of the non-grouped data by a maximum likelihood method. Here the vector $\mathbf{P} = (P_1, \ldots, P_k)^\tau$ is assumed to be specified, while the boundary points of the intervals are defined using the relations $x_i(\theta) = F^{-1}(P_1 + \ldots + P_i)$, $i = 1, \ldots, k-1$. The proposed statistic has the form [4]:

$$Y_n^2(\theta) = X_n^2 + n^{-1} a^\tau(\theta) \boldsymbol{\lambda}(\theta) a(\theta), \tag{9}$$

where X_n^2 is calculated using (1). For distribution laws that are determined only by shift and scale parameters,

$$\boldsymbol{\lambda}(\theta) = \left[\mathbf{J}(\theta) - \mathbf{J}_g(\theta)\right]^{-1}.$$

In the case of the normal distribution with parameter vector $\theta^\tau = (\mu, \sigma)$, the Fisher information matrix has the form:

$$\mathbf{J}(\theta) = \begin{bmatrix} 1/\sigma^2 & 0 \\ 0 & 2/\sigma^2 \end{bmatrix},$$

with the elements of the information matrix based on grouped data $\mathbf{J}_g(\theta)$ given by

$$J_g(\mu, \mu) = \sum_{i=1}^{k} \frac{1}{\sigma^2 P_i(\theta)} (f(t_{i-1}) - f(t_i))^2,$$

$$J_g(\sigma, \sigma) = \sum_{i=1}^{k} \frac{1}{\sigma^2 P_i(\theta)} (t_{i-1} f(t_{i-1}) - t_i f(t_i))^2,$$

$$J_g(\mu, \sigma) = J_\gamma(\sigma, \mu) = \sum_{i=1}^{k} \frac{1}{\sigma^2 P_i(\theta)} (f(t_{i-1}) - f(t_i))(t_{i-1} f(t_{i-1}) - t_i f(t_i)),$$

where

$$t_i = (x_i - \mu)/\sigma, t_0 = -\infty, t_k = \infty, f(t) = \frac{1}{\sqrt{2\pi}} e^{-t^2/2}$$

is the standard normal distribution. The elements of the vector $a^{\tau}(\theta) = [a(\mu), a(\sigma)]$ are given by

Table 6. Power of the Nikulin-Rao-Robson set with respect to hypotheses H_1, H_2 and H_3

n	$k_{\max}$	k_{opt}	α				
			0.15	0.1	0.05	0.025	0.01
H_1							
10	4	4	0.348	0.1	0.029	0.009	0.006
20	4	5	0.234	0.143	0.074	0.041	0.016
30	5	5	0.256	0.197	0.102	0.053	0.023
40	6	5	0.293	0.221	0.123	0.079	0.035
50	6	5	0.326	0.240	0.148	0.083	0.040
100	9	5	0.485	0.395	0.271	0.179	0.102
150	10	6	0.619	0.530	0.397	0.284	0.179
		7	0.641	0.539	0.383	0.261	0.148
200	11	9	0.713	0.616	0.464	0.339	0.214
300	13	11	0.872	0.810	0.695	0.573	0.420
H_2							
10	4	4	0.368	0.103	0.055	0.031	0.007
20	4	5	0.250	0.210	0.126	0.065	0.039
30	5	5	0.349	0.265	0.185	0.127	0.078
40	6	5	0.474	0.403	0.297	0.218	0.149
50	6	5	0.548	0.473	0.365	0.281	0.190
100	9	5	0.807	0.755	0.667	0.583	0.482
150	10	7	0.919	0.889	0.834	0.774	0.691
200	11	9	0.973	0.961	0.933	0.900	0.849
300	13	11	0.997	0.995	0.990	0.983	0.968
		13	0.997	0.995	0.990	0.983	0.968
H_3							
10	4	4	0.321	0.083	0.034	0.014	0.005
20	4	5	0.166	0.120	0.065	0.030	0.014
30	5	6	0.198	0.138	0.080	0.047	0.024
40	6	7	0.232	0.173	0.104	0.063	0.034
50	6	7	0.251	0.188	0.117	0.074	0.040
100	9	10	0.360	0.290	0.202	0.141	0.091
150	10	10	0.432	0.358	0.263	0.195	0.131
200	11	11	0.509	0.436	0.337	0.259	0.183
300	13	13	0.641	0.572	0.469	0.381	0.288

$$a(\mu) = \sum_{i=1}^{k} \frac{n_i(f(t_{i-1}) - f(t_i))}{\sigma P_i(\theta)},$$

$$a(\sigma) = \sum_{i=1}^{k} \frac{n_i}{\sigma P_i(\theta)} (t_{i-1} f(t_{i-1}) - t_i f(t_i)).$$

As in the case of the Pearson test, when testing for normality with MLE estimation of the parameters μ and σ based on the non-grouped data, Tables 2 and 3 can be used.

For calculating statistic (9), the boundaries separating the intervals for given k are found from the values of t_i in the corresponding row of Table 2 using the formula $x_i = \hat{\sigma} t_i + \hat{\mu}$, where $\hat{\mu}$ and $\hat{\sigma}$ are the MLE parameters found from the sample data. Then the number of observations n_i in each interval is counted. The probabilities $P_i(\theta)$ of falling into an interval when calculating statistic (9) are taken from the corresponding line of Table 3. The elements of the vector $\alpha(\theta)$ and matrix $\Lambda(\theta)$ are calculated using the tabulated data for t_i, P_i and the resulting estimates of $\hat{\sigma}$.

To decide on the test results for hypothesis H_0, the value of the statistic Y_n^{2*} is compared with the corresponding critical $\chi^2_{k-1,\alpha}$ or the achieved level of significance (p-value) $P\{Y_n^2 > Y_n^{2*}\}$ is found from the corresponding χ^2_{k-1} distribution.

To test for normality with MLE calculation of the parameters μ or σ separately on the basis of non-grouped samples, the required tables of AOG can be found in [15].

Estimates of the power of the Nikulin-Rao-Robson test for the competing hypotheses H_1, H_2 and H_3 for k_{opt} are given in Table 6. This test is generally more powerful than the Pearson test (for example, see its powers relative to the competing hypotheses H_2 and H_3). Here we often have $k_{opt} = k_{\max}$ for

$$\min_i \{nP_i(\theta)\} > 1,$$

However, this is not always so. In terms of its power relative to the "tricky" hypothesis H_1 it is inferior to the Pearson test, and k_{opt} in this case is considerably smaller than $k_{\max}$ with AOG.

6 Conclusion

The power of the Pearson χ^2 test and Nikulin-Rao-Robson test can be maximized by the optimal selection of the number of intervals and interval boundary points.

Combining the obtained results of the power analysis for the Pearson χ^2 test and Nikulin-Rao-Robson test with the results presented in [15–17], we can see that in regard to the competing hypothesis H_1, the Pearson χ^2 test shows very good results, yielding in power only to some special normality tests.

At the same time, in regard to competing hypotheses H_2 and H_3, the Pearson χ^2 test and Nikulin-Rao-Robson test inferior in power to most special normality tests and to nonparametric goodness-of-fit tests (Anderson-Darling, Cramer-Mises-Smirnov, Watson, Kuiper, Zhang, Kolmogorov tests).

This work is supported by the Russian Ministry of Education and Science (project 2.541.2014 K).

References

1. Chernoff, H., Lehmann, E.L.: The use of maximum likelihood estimates in χ^2 test for goodness of fit. Ann. Math. Stat. **25**(3), 579–586 (1954)
2. Lemeshko, B.Y., Postovalov, S.N.: Limit distributions of the Pearson chi 2 and likelihood ratio statistics and their dependence on the mode of data grouping. Ind. Lab. **64**(5), 344–351 (1998)
3. Nikulin, M.S.: χ^2 tests for continuous distributions with shift and scale parameters. Teor. Veroyatn. Primen. **XVIII**(3), 583–591 (1973)
4. Nikulin, M.S.: On χ^2 tests for continuous distributions. Teor. Veroyatn. Primen. **XVIII**(3), 675–676 (1973)
5. Rao, K.C., Robson, D.S.: A chi-squared statistic for goodness-of-fit tests within the exponential family. Commun. Stat. **3**(1), 1139–1153 (1974)
6. Lemeshko, B.Y., Chimitova, E.V.: On the choice of the number of intervals in χ^2-type goodness-of-fi t tests. Zavod. Lab. Diagn. Mater. **69**(1), 61–67 (2003)
7. Lemeshko, BYu.: Asymptotically optimum grouping of observations in goodness-of-fit tests. Ind. Lab. **64**(1), 59–67 (1998)
8. Voinov, V., Pya, N., Alloyarova, R.: A comparative study of some modified chi-squared tests. Commun. Stat. Simul. Comput. **38**(2), 355–367 (2009)
9. Denisov, V., Lemeshko, B.: Optimal grouping in estimation and tests of goodness-of-fit hypotheses. Wissenschaftliche Schriftenreihe der Technishen universitat Karl-Marx-Stadt. **10**, 63–81 (1989)
10. Lemeshko, B., Postovalov, S., Chimitova, E.: On statistic distributions and the power of the Nikulin χ^2 test. Ind. Lab. **67**(3), 52–58 (2001)
11. Lemeshko, B., Chimitova, E.: Maximization of the power of $\chi 2$ tests. Papers of Siberian Branch of Academy of Science of Higher Education, vol. 2, pp. 53–61 (2000)
12. Denisov, V., Lemeshko, B.: Optimal grouping in the analysis of experimental data. In: Measuring Information Systems, Novosibirsk, pp. 5–14 (1979). [In Russian]
13. Lemeshko, B.Y., Lemeshko, S.B., Postovalov, S.N., Chimitova, E.V.: Statistical Data Analysis, Simulation and Study of Probability Regularities. Computer Approach, 888 pp. NSTU Publisher, Novosibirsk (2011). [In Russian]
14. Greenwood, P.E., Nikulin, M.S.: A Guide to Chi-Squared Testing. John Wiley & Sons Inc., New York (1996)
15. Lemeshko, B.Y.: Tests for checking the deviation from normal distribution law. In: Guide on the Application. INFRA-M, Moskow (2015). doi:10.12737/6086
16. Lemeshko, B.Y., Lemeshko, S.B.: Comparative analysis of tests for verifying deviation of a distribution from normal. Metrologiya **2**, 3–24 (2005). [In Russian]
17. Lemeshko, B.Y., Rogozhnikov, A.P.: Features and power of some tests of normality. Metrologiya **4**, 3–24 (2009). [In Russian]

Homogeneity Tests for Interval Data

Stanislav S. Vozhov and Ekaterina V. Chimitova(✉)

Novosibirsk State Technical University, Novosibirsk, Russia
chimitova@corp.nstu.ru

Abstract. In many practical situations, we only know the upper bound Δ of the measurement error. It means that the precise measurement is located on the interval $(x - \Delta, x + \Delta)$. In other words, the data can be represented as a sample of interval observations. When performing statistical tests, ignoring this uncertainty in data may lead to unreliable decisions. For interval data, standard nonparametric and semiparametric methodologies include various modifications of the logrank test for comparing distribution functions. The statistics of the logrank homogeneity tests are based on comparing the nonparametric maximum likelihood estimates (NPMLE) of the distribution functions. In this paper, NPMLE is calculated by the ICM-algorithm (iterative convex minorant algorithm). The purpose of this paper is to investigate some homogeneity tests for interval data and to carry out the comparative analysis in terms of the power of tests for close competing hypotheses.

Keywords: Interval data · Generalized logrank test · ICM algorithm

1 Introduction

The basis of interval data analysis was initially laid in metrology, where an interval uncertainty is introduced naturally. However, interval data arise in various spheres. For example, in marketing research, observations obtained from a survey of the target group of consumers are usually interval. In reliability and survival analysis, lifetime data are often interval-censored. Observation interval length may be different. For example, a concept of progressing measurement error (an unpredictable error, slowly changing in time) is introduced in metrology. In sociological surveys, where it is necessary to specify the range of values of a certain numerical characteristic, each respondent can give his own range of values.

In this paper, we assume that every observation is a value measured by an instrument with an error. Thus, if the precise value of observed response is $\dot{x}$, measurement error is $e \in [-\Delta, \Delta]$, then the value $x = \dot{x} + e$ is observed. In this case, we deal with a usual complete sample $X_n = \{X_1, \ldots, X_n\}$. In fact, during the statistical analysis on the basis of complete samples, the value of the measurement error Δ is not taken into account. However, there is another approach, based on presenting observations in the form of intervals $(x - \Delta, x + \Delta) = (L, R)$. In this case, we obtain a sample of interval observations in the form $I_n = \{(L_1, R_1), \ldots, (L_n, R_n)\}$.

Interval observations are considered in many publications, see for example [1–7]. In these papers, the authors argued the reasonability of constructing new mathematical

R. Szewczyk and M. Kaliczyńska (eds.), *Recent Advances in Systems, Control and Information Technology*, Advances in Intelligent Systems and Computing 543, DOI 10.1007/978-3-319-48923-0_83

and statistical models and methods, according to which observations are not numbers, but intervals.

In this paper, we investigate the distributions of statistics and the power of tests of homogeneity of distributions for interval data. Different generalizations of the logrank homogeneity test for interval data are considered.

2 Homogeneity Tests

Let we have k independent interval samples of the form:

$$\mathrm{I}_{n_1}^1 = \left\{ \left(L_1^1, R_1^1\right), \ldots, \left(L_{n_1}^1, R_{n_1}^1\right) \right\}$$

$$, \ldots ,$$

$$\mathrm{I}_{n_k}^k = \left\{ \left(L_1^k, R_1^k\right), \ldots, \left(L_{n_k}^k, R_{n_k}^k\right) \right\}$$

The number of all observation is $n = \sum_{l=1}^{k} n_l$.

For testing the fact that all these samples belong to one distribution, it is necessary to test the hypothesis H_0: $F_1(x) = F_2(x) = \ldots + F_k(x)$, where $F_i(x)$ is the distribution function of i-th sample.

This hypothesis can be tested with homogeneity tests proposed in [11–14].

2.1 Generalized Logrank Test II (gLRT2)

In this test, k samples are combined into one common sample of size n. Let vector $e = e_1, \ldots, e_n$ indicates that observation (L_i, R_i) belongs to the l-th sample, $l = \overline{1, k}$. The test statistic has the form [11, 14]

$$S_2 = \frac{1}{n} U_{II}^{*\prime} V_{II}^{*-1} U_{II}^{*}, \tag{1}$$

where U_{II}^* is the vector of first $k-1$ components of vector $U_{II} = \sum_{i=1}^{n} e_i K_n(L_i, R_i)$, and V_{II}^* is the matrix that is derived by deleting the last row and column of matrix

$$V_{II} = \left(\begin{cases} \frac{n_l(n-n_l)}{n^2} \frac{1}{n} \sum_{i=1}^{n} K_n^2(L_i, R_i), \text{ if } l = r \\ \frac{-n_l n_r}{n^2} \frac{1}{n} \sum_{i=1}^{n} K_n^2(L_i, R_i), \text{ else} \end{cases} \right)_{k \times k}, \tag{2}$$

$$K_i(L_i, R_i) = \frac{\eta\left(\hat{F}(R_i)\right) - \eta\left(\hat{F}(L_i)\right)}{\hat{F}(R_i) - \hat{F}(L_i)}. \tag{3}$$

In this paper, we consider $\eta(u) = 1 - (1-u)\log(1-u)(1-u)^b x^c$, where b and c are some numbers from the interval [0, 1].

Statistic S_2 has the limiting χ^2 distribution with $k-1$ degrees of freedom.

2.2 Generalized Logrank Test III (gLRT3)

This class of generalized logrank tests was proposed by Zhao et al. in [12, 14].

Using the idea behind the gLRT2 test, they consider interval-censored data that may have both censored and exact observations (when we have a point instead of the interval). Let e_i be the indicator for an exact observation such that $e_i = 1$ if $L_i = R_i$ and 0 otherwise. For group l, let n_{l1} and n_{l2} be the numbers of exactly observed and interval-censored observations, correspondingly. Moreover, let $N_1 = \sum_{l=1}^{k} n_{l1}$ denotes the total number of exact observations and $N_2 = \sum_{l=1}^{k} n_{l2}$ is the total number of interval observations. The test statistic is given by

$$U_{III} = Diag\left(\frac{N_1}{n_{l1}}\right) \sum_{i=1}^{n} e_i K_i(L_i, R_i) + Diag\left(\frac{N_2}{n_{l2}}\right) \sum_{i=1}^{n} (1-e_i) K_i(L_i, R_i),\ l = \overline{1,k}, \tag{4}$$

where $K_i(L_i, R_i)$ was defined earlier in gLRT2. The covariance matrix can be consistently estimated by $V_{III} = \left(v^*_{lr}\right)_{k\times k}$, where

$$v^*_{lr} = \begin{cases} \frac{N_1}{n}\left(\frac{N_1}{n_{l1}}\right)P_{N_1} + \frac{N_2}{n}\left(\frac{N_1}{n_{l2}} - 1\right)Q_{N_2}, & \text{if } l = r, \\ -\frac{N_1}{n}P_{N_1} - \frac{N_2}{n}Q_{N_2}, & \text{otherwise.} \end{cases}, \tag{5}$$

$$Q_{N_2} = \frac{1}{N_2}\sum_{i=1}^{n}(1-e_i)\left(K_i(L_i, R_i)\right)^2,$$

$$P_{N_1} = \frac{1}{N_1}\sum_{i=1}^{n} e_i \begin{pmatrix} -\frac{d}{dx}\eta(x)|_{x=F(R_i)}, \text{if } F(L_i) = F(R_i), \\ K_i(L_i, R_i), \text{ otherwise.} \end{pmatrix}^2.$$

When there are no exact observations, U_{III} reduces to U_{II} in gLRT2. Otherwise, H_0 can be tested using the test statistic $S_3 = U^T_{III} V^{-1}_{III} U_{III}/n$, which asymptotically has the χ^2 distribution with k degrees of freedom.

2.3 Generalized Logrank Test IV (gLRT4)

The test statistic is defined as follows [13]:

$$S_4 = U_{n_1}^2 / U_{n_2}^2, \tag{6}$$

where

$$U_{IV} = (U_{n_1}, U_{n_2})^T = \left(\frac{1}{\sqrt{n_1}} \sum_{i \in S_1} K_{n_1,n_2}(L_i, R_i), \frac{1}{\sqrt{n_2}} \sum_{i \in S_2} K_{n_2,n_1}(L_i, R_i) \right), \tag{7}$$

$$K_{i,j}(L, R) = \frac{\eta(\hat{F}_i(R)) - \eta(\hat{F}_i(L))}{\hat{F}_j(R) - \hat{F}_j(L)}.$$

The distribution of S_4 can be approximated by the $F(1, 1)$ distribution. To implement the test procedure, one needs to determine $\hat{F}_{n_1}$ and $\hat{F}_{n_2}$ and to select the function η.

The nonparametric estimate of the distribution function by the interval data can be calculated using the ICM-algorithm.

3 ICM Algorithm

Let $X_1, X_2, \ldots, X_n$ is a sample of independent identically distributed random variables from $F(x)$. However, there are situations when exact values X_i, $i = \overline{1,n}$ are unknown, but it is known that they belong to the intervals (L_i, R_i), $i = \overline{1,n}$. Then, the original sample can be expressed as:

$$\mathrm{X}_n = \{(L_1, R_1), (L_2, R_2), \ldots, (L_n, R_n)\}.$$

The idea of building a nonparametric estimation of the distribution function from interval data is based on the maximization of log-likelihood function

$$\ln L(\mathrm{X}_n) = \sum_{i=1}^{n} \ln(F(R_i) - F(L_i))$$

by the values of the distribution function at the boundary points of observation intervals, subject to the conditions of monotonicity of the distribution function. However, the solution of the optimization problem by penalty functions requires large computational resources. Instead, it is advisable to use special algorithms, such as algorithm ICM.

For the estimation of the distribution function $F(x)$, consider a partition $0 < \tau_0 < \tau_1 < \ldots < \tau_m$, consisting of all non-repeating ordered boundary points L_i and R_i, $i = \overline{1,n}$.

The ICM-algorithm reduced the problem of maximizing the likelihood function to the problem of sequential computation of isotonic regression [8–10]. The distribution function is maximal at the points of left derivative of convex minorants, determined by the points

$$P_j = \left(G_j^{(k)}, V_j^{(k)}\right),$$

for $P_0 = (0,0)$. Minorant is a function whose value is not greater than the corresponding values of the function.

In ICM-algorithm, for each $j = \overline{1,m}$ and $i = \overline{1,n}$ the weights α_{ji} establish the correspondence between the elements of the sample (L_i, R_i) and elements of vector τ_j.

The iterative process of ICM-algorithm for an arbitrary step of algorithm $k \geq 1$:

1. For each $j = \overline{1,m}$ find the point $\left(G_j^{(k)}, V_j^{(k)}\right)$ in accordance with the following expressions:

$$G_j^{(k)} = G_{j-1}^{(k)} + \sum_{i=1}^{n} \frac{1}{\left|\alpha_{ji}\right| \cdot \left(\hat{F}^{(k)}(\tau_j) - \hat{F}^{(k)}(\tau_{j-1})\right)^2},$$

$$W_j^{(k)} = W_{j-1}^{(k)} + \sum_{i=1}^{n} \frac{1}{\alpha_{ji} \cdot \left(\hat{F}^{(k)}(\tau_j) - \hat{F}^{(k)}(\tau_{j-1})\right)},$$

$$D_j^{(k)} = D_{j-1}^{(k)} + \sum_{i=1}^{n} \frac{\hat{F}^{(k)}(\tau_j)}{\left|\alpha_{ji}\right| \cdot \left(\hat{F}^{(k)}(\tau_j) - \hat{F}^{(k)}(\tau_{j-1})\right)^2},$$

$$V_j^{(k)} = W_j^{(k)} + D_j^{(k)}.$$

2. Set $l = 0$.
3. The estimate of distribution function is equal to the left derivative of convex minorants:

$$\hat{F}^{(k)}(\tau_j) = \min_{l+1 \leq s \leq m} \left(\frac{V_s^{(k)} - V_l^{(k)}}{G_s^{(k)} - G_l^{(k)}}\right),$$

   where $j = \overline{l+1, s}$, s is the index of the corner points of the convex minorants.
4. Change $l = s$. If $l < m$ go to item 3.

The algorithm is repeated, until

$$\left|\hat{F}^{(k)}(\tau_j) - \hat{F}^{(k-1)}(\tau_j)\right| \leq 10^{-7}$$

is satisfied for all $j = \overline{1,m}$. The estimate of distribution function by the ICM algorithm is strictly consistent [5, 10].

Figure 1 illustrates the dependence of the mean distance between nonparametric estimate and the true distribution function $D_n = \sup_{|x|<\infty} \left|\hat{F}_n(x) - F(x)\right|$ on n for different lengths of observation intervals, equal to 1%, 3%, 5% or 10% from the length of the interval $(0, F^{-1}(0.99))$, where $F^{-1}(\cdot)$ is the inverse distribution function.

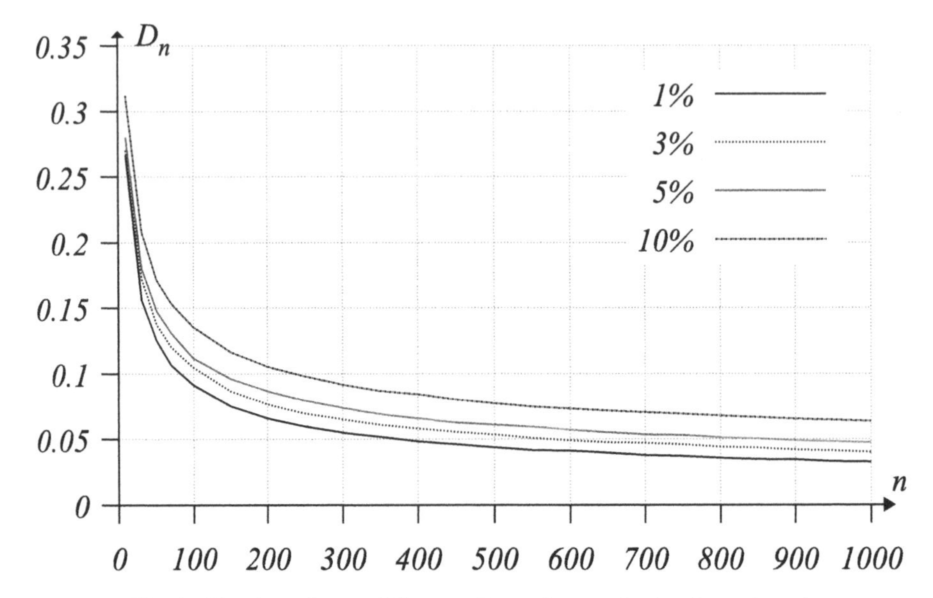

Fig. 1. The dependence of distance D_n on the sample size for various Δ

4 A Comparative Analysis of the Test Power

In this paper, we investigate the distributions of considered test statistics by computer simulation methods (Monte-Carlo method). Interval samples I_n, in which each element is an interval of length Δ, are generated according to the following algorithm.

1. Set the value of length Δ for the observation intervals.
2. Using the inverse function method, generate an observation X from the distribution $F(x)$.
3. Generate a measurement error e from a truncated distribution $F_e(x)$ with the definition range $[0, \Delta]$.
4. Calculate interval observation:

$$L = X - e, R = X - e + \Delta.$$

5. Repeating steps 2–4 n times, obtain the interval sample (L_1, R_1), …, (L_n, R_n).

The distribution of error e is the truncated normal distribution with the density function

$$f_e(x|0 < X \leq \Delta) = \frac{f(x)}{\Phi\left(\frac{\Delta-\mu}{\sigma}\right) - \Phi\left(\frac{-\mu}{\sigma}\right)},$$

where $f(x)$ is the density function of the normal distribution, with parameters $\mu = \frac{\Delta}{2}$, $\sigma = \frac{\Delta}{6}$.

Let us consider two pairs of competing hypotheses for investigating the power of considered homogeneity tests. H_0: $F_1(x) = F_2(x)$, H_1: the first sample was generated from the gamma-distribution $F_1(x) = \Gamma(5, 1, 1.5)$ and the second one – from $F_2(x) = (5, 1, 1)$.

H_0: $F_1(x) = F_3(x)$, H_2: the first sample was generated from the gamma-distribution $F_1(x) = \Gamma(5, 1, 1.5)$ and the second one – from $F_3(x) = \Gamma(5, 0.5, 3)$.

Samples were generated with the sizes $n_l = 100, 300, 500$, $l = 1, 2$ and the number of simulations $N = 16600$.

The power estimates of the considered homogeneity tests for competing hypotheses H_1 and H_2 in the two-sample case are given in Tables 1 and 2. The sizes of each sample are given in the column n_l, $l = 1, 2$, we considered the case when $n_1 = n_2$.

Table 1. The power estimates for $\Delta = 0.2$

	n	H_1			H_2		
		$\alpha = 0.1$	$\alpha = 0.05$	$\alpha = 0.01$	$\alpha = 0.1$	$\alpha = 0.05$	$\alpha = 0.01$
gLRT(2/3)	100	0.920	0.863	0.678	0.104	0.053	0.009
	300	0.999	0.999	0.996	0.150	0.087	0.021
	500	1	1	0.999	0.208	0.120	0.035
gLRT4	100	0.239	0.127	0.029	0.150	0.075	0.016
	300	0.687	0.512	0.174	0.247	0.152	0.046
	500	0.867	0.777	0.420	0.302	0.210	0.072

Table 2. The power estimates for $\Delta = 0.6$

	n	H_1			H_2		
		$\alpha = 0.1$	$\alpha = 0.05$	$\alpha = 0.01$	$\alpha = 0.1$	$\alpha = 0.05$	$\alpha = 0.01$
gLRT(2/3)	100	0.913	0.852	0.672	0.117	0.061	0.012
	300	0.999	0.999	0.996	0.178	0.104	0.030
	500	1	1	1	0.243	0.146	0.042
gLRT4	100	0.265	0.130	0.029	0.153	0.076	0.015
	300	0.572	0.363	0.079	0.253	0.141	0.033
	500	0.739	0.575	0.197	0.305	0.194	0.050

Table 1 illustrates the case, when the length of observation intervals $\Delta = 0.2$, and in Table 2 there are the power estimates for $\Delta = 0.6$.

As can be seen from Tables 1 and 2, for the competing hypothesis H_1, the generalized logrank tests II and III have much higher power than the generalized logrank test IV, and with the sample size growth, the tests power increases. Comparing the power estimates for different lengths of intervals, it is possible to conclude, that the less length of intervals, the higher power of tests. All three considered tests are not able to distinguish the crossing distribution functions, which correspond to the competing hypothesis H_2.

The power estimates of the considered homogeneity tests in the case of interval-censored data for the same competing hypotheses are given in Table 3. At the

Table 3. The power estimates for interval-censored data, $n_1 = n_2 = 100$, $\Delta = 0.6$

		H_1			H_2		
		$\alpha = 0.1$	$\alpha = 0.05$	$\alpha = 0.01$	$\alpha = 0.1$	$\alpha = 0.05$	$\alpha = 0.01$
40%, 20%, 40%	gLRT(2/3)	0.902	0.835	0.642	0.104	0.051	0.011
	gLRT4	0.098	0.049	0.011	0.099	0.049	0.010
30%, 40%, 30%	gLRT(2/3)	0.896	0.825	0.638	0.095	0.047	0.011
	gLRT4	0.111	0.057	0.011	0.111	0.057	0.011
15%, 70%, 15%	gLRT(2/3)	0.897	0.832	0.655	0.101	0.051	0.010
	gLRT4	0.174	0.082	0.013	0.151	0.072	0.013

first column, there are the numbers of left censored, interval and right censored observations in percents from the sample size.

As it is seen from Table 3, the test power decreases slightly with the censoring degree growth. The generalized logrank tests II and III have much higher power than the generalized logrank test IV, as in the previous case.

5 Conclusions

According to the obtained investigation results, the generalized logrank tests II and III can be recommended as the most appropriate among considered tests. However, all considered homogeneity tests are able to distinguish only competing hypotheses corresponding to the distribution functions without intersection. So, the development of the homogeneity tests for interval-censored data, which have the high power against competing hypotheses with crossing distribution functions, is the subject of our future research.

Acknowledgement. This work is supported by the Russian Ministry of Education and Science (project 2.541.2014 K).

References

1. Kreinovich, V.: Interval computations and interval-related statistical techniques: estimating uncertainty of the results of data processing and indirect measurements. In: Advanced Mathematical and Computational Tools in Metrology and Testing X, Advances in Mathematics for Applied Sciences; Vol. 86, pp. 38–49. World Scientific, Singapore (2015). doi:10.1142/9789814678629_0014
2. Lemeshko, B.Y., Postovalov, S.N.: On estimation of distribution parameters by interval observations. Comput. Technol. **3**(2), 31–38 (1998). (in Russian)
3. Lemeshko, B.Y., Postovalov, S.N.: On solving the problems of statistical analysis of interval data. Comput. Technol. **2**(1), 28–36 (1997). (in Russian)
4. Lemeshko, B.Y., Postovalov, S.N.: Statistical analysis of interval observations, No. 1, 3–12 (1996). Science Bulletin of the Novosibirsk State Technical University (in Russian)

5. Vozhov, S.S.: Investigation of the properties of nonparametric estimate for distribution function with interval data **1**(79), 33–44 (2015). Science Bulletin of the Novosibirsk State Technical University (in Russian)
6. Vozhov, S.S., Chimitova, E.V.: Investigation of maximum likelihood estimates and goodness-of-fit tests for data with known measurement error. In: Applied Methods of Statistical Analysis. Applications in Survival Analysis, Reliability and Quality Control – AMSA 2015, Novosibirsk, 14–19 Sept. 2015: proceedings of the Intern. Workshop, pp. 124–130. NSTU publ., Novosibirsk (2015)
7. Vozhov, S.S., Chimitova, E.V.: The proof of hypothesis of the form of distribution on interval data. Bull. TSU Manage. Comput. Sci. Inf. **1**(34), 35–42 (2016). (in Russian)
8. Groeneboom, P.: Asymptotics for interval censored observations. Technical Report 87-18, Department of Mathematics, University of Amsterdam, 69 p. (1987)
9. Groeneboom, P.: Nonparametric maximum likelihood estimation for interval censored data. Technical Report, Statistics Department, Stanford University, 87 p. (1991)
10. Groeneboom, P., Wellner, J.A.: Information Bounds and Nonparametric Maximum Likelihood Estimation, 126 p. Birkhauser Verlag, Basel (1992)
11. Sun, J., Zhao, Q., Zhao, X.: Generalized log-rank test for interval-censored failure time data. Scandinavian J. Stat. Vol. **32**, 49–57 (2005)
12. Zhao, X., Zhao, Q., Sun, J., Kim, J.S.: Generalized log-rank tests for partly interval-censored failure time data. Biometrical J. **50**(3), 375–385 (2008)
13. Zhao, X., Duan, R., Zhao, Q., Sun, J.: A new class of generalized log rank tests for interval-censored failure time data. Comput. Stat. Data Anal. **60**, 123–131 (2013)
14. Zhao, Q.: gLRT - A New R Package for Analyzing Interval-censored Survival Data, Interval-Censored Time-to-Event Data: Methods and Applications, pp. 377–396. CRC Press (2012)

On the Properties and Application of Tests for Homogeneity of Variances in the Problems of Metrology and Control

Boris Yu. Lemeshko(✉) and Tatyana S. Sataeva

Novosibirsk State Technical University, Novosibirsk, Russia
Lemeshko@ami.nstu.ru

Abstract. Distributions of test statistics of classical tests for homogeneity of variance (Neyman–Pearson, O'Brien, Link, Newman, Bliss–Cochran–Tukey, Cadwell–Leslie–Brown, Overall–Woodward Z-variance and modified Overall–Woodward Z-variance tests) are investigated including a case when the standard assumption of the normality is violated. The comparative analysis of power of the classical tests is carried out. Method of application of the tests of violation of the standard assumption that provides an interactive simulation of distributions of the test statistics is proposed and tested.

Keywords: Test homogeneity variances · Neyman–Pearson test · O'Brien test · Link test · Newman test · Bliss–Cochran–Tukey test · Cadwell–Leslie–Brown test · Overall–Woodward Z-variance test · Modified overall–Woodward Z-variance test · Test power

1 Introduction

Tests of homogeneity of variances are frequently used in various applications. The tasks of processing of measuring results are no exception. Perhaps, the most striking example of demand for tests of homogeneity of variances in the area of metrology is the task of comparison of laboratory tests.

The hypothesis of constant variances of m samples and the competing hypothesis have the form

$$\begin{aligned} H_0 &: \sigma_1^2 = \sigma_2^2 = \ldots = \sigma_m^2, \\ H_1 &: \sigma_{i_1}^2 \neq \sigma_{i_2}^2, \end{aligned} \tag{1}$$

where the inequality holds for at least one pair of indices i_1, i_2. Some tests can be used only for $m = 2$.

The quality of statistical conclusions that is carried out by the results of analysis is provided with correct application of corresponding tests that have the best power.

The standard assumption to determine the possibility of application of classical tests of homogeneity of variances is that the samples follow a normal distribution. This condition sharply limits the area of application of classical tests. This restriction is not imposed on nonparametric tests to test the hypothesis of equality of scaling parameters.

R. Szewczyk and M. Kaliczyńska (eds.), *Recent Advances in Systems, Control and Information Technology*, Advances in Intelligent Systems and Computing 543, DOI 10.1007/978-3-319-48923-0_84

However, on nonparametric tests the samples should belong to the same type of distribution.

Previously, in [1–5], the distributions of test statistics of number of classical (Bartlett [6], Cochran [7], Hartley [8], Levene [9], Fisher [10]) and nonparametric (Ansari–Bradley [11], Mood [12], Siegel–Tukey [13], Klotz [14]) tests used for testing the hypothesis of homogeneity of variances were studied and the analysis of power was conducted. The distributions of test statistics are investigated including a case when the standard assumption is violated. In this paper, the conclusions of [1–4] have been added to results of comparative analysis of number of classical tests for homogeneity of variances (Neyman–Pearson [15], O'Brien [16], Link [17], Newman [18], Bliss–Cochran–Tukey [19], Cadwell–Leslie–Brown [20], Overall–Woodward Z-variance [21] and modified Overall–Woodward Z-variance tests [22]). The purpose of the work is to study the distributions of the statistics of these tests, to extend the table of percentage points, to make the comparative analysis of the power of the tests, to realize the feasibility of using the tests when the standard assumption is violated.

A study of the distributions of the statistics and an estimate of the power of the tests with respect to various alternative hypothesis have been done using a method of statistical simulation in the framework of the Windows Controlled Interval Statistics (ISW) program system. The number of statistical experiments for simulation of samples of the statistics was $N = 10^6$. Than the difference between the true distribution of the statistics and the simulated empirical distribution usually is less than 10^{-3} in absolute value.

When the standard assumption of normality is violated the distributions of the test statistics are studied in a case where the simulated samples belong to a family with the density

$$f(x) = De(\theta_0) = \frac{\theta_0}{2\theta_1\Gamma(1/\theta_0)}\exp\left(-\left(\frac{|x-\theta_2|}{\theta_1}\right)^{\theta_0}\right) \tag{2}$$

for different values of the shape parameter θ_0. The distribution $De(\theta_0)$ includes the normal $\theta_0 = 2$ and the Laplace $\theta_0 = 1$ distribution as special cases.

2 Homogeneity Variance Tests

The Neyman-Pearson test. The statistics of the test [15] is a ratio between arithmetic mean of the all estimate of variances s_i^2 and geometric mean:

$$h = \frac{1}{m}\sum_{i=1}^{m} s_i^2 \Big/ \left(\prod_{i=1}^{m} s_i^2\right)^{\frac{1}{m}}, \tag{3}$$

where m is the number of samples, $s_i^2 = \frac{1}{n_i-1}\sum_{j=1}^{n_i}\left(x_{ij}-\bar{x}_i\right)^2$ is estimate of sample variances, $\bar{x}_i = \frac{1}{n_i}\sum_{j=1}^{n_i} x_{ij}$ is the mean of i-th sample, x_{ij} – j-th observation in the i-th sample. It is assumed that $n_1 = n_2 = \ldots = n_m = n$. The test is right-sided. The hypothesis H_0 is rejected when $h > h_{1-\alpha}$.

The distributions of test statistic (3) depend on n and m. In this work, the values of percentage points have been refined for distributions of statistic. The Neyman-Pearson test is very sensitive to any departures from normality (Fig. 1).

Naturally, that the test can be used for unequal n_i. However, in this case, the distribution of the statistics for true null hypothesis H_0 differs from the distribution with equal n_i.

The O'Brien test. Every raw score x_{ij} is transformed using the following formula [16]:

$$V_{ij} = \frac{(n_i-1.5)n_i\left(x_{ij}-\bar{x}_i\right)^2 - 0.5s_i^2(n_i-1)}{(n_i-1)(n_i-2)}, \tag{4}$$

where n_i is sample size, $\bar{x}_i$ is the mean, s_i^2 – the unbiased estimate of variance for i-th sample.

The test statistic is:

$$V = \frac{1}{m-1}\sum_{i=1}^{m} n_i\left(\overline{V_i}-\overline{\overline{V_i}}\right)^2 \Big/ \frac{1}{N-m}\sum_{i=1}^{m}\sum_{j=1}^{n_i}\left(V_{ij}-\overline{V_i}\right)^2, \tag{5}$$

where $\overline{V_i} = \frac{1}{n_i}\sum_{j=1}^{n_i} V_{ij}$, $\overline{\overline{V_i}} = \frac{1}{N}\sum_{i=1}^{m}\sum_{j=1}^{n_i} V_{ij}$, $N = \sum_{i=1}^{m} n_i$.

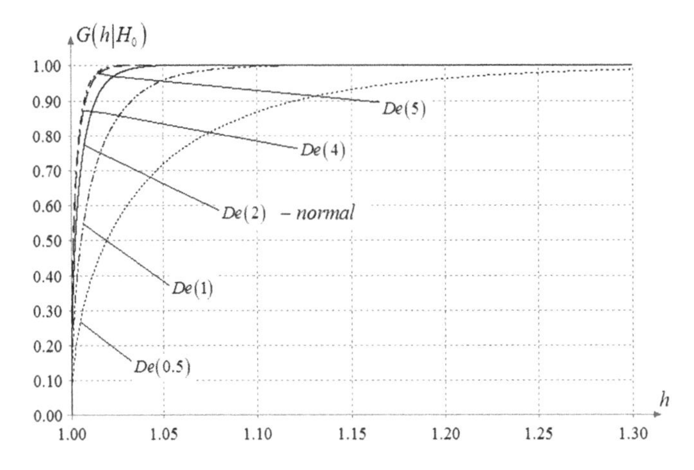

Fig. 1. Distributions of the statistic of the Neyman-Pearson test depending on the type of distributional law, for n = 100, m = 2

The test is right-sided. If the test statistic (5) exceeds the critical value the null hypothesis H_0 is rejected. When the null hypothesis is true the statistic of the O'Brien test has approximately $F_{m-1,N-m}$ – distribution [7]. Nevertheless, the study has shown that the distribution of statistic (5) converges quite slowly to $F_{m-1,N-m}$ – distribution. For example, in the case of $m = 2$, difference between real distribution $G(V|H_0)$ of statistic (5) and conforming $F_{1,N-m}$ – distribution can be neglected only when $n_1 = n_2 = n \geq 80$. For small sample sizes, the essential difference is for large values V therefore using of percentage points for the $F_{m-1,N-m}$ – distribution increases the probability of a Type 2 error in consequence of decrease of a set significance level α.

The upper critical values have been obtained for different number m of compared samples for $n_1 = n_2 = n \leq 80$ to provide the possibility of correct using of the test for small sample sized.

For $N - m \leq 80$, the distributions $G(V|H_0)$ for values V such that $1 - G(V|H_0) < 0.1$ are nearer to $F_{m-1,\infty}$ – than the $F_{m-1,N-m}$ – distribution. Therefore, in these situations, correctness of results can be increased using $F_{m-1,\infty}$ – distribution to estimate the achieved significance level (p_{value}) or choosing critical values $V_{1-\alpha}$ according to $F_{m-1,\infty}$ – distribution.

Distributions of statistic of the O'Brien test are quite robust to violation of the assumption of normality. If the tails of distribution are "easier" than tails of normal distribution, the test statistic doesn't change significantly. If the tails are "heavier", deviations are smaller than deviations for other classical tests. Only the modified Overall–Woodward Z-variance tests has the similar robustness to deviation from normality among the tests that are considered in this work.

The Link test. The Link test is analogue of Fisher test using only for analysis of two ($m = 2$). The test statistic is defined as [17]:

$$F^* = \omega_{n_1} / \omega_{n_2}, \tag{6}$$

where $\omega_{n_1} = x_{1,\max} - x_{1,\min}$, $\omega_{n_1} = x_{2,\max} - x_{2,\min}$ is ranges of samples.

The test is two-sided. The hypothesis is rejected if $F^* > F^*_{1-\alpha/2}$ or $F^* < F^*_{\alpha/2}$, where α is significance level, $F^*_{1-\alpha/2}$ and $F^*_{\alpha/2}$ is upper and lower critical values of statistic.

The distribution of test statistic depends essentially on sample sizes. The Link test is very sensitive to any violation of standard assumption. Upper and lower critical values of statistic (6) have been refined using the methods of statistical simulation.

The Newman test. The statistic of Newman test is defined as follows [18]:

$$q = \omega_{n_1} / s_{n_2}, \tag{7}$$

where $\omega_{n_1} = x_{1,\max} - x_{1,\min}$, $s_{n_2} = \sqrt{\frac{1}{n_2 - 1} \sum_{i=1}^{n_2} (x_{2i} - \bar{x}_2)^2}$.

As the previous test, the Newman test is two-sided. If the null hypothesis H_0 is true, the distribution of test statistics (7) depends on sample sizes and the law of distribution of samples. In this work, upper and lower critical values of statistic (7) have been refined.

The Bliss–Cochran–Tukey test. The test [19] was proposed as analogue of the Cochran test

$$c = \max_{1 \le i \le m} \omega_i \Big/ \sum_{i=1}^{m} \omega_i, \tag{8}$$

where m is number of samples, $\omega_i = \max_{1 \le j \le n_i} x_{ij} - \min_{1 \le j \le n_i} x_{ij}$ is range of i-th sample.

The test is right-sided. The distribution of test statistic highly depends on sample sizes. As and the Cochran test, the distributions of the test statistic are very sensitive to departure from normality.

The Cadwell–Leslie–Brown test. The test [20] is analogue the Hartley test:

$$K = \max_{1 \le i \le m} \omega_i \Big/ \min_{1 \le i \le m} \omega_i, \tag{9}$$

where m is sample sizes, ω_i is range of i-th sample.

The test is right-sided. The distribution of statistic of the Cadwell–Leslie–Brown test, as the distribution of Bliss-Cochran test-statistic, depends essentially on sample sizes and on law of distribution.

In this work, critical values $K_{1-\alpha}$ have been refined for different number of samples m and equal sample sizes $n_i = n, \ \ i = \overline{1,m}$. The study shows, that most of the tests significantly exceed the Cadwell–Leslie–Brown test on the power when the sample sizes are large.

The Overall–Woodward Z-variance test. The test statistic is written as [21]:

$$Z = \frac{1}{m-1} \sum_{i=1}^{m} Z_i^2, \tag{10}$$

where m is number of samples, $Z_i = \sqrt{\frac{c_i(n_i-1)s_i^2}{MSE}} - \sqrt{c_i(n_i-1) - \frac{c_i}{2}}$, $c_i = 2 + 1/n_i$, $MSE = \frac{1}{N-m} \sum_{i=1}^{m} \sum_{j=1}^{n_i} \left(x_{ij} - \bar{x}_i\right)^2$, n_i is size of i-th sample, s_i^2 – is the unbiased estimate of sample variances, $N = \sum_{i=1}^{m} n_i$.

If the null hypothesis H_0 is true and the samples obey normal law of distribution, the distribution of test statistic (10) has approximately $F_{m-1,\infty}$ – distribution and doesn't depend on sample sizes. However, for small sample sizes, distribution of statistic differs significantly from $F_{m-1,\infty}$ – distribution. The analysis has shown, that difference between real and $F_{m-1,\infty}$ – distribution can be neglected, if sample sizes $n \ge 50$. In case when assumption of normality was satisfied and sample sizes are

$n \leq 50$, upper critical values $Z_{1-\alpha}$ have been computed using methods of statistical simulation.

As with most of the classical tests for homogeneity of variance, distribution of Z-variance test statistic is very sensitive to violation of assumption of normality.

The modified Overall–Woodward Z-variance test

Overall and Woodward proposed modification of Z-variance test [22] to construct test that would remain stable when sample data deviate from normality. The new values c_i depends on the sample sizes and the mean of kurtosis indices:

$$c_i = 2.0\left[\frac{1}{K_i}\left(2.9 + \frac{0.2}{n_i}\right)\right]^{\frac{1.6(n_i - 1.8K_i + 14.7)}{n_i}}, \tag{11}$$

where $K_i = \frac{1}{n_i - 2}\sum_{j=1}^{n_i} G_{ij}^4$ – estimate of kurtosis index of i-th sample, $G_{ij} = (x_{ij} - \bar{x}_i) \Big/ \sqrt{\frac{n_i - 1}{n_i} s_i^2}$, $\bar{K}$ – the mean of kurtosis indices.

The studies were demonstrated, that the distribution of test statistic converges slowly to $F_{m-1,\infty}$ – distribution with increased sample sizes. Even when the sample sizes are large values, the distribution of test statistic differs from $F_{m-1,\infty}$ – distribution. However, in the range of large values of the statistic, difference between distribution of test statistic and $F_{m-1,\infty}$ – distribution is not significant. Critical values were found to apply the test correctly for small sample sizes.

At the same time, it should be noted, that the distribution of statistic of the modified Overall-Woodward Z-variance test really is more robust to departures from normality. The obvious difference between distribution of modified test statistic for symmetric laws of distributions and distribution of modified test statistic for normal law of distribution is only for heavy tails. The struggle for robustness led to a decrease of power.

3 The Comparitive Analysis of Power of the Tests

The analysis of power of the tests was made concerning the alternative hypotheses ($H_1 : \sigma_m = 1.1\sigma_0$, $H_2 : \sigma_m = 1.2\sigma_0$, $H_3 : \sigma_m = 1.5\sigma_0$). The estimates of power of classical Bartlett, Cochran, Levene, Hartley, Fisher tests [2] and nonparametric Mood, Ansari-Bradley, Siegel-Tukey tests [3] were included in the comparative analysis.

The obtained estimates of power of the tests for the case of normal samples for significance levels $\alpha = 0.1,\ 0.05,\ 0.01$ and number of samples $m = 2$ are shown in descending order of power in Tables 1, 2.

The Neyman-Pearson, Overall-Woodward Z-variance Bartlett, Cochran, Hartley and Fisher tests appear to be equivalent in power. Difference between modified Overall-Woodward Z-variance and O'Brien tests is noticeable only if the competing hypothesis is relatively distant (H_3). At the same time, both tests have an advantage in power over the Levene test. Note that the O'Brien, Levene and modified Z-variance tests are relatively robust to violation of normal assumption.

Table 1. Power of tests of homogeneity of variances relative to alternative hypothesis $H_1: \ \sigma_2 = 1.1\sigma_1$

Test	α	Sample sizes				
		$n = 10$	$n = 20$	$n = 40$	$n = 60$	$n = 100$
Bartlett, Cochran, Hartley, Fisher, Neyman–Pearson, Z-variance	0.1	0.112	0.128	0.157	0.188	0.246
	0.05	0.058	0.068	0.090	0.111	0.156
	0.01	0.012	0.016	0.023	0.032	0.051
O'Brien, Modified Z-variance	0.1	0.109	0.125	0.154	0.184	0.243
	0.05	0.056	0.066	0.087	0.108	0.153
	0.01	0.012	0.015	0.022	0.030	0.049
Klotz	0.1	0.109	0.123	0.151	0.181	0.236
	0.05	0.056	0.065	0.085	0.106	0.149
	0.01	0.012	0.015	0.021	0.030	0.047
Levene	0.1	0.110	0.123	0.150	0.176	0.228
	0.05	0.056	0.065	0.084	0.103	0.141
	0.01	0.012	0.014	0.021	0.028	0.044
Mood	0.1	0.111	0.120	0.143	0.166	0.211
	0.05	0.057	0.064	0.080	0.096	0.128
	0.01	0.012	0.014	0.020	0.026	0.039
Newman	0.1	0.111	0.123	0.143	0.159	0.186
	0.05	0.057	0.066	0.080	0.091	0.112
	0.01	0.012	0.015	0.020	0.025	0.033
Ansari-Bradley	0.1	0.101	0.125	0.135	0.154	0.190
	0.05	0.052	0.064	0.074	0.087	0.113
	0.01	0.011	0.014	0.019	0.023	0.033
Siegel-Tukey	0.1	0.106	0.121	0.135	0.154	0.190
	0.05	0.055	0.062	0.075	0.087	0.113
	0.01	0.011	0.010	0.018	0.023	0.033
Bliss–Cochran–Tukey, Cadwell–Leslie–Brown, Link	0.1	0.111	0.119	0.133	0.141	0.154
	0.05	0.057	0.063	0.072	0.078	0.087
	0.01	0.012	0.014	0.018	0.019	0.023

The Newman test is inferior to the Levene test in power with an increase in sample sizes. At the same time, the Newman test is superior to Bliss-Cochran-Tukey, Cadwell-Lesley-Brown and Link in power (except when $n = 10$). The last three tests are equivalent in power.

It should be noted, that for small sample sizes ($n = 10$), group of robust tests (modified Z-variance, O'Brien and Levene tests) is inferior to the Newman, Link, Bliss-Cochran-Tukey, Cadwell-Lesley-Brown tests in power, but with the increase of sample sizes n, has a significant advantage over these tests and nonparametric tests. In the robust group the O'Brien test has a slight advantage.

The Newman, Bliss-Cochran-Tukey, Cadwell-Lesley-Brown are more powerful than nonparametric tests only when sample sizes are small ($n = 10 \div 20$).

Table 2. Power of tests of homogeneity of variances relative to alternative hypothesis $H_3: \ \sigma_2 = 1.5\sigma_1$

Test	α	Sample size				
		n = 10	n = 20	n = 40	n = 60	n = 100
Bartlett, Cochran, Hartley, Fisher, Neyman–Pearson, Z-variance	0.1	0.312	0.532	0.806	0.926	0.991
	0.05	0.201	0.402	0.705	0.871	0.98
	0.01	0.064	0.182	0.463	0.692	0.924
O'Brien	0.1	0.266	0.49	0.783	0.917	0.99
	0.05	0.155	0.344	0.664	0.849	0.976
	0.01	0.039	0.127	0.379	0.628	0.903
Modified Z-variance	0.1	0.265	0.489	0.781	0.916	0.99
	0.05	0.158	0.348	0.666	0.849	0.976
	0.01	0.043	0.138	0.397	0.639	0.906
Klotz	0.1	0.258	0.463	0.754	0.9	0.987
	0.05	0.158	0.334	0.638	0.829	0.971
	0.01	0.047	0.137	0.379	0.619	0.892
Levene	0.1	0.269	0.471	0.746	0.888	0.981
	0.05	0.163	0.338	0.628	0.812	0.96
	0.01	0.045	0.131	0.364	0.59	0.866
Mood	0.1	0.255	0.425	0.688	0.841	0.964
	0.05	0.158	0.302	0.565	0.751	0.931
	0.01	0.045	0.121	0.319	0.518	0.802
Newman	0.1	0.296	0.473	0.682	0.796	0.901
	0.05	0.19	0.348	0.566	0.699	0.84
	0.01	0.06	0.153	0.326	0.473	0.667
Ansari-Bradley	0.1	0.242	0.393	0.608	0.768	0.926
	0.05	0.15	0.27	0.484	0.659	0.869
	0.01	0.041	0.104	0.254	0.413	0.693
Siegel-Tukey	0.1	0.246	0.383	0.609	0.768	0.926
	0.05	0.155	0.261	0.484	0.659	0.869
	0.01	0.043	0.056	0.251	0.414	0.693
Bliss–Cochran–Tukey, Cadwell–Leslie–Brown, Link	0.1	0.285	0.425	0.584	0.674	0.776
	0.05	0.181	0.305	0.458	0.554	0.671
	0.01	0.057	0.127	0.237	0.314	0.43

The Bartlett, Cochran, Hartley, Levene, Neyman-Pearson, O'Brien, Bliss-Cochran-Tukey, Cadwell-Lesley-Brown, Overall-Woodward Z-variance and modified Z-variance tests can be used when the number of samples is more than two. However, The Bartlett, Cochran, Hartley, Neyman-Pearson and Overall-Woodward Z-variance test are not equivalent in power anymore.

The Table 3 contains the obtained estimates of power of the multisample tests for the case of normal samples for significance levels $\alpha = 0.1, \ 0.05, \ 0.01$ and number of samples $m = 3$ and $m = 5$ relative to alternative hypothesis H_3.

Table 3. Power of multisample tests of homogeneity of variances relative to alternative hypothesis $H_3: \ \sigma_m = 1.2\sigma_1,\ n_i = 100,\ i = \overline{1,m}$

Test	α					
	0.1	0.05	0.01	0.1	0.05	0.01
	$m = 3$			$m = 5$		
Cochran	0.609	0.494	0.286	0.624	0.515	0.316
O'Brien	0.583	0.461	0.247	0.575	0.460	0.258
Z–variance	0.583	0.461	0.246	0.565	0.445	0.241
Neyman-Pearson	0.580	0.457	0.240	0.557	0.434	0.228
Bartlett	0.577	0.459	0.237	0.557	0.434	0.227
Modified Z-variance	0.574	0.449	0.232	0.554	0.433	0.228
Hartley	0.568	0.443	0.217	0.545	0.418	0.204
Levene	0.530	0.409	0.200	0.513	0.390	0.197
Bliss–Cochran–Tukey	0.359	0.187	0.068	0.262	0.170	0.061
Cadwell–Leslie–Brown	0.280	0.180	0.061	0.253	0.158	0.052

The Cochran test performed the best in power with clear advantage. The next best is the O'Brien test. However when the number of samples is three and the hypothesis is close competing, the test hasn't advantages over the Overall–Woodward Z-variance, Neyman–Pearson and Bartlett tests. At the same time, the O'Brien test is more powerful than modified Z-variance and Levene tests that are robust to violation of the standard assumption of normality.

When the normal assumption is violated the distributions of the test statistics are studied for the simulated samples which belong to the generalized normal distribution with density (2) for different values of the parameter θ_0.

In the case when the assumption of normal distribution of random variables does not hold and the two analyzed samples belong to some symmetric distribution law, the Bartlett, Cochran, Hartley, Fisher, Neyman–Pearson, Z-variance tests are practically equivalent in power.

Similarly, the Bliss–Cochran–Tukey, Cadwell–Leslie–Brown, Link tests are equivalent by power.

In the case when the samples belong to the law with tails "easier" than tails of normal distribution, the tests are ordered by power as in the case of normal law.

If the tails of distribution are "heavier" than the order of preference changes as following:

Mood ≻ Levene ≻ Siegel-Tukey ∼ Ansari-Bradley ≻ O'Brien ≻ Modified Z-variance ≻ ***group of tests*** *(Bartlett, Cochran, Hartley, Fisher, Neyman–Pearson, Z-variance) ≻ Newman ≻* ***group of tests*** *(Bliss–Cochran–Tukey, Cadwell–Leslie–Brown, Link)*

It should be noted, that the power of all classical tests decreases with growth of "heaviness" of tails.

Classical Bliss–Cochran–Tukey, Cadwell–Leslie–Brown, Link tests have less power than nonparametric tests for any laws of distributions. The Newman test has an advantage over nonparametric tests if the tails of distributions are very "easy" only. When the sample sizes are small, the nonparametric tests has not the advantage.

The situation has changed with the increasing number of samples ($m > 2$). The groups of equivalent tests are almost disappearing. Bartlett and Neyman–Pearson tests are equivalent in power in any situation. If the samples belongs to laws with "easy" tails (in comparison with tails of normal law), the tests are ordered by power as in the case of normal law (see Table 3). However, when the tails are "heavy", the tests are in descending ordered by power as follows:

Levene $\succ$ *O'Brien* $\succ$ *Modified Z-variance* $\succ$ *Bartlett* $\sim$ *Neyman–Pearson* $\succ$ *Z-variance Overall–Woodward* $\succ$ *Hartley* $\succ$ *Cochran* $\succ$ *Cadwell–Leslie–Brown* $\succ$ *Bliss–Cochran–Tukey.*

4 Problems of Application of the Tests

We should consider the following facts for choosing an applicable test of homogeneity of variances.

1. The standard assumption that the analyzed sample belongs to the normal distribution determines the possibility of using classical tests of homogeneity of variances. In the case of violation of the assumption, distributions of test statistics that are corresponding accuracy H_0 are changed considerably. The robust tests (O'Brien, Levene and Modified Z-variance Overall–Woodward) is an exception. However, in the case of these tests, dependence on the type of distribution law of the analyzed samples is also observed.
2. Even in the case when the standard assumption is fulfilled, the possibility of correct application of many classical tests is limited by the fact that statistic distributions are unknown and there are only tables of critical values. Therefore, it is impossible to estimate the current significance level p_{value} when hypothesis is tested.
3. For the limited sample sizes, distributions of statistics of classical tests often differ substantially from the known asymptotic distributions of these statistics.
4. The assumption of normality is not imposed on nonparametric tests of homogeneity of the measure of scatter. The nonparametric tests checked in essence a hypothesis about the equality of scale parameters. However, the implementation of at least a strong assumption of laws homogeneity of analyzed samples is required [3].
5. The distributions of normalized statistics of nonparametric tests (Ansari-Bradley [11], Mood [12], Siegel-Tukey [13]) are discrete and, for small sample sizes, significantly different from asymptotic standard normal [3].
6. The classical tests have the obvious advantage in power over nonparametric tests even when the samples belong to non-normal law of distribution. This raises a questions about possibility of correct application of the tests for the use in case of violation of standard.

Hence, for construction of correct statistical conclusion from testing of hypothesis, it is necessary to choose the most powerful test and to estimate the attained significance level p_{value} in accordance with the test. The making a decision on the basis of the estimate p_{value} always is more informative than on the basis of comparison between calculated statistics and some critical value.

When the test is right-sided, the attained significance level (the probability of exceeding the obtained value of statistic when the null hypothesis H_0 is true) is determined as

$$p_{value} = P\{S > S^* | H_0\} = 1 - G(S^* | H_0),$$

where $G(S|H_0)$ is probability distribution function of statistics of the test for true null hypothesis H_0.

In the case of two-sided test, critical region has two parts. The attained significance level is defined as

$$p_{value} = 2\min\{G(S^* | H_0),\ 1 - G(S^* | H_0)\}.$$

When the standard assumption is satisfied, in the case of application of the tests for which distribution of test statistics is unknown or differs from asymptotic distribution, the estimate p_{value} is a problem. However, the problem is solved if there is corresponding software to find the estimate p_{value} as a result of statistical simulation [23, 24].

5 The Applications of the Tests in Nonstandard Conditions

We illustrate the use of interactive mode of investigation $G(S|H_0)$ and accuracy of estimate p_{value} depending on number of experiments N of modeling empirical distributions of statistics including a case when the standard assumption of the normality is violated.

The number of experiments of simulation modelling N should be 16 600 in order that the estimation error p_{value} does not exceed 0.01 for the confidence probability 0.99. The number of experiments should be 1 660 000 in order that the error does not exceed 0.001 [4].

Example 1. The hypothesis of homogeneity of variances of two samples is tested. Assume that the samples belong to the normal law of distribution. The sample sizes is $n_i = 40,\ i = \overline{1,2}$:

0.205	0.232	−0.219	0.829	0.127	0.939	0.995	0.706	−0.450	−0.361
−0.364	−0.107	1.054	−0.095	−2.188	0.453	−1.052	0.640	−0.417	−2.144
−3.473	−0.857	−0.678	0.070	−1.139	0.574	0.409	0.206	0.184	1.273
−0.326	−1.245	0.227	0.185	0.383	0.126	0.255	1.110	−0.310	−0.178

0.269	−0.187	−0.013	−1.248	−0.247	−0.541	1.209	−2.814	0.575	−0.452
−0.427	0.337	1.138	−1.090	−0.858	−0.006	−1.212	−0.180	1.751	−0.485
−0.779	−0.752	0.342	−0.175	0.509	0.209	0.596	1.869	1.764	1.084
0.995	0.633	0.003	−0.642	−1.225	−0.115	−1.543	0.137	−1.290	2.189

Table 4 gives the values of statistics that have been computed. The table presents the estimates p_{value} that have been obtained by simulated distributions of the test

Table 4. The estimates p_{value} obtained by testing of homogeneity of first two samples (hypothesis H_0 is true)

Test	Statistic	p_{value}			
		For normal law			For Laplas law
		Theoretical	$N = 10^4$	$N = 10^6$	$N = 10^6$
Bartlett	0.268028	0.604658	0.605	0.6045	0.734
Cochran	0.541643	–	0.605	0.6045	0.734
Fisher	0.846236	0.604671	0.596	0.6045	0.734
Hartley	1.1817	–	0.605	0.6045	0.734
Neyman-Pearson	1.00349	0.607	0.605	0.6045	0.734
Z–variance	0.279266	0.597183	0.605	0.6045	0.734
Modified Z-variance	0.115111	0.734398	0.741	0.7348	0.732
O'Brien	0.162623	0.687856	0.702	0.6971	0.722
Levene	0.604953	–	0.454	0.4451	0.459
Newman	4.56411	–	0.780	0.7777	0.730
Link	0.948631	–	0.806	0.8079	0.877
Bliss–Cochran–Tukey	0.513181	–	0.814	0.8084	0.878
Cadwell–Leslie–Brown	1.05415	–	0.814	0.8084	0.878

statistics for number of experiments $N = 10^4$ and $N = 10^6$ when the assumption about normality has been satisfied. The theoretical estimates p_{value} is given in the table for tests that have known limit distributions of test statistics.

In fact, Both samples have been simulated in accordance with Laplace distribution with parameter $\sigma = 1$. Therefore, the last column of table presents the estimate p_{value} that has been obtained by simulated distributions of test statistics for number of experiments $N = 10^6$ when the assumption about membership of random variables to Laplace distribution law is satisfied.

As can be seen, the estimates p_{value} for Laplace and normal law of distribution are significantly different. At the same time, in the case of robust Levene, O'Brien and modified Z-variance tests, difference in the estimates is minimal.

Example 2. The hypothesis of homogeneity of variances is tested for three samples, two of which have been taken from previous example. The third sample is:

0.254	−0.254	−0.017	0.002	1.937	−2.476	−0.092	−0.543	2.588	1.970
1.869	0.453	−0.616	−2.806	2.382	0.476	0.641	−2.581	−0.659	−0.027
1.775	2.154	−1.801	−0.774	−0.522	1.413	−0.042	−0.175	−0.929	0.664
−0.298	0.409	0.040	0.418	0.478	−0.052	−4.354	1.521	−2.126	1.177

The sample have been simulated in accordance with the Laplace distribution, where $\sigma = 1.5$.

Table 5 presents values of test statistics that have been computed for checking homogeneity of variances of the three samples. In the case of assumption of normal distribution the table is given the theoretical values p_{value} and the estimates that are

Table 5. The estimates p_{value} obtained by testing of homogeneity of three samples (the ratio of standard deviations 1:1:1.5)

Test	Statistic	p_{value}		
		For normal law		For Laplas law
		Theoretical	$N = 10^6$	$N = 10^6$
Bartlett	9.72943	0.0077	0.0079	0.1198
Cochran	0.534165	–	0.0032	0.0667
Hartley	2.50172	–	0.0140	0.1508
Neyman-Pearson	1.08774	–	0.0079	0.1198
Z–variance	5.0054	0.0067	0.0065	0.1098
Modified Z-variance	2.31571	0.0987	0.0953	0.0897
O'Brien	3.2241	0.0434	0.0396	0.0336
Levene	2.82473	–	0.0661	0.0713
Bliss–Cochran–Tukey	0.415913	–	0.0861	0.3301
Cadwell–Leslie–Brown	1.46271	–	0.1899	0.5106

obtained from statistical simulation for number of experiments 10^6. In the case of assumption of Laplace distribution law of distribution, the estimates p_{value} is given for $N = 10^6$.

It is seen that, if the violation of assumption about normality is ignored, all tests (except modified Overall-Woodward Z-test and O'Brien test) have the values p_{value} less than true values obtained by Laplace distribution.

Hence, classical tests of homogeneity of variances have largest power and can be used correctly for the case of standard assumption about normality or the case of violation of this assumption. In either case, possibility of computation of estimates of attained significance level increases informativeness of statistical conclusions.

6 Conclusion

As follows from the studies, correctness of statistical conclusions, that is carried out at test of hypothesis using classical tests for homogeneity of variances, directly depends on knowledge of the law of distribution of statistic for true null hypothesis H_0. Frequently, even when the assumption of normal distributions holds, distribution of test statistic is unknown and differs from asymptotic distribution. As a result, p-value can not be estimated.

When the standard assumption of normality is violated and the samples have some other type of distribution, statistics of the tests, as a rule, are unknown. Consequently, conclusion of the results of testing hypothesis can not be drawn.

However, if assumptions about estimated type of law of distribution can be justified, problem is not unsolvable. Based on the methods of statistical simulation (and corresponding software), the distributions of test statistics can be found in the process of the analysis, as in [23, 24], including the interactive computing [24].

Acknowledgement. This work is supported by the Russian Ministry of Education and Science (project 2.541.2014K).

References

1. Lemeshko, B., Mirkin, E.: Bartlett and Cochran tests in measurements with probability laws different from normal. Meas. Tech. **47**(10), 960–968 (2004)
2. Lemeshko, BYu., Lemeshko, S.B., Gorbunova, A.A.: Application and power of criteria for testing the homogeneity of variances. Meas. Tech. **53**(3), 237–246 (2010). Part I. Parametric criteria
3. Lemeshko, BYu., Lemeshko, S.B., Gorbunova, A.A.: Application and power of criteria for testing the homogeneity of variances. Meas. Tech. **53**(50), 476–486 (2010). Part II. Nonparametric criteria
4. Lemeshko, B.Y., Lemeshko, S.B., Postovalov, S.N., Chimitova, E.V.: Statistical data analysis, simulation and study of probability regularities. Computer approach, 888 pp. NSTU Publisher, Novosibirsk (2011)
5. Gorbunova, A.A., Lemeshko, B.Y.: Application of parametric homogeneity of variances tests under violation of classical assumption. In: Proceedings, 2nd Stochastic Modeling Techniques and Data Analysis International Conference, Chania, Crete, Greece, 5–8 June 2012, pp. 253–260. http://www.smtda.net/images/1_SMTDA2012_Proceedings_D-J_119-338.pdf
6. Bartlett, M.S.: Properties of sufficiency of statistical tests. Proc. Royal Soc. Lond. Ser. A Math. Phys. Sci. **160**(902), 268–282 (1937)
7. Cochran, W.G.: The distribution of the largest of a set of estimated variances as a fraction of their total. Ann. Eugenics **11**(1), 47–52 (1941)
8. Hartley, H.O.: The maximum F-ratio as a short-cut test of heterogeneity of variance. Biometrika **37**(3/4), 308–312 (1950)
9. Levene, H.: Robust tests for equality of variances. In: Contributions to Probability and Statistics: Essays in Honor of Harold Hotelling, 278–292 (1960)
10. Bolshev, L.N., Smirnov, N.V.: Tables for Mathematical Statistics [in Russian]. Nauka, Moscow (1983)
11. Ansari, A.R., Bradley, R.A.: Rank-tests for dispersions. Ann. Math. Stat. **31**(4), 1174–1189 (1960)
12. Mood, A.: On the asymptotic efficiency of certain nonparametric tests. Ann. Math. Stat. **25** (3), 514–522 (1954)
13. Siegel, S., Tukey, J.W.: A nonparametric sum of rank procedure for relative spread in unpaired samples. J. Am. Stat. Assoc. **55**(291), 429–445 (1960)
14. Klotz, J.: Nonparametric tests for scale. Ann. Math. Stat. **33**, 498–512 (1962)
15. Kobzar, A.I.: Applied mathematical statistics for engineers and academic researchers, p. 816 (2006). M. : Fizmatlit
16. O'Brien, R.G.: Robust techniques for testing heterogeneity of variance effects in factorial designs. Psychometrika **43**(3), 327–342 (1978)
17. Link, R.F.: The sampling distribution of the ratio of two ranges from independent samples. Ann. Math. Stat. **21**(1), 112–116 (1950)
18. Newman, D.: The distribution of range in samples from a normal population, expressed in terms of an independent estimate of standard deviation. Biometrika **31**(1/2), 20–30 (1939)
19. Bliss, C.I., Cochran, W.G., Tukey, J.W.: A rejection criterion based upon the range. Biometrika **43**(3/4), 418–422 (1956)

20. Leslie, R.T., Brown, B.M.: Use of range in testing heterogeneity of variance. Biometrika **53** (1/2), 221–227 (1966)
21. Overall, J.E., Woodward, J.A.: A simple test for heterogeneity of variance in complex factorial design. Psychometrika **39**, 311–318 (1974)
22. Overall, J.E., Woodward, J.A.: A robust and powerful test for heterogeneity of variance. University of Texas Medical Branch Psychometric Laboratory (1976)
23. Gorbunova, A.A., Lemeshko, B.Y.: Application of variance homogeneity tests under violation of normality assumption. In: Proceedings of the International Workshop "Applied Methods of Statistical Analysis. Simulations and Statistical Inference" - AMSA'2011, Novosibirsk, Russia, pp. 28–36, 20–22 September 2011
24. Lemeshko, B.Y., Lemeshko, S.B., Rogozhnikov, A.P.: Real-time studying of statistic distributions of non-parametric goodness-of-fit tests when testing complex hypotheses. In: Proceedings of the International Workshop "Applied Methods of Statistical Analysis, Simulations and Statistical Inference" - AMSA'2011, Novosibirsk, Russia, pp. 19–27, 20–22 September 2011

E2LP Remote Laboratory: Evolution of the System and Lessons Learned

Rafał Kłoda(✉) and Jan Piwiński

Industrial Research Institute for Automation and Measurements PIAP,
Al. Jerozolimskie 202, 02-486 Warsaw, Poland
{rkloda, jpiwinski}@piap.pl

Abstract. Embedded Engineering Learning Platform (E2LP) was a FP7 funded project, which focused on Embedded Systems Education at University level and required a multidisciplinary approach, involving different technologies and system solution optimizations. The paper presents the results of Remote Laboratory (RL) services evolution for distance learning for Embedded Systems, developed under E2LP. The paper addresses advanced information technologies solutions in the integration stages along with novel hardware technologies involved. E2LP RL assembled hardware and the education materials, and delivered secure and open access e-learning portal, which allowed to create full course and provided alternative teaching methods through the real-time experiments. This paper reports also on Remote Laboratory evaluation results performed in Warsaw University of Technology, where we introduced the new learning model in Digital System Design course.

Keywords: E2LP · Remote laboratory · Curriculum integration · Embedded systems

1 Introduction

Recently, with advent and exploitation of computer and communication technologies, remote laboratories have been widely popular among many universities. They are built in order to enhance learning and minimize the gap between theory and practice. Remote laboratories provide on-line pervasive workbenches, which allow an interactive learning environment that maintains student attention.

As embedded software systems have grown in number, complexity, and importance in the modern world, a corresponding need to teach computer science students how to effectively engineer such systems has arisen [1].

Many undergraduate computer engineering programs still teach programming and design skills that are applicable to a general-purpose computer rather than to the more specialized embedded systems [2].

Early exposure to embedded computing systems is crucial for students to be prepared for the embedded computing demands of today's world. However, exposure to systems knowledge often comes too late in the curriculum to stimulate students' interests and to provide a meaningful difference in how they direct their choice of electives for future education and careers [3].

R. Szewczyk and M. Kaliczyńska (eds.), *Recent Advances in Systems, Control and Information Technology*, Advances in Intelligent Systems and Computing 543, DOI 10.1007/978-3-319-48923-0_85

Those aforementioned issues were a genesis to create an unified learning platform, customized to embedded systems curriculum and was the main goal of the E2LP project.

Embedded Computer Engineering Learning Platform (E2LP) is a European FP7 project of three years duration, started in September 2012 [4].

In E2LP project a Remote Laboratory is an experiment, demonstration and a process running locally to design and control an experiment board based on a FPGA device, but with the ability to be monitored and controlled over the Internet (E-learning portal).

In the base case, the RL can be an experiment board connected to a computer through a standard interface and with the host computer connected to the Internet, which provides a remote access. The client can be any computer connected to the Internet with an ability to see the same interface as the local host as well as has the same programs, interfaces and modules.

RL framework consists of three main elements:

1. E-learning portal. This part of RL provides an access to knowledge (on-line exercises, data sheets) as well as remote operations with E2LP main board through a web user interfaces.
2. Laboratory hardware. Main element is E2LP experimental board with programming cable device and other equipment to conduct remote learning process (E2LP server, digital card, serial port server).
3. Laboratory software. It includes the necessary software to programming board and other applications/services/interfaces based on several IT technologies, which provide proper functioning of the whole Remote Laboratory and their hardware components. Here there are also a number of communication ports, which provide flawless operation of specific applications and services in E2LP server as well as in several cases enable user to individually configure the communication with a given device.

2 Remote Laboratories in Learning Courses

As technology is increasingly being seen as a facilitator to learning, open remote laboratories are increasingly available and in widespread use around the world [5].

Remote laboratories are those laboratories that can be controlled and administrated online. They differ from the virtual simulated laboratories as they are interacting with physical instruments [6].

Virtual and remote laboratories (VRLs) are e-learning resources that enhance the accessibility of experimental setups providing a distance teaching framework which meets the student's hands-on learning needs [7]. They have been considered as one of the five major shifts in a century of engineering education, thanks to the influence of information and computational technologies [8].

An important study of the implementation of VRLs into learning courses was reported here [9]. This study presents the results of integrating the open remote laboratories into several courses, in various contexts and using various methodologies.

These integrations, all related to higher education engineering, were designed by teachers with different perspectives to achieve a range of learning outcomes.

In a traditional laboratory, the user interacts directly with the equipment by performing physical actions (e.g. manipulating with the hands, pressing buttons, turning knobs) and receiving sensory feedback (visual and audio). However, equipping a laboratory is a major expense and its maintenance can be difficult [10].

Since the experiments are performed in a laboratory that contains expensive equipment, the students must be supervised which limits the time they have. This also requires a class with many groups performing the experiment at the same time, and thus many instruments are required to support each group. Laboratory experiments are also a serious problem for distance learning students who may not have an access to the laboratory at all [11].

However a review of literature highlighted the lack of meaningful assessment tools for virtual laboratory environments in engineering education. Literature review also highlighted the shortcomings of traditional student lab work assessment practices, among engineering faculties. There are also the problems of time demands, bias, inconsistency, and increasing student numbers which make the traditional lab work assessment scheme impractical [12].

It should be stressed that Remote Laboratories cannot replace the classical education course. There are of course drawbacks of implementation such tools, mainly in lack of communication between student and course supervisor. This type of systems can isolate students and reduce their motivation in learning process. Furthermore, students could not receive instant feedback from their questions and cannot talk in real-time about results obtained in the learning activities with the teacher. Finally students are required to demonstrate their final projects on the actual hardware, the Massive Open Online Course (MOOC) platforms enable them to prepare for this at home, and then to be able to demonstrate valid hardware results in the laboratory [13].

3 E2LP RL Concept of Design and Project Research Objectives

In E2LP project a Remote Laboratory is a service, which enable students to access the laboratory equipment and execute remote operations to carry out exercises. The main goal of RL was implementation of instant feedback from remote E2LP board in a way that user would operate with the real board as if it was connected locally. This functionality was a purpose to develop the GUI web interface of E2LP board front panel that exactly reflects the real board, which has connections to real signals from the real board.

The main advantage of proposed E2LP RL framework (Fig. 1) is a possibility for students to interact with the real E2LP platform interfaces, implemented as a web services in Moodle [14] and work with software applications, on the same operational level like they are actually operating the same tools and instruments in classic lesson in laboratory.

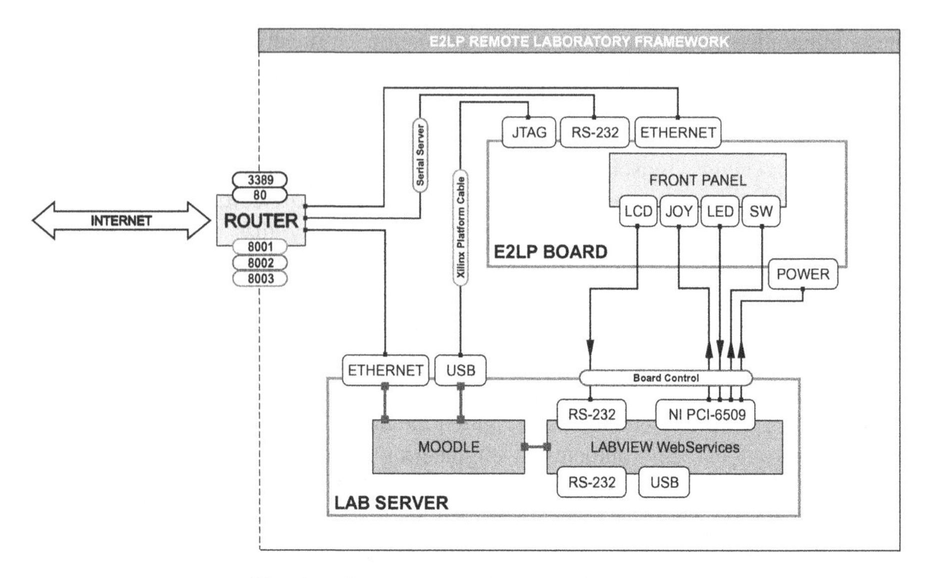

Fig. 1. E2LP Remote Laboratory framework

RL is a gate which provides an access to continuously refreshed interfaces and signals from the real board and enable users to remotely control and program the board directly from their computer at home, having instant visual feedback.

To achieve this, it is necessary to forward data directly to the server over common interfaces or over local network by using dedicated hardware solutions and specified proper router configuration.

The E2LP RL should allow users to do following actions over an Internet connection, which are the list of E2LP Remote Laboratory main functionalities:

1. Dedicated software and hardware solutions provide an access to laboratory equipment and enable students to set them up and operate them at the required level to carry out selected exercises.
2. Users could access the essential data sheets, tutorials and software tools, which are available on the E-learning portal as an introduction to the course. Each laboratory exercise is presented in transparent form to the user through tabs and such division is implemented into Moodle based platform for e-learning course (Basic information, Theoretical explanations, Instructions, Configure Platform, Feedback, Discussion on results questionnaire for lab evaluation).
3. After booking in a given time slot users could remotely program given set of exercises over the Internet and simultaneously, in real time, could monitor the evolution of the experiment on implemented dedicated Graphical User interface (GUI) of the Front Panel of the real E2LP board.
4. Automatic verification of course assignments will allow an advanced management of assignments and submissions together with feedback information mechanisms for both teachers and students, which will verify, whether the students designs are correct or not according to the specifications.

4 RL Development and Implementation Phases

Connection with the Remote Laboratory is provided via e-learning portal, which is based on Apache server, PHP and SQL server. It provides an access to knowledge (exercises, data sheets) and laboratory hardware through a web user interfaces. The second role of e-learning portal is management of users, which means enable them access to the laboratory hardware and software (booking functionality and authorization). In E2LP project the e-learning platform is based on Moodle Platform, which is one of the most popular open source learning management systems. The URL of the Remote Laboratory portal is [15].

RL presents fully operational and tested system, which is enriched with dedicated modules to E2LP Mother Board, which provide real-time remote control, monitoring and programming. Below we show the main advantages of the system:

- The final laboratory exercise on the web has sections (tabs) to enable user to have the full experience of working on the laboratory exercise. These are Digital System Design course exercises, which aim is to control Switches, JOY Push Buttons, LEDs, LCD output in the front panel of the E2LP board as well as RS-232 port are available for remote operations.
- Advance booking system, which enables to reserve a time slot for individual remotely tests of the solution for a given exercise. Booking functionality enables to access up to 4 remote E2LP boards.
- The fast bit file loading module enables remote configuration and immediate respond of the successful E2LP board configuration, without a requirement for users to have a specialized Xilinx [16] software to do it.
- The user friendly Graphical User Interface of the Front Panel, which reflects to the same panel on the real E2LP board, enables user to monitor and control remotely each switch, button, LED and LCD output. The GUI is enriched with the checking correctness of the solution module, which compares the students solution with a master, created by the teacher.
- Automatic verification module, which is based on regular expressions, checks the correctness of the users solution. The pattern for solution is prepared by the teacher or course creator. After comparison the user is informed visually about correctness of his solution.
- The 'Discussion on results' functionality module consist the output information from check correctness solution module, by showing the log records output from the E2LP board Front Panel and enable Teacher and user to exchange information about given exercise.

The whole environment is managed by powerful E2LP Server, controlled by National Instruments LabVIEW software, which is equipped with all common interfaces, which are essential for internal hardware and software compatibility. E2LP Server is connected via Ethernet interface to the local network, which is responsible for seamless data communication between environment's components. The crucial component of the remotely controlled environment is an experiment base board, which is controlled by programming device (Xilinx Platform HW-USB-II-G). This programming device provides integrated firmware to deliver high-performance, reliable and user-friendly configuration of the base board and enables user to program other Xilinx CPLD devices. This programming device is fully integrated and optimized for use with specialized Xilinx iMPACT software, which enable users to perform remote operations such as programming and configuring FPGA via JTAG interface (Fig. 2).

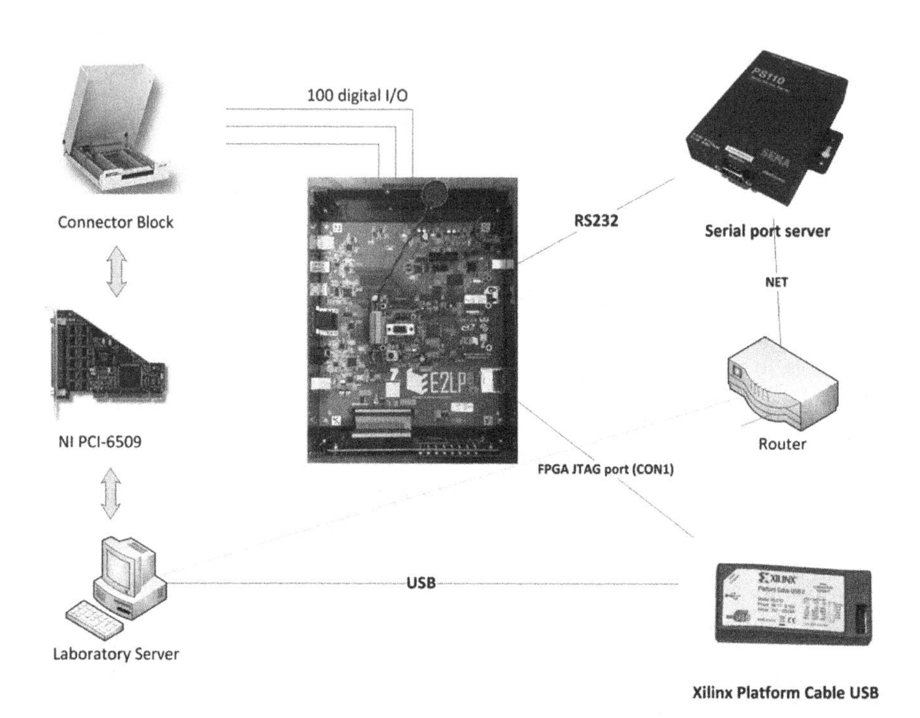

Fig. 2. Remote Laboratory concept of solution

The NI PCI-6509 digital card with 96 bidirectional I/O lines enable user by dedicated GUI interface (Fig. 4) to control each pin in the boards front panel interface and consequently enable him to control each led, switch and button.

Furthermore specific converter communicates with LCD pins on boards front panel interface and translates them into RS-232 ASCII chars. This converter was designed to enable remote characters reading from LCD 2X16 chars.

The chronological evolution of the RL system and main development objectives of the given phase presents the Fig. 3 below.

It should be pointed that all exercise could be done remotely, but feedback from some interfaces is not available. Table 1 below represents implemented necessary interfaces according to requirements.

To integrate physical layer (NI 6509 device) with application layer (user interface) the Web Service (WS) (Fig. 4) was developed using LabVIEW environment.

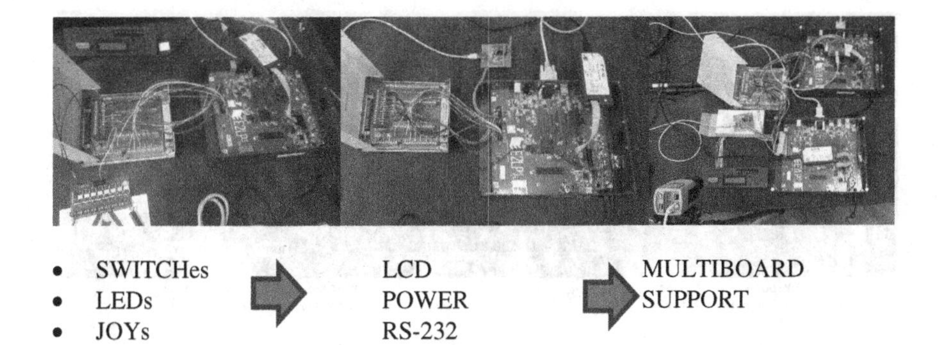

Fig. 3. Remote Laboratory evolution and development objectives

Table 1. Implemented interfaces in RL

Component		Implementation in RL	Access to the interface
Type	Direction (In/Out)		
LCD 16×2 Character Display	O/●	Full	GUI
Dip Switches	●/O	Full	GUI
Push Buttons	●/O	Full	GUI
LED	O/●	Full	GUI
RS-232	●/●	Full	Standard tool[a]
Power Supply ON/OFF	●/O	Full	GUI
[a]e.g. HyperTerminal, Putty, etc. GUI – web interface			

This Web Service has its own user interface (web application) that provide access to control and monitor physical signals on E2LP Board. This web application uses AJAX programming technique that quickly respond to user requests. AJAX enables JavaScript to communicate directly with the Web Service using the XMLHttpRequest object, which requests and updates only the required data instead of reloading whole interface. This concept also enables easy integration with any e-learning systems [6].

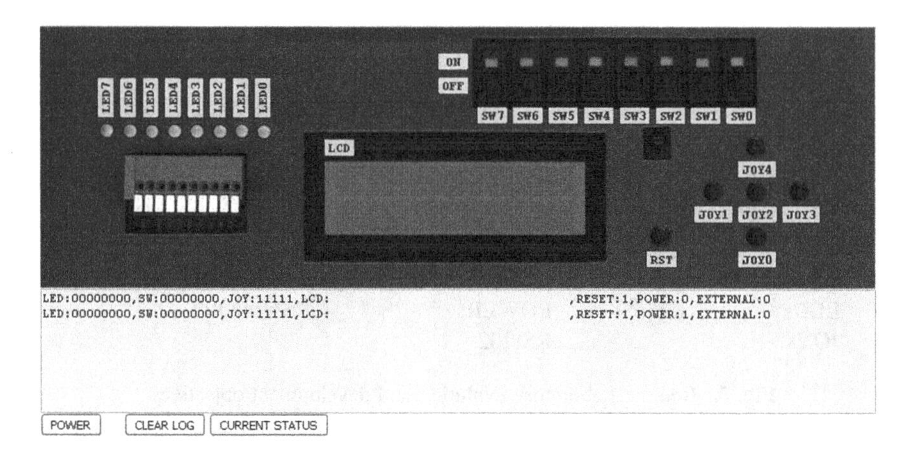

Fig. 4. E2LP board Front Panel web service

5 Scenario of RL Usage

For remote operations users are provided user manual guideline, dedicated software and an access to laboratory equipment, which enable students to set them up and carry out exercises. Our aim was to provide instant feedback to the board in a way that user would work with the real board as if it was connected locally. This functionality was a purpose to developed GUI web interface of front panel that exactly looks like the real board, which has connections to real signals from the real board. States of the physical signals are continuously refreshed at defined times (250 ms) and tests confirmed that it is sufficient to inform the user (by observing GUI) that his program – made during the lab exercise – works correctly. Below we described the scenario of RL usage during the exercise of design the logic circuits:

- User write code of his program according to exercise on local Xilinx ISE environment
- Generate bit file
- Remotely configure FPGA (setup the board, power up, connect the programming modules,)
- Test solution using GUI and standard tool

In the design of digital systems logic circuits are used as fundamental components. The NAND (Negated AND or NOT AND) logic table its truth table inherited from Boolean algebra. After setting the input voltages to the desired values (low or high), these circuits are capable of calculating values of Boolean functions they represent and present the results of Boolean functions on their outputs. and this exercise will not go into detail on how these circuits are implemented.

System entity represents the system as seen from outside. Imagine observing the system on (Fig. 5) from outside, without the possibility to see what is inside the system. The only things you will see are its input and output ports. Input ports are like input variables to a function, while output ports are like results of the functions.

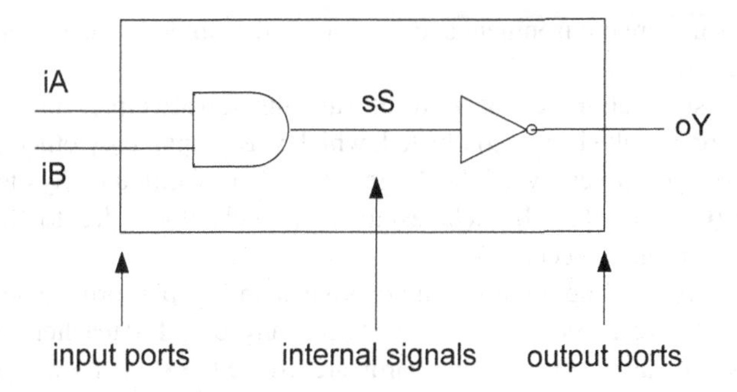

Fig. 5. An example of the digital system

Each output port in a combinational digital system represents the result of one Boolean function of input ports. If the system has N output ports and M input ports, it computes N Boolean functions of M variables, i.e. functions of the same input ports.

In order for the tool to know to which components on the board we want to connect inputs and outputs of our system, we need to specify which pin from FPGA we want to associate with which port of our system. Let us connect inputs iA and iB to two switches and output oY to a LED. You can always refer to the complete list of FPGA pins on E2LP platform and to which components they are connected. Pin assignment is done in a special tool for that, the Xilinx PlanAhead. It can be run from Xilinx ISE.

If configuration completes without errors, the FPGA_DONE diode on the board should be turned on, meaning that the FPGA is configured and working. If you change the state of the switches connected to inputs iA and iB, you should observe the corresponding change to the LED connected to output oY based on a Boolean function which the circuit implements (NAND).

6 Evaluation and Discussion on Results

This chapter presents the preliminary student's practical validation of developed E2LP Remote Laboratory, which was performed at Warsaw University of Technology (WUT) at Mechatronics Faculty during the evaluation stage of the E2LP project. The main purpose of performed evaluation was showing to students system capabilities and engaging them in contribution in testing the developed RL platform as a additional value to study programs in WUT.

A study was completed under "Intelligent Measurement Devices" – a new course in the Electronic Measurement Systems specialization on engineering degree.

During this course students gain comprehensive skills: knowledge about the intelligent sensors, measurements devices and systems operation rules, competence in signal processing and the methodology of novel apparatus construction.

One of the main purpose of the evaluation and the E2LP project was the enrich aforementioned skills with understanding the different digital logic circuits and their operation, implementation of Boolean functions using digital logic circuits, understand the Xilinx ISE software environment and tools as well as understand VHDL description of digital logic circuits.

To get the summative feedback from students towards presented system, the quantitative on-line analysis was conducted, which was prepared by other E2LP project partner Ben-Gurion University of the Negev, based on Computer Sys-tem Usability Questionnaire (CSUQ) [17]. It includes many aspects that refer to the usage RL platform and its and user acceptance.

Our E2LP Remote Laboratory is innovative learning platform, which easy customizes to any course needs and doesn't require any cost for teachers, namely they don't need any specialize software and hardware. Since the students were beginners in VHDL language and Xilinx environments we prepared separate set of easy exercises, which were in line with the curriculum of the subject.

Regarding the e-learning platform for FPGA, the users confirm, that proposed solution are powerful and efficiently improved by using RL. In this sense, students declare they somewhat agree with the idea that remote work is possible without the need to work with the real board.

Considering the aspects related to the user graphic interface, users proof that it is easy to use and its readability increase significantly. Moreover students stress that they are able to quick learning.

The biggest encountered problem was connected with using Xilinx ISE software, which was source of error messages and poor programming experience. The low mark might be a reason of very low student's initial knowledge level of FPGA systems.

7 Conclusions

This paper has discussed all the features provided by E2LP RL for its implementation and deployment in embedded systems engineering education along with feedback from the universities that had deployed it in their learning curricula. Robust RL portal enables users to access E2LP platform over the Internet, configure it compiling VHDL code and having the immediate feedback of solution on their own computer.

Results presented in the paper confirms that introduction of RL into curriculum and new learning model is challenging in the education of engineers in embedded systems. Student, who has never had any practice with Xilinx ISE environment and any FPGA board configuration needs really precise procedure what to do in current exercise.

Proposed solutions based on integrated together Remote Laboratory components and e-learning Moodle Platform enable student to acquire desired knowledge about digital systems and significantly support learning process.

During the evaluation it occurred that remote operations through real-time experiments stimulate the students curiosity and productivity.

Acknowledgments. E2LP Remote Laboratory development were performed in the Industrial Research Institute for Automation and Measurements PIAP. E2LP Remote Laboratory evaluation were made in the Institute of Metrology and Biomedical Engineering, Warsaw University of Technology.

References

1. Mattmann, C.A., Medvidović, N., Malek, S., Edwards, G., Banerjee, S.: A middleware platform for providing mobile and embedded computing instruction to software engineering students. IEEE Trans. Educ. **55**(3), 425 (2012)
2. Jackson, D.J., Caspi, P.: Embedded systems education: future directions, initiatives, and cooperation. SIGBED Rev. **2**(4), 1–4 (2005)
3. Benson, B., Arfaee, A., Kim, C., Kastner, R., Gupta, R.K.: Integrating embedded computing systems into high school and early undergraduate education. IEEE Trans. Educ. **54**(2), 197 (2011)
4. E2LP project website. http://www.e2lp.org
5. Marques, M.A., Viegas, M.C., Costa-Lobo, M.C., Fidalgo, A.V., Alves, G.R., Rocha, J.S., Gustavsson, I.: How remote labs impact on course outcomes: various practices using VISIR. IEEE Trans. Educ. **57**(3), 151–159 (2014)
6. Tawfik, M., Sancristobal, E., Martin, S., Diaz, G., Castro, M.: State-of-the-art remote laboratories for industrial electronics applications. In: Technologies Applied to Electronics Teaching (TAEE), pp. 359–364 (2012)
7. de la Torre, L., Heradio, R., Jara, C.A., Sanchez, J., Dormido, S., Torres, F., Candelas, F.A.: Providing collaborative support to virtual and remote laboratories. IEEE Trans. Learn. Technol. **6**(4), 312–323 (2013)
8. Froyd, J., Wankat, P., Smith, K.: Five major shifts in 100 years of engineering education. Proc. IEEE **100**(Special Centennial Issue), 1344–1360 (2012)
9. García Zubía, J., Alves, G.R. (eds.): Using Remote Labs in Education. Two Little Ducks in Remote Experimentation. Prize for Best Research UD – Grupo Santander. University of Deusto (2011)
10. Distance-Learning Remote Laboratories using LabVIEW, 06 September 2006. http://www.ni.com/white-paper/3301/en/
11. Nafalski, A., Machotka, J., Nedic, Z.: Collaborative remote laboratory NetLab for experiments in electrical engineering in using remote labs in education. Two Little Ducks in Remote Experimentation, pp. 177–199. University of Deusto (2011)
12. Achumba, I.E., Azzi, D., Dunn, V.L., Chukwudebe, G.A.: Intelligent performance assessment of students' laboratory work in a virtual electronic laboratory environment. IEEE Trans. Learn. Technol. **6**(2), 103–116 (2013)
13. Ackovska, N., Ristov, S.: OER approach for specific student groups in hardware-based courses. IEEE Trans. Educ. **57**(4), 242–247 (2014)
14. https://moodle.org/
15. E2LP Remote Laboratory, e-learning portal for E2LP. http://e2lp.piap.pl
16. www.xilinx.com/
17. Lewis, J.R.: IBM computer usability satisfaction questionnaires: psychometric evaluation and instructions for use. Int. J. Hum. Comput. Interact. **7**(1), 57–78 (1995)

Modern Business Internet Technology: Trends, Perspectives and Risks

Svetlana Kumova(✉) and Olga Toropova(✉)

Yuri Gagarin Saratov State Technical University of Saratov, Saratov, Russia
skumova@mail.ru, toropovaoa@inbox.ru

Abstract. The modern technology development affects the business. But, conversely, the business makes its own demands on the internet technologies development. SMM and SEO are the classical Internet promotion methods. These tools become popular around the world and in all sectors of business and entrepreneurship. However, the development of social networks and their integration into the business processes lead to an exacerbation of information security and reputational risks the problem. The aim of the article is to analyze such trends as the growth of cloud services, SMM services, development and deployment of commercial mobile applications, IT-outsourcing services, new trends in site building. The authors present the detailed analysis of trends, forecast the further development, as well as they identify the causes of the risks and threats that the Internet technologies provide to the modern business process. Based on these conclusions, the authors emphasize the need to improve the level of IT-education among workers of different levels.

Keywords: Internet · Information technologies · Social Media Marketing · Information security · SEO · Cloud services · IT-education

1 Introduction

Popularity of information technologies in business leads to continuous complication of instruments of internet marketing, expansion of its opportunities. At the same time introduction on the Internet market leads to increase in number of threats. As a result the accruing criticism of internet marketing and the analysis of Internet threats become a new tendency in modern analytics. Fast speed of changes in the Internet-services market promotes emergence of negative responses.

In this article authors carried out the analysis of new instruments of Internet advance, prospects and threats of their use. Thus not only marketing, but also information technologies are in focus of attention.

2 Main Tendencies of Modern Internet Technologies Development

The competition between advertisers for the right to be seen continuously grows. Many advertising technologies stop being profitable. Large advertisers monopolize search delivery and small business returns to offline instruments of promotion. [1]. Users prefer

R. Szewczyk and M. Kaliczyńska (eds.), *Recent Advances in Systems, Control and Information Technology*, Advances in Intelligent Systems and Computing 543, DOI 10.1007/978-3-319-48923-0_86

to trust their relatives and familiar people councils. Therefore reposts and responses are effective, and the paid announcements, on the contrary, don't achieve the objectives.

Responses and likes are the main driving force of marketing. In other words, classical SMM instrument are in the lead in a market-places and aggregators. As the result, the official site of the company stops being the basis in the digital environment. Also doesn't matter what type this site belongs to ("business card", corporative site, informational portal, promo-site). Now it became more effective to work with target audience through social networks, appendices and accounts at other platforms. For many companies the role of the corporative site was narrowed to a distribution to platforms which are more convenient for visitors. The development of mobile applications and/or adaptation of the corporative sites for mobile devices become a key tendency in the sphere of web development.

These are two different tendencies, but they have the general indicators of productivity. At the moment the majority of advertizing networks provides service of automatic placement of notifications and announcements to users of mobile devices. As result there are a lot of systems for mobile advertizing and analytics. But experience with such tools is still small. According to researches of the analytical company comScore the world mobile traffic in 2014 exceeded desktop- traffic and surely moved towards a further gain [2].

It, in fact, is the continuous movement to reduction of the habitual web environment and complete dominance of appendices over the sites. Such dynamics is characteristic not only for the countries of Europe and America. The Association of Communication Agencies of Russia (ACAR) published the results of the advertising market development in Russia for the first three quarters 2015. And though at the moment in Russia the desktop advertising traffic prevails, the tendency is traced with the same vector that around the world. Full falling of interest in the standard sites and growth of popularity of mobile versions and mobile applications can become the general tendency.

In Russia growth of number of mobile users proceeds: totally (mobile devices and tablets) the share of mobile users in the Runet for the end of 2015 made 32%, and there are bases to believe that by the end of 2017 this share will exceed a 50%.

Search engines started lowering positions of the sites which are badly displayed on mobile devices. So far it occurs only in mobile delivery. Customers don't have more sense to start web projects without adaptive imposition and mobile versions. Search engines started indexing applications and give out them in a priority order before the sites if the request is made on the mobile device. Specially developed ARC Welder expansion for the Google Chrome browser allows to open these applications without installation. ARC Welder uses the technology with Native Client open source code providing start of native applications in the browser [3].

In the short term it will force out the sites from search and will increase popularity of applications. Clients lose interest in development and use substitutes such as CRM or social networks.

Many modern companies choose social networks as the main instrument of Internet marketing. In this case well planned marketing strategy allows to capture large target audience. Modern social networks give the chance to receive detailed reports about a rating of use of links, articles, distributions of content on a web resource of the company, etc. The organization itself can trace popularity of the created community or group among users, actions of participants on the platform (for example, by means of Google Webmaster Tools, Facebook Insights) – to see that attracts target audience and that isn't present, to learn opinions on this or that product, etc. Finally it allows to create loyal base of consumers.

It is necessary to mention two main models of brand promotion. There are SMM and SMO. SMM (Social Media Marketing) represents promotion of the site or company services through social media marketing. This method allows to attract the user traffic on the site directly but not through search engines. SMM covers not only social networks but also such platforms as blogs, forums, network communities, different media resources assuming active communicative interaction of users.

The second model is SMO (Social Media Optimization) that means carrying out the internal technical works increasing efficiency of interaction of the site with social systems. The main actions for content and interface optimisation allow to integrate the site with one or several social networks.

The set of SMO-tools is rather limited (it is visible at least from comparison of 3–4 cases of different optimisers). Not the quantity and a variety, but quality of use matters. These are the most widespread SMO tools:

- Maintaining corporate blogs of the company and personal blogs of employees with possibility of open commenting;
- Development and conducting thematic forums;
- Existence of service of vote with public announcement of results;
- Existence of buttons of fast addition of records in services of social bookmarks (it facilitates distribution of content);
- Creation of various formats of content: PDF files, video and audio recordings. Export of such content will give to the site additional external references.

However, one of classical threats to business which proceed from social networks is reputation risk. The heads of the companies forget that not only official groups of their organizations but also personal pages of top-managers and directors come into the view. As a result the open accounts bearing personal photos and inscriptions "on walls" actively appear at inquiries in search engines. Especially it is actual in connection with emergence of such new technology as "Knowledge Graph".

Knowledge Graph is the technology of a semantic web and the knowledge base of the Google company allowing to find not only narrow information on the asked object or the phenomenon, but also various data concerning required object. Now at request for the company in Google (besides usual search delivery) there is a card on the right with information which is had semantic close to required object. It can be data on a personnel, news messages, etc.

Main objective of this technology is data collection and submission of the necessary information on the search delivery page, exempting the user from an excess waste of time on search on other sites. In certain cases, for example, in a situation of bad Internet communication, this technology will save a lot of time as the user should load smaller number of pages.

It is possible to assume that gradually the share of the sites on which not author's information is provided, will decrease steadily in a delivery top. Also the sites which can give the deep uncommon analysis of some fact or phenomenon, or at least a large number of the structured information on this question will succeed.

Analog of "Knowledge Graph" is service of other large searcher "Yandex. Islands" Search engines start giving out direct replies to the requests of users through "Islands" and "Knowledge Graph". They use data of the sites but don't suggest to pass to them. It reduces possibilities of the companies on attraction of a traffic by the sites and to interaction with visitors. And voice digital assistants (Siri, Cortana and Google Now) give to users the chance to ask questions by voice and to receive direct answers passing the sites and pages of delivery of search results.

It is impossible to ignore such tendency as development of ready CMS, and, in particular, emergence of visual CMS. The platforms which appeared not so long ago for assembly of the sites and the publication of content (Medium, Tilda, Squarespace, Webflow and others) are improved, acquire functionality, add templates and start taking away the market from developers of the sites.

The increasing number of clients prefer ready platforms for expansion of the sites and shops. Standard online stores completely go to this zone. There are ready industry solutions for car dealers, builders and other business.

Among web-services and tools there are a lot of "entertainment technologies" such as:

- the computer games which are at the same time involving many thousands of people interacting in real time with similar;
- the computer games having special sensors and devices for interaction with the real physical world and creations of high-realistic virtual (VR) and added reality (AR);
- mass distribution of Internet video cameras will provide not only the improved communication interaction, but also will become one of serious problems of preservation of confidentiality;
- active interaction of people of various cultures as result of introduction of machine translation systems;
- the video glasses placing the image just before the viewer's eyes, creating effect of presence;
- music and cinema on demand at any time and e-book readers.

3 Risks and Threats

It is very important to define the risks and dangers connected with development of Internet technologies. We will define some of those risks which can be predicted, proceeding from the listed above tendencies of development of Internet technologies.

The first problem is cybercrime and such phenomena as hacker attacks to business or websites for the purpose of stealing of information, establishment of control over remote system or removal of systems and computers out of operation. It breaks normal work of the organizations and yield huge losses.

The separate group is formed by cybercriminals. Object of their attacks are the devices controlling our existence, movement and work of a large number of services of providing life and comfort. Efforts of cybercriminals will be directed on breaking or theft of similar systems, free use by services, adjustment or removal of information on the order and volume similar activity.

With development of technologies roguish schemes of plunders of money on the Internet, such as phishing, will be improved. They steal numbers of credit cards, bank accounts and other confidential information. Typical tools of a phishing are the post messages using the methods of social engineering playing on psychology and weaknesses, and specially developed websites, intended only for that the victim herself brought the confidential data.

4 Conclusions

So, summing up the result of the carried-out analysis of key tendencies in the sphere of Internet technologies, we will note that in the closest years of development of new technologies will continuously stimulate information revolution. In the field of information technologies we can distinguish the following from the most significant events:

- transition from web resources, desktop to mobile versions;
- convergence of various applications on the basis of the Internet protocol;
- deep integration of data, voices and video both in global, and in national networks, active introduction of VR and AR-applications.
- introduction and development of cloudy services and systems of safety of information stored in them;
- further development of a semantic web;
- complex revision of content policy and personalization of Internet communications in connection with growth of influence of SMO technologies.

Thus, active development of Internet technologies shows the actuality of increase IT literacy and IT education level. It`s important not only for the employees who are responsible for the content and technical maintenance of web resources but also for directors, managers, marketing specialists, advertisement makers.

References

1. The letter in edition: The market of the sites development dies as a video hire shop (2002). https://vc.ru/p/digital-is-dying
2. Tokovinin, V.: QSOFT+AIC: The market of web development goes to a bottom, it is necessary to work resolutely. https://roem.ru/22-07-2015/200937/webdev-is-dying/
3. Marshal, J.: Mobile Isn't Killing the Desktop Internet. Electronic resource. http://blogs.wsj.com/cmo/2015/05/26/mobile-isnt-killing-the-desktop-internet/
4. Android-applications became possible to be started in Windows, Mac, Linux and Chrome OS. Instruction. http://www.cnews.ru/news/top/androidprilozheniya_stalo_vozmozhno_zapuskat
5. Social networks in Russia, winter of 2015–2016. Figures, trends, forecasts. Electronic resource. http://br-analytics.ru/blog/socialnye-seti-v-rossii-zima-2015-2016-cifry-trendy-prognozy/

6. Electronic resource. https://ru.wikipedia.org/wiki/Knowledge_Graph
7. Electronic resource. https://ru.wikipedia.org
8. Scherbakova, T., Vergelis, M., Demidova, N.: Spam and phishing in the third quarter 2015: quarterly report. Electronic resource. https://securelist.ru/analysis/spam-quarterly/27294/spam-i-fishing-v-tretem-kvartale-2015
9. Virin, F.: Internet marketing: full collection of practical tools / Fedor Virin. – 2 e prod. – M.: EKSMO (2012)
10. Koshik, A.: Web-Analytics 2.0 in Practice. Subtleties and the Best Techniques. Dialektika Publishing house (2011)
11. Ashmanov, I., Ivanov, A.: Optimisation and advance of the sites in search engines. SPb. St. Petersburg (2011)

Quaternion Based Dynamics for Servicing Satellite with Mission Protocol Description

Elżbieta Jarzębowska[1(✉)] and Michał Szwajewski[2]

[1] Institute of Aeronautics and Applied Mechanics, Warsaw University of Technology, Warsaw, Poland
elajarz@meil.pw.edu.pl

[2] Aerospace Section Poland Division, SENER Sp. z o.o., Warsaw, Poland
michal.szwajewski@sener.pl

Abstract. The paper presents the dynamics of the servicing satellite. The model includes attitude and orbital dynamics of the satellite, with the goal to dock to the target satellite. The dynamics model is free of the singularities because of quaternion use. Orbital dynamics was defined in the internal reference frame with disturbances. To the satellite dynamic model the control system was proposed and the task was defined for the system was defined. Control algorithm was proposed based on Computed Torque control law. The new approach was to defined step by step procedure of the servicing mission of the telecommunication satellites.

Keywords: Attitude dynamics · Attitude control system · Dynamics in quaternions

1 Introduction

The motivation to of the work based on analysis of already planned ESA (European Space Agency) missions. The main area of investigation is a removal of the space debris on low earth orbit or formation flying missions. Two main examples are E-deorbit and PROBA-3 mission. The goal of the E-deorbit is to deorbiting the Envisat satellite, which at this time is not operating satellite, with the use of robotic arm on deorbiting satellite. In mission PROBA-3, two satellites launched in connected configuration have to disconnect on highly elliptical orbit and create a scientific instrument coronagraph and flight in formation.

The new idea of the mission is to investigate possibility of servicing satellites on geostationary orbits. Main object located there are telecommunication satellites, which are most expensive. The extension of their operations will be an advantage in comparison sending new such satellite for replacement. The work based on this approach is investigated by companies acting in space market. In the paper presented by the authors in [1] more details about the current status of the mission is described.

In the paper main focus was oriented on control system of the attitude dynamics. The model presents the complex description of the satellite dynamics including equipment used for reorientation and orbit changing as for example reaction wheels and thrusters. In the control system the guidance navigation system had to be implemented

R. Szewczyk and M. Kaliczyńska (eds.), *Recent Advances in Systems, Control and Information Technology*, Advances in Intelligent Systems and Computing 543, DOI 10.1007/978-3-319-48923-0_87

to define the path, which the servicing satellite has to follow. In the previous paper [1] main focus was placed on mathematical modeling of the satellite, but with limited focus of control system. The control system development is presented via set of simulations, which are defined in inertial reference frame for easier understanding of the problem.

2 Phases of the Servicing Mission

Mission planning is planned in the point where servicing satellite in located on the target satellite orbit. The reaching define orbit is well known task which is out of the scope this paper. In the rendezvous mission planning coordinate system is fixed in target satellite and defined as LVLH – Local Vertical Local Horizontals coordinate system. In detail described in [2] and presented in Fig. 1.

In Table 1. The mission protocol was presented. The time values of each maneuver depends on the equipment installed on the satellite. If effectors of the satellite can assure high data processing and accurate maneuvers this period can be decreased. The detailed calculation performed were presented in [1] and [3].

The phases of the mission were planned to receive the maximum required velocity change in each maneuvers. Based on this data preliminary thrusters selection can be performed. For such analysis Clohessy Wiltshire equation were used.

The total velocity change is defined as in Eq. 1.

$$\Delta v_{total} = ||\Delta v_0|| + ||\Delta v_f|| \tag{1}$$

where

$$\Delta v_0 = \delta v_0^+ - \delta v_0^- = \begin{pmatrix} \delta u_0^+ \\ \delta v_0^+ \\ \delta \omega_0^+ \end{pmatrix} - \begin{pmatrix} \delta u_0^- \\ \delta v_0^- \\ \delta \omega_0^- \end{pmatrix} \tag{2}$$

and

$$\delta v_0^+ = -\Phi_{rv}(t_f)^{-1}\Phi_{rr}(t_f) * r_0 \tag{3}$$

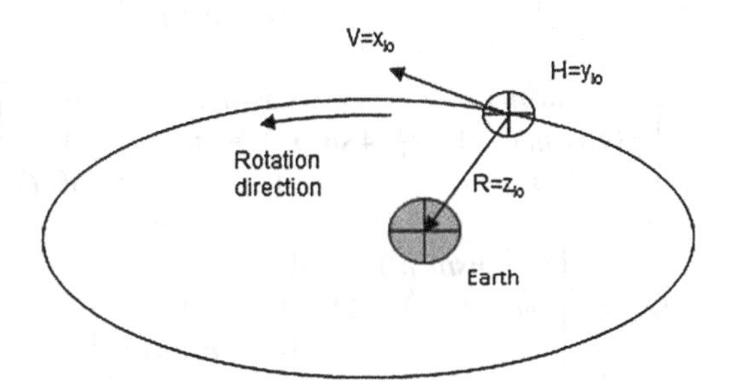

Fig. 1. LV/LH coordinate system

Table 1. Mission protocol description

Control Points	Duration	Description
A1 (+35 km on V bar)	6 h	The satellite is on V bar and is searching the target. The GNC system is defining the orbit of target and servicing satellites
A1 to A2	12 h	Transferring to A2 point. Approaching to the target
A2 (+2 km on V bar)	1 h	Approaching to the target at 2000 m Searching and defining the satellite motion
A2 to A3a (+400 m on V bar)	2 h	Preparation to a spiral approach
A3a to A3b (−200 on V bar)	1 h	Approaching and orbit determination
A3b to A3 (+100 on V bar)	1 h	Motion form V bar to R bar
A3 to A4 (−500 m on R bar)	6 h	Approaching on R bar
A4 do A5 (−20 m on R bar)	10 min	Satellite is about 20 m from the target
A5 do A6 (7 m from the target)	10 min	At A6 point servicing satellite is at a distance of about 7 m from the target
A6 (−7 m on R bar)	10 min	Servicing satellite is starting docking
A6 do A7	20 min	Servicing satellite is ready to dock

$$\delta v_f^- = \phi_{vr}(t_f)\delta r_0 + \phi_{vv}(t_f)\delta v_0^+ \tag{4}$$

$$\Delta v_f = \delta v_f^+ - \delta v_f^- \tag{5}$$

The notation in Eqs. (3)–(5) is: δv_0^+ is the velocity to initiate the start of motion and δv_f^- is the velocity to stop the rendezvous maneuver. Based on above equation and with relations presented by Clohessy and Wiltshire presented below the total change of velocity can be easily defined.

$$\Phi_{rr} = \begin{bmatrix} 4 - 3\cos(nt) & 0 & 0 \\ 6(\sin(\mathrm{nt}) - \mathrm{nt}) & 1 & 0 \\ 0 & 0 & \cos(nt) \end{bmatrix} \tag{6}$$

$$\Phi_{rv} = \begin{bmatrix} \frac{1}{n}\sin(nt) & \frac{2}{n}(1 - \cos(nt)) & 0 \\ \frac{2}{n}(\cos(nt) - 1) & \frac{1}{n}(4\sin(nt) - 3nt) & 0 \\ 0 & 0 & \frac{1}{n}\sin(nt) \end{bmatrix} \tag{7}$$

$$\Phi_{vr} = \begin{bmatrix} 3nsin(nt) & 0 & 0 \\ 6n(\cos(nt) - 1) & 0 & 0 \\ 0 & 0 & -nsin(nt) \end{bmatrix} \tag{8}$$

$$\Phi_{vv} = \begin{bmatrix} \cos(nt) & 2\sin(nt) & 0 \\ -2\sin(nt) & 4\cos(nt) - 3 & 0 \\ 0 & 0 & \cos(nt) \end{bmatrix} \tag{9}$$

where n is the angular velocity for a circular orbit and it is expressed as

$$n = \frac{v}{r} = \sqrt{\frac{\mu}{r_0^3}} \tag{10}$$

Detailed scheme of calculation can be found in [1], where the analysis was presented with assumed data for circular geostationary orbit.

3 Attitude Dynamics in Quaternions

This section is presented description of any rotation of the spacecraft. The main effector in the model are reaction wheels acting on main axis of the spacecraft. In the model the thrusters easily can be implemented. The model was defined in the purpose of any modification to be implement on each step of the project.

Rotation of the satellite is described by Euler equations, i.e. that the time rate of change of an angular momentum is equal to the sum of external moments applied to a body. The upper case at each variable indicates the reference frame, where a letter ***I*** stands for the inertial frame and ***B*** specifies a body fixed reference frame.

$$I^B \frac{d\omega^{BI}}{dt} = \sum \mathrm{M_B} \tag{11}$$

Following the Euler angle notation Eq. (11) can be expressed as follows below. The equation is expressed in body reference frame.

$$\frac{d\omega^B}{dt} = (I^B)^{-1}\left[(-\Omega^B)(I^B)\omega^B + m_B\right], \tag{12}$$

Euler angles notation is inconvenient when the rotation exceeds or pass right angle. To omit the singularities the quaternion notation is proposed.

The quaternion formulation is described as follows, i.e.

$$q = q_0 + iq_1 + jq_2 + kq_3 \tag{13}$$

Quaternions have a specific property, that norm of the quaternion expresses in Eq. (14) has to be equal to one during whole numerical integration and corrected.

$$q_0^2 + q_1^2 + q_2^2 + q_3^2 = 1 \tag{14}$$

Body fixed velocities are related with the quaternions as presented in Eq. (15).

$$\begin{bmatrix} \dot{q}_0 \\ \dot{q}_1 \\ \dot{q}_2 \\ \dot{q}_3 \end{bmatrix} = \frac{1}{2} \begin{bmatrix} 0 & -p & -q & -r \\ p & 0 & r & -q \\ q & -r & 0 & p \\ r & q & -p & 0 \end{bmatrix} \begin{bmatrix} q_0 \\ q_1 \\ q_2 \\ q_3 \end{bmatrix} \tag{15}$$

As mentioned in previous paragraph during integration the norm of the quaternions has to be monitored. The additional element to the equation has to be implemented to the norm on the quaternion and corrected value of the satellite reorientation. Below relation with additional element is presented

$$\begin{bmatrix} \dot{q}_0 \\ \dot{q}_1 \\ \dot{q}_2 \\ \dot{q}_3 \end{bmatrix} = \frac{1}{2} \begin{bmatrix} 0 & -p & -q & -r \\ p & 0 & r & -q \\ q & -r & 0 & p \\ r & q & -p & 0 \end{bmatrix} \begin{bmatrix} q_0 \\ q_1 \\ q_2 \\ q_3 \end{bmatrix} + K\lambda \begin{bmatrix} q_0 \\ q_1 \\ q_2 \\ q_3 \end{bmatrix} \tag{16}$$

where $\lambda = 1 - \left(q_0^2 + q_1^2 + q_2^2 + q_3^2\right)$ the correcting factor and the K is a constantan which is defined based on numerical experiments.

For expressing the quaternions in internal reference frame the rotations matrix expresses below can be used.

$$\mathrm{M}^{BE} = \begin{bmatrix} q_0^2 + q_1^2 - q_2^2 - q_3^2 & 2(q_1 q_2 + q_0 q_3) & 2(q_1 q_3 - q_0 q_2) \\ 2(q_1 q_2 - q_0 q_3) & q_0^2 - q_1^2 + q_2^2 - q_3^2 & 2(q_2 q_3 + q_0 q_1) \\ 2(q_1 q_3 + q_0 q_2) & 2(q_2 q_3 - q_0 q_1) & q_0^2 - q_1^2 - q_2^2 + q_3^2 \end{bmatrix} \tag{17}$$

4 Orbital Dynamics

General equation of motion around the point mass understood as a planet is given in the internal reference frame can be expressed as below:

$$\frac{d^2 r}{dt^2} = -\frac{\mu}{r^3} \boldsymbol{r} \tag{18}$$

where, $\boldsymbol{r}$ is the position vector of the satellite, μ is the gravitational constant and t is time. To the above equation the additional element as *ap* acceleration vector can be added, which express the perturbation vector acting on the satellite.

$$\frac{d^2 r}{dt^2} = -\mu \frac{\boldsymbol{r}}{r^3} + ap \tag{19}$$

The above equation were solved with a use of direct Gauss–Jackson numerical integration method [4]. The Gauss–Jackson method is a predictions step method what allow to maintain the energy in systems. In such method the increase of energy is controlled in each step of integration.

5 Control System of the Attitude Dynamics

The main effectors which acting on the satellite to change the orientation of the spacecraft are reaction wheels and the thrusters. For defining the orientation the system require an information which path of the movement or orientation change has to be executed. For that reason input information is required. In the spacecraft this information is provided by sensors as star-trackers and suns sensor. In simulation study such information was obtained based on geometrical calculation. Based on them the reorientation of the satellite was planned. To the control system was defined a task "follow by the defined path". The satellite had to make a reorientation in such way that the camera has to follow defined trajectory.

For this task the Computed Torque Control Technique was defined. This Technique require a definition a liner differential equation for error command [5]. The equation then is used in dynamic equation of the satellite.

The Eq. (12) can be expresses in general form as below:

$$m_B = D(q) * \ddot{q} + H(g, \dot{q}) \tag{20}$$

where **q** is a vector of variable, which is vector of the orientation of the spacecraft in body reference frame. Into the Eq. (20) is introduced the computed torque control law with an error vector, so the dynamic equation with the control law can be expressed as follows:

$$m_b = D(q) * (\ddot{q}_d - k_d * \dot{e} - k_p * e) + H(g, \dot{q}) \tag{21}$$

$$e = q - q_d \tag{22}$$

where q_d is defined path vector and kp and kd are constant.

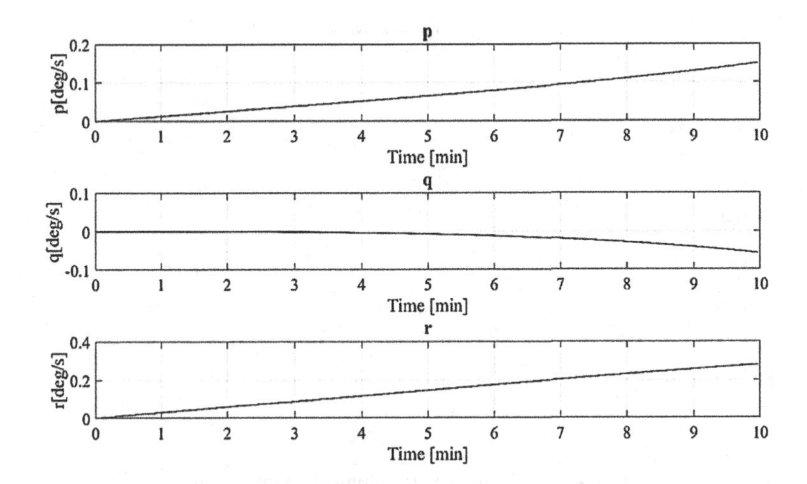

Fig. 2. Changes of the satellite angular velocities under the action of reaction wheels

6 Simulation Study – Attitude Orientation and Execution of the Task "Follow the Path"

The section presents the simulation studies. The results illustrating the orientation change of the spacecraft with introduced computed torque control law. The servicing satellite moments of inertia are Ixx = 2645, Iyy = 698, Izz = 2355 kg/m^2.

The first simulation were presented in case where the control law was not introduced (Fig. 2). In the simulation low torques on reaction wheels were applied during 10 min time simulation (Fig. 3). The quaternion norm were checked and presented (Fig. 4).

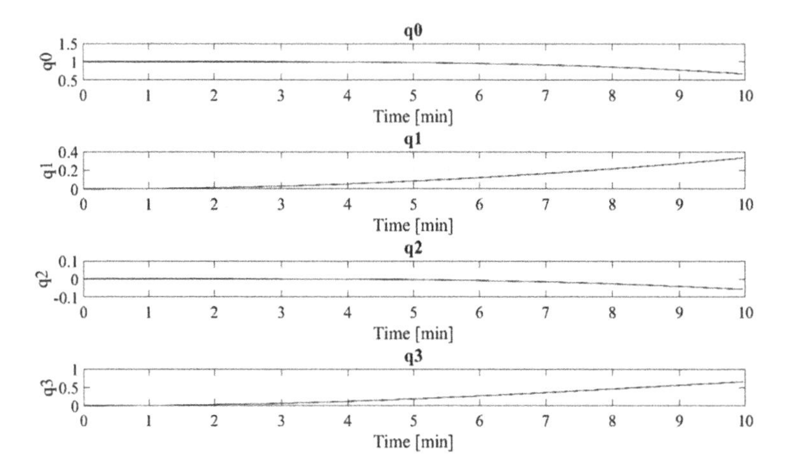

Fig. 3. Euler parameters changes in time

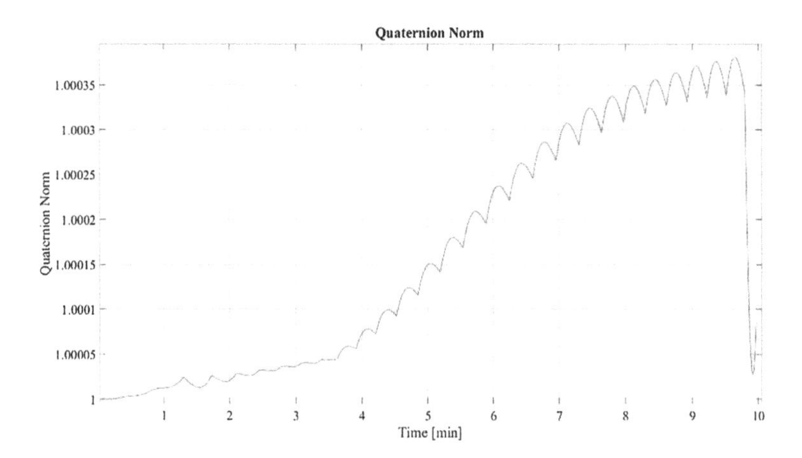

Fig. 4. Quaternion norm along the simulation

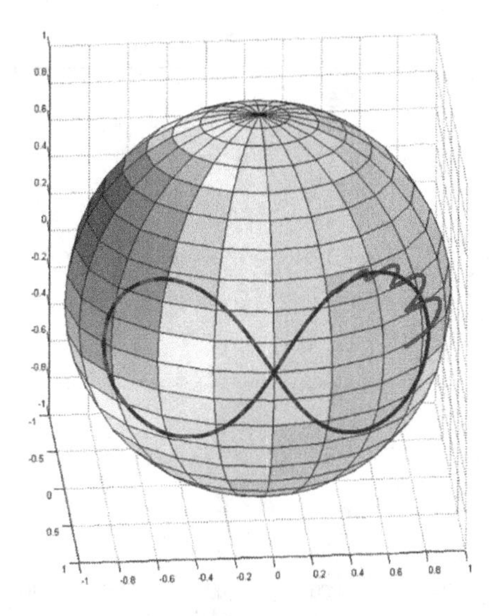

Fig. 5. The artificial surface on which the defined and real path of movement is presented (Color figure online)

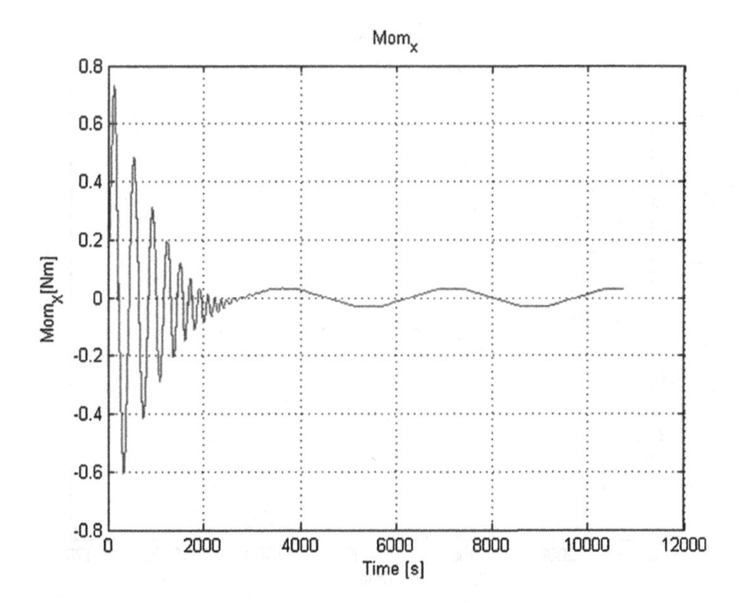

Fig. 6. Torque of X axis

After numerical check of the quaternion norm the analysis with introduced control law were executed. In the Fig. 5 is shown the sphere with the red and blue path on it. The sphere is an illustration on the surface of the movement of the artificial optic of the camera. The blue path is the required path to follow, defined based on proposed

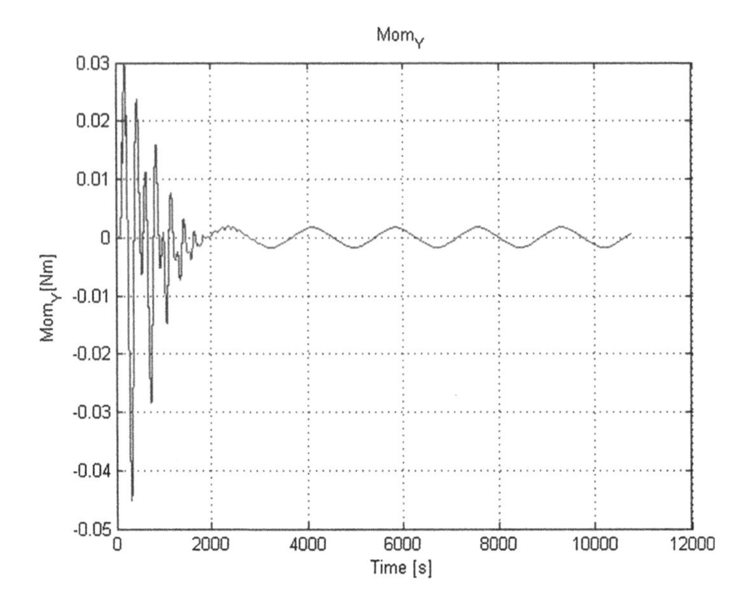

Fig. 7. Torque on Y axis

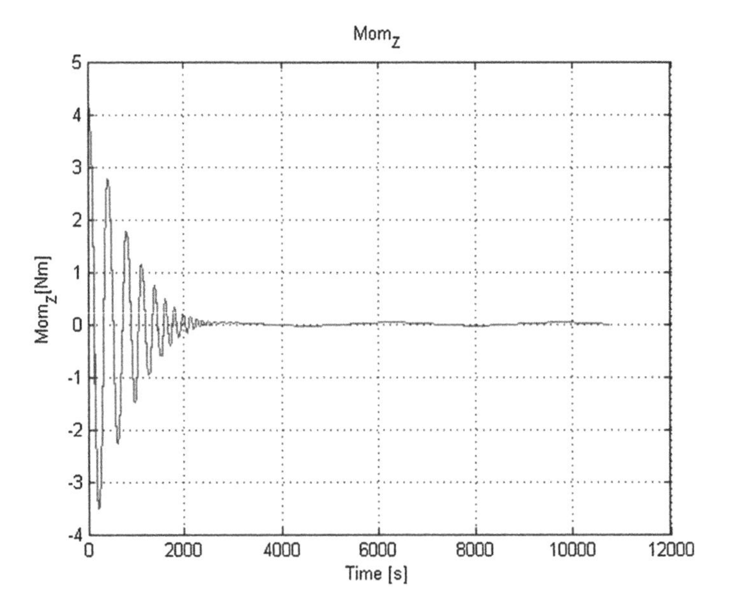

Fig. 8. Torque on Z axis

movement of the target satellite. The red path is the path which follows the artificial optics of the camera reproduce the movement of the target satellite. This maneuver is executing when the satellite is docking to the target spacecraft. In reality the changes of the orientation are far smaller but for simulation purpose was presented bigger changes.

This approach allows also to estimate the setting in control law in case there is require quick change in limited period of time.

For presented above trajectory the graphs of the torques on reaction wheels were obtained (Figs. 6, 7, 8).

The oscillation of the toques on the graphs are torques required to follow the path, so reproducing the torques which act on the target satellite.

7 Conclusions

In the paper description of the mission planning was presented. The development of dynamics in quaternion was described with simulation study. The control algorithm was presented with simulation study of the docking movement reproduction. The satellite model presented in the paper is complex and allow to make a simulation studies of the whole servicing missions.

The future planned work is implementation of other control system algorithms with validation of energy consumption during the simulation, as a most important parameter for implementation on real spacecraft.

References

1. Jarzębowska, E., Szwajewski, M.: A Docking Maneuver of Servicing Satellite – Quaternion Based Dynamics and Control Design
2. Curtis H.D.: Orbital Mechanics for Engineering Students. Elsevier (2014)
3. Wigbert, F.: Automated Rendezvous and Docking of Spacecraft, Cambridge Aerospace Series, November 2008. ISBN 9780521089869
4. Berry, M.M., Liam, M.H.: Implementation of Gauss-Jackson integration for orbit propagation. J. Astronaut. Sci. **52**(3), 331–357 (2004)
5. Jazar, R.N.: Theory of Applied Robotics. Springer, ISBN 978-1-4419-1749-2

Author Index

B
Bączyk, Anna, 610
Balyuk, Vladislav, 734
Bezvesilna, Elena, 43
Bezvesilna, Olena, 481
Bieńkowski, Adam, 682, 689
Blicharz, Bartosz, 345
Bojar, Konrad, 163
Boyko, Regina, 59, 189
Bučinskas, Vytautas, 64, 601
Budny, Eugeniusz, 172
Buliński, Damian, 396
Burdega, Mariana, 567

C
Cader, Maciej, 312, 327, 345
Charubin, Tomasz, 593
Cherepanska, Irina, 43
Chilchenko, Tetyana, 481
Chimitova, Ekaterina V., 760, 775
Czubaczyński, Filip, 368, 396

D
Dąbek, Przemysław, 409
Danik, Yuriy G., 20, 263
Davydenko, Liudmyla, 196
Davydenko, Nina, 196
Davydenko, Volodymyr, 196
Dorozhovets, Mykhaylo, 567
Dróżdż, Tomasz, 211
Dunaj, Jacek, 288, 358
Dupelich, Sergey, 20
Dzedzickis, Andrius, 64, 601
Dziekoński, Cezary, 495

F
Fedushko, Solomia, 104
Fraś, Jan, 368

G
Gladka, Miroslava, 59
Główka, Jakub, 368
Glukhenkyi, Aleksandr I., 535
Gordiichuk, Svetlana, 751
Goszczyński, Tadeusz, 427
Grabar, Ivan, 464
Granovskii, Valerii A., 624
Grygoruk, Mateusz, 610

H
Hajduk, Mikulas, 387
Hamela, Marek, 584
Holub, Zoriana, 111
Hots, Nataliya, 556
Hristoforou, Evangelos V., 647
Hryshchuk, Ruslan, 34, 504

I
Idźkowski, Adam, 524, 740
Ilchenko, Andrii, 27, 734
Iljin, Igor, 64, 601

J
Jackiewicz, Dorota, 263, 577, 689
Janičkovič, Dušan, 705, 713
Janotová, Irena, 705, 713
Jarzębowska, Elżbieta, 816
Juś, Andrzej, 144, 448, 473, 488, 495, 584

K
Kachniarz, Maciej, 481, 495, 682, 689, 698
Kaliczyńska, Małgorzata, 211
Kalinowski, Mateusz, 172
Kamiński, Marcin, 189, 243, 584
Karachun, Volodimir, 459
Kasprzyczak, Leszek, 327
Kataja, Juhani, 172
Kharlamenko, Vadim, 227

R. Szewczyk and M. Kaliczyńska (eds.), *Recent Advances in Systems, Control and Information Technology*, Advances in Intelligent Systems and Computing 543, DOI 10.1007/978-3-319-48923-0

Klimasara, Wojciech J., 273, 358
Kłoda, Rafał, 155, 610, 799
Kolpakova, Tetiana, 3
Korobiichuk, Igor, 20, 189, 227, 243, 263, 464, 481
Korobiichuk, Valentyn, 653
Korzh, Roman, 104
Kosheva, Larysa A., 524, 740
Kosnitskaya, Neonila, 734
Kostro, Sylwester, 512
Kovalenko, Yulia, 234
Kowalski, Grzegorz, 396
Krześniak, Magdalena, 577
Ktena, Aphrodite, 647
Kubisa, Stefan, 630
Kucheruk, Volodymyr, 435
Kudryavtsev, Mikhail D., 624
Kulakov, Pavel, 435
Kulik, Natasha, 674
Kumova, Svetlana, 810
Kuzmych, Lyudmyla, 659
Kvasnikov, Volodymyr, 52, 659
Kyshenko, Vasil, 220

L
Ladanyuk, Anatoliy, 189, 220
Lemeshko, Boris Yu., 760, 784
Lendiel, Taras, 243
Lis, Stanisław, 211
Lomakin, Volodymyr, 27
Lovkin, Valerii, 3
Lysenko, Vitaliy, 243

M
Maciaś, Mateusz, 368
Mako, Igor, 713
Mangiorou, Eleni, 647
Marchenkova, Svitlana, 281
Marcin, Jozef, 705, 713
Matko, Igor, 705
Mel'nick, Viktorij, 459
Mikhal, Aleksandr A., 535
Missala, Tadeusz, 584
Molodetska, Kateryna, 34
Mores, Stelios, 647

N
Niewiatowski, Jerzy, 203
Nowak, Aleksandra, 155
Nowak, Paweł, 75, 203, 448, 473, 488
Nowicki, Michał, 20, 82, 135, 512, 593

O
Obrzut, Łukasz, 252
Oleksyj, Samchyshyn, 20
Oliinyk, Andrii A., 11, 88, 97
Owczarek, Stefan, 144

P
Peleshchyshyn, Andrij, 111
Peleshchyshyn, Andriy, 104
Pełka, Jakub, 135
Petruk, Oleg, 227, 464, 682
Pilat, Zbigniew, 273, 358, 387
Piwiński, Jan, 155, 610, 799
Pogorzelski, Dawid, 584
Pozdniakov, Pavlo, 263
Praczukowska, Alicja, 135

R
Råback, Peter, 172
Radzikowska-Juś, Weronika, 144, 448, 473
Reshetiuk, Volodymyr, 189, 243
Rozen, Viktor, 196
Ruban, Sergii, 227
Ruokolainen, Juha, 172
Rzeplińska-Rykała, Katarzyna, 82, 512

S
Safinowski, Marcin, 118, 495, 584
Salach, Jacek, 577, 698
Sałek, Paweł, 368
Sataeva, Tatyana S., 784
Sazonov, Artem, 43
Šešok, Nikolaj, 64, 601
Shkolna, Olena, 220
Shumyhai, Dmitriy, 59
Shumyhai, Dmytro, 189
Shygonska, Natalia, 127
Shygonskyy, Valerij, 723
Skliarov, Vladimir, 664
Škorvánek, Ivan, 705, 713
Skrupsky, Stepan, 88
Słowikowski, Marcin, 273
Smater, Michał, 273
Sprońska, Agnieszka, 396
Stashynsky, Oleksandr, 52
Storozhuk, Natalia, 435
Subbotin, Sergey A., 11, 88, 97
Šutinys, Ernestas, 64, 601
Švec Sr., Peter, 705, 713
Švec, Peter, 705, 713
Syerov, Yuriy, 104
Szałatkiewicz, Jakub, 172, 252
Szałatkiewicz, Marzena, 252

Szewczyk, Roman, 75, 82, 118, 172, 448, 473, 488, 495, 512, 584, 682, 689
Szudarek, Maciej, 118, 144
Szwajewski, Michał, 816

T
Tkachuk, Andrii, 481
Tomasik, Marcin, 211
Toropova, Olga, 810
Trojnacki, Maciej, 378, 409
Trzcinka, Krzysztof, 495

U
Urbański, Michał, 593

V
Varga, Jozef, 387
Volodarsky, Eugenij, 524, 740
Vourna, Polyxeni, 647
Vozhov, Stanislav S., 775

W
Warsza, Zygmunt L., 524, 535, 567, 630, 740
Winiarski, Wojciech, 118, 495, 584
Wołoszczuk, Adam, 396
Wyszyński, Dominik, 312

Y
Yehorov, Serhii, 546

Z
Zalohin, Maxim, 664
Zavada, Andriy, 504
Zieliński, Jacek, 273
Zielono, Grzegorz, 252
Zigo, Juraj, 705, 713

Printed in the United States
By Bookmasters